A Joint Conference on Materials in the Space Environment (ISMSE 15 & ICPMSE 13 2022)

IOP Conference Series: Materials Science and Engineering
Volume 1287

Leiden, Netherlands
18-23 September 2022

Editors:

Malgorzata Holynska
Adrian Tighe
Elisabeth Laurent
Jacob Kleiman
Sophie Duzellier
Timothy Minton

ISBN: 978-1-7138-7727-1
ISSN: 1757-8981

Printed from e-media with permission by:

Curran Associates, Inc.
57 Morehouse Lane
Red Hook, NY 12571

proceedings.com

Some format issues inherent in the e-media version may also appear in this print version.

Institute of Physics
Dirac House, Temple Back
Bristol BS1 6BE UK

Phone: 44 1 17 929 7481
Fax: 44 1 17 920 0979

techtracking@iop.org

Additional copies of this publication are available from:

Curran Associates, Inc.
57 Morehouse Lane
Red Hook, NY 12571 USA
Phone: 845-758-0400
Fax: 845-758-2633
Email: curran@proceedings.com
Web: www.proceedings.com

TABLE OF CONTENTS

cnes
ONERA
ISMSE15
ICPMSE13
esa

IOP Conf. Series: Materials Science and Engineering 1287 (2023) 011001 doi:10.1088/1757-899X/1287/1/011001

Preface

The 15th International Symposium on Materials in the Space Environment (ISMSE) and 13th International Conference on Protection of Materials and Structures from the Space Environment (ICPMSE) took place 18-23 September 2022 in a special venue, the Naturalis Biodiversity Center in the beautiful city of Leiden (the Netherlands). This joint conference was attended by about 210 participants from 25 different countries worldwide.

We are now happy to present to you the final version of the conference proceedings. This unique collection includes 42 original scientific papers in a variety of fields, including:

- Contamination modelling, applications and monitoring,
- Low / Very low Earth orbit (LEO/VLEO),
- Lunar environment and dust,
- Clean Space and sustainability,
- Flight experiments,
- Radiation and synergistic effects,
- Innovative materials and processes,
- Ground facilities and testing,
- Long-term storage.

The papers reflect well the topics discussed during the conference.

To maintain the high quality of the papers, we used a single-blind peer-review process, established within the framework of the Morresier platform provided by IOP. Moreover, all papers are now assigned DOIs and published open access online by IOP.

The papers resulted from at least 4 years of work since the last ISMSE conference. Some authors are contributing to several papers as they actively work in several research subgroups and supervise multiple student projects. In each case all authors have contributed to all mandatory aspects: substantial contributions to the conception or design of the work; or the acquisition, analysis, or interpretation of data for the work; and drafting the work or revising it critically for important intellectual content; and final approval of the version to be published; agreement to be accountable for all aspects of the work in ensuring that questions related to the accuracy or integrity of any part of the work are appropriately investigated and resolved.

IOP Conf. Series: Materials Science and Engineering 1287 (2023) 011001 doi:10.1088/1757-899X/1287/1/011001

We hope you find the proceedings informative and we look forward to more exciting high-quality research publications in the future as part of the forthcoming ISMSE-16/ICPMSE-14 event.

IOP Conf. Series: Materials Science and Engineering 1287 (2023) 011002 doi:10.1088/1757-899X/1287/1/011002

Peer Review Statement

All papers published in this volume have been reviewed through processes administered by the Editors. Reviews were conducted by expert referees to the professional and scientific standards expected of a proceedings journal published by IOP Publishing.

1. Type of peer review: Single anonymous
2. Conference submission management system: Morressier
3. Number of submissions received: 43
4. Number of submissions sent for review: 43
5. Number of submissions accepted: 42
6. Acceptance Rate (Submissions Accepted / Submissions Received × 100): 97.7 %
7. Average number of reviews per paper: 1
8. Total number of reviewers involved: 27
9. Contact person for queries:
   Name: Malgorzata Holynska
   Affiliation: ESA / ESTEC
   Email: Malgorzata.Holynska@esa.int

* means value has been edited

Compliant mechanisms for dust mitigation in Lunar hardware development: technology and material considerations

D H Budzyn[1], H Zare-Behtash[1], A Cowley[2] and A Cammarano[1]

[1] University of Glasgow, G12 8QQ Glasgow, United Kingdom
[2] European Astronaut Centre - ESA, Linder Hoehe, 51147 Köln, Germany

d.budzyn.1@research.gla.ac.uk

Abstract. During the Apollo missions, astronauts observed negative impact of Lunar dust on the surface hardware. The characteristics of Lunar environment and the regolith properties accelerate the contamination and promote the abrasion and clogging of different components in the equipment. To protect the hardware from damages in the future Lunar missions, several mitigation technologies must be adapted. In this work, we propose to consider application of solutions that are naturally dust resilient. Such solutions, called implicit dust mitigation technologies, include usage of compliant mechanisms. Compliant mechanisms use elastic deformation to achieve motion and can replace rigid-body mechanisms that suffer increased friction and jamming due to dust accumulation in the inter-element gaps. Material selection for compliant mechanisms needs to be considered very early in the design process, and as demonstrated in our work, it is crucial to the final mechanism performance.

1. Introduction

Apollo astronauts reported multiple dust-induced problems, which were classified by Gaier [1]. All recorded problems fall into one of the following categories: dust coating and contamination, seal failures, abrasion, clogging of the mechanisms, thermal control problems, false instrument readings, loss of traction, inhalation and irritation, vision obstruction. The range of the problems is very broad and therefore multiple dust mitigation strategies should be developed for the future Lunar exploration [2]. Furthermore, understanding of the Lunar environment and dust characteristics can help to identify useful mitigation strategies.

Our Moon does not have atmosphere or magnetosphere that could protect it from the solar wind plasma, UV rays and X-rays [3]. As a result, exposed Lunar surface develops electrostatic charge on its surface [4–6]. The electrostatic potential changes with the day-night Lunar cycle [7]. Such electrostatic behaviour enhances the adhesion of Lunar dust particles to surfaces of scientific hardware and makes it hard to clean. The difference of potentials is dependent on the work function of materials used in the equipment as compared to work function of Lunar regolith [8]. Furthermore, Lunar regolith has very wide distribution of particle sizes, with the smallest grains at around 2μm [9]. Such small particles can easily penetrate the gaps in rigid-body mechanisms as well as the gaps in the woven fabrics of spacesuit outer layers. The grains have sharp edges [10] and are made of minerals with hardness that often exceeds the hardness of 6 and above on the Mohs scale [11] – this is higher value as compared to hardness of frequently used engineering materials that build the elements of

Lunar hardware. As a result, Lunar dust particles are abrasive agents that can cut through fabrics and seals and etch equipment surfaces.

In this work, we focus on rigid-body mechanisms dust protection. Rigid body mechanisms (e.g., hinges, sliders etc.) are built with multiple components, that have gaps in between them. Those gaps can get penetrated by the Lunar dust which increases the internal friction of the mechanisms and can ultimately lead to jamming. There are multiple strategies to deal with the dust exposure and contamination [4,12]. Most of them belong to one of the following categories: active dust mitigation that uses external force to clean off the dust (mechanical, fluidal, electrodynamic etc.) and passive dust mitigation which mitigates the dust attraction without the use of external forces (coatings, modified surfaces, seals etc.). Another approach is to design hardware in a way that it is naturally dust resilient. In this work we focus on solving the problem of rigid-body mechanisms getting damaged by dust. The root cause of the problem is dust particles penetrating the inter-element gaps of mechanisms and clogging them [10]. As such, designing mechanisms without those gaps would create naturally dust resilient solutions – implicit dust mitigation. Compliant mechanisms can perform their kinematic functions, without the need of inter-element gaps providing capability to mitigate the dust impact on the hardware at the design level. In this work we explore the intricate relations existing between design, material and performance of compliant mechanisms, highlighting some key aspects that need to be considered in the material selection.

2. Implicit dust mitigation: compliant mechanisms

Compliant mechanisms are mechanisms that use flexible deformation to achieve motion [13]. As such they move without inter-element friction and do not require lubrication and planning for tribological wear [14]. The lack of tribological surfaces in contact with Lunar dust makes compliant mechanisms suitable solution to avoid the dust damage in moving components. Furthermore, compliant mechanisms can be constructed as monolithic pieces which can speed up manufacturing and assembly process as compared to rigid-body mechanisms. Compliant mechanisms deform elastically which also means they store elastic (strain) energy [15]. As such after the removal of input force compliant mechanisms spring back to their original shapes – this is behaviour that is desirable in multiple hold down and release mechanisms in the space industry. Compliant mechanisms have also some disadvantages, including limited motion range (due to elastic range of the deformation), non-linear effects with big displacements, non-intuitive design methodology and challenging loading conditions for some materials (e.g., polymers being susceptible to creep and stress relaxation). Nonetheless, the advantages and natural dust resilient make compliant mechanisms interesting candidates to be used in the Lunar hardware.

3. Compliant mechanisms synthesis

As already mentioned, compliant mechanisms can be less intuitive to design than rigid body mechanisms and the material selection usually poses a challenge, even though there is a wide range of materials applicable for compliant mechanisms. The sections below provide some discussion about those topics.

3.1. Design methodologies

There are multiple methods used to synthesize compliant mechanisms. They are divided into two main categories: analytical design and topology optimisation [14,16]. Analytical methods usually use a library of compliant kinematic pairs or building blocks that fulfil a certain kinematic function. Combination of those elements yields the final design, and from the designer point of view this method resembles the design of more traditional mechanisms. Topology optimisation is a design method that searches for optimum material distribution for a given problem defined by boundary conditions, objective function, and optimisation constraints. This method is popular in the design of static structures. For compliant mechanisms design the definition of objective function is still an open research question and there are multiple formulations that support the synthesis of compliant mechanisms [17].

Adjusting the stiffness of a compliant mechanism can be done in few ways. First one is to change the topology – e.g., instead of using one slender beam as a flexure two beams can be used to increase the stiffness. Next method changes the geometry – increasing the thickness of the flexure leads to higher stiffness. Another method is the change of material – for a given geometry of a flexure usage of material with lower Young's modulus will result in bigger deflections under lower stress.

3.2. Material selection
Selection of a material for compliant mechanism needs careful consideration, and as mentioned above it has big impact on the stiffness characteristic of a system. Furthermore, compliant mechanisms can present lumped compliance where the topology consist of structural (stiff) areas and flexible elements (flexures) that deform elastically. It is also possible to achieve distributed compliance, where most of the topology deforms elastically [16]. Nonetheless, the lumped compliance is more prominent in the field and the flexures (areas with most elastic deformation) are most prone to fatigue damage.

Polymers are generally susceptible to stress relaxation and therefore, for compliant mechanisms which might be required to remain in deformed configuration for extended periods, a plastic (permanent) deformation can occur even below yield stress. Another danger of using polymers is creep, that can plastically deform flexures under constant stress. Nevertheless, polymers are quite popular in consumer products which feature compliant mechanisms: flexible lids for cosmetic bottles, sealing clips, lids of food containers etc. One of the reasons for using polymers in these applications is their price but also high ratio of yield strength to Young's modulus. High yield strength means that the element can undergo higher stress before plastic deformation develops and low Young's modulus ensures higher range of flexible deformations. Metals are also suitable candidates for use in compliant mechanisms. They are considerably stiffer and therefore might require more force to achieve expected deformations but can offer good choice for precision applications. Another way to achieve bigger deformations, as already mentioned, is to change the geometry and therefore metal compliant mechanisms often present very thin long flexures that might be challenging to manufacture. As such, geometry, material, and manufacturing method selection need to be very closely considered together when designing compliant mechanisms.

3.3. Application examples
In this section the impact of material and design method on the performance of compliant gripper will be presented. Two design methods are demonstrated: topology optimisation utilizing a simple and versatile formulation from Koppen et al. [18] implemented in commercial software HyperWorks OptiStruct and analytical method utilizing the instant canter approach [19]. The grippers presented here are displacement driven; input displacement pushing on the input port results in the jaws of gripper closing. Furthermore, the grippers are prototypes, and the aim was to develop and manufacture them relatively quickly. Therefore, 3D printing Fused Filament Fabrication (FDM) was selected as a manufacturing method. In future space missions, 3D printing could allow manufacturing by astronauts in space to produce elements of hardware when needed. In this project, two different 3D printing materials were evaluated: Polyactic acid (PLA) and Thermoplastic Co-Polyester (TPC). Some representative values of mechanical properties of these materials are presented in Table 1. The exact values usually differ with the brand of filament, 3D printer used and multiple parameters like the orientation of specimen. Therefore, literature presents a range of the values differing from one publication to another [20–22]. Nevertheless, it is undeniable that PLA is stiffer than TPC, which will have impact on the resulting designs.

Table 1. Selected mechanical properties of PLA and TPC.

	PLA **Polylactic acid**	**TPC** **Thermoplastic Co-Polyester**
Young's modulus [GPa]	3.8	0.029
Poisson's ratio	0.33	0.48
Elongation at Yield [%]	2	35

The first gripper design presented here utilises instant centre approach. This method works the same way regardless of material selected. It follows a few different steps to produce a sketch with desired topology, details on the method can be found in literature [19,23]. The drawing of produced topology is presented in Figure 1. As visible there are long flexures present in the design. The thinner the flexures the smaller stiffness they will have, and the width of 0.8 mm was selected here.

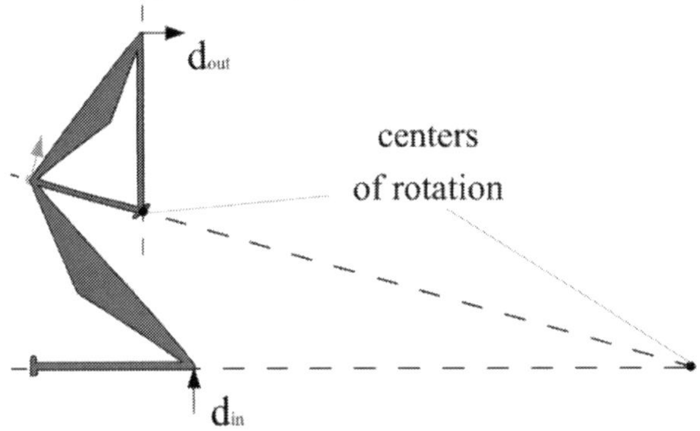

Figure 1. Half of the gripper conceptual design produced with instant centre approach method (analytical design).

The First gripper based on the design discussed above was printed using PLA. It is visible in Figure 2 in its original configuration and fully deformed state with closed jaws. The expected kinematic behaviour of closing induced by linear input force was achieved, and no problems concerning this gripper were noted.

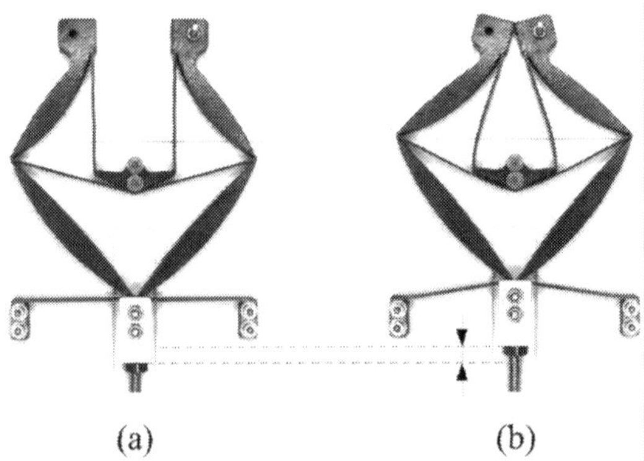

Figure 2. PLA 3D printed gripper based on the analytical design: (a) default position and (b) closed jaws position; dimension marked with arrows represent input displacement.

The same gripper topology (shape) was printed with TPC. This gripper is presented Figure 3. As already mentioned, TPC is more compliant material, and this has an impact on the behaviour of the gripper. It opens and closes as demonstrated in Figure 3 (a) and (b), but it differs from the PLA gripper when it comes to overall output stiffness. As demonstrated in the Figure 3 (c) it is possible to force the output to closed jaws position even though the input is fixed in the original (open) position. Such behaviour is characteristic of underconstrained compliant design, but this behaviour was not observed in the PLA gripper presented above.

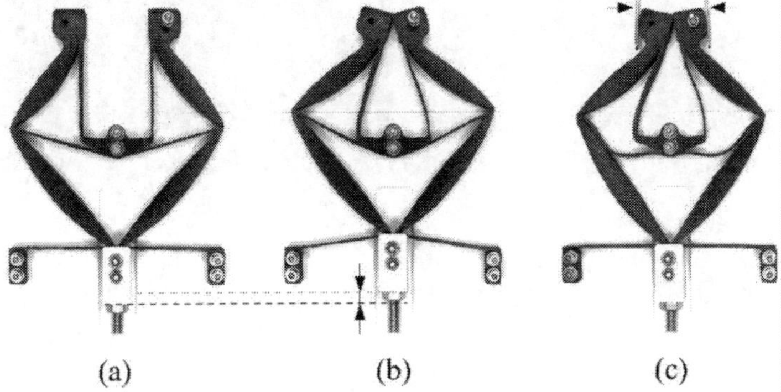

Figure 3. TPC 3D printed gripper based on the analytical design: (a) default position, (b) closed jaws position, (c) forced output closure with arrows demonstrating the external force; dimension marked with arrows (between (a) and (b)) represent input displacement.

The unexpected output behaviour is the result of the lack of appropriate stiffness of the flexures. Flexures provide degrees of freedom (DOFs), but they also provide degrees of constraint (DOCs) – Figure 4 demonstrates the theoretical DOFs and DOCs for a flexure that was used in the discussed gripper. More information about accessing the DOFs and DOCs for compliant elements can be found in the literature[24–26]. Here the problematic DOC is the longitudinal one. As visible in the Figure 3 (c) the flexures buckle with minimal force at the output. This lack of stiffness (as compared to PLA design) results from the application of TPC with the giver geometry (mainly width) of flexure.

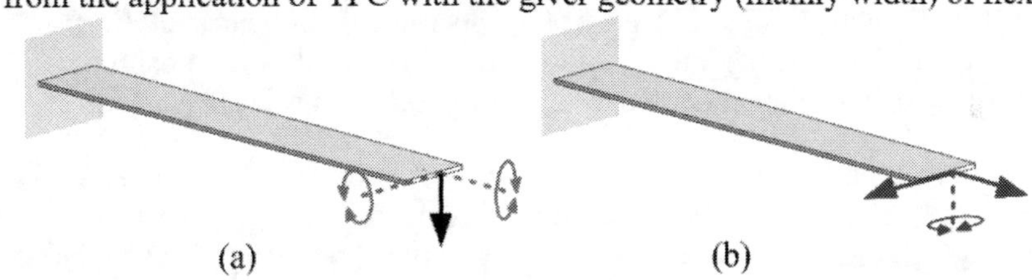

Figure 4. Flexure (a) degrees of freedom (DOFs), (b) degrees of constrained.

Topology optimisation of the next grippers was performed in commercial software HyperWorks utilizing the displacement driven 'simple and versatile topology optimisation formulation for flexure syntheses' developed by Koppen et al. [18]. This formulation was adapted to use for the gripper design, the representative selected iterations of optimisation are presented in Figure 5. Simulation was set up for maximum of 100 iterations but here it reached the convergence after 50 iterations. Half of the design domain was modelled assuming gripper's symmetry. As visible in the last step, most of the intermediate densities (yellow and green elements) are removed from the design domain leaving an almost binary design (red and blue). In post processing all elements with densities below 0.75 were removed and the design was finalised to be 3D printed.

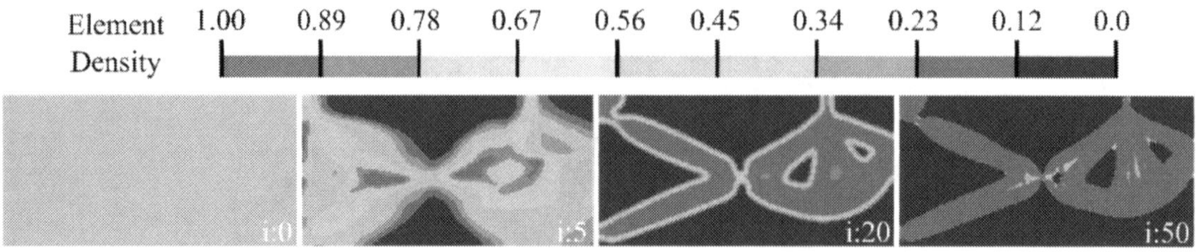

Figure 5. Topology optimisation progression for TPC compliant gripper; 'i' – number of iterations.

Figure 6 presents TPC 3D printed gripper in its default and deflected state. As visible, the gripper has expected kinematic behaviour – pushing on the input force closes the jaws. No problems were noted with this design and material combination.

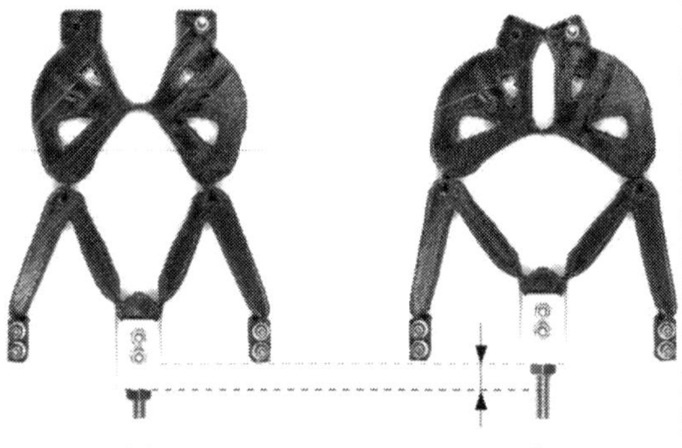

Figure 6. TPC 3D printed gripper based on the analytical design: (a) default position and (b) closed jaws position; dimension marked with arrows represent input displacement.

Topology optimisation was also performed for PLA gripper. As compared with the previous optimisation, the only change in the simulation was the material. Figure 7 shows the evolution of the topology. Even though this optimisation used more iterations, the green intermediate densities are still visible in the last step (i:100). To have continuous material in the design domain the densities above 0.35 were kept (which means they were shifter to density of 1). This number is much lower than for the TPC topology optimized design. It is important to note that the formulation used for the optimisation has a constrained on the global stiffness/compliance and both topology optimisation simulations (TPC and PLA) had exactly same constraints. As lower densities will make it to the final PLA design, it is evident that this design will exhibit higher stiffness.

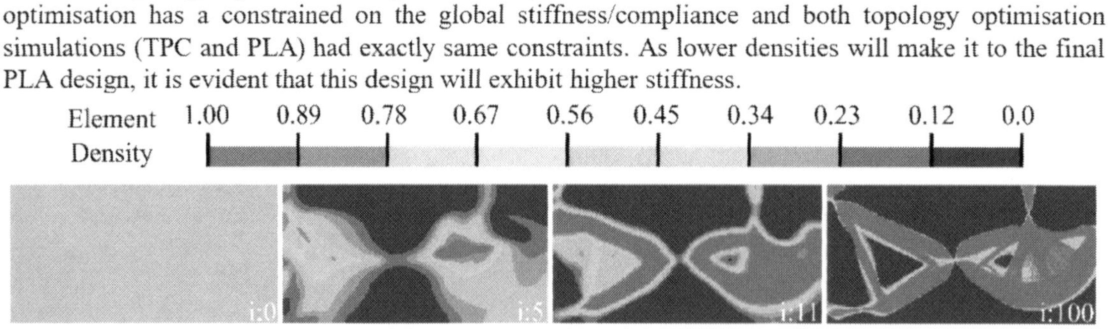

Figure 7. Topology optimisation progression for PLA compliant gripper; 'i' – number of iterations.

3D printed PLA gripper is presented in the Figure 8 in its default and deflected position. The deflection presented is very small since FEM analysis (Figure 8 (c)) show excessive stress occurring in the localised compliant hinge produced by the topology optimisation. To avoid damage of the hinge only small displacements induced performed. It is worth noting that adjusting the topology optimisation parameters or performing additional shape optimisation could immensely improve this design, but the scope of this paper is to show the impact of material selection combined with the design methodology.

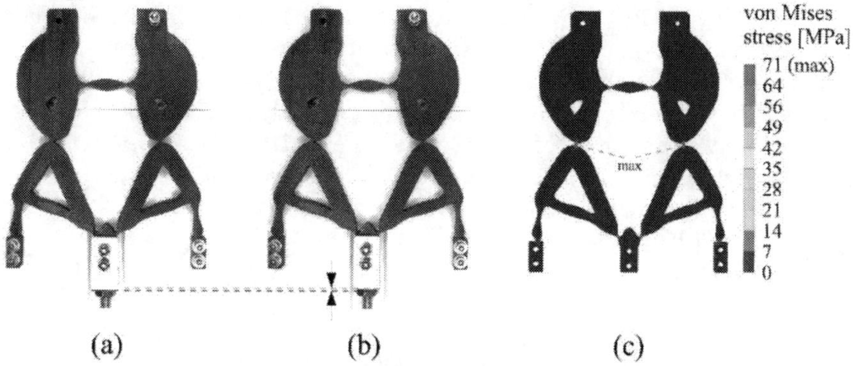

Figure 8. TPC 3D printed gripper based on the analytical design: (a) default position, (b) closed jaws position, (c) forced output closure with arrows demonstrating the external force; dimension marked with arrows (between (a) and (b)) represent input displacement.

4. Discussion

Lunar dust posed challenges in Apollo exploration operations and an effort should be made to mitigate its hardware impact in the future. No single dust mitigation solution will be capable of protecting all types of systems. Therefore, development of different approaches that can work together is crucial. One of the solutions to this problem is the usage of naturally dust resilient compliant mechanisms. As outlined in this work, complaint mechanisms have significant advantages in the Lunar environment. They also bring some design challenges, some of which were discussed in this paper.

We explored two different design approaches and two materials for 3D printing of compliant mechanisms. Topology optimized TPC gripper and analytically design PLA gripper exhibited satisfactory kinematic behaviour. As opposed to that, TPC analytically designed gripper failed in providing enough DOCs in the flexures, resulting in flexures buckling and gripper displaying underconnstrained behaviour. The PLA topology optimisation also led to unsatisfactory solution with high stiffness, and as demonstrated, the desired displacements cannot be supported by the flexural hinges present in this design. This shows that selection of the material and design have considerable impact on the final performance and should be selected simultaneously. The authors acknowledge that the materials presented here would not be suitable for the Lunar use. Nonetheless, there are materials (PEEK, PEKK) suitable for 3D printing that could be included in the hardware on the Lunar surface. Future efforts should focus on finding a range of Moon-suitable materials that could fulfil different functions and work with different design methods for compliant mechanisms.

References

[1] Gaier JR. The Effects of Lunar Dust on EVA Systems During the Apollo Missions. NASA/Tm-2005-213610/Rev1 2007.
[2] International Space Exploration Coordination Group / Technology Working Group. Global Exploration Roadmap Critical Technology Neends 2019.
[3] Stubbs TJ, Halekas JS, Farrell WM, Vondrak RR. Lunar surface charging: A global perspective using lunar prospector data. ESA - Workshop on Dust in Planetary Systems 2007:181–4.
[4] Budzyń D, Tuohy E, Garrivier N, Schild T, Cowley A, Cruise R, et al. Lunar Dust: Its Impact on Hardware and Mitigation Technologies. 46th Aerospace Mechanisms Symposium 2022:287.

[5] Halekas JS, Lin RP, Mitchell DL. Large negative lunar surface potentials in sunlight and shadow 2005;32. https://doi.org/10.1029/2005GL022627.
[6] Colwell JE, Robertson SR, Horányi M, Wang X, Poppe A, Wheeler P. Lunar dust levitation. J Aerosp Eng 2009;22:2–9. https://doi.org/10.1061/(ASCE)0893-1321(2009)22:1(2).
[7] Colwell JE, Batiste S, Horányi M, Robertson S, Sture S. Lunar surface: Dust dynamics and regolith mechanics. Reviews of Geophysics 2007;45:1–26. https://doi.org/10.1029/2005RG000184.
[8] Sternovsky Z, Robertson S, Sickafoose A, Colwell J, Horányi M. Contact charging of lunar and Martian dust simulants. Journal of Geophysical Research E: Planets 2002;107:15–1.
[9] Carrier WD. Particle Size Distribution of Lunar Soil. Journal of Geotechnical and Geoenvironmental Engineering 2003;129:956–9. https://doi.org/10.1061/(asce)1090-0241(2003)129:10(956).
[10] Levine JS, Winterhalter D, Kerschmann RL. The Impact of Lunar Dust on Human Exploration. 2020.
[11] Rickman D, Street KW. Some expected mechanical characteristics of lunar dust: A geological view. AIP Conf Proc 2008;969:949–55. https://doi.org/10.1063/1.2845062.
[12] Pirrotta S, Lefebvre D, Co-chair MW, Wong H, Buffington J, Gaier JJ. Dust Mitigation Gap Assesment Report 2016.
[13] Howell LL, Magleby SP, Olsen BM. Handbook of Compliant Mechanisms. 2013. https://doi.org/10.1002/9781118516485.
[14] Gallego JA, Herder J. Synthesis methods in compliant mechanisms: An overview. Proceedings of the ASME Design Engineering Technical Conference 2009;7:193–214. https://doi.org/10.1115/DETC2009-86845.
[15] Kuppens PR, Bessa MA, Herder JL, Hopkins JB. Compliant Mechanisms That Use Static Balancing to Achieve Dramatically Different States of Stiffness. J Mech Robot 2021;13:1–6. https://doi.org/10.1115/1.4049438.
[16] Zhu B, Zhang X, Zhang H, Liang J, Zang H, Li H, et al. Design of compliant mechanisms using continuum topology optimization : A review. Mech Mach Theory 2020;143. https://doi.org/10.1016/j.mechmachtheory.2019.103622.
[17] Saxena A, Ananthasuresh GK. On an optimal property of compliant topologies. Structural and Multidisciplinary Optimization 2000:36–49.
[18] Koppen S, Langelaar M, van Keulen F. A simple and versatile topology optimization formulation for flexure synthesis. Mech Mach Theory 2022;172:104743. https://doi.org/10.1016/j.mechmachtheory.2022.104743.
[19] Kim CJ, Arbor A, Moon Y. An Instant Center Approach Toward the Conceptual Design of Compliant Mechanisms 2006;128. https://doi.org/10.1115/1.2181992.
[20] Travieso-Rodriguez JA, Jerez-Mesa R, Llum J, Traver-Ramos O, Gomez-Gras G, Josep J, et al. Mechanical Properties of 3D-Printing Polylactic Acid Parts subjected to Bending Stress and Fatigue Testing n.d.
[21] Scallan P. Comparative Study of the Sensitivity of PLA, ABS, PEEK, and PETG's Mechanical Properties to FDM Printing Process Parameters. Process Planning 2003:219–50. https://doi.org/10.1016/b978-075065129-5/50007-6.
[22] Naveed N. Investigate the effects of process parameters on material properties and microstructural changes of 3D-printed specimens using fused deposition modelling (FDM). Materials Technology 2021;36:317–30. https://doi.org/10.1080/10667857.2020.1758475.
[23] Kim CJ, Kota S, Moon YM. An instant center approach to the conceptual design of compliant mechanisms. Proceedings of the ASME Design Engineering Technical Conference 2004;2 B:993–1005. https://doi.org/10.1115/detc2004-57388.
[24] Hopkins JB, Culpepper ML. Synthesis of multi-degree of freedom, parallel flexure system concepts via Freedom and Constraint Topology (FACT) - Part I: Principles. Precis Eng 2010;34:259–70. https://doi.org/10.1016/j.precisioneng.2009.06.008.

[25] Hopkins JB, Culpepper ML. Synthesis of multi-degree of freedom, parallel flexure system concepts via freedom and constraint topology (FACT). Part II: Practice. Precis Eng 2010;34:271–8. https://doi.org/10.1016/j.precisioneng.2009.06.007.

[26] Hopkins JB. Designing hybrid flexure systems and elements using freedom and constraint topologies. Mechanical Sciences 2013;4:319–31. https://doi.org/10.5194/ms-4-319-2013.

A predictive model of Lunar Gateway molecular outgassing and plume-induced contamination

W A Hoey [1*], M G Martin [1], C A Steagall [2], C E Soares [1], G S Shallcross [1] and E S Worthy [3]

[1] Jet Propulsion Laboratory, California Institute of Technology; 4800 Oak Grove Dr. Pasadena CA 91109 (USA)

[2] Jacobs, 2224 Bay Area Blvd. Houston TX 77058 (USA)

[3] NASA Johnson Space Center, Houston TX 77058 (USA)

[*] Corresponding author: william.a.hoey@jpl.nasa.gov

Abstract. The Gateway space station will operate both as a lunar outpost and a hub for space exploration missions in support of a long-term human presence on the Moon, and is under development by NASA in collaboration with ESA, CSA, JAXA, and international commercial partners. Gateway will experience induced molecular contamination from sources including materials outgassing, venting, and the nominal operation of its chemical and electric propulsion systems. The degradation effects of molecular contamination on spacecraft are well-documented and understood, and predictive modeling tools have been developed to aid in the mitigation of such contamination effects in the mission design phase. Therefore, the Gateway Induced Environments and Thermal teams have worked with JPL Contamination Control to develop modular simulation tools for the prediction of molecular contaminant depositions across the mission life, and for use in requirements derivation and sensitivity assessments. Free-molecular transport and deposition of outgassing products onto sensitive Gateway receiver surfaces is calculated using a viewfactor matrix approach, while International Space Station (ISS)-heritage bipropellant plume models are used to evaluate contaminant fluxes generated by the operation of thrusters on Gateway modules and the Orion spacecraft. The alteration of optical properties, e.g. the solar absorptance, of contaminated surfaces is likewise calculated using an ISS-heritage semi-empirical model premised on flight and laboratory testing data. This framework is intended to support early identification of potential integration issues and establish a baseline for incorporating improved analysis and test data as the Gateway design matures.

1. Introduction

Each spacecraft experiences and contributes to its own unique *contamination environment* as its thruster operations, materials outgassing, and other nominal processes disperse unwanted material that may interfere with spacecraft or instrument operations. Space stations can experience particularly diverse and extreme contamination vectors in comparison to other types of spacecraft given factors including their long mission lifetimes; relatively large exposed areas of contamination source materials; repeated docking and undocking maneuvers requiring proximal thruster operation; and repeated venting of unwanted gases and liquids [2]. Harmful effects of molecular and particulate contamination relevant to space stations include the performance degradation of radiators and solar arrays; spurious instrument measurements; and material damage including chemical and mechanical erosive processes. Examples of such effects experienced during International Space Station (ISS) operations are shown in figure 1.

Figure 1. ISS Contamination Processes and Effects. From left to right: outgassing-induced molecular contaminant deposition on external ISS surfaces; ATV thruster firing effects proximal to ISS; and plume-induced mechanical erosion of a sample coupon. Image credits: NASA. [2,5,6]

Contamination control (CC) is the spacecraft engineering discipline concerned with identification, characterization, and mitigation of such contaminant threats throughout the spacecraft mission lifecycle. NASA CC engineers support flight projects across all mission phases from concept generation through operation and end-of-mission, with physics-based modeling, experimental testing of materials and simulation of space conditions, materials screening, and cleanroom support. Particularly in early mission design phases, tools for physics-based modeling and analysis are a key component of spacecraft contamination control strategies that enable proactive identification and mitigation of threats. The CC discipline at NASA's Jet Propulsion Laboratory (JPL) actively develops and implements physics-based modeling practices across the spectrum of multi-regime mass transport problems relevant to spacecraft contamination. Driving questions for this discipline include:

- How are molecular and particulate contaminants generated, dispersed, and transported to contamination-sensitive surfaces and instruments across spacecraft environments?
- How do such contaminants degrade spacecraft and instrument performance?
- How can spacecraft engineers act to ensure that mission performance is not compromised?

1.1. Contamination models for early project phases

NASA contamination analysis teams must maintain a robust set of tools and techniques for multi-regime and multi-phase flows, as mission contamination sensitivities and operational environments vary widely across NASA's range of spacecraft and space station projects. However, modeling approaches also need to be flexible within individual projects across design phases and therefore responsive to the quality and availability of critical inputs. Influencing project decisions in early design phases can be critical to meeting contamination requirements and ensuring performance across mission lifetimes – to that end, contamination models can help to inform a project's material selections, outgassing rate requirements, materials characterization and bakeout programs, mechanical design, propulsion system design and planned operations, and so on. Contamination models suited to such early project phases (i.e. 'early-look' models) must start from relatively few available inputs, and be robust to expansion as designs mature; must be useful for parametric studies and requirements derivation purposes; and must be lightweight – distributable to and usable by diverse systems engineering teams. A template for such a contamination model was outlined by JPL CC for NASA's Psyche Asteroid Mission [3]. The Psyche materials outgassing and external molecular contamination model was kept up-to-date throughout that project as spacecraft geometries evolved; as planned mission operations and thermal predictions became well-defined; and as anticipated outgassing rates became superseded by as-measured materials and hardware characterization data. This model framework was used to track mission-life outgassing budgets and molecular deposition effects, and also found utility in e.g. selecting optimal MLI vent locations and in assessing relative instrument susceptibilities across the range of solar array rotations.

1.2. *The scope of an early-look model for Gateway external molecular contamination*
The Gateway Program is led by NASA's Johnson Space Center (JSC) in a collaboration with space agencies including ESA, CSA, and JAXA and international commercial partners. As a part of NASA's Artemis program, Gateway will serve as a multi-purpose outpost orbiting the Moon to support the long-term human return to the lunar surface and serve as a staging point for deep space exploration [1]. The Gateway Induced Environments team has collaborated with the Gateway Thermal team and the JPL Contamination Control Engineering modeling and analysis team to develop an integrated, early-look modeling framework for the assessment of materials outgassing, transport, deposition, and optical property degradation processes. The key objectives of this external contamination model framework are to inform Gateway's system-level and element-level requirements (and associated data requirements); to aid in the proactive identification of any integration issues or risks; and to aid in the formulation of long-term strategies for evolving external contamination tools, processes, and verification steps. This paper will outline the early-look framework developed for Gateway by JPL CC in response to these objectives including review of the model workflow, its necessary inputs and assumptions, and its associated methodologies and example results.

2. Outgassing Model Workflow and Methodology
JPL CC has generated a model of molecular contamination and optical property degradation for Gateway module sensitive surfaces as a function of molecular outgassing, transport, and deposition from designated source groups. Boundary conditions and unique inputs for this model include Gateway module CAD geometries and meshes; the user identification of N_S outgassing source surfaces and associated outgassing requirements, module arrival times, etc.; and the user identification of N_R sensitive receiver surfaces and associated optical properties, equivalent sun hour (ESH) exposure, etc. For a given geometric configuration of Gateway, JPL CC pre-calculates matrices of free-molecular transport coefficients between all N_S source and N_R receiver groups for cases of both *direct* and *direct + reflective* flux using a viewfactor approach. Here, we define direct molecular flux as transport along the first line-of-sight from a source to a receiver, i.e. the case in which all elements stick all incident mass. Reflective flux occurs if non-receiver elements instantaneously re-emit all incident mass from a cosine angular distribution, and only receiver elements stick incident mass.

The model is packaged into a set of Excel spreadsheet workbooks for delivery which include expressions for outgassing rate time dependence and optical property degradation. For a given geometry, these workbooks can be tuned by users to parametrically evaluate the effects of material properties, beginning-of-life (BOL) source outgassing rate requirements, BOL optical properties, and UV exposure on contaminant deposition and optical performance. This approach decouples the relatively expensive geometric model processing and viewfactor transport steps performed by JPL CC from the set of lightweight workbook products that are distributed to and used by systems engineering teams. Of course, significant assumptions and simplifications are implied by this decoupling that are most appropriate for an 'early-look' – before designs, materials selections, and temperature predictions are finalized. In later project stages, and particularly for cases of sensitive scientific instrumentation, contamination analysts should not necessarily decouple outgassing and transport models – examples of such models are outlined by JPL CC for e.g. the Europa Clipper fly-by mission [4] and the cryogenic SPHEREx telescope [9].

2.1. *Framework for an integrated, early-look outgassing model*
The early-look Gateway molecular contamination model workflow is described in figure 2 in terms of its inputs, methodologies, and results and deliverable products. The model comprises a series of steps that *take in* information about Gateway module geometries and outgassing source and sensitive receiver surface definitions, and *deliver* workbooks that calculate molecular contaminant deposition effects onto receiver surface as functions of user inputs. Each sequential set of inputs, methods, and results will be described in the following subsections of this paper.

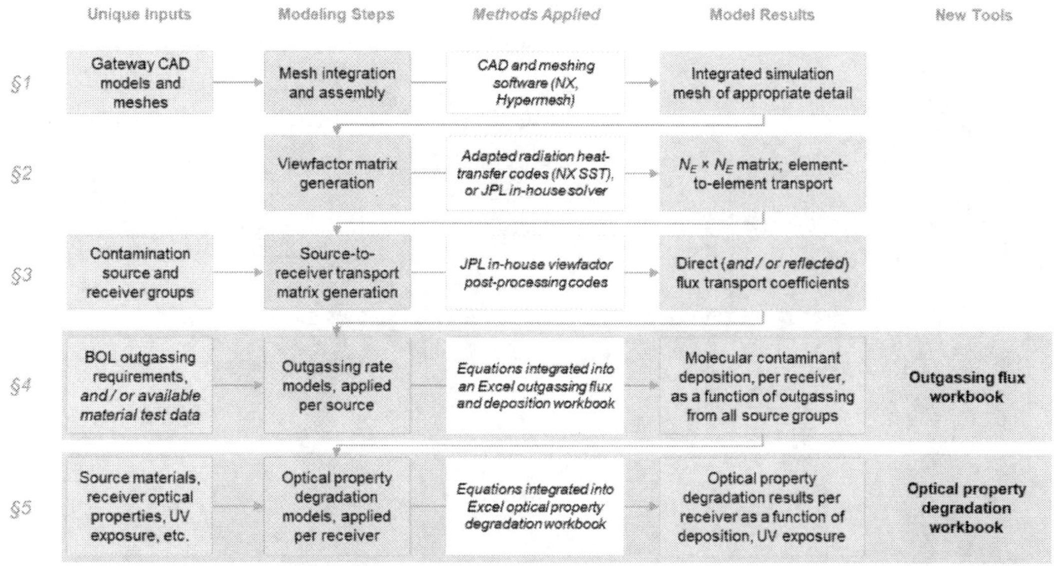

Figure 2. Model Workflow. Provided Gateway module geometries and source and receiver groupings, the JPL CC team develops workbook products (highlighted in purple) that enable rapid calculations of Gateway external molecular contaminant deposition and related optical property degradation effects.

2.1.1. Mesh integration and assembly. The first set of unique inputs for early-look contamination analyses are geometries for each Gateway module, combined with a specified analysis configuration defining module connectivity. Geometric inputs may include low- and high-fidelity CAD models and meshes developed for other spacecraft design analyses, e.g. structural NASTRAN meshes. These separate inputs must be assembled into an integrated simulation mesh in the analysis configuration of interest that retains important geometric features, but is not 'too detailed' for viewfactor calculations.

Outgassing source and sensitive surface receiver groups must be represented accurately in shape and area, as must intermediate shadowing and reflective surfaces. Total mesh element count (N_E) can be a significant constraint – viewfactor matrix storage size scales with $N_E^{\,2}$, while calculation and post-processing of the matrix tends to scale between $N_E^{\,2}$ and $N_E^{\,3}$. Other mesh constraints may include the enforcement of mesh *water-tightness* and of appropriate mesh size and angular resolution for results visualization and interpretation. For this purpose, a mesh is said to be *water-tight* if it forms a set of closed 3-D volumes without element self-intersection and without non-manifold or unmatched edges – this condition avoids spurious transport of molecular contaminant. Given input geometries for each Gateway module, to meet such criteria the JPL CC team may opt on a case-by-case basis to either *(1)* generate new reduced CAD models and meshes from scratch, *(2)* mesh input models or re-mesh input meshes directly, *or (3)* use input meshes as-is. A detailed description of geometric model reduction and meshing steps for free-molecular transport applications is provided by Martin et al. (2020) [3].

The integrated simulation mesh for the representative Gateway analysis addressed in this paper is shown in figure 3 alongside a CAD model that describes the component modules and configuration of interest. In this example a mesh of $N_E = 30{,}522$ triangular elements was generated from a combination of CAD model and NASTRAN mesh inputs. New, custom models were developed from high-fidelity reference models for the dominant source and receiver group modules HALO (the *Habitation and Logistics Outpost*) and I-HAB (the *International Habitation Module*). Meshes for the remaining modules – including PPE (the *Power and Propulsion Element*), ESPRIT-RM (the *European System Providing Refueling, Infrastructure and Telecommunications - Refueling Module*), and a representative airlock module – were developed by re-meshing structural NASTRAN meshes to refine size resolution and to enforce water-tightness. Orion and Dragon XL docking vehicles were also assessed. More information on Gateway, its modules and configurations is available at NASA's Gateway Overview [1].

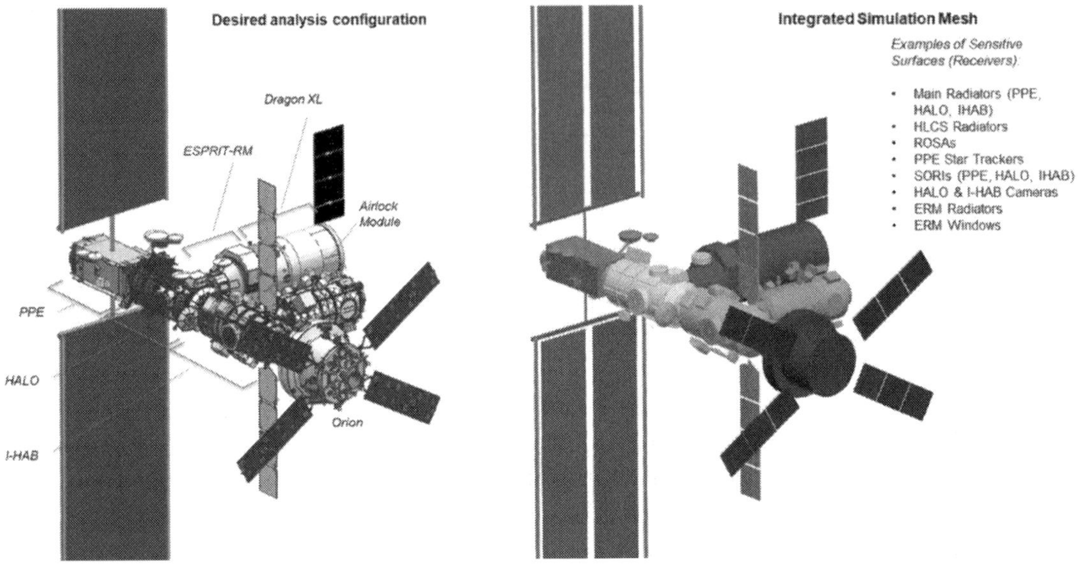

Figure 3. An Integrated Simulation Mesh for Gateway. At left, an input set of CAD models for Gateway in the representative analysis configuration addressed throughout this paper; at right, a corresponding integrated simulation mesh which represents essential outgassing source and sensitive receiver group surfaces and satisfies size resolution and water-tightness constraints.

2.1.2. Viewfactor matrix generation. Predictions of molecular contaminant deposition onto Gateway's sensitive surfaces require an understanding of free-molecular transport processes in the space station's induced environment. Ambient pressures experienced by Gateway on orbit will be sufficiently low for outgassing transport to be nearly-collisionless, with mean-free-paths between inter-molecular collision greater than 1 km and Knudsen numbers *(Kn)* greater than 100 relative to Gateway's length. Neutral molecular contaminant self-scatter and ambient-scatter processes are therefore treated as negligible for the purposes of this analysis in comparison to free-molecular direct and reflective transport vectors.

Treatments for rarefied gas dynamic transport processes were developed for the ISS program [10] and have more recently been applied by JPL CC to predict ambient-scatter return flux processes for the NASA Europa Clipper mission during Europan exospheric encounters [11]. If needed, these tools could also be applied to Gateway; however, and consistent with our assumptions for Gateway, ISS ambient-scatter return flux analyses found self- and ambient-scatter vectors to be negligible in comparison to free-molecular direct and reflective vectors for that space station [10]. We conclude that over short distances in the vicinity of Gateway, the transport of neutral outgassing products is well-modeled as free-molecular and ballistic. The effects of molecular contaminant ionization and transport under electric and magnetic field body forces have been assessed separately by JPL CC and were also found to be second-order for Gateway in comparison to neutral direct and reflective vectors [12].

In this model, which decouples transport from surface adsorption and desorption processes, each surface interaction is treated *either* as a permanent molecular adsorption onto a designated sensitive surface, *or* an instantaneous re-equilibration to and reemission from a surface at a cosine-distributed angle. This early-look model framework must neglect time-dependent adsorption and desorption processes which require detailed input information about material properties, outgassing rates, and time-dependent surface temperature predictions throughout a mission life. In later mission design phases, it may become important to consider such effects – recent such applications by JPL CC include Europa Clipper [4] and the SPHEREx telescope [9]. Anticipating contamination sensitivities and the pending availability of Gateway materials data, preparatory work is already underway to evolve the early-look model framework described in this paper to align with the multi-species diffusion-driven outgassing models detailed by Alred et al. (2023) [13].

Free-molecular transport can be computed on a geometric mesh using viewfactors, analogous to models for radiative heat transfer. In this approach the fraction of mass transported from each geometric element to each other element in the Knudsen sense (cosine angular emittance) may be computed and stored in a viewfactor matrix of size $N_E \times N_E$. The viewfactor matrix can be used to calculate full-spacecraft direct and reflective (i.e. re-emitted through surface equilibration and desorption) molecular contaminant transmission fractions [4]. Radiative heat transfer codes can be used to generate viewfactor matrices (e.g. NX SST); JPL CC has also developed in-house HPC viewfactor matrix calculation and post-processing codes optimized for outgassing transport problems [3,4]. A key advantage of the viewfactor scheme is that, once computed and stored, a viewfactor matrix can be used to calculate transport around a static geometric configuration for any combination of source and receiver elements, or for groups of elements. A disadvantage is that a new matrix is needed for each geometric configuration. An important consideration is matrix storage size: for example, the $N_E = 30{,}522$ matrix corresponding to the integrated mesh of figure 4 has over 930 million possible values. While such external transport matrices tend in practice to be no more than 10 % full, storage of one such text matrix requires on the order of 1 gigabyte.

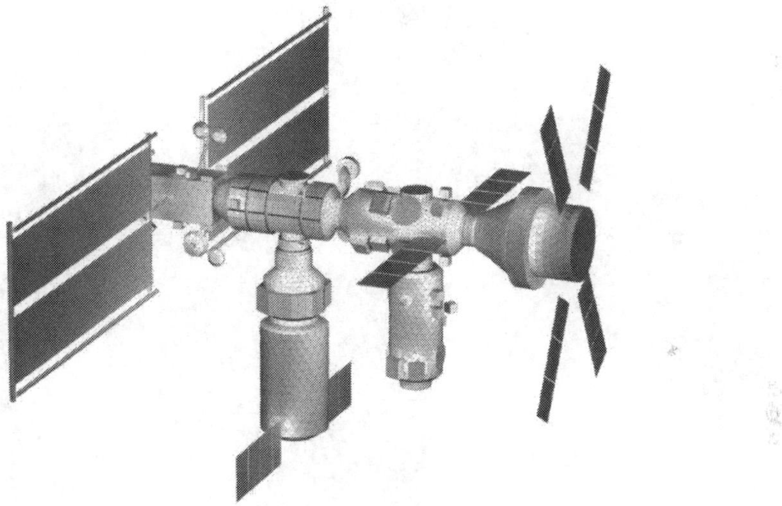

Figure 4. Viewfactor Result, Direct Flux. In this example all surfaces outgas at a uniform rate per area and adsorb all incident contaminant (nominal colormap). This is a representation of the full $N_E \times N_E$ viewfactor matrix that describes mass transport from each facet element to all other elements, and is used to compute mass transport from each outgassing source to each sensitive surface receiver.

2.1.3. Source-to-receiver transport matrix generation. The Gateway molecular contamination model enables users to distinguish the effects of hundreds of distinct contamination sources onto hundreds of distinct contamination receivers distributed across Gateway's range of modules and docking vehicles. The model framework can therefore incorporate an arbitrary number of source and receiver groups – in the example of this paper, $N_S = 144$ source element groups and $N_R = 146$ receiver element groups are tracked for a total of $N_S \times N_R = 21{,}024$ source-receiver group transmission coefficients.

Source element groups represent unique sources of outgassed and desorbed molecular contamination that may represent some combination of a uniform set of materials, a uniform geometric region, or a uniform outgassing rate requirement. For instance, each side of the PPE ROSA (*Roll Out Solar Array*) is identified as a source – one such group, the PPE ROSA +Z +X face, is highlighted green in figure 5.

IOP Conf. Series: Materials Science and Engineering 1287 (2023) 012002 doi:10.1088/1757-899X/1287/1/012002

Likewise, *receiver element groups* each represent sensitive surfaces for which it is important to track molecular contaminant deposition so as to characterize a performance degradation mode. For example, Gateway module and docking vehicle radiator and solar array panel faces are each tracked as unique receiver groups in this model, as are all distinct faces of each Gateway external payload box.

In figure 5, the PPE ROSA +Z +X face source group is shown to outgas and to deposit a normalized proportion of its outgassed mass onto all Gateway facet elements in its line-of-sight, i.e. as direct flux. The source group has a nominal uniform outgassing rate per unit area, and therefore receiver element colors represent the fraction of total mass outgassed from the PPE ROSA +Z +X face that arrives at and adsorbs onto each receiver element as direct flux, per unit receiver area (and per unit time, for a given outgassing time interval). Therefore, for each gram of mass outgassed per second across the PPE ROSA +Z +X source group, the colorbar in figure 5 would range from 10^{-5} to 10^{-2} g/m^2/s. A comparable result exists for each of the other 143 Gateway molecular contamination source groups treated in this example.

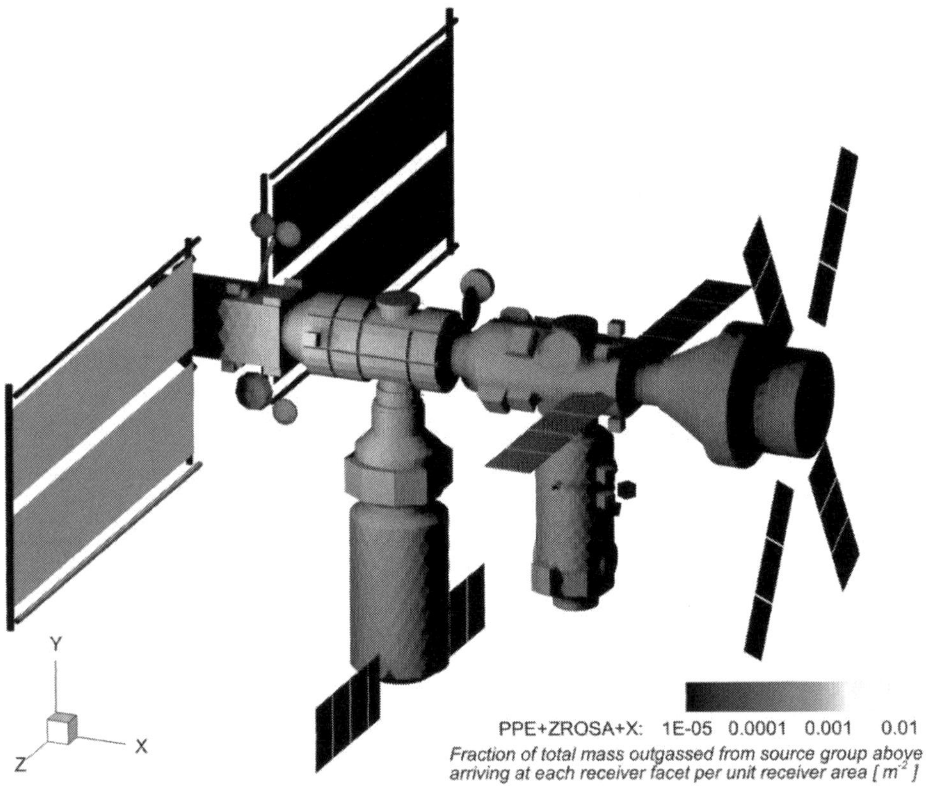

Figure 5. Visualization of Direct Flux from a Single Outgassing Source. In this example, a source group representing the PPE ROSA +Z +X face (highlighted in green) outgasses and deposits mass in the direct sense onto all Gateway elements in its line-of-sight.

Figure 5 shows the normalized free-molecular transmission of outgassing product *from* one set of source elements *to* all possible receiver elements for a mesh with $N_E = 30{,}522$. Receiver element colors represent the fraction of total mass outgassed from the source that arrives at each element per unit area in units of m^{-2}. Receiver element groups are collections of elements across which a normalized mass transmission from each source is averaged. In this fashion, an $N_E \times N_E$ viewfactor matrix may be processed into an $N_S \times N_R$ matrix of *transmission coefficients*, each of which describes the ratio of outgassed or desorbed mass transmitted from a source element group onto a receiver element group.

The mathematical details of this processing are not novel, and the treatments for direct and reflective flux applied for Gateway analyses are described in Anderson et al. (2023) [4]. In the example of this paper the source-receiver transmission coefficient matrix holds 21,024 possible entries – sufficiently compact to be stored and operated on within an Excel workbook, particularly by comparison to the corresponding viewfactor matrix of over 930 million possible entries. An important consideration in selecting source and receiver element groupings emerges from the compression of information inherent in processing the viewfactor matrix into the transmission coefficient matrix – each source-receiver pair of arbitrary geometric complexity, and with arbitrary intermediate shadowing surfaces, is reduced to a single coefficient value. This makes it straightforward to calculate incident molecular fluxes from all sources averaged across each receiver, but precludes the calculation of local minimum and maximum flux values. Receiver definitions and element groupings must be chosen carefully such that summing incident molecular fluxes from all sources averaged across each receiver is reasonable – therefore, receivers ought to be uniform in geometric orientation and in shape, and relatively small. For similar reasons, source groups share these considerations and should also be uniform in outgassing rate (implying uniformity in materials selection and in temperature). If an optical property degradation model is applied to a receiver, that group should have uniform initial properties and solar UV exposure.

2.1.4. Outgassing rate models, per source. In the Gateway early-look molecular contamination model, each source group releases a total amount of outgassed mass corresponding to an outgassing rate or rate requirement integrated over a time representative of module and docking vehicle arrivals and departures. Time- and temperature-dependent outgassing rate decay models are a rich area of study [4,13] that could inform future versions of a Gateway molecular contamination model. However, the Gateway program does not yet have many detailed or finalized inputs available for materials characterization or thermal environment predictions necessary to populate such models, and therefore the early-look model incorporates only a simple form of outgassing time decay dependent on an assumed decay limit and a set of bakeout requirement rates and times.

A conservative, diffusion-limited form of outgassing time decay may be obtained from a one-dimensional diffusion model [14] in which outgassing rate scales with the inverse square root of time: $r \propto t^{-1/2}$. For long times at high temperatures outgassing will approach an exponential decay. Without more specific material properties including a diffusion coefficient D, contaminant concentration C_0 and material thickness L, it is only possible to assume a generic functional form for outgassing rate. A conservatism inherent in the $t^{-1/2}$ formulation emerges from an implicit assumption that source material acts as an infinite source of contaminant – while the rate decays in time, its improper integral diverges.

Assume an outgassing rate of form $r(t) = c\, t^{-1/2}$ with constant c dependent on material properties. To extrapolate in time, it is necessary to know when an outgassing rate is measured: for instance, at the end of a thermal bakeout, t_{bake}, a requirement outgassing rate r_{req} may be met. With this information, it is possible to solve for c in terms of the assumed requirement rate measurement and to derive a general form for outgassing as $r(t) = r_{req}\, {t_{bake}}^{1/2}\, t^{-1/2}$. Thus, the total outgassed mass for a source material from end-of-bakeout to end-of-mission can be estimated by integrating as:

$$m = \int_{t_{bake}}^{t_{mission}+t_{bake}} r(t)\, dt = 2\, r_{req}\sqrt{t_{bake}}\left(\sqrt{t_{mission}+t_{bake}} - \sqrt{t_{bake}}\right). \quad (1)$$

With the total outgassed mass calculated per-source-group, the amount of mass incident upon each receiver group can be attained by multiplying through by the transmission coefficient between each source-receiver group pairing. Incident masses from each source group are then summed to calculate a total mass incident onto each unique receiver group. That total mass can be divided by an assumed contaminant density and a receiver group surface area to estimate a *molecular contamination deposit thickness*, a key deliverable result which can also be input into an optical property degradation model.

2.1.5. Optical property degradation, per receiver. A physics-based optical property model for absorptivity change as a function of molecular contaminant deposition and ultraviolet (UV) radiation exposure was developed for the ISS program [7,8] from spaceflight experiments and ground-based test data, including data from the Shuttle-Mir Phase-I Risk Mitigation Experiment Program and extensive laboratory testing conducted by the Boeing company and NASA. This model calculates surface absorptivity as a function of initial solar adsorption, α_0; solar adsorption limit, α_∞; UV exposure in equivalent sun hours, *ESH(t)*; total contaminant deposition, *d(t)*; and a contaminant-mixture-dependent solar adsorption rate change function, K_{mix}:

$$\alpha(t) = \alpha_\infty \left(1 - e^{-2\,K_{mix}\sqrt{ESH(t) \times d(t)}}\right) + \alpha_0\, e^{-2\,K_{mix}\sqrt{ESH(t) \times d(t)}}. \quad (2)$$

This predictive model has been extensively used by the ISS Program in support of ISS operations, science and life extension studies, and its predictions have been correlated with flight observations and operating temperature measurements from ISS thermal control systems and have shown good agreement with flight data [15]. In the Gateway analysis framework this optical property model is implemented per receiver group as a function of total molecular contaminant depositions (calculated as described in the prior section of this paper) as well as assumed input values for receiver beginning-of-life absorptivity, contaminant mixture fraction, and total UV exposure in ESH. Tables 1 and 2 provide an example of a representative calculation performed with the outgassing and optical property workbooks.

2.2. *An example of early-look outgassing model results for Gateway*

Early-look materials outgassing studies are very useful to inform requirement development, identify potential contamination drivers, and assess end-of-life molecular contamination levels. Gateway has successfully used the outgassing flux workbook for multiple early-look outgassing studies in which outgassing rates are applied to Gateway sources (e.g. element surfaces) to evaluate potential molecular deposition levels to various Gateway receiver surfaces of interest (e.g. radiators). These early-look studies consider time-phased element arrivals based on the Gateway assembly timeline and account for outgassing rate decay with time. Outgassing rates are currently applied parametrically, or with reference to heritage characterizations from past spacecraft. Actual outgassing rate source terms will be developed as the Gateway design matures and element-level contamination definition data becomes available.

An example of an early-look outgassing result for generic source and receiver groups is shown in table 1. Early-look materials outgassing results can also be used in the optical property model to assess the effect of material outgassing deposition on absorptivity. Table 2 shows a representative optical property result for receiver groups based on deposition results and estimated sun exposure.

Table 1. Example of Molecular Outgassing Contaminant Deposition from Multiple Sources [Å].

	Gateway Module 1	Gateway Module 2	Gateway Module 3	Gateway Module 4	Gateway Module 5	Visiting Vehicle 1	Visiting Vehicle 2	***Total***
Receiver 1	120	274	29	0	0	80	0	*503*
Receiver 2	120	262	3	199	79	182	138	*983*
Receiver 3	311	100	3	0	0	32	11	*457*

Table 2. Example of Contaminant-Induced Absorptivity Changes.

	Beginning-of-Life Absorptivity [α_0]	Total Molecular Deposition [Å]	Equivalent Sun Hour Exposure [ESH]	***End-of-Life Absorptivity [α]***
Receiver 1	0.09	503	5106	*0.14*
Receiver 2	0.09	983	832	*0.12*
Receiver 3	0.09	457	5716	*0.14*

2.3. *Outgassing model conclusions*

Multiple Gateway system studies have been performed using these contamination and optical property degradation workbooks. Workbooks have been updated to incorporate design changes, and to add new sensitive surface and outgassing source groups. As materials selection and characterizations develop, outgassing rate and species information can be incorporated into this framework. However, if e.g. transient adsorption and desorption processes; multispecies outgassing; or thermal effects become important to model, techniques that do not decouple transport and surface interactions can be applied.

The representative analysis described in this paper is inherently conservative in its assumption of maximal sticking coefficient (i.e. 100% sticking of incident contaminant molecules) onto receiving surfaces in its direct flux mode, and also in its assumption of complete re-emittance of contaminant (i.e. 0% sticking) from non-receiver surfaces in its direct+reflective flux mode. Such sticking coefficients can be set freely on an element-by-element or group-by-group basis during the calculation of the source-to-receiver transport matrix (§2.1.3) and could be empirically motivated by e.g. QTGA (*Quartz crystal microbalance Thermo-Gravimetric Analysis*) data for a known outgassing material and combination of source and receiver temperature. However, this would require that a distinct source-to-receiver transport matrix be generated for each set of sticking coefficients mapping to every interesting combination of source and receiver element temperatures – which is incompatible with the format of the early-look model workbooks, each of which operates on a single source-to-receiver matrix.

At this point it is necessary to consider more sophisticated and expensive analysis capabilities that do not decouple the transport matrix solution from the application of per-element temperature boundary conditions and temperature- and time-dependent outgassing rates. Anderson et al. (2023) describe a viewfactor-based capability developed at JPL that uses mission thermal predicts and outgassing test data to set outgassing rates and surface sticking coefficients over time across the duration of a spacecraft mission [4]. Likewise, Alred et al. (2023) describe a JPL application of a COMSOL viewfactor-based method to the solution of temperature- and time-dependent water outgassing and condensation processes during the early-mission commissioning phase of the SPHEREx telescope [9]; this model incorporates a vapor-pressure-derived residence time for water, but the method could be extended to multiple species. Brieda et al. (2022) describe a similar application of the Lagrangian CTSP analysis software to temperature- and time-dependent water outgassing and condensation problems for the James Webb Space Telescope (JWST) during its deployment sequence [16]. As the Gateway design matures and test data becomes available it may be necessary to evolve from the simplified early-look workbook model to such a fully-coupled method, and JPL is prepared for this eventuality.

3. Plume Effects Modeling

The empirical bipropellant plume contamination models of Soares et al. (2002, 2015) developed for ISS report permanent molecular contaminant deposition, droplet surface fluxes, and surface erosion effects [5,6]. These models were intended to inform engineering design, e.g. to influence trades on feathering angles for rotatable ISS sensitive surfaces, and to quantify liquid droplet distribution and physics-based properties in plumes. Over time, improvements have been implemented to the initial erosion and particle flux models, informed by flight hardware experiments like PIC (*Plume Impingement Contamination*) and SPIFEX (*Shuttle Plume Impingement Flight EXperiment*) and by computational analysis. Additional work was performed to generate particle induced impact damage data with the SPHINX Smooth Particle Hydrodynamics code from Los Alamos National Laboratory (LANL) [6].

Gateway Induced Environments and JPL CC have also worked together to perform parametric thruster pluming assessments for Gateway module thrusters. JPL CC conducted parametric thruster plume analyses on the Gateway system mesh using heritage ISS plume models. Deposition results are provided in workbook format by thruster and receiver surface to permit scaling of thruster on-time assumptions for analysis scenarios of interest. This workbook complements the outgassing workbook, in that it shares its simulation mesh and receiver groupings. Future work may address alternate pluming scenarios, alternate Gateway configurations, and environmental effects that may influence deposition and material transport during thruster operations. Figure 6 shows a normalized result for permanent molecular contaminant deposition induced by a single plume firing during a hypothetical Orion docking.

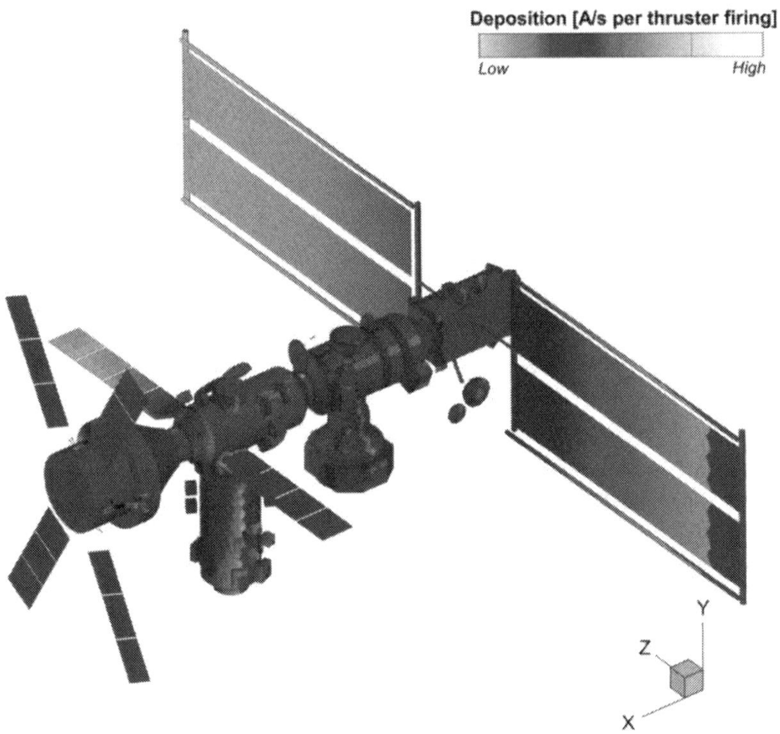

Figure 6. Empirical Plume Modeling for Gateway. A normalized result for molecular deposition.

4. Conclusions: The Status and Future of Gateway Molecular Contamination Modeling

The Gateway Induced Environments Team has collaborated with the Gateway Thermal Team and JPL CC on 'early-look' models for outgassing, molecular deposition, and optical property degradation. Induced Environments and JPL CC have also worked together on thruster pluming assessments with ISS heritage models. Analysis results have been integrated into workbook products which have been used in multiple Gateway system studies and revised to keep up-to-date with Gateway's design.

Spacecraft and space stations require sophisticated physics-based models to track and mitigate contaminant generation, transport and deposition processes and harmful contamination-induced effects across mission lifecycles [2]. Models developed for the International Space Station (ISS) continue to have relevance to NASA's Gateway program – particularly for problems of optical property degradation [7,8,15], neutral contaminant self- and ambient-scatter return flux [10], and thruster-plume-induced contaminant deposition and erosion [5,6]. JPL CC has leveraged this ISS heritage and an analysis framework developed for the Psyche Asteroid Mission [3] to provide 'early-look' models of outgassing and thruster pluming contaminant deposition effects for Gateway.

As Gateway's design matures and analysis boundary conditions become available, more detailed modeling approaches may become possible and beneficial. Future work for Gateway may incorporate recent advances by JPL CC in processing materials testing data into multispecies diffusion-driven outgassing models [13]; in assessing contaminant ionization and natural environment effects [12]; in modeling coupled outgassing and free-molecular transport processes as functions of spacecraft thermal predictions [4]; and in simulating thruster plume effects in rarefied gas dynamic regimes.

Acknowledgements
This work was performed at the Jet Propulsion Laboratory, California Institute of Technology, under a contract with the National Aeronautics and Space Administration.

High Performance Computing (HPC) resources used in this investigation were provided by funding from the JPL Information and Technology Solutions Directorate. The authors would like to thank Drs. John R. Anderson and John M. Alred of JPL CC Engineering (353D) for their development of models and techniques applied here to Gateway.

References

[1] NASA Gateway Overview https://www.nasa.gov/gateway/overview 2021
[2] Soares C E and Mikatarian R 2003 Understanding and control of external contamination on the International Space Station *Proc. 9th Int. Symp. on Materials in the Space Environment*
[3] Martin M G, Hoey W A, Alred J M and Soares C E 2020 Novel contamination control model development and application to the Psyche Asteroid Mission *Proc. IEEE Aerospace*
[4] Anderson J R, Alred J M and Hoey W A 2023 Modeling molecular outgassing transport and deposition for spacecraft operating in vacuum *Proc. 32nd Int. Symp. Rarefied Gas Dynamics*
[5] Soares C E, Barsamian H and Rauer S 2002 Thruster plume-induced contamination measurements from the PIC and SPIFEX flight experiments *Proc. SPIE* vol 4774 pp 199-209
[6] Soares C E, Olsen R, Steagall C A, Schmidl W D, Myers B, Mikatarian R, Koontz S and Worthy E 2015 Improvements in modeling thruster plume erosion damage to spacecraft surfaces *Proc. 13th Int. Symp. on Materials in the Space Environment*
[7] Alred J W and Soares C E 2001 Solar absorptivity as a function of spacecraft external contamination *Proc. 33rd SAMPE Technical Conference*
[8] Huang A Y, Schmidl W D and Soares C E 2008 Characterizing the optical performance of AZ93 with a fluoropolymer overcoat under ultraviolet exposure *Proc. SPIE* vol 7069
[9] Alred J M, Moore B D, Susca S, Ricchiuti V, Rocca J M and Soares C E 2023 Modeling contaminant outgassing and free molecular transport processes for the cryogenic SPHEREx observatory *Proc. 32nd Int. Symp. Rarefied Gas Dynamics*
[10] Roussel J F, Soares C E and Schmidl W D 2002 Development and validation of a new return flux model for the International Space Station *Proc. SPIE* vol 4774 pp 172-183
[11] Soares C E, Hoey W A, Anderson J R and Ferraro N W 2019 Spacecraft return flux considerations for missions targeting detection of organics with mass spectrometers *Proc. 70th Int. Astronautical Congress*
[12] Anderson J R, Soares C E and Steagall C A 2023 Assessment of contamination ionization due to interaction with the natural environment for Gateway *Proc. IEEE Aerospace*
[13] Alred J M, Roussel J F, Soares C E, Lansade D, Anderson J R, Sinha N, Faye D and Guillaume R 2023 Development of multispecies model for comprehensive characterization of outgassing molecular constituents *Proc. 15th Int. Symp. on Materials in the Space Environment*
[14] Liu C K and Glassford A P M 1977 Kinetics data for diffusion of outgas species from RTV 560 *Proc. NASA GSFC 9th Conf. on Space Simulation*
[15] Soares C E, Mikatarian R, Schmidl W D, Boeder P A and Alred J W 2006 Optical characteristics of materials in a space environment *Proc. 10th Int. Symp. on Materials in the Space Environment*
[16] Brieda L, Laugharn M, Woronowicz M, Henderson-Nelson K, May C and Wooldridge E 2022. Numerical study of water ice and molecular contamination build up during JWST deployment *Proc. SPIE* vol 12224

Metamaterial-based smart and flexible Optical Solar Reflectors

M Gaspari[1], S Mengali[1], M Simeoni[1], A Urbani[1], O L Muskens[2a], K Sun[2a], I Zeimpekis[2b], C H de Groot[2c], A Bialy[3], I Czolkos[3], L Kildebro[3], B Alpat[4], G Bartolini[5], M Jamalipour[5], J Frolec[6], T Kralik[6], F Tessarin[7], M Gottero[7] and T Schillaci[7]

[1]Consorzio CREO, SS 17 località boschetto, 67100 L'Aquila (Italy)
[2a]Physics and Astronomy, University of Southampton, University Road, SO17 1BJ Southampton (UK)
[2b]Optoelectronic Research Center, University of Southampton, University Road, SO17 1BJ Southampton (UK)
[2c]Electronics and Computer Science, University of Southampton, University Road, SO17 1BJ Southampton (UK)
[3]NIL Technology, Haldor Topsøes Allé 1, 2800 Kgs. Lyngby (Denmark)
[4]INFN sezione di Perugia, Via Alessandro Pascoli, 06123, Perugia, Italy
[5]Beamide srl, Via Cristoforo Colombo 19/I 06127 Perugia (Italy)
[6]Institute of the Scientific Instruments of the CAS, Kralovopolska 147, 612 00 Brno (Czech Republic)
[7]Thales Alenia Space, Strada Antica di Collegno 253, 10146 Torino (Italy)

sandro.mengali@consorziocreo.it; mirko.simeoni@consorziocreo.it

Abstract. In the frame of projects funded by the European Commission and the ESA, we developed a new type of Optical Solar Reflector (OSR) that combines the flexibility and easy handling of Second Surface Mirrors with the temperature-variable emittance necessary to ensure both effective radiative cooling in the hot phase and reduced heat losses in the cold phase. The new smart OSR consists of a metamaterial coating deposited on Kapton film. The coating is made of two functional blocks, namely a variable emitter topped by a dielectric solar reflector. The variable emitter is a Metamaterial Perfect Absorber designed for strong and broadband plasmonic resonance absorption in the thermal IR. It consists of a metal back-reflector, a dielectric spacer, and an array of doped VO_2 thermochromic micro-antennas that are switched-off when the temperature drops below the metal-to-insulator transition point. The solar reflector is a wideband dielectric filter made of materials that are transparent across the entire spectrum from the VIS to the thermal IR. All the layers of the two blocks are deposited by standard vacuum techniques, while the array is patterned by Nanoimprint Lithography, a technique that is often performed at the wafer level but allows for up-scaling via roll-to-roll or roll-to-plate production setups. The paper reports on the characterization and testing of samples of size up to 100 mm x 100 mm, at the Beginning of Life and after thermal, humidity, irradiation and handling tests.

1. Introduction

Optical Solar Reflectors (OSRs) are materials in the form of tiles or films that, glued onto radiator panels, are used to regulate the heat exchanged radiatively between the spacecraft and the outer space.

At a good extent, the behavior of an OSR is described by two thermo-optical parameters: Solar Absorptance α and InfraRed Emittance ε. Traditional OSRs have low α (≤ 0.20) and high ε (≥ 0.80), at *any* temperature. However, high ε is beneficial when the spacecraft is hot, because it prevents overheating, but detrimental when the spacecraft is cold, because it causes unwanted heat losses. The main objective of this research is to develop a smart OSR that features temperature-variable emittance, that is:

- High ε (together with low α) when the spacecraft is 'hot';
- Low ε when the spacecraft is 'cold';
- Transition from the hot to the cold state at ≈ 25 °C.

A second objective is to combine smart behavior as defined above with mechanical flexibility, to ease the handling and application of the OSR, particularly in the case of non-flat radiator panels.

The study of smart surfaces with temperature variable emittance has been an active field of research for the last 20 years. In the space domain, the advantage of a smart-OSR can be better appreciated at the system than at the component level: reduced heat losses in the cold phase would translate into reduced electric budget at the heaters, and reduced masses deriving from the integration of louvers. Until recently, smart-OSRs appeared most attractive for small satellites, that cannot be equipped with massive and power-hungry emittance-control systems. In the last few years, the relevance of temperature variable emittance has emerged also for telecom satellites which utilize electric propulsion: the transition to the final GEO orbit is a cold case that lasts several months, and in which the available electrical power would preferably be dedicated to electrical propulsion rather than to heating the satellite.

Initially, the quest for variable emittance surfaces concentrated on perovskite manganites [1], primarily in the form of bulk material. However, it was soon realized that manganites have very high crystallization temperatures above 800 °C, and a bit too low transition temperature (T_{MIT}) below – 20 °C. The focus was then shifted to continuous-thin-film coatings based on Vanadium Dioxide (VO_2)[2], which requires lower crystallization temperatures, and can provide transition at room temperature if properly doped with one of several metals, of which Tungsten (W) is the most popular [3][4]. One striking characteristic of most studies in this field, however, is that they focus on achieving strong emittance contrast $\Delta\varepsilon$ between the hot and the cold state, but pay little attention to solar absorptance, which is generally very high. A rare exception is a recent work by Dudon *et. al.*[5], who developed a dielectric solar reflector on top of a VO_2-based variable emitter and fabricated samples on 40 mm x 40 mm Silicon tiles having $\Delta\varepsilon = 0.38$, $\alpha = 0.44$, and $T_{MIT} < 60$°C. Starting from 2016, our research team began to explore the potential advantages in terms of α and $\Delta\varepsilon$ deriving from the use of metamaterial coating architectures described in previous articles[6][7][8], of which this paper is an experimental followup.

2. Proposed solution

The architecture of the proposed OSR is shown in Figure 1. It consists of a 75 μm poly-imide film and of a coating grown by Physical Vapour Deposition, having total thickness ≈ 3 μm and being made of two functional blocks: a temperature variable emitter and a dielectric solar reflector.

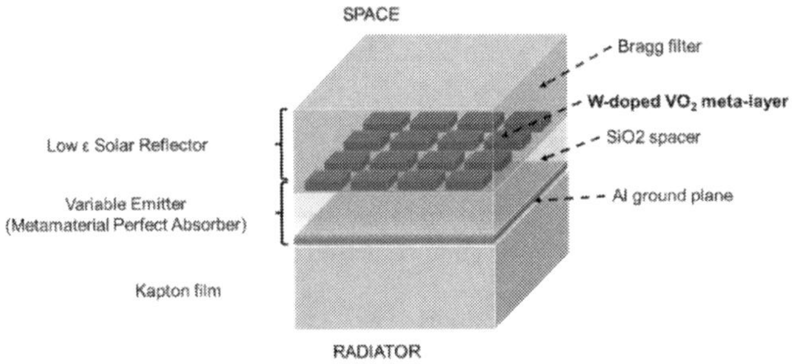

Figure 1. OSR architecture.

The first block comprises three layers: an Aluminum (Al) ground plane, a Silicon Dioxide (SiO_2) spacer, and a so-called 'meta-layer' of micron-size W-doped VO_2 (WVO_2) square islands and gaps. In the jargon of plasmonics, this three-layer structure is called a Metamaterial Perfect Absorber (MPA), the operating principle and detailed design of which are described elsewhere [7]. Here, it is worth to note that our MPA is very similar to a classical VO_2-based Fabry Perot cavity as the one for instance utilized in [5], the main difference being the presence of a meta-layer in the place of a continuous one. The meta-layer may have several beneficial effects on thermo-optical performance, of which reduced solar absorptance is the most important. This effect is easy to understand considering that WVO_2 has a strong absorption over the entire UV-VIS spectrum. Consequently, creating a meta-layer means removing high-α material from the stack. Experimentally, it was verified that smart OSR samples built with the meta-layer have 0.05÷0.10 lower α than analogous samples built with a continuous WVO_2 layer of the same thickness.

The second functional block is a broadband Bragg filter made of about 30 layers and designed to reflect solar radiation in the range from 400 to 1000 nm wavelength. The final choice of materials and growth conditions for the filter was the result of a long process of trials and trade-offs: on the one hand materials must have low emittance, otherwise they will dampen the emittance contrast generated by the MPA; on the other hand, they must have good resistance to bending and humidity, whereas most transparent low emittance materials are fragile and hygroscopic.

3. Test samples and thermo-optical properties at the Begin of Life

In the course of the study, tens of samples were built in two distinct formats: smaller samples 25 mm x 25 mm for ageing tests, and larger samples 100 mm x 100 mm for handling tests and demonstration, as the one in Figure 1, left. The surface appears uniform to the naked eye, but the matrix of square WVO_2 micro-antennas becomes visible when viewed under a microscope (Figure 2, right). The pattern is created by Nanoimprint Lithography, a technique here implemented on the wafer scale, yet scalable to larger areas at fair costs for instance via roll-to-plate set up [10]. Fabrication of the meta-layer proceeds in four steps: first imprint of the photoresist mask onto the surface of the dielectric spacer, then growth of the WVO_2 layer at room temperature, lift-off, and short treatment in vacuum at high temperature. Annealing temperature and time are trade-off values that allow the formation of WVO_2 micro-crystals and the activation of thermochromism while avoiding damage to the poly-imide substrate.

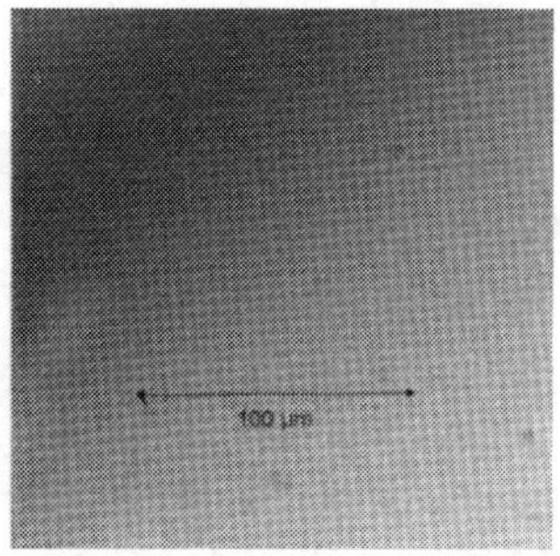

Figure 2. 10 cm × 10 cm test sample and its surface observed with optical microscopy.

The thermo-optical (T.O.) properties of the samples at the Beginning of Life (BoL) are given in Table 1. Figure 3 shows on the left a typical plot of normal emittance ε_N(T) and, on the right, examples of normal IR reflectance spectra taken at different temperatures and used to calculate ε_N(T). As typical of VO_2 material systems, emittance has some hysteresis in the transition zone. Therefore, the values of T_{MIT} and of the transition width ΔT_{MIT} are determined by taking the flex points FP1 and FP2 along the two branches of the plot, then calculating $(FP1 + FP2)/2$ and $|FP1 - FP2|$, respectively. On a few samples, the hot branch of the emittance curve was also measured by a calorimetric method [9]. Calorimetric emittance $\varepsilon_C(T)$ has a S shape similar to normal emittance $\varepsilon_N(T)$ yet slightly shifted to lower values. Unfortunately, $\varepsilon_C(T)$ could not be used to determine ε_{HOT} and $\Delta\varepsilon$, because the calorimeter cannot be heated to temperatures above 45°C and therefore cannot reach the hot end-point of the transition.

Table 1. T.O. properties BoL calculated on a set of 8 samples. Subscript HOT refers to quantities in the hot state. Subscript N indicates that ε values are derived from normal reflectance spectra in the range 3-21 µm as per ECSS.

	Average value	*Variance*
α_{HOT}	0.24	0.02
$\varepsilon_{N\ HOT}$	0.77	0.03
$\Delta\varepsilon_N$	0.32	0.02
T_{MIT}	27°C	8°C
ΔT_{MIT}	26°C	11°C

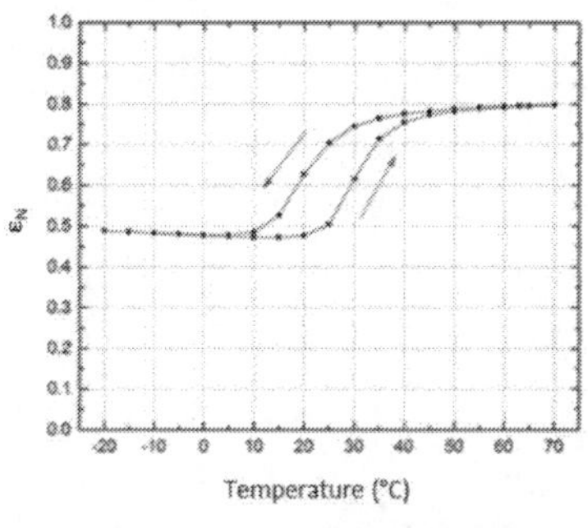

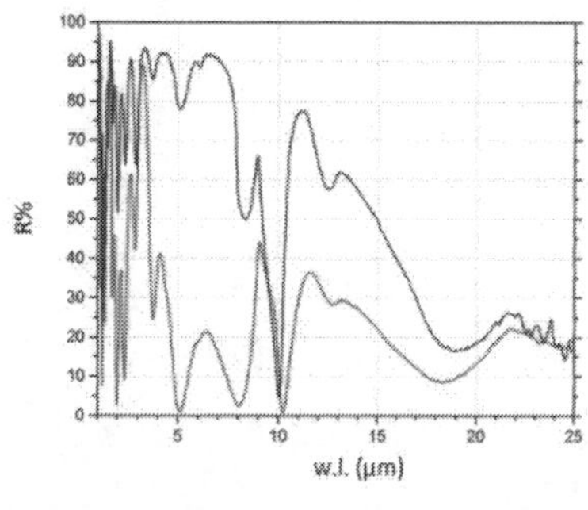

Figure 3. Left: ε_N (*T*), with arrows to indicate the hot (red) and the cold (blue) branches of the curve. Right: IR normal reflectance spectra at 55°C (red) and at 5°C (blue).

4. Testing

Tests were performed along two main paths, one aimed at assessing thermo-optical stability, the other focussing on mechanical durability and coating adhesion.

4.1. Thermo-optical stability

Four samples 25 mm x 25 mm were tested against the sequence described in Table 2, that comprises both thermal tests and irradiation tests with protons at two distinct energies and total dose equivalent to about two years in a GEO environment. A fifth sample was tested with a heavy dose of electrons ($1.35\times10^{16}/cm^2$ at 200 keV), corresponding to 15 years in a GEO environment.

Table 2. Test sequence to assess T.O. stability.

Test type	*Conditions*
TVAC	-70°/+180°C: 100 cycles
Extended TVAC	-180°C/+180°C: 8 cycles
Thermal Ageing	90°C in N_2: 350 hrs
Proton ageing	1×10^{15}p+/cm^2 at 50 keV
	2×10^{14}p+/cm^2 at 250 keV

The effects of the sequence are shown in Table 3 and in Figure 4, left. Solar absorptance remains essentially unchanged, as happens with transition position and width, while emittance values and contrasts are all slightly reduced. Subsequent studies have shown that ε(T) degradation occurs in the first 3-4 hot cycles of the TVAC test, then the values stabilize. This phenomenon is probably the effect of chemo-physical interactions at the interface between WVO_2 and neighbouring layers, promoted by the high temperature of the test (+180°C).

The effects of electron irradiation were found negligible both on α (Δα = - 0.01) and on ε and Δε (Figure 4, right).

Table 3. T.O. properties at the end of the test sequence of Table 2. In brackets variations from BoL values.

Sample ID	α_{HOT}	$\varepsilon_{N\,HOT}$	$\Delta\varepsilon_N$	T_{MIT} (°C)	ΔT_{MIT}(°C)
#1	0.25±0.01	0.78±0.01	0.29±0.02	24±2	30±5
	(--)	(--)	(-0.03)	(+1)	(--)
#2	0.24±0.01	0.77±0.01	0.30±0.02	32±2	20±5
	(+0.01)	(-0.02)	(-0.04)	(+3)	(--)
#3	0.24±0.01	0.73±0.01	0.31±0.02	32±2	20±5
	(+0.01)	(-0.03)	(-0.04)	(--)	(--)
#4	0.23±0.01	0.75±0.01	0.33±0.02	30±2	20±5
	(+0.01)	(-0.03)	(--)	(--)	(--)

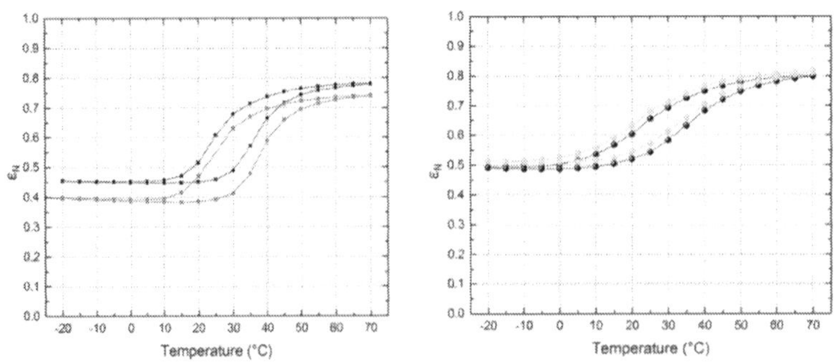

Figure 4. Effects of tests on emittance. Left: ε_N (T) before (black) and after (red) the test sequence of Table 2. Right: ε_N (T) before (black) and after (green) electron irradiation.

4.2. Mechanical stability and coating adhesion

Several samples were tested for adhesion with adhesion tape according to ECSS at the BoL and after one or the other of the tests listed in Table 4. No adhesion failure was observed.

Table 4. Single tests and short sequences to assess mechanical stability and coating adhesion.

Test type	*N° samples / type*	*Conditions*
Humidity	6 / 25 mm x 25 mm	45°C and 95% R.H.: 24 hrs
Bending sequence	1 / 30 mm x 100 mm	Rolling around cylinders of decreasing diameter
Cutting	1 / 100 mm x 100 mm	With scissors and with scalpel
TVAC + Thermal ageing	1 /100 mm x 100 mm	TVAC -180°C/+180°C: 8 cycles Ageing at 90°C in N_2; 350 hrs

The bending sequence was performed by rolling up a 30 mm x 100 mm strip around stainless-steel cylinders of decreasing diameter, from ø = 30 mm to ø = 4 mm (Figure 5). Cutting experiments were done both with scissors and with scalpel. Cutting edges show no major coating cracking at the microscope (Figure 6).

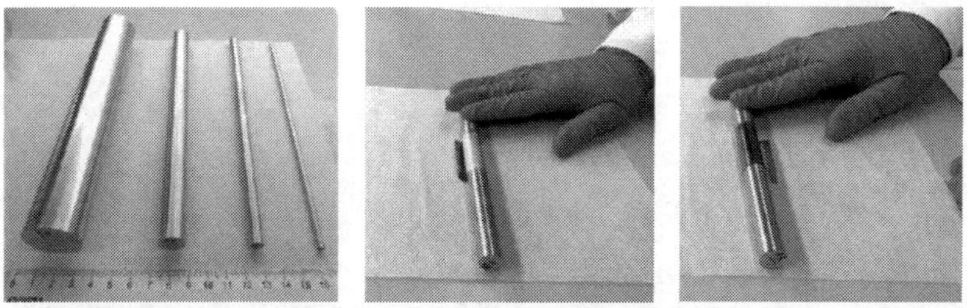

Figure 5. Rolling tests on cylinders of decreasing diameter. For each cylinder, the strip is rolled up 10 x with face up and 10 x with face down.

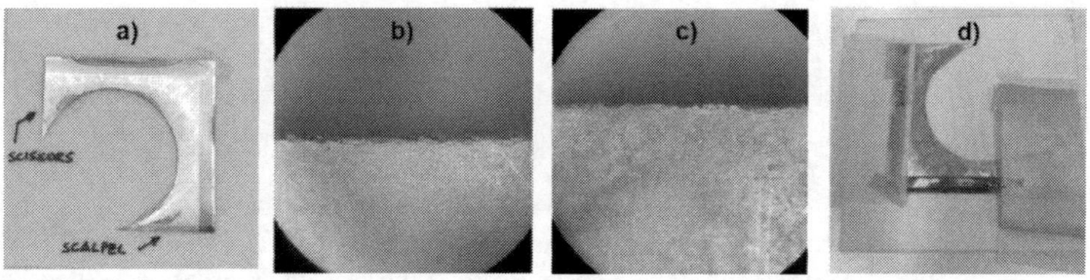

Figure 6. Cutting test. Cutting edges at the naked eye (a) and at the microscope: scissors (b) and scalpel (c); adhesion tape after cutting (d).

5. Conclusions and outlook

The study led to the realization of flexible samples up to 100 mm x 100 mm in size, having $\alpha \approx 0.24$, $\varepsilon_{HOT} \approx 0.80$, $\Delta\varepsilon \approx 0.30$, and $T_{MIT} \approx 25°C$ at the Beginning of Life. Samples are mechanically robust and show no coating adhesion failure after TVAC, thermal ageing, humidity, bending and cutting tests. Thermo-optical properties and transition point are all stable against irradiation with electrons and protons, while TVAC causes only a slight decline in emittance, which occurs in the first 3-4 hot cycles, and subsequently stabilizes.

Thermal analysis performed by Thales Alenia Space shows that, although thermo-optical performance still needs some improvement, efforts should be aimed primarily at reducing α below 0.20. As for the emittance contrast, $\Delta\varepsilon = 0.3$ is already an interesting value since, based on simulations for

LEO, MEO, and GEO missions, it would lead to savings around 40% of power budget at the heaters in the cold phase compared to a standard coating $\varepsilon = 0.8$. Clearly, savings would be even greater for larger contrasts, reaching 50% for $\Delta\varepsilon = 0.35$ and approaching 60% for $\Delta\varepsilon = 0.4$. These savings have real impact at the system level, since they translate into savings in terms of mass, complexity (of thermal control system, power distribution unit, software, and onboard computer), and time to reach the final orbit for electrical propulsion missions.

Funds pending, the authors have already outlined an evolution plan for the technology, and identified the main topics for future R&D work, such as new meta-layer designs with lower fill factor (to reduce α), and new materials for the dielectric spacer (to increase the contrast). In parallel, the authors are planning additional ageing tests for the existing configuration, from exposure to UV (ongoing), to an experiment in space.

Authors contribution

M.G., S.M., M.S., A.U. (CREO): concept and overall design, materials and breadboards development and characterization.

O.M., K.S., I.Z. C.G. (Uni. Southampton): concept, modelling and optical design, study of alternative methods for materials growth.

A.B., I.C., L.K. (NILT): nanoimprinting.

B.A., G.B., M.J. (INFN/Beamide): ageing tests.

J.F., T.K. (CAS Institute): calorimetric measurements.

F.T., M.G., T.S. (Thales Alenia Space): thermal analysis.

Acknowledgements

The study received funds from the EC Horizon 2020 Programme (contract n° 821932 "SMART-FLEX") and from the ESA Innovation Triangle Initiative (contract n° 4000124544/18/NL/CRS "HIPER-OSR").

TAS-Cannes (Jean-Paul Dudon and Pierre Jouanne) and AIRBUS UK (Elizabeth Abelho) supported the definition of requirements and test plan and the final assessment of the technology.

CNR-IFN Roma (Ennio Giovine) fabricated initial breadboards of the WVO_2 meta-layer by Electron Beam Lithography.

University of L'Aquila (Luca Lozzi) characterized the structure and composition of the coatings.

References

[1] Tachikawa S, Ohnishi A, Nakamura Y and Okamoto A 2008 Design and optical performance evaluation of smart radiation device with multi-layer coating *SAE Technical Paper* 2008-01-2152

[2] Haddad E, Kruzelecky R, Wong B, Jamroz W. et al. 2006 Multi-function tuneable emittance smart coatings for thermal control in harsh space environment *SAE Technical Paper* 2006-01-2263

[3] Junwei Y, Li Z, Fengjuan L, Ji Q, Weitao G, Yuan L and Guiling N 2010 Preparation, characterization and properties of thermochromic tungsten-doped vanadium dioxide by thermal reduction and annealing *J. Alloys Compd* **504** 2 503-7

[4] Ping J and Sakae T 1995 Relationship between transition temperature and x in V1-xWxO2 films deposited by dual-target magnetron sputtering *Jpn. J. Appl. Phys.* **34** 2459

[5] Dudon JP, Marcel C, Dubost L, Ravaux A, Aubert PH, Duzellier S, Remaury S and Divay L 2021 Development of variable emissivity coating for thermal radiator ICES-2021-63

[6] Sun K, Riedel CA, Wang Y, Urbani A, Simeoni M, Mengali S, Zalkovskij M, Bilenberg B, de Groot CH and Muskens OL 2018 Metasurface optical solar reflectors using AZO transparent conducting oxides for radiative cooling of spacecraft *ACS Photonics* **5** 2 495-501

[7] Sun K, Riedel CA, Urbani A, Simeoni M, Mengali S, Zalkovskij M, Bilenberg B, de Groot CH and Muskens OL 2018 VO2 thermochromic metamaterial-based smart optical solar reflector *ACS Photonics* **5** 6 2280-86

[8] Sun K, Xiao W, Wheeler C, Simeoni M, Urbani A, Gaspari M, Mengali S, de Groot CH and Muskens OL 2022 VO2 metasurface smart thermal emitter with high visual transparency for passive radiative cooling regulation in space and terrestrial applications *Nanophotonics* **11** 17 4101-14

[9] Králík T, Musilová V, Hanzelka P and Frolec J 2016 Method for measurement of emissivity and absorptivity of highly reflective surfaces from 20 K to room temperatures *Metrologia* **53** 743

[10] Lee H, Lee HE, Wang HS, Kang SM, Lee D, Kim YH, Shin JH, Lim YW, Lee KJ, Bae BS 2020 Hierarchically surface-textured ultrastable hybrid film for large-scale triboelectric nanogenerators *Adv. Funct. Mater.* **30** 2005610

IOP Conf. Series: Materials Science and Engineering 1287 (2023) 012004 doi:10.1088/1757-899X/1287/1/012004

Contamination assessment of a freely expanding green propellant thruster plume

L J Buntrock[1], **M Grabe**[1] **and H Fischer**[2]

[1] *DLR Institute of Aerodynamics and Flow Technology, Bunsenstraße 10, 37073 Göttingen (Germany)*
[2] *European Space Agency (ESA), ESTEC, Keplerlaan 1, PO Box 299 Noordwijk (The Netherlands)*

leonie.buntrock@dlr.de

Abstract. A number of propellant/thruster combinations are under development in recent years that aim to replace the prevailing hydrazine-driven reaction control thrusters with less hazardous substances ("green propellants"). With some of these systems already in orbit, characterizing their contamination potential in a space environment becomes relevant. In this paper we discuss experiments on plume induced contamination from a novel propene/nitrous oxide bipropellant thruster, including high-speed imaging, SEM-EDS analysis, QCM measurements and in-situ mass spectrometry. Additional measurements of combustion chamber pressure complement the overall characterization of the thruster performance in a high vacuum environment. The main findings of this exploratory study are strong indications of solid and liquid phase particles being ejected from the nozzle, which will be investigated in a subsequent phase of the activity.

1. Introduction

Spacecraft contamination due to reaction control thruster plume impingement is a subject of concern since the early days of space exploration, especially for long-term and manned missions, but also for spacecraft bearing sensitive instrumentation. Exhaust plume constituents of chemical propulsion systems are known to affect the properties and performance of functional spacecraft materials. In particular, they may alter the thermo-optical properties of thermal control surfaces, and mechanically or chemically erode protective or functional coatings. While the pertinent data gathered in space and on ground is sparse and hardly generalizable for the common hydrazine-based mono- and bipropellant propulsion systems [1], it is practically non-existent for either of the many alternative propellant/thruster combinations developed since about the mid-1990s in response to the imminent threat of a hydrazine ban under the European Regulation No 1907/2006 ("REACH"). This absence of data makes reliable plume contamination predictions during the design phase impossible, and potentially hinders the adoption of the novel propellant/thruster combinations for contamination sensitive missions. ESA has thus initiated a research activity with a particular green propellant/thruster combination, in which the composition of the freely expanding plume and its impact on properties of impinged representative spacecraft materials is evaluated in order to derive applicable input parameters for industry-standard numerical prediction software.

This paper presents the results of preliminary experiments with a green bipropellant propene/nitrous oxide reaction control thruster, dedicated to determining its contamination potential and general behaviour. Insights gained from these experiments will be used to optimize the aforementioned experiments regarding plume impingement and composition analysis.

2. Theory

Throughout this study we concern ourselves with the construction and operation of the particular thruster only insofar as it affects the plume gas expansion and production and distribution of potential non-gaseous contaminants. The main features of a thruster plume expanding into high vacuum ("free expansion") from a gas dynamic perspective are the following: Emerging from the combustion chamber, the propellant gas reaches sonic speed at the nozzle throat and continues to expand supersonically. Friction and heat transfer near the nozzle walls lead to the formation of a non-isentropic zone, the boundary layer. Boundary layer gas is much slower than the flow near the nozzle axis and thus tends to extend well upstream the nozzle exit plane ("backflow") once it reaches the nozzle lip, forming a potential source of spacecraft self-contamination. The supersonic core expands at an even greater rate after leaving the exit plane, associated with a decrease in thermal energy, which may promote condensation. A narrow zone around the nozzle axis, extending a few exit diameters downstream, is dense enough to be treated with the methods of continuum gas dynamics. Far away from the nozzle the plume gas is so rarefied that molecules practically do not interact with each other ("free molecular flow") and travel on straight paths unhindered.

Studying free plume expansion experimentally in a vacuum chamber is not trivial, as the chamber walls and the pumping system must be designed so as to prevent a build-up of background gas during thruster operation. The only way to achieve this in practice is to use large-surface cryopumps. The high-vacuum plume test facility for chemical thrusters at DLR Göttingen (STG-CT) [3] makes use of a liquid-helium cooled cryopump to enable plume analysis of actual mono- and bipropellant reaction control thrusters. Refrigeration to below 4.7 K becomes mandatory when hydrogen gas is a plume constituent and background pressure must be maintained below 10-5 hPa. While STG-CT provides this space-like vacuum for the thruster plume that will be required for QCM (quartz crystal microbalance) measurements, in-situ mass spectrometer species characterization, as well as material exposition in a later phase of the project, it is comparatively costly to run, and optical access is limited due to the cryopump necessarily enclosing the test section. However, much can be learned about the contamination potential of the thruster by conducting exploratory tests in a sufficiently large conventional, mechanically pumped vacuum chamber with better optical access, such as the DLR Contamination Chamber Göttingen (CCG). In such a chamber, background pressure will quickly rise during firing, providing a hindrance for the plume gas to expand. It is well known that in such a case a barrel-shock system will form, that will separate the supersonic dense gas ultimately downstream of the nozzle exit from the background gas. The axial extent of the barrel shock system is proportional to the square-root of the ratio of stagnation- to background pressure [4, 5]. Information from the background cannot propagate upstream past the shock system and thus all measurements and observations inside the domain enclosed by the barrel shock are perfectly representative of the free plume expansion.

3. Methods

Measurements are conducted at the CCG as well as the STG-CT test facility of DLR Göttingen. The bipropellant thruster under investigation is the Dawn Aerospace B20 model (6.6 N to 19.3 N), additively manufactured from Inconel 718. Propene (C_3H_6) as the fuel and nitrous oxide (N_2O, “laughing gas”) as the oxidizer are the two self-pressurizing propellants, each actuated by two solenoid valves (isolation and firing). Control electronics are built inhouse at DLR to trigger valves, igniter and measurement equipment in the required sequence. The thruster features a built-in ratiometric combustion chamber pressure sensor, along with a type-K (NiCr-Ni) thermocouple installed on the outside of the combustion chamber.

CCG is a 12 m^3 stainless steel vacuum chamber. The pumping system consists of two strands designed for high throughput, each of which is identically equipped with three pumps in series (Leybold SOGEVAC SV300B, RUVAC WH2500 and RUVAC WH7000), with a total effective pumping speed in excess of 12000 m^3/h. A background pressure in the low 10^{-3} hPa range is ultimately achieved, which rises to less than 0.2 hPa when firing a typical 250 ms pulse with the thruster described here, but returns to its original value within 30 s. The B20 thruster is shown in figure 1(a) mounted inside CCG, together with the Pitot probe used to characterize the expanding beam profile. A high-speed camera, seen in the background, is employed to detect potential liquid or solid contaminants ejected from the nozzle, provided they are sufficiently large. Possible solid contaminants are captured with double-sided adhesive carbon tape attached to both the Pitot probe support and the chamber wall facing the thruster. Samples are analysed using a SU3900 scanning electron microscope (SEM) from Hitachi High Technologies. The latter is equipped with an energy-dispersive x-ray (EDS) detector (Oxford Instruments Ultim Max).

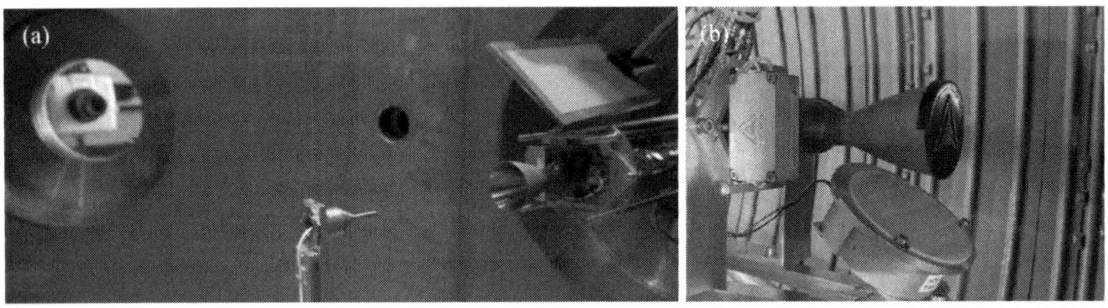

Figure 1. Dawn Aerospace B20 thruster inside CCG vacuum chamber, together with a Pitot probe and a high-speed camera (a). B20 thruster installed in STG-CT with LED beneath the nozzle for shadowgraphy imaging (b).

As described above in section 2, STG-CT is a cryo-cooled high-vacuum plume test facility which allows to maintain a vacuum level below 10^{-5} hPa while firing small chemical thrusters (mass flow rates at the order of g/s) with hydrogen as a plume constituent. The cylindrical test section with a volume of 10 m^3 allows for a plume expansion well into the highly rarefied flow regime including the backflow region. Figure 1(b) shows the thruster mounted at STG-CT with a LED sitting right below the nozzle which is used for shadowgraphy high-speed imaging. A quadrupole mass spectrometer (MS, Hiden Analytical Ltd., model "HAL PIC 101") faces the nozzle exit 3 m downstream on the plume axis for in-situ measurements. A spectral scan of integer increments of mass-to-charge ratio (m/z from 3 to 50) is used, so that both spectral and temporal information about the gaseous plume species can be obtained at reduced resolution compared to the attainable step size of 0.01. Additionally, a cryogenic QCM (CQCM, QCM Research, Type MARK 18) is mounted to the cryopump, facing the thruster under an angle of 30° to the plume axis at a distance of about 1,6 m in order to monitor condensable plume constituents just below 80 K. A built-in heater is used at the end of the thruster activity to conduct a thermogravimetric analysis (TGA) by heating the crystal surface at a rate of 2 K/min.

3.1. Thruster operation

To achieve nominal operation pressure in the feed lines (fuel and oxidizer are driven by their vapor pressure), the gas bottles are kept at room temperature. At CCG the latter varied between 21°C and 22.5°C from day to day. The resulting feed line pressure is documented for the length of each firing and varies between 9.25 and 9.58 bar (fuel) and 51 to 51.9 bar (oxidizer). An oxidizer-rich off-design operational point targeted for some firings is obtained by raising the N_2O gas bottle by 5 K, resulting in a pressure increase of 6.6 to 6.8 bar. In addition, operation at CCG allowed for the testing of varying pulse length as well as pulse trains. While testing at STG-CT, the maximum room temperature was 25°C with a resulting maximum feedline pressure of 11 bar (fuel) and 53.5 bar (oxidizer). In each case, a single pulse length of 250 ms was chosen.

3.2. *High-speed imaging*

A high-speed camera (Phantom UHS Series, Model v1210) equipped with a tele-lens (AF-S DX NIKKOR 55–200 mm) is set up outside an angled view port of CCG to observe the thruster and a downstream region of about 30 cm during firing, figure 1. Images are taken at a resolution of 1280 by 800 (approx. 400 µm per pixel) with frame rates ranging from 100 frames per second (fps) to 4000 fps. At STG-CT, a high-speed camera (ix Cameras, i-speed 726, resolution 2048x1536) observes the thruster nozzle with the same tele-lens from a top-view under an angle of 55°. In addition to indirect lighting, a LED with a diffuser plate serves as background light underneath the nozzle for shadowgraphy imaging. Images were taken at 200 to 1000 fps.

4. Measurements and results

4.1. *Thruster performance*

At CGG a variation of single thruster pulses and pulse trains were fired in the course of this investigation, with a focus on a nominal on-time per pulse of 250 ms. The overall behaviour of the thruster is very reproducible, as deduced from the high pulse-to-pulse repeatability of the combustion chamber pressure. Figure 2 shows a plot of the median combustion chamber pressure (*pcc*, in dark blue) for 30 representative 250 ms pulses. The pulse-to-pulse variation in the measurements is indicated by the 5th percentile range shown in light blue. A delay of about 15 ms from command to actuation is discernible for both, start-up and shut-down, with the combustion chamber pressure increasing only slightly during the duration of the pulse. Shown on the same time scale is the chamber background pressure, *pb*, recorded by a capacitance gauge mounted to the chamber wall. The background pressure repeatably remains below 0.2 hPa for duration of the pulse.

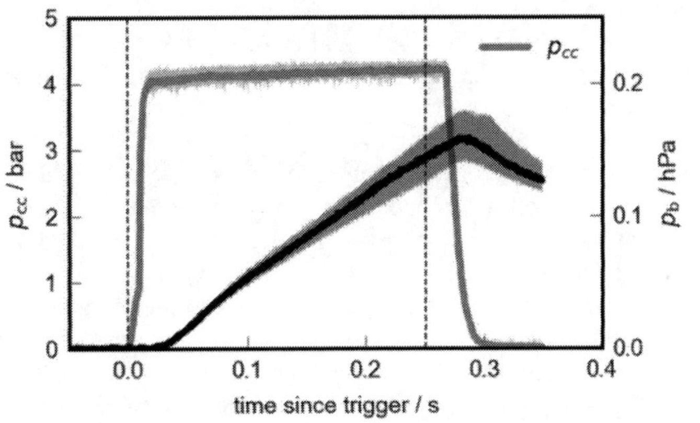

Figure 2. Combustion chamber pressure (box profile) and background chamber pressure for 30 pulses of 250 ms duration. Vertical dashed lines mark pulse start and end trigger.

At STG-CT about 30 pulses were triggered in order to familiarize with the thruster behaviour at the cryo-cooled facility. The thruster was first operated at a starting temperature slightly below the specified operational temperature and the first 15 pulses of 250 ms length were fired to raise the combustion chamber temperature to about 90°C. Before firing five nominal pulses (compare behaviour at CCG, figure 2) at the end of the test day, the temperature of the propellants had to be lowered as an increase of the feedline pressure greater than the specified operational pressure had presumptively led to a misfunctioning of the oxidizer valve.

4.2. Imaging

Examination of the thruster exit plane and near field using a high-speed camera at CCG, exposed a number of phenomena potentially relevant to contamination. Figure 3 displays what will be referred to as streaks of apparently glowing or reflecting solid particles which follow straight trajectories (a) and elastically bounce off of the Pitot probe structure (b). The number of streaks appears to change inconsistently between few and numerous from pulse to pulse, and typically no or very few streaks are observed during the second half of a nominal pulse (250 ms on-time). The same applies for pulse trains of several 250 ms pulses, where the sequence of observed events does not follow a discernible order. When evaluating a series of twelve oxidizer rich pulses, a slight decrease of streaks is observed on average only. No change compared to initial standard operational points can be observed when returning to these settings after the oxidizer rich runs. Overall, solid particles were detected leaving the thruster under angles no greater than 17° to the nozzle axis, with velocities ranging from approximately 175 m/s to 850 m/s (estimated from the streak lengths) and particle sizes in the range of 200 to 800 µm (estimated from the streak width). Smaller particles or droplets may well be present but not detectable with this technique. Further observations reveal that light is scattered in distinct areas which indicates the presence of small droplets.

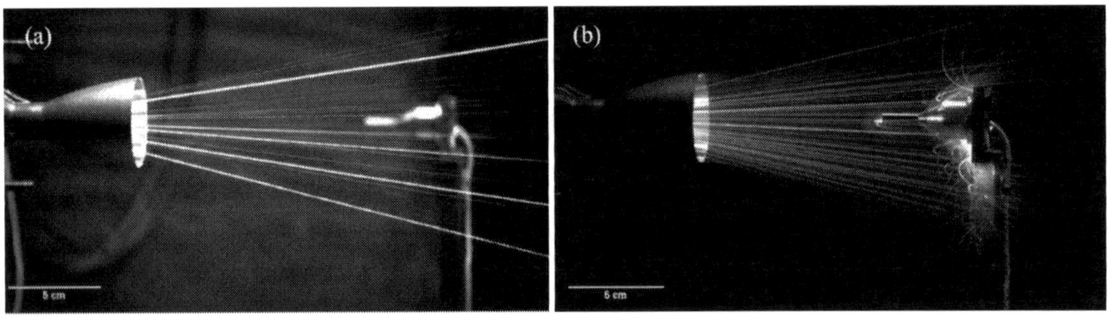

Figure 3. High-speed camera images of near field with bright streaks observed at 400 fps (a) and 1000 fps (b).

At STG-CT, the high-speed camera did not reveal as many details as in the CCG chamber, as the optical access is highly limited but a bright, striped cone emitted from the nozzle was visible at the beginning of most pulses as well.

4.2.1. Particle analysis. SEM inspection of adhesive carbon tape, which had been installed at CCG to capture ejected particles, revealed several beads with sizes on the order of 100 µm (figure 4(a)). An EDS analysis at 15 keV incident electron energy (table 1(a)) points to the composition of soda-lime-glass [6, 7]. The carbon signal, as well as about 14% of the oxygen signal, are due to the carbon tape itself. The beads appear transparent under the stereomicroscope and charge up lightly under the high-energy electron beam during EDS analysis, supporting the assumption of them being glass beads. Several even smaller spheres (figure 4(b)), with sizes ranging from 5 to 10 µm, yield EDS signals that strongly suggest Inconel 718 [8] (table 1(b)), the material the thruster nozzle is made of. Particles were captured in a pouch of adhesive carbon tape, and here too the oxygen signal as well as part of the carbon signal originate from the background. Both kinds of particles (glass and metal alloy) are so sparsely present, that no average population per surface area can be inferred from the findings.

4.3. Mass spectra

The mass spectrometry analysis conducted at STG-CT revealed the signature of both unburnt propellants and signals of exhaust gases nitrogen and water when the thruster was cold started below its operational temperature. With each following pulse the thruster warmed up and the amount of unburnt propellant decreased in comparison to the expected exhaust gases. The analysis yielded rather clean MS spectra at an initial combustion chamber temperature of 90°C with hydrogen as the main constituent, a fraction of water below 20% and an unknown peak of over 40% intensity at a mass-to-charge ratio of 29. The temporary misbehaviour of the oxidizer valve due to an out of range increase in propellant temperature was verified by a distinct spectrum of unburnt propene.

Table 1. EDS elemental analysis results of glass beads (a) and smaller spheres (b).

(a) Element	A / wt%	B / wt%
O	49.2	47.4
C	22.8	23.9
Si	17.6	15.1
Na	6.1	7.6
Ca	2.9	3.7
Mg	0.6	1.8
Al	0.5	0.4

(b) Element	A / wt%	B / wt%	C / wt%
Ni	34	34.9	47.3
Cr	28.5	14.8	17.6
Fe	14.2	14.9	19.7
C	18.2	26.2	10.8
Nb	5.7	3.6	1.8
O	3.8	3.9	0.6
Ti	2.9	1.4	0.7
Mo	1.6	-	1.3
Al	1.1	0.4	0.1

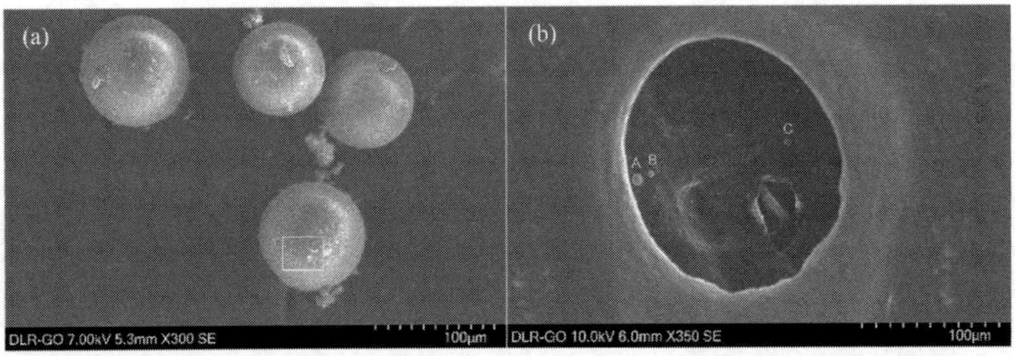

Figure 4. SEM images of beads (a) and spheres (b) captured on adhesive carbon tape during thruster testing.

4.4. *QCM analysis*
The QCM-TGA measurement conducted at the end of the test day at STG-CT exhibited three major desorption events (figure 5) of which the first one started just above 80 K exhibiting distinct rippling features. It is most likely caused by CO_2, one of the three main exhaust gases, while the unburnt oxidizer N2O could also be a source as a result of cold starting the thruster [9]. The second desorption peak occurs at about 150 K which may be related to the unburnt propellant propene. This correlation is based on the chemical species available from propellants and exhaust gases after ruling out all non-condensable species at the given QCM temperature. Propene impacted the QCM due to both the cold start of the thruster and one misfiring without the oxidizer. The 0th order desorption peak starting above 160 K is most likely due to evaporating water. The desorption temperature matches the corresponding vapour pressure at an ambient pressure of 10^{-6} mbar according to the low temperature Arden Buck equation [10].

5. Discussion and outlook
Actual contamination prediction for spacecraft is inherently complex, as it depends not only on the source of contamination (here a reaction control thruster), but also on the properties of the impinged surface and the environment it is subjected to. In this first part of the activity, we studied the plume of a 20 N bipropellant propene/nitrous oxide reaction control thruster supplied by Dawn Aerospace.

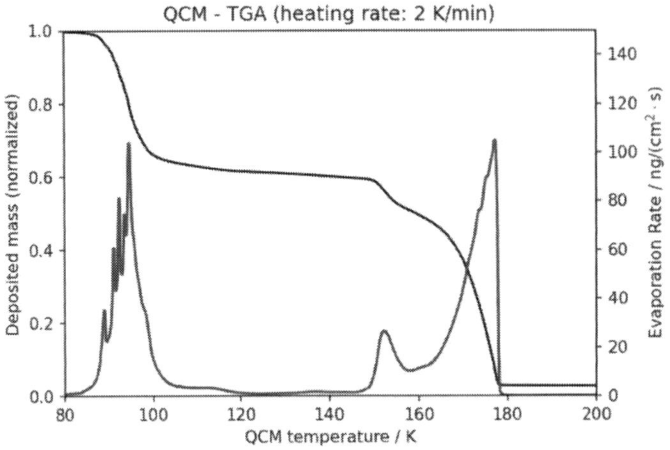

Figure 5. QCM-TGA plot of decreasing deposited mass and three major desorption events (evaporation rate).

The set of exploratory tests described in this paper provides the opportunity to familiarize with thruster operation, characterize its performance, as well as to verify and adapt plume diagnostics. The aforementioned findings help to design and improve subsequent sophisticated testing in the STG-CT facility. Operating the thruster inside the CCG test facility proved successful. With regard to the main objective of this study, the contamination behaviour of a green propellant thruster, potential solid and liquid particles ejected from the nozzle were captured with a high-speed camera and collected for chemical analysis. Operation of the thruster proved to be reliable and repeatable for the operating conditions foreseen for the subsequent in-depth STG-CT test campaign. A first test of the thruster at STG-CT pointed out which operational conditions must be followed for a reliable operation, while exploratory MS and QCM measurements proved successful. High-speed camera imaging was tested and will be optimized for future tests.

The observation of the thruster plume with a high-speed camera at CCG revealed particles being ejected from the nozzle, following linear trajectories with velocities of up to 850 m/s. As the combustion chamber emits light during a pulse, the particles could either scatter this light or could glow inherently. The original assumption of soot being the origin of the observed light streaks is contradicted by the ballistic behaviour of the particles as well as a rather ineffective shift to an oxidizer rich operational point of the thruster, aiming at reducing the tendency of soot formation. In order to identify expelled particles, surface samples positioned in the plume at CCG were inspected with SEM-EDS analysis. Generally, only very few noticeable objects were identified on the few sampled surfaces. Several of the identified spherical particles likely originate from the additive manufacturing process of the thruster body, as they exhibit EDS spectra resembling the composition of the material reportedly used. The analysis of another set of spheres is in good agreement with the elemental signature of soda-lime glass. The glass beads found by EDS analysis are likely the source of the bright streaks observable in the high-speed camera images, supported by their reflection of light and them elastically bouncing off of the metal structure of the Pitot probe. We suspect the glass beads to be remnants from the production process as they do not originate from the thruster body material itself. Both glass and metal spheres may have been trapped inside the combustion chamber, from where they gradually get expelled during firing. The 7 μm filters installed into the feedlines right in front of the valves preclude particles of this size or bigger to originate from upstream the thruster. Additionally, images taken with the high-speed camera indicate the presence of fine droplets, as light must be scattered off of particles condensed in the mentioned region. The phenomenon was observed multiple times during the ramp-up phase of a pulse and will be investigated further in the course of the project.

Exploratory testing at STG-CT provided first information about the freely expanding plume composition through mass spectrometry on the plume centreline and a QCM-TGA analysis at 30° off-axis. Both measurement techniques proved successful and will be repeated at a later in-depth experiment at the facility. Overall, MS data revealed a changing plume composition depending on the initial thruster temperature with dissipating unburnt propellant fractions after sufficient heating, but a general conclusion is not reasonable yet, as too many thruster pulses were fired outside of the applicable specification window. For the interpretation of mass spectra captured at low resolution as it is the case here due to thruster pulse length limitations, it is important to keep in mind that the mass-to-charge ratio of 44 might be misinterpreted as N_2O if the exhaust gas CO_2 is present, or vice versa. The QCM-TGA analysis delivered three distinct temperature ranges where desorption of one species or a mixture of soluble species took place, most probably caused by water, CO_2, propene and/or N_2O, as they were detected by the mass spectrometer and are condensable at the provided QCM temperature of 80 K. Generally, no definite attribution of chemical species can be made for a QCM-TGA when multiple unknown species are present, unless combined with a continuous MS scan during desorption. However, an analysis like the one conducted in this study, can provide information on how much mass is deposited on surfaces with at a certain temperature and how much heating is necessary to evaporate the deposited matter.

It is well known that the absence of an appreciable atmosphere in space allows for a wide expansion of the thruster plume, and it is necessary to replicate that high-vacuum environment in a ground test to obtain the distribution of the gaseous plume species that would be found in the vacuum of space. Additional measurements will thus be conducted in the DLR high-vacuum plume test facility Göttingen for chemical thrusters (STG-CT), using a mass spectrometer to record the gas phase composition at various angles from the plume axis. Non-gaseous effluents, i.e. droplets and particulates, will again be recorded with a high-speed camera during the STG-CT test campaign, providing information on size, velocity and trajectories if present and resolvable. The impact of plume contamination on the properties of relevant surface coatings and materials will be studied in a later phase of the activity.

6. Acknowledgements

This work is funded under ESA contract ESA Contract No.4000132279/20/NL/KML. The authors thank Dawn Aerospace for helpful discussions and engineering support, and gratefully acknowledge technical contributions from G v. Roden (DLR) and Dr. G Dettleff (DLR-retired).

References

[1] Grabe M and Soares C E 2019 Status and Future of Research on Plume Induced Contamination *Proceedings of the 70th International Astronautical Congress*, Washington D.C.
[2] Dettleff G and Grabe M 2011 Basics of Plume Impingement Analysis for Small Chemical and Cold Gas Thrusters *Models and Computational Methods for Rarefied Flows*, RTO/NATO
[3] Grabe M 2016 STG-CT: High-vacuum plume test facility for chemical thrusters *Journal of Large-Scale Research Facilities*, **2**
[4] Ashkenas H and Sherman F S 1966 The Structure and Utilization of Supersonic Free Jets in Low Density Wind Tunnels *Rarefied Gas Dynamics*
[5] Driftmyer R T 1972 A Correlation of Freejet Data *AIAA Journal*, **10**(8), pp. 1093–1095
[6] Martin J *Materials for Engineering,* Woodhead Publishing (Abington), 133, 2006
[7] Polysciences Inc. 2016 Glass Beads, Technical Data Sheet 857
[8] Timet Powder Metals 2012 Inconel 718 Powder, Material Safety Data Sheet
[9] Honig R E and Hook H O 1960 Vapor pressure data for some common gases *RCA review* **21**.3: 360-368
[10] Buck A L 1996 Updated equation from Buck, A L, *Buck Research Manual*; New equations for computing vapor pressure and enhancement factor *J. Appl. Meteorol.*, **20**, 1527-1532, 1981

Evaluation of the Material Properties of Solid Rocket Motor Slag

Yoshiki Matsuura,

IHI Aerospace Co., Ltd., 900 Fujiki, Tomioka-shi, Gunma, 370-2398, Japan
Yoshiki-matsuura@iac.ihi.co.jp

Kumi Nitta, Hirohide Ikeda, Masahiro Kinoshita, Kyohichi Ui

Japan Aerospace Exploration Agency, 2-1-1 Sengen, Tsukuba-shi, Ibaraki, 370-8505, Japan
nitta.kumi@jaxa.jp, ikeda.hirohide@jaxa.jp@jaxa.jp, kinoshita.masahiro@jaxa.jp, ui.kyohichi@jaxa.jp

Abstract. A considerable amount of aluminum oxide particles are generally produced during the Solid Rocket Motor (SRM) combustion. A project determining the damage caused by slag impact on spacecrafts in orbit is drafted to acquire physical property data of SRM slag by JAXA. After the grand firing test of small SRMs, all slag was collected and analyzed. This study describes the investigation results regarding the characteristics of SRM slag and the current possible scenario of the slag particle formation process.

1. Introduction

With the increasing number of satellites being launched into the low earth orbit, the problem of space debris is becoming more important over time. Therefore, monitoring this problem from the ground is very important. With tracking capabilities improving continuously, space debris with a diameter of 2 cm or more can be monitored from the ground [1]. In the near future, 1 cm class space debris is expected to be monitored as well.

According to the debris flux estimation around the International Space Station (ISS) orbit calculated by ESA/MASTER [2], "SRM slag" and "Explosion Fragments" are stated to constitute most of the debris source population (see Figure 1). As seen from the chart in Figure 1, the light blue line is derived from SRM, and the "1 mm to 1 cm region" shows a larger total volume than the trackable debris. Under such circumstances, reducing the amount of debris between 1 mm and 1 cm has become important. Based on debris flux predictions from the Orbital Debris Engineering Model (ORDEM) that is produced from NASA, new satellite constellations will face more than 10,000 orbital debris impacts of 1mm size or lager each year [3]. Against the background of such research results, ISO Space debris mitigation requirements have included a new regulation for SRM slag ejection larger than 1mm in size [4].

The MASTER model is based on the investigation results of SRM slag generation confirmed on solid rocket booster static fire tests published in the past [5]. However, as the results of the orbital stage rocket motors made in Japan were not available in the public domain, they have not been referred sufficiently in the published studies.

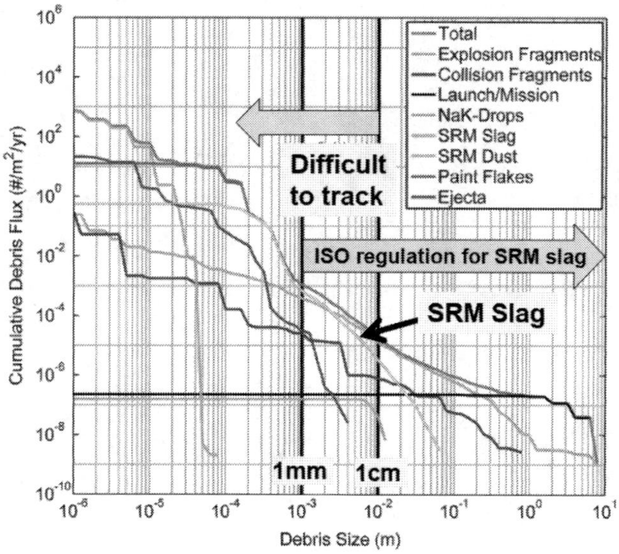

Figure 1. MASTER-2009 debris source population fluxes for the ISS orbit in 2014.

Generally, a large amount of slag remains inside the combustion chamber of a solid rocket booster and 1st stage SRM. The composite propellants currently used for SRMs contain approximately 20% aluminum powder by mass fraction. These aluminum powder change to alumina (AL_2O_3) by chemical reaction with oxidants contained in the propellant as ammonium perchlorate (AP), and mass fraction of combustion products alumina reaches to approximately 38%. Alumina, which mainly exists in the liquid phase in the combustion gas inside a SRM, has the possibility of being trapped in a stagnant flow field around the submerged nozzle or collide with the aft-dome region. Even though a small fraction of alumina is trapped, a significant amount of slag remains because of the abundance of aluminum in the composite propellant. The term "slag pool" is often used to describe the slag that remains in conventional solid rocket booster and 1st stage SRM. The situation where 116 lbm (52 kg) of slag remains with 11 tons of propellant and 4064 lbm (1843 kg) of slag remains with 300 tons of propellant, as introduced in the NASA-TP [6], can be described is a "slag pool". As alumina slag causes an increase in the inertia inside the motor, it leads to a decrease in performance. This effect is a minor issue for the booster and 1st stage SRM but a major problem for the orbital stage. Considering the discharged slag near the end of the motor combustion, the chamber pressure declines sharply and then the SRM-slag spreads out from the slag pool and is ejected through the nozzle. As the source is a large slag pool, ejected SRM-slag tends to be large, more than several centimetres.

Japan has focused on addressing this issue because of its history of pursuing efficient launches of satellites and probes using solid rockets. Improvements in the combustion efficiency of aluminum in upper stage SRM have been ongoing in Japan for decades. Figure 2 shows the ratio of slag to propellant volume in major Japanese SRMs. In Japan, propellants are labeled as the BP series. The latest propellant used for the upper-stage rocket motor is BP-211J, which is a successor to BP-205J. With the improvement in the combustion efficiency of aluminum, the amount of slag can be significantly reduced. For example, the amount of alumina slag for M-35, Epsilon's second-stage SRM, is 31 g, which is 0.0006% of the initial propellant volume of 15 tons with BP-211J. KM-V2, Epsilon's third-stage motor, has a slag amount of 20 g, which is 0.0002% of the initial propellant volume of 2.5 tons with BP-205J. Considering the data above, the discussion around space debris issues should be evaluated with the orbital stage and not the booster/1st stage SRM.

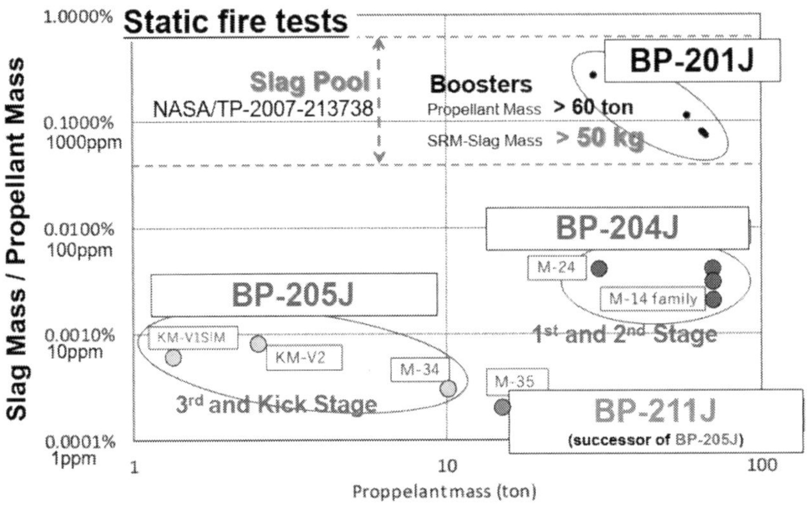

Figure 2. Ratio of alumina slag to propellant in Japanese major solid rocket motors.

2. Research procedures

To contribute to the orbital environmental conservation and sustainability of space development, it is important to correctly estimate the actual impact of "SRM slag in the 1mm to 1cm range" and continue research on reducing the amount of SRM slag generated in orbital stage. Therefore, further investigation has been performed, the implications of which are reported in Steps 1 and 2. The proposed future works are given in Steps 3 and 4 .

Step 1 Hot fire test with several solid rocket propellant for orbital stage.

- Measure the amount of slag that accumulates inside the motor.
- Collect all the slag from SRM for the next step.

Step 2 Investigate the properties of the SRM slag

- Determine the relationship between size and weight.
- Extract other features

Step 3 Carry out collision tests using the real "SRM slag in the 1mm to 1cm range"

Step 4 Build a numerical calculation model for evaluating the effect of "SRM slag in the 1mm to 1cm range" collisions

3. Details of experiment and research

3.1. Hot fire test

To obtain test data for multiple propellants, a hot-fire test was conducted using a standard test motor (see Figure 3). The inner diameter of the test motor was 250 mm, and the propellant mass was 10kg. The materials of the structural parts inside this standard test motor are the same as those of the actual motors used for flight, which are heat-resistant rubber, carbon/carbon composite, and carbon FRP. In this test campaign, two types of propellants were selected. One is the newest propellant for orbital stage that is currently in practical use is "BP-211J". The other is called "ISP-DT-24 (**I**mproved **S**olid **P**ropellant **D**eveloped at IHI Aerospace **T**omioka-Plant)" which is currently under development, further improves the aluminum combustion efficiency.

Figure 4 shows the compact low-pressure test facility owned by JAXA. The standard test motor was mounted horizontally on the test stand. Several static fire tests were conducted in a low-pressure environment at approximately 5 kPa.

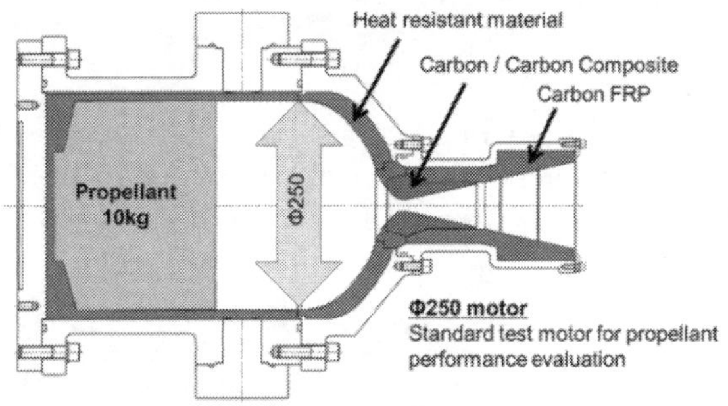

Figure 3. Motor design for the standard test Motor.

Figure 4. Compact low pressure test facility owned by JAXA.

3.2. Test results: at a glance

Figure 5 shows the aft dome of the standard test motor after the static fire test. Glossy aluminum combustion residue is seen remaining discretely on the insulation rubber. The residue comprised mostly single particles and there are no large agglomerates. The results with BP-211J demonstrate a significantly low amounts of SRM slag (no slag pool); moreover the improved propellant ISP-DT-24 is achieving even lower amounts.

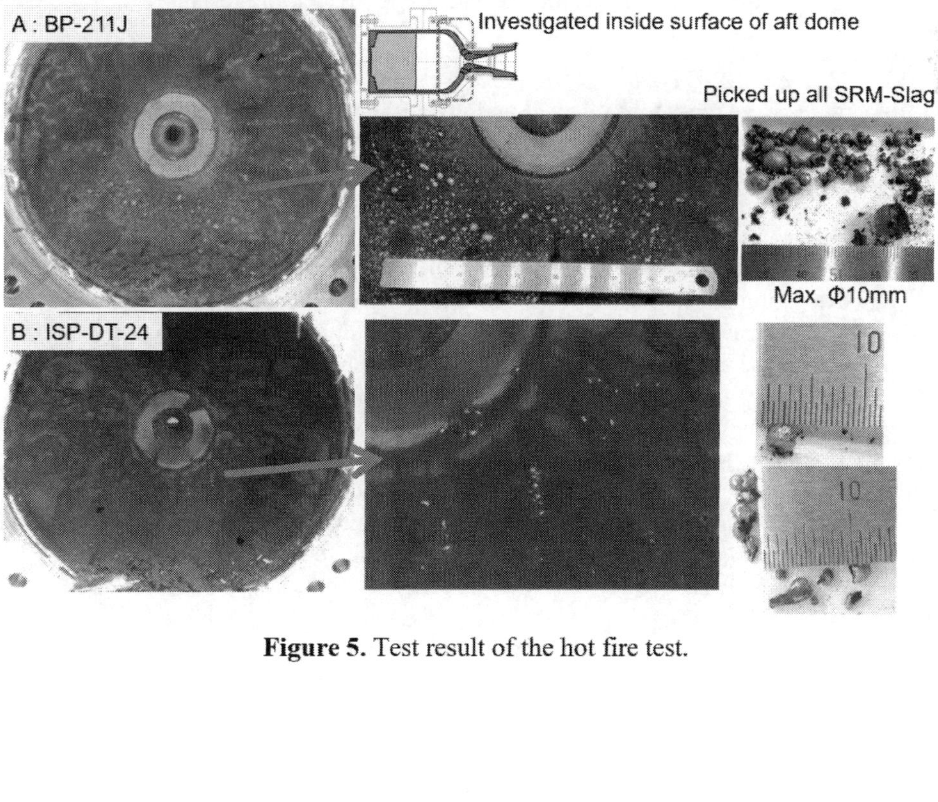

Figure 5. Test result of the hot fire test.

3.3. Test results: detailed investigation of the SRM slag particle

All the slag particles were picked and sorted by equivalent diameter (Figure 6).

1) BP-221J

The total number of SRM slag particles is over 1000, and the total weight is approximately 7g. As for the equivalent diameter of the particles, the largest ones are 5mm or more, with a maximum of approximately 10mm. They are widely distributed from approximately 0.5mm to 10mm, with a higher concentration of approximately 2mm.

2) ISP-DT-24

The fire test was operated three times. The total number of slag particles is less than 30 for each test. The total weight is less than 0.2 g for each. Also, the slag-to-propellant ratio was 0.1 ppm for ISP-DT-24, which is less than 50 times that of the BP-211J. If ISP-DT-24 is selected for the orbital stage, SRM slags could be reduced significantly.

No.	Propellant	**Number** of SRM-Slag								
		~ Φ5.0	Φ2.0~ Φ5.0	Φ1.4~ Φ 2.0	Φ1.0~ Φ1.4	Φ0.71~ Φ1.0	Φ0.5~ Φ0.71	Φ0.425~ Φ0.5	Φ0.425~	Total
1	BP-211J	11	111	743	188	156	306	226	0	1741
2	ISP-DT-24	0	0	11	8	4	0	0	0	23
3	ISP-DT-24	0	0	10	10	6	0	0	0	26
4	ISP-DT-24	0	1	1	2	7	0	0	0	11

No.	Propellant	**Weight** of SRM-Slag (g)									(ppm)
		~ Φ5.0	Φ2.0~ Φ5.0	Φ1.4~ Φ 2.0	Φ1.0~ Φ1.4	Φ0.71~ Φ1.0	Φ0.5~ Φ0.71	Φ0.425~ Φ0.5	Φ0.425~	Total	Slag to Prop. Ratio
1	BP-211J	1.581	2.254	2.677	0.354	0.097	0.086	0.022	0.161	7.232	7.2
2	ISP-DT-24	0.000	0.000	0.110	0.017	0.004	0.000	0.000	0.000	0.131	0.1
3	ISP-DT-24	0.000	0.000	0.078	0.019	0.004	0.000	0.000	0.000	0.101	
4	ISP-DT-24	0.000	0.012	0.004	0.002	0.004	0.005	0.000	0.000	0.027	

Less than 1/50

Figure 6. Test results of standard motor hot fire tests.

3) Both of BP-211J and ISP-DT-24

A total of 40 particles were randomly selected and investigated with the X-ray computed tomography (X-CT) (Figure 7). Every particle is hollow, with a large number of small voids distributed in the outer shell that constitutes the slag particle. The larger ones are fragile, while the smaller ones are brittle.
energy-dispersive X-ray spectroscopy (EDX) analysis was performed on the inner surface of the particle shell; carbon, oxygen, and aluminum were detected (Figure 8). These chemical elements were in abundance. Carbon was detected on all measurement surfaces, then combustion gases are also estimated to play a role in expanding the slag particles and forming internal voids.
Figure 9 shows the porosity and bulk density estimation results. All particles were assumed to be ellipsoid from the representative dimensions of their outer shape, which set the equivalent particle diameter for each. Porosity data for each particle are estimated from X-CT analysis. This analysis reveals that the larger the size, the higher the porosity. As a result, the outer shell must be thinner. For this reason, the larger particles tend to be fragile. The bulk density is estimated from the measured weight and equivalent particle diameter for each; hence, the larger the diameter, the lower is the bulk density. Crashing equations for hypervelocity particles impacting monolithic structures are given in [7], in which the density rate between projectile and structure is important to estimate penetration depth. In that equation, larger projectile density leads to deeper penetration depth. Based on the relationship between slag diameter and bulk density, a slag with a larger diameter tends to exhibit a lower bulk density, therefore a larger diameter does not imply a large collision impact.

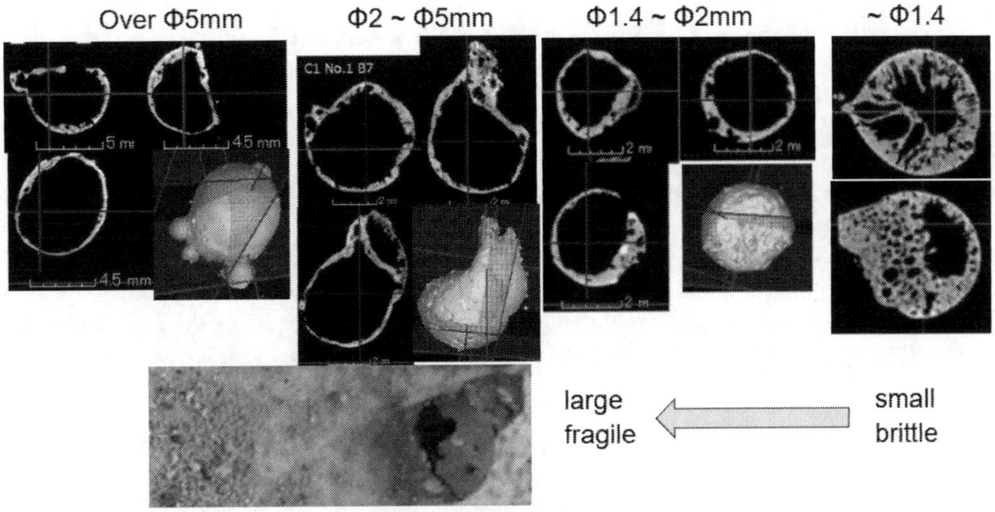

Figure 7. Results of X-ray computed tomography observations of slag particles.

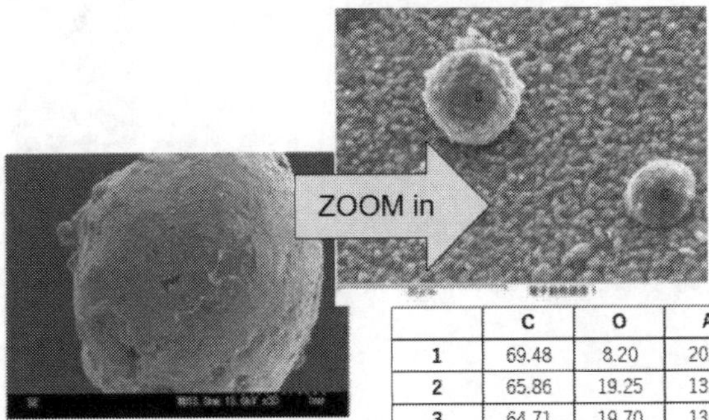

Mass concentration (%)

	C	O	Al	Si	S	Other	Total
1	69.48	8.20	20.62	0.19	0.08	1.42	99.99
2	65.86	19.25	13.35	0.24	0.05	1.25	100.00
3	64.71	19.70	13.89	0.22	0.05	1.43	100.00
4	71.17	12.52	14.93	0.37	0.09	0.91	99.99
5	68.01	13.41	17.25	0.17	0.07	1.09	100.00
Ave.	**67.85**	**14.62**	**16.01**	**0.24**	**0.07**	**1.22**	100.00

Figure 8. Energy-Dispersive X-ray spectroscopy investigation result.

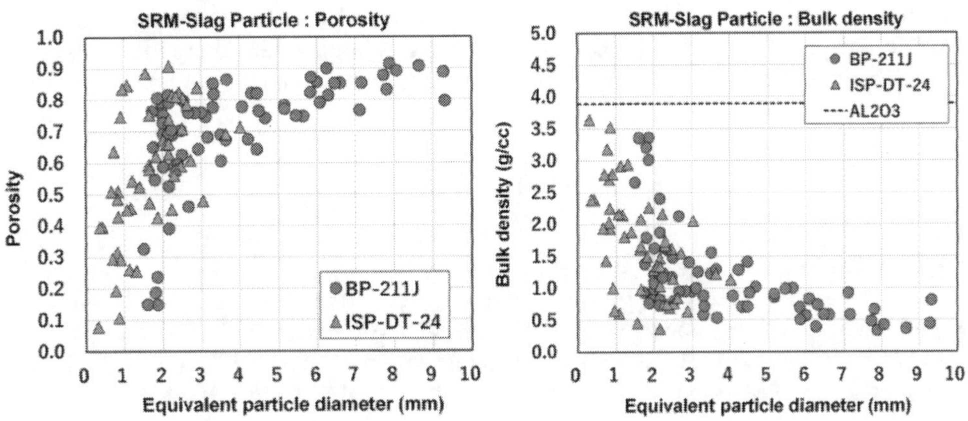

Figure 9. Porosity and bulk density estimation result.

4. Current possible mechanism for the formation of SRM slag particles

Based on the investigation results, the following scenario is assumed for the formation process of hollow slag particles with several voids inside a thin surface shell.

Step 1: Aluminum particles flying in the SRM-chamber grow into alumina particles.
Step 2: Many of them are ejected from the nozzle along with the combustion gas flow and are subdivided by a strong shearing force near the throat. However, several particles that are unable to follow the flow, collide with the nozzle and aft dome, and remain inside the motor. If the remaining amount is extremely small, it will be limited to slag particles (not growing to a slag pool).
Step 3: The remaining slag particles expand with the gas trapped inside during the final decompression. Consequently, slag particles are formed with a thin outer shell and internal voids, resulting in larger sizes that are fragile.

This estimated scenario is applicable to all orbital-stage SRMs with especially low slag deposits.

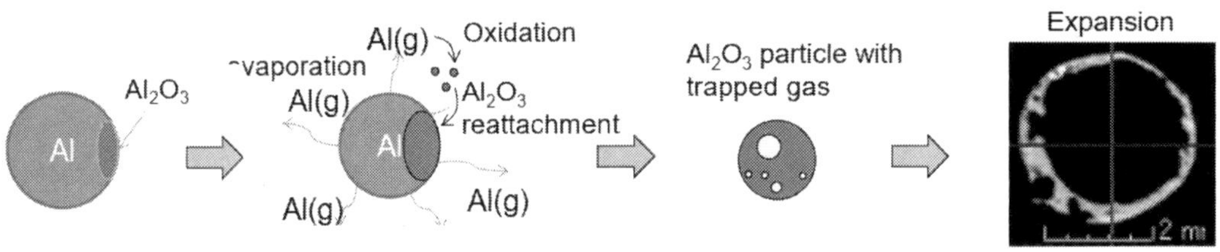

Figure 10. Formation process of hollow slag particles.

5. Ongoing activity

Japan is tackling this research to contributes to the orbital environmental conservation and sustainability of Space Development with newest SRM which capable of sufficiently reducing the probability of slug collision with orbital missions. One of such research projects is conducting risk assessments of small SRM slag particles that are difficult to track. Slag generation could clearly be minimized through the use of ISP-DT-24 propellant, however even then it is estimated that a little amount of slag particles with thin outer shells and internal voids will be deposited inside the SRM. An investigation of the data from Real Size SRMs with BP-211J and ISR-DT-24 are currently in progress in order to use it as data for risk assessment. These studies will be a meaningful action for sustainable space activities, and will be reported in the near future.

References

[1] Makena Y., "Trends in Global Space Situational Awareness", Advanced Maui Optical and Space Surveillance Technologies Conference, 2021
[2] Krisko, P. P., Flegel, S., Matney, M. J., Jarkey, D. R. & Braun, V. ORDEM 3.0 and MASTER-2009 modeled debris population comparison. Acta Astronautica 113 (2015), pp. 204-211.
[3] Joel W., Daniel P., Asha B., Stephen O., "Characterizing the Orbital Debris Environment Using Satellite Perturbation Anomaly Data" First Int'l. Orbital Debris Conf. (2019)
[4] ISO 24113 Space systems – Space deburis mitigation requirements
[5] Akiba R., Ishii N., Inatani Y.,Behavior of Alumina Particle Exhaustedby Solid Rocket Motors, AIAA/NASA/DoD Orbital Debris Conference: Tech-nical Issues & Future Directions, Baltimore MD, April 16-19 1990

[6] Mark M., "An Assessment of the Role of Solid Rocket Motors in The Generation of Orbital Debris", NASA/TP-2007-213738, 2007.
[7] Burton G. Cour-Palais, "A Career in Applied Physics: Apollo Through Space Station", International Journal of Impact Engineering 23 (1999) pp.137-168

Annealing effects on Al/polyimide adhesion in flexible optical solar reflectors

MJ Cordill[1], P Kreiml[1], A Lassnig[1], DD Gebhart[1], S Zak[1], C Mitterer[2], T Griesser[3], G Milassin[4]

[1] Erich Schmid Institute of Materials Science, Austrian Academy of Sciences, Leoben, AT
[2] Department of Materials Science, Montanuniversität Leoben, Leoben, AT
[3] Chair of Chemistry of Polymeric Materials, Montanuniversität Leoben, Leoben, AT
[4] European Space Research and Technology Centre, Noordwijk, NE

E-mail: megan.cordill@oeaw.ac.at

Abstract. Flexible optical solar reflectors are made of single and multi-layered metal thin films on polymer substrates and will encounter around 6000 thermal cycles exceeding +/-100°C during one year of operation in low earth orbit. The candidate thin film system of Inconel/silver (Ag)/Teflon (FEP) recently demonstrated early damage formation (cracks and voids) after only a few thermal cycles, most likely due to the poor interfacial properties between Ag and FEP. An alternative material system that could be used is colourless polyimide, Tormed, instead of FEP. Additionally, aluminium (Al) has demonstrated very good interfacial properties with polyimide even after thermal cycling and has suitable optical properties. The adhesion of the Al/Tormed and Ag/Tormed interfaces were evaluated with tensile induced delamination. Generally, Al/Tormed has a much higher interface adhesion energy compared to Ag/Tormed, and there is no significant degradation after bake-out in vacuum (200°C, 10^{-6} mbar) for 10 and 24 hrs. Thus, the Al/Tormed system could be a more robust coating system for future flexible solar reflectors.

1. Introduction

Multilayer insulators (MLIs) and optical solar reflectors (OSRs) are integral parts to space flight that are designed and manufactured for thermal control of satellites and spacecraft in orbit. During their lifetime, satellite OSRs and MLIs must withstand the harsh space environment including thermal cycling (exceeding +/- 100°C) during low earth orbit (LEO). In order to address these varied environmental conditions, innovative material solutions are needed. The currently established and widely used material system used for radiator surfaces and flexible OSRs is Inconel/Ag/Teflon (FEP). However, when the Inconel/Ag/FEP system is thermally cycled under simulated LEO conditions, cracks form in the Inconel layer and void growth in the Ag layer occurs, greatly reducing the reflectivity of the system (about 20% of the total reflectivity) [1]. The decrease in reflectively could be critical and stray light could enter optical instruments or heat up surfaces that should remain in shadow [1]. From these laboratory experiments it was observed that the cracks start to form after only 6 thermal cycles of +/- 150°C in a nitrogen atmosphere [1]. The Inconel/Ag/Teflon (FEP) system has also been mechanically stretched (strained) to evaluate its mechanical and interfacial strengths. With only a small amount of strain (approximately 0.5% applied strain), the system begins to fail [2]. Cracks in the Ag layer form at only

2% applied strain which will also degrade the reflectivity of the layer. These applied strain levels can easily be reached by bending, rolling or folding, as these loading conditions are realistic for the assembly, integration and testing (AIT) phase, but for foldable structures also transport, launch and deployment in space would be relevant. Further investigations have found that the interface strength (adhesion) of the Ag/FEP measured with tensile induced delamination [1,3] is quite low compared to other metal-polymer interfaces measured with the same method [3–7]. Compared to the Inconel/Ag/FEP system, the Al/polyimide (PI-Upilex) system is mechanically very robust even after 200 thermal cycles of +/- 150°C [8,9] due to a naturally forming amorphous interface layer between the Al and PI substrate. This system can be strained up to 40% without the Al film delaminating and with only short cracks forming through the film thickness. The adhesion energy of the Al/PI interface was also found to be very stable even after thermal cycling and long annealing times [9,10]. One drawback to Upilex polyimide is that it shows some absorption in the visible wavelengths, having a yellow hue. Recently, a transparent polyimide called Tormed has been introduced commercially and it should be evaluated for space applications. In comparison to its yellow/brown counterparts, the polymerization process is slightly different, where the colourless polyimide is not fully aromatic like regular polyimide (Upilex). Thus, for gaining the trait of transparency its mechanical and thermal stability are inferior [11]. However, Tormed is still an attractive alternative to FEP because it has been demonstrated that metals adhere better to polyimides [8] and has a larger stable temperature regime (up to 400°C). Higher temperature applications can be envisioned with Tormed that are not possible with FEP. The optical properties of Tormed are good with 90% transmittance [12]. Compared to FEP, the transmittance of Tormed is slightly worse but still should be considered a candidate substrate for flexible OSRs.

The above results indicate that the adhesion between a reflective metal layer and the polymer substrate needs to be further addressed for flexible OSR and MLI applications. In order to minimize or avoid multilayer cracking or void formation due to thermal cycling, the layer structure must be carefully designed to ensure a thin film system that is thermally, mechanically, and environmentally stable for LEO. A possible solution for improving the thermal cycling resistance is to use a polymer substrate, which has better interfacial properties with Ag or change the reflective layer to a different metal, such as Al. The change of substrate and metal would allow for the investigation of interface stability, in terms of adhesion and potential voiding caused by diffusion mechanisms during thermal cycling. The adhesion and thermal stability between FEP and metals is poor because strong chemical bonds cannot form at the metal-polymer interface as they can with polyimide [10]. The adhesion will be evaluated with tensile induced delamination, a method based on fragmentation testing [3,13–15]. Fragmentation testing has been in use for that last decade to investigate the mechanical and electrical behaviour of metal films on polymer substrates for flexible applications [16–18] and is an ideal method to apply to flexible OSR and MLI material systems. Under a uniaxial load, the film will form cracks perpendicular to the loading direction and buckle delaminations parallel to the loading direction. The film delaminates due to a high transversal compressive stress that builds up during straining. The size of the buckles and the elastic properties of the film system are used to evaluate the interface strength (adhesion energy) [3,19]. Generally, brittle film systems can easily delaminate, but ductile films will first plastically deform. In order to help induce buckling, a hard overlayer has been used to confine the plastic deformation of the ductile film and promote delamination [2,4,20].

It is proposed here to study two film systems on a colourless polyimide substrate to investigate the interfacial strengths of the different metal-polymer interfaces. The reflective layers are Al or Ag and the polymer substrate to be utilized is Tormed polyimide. Ag is a more mobile atom [21,22] which is one reason why voids and hillocks form during simulated LEO thermal cycling [1]. The other reason is the poor interface that forms between Ag and FEP with a lower adhesion energy [1]. Al is not as mobile as Ag and diffusion is not expected to be as detrimental as with Ag. Al is more reflective in the UV than Ag and slightly less reflective than Ag in the visible spectrum [23]. Additionally, a top layer on both Al and Ag will be used to better induce delamination of the ductile Al and Ag films as well as to compare to the Inconel/Ag/FEP system. Reactive magnetron sputtering with oxygen was applied to fabricate Mo-based oxides (MoO_x) layers [24,25] which are transparent and have a high hardness. MoO_3 films are of interest due to their atomic oxygen resistance and were used as the capping layer.

2. Methods and Materials

Films were deposited using reactive ($Mo+O_2$) and non-reactive (Al, Ag, Mo) magnetron sputtering onto polymer substrates of Tormed (25 μm thick, IST Corporation, Osaka, Japan). Two different film systems were fabricated [24,26]: 220 nm MoO_3/300 nm Al/25 μm Tormed and 50 nm Mo/250 nm Ag/25 μm Tormed. MoO_3 is transparent and non-conductive [24,26]. A summary of the deposition conditions for all samples are listed in Table 1. These film architectures have not yet been applied to any space applications, however both have been tested for atomic oxygen resistance and are part of the chosen samples for the European Materials Aging program (set to launch in May 2024). Samples chosen for this program will be sent to the International Space Station and mounted on the Bartolomeo Module (ram facing) for 6 to 12 months. MoO_3 and Mo are used to study how these materials would react to space environmental conditions. Before deposition, polymer substrates were pre-cut into tensile strips and notched strips (figure 1a) using a Cricut Explore Air 2 and ultrasonically cleaned in an ethanol bath for 10 min. Pre-cutting is helpful to reduce defects induced through the cutting process and to always have the same sample dimensions for comparison. The strip sample is a commonly used shape for fragmentation experiments; however, it was learned that the notched geometry allows for locally more strain to be induced, thus more compressive stress in the films to help films delaminate. Finite element method (FEM) was utilized to determine the amount of strain applied to the notched sections. From the FEM simulations it is evident that the strain changes from the edge of the notch to the middle of the sample (figure 1b). Here, the highest strain at the edge of the notch will be reported. After deposition, long term annealing experiments, to determine if interface strength degradation occurs, were performed on the MoO_3/Al/Tormed at a maximum temperature of 200°C for times of 10 and 24 hrs, and vacuum conditions of approximately 10^{-6} mbar. The adhesion energies in the as-deposited state will be compared to the heat treated samples using the tensile induced delamination method [3].

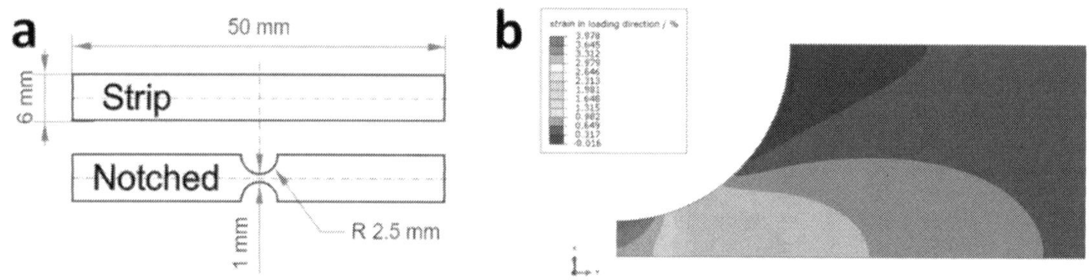

Figure 1. (a) Two sample geometries, the strip and notched, were used to induce delamination of the film systems. (b) FEM model of ¼ of the sample illustrating the strain gradient at the thinnest section of the sample. Colour online.

Table 1. Summary of deposition conditions and the deposition chambers used to create the MoO_3/Al/Tormed and Mo/Ag/Tormed.

Sample	Targets	Sputter power [W]	Pressure [mbar]	Substrate	Deposition time [sec]
220 nm MoO_3 / 300 nm Al / 25 μm Tormed	Planar Al (600 × 125 mm, 99.9995% purity)	3500	Ar: 5.2×10^{-3}	Grounded, no external heating	76
	Rotary Mo (Ø 125 × 600 mm, 99.97% purity)	2000	O_2: 1×10^{-3}	Grounded, no external heating	210
50 nm Mo / 250 nm Ag / 25 μm Tormed	Planar Ag (Ø 50.8 mm, 99.99% purity)	166	Ar: 3×10^{-3}	Grounded, no external heating	87
	Planar Mo (Ø 50.8 mm, 99.97% purity)	156	Ar: 3×10^{-3}	Grounded, no external heating	86

X-ray photoelectron spectroscopy (XPS) was used to determine the composition of the MoO_3 film. With reactive sputtering, different oxides can form depending on the deposition parameters. X-ray Photoelectron Spectrometer (XPS) spectra were recorded with a monochromatic XPS system (Thermo Fisher Scientific) with an Al Kα X-ray source (1486.6 eV) operating at a base pressure in the range of 10^{-10} to 10^{-8} mbar. Survey scans up to 1350 eV were recorded at a step size of 1 eV and a pass energy of 200 eV and high resolution Mo 3d core level spectra were taken with a step size of 0.1 eV and a pass energy of 50 eV for the range of 220 to 240 eV. All scans were taken at two different locations on the sample surface with a spot diameter of 400 μm. It was found that the reactive sputtering conditions with oxygen did result in a MoO_3 film as planned (figure 2) [27].

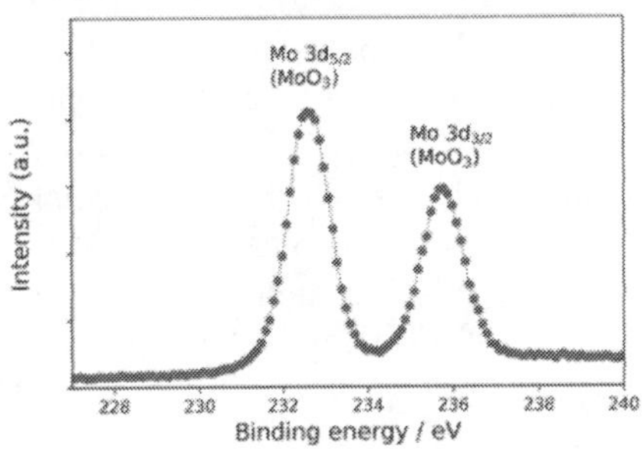

Figure 2. XPS Mo 3d core level spectrum of the MoO_3 layer indicates only the presence of chemical bonds characteristic for MoO_3.

When the films delaminate from the polymer substrates (form buckles) due to the tensile induced compressive stress that arises during the tensile straining, the adhesion energy of the interface can be evaluated [3]. Adhesion is calculated from the resulting delamination buckles that are imaged with 3D surface techniques, such as atomic force microscopy or confocal laser scanning microscopy (CLSM). In figure 3, an example of an image of a buckle and the corresponding buckle profile with the dimensions of buckle width, $2b$, and height, δ, indicated. The buckle dimensions are normalized by the total film thickness, h, and plotted as (b/h) vs. $(\delta/h)^{(1/2)}$. From this plot, Eqn. (1) is fit to the lower buckle data points where α is a fitting parameter,

$$\sqrt{\frac{\delta}{h}} = (2\alpha)^{\frac{1}{4}} \frac{b}{h}\left(1 + \sqrt{1 + \frac{3}{4}\alpha\left(\frac{b}{h}\right)^4}\right)^{-\frac{1}{4}}. \quad (1)$$

The fitting of the lower data has been shown to describe the buckling without additional influences from film cracking, which can occur at the buckle apex (top) or at the bases. Using the α parameter the adhesion energy, Γ, is calculated with Eqn. (2),

$$\Gamma = \frac{\alpha\, h\, E_f'}{4}\left(\frac{\pi}{2}\right)^4, \quad (2)$$

where the modified elastic modulus, $E' = E_f / (1-v_f^2)$, is thickness weighted for a multilayer system. The reader is referred to the original work for more details [3].

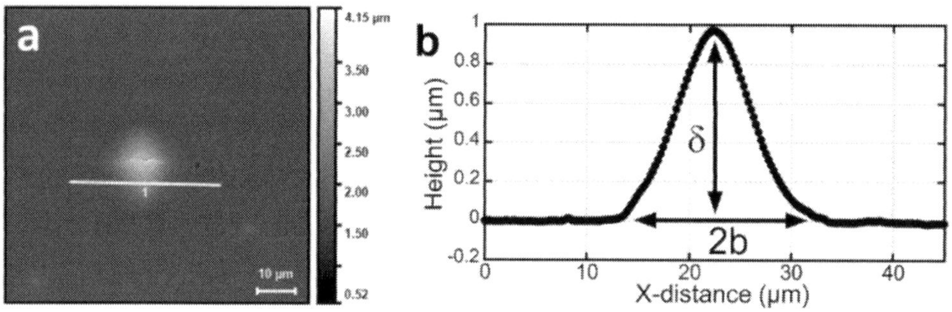

Figure 3. (a) CLSM height image of buckle in the MoO_3/Al/Tormed. (b) Buckle profile indicated in (a) with the methodology to measure the buckle dimensions to calculate adhesion.

3. Results and Discussion

As-deposited MoO_3/Al/Tormed samples did not produce any observable cracking and deformation during testing with in-situ CLSM straining of the strip sample (figure 4a). It should be noted that Tormed strip samples can only be strained to approximately 3% before complete fracture of the sample near the grips. Thus, the two different sample geometries were used for the straining experiments in order to increase the compressive stresses in the films and induce delamination. This is especially important for the ductile Al and Ag films which will first plastically deform before allowing delamination [28]. Figure 4b illustrates that using the notched sample increases the possibility to induce delamination due to the application of about 15% strain compared to the strip geometry. The Mo/Ag/Tormed system on the other hand, formed cracks and buckles well before the Tormed substrate broke (figure 4c). Buckles started forming at 2.3% applied strain. Furthermore, it was not necessary to use the notched sample shape to induce buckle delaminations. Qualitatively, the observation that the Mo/Ag film system delaminated so easily is an indication that Ag does not adhere well to the Tormed substrate. It is known that Al/Tormed and Ag/Tormed interfaces delaminate by examining the films through the transparent Tormed. A similar procedure was used by Putz et al. [2] to determine how the Ag film cracks during straining.

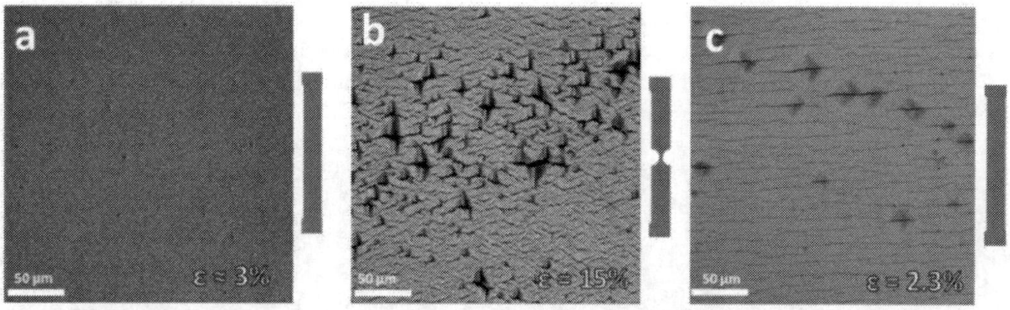

Figure 4. Two sample geometries were used for the straining experiments on the Mo₃/Al/Tormed samples. The strip geometry (a) did not lead to cracking and delamination of the film system. A notched geometry (b) allowed for 15% strain to be applied at the notch that causes cracking and delamination at the Al/Tormed interface. (c) Cracks and buckles formed on the Mo/Ag/Tormed sample on the strip sample, thus the notched geometry was not used. All images are CLSM laser intensity images.

To calculate the adhesion energy, Γ, of the delaminated interfaces, at least 20 buckles on each film system were measured. For the as-deposited MoO_3/Al/Tormed sample Γ = 55 Jm^{-2} was measured with an α parameter of 0.0015 (figure 5a). The adhesion energy for the as-deposited Mo/Ag/Tormed sample was found to be significantly lower at Γ = 0.7 Jm^{-2} (figure 5b). These results agree with the work of others where the adhesion energy of the Al/PI was measured to be near 70 Jm^{-2} [8] due to a strong interlayer that formed between the Al and PI substrate [9]. In the as-deposited state, it has been shown that Al creates a stronger interface to Tormed compared to Ag because the Al-Tormed system has significantly higher interface strength.

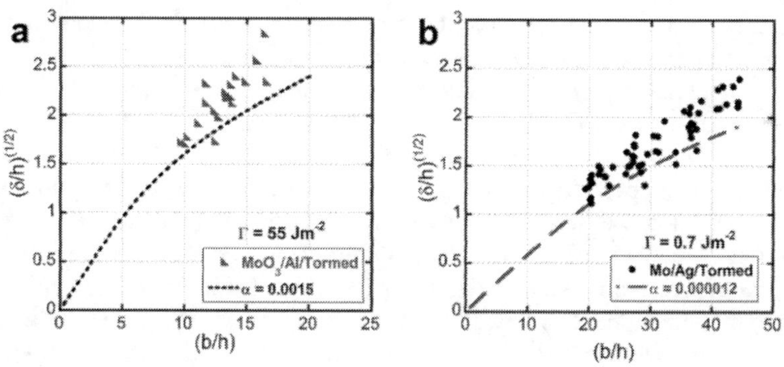

Figure 5. (a) The adhesion energy of the MoO_3/Al/Tormed system was found to be significantly higher than the Mo/Ag/Tormed (b) using the tensile induced delamination method.

Due to the fact that the as-deposited Mo/Ag/Tormed system had a low interface strength, annealing in vacuum at 200°C for 10 and 24 hrs was only performed on the MoO_3/Al/Tormed film system. Straining of the annealed MoO_3/Al/Tormed was performed with only the notched geometry to examine the fracture and delamination behaviour is shown in figure 6. What is observed is that the as-deposited sample (figure 6a) has cracks and buckles, but with increasing annealing times, only cracks at the notched areas where the highest level of strain is applied were observed with no buckling (figure 6b-c). This qualitatively indicates that the adhesion energy of the Al/Tormed interface is improved with annealing. Previous studies have shown that the strong mixed interlayer of Al and PI grows in thickness with annealing at 300 °C, but no increase in adhesion was observed due to measurement uncertainties [8,10]. The qualitative increase in adhesion observed here with annealing at 200 °C substantiates the assumption that a similar interlayer has formed during sputter deposition of Al and is the reason for the improved adhesion over Ag to the same substrate. Additionally, between 11 and 15% strain was applied to the Al/Tormed system to achieve the amount of damage visible in figure 6, compared to less than 3% applied strain on the Ag/Tormed system.

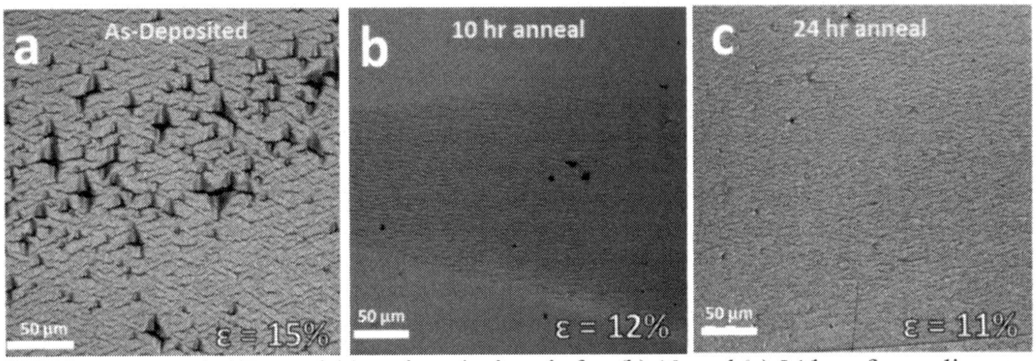

Figure 6. MoO_3/Al/Tormed (a) as-deposited, and after (b) 10, and (c) 24 hrs of annealing at 200°C in vacuum. Straining of the notched samples only revealed cracking at the applied strains and no buckles. Qualitatively the adhesion is improved.

4. Conclusions

Al and Ag films were sputtered deposited onto colourless polyimide called Tormed and evaluated in terms of adhesion energy using the tensile induced delamination method. These films could be used as future materials for flexible OSRs or outer layer MLI foils. From the resulting buckle delaminations, it was found that the Al/Tormed system had a higher adhesion energy compared to the Ag/Tormed film system. The adhesion energy of Al/Tormed was found to be qualitatively improved by annealing at 200°C for 10 and 24 hrs in vacuum. The colourless polyimide substrate (Tormed) can withstand higher temperatures than the currently used FEP substrates and metals, such as Al, adhere better to polyimide-based substrates. Annealing also illustrated that the Al/Tormed interface is stable and does not degrade at elevated temperatures for extended periods. Simulated LEO thermal cycling would be the next step to further evaluate the material system for future flexible OSR applications. The studied MoO_3/Al/Tormed system is part of the European Materials Aging program and once samples return from the mission, vital information for new possible protection coatings and new flexible substrates to use for MLIs and OSRs in LEO will be provided. With positive results, it can be envisioned to have the ability to design material combinations having the optimum protection against atomic oxygen and being robust against extreme temperature cycles, with mechanical and interfacial reliability necessary for the LEO environment. The new knowledge gained about the proposed material systems could also be applied to rigid substrates and applications, such as cover glasses and optical coatings on lenses and first surface mirrors, as the Mo-based oxide films can also be deposited onto glass or metal substrates.

5. Acknowledgements

This work was supported by the European Space Agency under contract: 4000129515/19/NL/KML7rk

References

[1] Putz B, Wurster S, Edwards TEJ, Völker B, Milassin G, Többens DM, Semprimoschnig COA and Cordill MJ 2020 *Acta Astronaut.* **175** 277
[2] Putz B, May-Miller C, Matl V, Völker B, Többens DM, Semprimoschnig C, and Cordill MJ 2018 *Scr Mater* **145** 5
[3] Cordill MJ, Fischer FD, Rammerstorfer FG, and Dehm G. 2010 *Acta Mater* **58(16)** 5520
[4] Cordill MJ, Paulitsch M, Katsarelis C, Putz B, Lassnig A, Kennedy MS 2022 *Thin Solid Films* **742** 139051

[5] Jörg T, Cordill MJ, Franz R, Kirchlechner C, Többens DM, Winkler J, and Mitterer C 2017 *Mater. Sci. Eng.* A **697** 17
[6] Cordill MJ, Kleinbichler A, Völker B, Kraker P, Economy DR, Többens D, Kirchlechner C and Kennedy MS 2018 *Mater. Sci. Eng.* A 735 456
[7] Kreiml P, Rausch M, Terziyska VL, Köstenbauer H, Winkler J, Mitterer C, and Cordill MJ 2020 *Materialia* **12** 0
[8] Putz B, Völker B, Semprimoschnig C, Cordill MJ. 2017 *Microelectron. Eng.* **167** 17
[9] Putz B, Milassin G, Butenko Y, Völker B, Gammer C, Semprimoschnig C, and Cordill MJ 2017 *Surf. Coat. Technol.* **332** 368
[10] Putz B, Milassin G, Butenko Y, Völker B, Gammer C, Semprimoschnig C, and Cordill MJ 2018 *Surf. Interface Anal.* **50** 579
[11] Tapaswi PK, Ha CS 2019 *Macromol. Chem. Phys.* **220(3)** 1
[12] Thermo-optical measurements of Tormed samples - Transparent polyimide. 2020.
[13] Kelly A, Tyson WR. 1965 *J. Mech. Phys. Solids* **13(6)** 329
[14] Agrawal DC, Raj R. 1989 *Acta Metall.* **37(4)** 1265
[15] Ho PS, Faupel F. 1988 *Appl. Phys. Lett.* **53(17)** 1602
[16] Cordill MJ, Kreiml P, Mitterer C. 2022 *Materials* **15** 926
[17] Lacour SP, Wagner S, Huang Z, Suo Z. 2003 *Appl. Phys. Lett.* **82(15)** 2404
[18] Olliges S, Gruber PA, Auzelyte V, Ekinci Y, Solak HH, Spolenak R. 2007 *Acta Mater.* **55** 5201
[19] Toth F, Rammerstorfer FG, Cordill MJ, Fischer FD. 2013 *Acta Mater.* **61** 2425
[20] Cordill MJ, Marx VM, Kirchlechner C. 2014 *Thin Solid Films* **571** 302
[21] Berry WB, Emery KA, Swartzlander AB, Nelson AJ. 1988 *Conf Rec IEEE Photovolt Spec Conf.* **1(February)** 262
[22] Agrawal PM, Rice BM, Thompson DL. 2022 *Surf. Sci.* **515(1)** 21
[23] Correia JH, Bartek M, Wolffenbuttel RF. 2000 *IEEE Trans Electron Devices* **47(3)** 553
[24] Pachlhofer JM, Martín-Luengo AT, Franz R, Franzke E, Köstenbauer H, Winkler J, Mitterer C. 2017 *J. Vac. Sci. Technol.* A **35(2)** 021504
[25] Pachlhofer JM, Tarazaga Martín-Luengo A, Franz R, Franzke E, Köstenbauer H, Winkler J, and Mitterer C. 2017 *Surf. Coatings Technol.* **332** 80
[26] Pachlhofer JM, Jachs C, Franz R, Franzke E, Köstenbauer H, Winkler J, and Mitterer C 2016 *Vacuum* **131** 246
[27] Scanlon DO, Watson GW, Payne DJ, Atkinson GR, Egdell RG, Law DSL. 2010 *J. Phys. Chem.* C. **114(10)** 4636
[28] Kreiml P, Rausch M, Terziyska VL, Köstenbauer H, Winkler J, Mitterer C, and Cordill MJ 2018 *Thin Solid Films* **665** 131

IOP Conf. Series: Materials Science and Engineering 1287 (2023) 012007 doi:10.1088/1757-899X/1287/1/012007

Assessing the effects of critical parameters in standoff bonding

O Batzilis[1,2,4] , S Rodriguez[1], N Blasakis[3], A Baltopoulos[3], P Janik[1] and A Stimoniaris[2]

[1] European Space Agency ESTEC, TEC-QEE, Keplerlaan 1, 2201AZ Noordwijk, The Netherlands

[2] University of Western Macedonia, Department of Chemical Engineering, Kila 50100 Kozani, Greece

[3] Adamant Composites Ltd., Agias Lavras & Stadiou Str., Platani-Patras,26504, Greece

Abstract. In this contribution we present the effect of several parameters that are critical for bonding standoffs onto the surface of spacecrafts' panels with polymeric adhesives. Standoffs are an integral part of the thermal management of space satellites since they ensure that the MLI blankets are mounted and maintained in the correct position during the whole mission duration, including launch. In the recent years, there have been several de-bonding problems during qualification phases, leading to long and costly delays, oftentimes caused from poor design and preparation[1]. For these reasons, we have performed a study of the bond strength of Torlon® 5030 rods bonded over sandwich panels made of an aluminium honeycomb core and CFRP skins. The bond strength was determined by mechanical tests, conducted in tensile and shear modes, both prior and after thermal cycling. Thermal analysis was also performed on the main components under study. The study consists on the following steps: first, different surface treatments of the CFRP skins have been investigated: (i) bonding standoffs with the surface as manufactured, (ii) cleaned with a solvent and (iii) mechanically sanded before cleaning. The best performing treatment was selected and used systematically in the rest of the study. Afterward, different types of adhesives (epoxy, graphene nano-enabled epoxy, polyurethane) and bond-line thicknesses were studied. The findings of this study can help to steer the industry towards better practices regarding the standoff bonding. This will improve the integrity of future space missions.

1. Introduction

Standoffs are used in spacecrafts to mount in place Multi-layer Insulation (MLI) blankets. MLI is used as an insulator in vacuum environments, either for space missions or cryogenics. Common insulation materials, such as the ones used in home construction function by reducing heat transfer mainly by means of conduction and convection. In a vacuum environment, for example between the spacecraft and the space environment, conduction and convection mechanisms are not present and only radiation occurs. Therefore, to reduce radiation heat loss to or from space, a radiative material such as MLI must be used. If MLI was directly wrapped in contact with the structure of the satellite, it would create significant heat transfer paths in which conduction would play a leading role. To avoid these heat

[4] Author to whom any correspondence should be addressed at orfeas.batzilis@gmail.com

IOP Conf. Series: Materials Science and Engineering 1287 (2023) 012007 doi:10.1088/1757-899X/1287/1/012007

transfer channels, standoffs are used to mount MLI in place eliminating the points of contact between MLI and the spacecraft. Standoff bonding currently is a common cause of non-conformances occurring often as a result of poor design and preparation. Issues with adhesive selection, curing or application procedures emerge during the qualification campaign. Further investigations usually consider different materials or processes, which imply additional delta-qualifications and therefore costly delays. These unexpected steps could be avoided if more data were openly available. Current open-source research, while adequate, only examines epoxies and specifically in cryogenic temperatures[2]. In this contribution we present information of room-temperature tests on epoxies and polyurethane adhesives. To obtain this data, the bonding strength was measured using a tensile machine, pulling in shear and tension modes. The effects of adhesive type, bondline thickness and surface preparation were assessed and evaluated. We have also considered the effects of temperature cycling to conclude how and when critical parameters affect the bonding strength. Residual stresses in the adhesives are expected to have an effect on the bonding strength of pristine versus thermally cycled samples. The effect of these stresses can be magnified by the bondline thickness, and therefore the amount of adhesive material.

2. Materials

Two main test types were carried out. Initially, a pre-screening test was performed in pristine samples in shear mode. This pre-screening was performed using 2216 epoxy adhesive from 3M, and Solithane 113 adhesives and a 120 μm spacer wire, -made out of INVAR- to achieve a specific bondline thickness. These adhesives were selected as they are used to a great extent in space missions, especially 2216 epoxy is often used for standoff bonding applications. The standoffs bonded were made from Torlon® 5030 HF, a thermoplastic material often used in these types of applications as it shows high strength and stiffness. The substrate material was CFRP/Aluminum sandwich structure, consisting of an aluminum honeycomb core with CFRP skins bonded on each side. The purpose of pre-screening was to identify which surface preparation was better, for which 3 standoffs were bonded for each adhesive.

After the pre-screening test, the best surface treatment was selected and tests in shear and tension were performed, both for pristine and thermally cycled samples. For these tests, in addition to 3M 2216 and Solithane 113, a graphene nano-enabled epoxy was tested. For each adhesive, three different bondline thicknesses were selected, 120 μm, 250 μm and 500 μm. For each set of parameters, three standoffs were bonded and tested.

3. Methodology

The first step in the adhesive bonding process was to study the effect of surface preparation of sandwich panels and standoffs. All standoffs were cleaned with isopropyl alcohol (IPA). For the sandwich panels the CFRP had 3 different preparations: (i) no treatment, (ii) IPA cleaning and (iii) sanding using fiber glass pen and then IPA cleaning to remove any sanding debris.

After the surfaces were prepared, they were laid on top of clean poly-wipe. Then the mixing of the adhesives took place. An empty plastic cup container was weighed, and then parts A and B were poured in accordance with the weight ratio suggested by the manufacturer. They were then carefully mixed with a clean wooden spatula, aiming to minimize entrapped air bubbles. After the adhesive was thoroughly mixed for at least 30 s, the cup was placed inside a vacuum chamber, and it was cycled 3 times between atmospheric pressure and vacuum to remove any air pockets introduced during the mixing. Then, it was carefully inserted in syringes to help the adhesive application. Figure 1 depicts the nano-enabled adhesive in the syringe, right before the application.

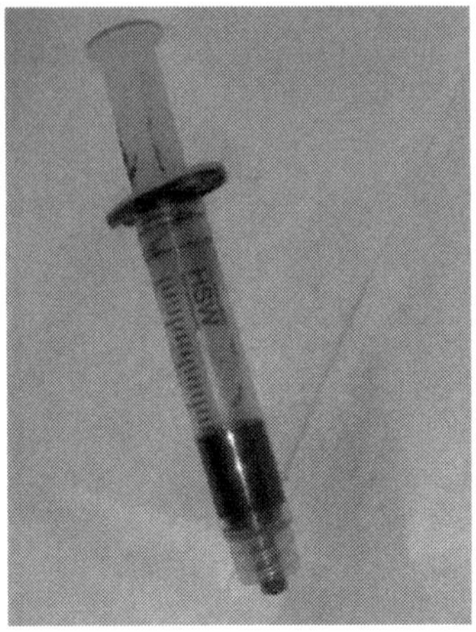

Figure 1. Syringe filled with nano-enabled adhesive.

With the adhesive and surfaces ready, the bonding procedure started. At first, a small amount of adhesive in form of a blob was laid in the sandwich panel. Then the 3 spacers that would secure the bondline thickness were laid as evenly as possible separated 120 degrees from one another. After the spacers were in place, a small amount of adhesive was applied to the bottom of the base of the standoff and, using tweezers, it was pressed firmly to the bonding area for a couple of seconds. While placing the standoff onto the CFRP it was important to pay attention to the spacer wires as they would often move resulting in a not ideal placement or even being removed completely. To avoid this, it was important to try to keep them in place by holding them while also pressing down the standoff. After this procedure was performed for one adhesive, the procedure was repeated for the other ones, then the samples were left to cure for at least 7 days.

To compare pristine samples with thermally cycled samples, a thermal cycling vacuum (TVAC) chamber was used. The samples were kept for a about a day just in vacuum, to reach high vacuum, as representative as possible for space applications. In total, the samples were exposed to 50 heat-cool cycles between 90 °C and -100 °C. The heating rate was 3 K/min and a dwell time after heating and cooling of 10 minutes. Finally, a thermocouple was attached to the top surface of each panel to have a good reading of the actual temperature that was present in each sample (Figure 2).

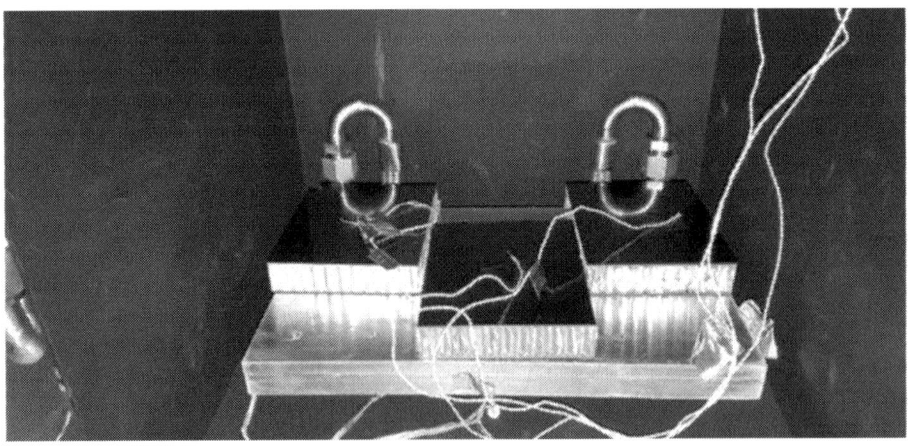

Figure 2. Thermal cycling vacuum chamber with dummy samples for fit check.

To measure the strength of the bonding the LR5K Plus tensile instrument by AMETEK, as it can be seen in the Figure 3, was used. The following parameters were chosen: pulling rate of 5 mm/min and 0.5 N pre-force. If the instrument sensed a drop of more than 10% it would stop. This limit was sufficient to detect when the standoffs had debonded.

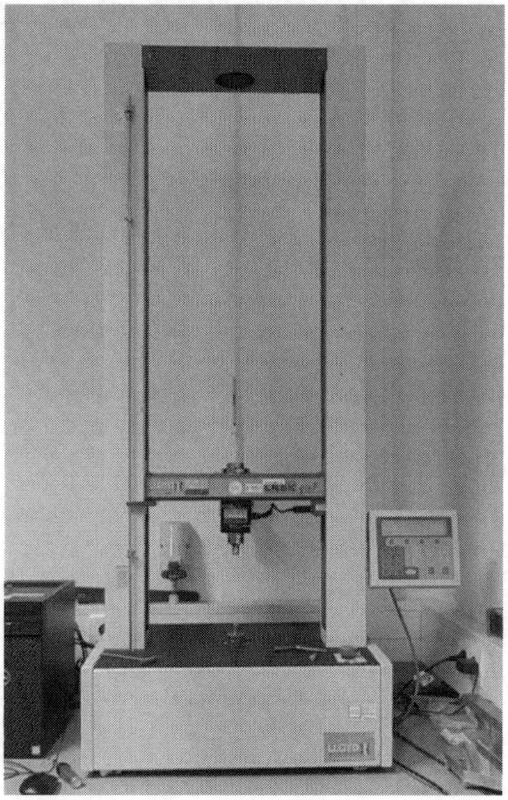

Figure 3. Lloyd LR5K Plus tensile instrument available in ESTEC labs.

As the sample geometry was not the standard lap-shear samples used in this instrument, the available jigs were not able to clamp the samples correctly. For this reason, it was necessary to design and manufacture dedicated custom jigs, made from stainless steel and aluminum, for the tension and shear tests.

In Figure 4 a standoff being bent can be seen during shear testing for the surface preparation study. In that figure some fracture modes are also visible. In particular, two out of three fractures while having no surface preparation are non-cohesive, and for IPA cleaning surface preparation the only fracture that had been performed at that moment was cohesive.

Figure 4. Bent standoff during surface preparation shear testing.

4. Results

As mentioned previously, the first step in the standoff bonding was selecting the surface treatment. Graph 1 shows the results of the pre-screening test using both 3M 2216 and Solithane 113, between CFRP/Aluminum honeycomb and standoffs, and pulled in shear.

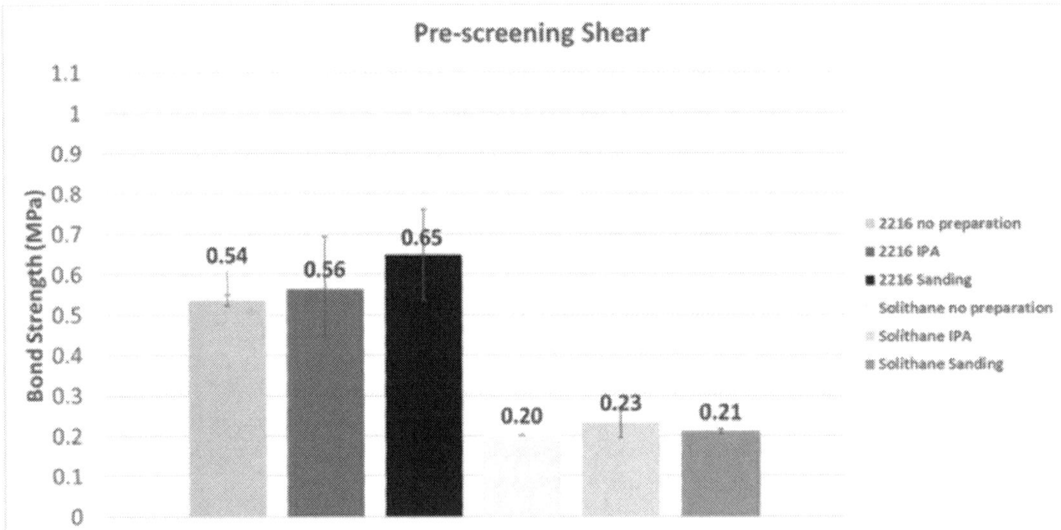

Graph 1. Bonding strength in pre-screening surface treatment shear samples.

Sanding was considered to be the best surface treatment, as it showed 17% higher bond strength as compared to the lowest value in the case of the epoxy. This trend was not confirmed in the case of the polyurethane adhesive, but in that case the differences are less obvious. After the pre-screening, the CFRP was sanded and IPA-cleaned, and then the rest of the samples were bonded and tested in pristine and thermal vacuum cycling configurations. The graphs below show the bonding strength of pristine shear, pristine tension, thermally cycled shear and thermally cycled tension tests. Graph 2 and

Graph 3 display the results of shear testing, with pristine and thermal cycled samples respectively. It can be noted that thermal cycling benefits the bonding strength of all epoxy-based adhesives, while not having a clear trend regarding the polyurethane based adhesive. In epoxy 2216, when pristine, the 500 μm bondline thickness is performing better than the rest thicknesses but after thermal cycling it gets outperformed by the 120 μm bondline thickness. For nano-enabled epoxy, when pristine, the 250 μm thickness is the option with the highest bonding strength but after thermal cycling it is the 500 μm. Finally, for Solithane 120 μm and 250 μm are performing identically and 500 μm is performing the worst. After thermal cycling, 500μm outperforms the rest thicknesses, with 120 μm being the second best and 250μm being the worst.

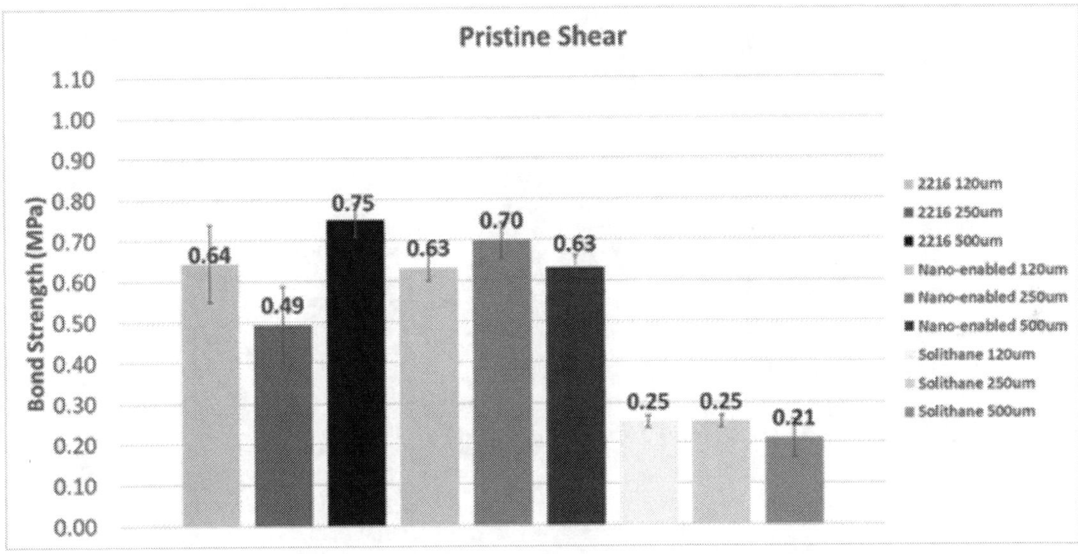

Graph 2. Bonding strength in pristine shear samples.

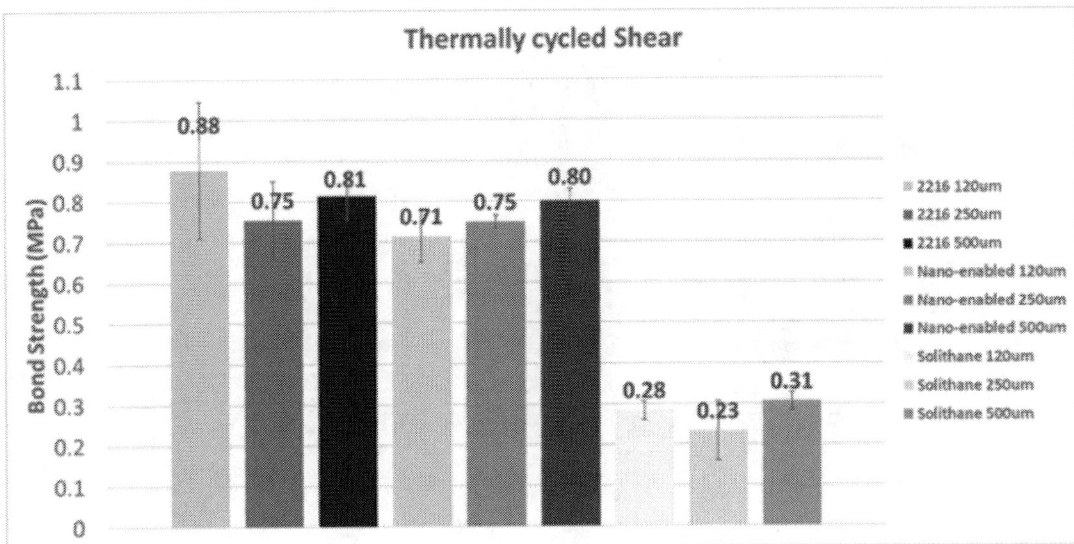

Graph 3. Bonding strength in thermally cycled shear samples.

Graph 4 and Graph 5 show similar trends with Graph 2 and Graph 3. In tension, epoxy 2216 and nano-enabled epoxy have the same best thickness as in shear for both pristine and thermally cycled samples. Solithane is different as the identical performing thicknesses are now the 250 μm and 500 μm, with the 250 μm being the best after thermal cycling, contrary to thermally cycled shear where it was worst.

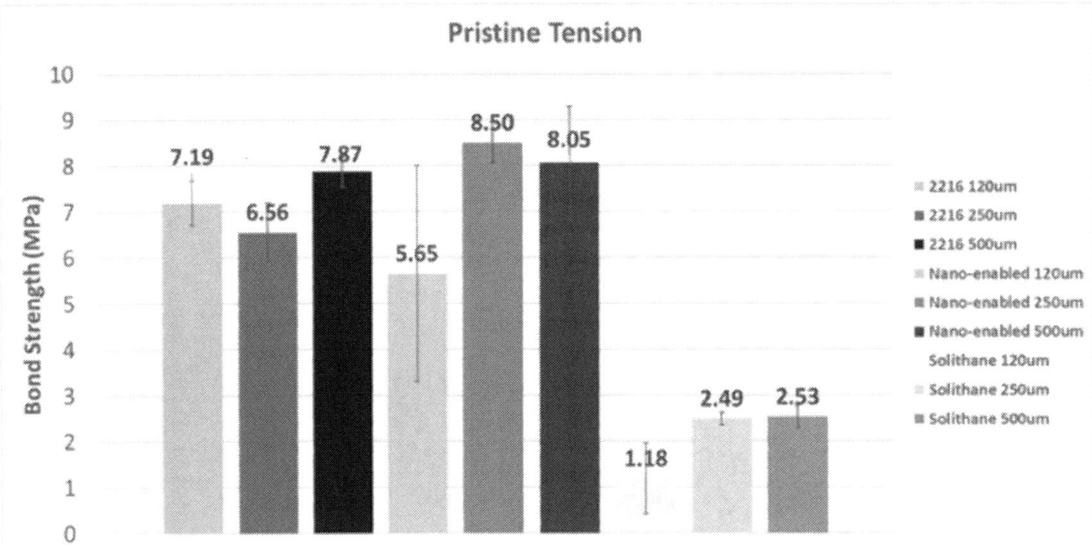

Graph 4. Bonding strength in pristine tension samples.

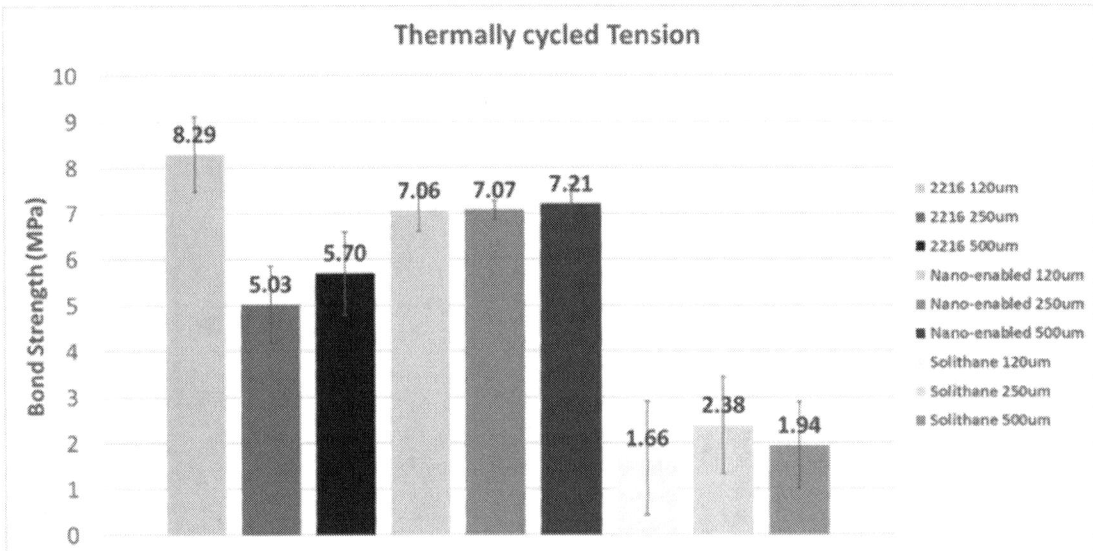

Graph 5. Bonding strength in thermally cycled tension samples.

5. Conclusions

Sanding the CFRP as a surface treatment, and cleaning afterwards with isopropyl alcohol, provided an 17% increase in the bond strength for epoxies. Polyurethane was not affected significantly by the surface preparations, as the data values were in the error limits. Epoxy adhesives, as expected, showed a higher bond strength than polyurethane. In particular, 2216 epoxy (Gray) results in at least 96% higher bond strength than Solithane 113. The nano-enabled adhesive showed similar bonding strength as the 2216 epoxy (Gray). In pristine 2216 samples, 500 μm proved to be the best bondline thickness for shear and tension, while for thermally cycled in vacuum 2216 samples, 120 μm bondline thickness outperformed the rest thicknesses. Pristine shear and tension samples present similar trends in the effect of bondline thickness of the epoxies, with 250μm thickness being the worst for 2216, while the best for the nano-enabled epoxy. Especially in tension, as it can be seen in the Figure 5 the adhesive fracture joint mode was mostly cohesive for Solithane 113, non-cohesive for 2216 as the CFRP partially delaminated staying bonded on to the standoff, and a mix of those for the nano-enabled adhesive. In shear mode, as can be seen in Figure 6, adhesive fracture joint modes did not follow a clear trend. Solithane 113 was cohesive, nano-enabled epoxy had some standoff fractures as its base broke leaving a part of it bonded on to the CFRP/Aluminum sandwich structure, and 2216 epoxy had only one standoff fracture in the same manner as before. Lastly, the effect of thermal cycling under vacuum was apparent as in most cases there was an increase in the bonding strength for epoxies but not for polyurethane. In Figure 6 where the Solithane 113 samples were bonded, in the right top there was no sample bonded, and this explains why there is no trace of adhesive.

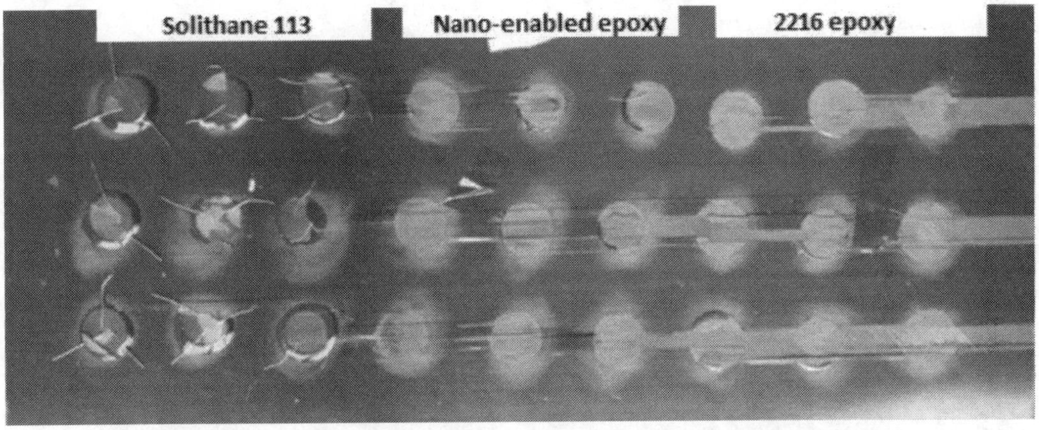

Figure 5. Tension CFRP/Alu TVAC cycled samples after testing.

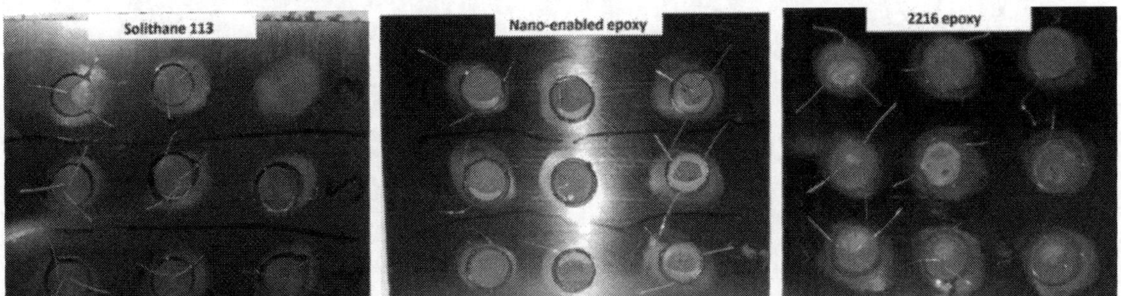

Figure 6. Shear CFRP/Alu TVAC cycled samples after testing.

To conclude, for 3M 2216 epoxy, if the sample will undergo thermal cycling, it is clear that the best option is to bond with the smallest bondline thickness. For the same conditions, the nano-enabled epoxy bonds stronger in shear with the larger bondline thickness, while in tension the bondline thickness does not show a clear trend as the value ranges for the different thicknesses seem to overlap. In order to identify a trend, regarding the bondline thickness effect, more samples should be tested for each set of parameters. Finally, Solithane 113 did not show a clear trend for both shear and tension modes.

The standoff bonding study could benefit from a further examination of more adhesive families and including a larger number of samples for a better statistical distribution of results. Conditioning of the samples could also be performed e.g., vibrations, humidity. Finally, a fracture observation using SEM would also provide valuable insights regarding the mode the adhesive fractured.

Acknowledgements
The authors would like to thank the Materials' Physics and Chemistry laboratory personnel at ESA ESTEC, and in particular the head of section Riccardo Rampini, for their support to conduct this research.

References

[1] Janik P, Olive M, Holynska M and Semprimoschnig C 2018 Proc. *Int. Symp. on Materials in the Space Environment* (Biarritz, France)
[2] Alberts S, Dohene C and Johnson W 2017 Proc. *Cryogenic Engineering Conf.* (Madison, USA)

Space Silicone Adhesive Monitoring Via Minimally Intrusive Optical Fibre Sensing During Vacuum UV/VUV Radiation Exposure

L Fazzi[1,2], N Dias[2], M Holynska[2], R Rampini[2] and R M Groves[1]

[1] Structural Integrity and Composites Group, Faculty of Aerospace Engineering, Delft University of Technology, Delft, 2629 HS, The Netherlands.
[2] ESA-ESTEC, Keperlaan 1, 2201 AZ Noordwijk, The Netherlands.

l.fazzi@tudelft.nl

Abstract. In this work a technique to monitor the health state of space adhesives during environmental testing is presented. Specifically, minimally intrusive Tilted Fibre Bragg Grating (TFBG) sensor was embedded in silicone adhesive Nusil CV16-2500 used to bond two thin cover glasses. The resulting solar cell assembly was then exposed to the UV/VUV radiation in a high vacuum chamber (less than 10^{-5} mbar). The sensor is demonstrated that it was able to measure simultaneously the strain, temperature and refractive index of the adhesive. The three parameters measured with the TFBG, can be used to detect the thermomechanical and refractometric state of the silicone adhesive.

1. Introduction

Space technologies widely use silicone polymers as adhesives, encapsulants and sealants. Indeed, once cured, these polymers have high transparency and resistance to chemicals and harsh environmental conditions. Furthermore, they remain highly flexible even at cold temperatures and keep their mechanical performance in a wide temperature range as well as adhesion to several materials [1]. A common application is for solar arrays. Specifically, a silicone sticking layer bonds the solar cells with an overlying protective cover glass. Another main application is for docking systems, while smaller amounts are exploited in LEDs, electronic and optical components and sensors [2]. Despite these excellent features, the appreciated properties of silicone adhesives may be negatively affected by some phenomena occurring under space environment. In particular, when considering the operational conditions of the space solar arrays, the combination of long high vacuum and UV radiation exposure can affect the original dimensional stability and composition of the silicone adhesive. Indeed, the vacuum space environment causes the outgassing of low-molecular weight residues and additives of the silicone, which increase contaminants' levels. The exposure of the contaminant layers to UV radiation generates the fixation of these layers on the surfaces, causing, in this way, severe efficiency and operational issues. Another drawback regards the absorption of the UV radiation by organic molecules. This leads to photochemical reactions which cause the rupture of chemical bonds and the generation of radicals. The final result is then a change in colour of the polymer, a variation in material properties, micro-cracking, loss of efficiency and premature failures [3].

IOP Conf. Series: Materials Science and Engineering 1287 (2023) 012008 doi:10.1088/1757-899X/1287/1/012008

These undesirable consequences can affect structurally fundamental parts of the spacecraft and/or reduce the efficiency of its components during the mission. Considering that in the space technological field, each issue may be crucial for the achievement of the mission. However, some vital parts, for example the solar arrays, are permanently exposed to high-vacuum and UV radiation. A sensing technology for the monitoring of the silicone health state during the spacecraft operational life or testing phase could provide important information and improve the safety level of the components. This apparatus should be physically minimally intrusive, light and resistant, but also provide a complete material state, accurate and real-time remote measurements and have a simple acquisition system.

With this aim, herein, a weakly Tilted Fibre Bragg Grating (TFBG) is embedded inside the silicone layer of a solar cell assembly to monitor its state during UV/VUV exposure in high vacuum. This sensor matches all the embedding requirements as it is minimally intrusive (Ø = 125 μm), lightweight (60-70 g/km) and easy to interrogate. Each single TFBG can perform in-situ accurate multi-parameter measurements without the support of other sensors [4,5], also in real-time and remote way without interfering with the surrounding material. Specifically, here, the TFBG sensor is used to obtain simultaneous information on the thermomechanical and refractometric states of the Nusil CV16-2500 silicone adhesive. The strain-temperature measurements give an indication on the dimensional and physical condition of the material. At the same time, the variations of its refractive index (RI) are linked to the chemical changes in the polymer [6,7]. Hence, since some chemical reactions may be induced by UV/VUV radiation, the silicone RI may undergo a variation detectable by the TFBG sensor.

2. TFBG multi-sensing techniques

TFBGs are a special kind of FBG sensor where the short-period RI modulation in the core layer of the optical fibre (OF) is engraved by imposing a tilt angle (θ) with respect to the optical axis of the OF (Fig.1). In this way, the tilted Bragg gratings generate a coupling system of the modes between the core forward-propagating mode and the backward-propagating core and the cladding modes.

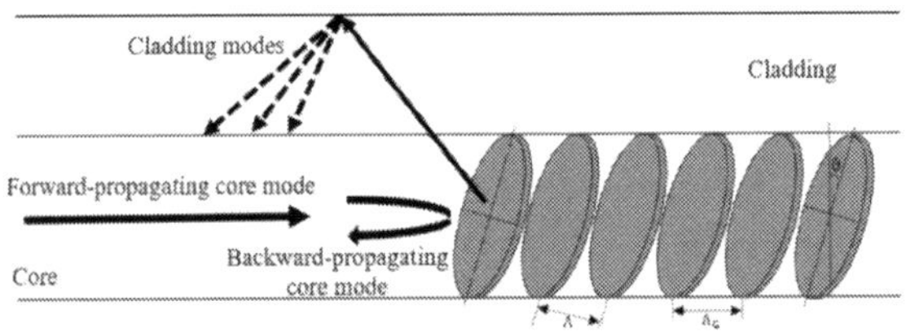

Figure 1. Schematic of TFBG internal structure.

In Fig.1, Λ_G and Λ are the nominal spacing and the period of the grating, respectively. The result of this coupling-mode system is the enhancement of the cladding resonance peaks in the transmission spectrum. Specifically, for weakly TFBG ($\theta < 10°$), the spectrum is composed of three different kinds of peaks: Bragg, ghost and cladding peaks (Fig.2).

These three kinds of peaks can be used to perform, simultaneously and separately, strain, temperature and RI measurements. Indeed, the Bragg and the ghost peaks are sensitive to thermomechanical perturbations but not to external RI variations. Therefore, a demodulation technique based on the wavelength shifting of these two peaks can be applied to obtain the strain ($\Delta\varepsilon$) and the temperature (ΔT) variations [5]. This is allowed by using the following matrix system:

$$\begin{pmatrix}\Delta\varepsilon \\ \Delta T\end{pmatrix} = \begin{pmatrix}K_{\varepsilon,Bragg} & K_{T,Bragg} \\ K_{\varepsilon,Ghost} & K_{T,Ghost}\end{pmatrix}^{-1} \begin{pmatrix}\Delta\lambda_{Bragg} \\ \Delta\lambda_{Ghost}\end{pmatrix}. \quad (1)$$

IOP Conf. Series: Materials Science and Engineering 1287 (2023) 012008 doi:10.1088/1757-899X/1287/1/012008

K_ε and K_T are the sensitivity coefficients of the selected peaks for unitary temperature and strain variation. These coefficients are obtained by calibration of the TFBG. The sensor here used was calibrated thermally and mechanically as described in [5].
The values found for the coefficients are:

- $K_{\varepsilon,Bragg} = 1.255 \pm 0.004$ pm/με,
- $K_{\varepsilon,Ghost} = 1.255 \pm 0.006$ pm/με,
- $K_{T,Bragg} = 8.9593 \pm 0.06$ pm/°C,
- $K_{T,Ghost} = 9.5303 \pm 0.1$ pm/°C.

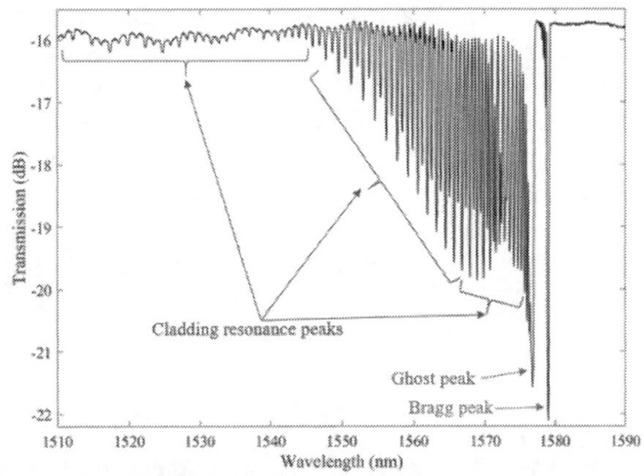

Figure 2. 3° TFBG transmission spectrum.

The measurements performed were interpreted calculating with eq.1 by respecting the thermal resolution of the TFBG. This is given by the ratio between the scanning wavelength resolution (*swR*) of the sensor interrogation system and the absolute difference between $K_{T,Bragg}$ and $K_{T,Ghost}$. For this research, National Instruments NI-PXIe 4844 FBG interrogator was used, which has an $swR = 4 \pm 1$ pm and minimum power detection of 6.103×10^{-4} dBm. The resulting *TR* is ~7 °C .

At the same time, the cladding resonance peaks are strongly affected, in wavelength and amplitude, by the values of the surrounding RI. Hence, the area obtained from the envelope of the upper and lower cladding resonance peaks can be exploited to find the variation of the external RI [8]. For this scope, through a preliminary refractometric calibration of the TFBG, the envelope area is correlated to each external RI. A correlation function is then extracted, which is able to provide the RI value from the normalised area found in the spectrum. A novel demodulation technique based on the Delaunay triangulation technique was used here to obtain the envelope area. The details can be consulted in [4]. For the here used TFBG, the correlation between envelope areas and RIs is shown in Fig.3.

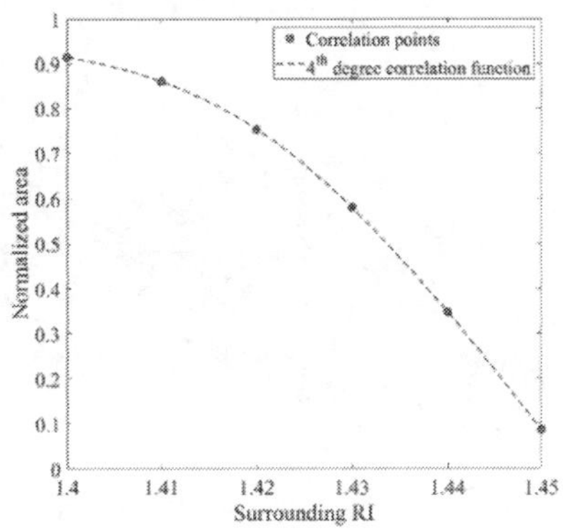

Figure 3. RI correlation points and fitting function in the range of 1.4-1.45.

The RI range in which the TFBG should be calibrated is usually based on the application of the sensor. In this case, since the Nusil CV16-2500 silicone has an initial RI of around 1.43 [9], the chosen RI range is 1.4-1.45. The R^2-error of the fitting is 0.9999 when using a 4th degree polynomial function, with a maximum normalised area standard deviation of 0.0016 and a mean RI resolution of 2×10^{-7}.
At this point, the TFBG sensor is ready to be embedded inside the silicone adhesive and measure simultaneously and separately the three parameters.

3. Methodology

The TFBG used for the in-situ measures possessed a tilt angle of 3° and a length of 4 mm. The sensor was written in PS1250/1500 OF and was not re-coated after the Bragg grating inscription. The tested sample was composed of two Cerium Doped Microsheets (CMX) cover glasses (20×20×0.05 mm) bonded by Nusil CV16-2500 silicone adhesive. This adhesive is a space qualified elastomer, specific for low outgassing and minimal volatile condensables under extreme operating conditions. The waveguide containing the TFBG was embedded inside the silicone adhesive with the sensor in the centre of the solar cell. A k-type thermocouple (k-TC) was also immersed in the polymer and placed as close as possible to the TFBG. In this way, the comparison of the measurements of the two sensors can be performed. The sample configuration and dimensions are shown in Fig.4.

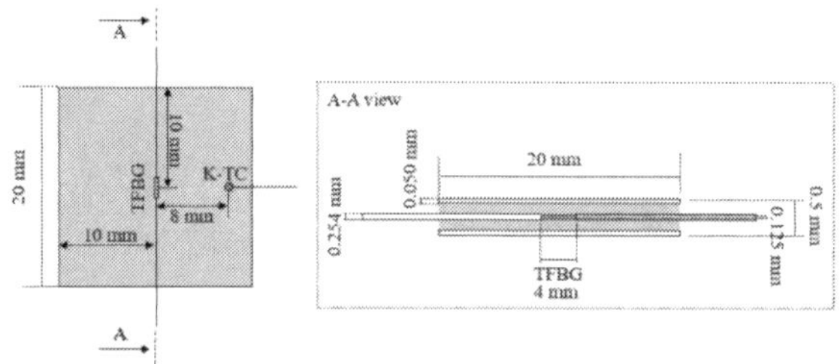

Figure 4. Schematic of the sensorised solar cell assembly sample.

Once the sensors were placed inside the silicone, the assembly was completed by adding the second cover glass on top. At this point, the element was put inside an oven to cure the silicone at the temperature profile provided by the manufacturer [9] (4 hours at 65 °C).
After the manufacturing step, the sample was prepared to be exposed inside a vacuum chamber under the light of UV/VUV lamps.
Fig.5 is a representation of the test sample plate where the specimens are placed to be irradiated. The sample plate is composed of several layers. The bottom part of the plate is the cooling section where liquid nitrogen (LN_2) conducts are present. On top of this layer, the plate has a section containing a thermocoax element able to heat the plate up to 250 °C. The heating and cooling systems were used during the experiment to keep a temperature as stable as possible. The last element of the test plate is a metallic support plate with several threaded holes, on this surface the sample was mounted.

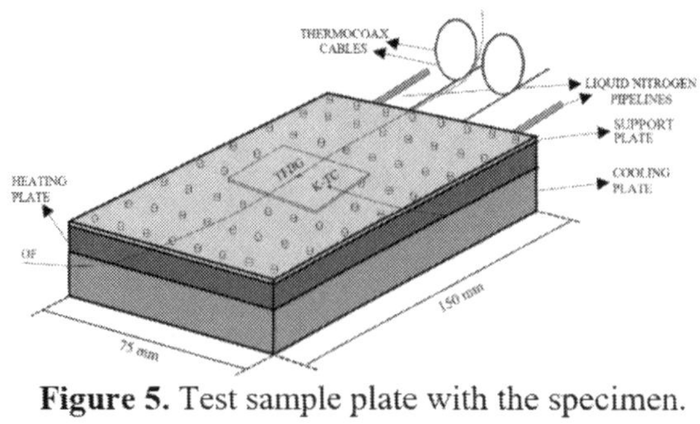

Figure 5. Test sample plate with the specimen.

The wires of the embedded k-TC were connected with the data acquisition system (DAQ) of the vacuum chamber. On the other hand, the ends of the OF containing the embedded TFBG were connected to the external FBG interrogator through vacuum OF feedthroughs.
The intensity emitted by the UV lamps was measured in solar constants (SC). The intensity was measured before mounting the sample on the test plate and in the position planned for the exposure. Solar constants were calculated based on the obtained UVA, UVB and UVC ranges. The intensities in these ranges were automatically calculated with a UV radiometer pad sensor. The value of the SC is important as it is used to know the equivalent solar hours (esh) exposure of the experiment.
Once all the elements were ready for the experiment, the chamber was closed and the pumps were activated to pump down the chamber. The DAQ system of the TFBG and TC started before turning-on of the pumps. Since the TFBG used here can operate until a maximum temperature of 85 °C, the cover glass sandwich needed to be cooled during the exposure. To avoid over-heating of the sensor due to the radiation, the cover glass sandwich was cooled in advance, starting from when the internal pressure of the chamber reached 10^{-4} mbar. The cooling plate was kept on for the entire exposure time. The UV and VUV lamps were switched-on, once the embedded TC detected a stable temperature around -40 °C.

This approach allowed the operational average temperature to be kept around 80 °C, with a range of fluctuation between 78 °C and 84 °C. The exposure was performed for 565.77 hours at 3.827 SC, which means 2165.21 esh of exposure. After the temperature was stabilised and the final registered pressure was 1.2×10^{-6} mbar.

4. Results

As eq.1 defines, the strain-temperature measurements are performed by detecting the wavelength variations of the Bragg and ghost peaks. Their values along with the exposure time can be substituted in eq.1 and, $\Delta\varepsilon$ and ΔT can be calculated from the starting point. The origin of the calculation is usually considered as $\Delta\varepsilon=0$ and $\Delta T=0$, hence, here, it is the beginning of the DAQ. The trends of the selected peaks wavelength variation are plotted in Fig.6 along with the TC temperature.

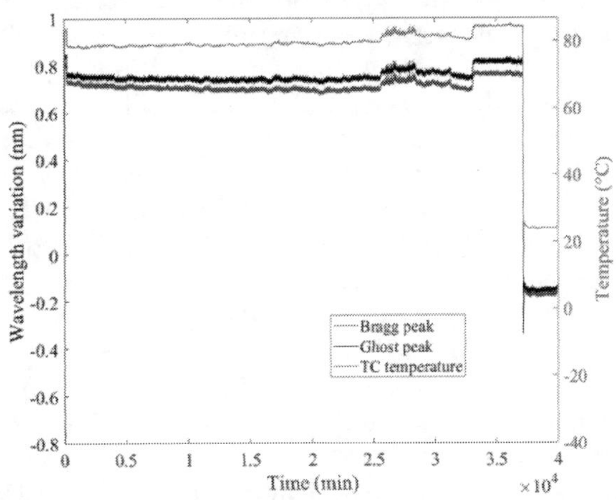

Figure 6. Bragg and ghost peak wavelength variation and TC temperature trends during the exposure time.

Both the trends start from 0 and follow the temperature trend detected from the TC. As expected from the peak thermal coefficients, $\Delta\lambda_{Ghost}$ is bigger than $\Delta\lambda_{Bragg}$ when the temperature reaches the plateau. However, after the shut-down of the lamps, they do not come back to a neutral value but are both negative. Since the temperature given by the TC is close to the initial one, this could indicate a strain level obtained in the silicone due to the radiation in high vacuum. Considering the TFBG spectra acquired, after that the shut-down of the lamps, the sensor was measuring a strain of -138.22 $\mu\varepsilon$ at the same initial temperature of 24.4 °C (confirmed also by the TC measured results). This indicates a negative deformation of the silicone across the TFBG. It may be caused by a shrinking effect caused by an over-curing or degradation under the UV radiation, and/or the volatiles induced by vacuum which can dimensionally influence the material.

At the same time, the envelope area of the cladding resonance peaks can be observed to find some information regarding the refractometric characteristics, and hence, the chemical state of the silicone. In Fig.7, the trend of the envelope area is reported during the exposure time.

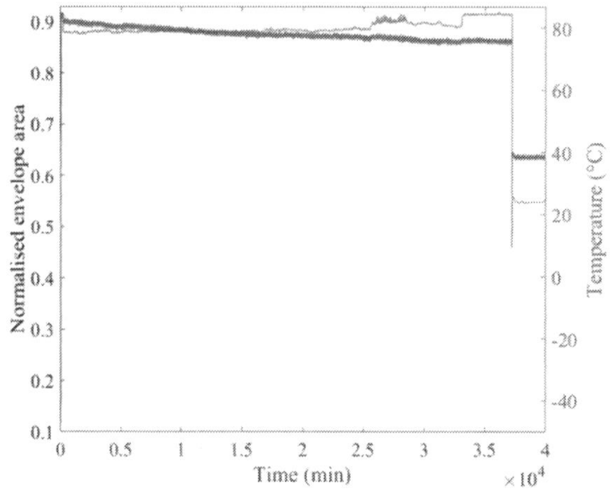

Figure 7. Normalised envelope area and TC temperature during exposure.

As in the case of the Bragg and ghost peaks, the envelope area trend is a function of the temperature. Indeed, as expected, the silicone RI decreases with the increasing of the temperature, while it grows when ΔT is negative. Therefore, as observed from the calibration curve in Fig.3, since the envelope area is inversely and quartically proportional to the RI (in this refractometric range), its value grows during the heating and decreases during the cooling step. However, the area trend is observed to have a slight decrease after the temperature is stable or even it slightly increases. The reason for this lies in the effect of the UV radiation on the silicone adhesive. The photochemical reactions induced from the radiation are able to modify the refractometric conditions of the silicone. Specifically, with increase in exposure time, the RI is noted to increase and it can be calculated at any moment of the test through the envelope area of the TFBG. In particular, the initial value, before exposure, is 1.4248241 while, once the lamps are off, the RI is 1.4270879 at the same temperature. This means the radiation generated a difference of $\Delta RI = 2.2638\times10^{-3}$. This variation depends, of course, on the UV effect on the chemical composition of the silicone, but also may depend on the new physical state due to the high-vacuum environment.

In conclusion, a single embedded TFBG has provided important information on the state of the silicone adhesive before, during and after the testing activity. The strain-temperature-RI measurements in the silicone adhesive can be performed along the entire time duration of the exposure, even in real-time. This allows rapid information to be obtained without further post-tests or processing.

In order to confirm the changes in the chemical composition of the silicone, a further investigation was performed by testing the percentage of light transmission through the cover glass sandwich assembly before and after the UV exposure. This was tested in a UV-Vis-NIR Agilent® Cary 5000 spectrophotometer where 3 spectra were obtained at each step. The resulting spectra, which comes from the average of the 3 measurements, are reported in Fig.8.

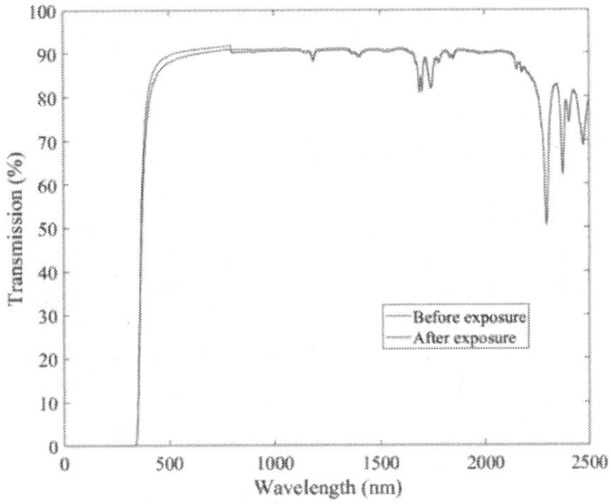

Figure 8. Transmission spectrum of the light through cover glass sandwich before and after the test.

The percentage of transmitted light through the sample is approximately the same in the range from 800 nm to 2500 nm. After a small jump at 800 nm (due to the changing of the light detector), the curves start to deviate. Specifically, the greatest difference is obtained in the UV wavelength range. Here it is usually the interval where most of the silicone absorption coefficient changes due to the UV radiation. As a consequence, this may confirm some variations in the chemical composition of the silicone.

5. Conclusions

The thermomechanical and refractometric state of the Nusil CV16-2500 silicone adhesive was monitored through a single embedded TFBG sensor. This adhesive was used to bond two thin cover glasses and the cover glass sandwich was exposed to light from UV and VUV lamps. The wavelength shifting of the Bragg and ghost peaks in the TFBG transmission spectrum, was used to find the strain-temperature variations in the silicone adhesive. Specifically, after the high vacuum UV/VUV exposure, the TFBG measured a negative strain of ~138 $\mu\varepsilon$ at the same initial temperature. This indicates a change in the state of the tested silicone due to the testing conditions. Specifically, the outgassing and the UV light, which also generate heat, caused a shrinkage of the material. Therefore, the dimensional stability of the silicone could influence the results. Furthermore, the silicone refractometric condition was also monitored through the envelope area generated from the upper and lower cladding resonance peaks. This allowed an RI variation of 2.2638×10^{-3} to be calculated, which can be linked with a change of the chemical composition of the material due to the radiation. This aspect was also confirmed by measuring the transmission of the light through the sample, which has shown a decrease in the experiment. The three-parameter TFBG measurements can be performed at any moment of the test, in real-time and in remote mode. This allows a considerably improvement in the investigation on the degradation of the polymers, and also reduces the testing steps and time.

6. Acknowledgments

The authors want to thank Adrian Tighe (ESA TEC-QEE) for the support and suggestions given during the execution and writing of this research. This research was supported by the ESA/ESTEC (visiting researcher programme) and the Operationeel Programma Zuid-Nederland (Op-Zuid) Project as part of the Dutch Composite Maintenance Centre (DCMC), supported by the Europees Fonds voor Regionale Ontwikkeling (EFRO) and the North Brabant province of the Netherlands. The views expressed herein can in no way be taken to reflect the official opinion of the European Space Agency (ESA) and are not intended to endorse particular technologies, companies, or products.

References

[1] White C, Tan K, Wolf A and Carbary L, 2010 *Advances in Structural Adhesive Bonding Woodhead Publishing Series in Welding and Other Joining Technologies*, ed David A. Dillard (Amsterdam/American Elsevier) pp 66-95.

[2] de Groh III HC, Daniels CC, Dever JA, Miller SK, Waters DL, Finbeiner JR, Dunlap Jr P.H and Steinetz BM, 2010 *NASA/TM* **216332**.

[3] Fischer HR, Semprimoschnig C, Mooney C, Rohr T, van Eck ERH and Verkuijlen MHW 2013 *Polymer Degradation and Stability* **98**.

[4] Fazzi L and RM Groves 2020 Demodulation of a tilted fibre Bragg grating transmission signal using α-shape modified Delaunay triangulation *Measurement* **166** 108197.

[5] Fazzi L, Valvano S, Alaimo A and Groves RM 2021 A simultaneous dual-parameter optical fibre single sensor embedded in a glass fibre/epoxy composite *Composite Structures* **270** 15.

[6] Gu QG and Zhou QL 1998 Preparation of high strength and optically transparent silicone rubber *European Polymers Journal* **34** 11.

[7] Fazzi L, Dias N, Holynska M, Tighe A, Rampini R and RM Groves 2022 Monitoring of silicone adhesive in space solar cells with an embedded multi-parameter TFBG sensor in a simulated space environment *Measurement Science and Technology* **33** 085108.

[8] Laffont G and Ferdinand P 2001 Tilted short-period fibre-Bragg-grating-induced coupling to cladding modes for accurate refractometry *Measurement Science and Technology* **12** 765.

[9] https://www.avantorsciences.com/assetsvc/asset/en_US/id/29017340/contents/en_us_tds_nusicv 16-2500.pdf last access 07/02/2022.

IOP Conf. Series: Materials Science and Engineering 1287 (2023) 012009 doi:10.1088/1757-899X/1287/1/012009

Solar Absorptivity Degradation of Spacecraft Materials Due to UV and Charged Particles in the Gateway Environment

A Zinecker[(*)1], **B Spivey**[1], **G Jayne**[2], **M Hasegawa**[2]

[1] *Jacobs, 2224 Bay Area Blvd, Houston, TX 77058 (USA)*
[2] *NASA Goddard Space Flight Center, 8800 Greenbelt Rd, Greenbelt, MD 20771 (USA)*
[(*)] abigail.a.zinecker@nasa.gov

Abstract. Gateway will be an outpost in cislunar space designed to survive a 15-year mission to support human Moon landings and future deep space exploration. The Gateway Passive Thermal Control System (PTCS) Group has identified charged particle and UV degradation of potential Gateway materials as a knowledge gap for the Near Rectilinear Halo Orbit (NRHO) and transit environment, especially the slow spiral transit planned for the first two modules. Solar absorptivity degradation is important to quantify because the end-of-life absorptivity impacts thermal performance of the system. To address this, ground testing has been completed at NASA Goddard Space Flight Center in which potential Gateway materials, including radiators, multi-layer insulation (MLI), and structural materials are subjected to the expected transit plus 15-year on-orbit charged particle fluence (9.86×10^{15} protons/cm^2 at 2.5 keV and 3.1×10^{16} electrons/cm^2 at 10 keV) and 5093 Equivalent Solar Hours (ESH). Reflectivity of a single sample of each material was measured in atmosphere and in vacuum inside the chamber after exposure to incremental levels of ESH and charged particle fluence. Increase in absorptivity of most samples were seen throughout the test including large differences in absorptivity of materials of the same category. Future planned testing will validate these results and include additional materials of interest.

1. Introduction
Traditionally, most of the data accumulated for material optical property degradation in space has been in the Low Earth Orbit (LEO) environment [2,3,5,8]. However, Gateway is a new space station that will be assembled and flown in Near Rectilinear Halo Orbit (NRHO) for 15 years. Additionally, some Gateway modules will have a slow spiral transit from Earth to NRHO, taking on the order of hundreds of days to pass through the Van Allen Belts, which contain damaging charged particles [7]. Due to the long duration transit, significant damage is expected occur before the NRHO portion of the mission begins. Over 95% of the charged particle exposure for the 15 year mission occurs during the transit portion of the environmental bounding mission for these modules. Therefore, this test emphasises materials for these modules. Degradaton of materials through the environments Gateway modules will encounter through the mission lifetime are not well characterized. Environmental factors such as charged particles and ultraviolet (UV) radiation have been shown to significantly increase solar absorptivity (α) for some thermal control materials and coatings [3,4,5,8]. Since α will impact thermal performance over the mission, the Gateway Passive Thermal Control System (PTCS) Group has identified charged particle and UV degradation of Gateway materials as a knowledge gap. Candidate materials and coatings were

subjected to expected mission charged particle environments and 5093 Equivalent Solar Hours (ESH) of UV at NASA Goddard Space Flight Center (GSFC). In situ change in reflectivity was observed at specified intervals during testing such that change in α could be plotted as a function of time of solar exposure.

2. Methodology

The Solar Wind Facility at GSFC was chosen for this testing due to the expertise of the group and their previous experience with similar testing [3,4]. This facility is decribed by Kauder in 2009 [6]. Twelve potential Gateway materials and coatings samples were tested, including radiator coatings, external MLI layers, and structural materials. The materials chosen are shown in Table 1; special focus was given to electrically dissipative coatings due to the charging environment expected in NRHO and transit, however electrical dissipation was not measured before or after exposure.

A passive contamination monitor, vapor deposited aluminum (VDA) thin film on Kapton-HN, was placed in one of the sample slots. This material was chosen because it is known to be stable in UV and charged particle exposures and any degradation could be attributed to contamination. Changes in the contamination monitor were found to be within the uncertainty of the instrument, indicating the contamination environment in the chamber was not a factor impacting test results. Eight of the 12 samples were measured in-vacuo; those shown in Table 1 with a * were not measured in-vacuo due to chamber restrictions.

Table 1. List of Material Samples

	Material Name	Description
1	Aptek 2719-A/B	Silicone based white paint radiator coating
2	Aptek 2711-1	Silicate based white paint radiator coating
3	Alion DS13N:6N/LO-HP	Silicone based white paint radiator coating
4	Alion Z93C55	Silicate based white paint radiator coating
5	Optical Solar Reflector (OSR)	6 mil CMX UV R, radiator material
6	Sulfuric Acid Anodize (SAA) with Indium Tin Oxide (ITO)	6061-T6 with nickel acetate seal, structural material
7	Germanium Black Kapton (GBK)	Electrically dissipative external MLI layer
8	Astroquartz	High performance pure silica fabric, external MLI layer
9*	Stamet Black Kapton	Electrically dissipative external MLI layer
10*	Contamination Monitor	VDA thin film on Kapton-HN
11*	VDA Kapton with scrim	Traditional external MLI layer
12*	SAA; no ITO	6061-T6 with nickel acetate seal, structural material

The samples were 19 mm diameter discs which were mounted inside the chamber and exposed to simultaneous proton, electron, and UV up to 5093 ESH while maintaining approximately $2x10^{-5}$ Torr vacuum. UV exposure was applied using an Energetiq EQ400 and Hamamatsu 100W Deuterium lamp which had a minimal infrared (IR) component to minimize heating of samples and sample carousel during the test. Temperature of the samples was not monitored during the test due to test equipment limitations, but preliminary testing of similar samples suggested the samples did not go above 40 C. Slight excursions of this may be possible. Due to the UV sources having a small IR component, it is not expected at this time that results are impacted and any degradation due to temperature is not expected to be measurable. The solar constant was ~2.6 equivalent suns, which was averaged over a wavelength range of 200 to 400 nm.

Total Ionizing Dose (TID) was applied with a 2.5 keV proton beam and a 10 keV electron beam and was calculated using NOVICE, a radiation modeling software. The dose depth profile was determined by the worst-case Gateway and transient environment particle fluences. Looking at the surface dose, the TID was converted into a particle fluence based on the available particle energies. The dose was calculated to be representative for these specific samples and was a fluence of approximately 9.86×10^{15} protons at 2.5 keV and 3.10×10^{16} electrons at 10 keV.

Solar reflectivity of each sample was measured at ambient pressure, vacuum after an overnight pump down, 13 intervals of ESH, and after return to ambient pressure. TID was applied at two of those intervals to get approximate 45% and 100% dose. Sensors were placed next to each sample in the chamber to monitor the approximate proton and electron dose received. Each sample was mounted in the chamber on an individual pawn and samples 1-8 were measured in-vacuo using a set of manipulator arms which would grab each pawn and bring it to the measurement device to be measured. All sample reflectivities were measured before and after the test in ambient atmosphere conditions.

During testing, there were facility issues after 1000 ESH of UV exposure which caused testing to have to enter a safe-hold while UV sources were being repaired. The test chamber remained under high vacuum ($<2 \times 10^{-5}$ Torr or lower) without active UV or charged particle exposure for a period of nearly 100 days. During this time, some samples experienced a decrease in α. This phenomena of recovery in α is theorized to be a bleaching effect caused by reaction with the small amount of oxygen present in the vacuum. Further testing is required to determine the validity of this theory. After UV was applied again, damage was rapidly recovered, much more quickly than it originally took to cause the damage. Due to this quick recovery, it is deduced that overall test results are not impacted by the idle time.

Reflectivity measurements were taken at wavelengths ranging from 250 to 2500 nm. Total α was calculated from these measurements by integrating 1-%R with respect to the solar spectral irradiance as defined by ASTM E490 [1], assuming a transmittance of zero. %R is defined as the reflectance measured at a given wavelength.

3. Results

Figure 1 and Figure 2 show the pre- and post-test images of the sample plate. The numbers next to each sample correspond to the those shown in Table 1. Looking closely, degradation occurs visibly in samples 1 and 3, corresponding to the silicone based white paint radiator coatings. Figure 3 plots α of each sample during the charged particle and UV testing over ESH. The samples which were not measured in situ are omitted from Figure 3. The ambient α of Stamet Black Kapton, VDA Kapton with Scrim, and SAA without ITO increased by 0.04, 0.08, and 0.10, respectively. The two vertical dashed lines identify the intervals when the charged particle exposures occurred. Shortly after the second dashed line is when the idle time discussed in the previous section occurred. Absorptivity at each ESH step for each sample as well as reflectance data are will be published at a later date.

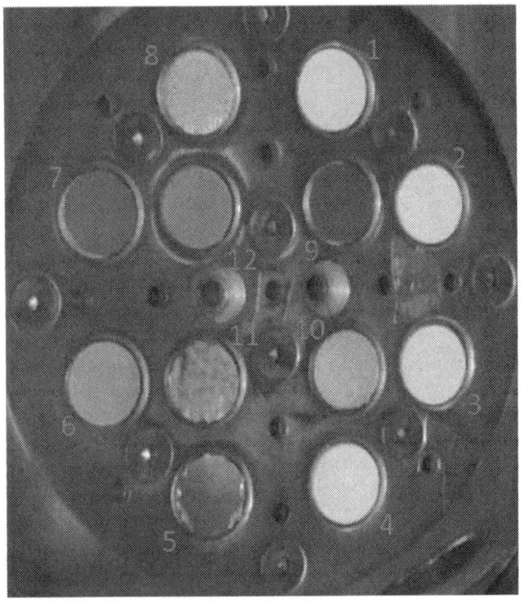

Figure 1. Sample plate pre-test

Figure 2. Sample plate post-test

Table 2. Absorptivity degradation broken down by source

	Sample	Δα due to TID	Δα due to UV	Total Δα (Vacuum)
1	Aptek 2719-A/B	0.09	0.19	0.27
2	Aptek 2711-1	0.11	0.00	0.11
3	DS13N:6N/LO-HP	0.13	0.28	0.40
4	Z93C55	0.04	0.15	0.20
5	OSR	0.01	0.09	0.10
6	SAA, ITO	0.04	0.03	0.08
7	GBK	0.01	0.01	0.02
8	Astroquartz	0.01	0.19	0.19

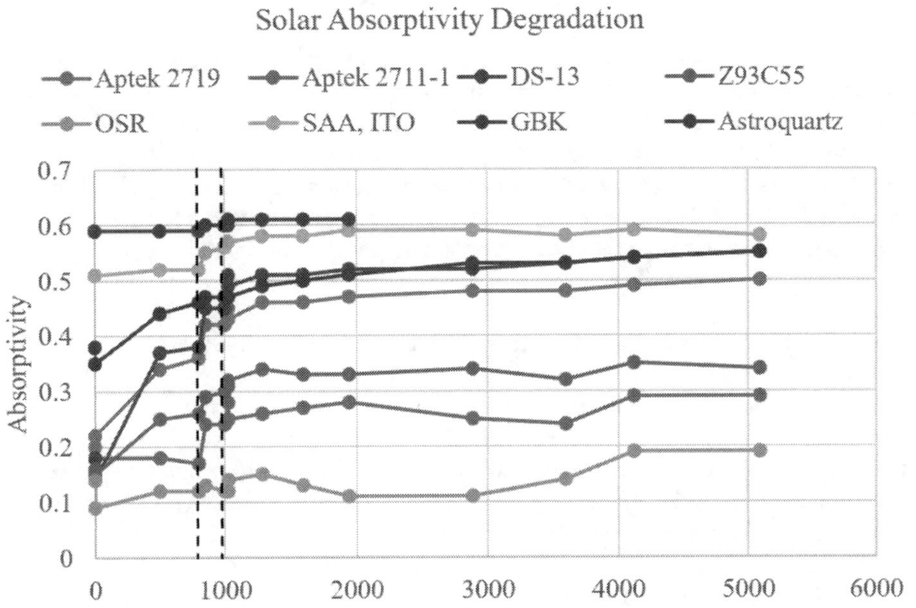

Figure 3. Solar Absorptivity degradation of 8 in-vacuo samples plotted over ESH. Dashed lines represent the intervals when the charged particle exposures occurred.

4. Conclusions and Future Work

This study shows a helpful comparison of solar absorptivity degradation of different types of materials such as white paint radiator coatings, OSR, several different MLI outer layers, and even strucutral materials with and without an electrically dissipative coating. The radiator material with the best α resistance to UV and charged particles is OSR while the white paint with the best resistance to UV and charged particles is Aptek 2711-1. The white paint with the best resistance to charged particles alone is Z93C55. Silicate-based white paints degraded less than silicone paints, especially in UV. In post-test ambient conditions the SAA with ITO degraded less than the SAA without ITO. Testing results have reduced uncertainty in material degradation in the Gateway environment by quantifying the charged particle and UV induced optical property degradation for critical thermal surfaces. Results from this test will be integrated with degradation resulting from other sources such as molecular contamination and lunar dust coverage.

After 1941 ESH of UV exposure, the pawn holding the GBK sample became stuck and the team was unable to take further measurements. However, based off of the previous data points before the incident, it is concluded that the sample was unlikely to degrade further. Stamet Black Kapton is expected to behave similarly, but further testing is needed to validate this.

In this test, only one sample of each material was able to be included. Future tests will aim to include more samples of the same materials to increase statistical significance. There is some uneven distribution of charged particles inside the chamber such that some samples received more or less dose than others. The samples received anywhere from 101% to 136% of the targeted proton exposure and 100% to 128% of the targeted electron exposure. This is due to samples in the center of the plate receiving higher charged particle flux than those at the edges. One more consideration is that the energy spectrum from the Gateway environment was consolidated from a wide range of energetic particles to a single energy level for each protons and electrons. This is not truly flight-like, but it is a ground testing limitation and it is representative of the flight environment based on the step profile. Measurement location on each sample varies slightly due to the arm manipulators' spatial accuracy which leads to a small measurement uncertainty.

Future work on this project will test more Gateway materials and retest some samples that were shown here. The radiation environment affected the OSR surficially, and changes that are evident in the latter UV exposure times are most likely due to the cracked edges of the sample and its rotation during manipulation for reflectance measurements. The team expects actual degradation of OSR in this environment to be less than what is recorded here. Future testing aims to resolve this anomaly. More testing is also needed for materials under different environmental transit assumptions. Another round of materials have been selected for the next round of testing and procurement is underway.

References

[1] ASTM, Standard Extraterrestrial Reference Spectrum, ASTM E490-00a, 2019.

[2] Finckenor, M. M., Golden, J. L., O'Rourke, M. E., Kravchenko, M., "Results of International Space Station Vehicle Materials Exposed on MISSE-7B" National Space and Missile Materials Symposium, Tampa, FL, June 2012.

[3] Hasegawa, M., Babel, H., "Estimation of the end-of-life optical properties of the Z-93 thermal control coating for the Space Station Freedom" Materials Specialist Conference – Coating Technology for Aerospace Systems, Dallas, TX, April 1992. AIAA-92-2168-CP.

[4] Hasegawa, M., Freese, S., Kauder, L., Triolo, J., "End-of-Life Optical Property Predictions of White Conductive Thermal Control Coatings through Analysis of On-Orbit and Ground Based Testing Data" AIAA Atmospheric and Space Environments Conference, Honolulu, HI, June 2011.

[5] Jaworski, D. A., Kline, S. E., "Review of End-of-Life Thermal Control Coating Performance" April 2008.

[6] Kauder, L., Space "Environmental Testing at GSFC", Contamination, Coatings and Materials Workshop, Greenbelt, MD, July 2009.

[7] McGuire, M. L., McCarty S. L., et al., "Overview of the Lunar Transfer Trajectory of the Co-Manifested First Elements of NASA's Gateway" AAS/AIAA Astrodynamics Specialist Conference, Big Sky, Virtual, Aug 2021. AAS 21-697.

[8] Soares, C. E., Mikatarian, R., Schmidl, W., Boeder, P., Alred, J., "Optical Characteristics of Materials in a Space Environment" 10th International Symposium on Materials in a Space Environment, Collioure, France, June 2006.

IOP Conf. Series: Materials Science and Engineering 1287 (2023) 012010 doi:10.1088/1757-899X/1287/1/012010

Interaction mechanisms between Atomic Oxygen and materials: investigation on reflected beam

D Leveque[1], S Duzellier[1], M Vilaranda Fernandes[2], G Chanteperdrix[2], D Nguyen Van Sang[3], V Perrin-Bailly[3], P Jouanne[3], E Laurent[4]

[1] ONERA/DPHY, Université de Toulouse, F-31055 Toulouse - France
[2] AIRBUS DEFENCE AND SPACE, 31 Rue des Cosmonautes, 31400 Toulouse, France
[3] THALES ALENIA SPACE, 5 Allée des Gabians, 06150 Cannes, France
[4] Sous-direction Qualité Expertise, Service Technologies, Matériaux et Procédés, CNES, 18 avenue Edouard Belin, 31401 Toulouse Cedex 9, France

E-mail: david.leveque@onera.fr

Abstract. Atomic Oxygen (AO) is the main component of the residual atmosphere present at low earth orbit. Multiple reflections can locally enhance or lower erosion rate in case of complex geometry. The need for precise estimate of hyperthermal AO flux is thus crucial at critical target embedded into realistic assembly. Industrial modeling tools (example Atomox module in ESABASE2 and SYSTEMA toolboxes) may account for multiple reflections with simplistic approach (ray-tracing) and a set of semi-empirical parameters. In this context, an original geometrical setup was designed in order to expose target surfaces with normal and tilted incidences and after one reflection on a set of materials of interest (black conductive thermal coatings, coverglass, surface treatment…). Candidate materials have been selected in view to compare numerical results and experimental ones. Results of two test campaigns highlight the effect of incidence angle on erosion rate (higher erosion with tilted beam) and provide an estimate of residual fluence after one reflection (in the 4-10 % range depending on material type). First computations with engineering codes have been successfully conducted, the erosion patterns being rather well described with some basic hypotheses on material parameters (absorption, specular/diffuse reflection) and taken into account the reflection on adjacent metallic surfaces.

1. Introduction

Atomic oxygen (AO) is the main component of the residual atmosphere present at low earth orbit [1]. AO is considered as one of the most serious hazards to spacecraft materials. The main consequence is oxidation and erosion of surface materials. Some classes of materials are stable under AO attack (oxides for instance) and incoming flux is then re-emitted with specific angular and energy distributions. The response to AO of space materials is known via measurement of the erosion yield. Multiple reflections can locally enhance or lower erosion rate in case of complex geometry.

The issue of materials erosion by AO is mainly considered in a worth case manner i.e. with an AO flux fully directed towards exposed surface. Nevertheless in some critical applicative cases, the local geometry of sub-system may be complex (for instance: optical devices with baffles, surfaces in sight of solar panels …) where phenomena of oxygen atoms diffusion/reflection/absorption can take place, and can lead to erosion of surfaces without direct exposure to space environment [2].

The need for precise estimate of hyperthermal AO flux is crucial for critical target embedded into a complex geometry. Depending on the configuration, target can be partially masked or exposed to additional AO flux due to scattering. Scattering mechanisms have been reported for instance in flight data [3][4] and at ground studies [5].

Modeling tools (example Atomox module in ESABASE2 and SYSTEMA industrial frameworks) may account for multiple reflections with simplistic approach (ray-tracing) and a set of semi-empirical parameters.

In this context, an original geometrical setup was designed in order to expose target surfaces with normal incidence or after one reflection. Experimental investigation was carried out to study the effect of hyperthermal AO in case of complex geometry and estimate the "secondary" AO flux after reflection on selected set of targets. At last a comparison was made between experimental observations and calculations of AO erosion in case of such scattered AO mechanisms. Preliminary results presented in [6] are completed and detailed in this paper.

2. Test approach and ground experiments

2.1. Geometry design

Ground experiment of AO multi-reflection effect can be found in the literature and address different thematic and objectives:

- Investigation of local erosion enhancement linked to chamfered geometries (edge effect, [7]),
- Develop mitigation and attenuation approach to secondary thermal AO beams [8],
- In [5], reflection jig was used to measure the ratio of reflected component at different angle.

Similarly, we designed and developed an original setup to expose target surfaces with normal incidence or after one reflection (Figure 1). First, the geometry was designed in order to be able to expose at the same time different materials in normal (position 1 to 3) and inclined positions (position 4). The angle of inclination in our case is 65° between the main flux direction and the normal of the surface. At position number 5, a detector is placed in such manner it is hidden from the main flux but receive the reflected flux from the inclined material. Positions 3 and 4 are fitted with candidate materials whose erosion yields at normal and tilted incidence can be measured based on mass loss measurement. The configuration of the assembly also generates AO scattering on part 4 and detection of scattered flux at position 5 (reference kapton foil with known erosion yield, or active Resistack detector [9]). At position 5, it comes an estimate of the ratio of reflected AO flux received by the detector. Lastly, reference Kapton foil (position 1) is used for the measurement of the fluence level in primary (normal) beam. In practise, the Kapton erosion yield is independent of the macroscopic incident angle, using the cos(θ) correction [10]. Position 2 is used for any additional material.

IOP Conf. Series: Materials Science and Engineering 1287 (2023) 012010 doi:10.1088/1757-899X/1287/1/012010

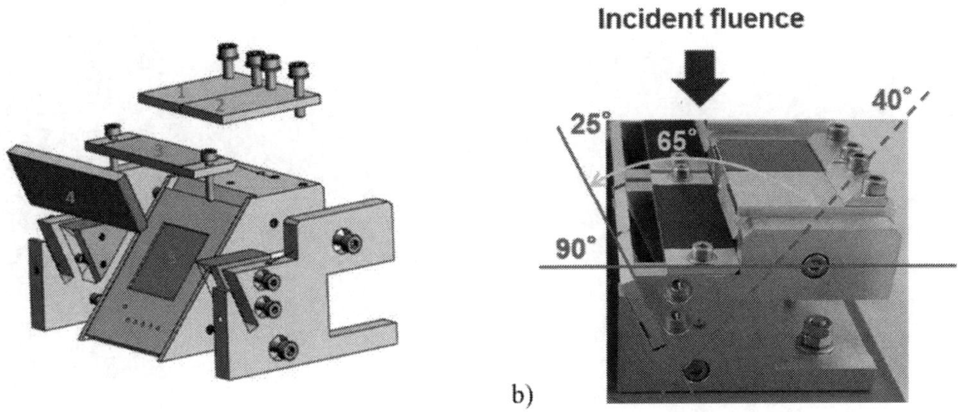

a) b)

Figure 1. a. geometry design (exploded view). **b.** assembled geometry with orientation references.

2.2. Materials choice

In terms of materials choice, Kapton HN film is used as a test-fluence standard, like proposed by Minton [11]. At position 1, the local incident fluence is estimated for each geometry. At position 5, the residual fluence after reflection is also estimated by measuring the mass loss and knowing the erosion yield of Kapton.

The materials under test are composed of black conductive thermal coatings (BCTC1, BCTC2 and BCTC3), a plasma chemical coating (PCC), a polyimide film like kapton (KAP), a coverglass (CVG) and an aluminized Kapton films too. Due to confidentiality purpose, only the generic family is indicated for each material.

All materials under test (MUT) samples have their witness samples, i.e. samples of same geometry and same nature but that are kept safely in a dry and neutral environment, used as a reference for mass loss measurement.

2.3. Test setup

Two Atomic Oxygen exposure campaigns were conducted at LEOX facility in the ESA-ESTEC laboratory (figure 2a), with a first screening test campaign (selection of materials) with a low fluence (~2×10^{20} atoms/cm²), and a second test campaign with higher fluence (~2×10^{21} atoms/cm²). The LEOX facility has a source concept based on the Laser Pulse Induced Breakdown (LPIB) principle with an AO hyperthermal energy of ~ 5 eV (relative velocity of about 8 km/s). Figure 2b depicts the sample holder of this second test campaign with the associated selected materials shared between six different geometries.

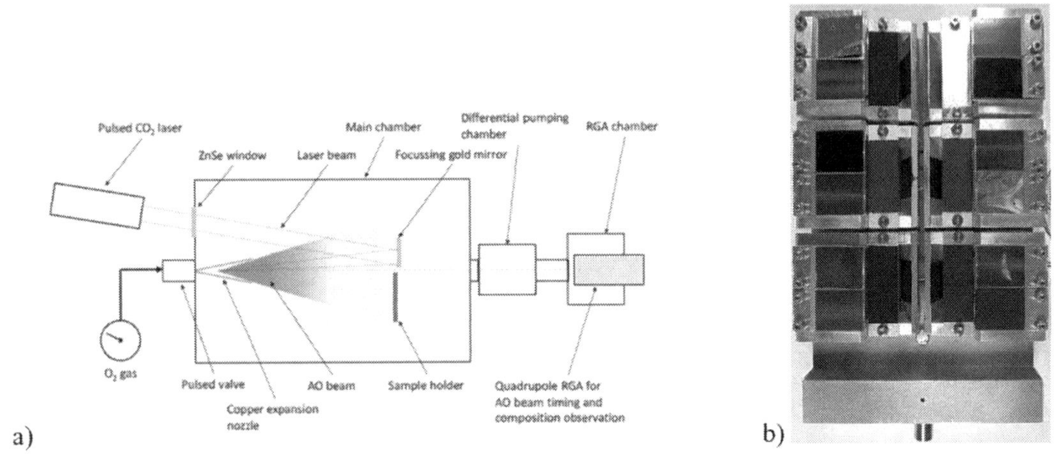

Figure 2. a. Schematic of the LEOX facility (ESA-ESTEC). **b.** LEOX2 sample holder before AO exposure test.

Before and after each test campaign, all the material samples are dried and weighted following the procedure detailed in [11]. From the mass loss, one can easily determine the erosion yield knowing the exposed surface area, the MUT density and the local incident fluence. The local incident fluence is estimated on each geometry thanks to the reference Kapton foil (located at position 1) from its mass loss and its well-known erosion yield (3.0 10^{-24} cm3/atom). Due to the wide discrepancy of the incident fluence level over the sample holder from one side to another, the fluence level was interpolate on all the exposed surfaces in order to better estimate the incident fluence associated with each MUT sample. Moreover, from this mapping the mean global fluence can be estimated over the whole holder plate at ~$1.7x10^{20}$ atoms/cm² for LEOX1 and ~$2.4x10^{21}$ atoms/cm² for LEOX2 test campaigns.

3. Test preliminary results

3.1. Erosion rates

Figure 3 discloses the normal-to-tilted erosion yield ratio for the set of materials under test. This is the ratio of the normal erosion yield to the erosion yield measured with an impingement angle θ ($\theta = 65°$ in our case). For these estimations, the effective fluence was used, *i.e.* corrected by the cos(θ) factor. When this ratio is equal to unity, the erosion yield is independent on the impingement angle. The standard deviation estimated on each ratio value includes the standard deviation due to the measurement of mass loss and exposed area surface (erosion yield calculation) as well as the incident fluence estimation (residual of fluence mapping).

At first, it is confirmed the absence of angle influence on the erosion yield of Kapton HN considering the proposed cos(θ) correction [10]. For the other MUT samples, this ratio is much lower than 1 showing higher erosion yield for all tilted samples whatever the material family and type tested in this study. It should not be always the case, as some other works have measured a strong material-dependent erosion yield ratio, this ratio could be even greater than 1 [12].

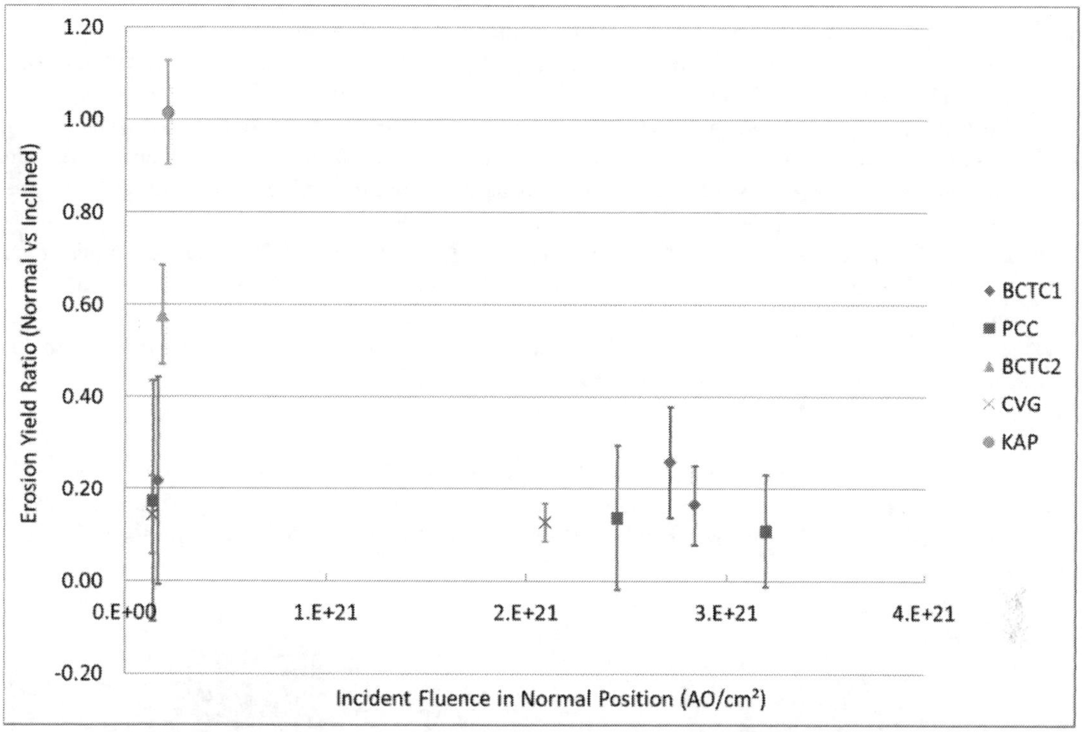

Figure 3. Erosion yield ratio (ratio of normal erosion yield to tilted erosion yield).

3.2. Residual fluences after reflection

The values reported in Table 1 correspond to the proportion (in percent) of incident fluence resulting in erosion of Kapton HN foils (used in the detection plane) after reflection on the MUT surfaces. Secondary-to-primary flux ratio ranges in the 4-10 % values that are of the order of magnitude of data reported in [5] or [8] (at a different reflection angle and for different materials). This secondary fluence estimation is consistent between LEOX1 and LEOX2 campaigns (BCTC1, CVG and PCC comparison in Table 1) even with a lower fluence level (one order of magnitude) in the first case.

Table 1. Secondary fluence comparison.

	Secondary fluence (%)	
Reflector material	LEOX1	LEOX2
BCTC1	4	5
BCTC2	7	-
CVG	9	10
KAP	6	-
PCC	7	8
BCTC3	-	10
AL	7	-

3.3. Erosion observations

In terms of AO erosion analysis, several observations were conducted. The first one is a visual inspection (optical) of the Kapton film detectors. By optical contrast, we can see interesting erosion feature as disclosed in Figure 4b for PCC sample and in Figure 5 for all samples:

- Erosion area can be observed and measured. The area between the two yellow dash lines in Figure 5 corresponds to the theoretical specular zone, *i.e.* with same incident and reflected angles (see Figure 4a), well defined geometrically speaking (5.3 mm in height in our case; Figure 4a).
- Clearly both specular and diffuse components add to the overall erosion of Kapton HN. Specular and diffuse response surface is lower than theoretical total exposed surface (not masked, see Figure 4b).

Observation of Kapton detectors for all MUT samples shows specific material-dependent signature. For instance, similar ratios of secondary fluence were measured for PCC and CVG (resp. 8 and 10 %; Table 1) but the feature of visible eroded areas are very different with (Figure 5):

- A total eroded zone much larger for PCC,
- Eroded zone limit is rather concave for CVG, straight for PCC.

For BCTC1 reflector material, the eroded area limit is rather concave like CVG material; for the BCTC3 material, the limit is rather straight and the diffuse erosion seems more extended like in the PCC case.

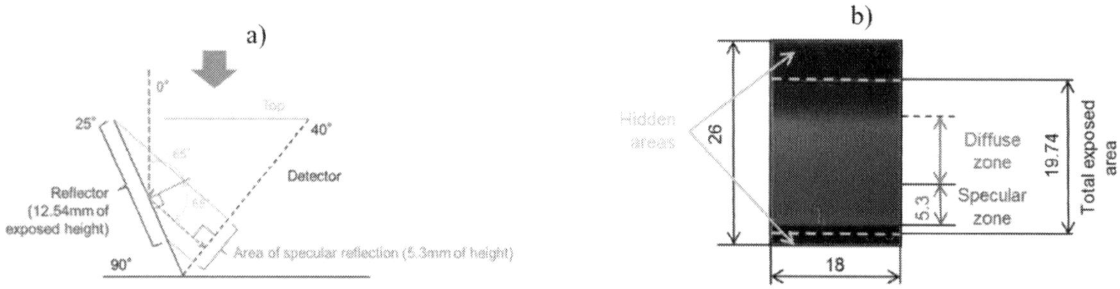

Figure 4. a. Reflection angles inside a geometry. **b.** Optical observation of erosion areas for PCC sample.

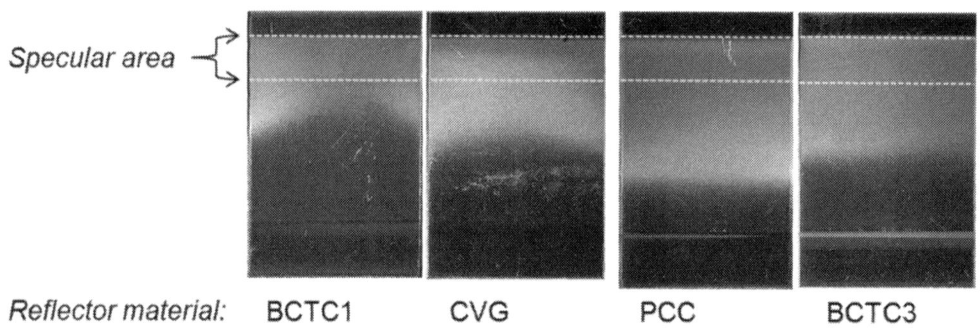

Figure 5. Optical observation of LEOX2 Kapton film detectors exposed to MUT scattered beam.

The second observation corresponds to the investigation of the eroded surfaces by SEM. The Figure 6 depicts the particular case of the Kapton film detector exposed to the BCTC3 reflector material. The total theoretical exposed area is defined between the two dash lines. The Z1 area corresponds to the specular reflection, with a canyon-like eroded facies but with rather smooth peaks (Figure 7). The highest peaks under localized dust give an idea of the erosion depth. In the pure diffuse area Z2, the erosion depth is weaker and the Z3 area, without optical contrast, seems not affected by

IOP Conf. Series: Materials Science and Engineering 1287 (2023) 012010 doi:10.1088/1757-899X/1287/1/012010

the erosion. One should note in the extreme border Z1' the erosion is well marked with higher peaks, probably due to a concentration effect (see next paragraph), as suggested by Banks and Miller [7].

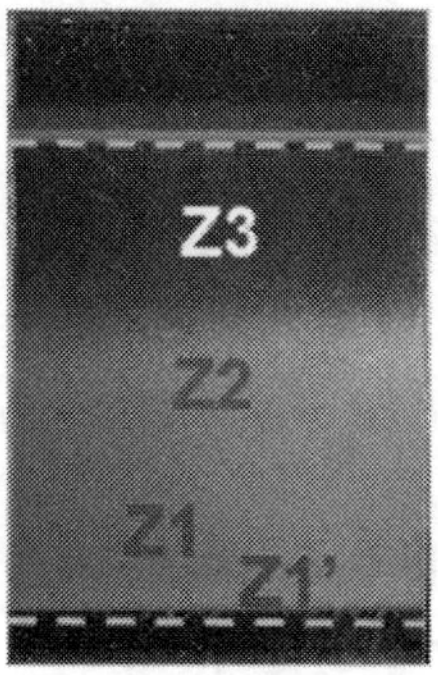

Figure 6. Kapton film detector exposed to reflected flux of BCTC3 sample: definition of the different zones of SEM observations (Figure 7).

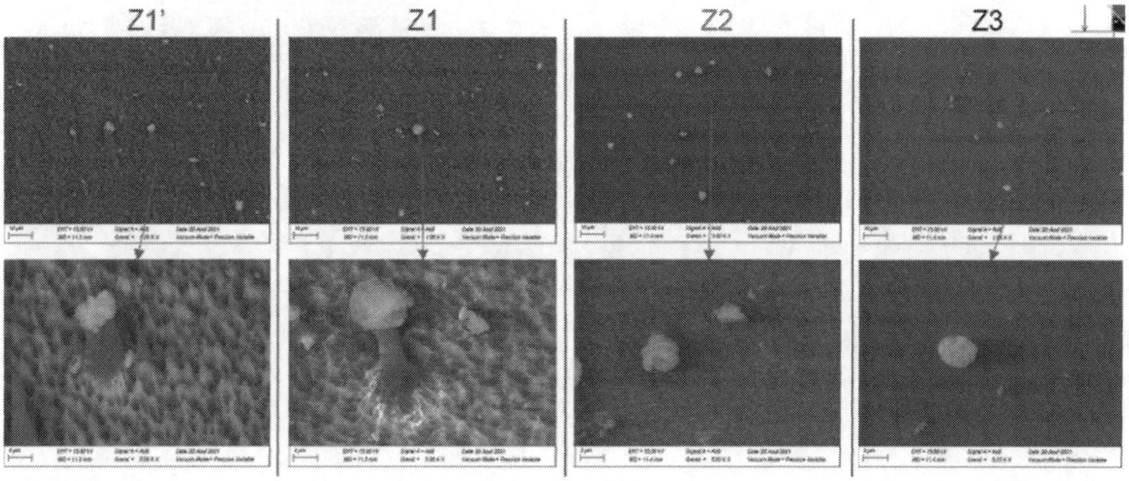

Figure 7. Different zones of SEM investigation on Kapton film detector surface (BCTC3 reflector).

These observations were followed by complementary profilometry measurements conducted with a 3D optical profilometer. For the same film detector of Figure 6, the profilometry measurement is compared to the optical observation at Figure 8. The optical observation is reported here with an amplified length scale and compared to the profilometry realized along the median line (in white). In the specular area (at left of the specular/diffuse border) the profile of erosion is more perturbed and in the diffuse area the profile is smoother. We retrieve the amplified erosion depth located at the two extremities of the exposed surface (arrows), as suggested by Banks and Miller [7] it's identified as a sample holder edge effect.

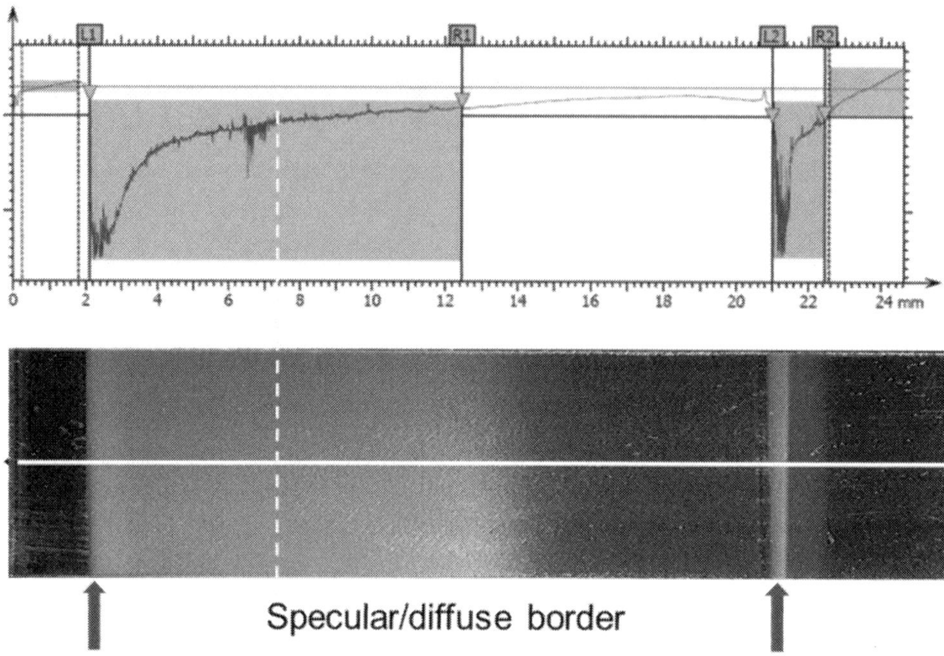

Figure 8. Optical observation vs profilometry: KAP-F-2 detector (BCTC3 reflector).

4. Modelling with engineering tools

4.1. Background and inputs for calculations

ESABASE2 and Systema-Atomox is using the ray-tracing technique which is well adapted for treating multiple-reflections. The method is summarized by the following. For a ray carrying N number of particles (Figure 9):

- Part of this number is absorbed by the surface, such as $N_a = \Omega N$,
- Part of this number is specularly reflected, such as $N_s = S_r N$, S_r being the specular reflection coefficient,
- Part of this number is diffusely reflected in a defined number of rays Lambert's cosine law, such as $N_d = S_d N (1\text{-}R)$, S_d being the diffuse reflection coefficient and R the accommodation ratio.

These coefficients are linked by the relationship: $\Omega + S_r + S_d = 1$.

To use the ATOX module, the user has to define a set of 4 parameters for each material:

- Ω, the absorption coefficient,
- The specularity ratio S, such as $S = \frac{Sr}{Sr+Sd}$,
- The accommodation ratio R in case of diffuse reflection,
- The (known) erosion rate.

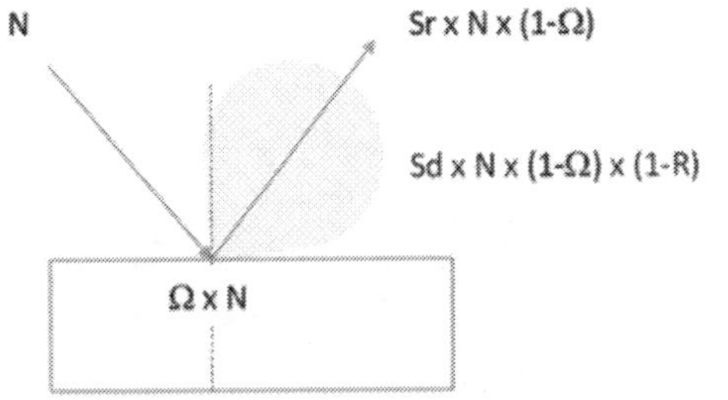

Figure 9. Main parameters of ATOMOX model.

Due to the complexity of the actual geometry and non-perfect material surfaces, several hypotheses are here considered:

- Hard-cube model, i.e. the reflected angle is the same than the incident one,
- No roughness of the target plane taken into account,
- No thermal motion,
- No accommodation, hence r = 0,
- Using this method, one can deduce that for a ray carrying N number of particles and hitting a surface, the particle conservation gives $N = N_a + N_s + N_d$,
- Multiple-reflection authorized but primary reflection results mainly observed,
- Estimate of the parameters for Kapton film detector can be deduced from the literature [13][14], with $\Omega = 0.15$ and $S = 0.5$.

Figure 10. Geometrical model.

Figure 10 is a representation of the model of the experimental geometry used. It has been decided to check if the eroded patterns of Figure 5 could be numerically reproduced with some basic hypotheses.

4.2. Boundary cases simulations

It is first important to check if the specular zone's height of 5.3 mm is reproduced numerically.

To do so, let us consider a fully specular material. With the engineering tool, the eroded zone on the sensor is shown in Figure 11. Here the specular zone corresponds to the red rectangle. Its height is about 5.4 mm.

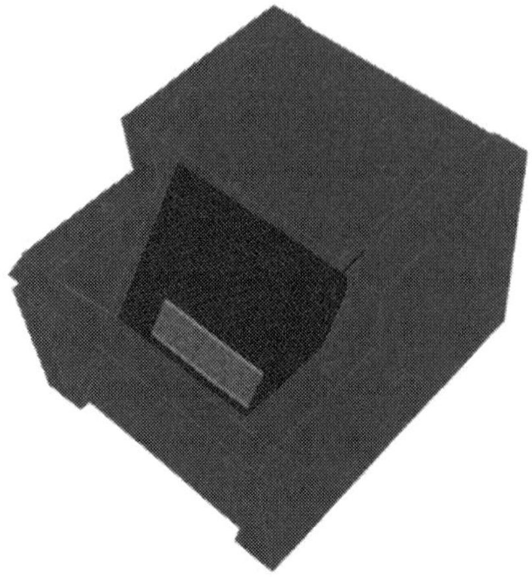

Figure 11. Eroded area in case of a fully specular material.

After the specular zone check, a fully diffusive material has been computed to see if the different eroded patterns of Figure 5 could be explained by a diffuse material. The eroded zone on the sensor is shown at Figure 12. According to the pattern of this fully diffusive material, the experimental eroded patterns which show higher erosion levels on the sides than at the center (especially BCTC1 and CVG) cannot be fully explained by diffusive materials in simulations. When examining the test geometry, one can observe that the samples don't fully cover the support (Figure 13a). One explanation of the eroded pattern could then be the diffusive reflection of the metallic support (aluminum material).

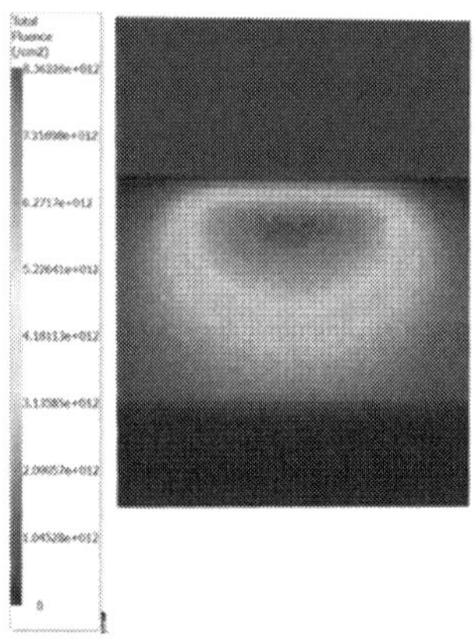

Figure 12. Eroded area in case of a fully diffuse material.

According to a simulation, in case of diffusive metallic surfaces of the side of the samples, the eroded pattern corresponding to these surfaces alone are as shown below (Figure 13b). It is then suggested that the eroded patterns obtained are due to the diffusive reflection of the metallic surfaces which is added to the AO reflections on the MUT samples.

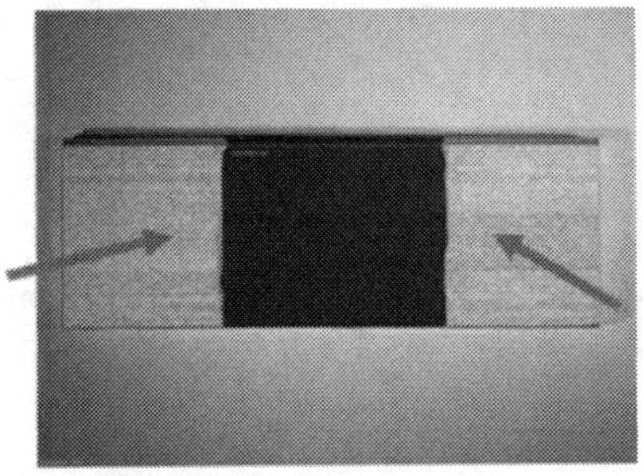

a)

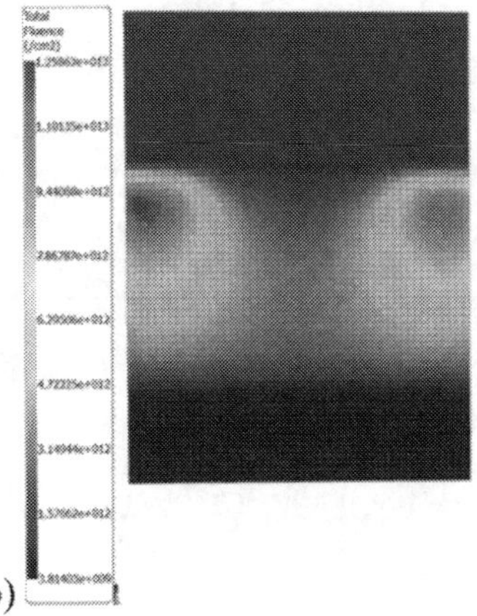

b)

Figure 13. a. Inclined MUT sample with aluminium support. **b.** Computed eroded pattern due to metallic sides diffusion.

4.3. First comparison with experimental results

Because of the supposed contribution of metallic parts, a new set of parameter has to be determined. To do so, it is decided to suppose that the BCTC1 is a specular material and thus the erosion observed on the sides (Figure 14) should mainly be due to the metallic parts. Indeed, to determine a set of parameters thanks to analytical equations (estimation of reflected fluences), points of control have to be defined (see Figure 14). It is first arbitrary chosen to put the eroded level of the "white" blurry line to a corresponding erosion depth of Kapton of 2 μm. With this hypothesis, one can decouple the analytic equations of the two different kind of materials. Using this method, metallic sides parameters are put at $\Omega = 0.12$ and $S = 0.08$ in our simulations.

Figure 14. Points of control used for BCTC1 material.

Using the same reasoning, one can approximate the samples' set of parameters (see Table 2) using this point of control considering the hypothesis on the eroded level at the "white line" at 2 μm. To numerically recreate the eroded pattern, the calculated points of control have been taken in the middle of the top side (Figure 14) because according to Figure 5, erosion levels coming from metallic surfaces are lower here, so less perturbation is introduced.

Table 2. Estimation of MUT samples' set of parameters.

MUT sample	BCTC1	CVG	PCC	BCTC3
Ω	0.85	0.82	0.73	0.75
S	0.52	0.45	0.3	0.37
S_r	0.078	0.081	0.081	0.0925
S_d	0.072	0.099	0.189	0.1575

With this set of parameters, computed eroded patterns are compared to the experimental observations for a rather specular material (BCTC1) at Figure 15 and for a rather diffuse material (PCC) at Figure 16. One should note in the following simulations, the incident fluence has been set to the same value of 2.47 10^{21} at/cm² for all geometries.

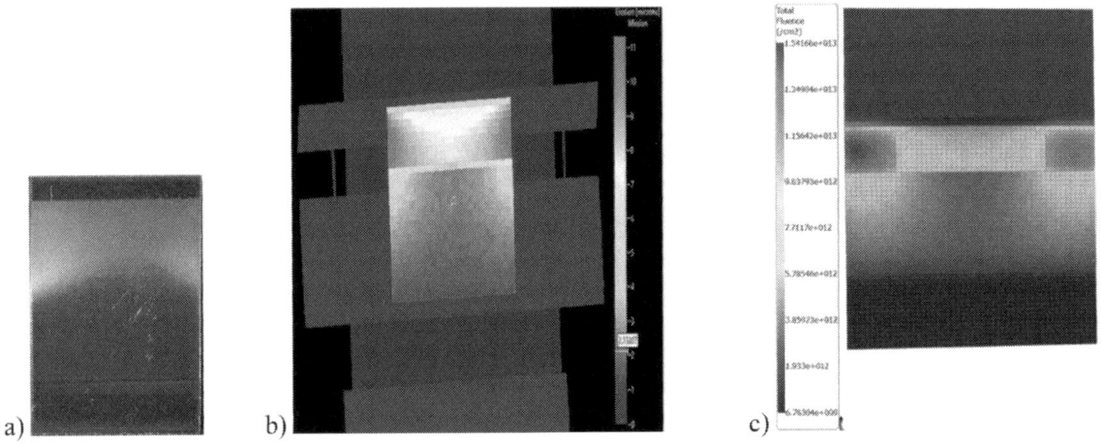

Figure 15. a. Optical eroded pattern (BCTC1 reflector). Computed eroded pattern (**b**: ESABASE2; **c**: SYSTEMA).

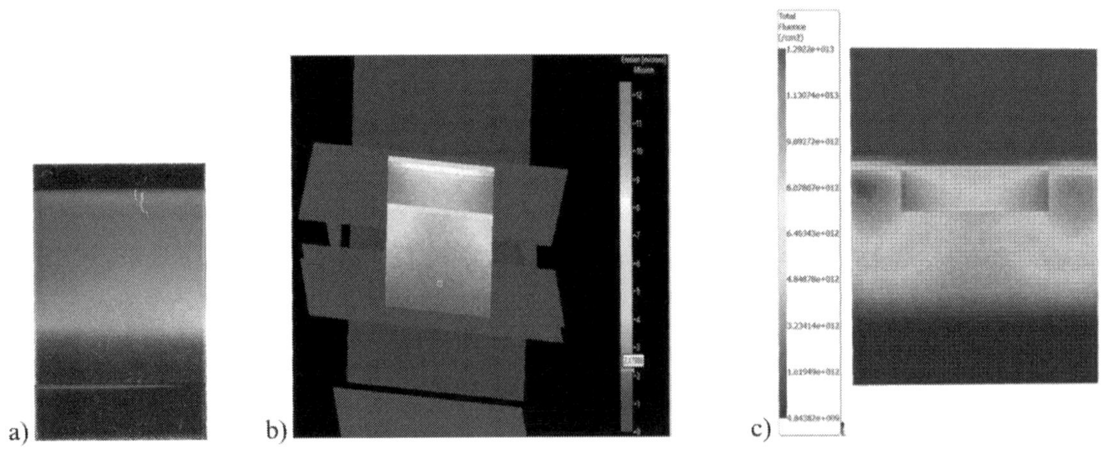

Figure 16. a. Optical eroded pattern (PCC reflector). Computed eroded pattern (**b**: ESABASE2; **c**: SYSTEMA).

It can be concluded that eroded pattern could be fairly well reproduced by simulation tools. Nonetheless, this work only describes a rough reproduction of the experimental results. A better methodology should be thought of to improve the accuracy of the simulations. In particular another check point concerns the total mass loss whose discrepancy between computed and experimental values remains in the range 8 to 40 % in this first approach.

5. Conclusions and perspectives

In this work, an original geometrical setup was designed in order to expose target surfaces with normal incidence or after one reflection. Candidate materials have been selected in view to compare numerical results conducted with usual modeling tools and experimental results. As already mentioned in the literature, Kapton HN was confirmed as a reference material to evaluate the level of fluence during the test campaign. Its erosion yield was confirmed to be independent of the impingement angle θ, by using the effective fluence ('cos(θ)' correction factor). Concerning the geometrical effect, all the materials tested with a 65°-impingement angle seem more sensitive to inclined AO erosion than normal-oriented erosion. After reflection on the inclined-MUT samples, the residual fluence coming on the detector surface is about one tenth of the direct incident fluence. The Coverglass is the more reflective material with ~10 % of secondary fluence, other materials are between 4 and 10 %. From the optical observation of eroded Kapton films used as detectors, it's clear the erosion area includes the area of specular reflection as well as a more or less large diffuse reflection area, the size and the form being dependent on the reflective materials nature. First computations with industrial codes implementing AO erosion process with multiple reflections have been successfully conducted, the erosion patterns being rather well described with some basic hypotheses on material parameters and taken into account the reflection on metallic sides, in order to reproduce the concave shape between eroded and non-eroded areas.

The first perspective will be to improve the modeling method with a parameters optimization based on the total mass loss and the use of multiple points of control with extracted erosion depths from the profilometry measurements. Moreover a near future goal is to apply this modeling to SESAME geometries (being part of the ESA Bartolomeo platform for material space experiment onboard the ISS) and compare ground vs flight data.

Acknowledgments
The authors would like to acknowledge the ESA-ESTEC technical staff for their helpful work at LEOX facility and the financial support from CNES R&T program and ADS and TAS co-funding.

References

[1] Crisp N H *et al.* 2021 System modelling of very low Earth orbit satellites for Earth observation *Acta Astronautica* **187** pp 475–491
[2] Banks B A, Miller S K R, de Groh K K and Demko R 2003 Scattered Atomic Oxygen Effects on Spacecraft Materials *Proceedings of the 9th International Symposium on Materials in a Space Environment* ESA SP-540 Noordwijk The Netherlands pp 145–152
[3] Banks B A, de Groh K K, Miller S K and Waters D L 2008 Lessons Learned From Atomic Oxygen Interaction With Spacecraft Materials in Low Earth Orbit *NASA Technical Memorandum* NASA/TM—2008-215264
[4] Bourassa R J and Gillis J R 1991 LDEF Atomic Oxygen Fluence Update LDEF Materials Workshop '91 NASA Conference Publication 3162 Part 1 pp 59-69
[5] Kimoto Y, Shimabukuro T, Sakai M and Koizumi T 2015 Simulation of Atomic Oxygen Behavior Using the NC-DSMC Method in Combined Space Effects Test Facility *Proceedings of the 13th ISMSE*, Pau, France, 22-26 June 2015
[6] Duzellier S, Lévêque D, Pons C, Soonckindt S, Laurent E, Chanteperdrix G, Vilaranda-Fernandez M, Jouanne P, Perrin-Bailly V and Nguyen Van Sang D 2021 Ground investigation of AO effect in case of complex geometry *Proceedings of the 1st International Symposium on very low earth orbit missions and technologies*, Virtual Event, 28-29 June 2021

[7] Banks B A and Miller S K 2018 Effects of Sample Holder Edge Geometry on Atomic Oxygen Erosion Yield of Pyrolytic Graphite Exposed in Low Earth Orbit *NASA Technical Memorandum* NASA/TM—2018-219910
[8] Banks B A, Seroka K T, McPhate J B and Miller S K 2011 Attenuation of Scattered Thermal Energy Atomic Oxygen *NASA Technical Memorandum* NASA/TM—2011-217028
[9] Duzellier S, Soonckindt S, Chardon J P, Pons C, Laurent E, Remaury S, Minton T and Murray V 2018 Validation of Atomic Oxygen Active Detector Resistack *Proceedings of the 14th ISMSE & 12th ICPMSE* Biarritz France 1 to 5 October 2018
[10] Yokota K, Tagawa M, and Ohmae N 2002 Impingement Angle Dependence of Erosion Rate of Polyimide in Atomic Oxygen Exposures *J Spacecraft* **39** NO.1: Engineering Notes pp 155-156
[11] Minton T K 1995 Protocol for Atomic Oxygen Testing of Materials in Ground-Based Facilities Version Number 2 *JPL Publication* 95-17
[12] Banks B A, Waters D L, Thorson S D, de Groh K K, Snyder A and Miller S 2006 Comparison of Atomic Oxygen Erosion Yields of Materials at Various Energy and Impact Angles *NASA Technical Memorandum* NASA/TM—2006-214363
[13] Banks B A et al. 1992 Monte Carlo Modeling of Atomic Oxygen Interaction with Protected polymers for Projection of Materials Durability in Low Earth Orbit *Mat. Res. Soc.* **278**
[14] Chen L and Lee C-H 2006 Interaction potential between atomic oxygen and polymer surfaces in low earth orbit *Journal of Spacecraft and Rockets* **43** n°3

Investigation of long-term storage of space materials for future constellation missions: study of Braycote® 601 EF lubricant

T Henry[1a] and M Holynska[1b]

[1]European Space Agency (ESA), European Space Research and Technology Centre (ESTEC), Materials' Physics and Chemistry Section (TEC-QEE), Keplerlaan 1, Postbus 299, 2200 AG Noordwijk, The Netherlands

[a]Theo.Henry@esa.int; [b]Malgorzata.Holynska@esa.int

Abstract. The long-term storage (LTS) of space materials is becoming more and more critical for the next missions. For instance, for MTG (Meteosat Third Generation) mission, 5-year assemble, integration and testing (AIT) time on ground, up to 17.5-year storage for recurring models plus at least 8.5 years in orbit operation are planned. Hence, a good understanding of materials' aging properties is desirable to ensure that materials' specifications are still suitable after LTS, and, eventually, to anticipate the potential non-conformities. For this purpose, the European Space Research and Technology Centre (ESTEC) is investigating these problematics thanks to its wide capabilities from aging with environmental, climatic and vacuum chambers up to physico-chemical characterizations by numerous instruments such as microscopes, spectroscopes, thermal, mechanical or electrical analyzers. An overview is presented herein through the study of a fluorinated based grease lubricant named Braycote® 601EF widely used in mechanisms in space sector such as the US space shuttles actuator or the basic end effectors of The European Robotic Arm.

1. Introduction

Space projects often last for several years or decades from the mission definition up to the launch. Between the constant increase of scientific requirements of space scientific missions and the development of numerous satellites to fly in constellations, ensuring the durability of materials and satellites parts between their procurement and the launch becomes necessary; either for a voluntary long-term storage (for example if satellites are built in series) or involuntary one (unplanned issue, difficulty to solve new cutting-edge scientific challenges). For example, ADM-ALADIN launched more than 10 years after it was initially programmed. On-ground long-term storage (LTS) can also be foreseen for satellites launched in constellations. For instance, for the MTG (Meteosat Third Generation) mission, 5-year assembly, integration and testing (AIT) time on ground, up to 17.5-year storage for recurring models plus at least 8.5 years in orbit operation are planned [1]. Another benefit of LTS is the cost reduction through reuse of spacecraft parts from the prototype flight models.

During this storage, even in cleanroom environment, space materials are subjected to aging and various degradation patterns may occur. As a proof of interest of the space community to study the long-term storage, a dedicated workshop conference was held by the European Space Agency (ESA) in 2016 [2] and conferences focusing on space applications now include session on this topic (e.g. ISMSE 15th / ICPMSE 13th, ECSSMET 17th). Wide studies led by ESA [3] or Airbus [4], including hundreds of samples stored for years, can also be pointed out. This LTS topic typically applies to all types of materials, from coatings and mirror which may degrade [5–7], to adhesive tapes which may lose adhesion or debond as a result of adhesive embrittlement, passing by CFRP structures [8], connectors [9], pyrotechnic igniter [10] and lubricants [11]. Among these last materials, Perfluoropolyether (PFPE) type lubricants and greases are common in space mechanism applications owing to their high thermal and chemical stability, as well as low outgassing properties. Therefore, they have adequate performances for less demanding space applications [12]. Braycote® 601 EF from Castrol is one of these PFPE greases. It was used for example in the United States (US) space shuttle body flap actuators [13] and the European Robotic Arm (ERA) of the International Space Station (ISS) [14].

Braycote® 601EF grease combines a Z-type perfluoropolyether base oil named Brayco® 815Z (refined version of Fomblin Z25 according to Prat et al. [15]) and a polytetrafluoroethylene (PTFE) thickener (≈ 23 wt%) [15]. It also contains $NaNO_2$, bentonite-like and $C_{20}H_{44}ClN$-like additives [13]. According to the manufacturer Castrol, it has exceptional chemical resistance, extremely low volatility, and a temperature service range from -80 °C to 204 °C. The manufacturer assesses a shelf-life of 10 years. From the literature, it was reported in the NASA's technical bulletins of 2007 [16] that Braycote® 601 retains tribological properties for 17 years in controlled storage and shows no detrimental effect on lubricity nor significant decrease in performance due to static controlled storage according to Thermo-Gravimetric Analysis (TGA) results. However, the detailed study has not been released. Additionally, an article [13] was published focusing on the Braycote® 601 EF used in the US space shuttles actuators during 19 years. Based on Fourier-Transform InfraRed spectroscopy (FTIR), Size Exclusion Chromatography analysis (SEC), gravimetric analysis and Inductively Coupled Plasma (ICP) analysis, the authors concluded that there were no significant differences in comparison to pristine samples and the grease coloration was related to metallic iron and oxidation of metallic wear particles instead of the grease alteration. These observations are coherent with the high stability of the grease. From a thermal point of view and in the absence of catalyst, Z-PFPE thermal degradation has been reported in the range 250 °C to 390 °C with the main decomposition occurring around 370 °C [17–19]. The decomposition path is suggested to be driven by successive end group reactions owing to radical formation [19–21]. In contact with a catalyst (e.g. Al_2O_3, FeF_3), the thermal degradation is hugely accelerated and the decomposition temperature is decreased. [17,22,23]. It is considered to be induced by Lewis acids, causing chain scissions of PFPEs. This might happen if wear particles are introduced or generated in the mechanisms. Nevertheless, according to Buttery et al. [11], there is limited experimental evidence of PFPE auto-catalytic degradation from tribological experimentations. Tribologically, the grease lubrications performances can be attributed to various factor depending on the environmental conditions (ambient pressure, relative humidity, atmosphere composition, temperature mainly) the tribological conditions (stress, velocity, amount of the grease deposited, finishing of the mechanical parts etc.) and the lubricants properties (e.g. viscosity, thermal and chemical stability) [11,24,25]. Therefore, in an environment appropriate for a given lubricant, the lubrication failure is often related to its chemical alteration. This alteration is evidenced by grease bleeding, reflow and viscosity changes, oxidation of the lubricant, bond-splitting and reduction of the averaged chain lengths or the molecular weight distribution [17].

In order to verify the previous observation suggesting a high stability of the Braycote® 601 EF grease after being used over years or stored for years and its probable suitability in the frame of LTS topics, it was decided to investigate its chemical and thermal behaviour when it is submitted to stress under hygrothermal conditions. After introducing briefly, the main accelerating aging models, this article will present various thermal and chemical analyses of the pristine and aged greases under hygrothermal conditions. In addition, an old grease batch manufactured 17 years ago which remained stored has been analysed. The results confirm the high grease stability, both from thermal and chemical changes even tough small features can be distinguished presumably related to the effect of moisture. Finally, tribological measurements are suggesting few to no changes in the grease friction properties due to aging when the friction coefficient is measured as a function of velocity.

2. Accelerated aging models: a short overview

2.1. General introduction to accelerating aging

Accelerating aging test aims at artificially reproducing the degradation of materials under a given environment (e.g. weather, stress, conditions of use etc.) with laboratory experiments at a faster rate. As an example, accelerated weathering experiment would take into account the cycled temperature over days and nights, the illumination, the humidity of the environment, the pollutants, eventually salt exposure depending on the reference location, the effect of sand, rain or other factors. Hence, this is a broad field which may include the estimation of lifetime of materials, products or assemblies, their shelf-life or their sustainability / viability under given conditions. The related scientific literature is relatively abundant since it gathers various fields of applications. Several reviews or books focusing on accelerated aging can be found either orientated toward engineering or with more fundamental approaches [26–36]. Despite the relative abundance of literature on accelerating aging, the high complexity to determine and simulate materials' behaviors while being aged brings difficulties to obtain generalized models. Indeed, the aging conditions are solely identical, and the alteration paths of materials can vary strongly and be synergistic even if the chemically properties differences are limited. The lack of real aging data and correlations with the acceleration conditions also lead to difficulties to assess the pertinence of the aging conditions. As a consequence, studies are often led with empirical or semi-empirical approaches considering aging models which are, in main cases, simplified approaches with the objective to obtain screening tests more than purely quantitative tests (this last comment may not apply to specific cases, for example for fatigue studies). This is the case for a large part of the available standards.

In the frame of LTS, the aging parameters for materials can be reduced to temperature, relative humidity, light (if lights are on during the storage) and atmosphere composition (dry air, N_2 atmosphere). In specific cases, the gravity may also influence the aging few materials. Because in the present study the investigated grease is considered to be hidden from light, this parameter was excluded from this investigation. The subject of the study is grease aging, only a part of the models including temperature and relative humidity are introduced. For more information about these models, the reader can refer to the articles and books referred previously in this section.

The main models are based on the Arrhenius law which describes the temperature influence on the rate of a simple chemical reaction present over a temperature range. In case several chemical processes or phenomena occur, Arrhenius law is not suitable for modelling except if one behavior is dominant and rate-limiting. In most cases, it would be more appropriate to refer to a quasi-activation energy [34] which encompasses all underlying phenomena contributing to the aging process. Similar as the Arrhenius model, the Eyring model displays a temperature-dependent pre-exponential factor. It is based on quantum mechanics involving free energy, heat and entropy of activation [31]. The "Q_{10} method" also known as the "10-degrees rule" is another approach derived from the Arrhenius law [26]. It is often used in medical devices' industry among others. It has been used now as a standard method for accelerated aging and is described in the ASTM F-1980 standard [37]. It uses a unitless factor Q_{10} describing how fast a reaction rate changes with temperature. For many cases and when the kinetics is unknown, a $Q_{10} = 2$ is used; meaning the reaction rate would double every 10 degrees. Finally, the Peck-Hallberg model initially used for epoxy studies [38,39] includes another parameter to take into account the relative humidity in its equation in addition to temperature. In Peck-Hallberg approach these two parameters are not synergistic. The related kinetics laws and acceleration factors are reported in Table 1.

Table 1. Four common models and their acceleration factors used in accelerated aging studies.

Model	Kinetics law	Acceleration factor $A_f = k_{test}/k_{use}$	Nomenclature
Arrhenius	$k = A \cdot e^{-\frac{E_a}{k_b T}}$	$e^{\frac{E_a}{k_b}\left(\frac{1}{T_{use}}-\frac{1}{T_{test}}\right)}$	A_f Acceleration factor
Eyring	$k = A \cdot T^m \cdot e^{-\frac{E_a}{k_b T}}$	$\left(\frac{T_{test}}{T_{use}}\right)^m \cdot e^{\frac{E_a}{k_b}\left(\frac{1}{T_{use}}-\frac{1}{T_{test}}\right)}$	A Pre-exponential factor E_a Activation energy (eV) k Rate constant
Q_{10} method	$Q_{10} = \left(\frac{k_{test}}{k_{use}}\right)^{\frac{10}{T_{test}-T_{use}}}$	$Q_{10}^{\frac{T_{test}-T_{use}}{10}}$	k_b Boltzmann constant n Power factor
Peck-Hallberg	$k = A \cdot (RH)^n \cdot e^{\frac{E_a}{k_b T}}$	$\left(\frac{RH_{use}}{RH_{test}}\right)^{-n} \cdot e^{\frac{E_a}{k_b}\left(\frac{1}{T_{use}}-\frac{1}{T_{test}}\right)}$	RH Relative humidity (%) T Temperature (K)

2.2. The activation energy inconsistency in the accelerating factor

In the formulas introduced above, it is important to note an inconsistency between the kinetics and the acceleration factors. From the Arrhenius approach, the activation energy can be seen as a minimum energy required by the reactants to result in a reaction. At a given temperature, if the activation energy is lowered, the reaction rate coefficient k would be higher (if $E_a \rightarrow 0$, $k \rightarrow A$), and inversely (if $E_a \rightarrow \infty$, then $k \rightarrow 0$). Considering now the acceleration factor, if $E_a \rightarrow 0$, then $A_f \rightarrow 1$ (meaning no aging); while if $E_a \rightarrow \infty$, then $A_f \rightarrow \infty$ (meaning an instantaneous failure or infinite aging). In consequence the acceleration factor acts in the opposite way of the physical signification of Arrhenius-based laws. This inconsistency was already addressed earlier by Kececioglu and Sun [40]. To avoid it, the activation energy in accelerated aging should be seen as representative of the aging test conditions for a given material and a given property that is measured (e.g. performance of the functional property of the material); instead of picking a value of a chemical or physical phenomenon which may not represent well enough the material alteration in its entirety. As a consequence, it needs to be determined experimentally to exploit the acceleration factor properly. For non-exhaustive examples, the reader can refer to the next references [41–43].

2.3. Aging conditions used

In the present case, along the grease LTS, its aging is considered to be potentially induced by the temperature and the relative humidity. Therefore, it was decided to impose environmental conditions implying stress on these two variables with higher values than the nominal storage conditions. In the absence of well-known and established degradation mechanisms and kinetics for the Braycote® 601 EF, or a specific parameter to trigger its performances, it was decided to impose fixed environmental conditions without determining the proper activation energy; the objective being first to spot the potential changes of properties. Test values were chosen by considering the heritage of accelerated aging test and standards. A temperature of 85 °C and a relative humidity of 85 %RH were considered within up to 1900 h of aging. These parameters are, for example, currently in use to characterize the concentrator photovoltaic modules or components (e.g. IEC 62108, IEC 60068). These conditions were preferred to milder conditions used for plastics or medical devices (e.g. ASTM F1980, EN 2243-5) because the grease is recognized as very stable.

2.4. A rough estimation of the acceleration factor

Even if the current investigation is not designed to obtain a very reliable acceleration factor, it is interesting to have an idea of "how much the material is aged". Ideally, a model considering both the temperature and the relative humidity as acceleration parameters should be considered. On such a case, a Peck-Hallberg model appears to be an acceptable compromise owing to its apparent simplicity.

Nevertheless, it is not pertinent to consider a specific model because the activation energy of the tested material is complex and not known precisely, this applies also to its degradation processes over accelerated aging test. Then, a multi-approach is preferred to roughly estimate the aging by crossing three methods: an Arrhenius law, a Peck-Hallberg law, and the 10-degrees rule. The models' parameters were selected as follows: $E_a = 0.7\,eV$, $Q_{10} = 2$ and $n = -2$. For E_a and Q_{10}, values suggested by common practices (e.g. AECTP 300 meth., ASTM F1980) were considered while the n value for the Peck-Hallberg model was chosen to limit the impact on the acceleration factor. Applying the nominal cleanroom conditions (22 °C and 55 %RH) and test conditions into the models leads to the acceleration factors and the corresponding aging times reported in Table 2. A quick comparison shows evident differences between these approaches: after 1900 h of accelerated test, it might correspond to at least 17 years of real time storage up to tens of years. The large value range of acceleration factors makes sense since the models consider different approaches and the variables used for calculation were simple postulates. Even though it does not take into account the relative humidity, the most robust approach to consider would probably be the most restrictive method to extrapolate the time, here being the Q_{10} method which estimates 17.1 years after 1900 h of aging. Once again and as introduced above, the reader should be aware that this approach is an approximation based on assumptions: the value might be largely over- or underestimated. By the propagation of uncertainties, the uncertainty of the acceleration factor can also very quickly reach folded values, especially because the cleanroom conditions (storage conditions) have a relatively high latitude versus the nominal value (± 3 °C and ± 10 %RH).

Table 2. Acceleration factors and equivalent aging times for three different approaches considering the test and conditions used for this study with $E_a = 0.7$ eV, n = -2 and $Q_{10} = 2$.

Model	Q_{10}	Arrhenius	Peck
Acceleration factor	78.8	126.7	302.5
Equivalent years after 240 h	2.1	3.5	8.1
Equivalent years after 1000 h	9.0	14.5	34.5
Equivalent years after 1900 h	17.1	27.5	65.6

3. Experimental

3.1. Samples

The Braycote® grease used for hygrothermal aging test comes from an unused tube with a date of manufacturing of 27/01/2020. From its reception to the aging experiment, the tube has remained sealed and stored in a drawer. It was fully protected from light. The nominal storage room conditions were 22 °C and 55 %RH. For the test, approximately 5 g of the grease was deposited (in wire shapes) onto 10 different soda-lime petri dishes. One of the 10 samples was kept as a witness. The 9 other samples were aged for different durations. After aging, the samples remained stored in their petri dishes (closed) and shaded from light in a drawer with the same environmental conditions as for the ones introduced above.

A Braycote® 601 EF tube with a date of manufacturing of 05/2005 was also investigated to obtain a reference value of real aging time. Its precise history is unknown. The tube was already open before these experimentations. It was assumed that this grease was stored in cleanroom-like and dark conditions.

At the same time, PFPE base oil Brayco® 815Z were analyzed in parallel to the grease. A Braycote® 601 EF grease was also placed in an oven for 524 h at 85 °C in order to qualitatively determine the humidity effect by visual inspection and weight measurements. 524 h of aging were performed at 85 °C.

Table 3 summarizes the samples investigated in the current article.

Table 3. Samples of Braycote® 601 EF used in this study, their corresponding aging time and year of manufacturing.

Sample	Aging time under 85 °C and 85 % RH (h)		Year of manufacturing
	Exact	Rounded	
A	0	0	2020
B	25	24	2020
C	51	50	2020
D	80	80	2020
E	151	150	2020
F	357	350	2020
G	645	650	2020
H	997	1000	2020
I	1476	1500	2020
J	1900	1900	2020
DOM 2005	Non applicable, real aging time		2005
Oven	85 °C for 524 h, no humidity control		2020

3.2. Cleanroom storage conditions
The cleanroom storage conditions are under dark, 22 ± 3 °C and 55 ± 10 %RH. The corresponding nominal dew point and absolute humidity are 12.5 °C and 10.68. g·m^{-3}, respectively.

3.3. Accelerated aging
Accelerated aging were performed with a climatic chamber ClimEvent C/270/70/15/M from Weiss Technik at constant temperature of 85 ± 1 °C and constant relative humidity 85.0 ± 0.1 %RH. The indicated uncertainties are experimental. The corresponding nominal dew point and absolute humidity are 80.9 °C and 297.56 g.m^{-3}, respectively. When samples were removed from the chamber, overshoot (max +2 °C and +2 HR % versus the nominal conditions) has been observed for a short period of time in order to reach back the stabilization conditions. It is assumed this has no impact on the aging process. For the hygrothermally aged samples, all the analyses were performed at least two weeks after the sample removal from the environmental chamber. This comment does not apply to weighting, optical inspection and when it is notified explicitly in the text.

3.4. Weighting
Mass measurements were performed with a Sartorius BP211D microbalance. The balance precision is 0.01 mg. Experimentally and for the present study, the reproducibility mass measurement error is estimated at 0.1 mg.

3.5. Optical inspections
A digital microscope Keyence VHX 6000 was used to inspect the samples with a VH-Z20UR/W/T objective.

3.6. UV-VIS, FTIR and Raman spectroscopies
Spectroscopic measurements at wavelengths from 250 nm to 800 nm were performed on the grease in reflection mode with a Cary spectrometer from Agilent using an integrating sphere of 110 mm and an incidence angle of 8°. To deposit the grease, an in-house-made sample holder was manufactured from PTFE with an empty cylinder shape (1 mm deep) that can be filled with grease. Data were treated

considering a 99 % diffuse reflectance reference from Labsphere. The deposition method of the grease into this sample holder is considered to be qualitative and semi-quantitative for a grease coming from the same batch. Purely quantitative characteristics were not assessed.

Fourier transform infrared (FTIR) spectroscopy measurements were performed on a Bruker Optics VERTEX70v vacuum FTIR device at 485 – 4000 cm^{-1} wavenumber range with a resolution of 4 cm^{-1} with a RT-DLaTGS detector and a KBr beam splitter. Each sample was placed with a small plastic tip onto a zinc selenide (ZnSe) window and spread as a thin layer over the Infra-Red (IR) beam area. Subsequently the ZnSe window was measured in transmittance. As a reference, the clean ZnSe window was measured prior to the sample analysis. Data treatment includes a normalization of the contaminated window by data obtained for the clean window followed by subtraction of a baseline made manually, and a normalization.

Raman spectra were acquired with a WITec's Raman microscope alpha300 R using a 532 nm laser excitation (green laser). Measurements were performed in the confocal mode with an objective Zeiss LD EC Epiplan-Neofluar HD Dic 50x / 0.55 between 37 cm^{-1} and 3900 cm^{-1} with a resolution ranging from 3 cm^{-1} for lowest wavenumbers to 2 cm^{-1} for highest wavenumbers. The integration time was 1 s. Raman spectroscopy was used to characterize the TGA residue of the grease. The power and the number of accumulations were 5 mW with 231 accumulations for the TGA residue obtained under air, and 1 mW with 120 accumulations for the residue obtained under N_2.

3.7. SEM/EDX analyses

An FEI TENO SEM was used for the TGA residues analyses. They were performed under secondary vacuum at 20 kV with a current of 0. 82 nA. EDX analyses were performed with an Ultim Max 65 EDS detector from Oxford Instruments. The powder residues were deposited on a black carbon tape and analyzed as raw without being coated.

3.8. TGA and DSC analyses

Thermo-Gravimetric Analyses (TGA) were carried out with a Mettler-Toledo device: TGA/SDTA 851e/LF1600/robot module operating with the STARe software using a heating program from 25 °C to 700 °C with a 10 $K \cdot min^{-1}$ heating rate. The furnace was continuously purged by nitrogen gas or air at a flow rate of 50 $mL \cdot min^{-1}$. Buoyancy and baseline corrections were made using an equivalent empty pan run using the same method. Alumina crucibles of 70 µL without lids were used for the TGA experiments. Typical sample mass ranged between 15 mg and 20 mg.

Differential scanning calorimetry (DSC) was used to study possible phase transitions. DSC was conducted on a Mettler-Toledo device: DSC 822e/700/robot module operating with the STARe software. A heat-cool-heat-cool program from -150 °C to 450 °C was used with a heating and cooling rate of ± 10 $K \cdot min^{-1}$. The DSC furnace was continuously purged with nitrogen at a flow rate of 50 $mL \cdot min^{-1}$. Aluminum crucibles of 40 µL with pierced lids were used for the experiments with typical sample mass between 10 to 15 mg.

3.9. GCMS analyses

Gas Chromatograph/Mass Spectrometry (GC-MS) analyses were conducted using an Agilent bench-top model GC-MS, consisting of a 7890B GC and a 5977B MS. The GC is fitted with a Multimode Inlet injection port. Samples were analyzed by thermo-desorption technique by placing a staple into a liner of thermo-desorption injector port. The liner temperature was set on 40 °C and ramped at 60 $°C \cdot s^{-1}$ to 300 °C after the injection. Flow rate of He as a carrier gas was set at 1.5 $mL \cdot min^{-1}$ in 1:10 split mode. The GC oven temperature program started at 40 °C with a hold time of 10 minutes and increased at a rate of 10 $°C \cdot min^{-1}$ to 300 °C, where it was held for 60 minutes. Mass spectrum was acquired masses from 29 to 500 amu. Analyses were performed on capillary column J&W DB-5ms 60m/0.25mm/0.25 µm with a source and quadrupole temperature of 230 °C and 150 °C, respectively. The presented graphs were normalized according to the main peak after the baseline was subtracted.

3.10. Rheology

The viscosity measurements were performed with the aid of TA Instruments (Waters) DHR-2 Rheometer at 20.0 °C. 25 mm aluminum plates were mounted on the Peltier temperature control system. Before each sample measurement, the upper plate was renewed, and the lower plate cleaned with Galden SV 55 from Solvay and then with IPA. Before each sample measurement, and after having installed the upper plate, the instrument was recalibrated (inertia, friction, and rotational mapping in fine mode with two iterations). Upon calibration, each sample was mounted on the lower plate and trimmed off at the trim gap of 115 µm. The gap was finally applied at 100 µm. Measurements were performed between 10^{-2} s^{-1} and $5\cdot10^{3}$ s^{-1} with 5 s of stabilization and 10 s of measurement time. Prior to the measurement, temperature stabilization dwell time of 3 min was applied, followed by a peak hold at 100 s^{-1} for 3 min.

3.11. Tribology

The tribological measurements were performed with the aid of TA Instrument DHR-2 Rheometer equipped with 4 ball geometry tribo-rheometer tool. Measurements were run at room temperature with 440C stainless steel ball of 12.7 mm diameter. Each measurement was performed with new balls to ensure similar surface finishing. Before the measurements, the balls were cleaned with IPA and a cleanroom tissue. The applied axial load was 40 N giving an Hertzian contact pressure reaching 1.55 GPa considering Young modulus and Poisson coefficient of 200 GPa and 0.285 for the stainless steel balls, respectively. A small amount of the grease was deposited by hand on the top ball only. The deposited mass was about 50 mg (not controlled precisely). This large mass was applied for three reasons: 1) the balls are fixed (i.e. they cannot roll), so the friction points remained unchanged and the grease amount implied in the friction is a small fraction on the deposited grease; 2) when a small quantity (0.1 mg to 0.5 mg) was spread over the balls during preliminary experiments (with quantitative reproducible method), a rapid failure at intermediate speed under the test conditions was observed; 3) on several mechanisms, the operators deposit the grease by hand without controlling precisely the lubricant concentration with a dedicated process.

After applying the force, a relaxation time of 1 min was held. Then a ramp flow from 0.01 $rad\cdot s^{-1}$ to 5 $rad\cdot s^{-1}$ with a duration of 1 min was performed. The velocity was increased and then decreased. These cycles were repeated four times. Finally, a ramp flow in an extended range from 0.01 $rad\cdot s^{-1}$ to 30 $rad\cdot s^{-1}$ with a duration of 3 min was performed. The velocity was increased and then decreased. These cycles were repeated two times. The first sweeps in a limited velocity range were carried out to highlight the properties under stables conditions because it was noticed that the grease lubrication properties are modified once reaching approximately 10 $rad\cdot s^{-1}$.

4. Accelerated aging of the grease: results and discussions

4.1. Grease coloration and moisture absorption

Figure 1 shows the general characteristics observed during optical inspection before and after aging. While the grease has a whitish color in its pristine state, after aging a yellowing is observed. This coloration was evident, even after 24 h of aging. The surface seems also rougher and drier. In addition, particle-like shapes of sizes between 1 – 5 µm to 200 – 300 µm are appearing on the grease surface after aging. These particle-like shapes revealed to be liquid aggregates such as bubbles or oozing when the magnification was increased. Although they might be local oil bleeding, no generalized oil separation is observed. These images are representative to all the aged samples where the observed oozing-like feature becomes more evident upon increase in the aging time.

This coloration was characterized by UV-VIS spectroscopy. Spectra are shown in Figure 2 a). The strongest absorption change occurs in the 400 – 500 nm range with an absorption increase comprised between 15 % and 20 %. The absorption centre undergoes a blue shift with the aging time (inset of Figure 2 a)). The chemical origin of this colouration is unexplained but might be related to the oxidation of iron either added as additive or already present in the used bentonite additive (see section 4.2).

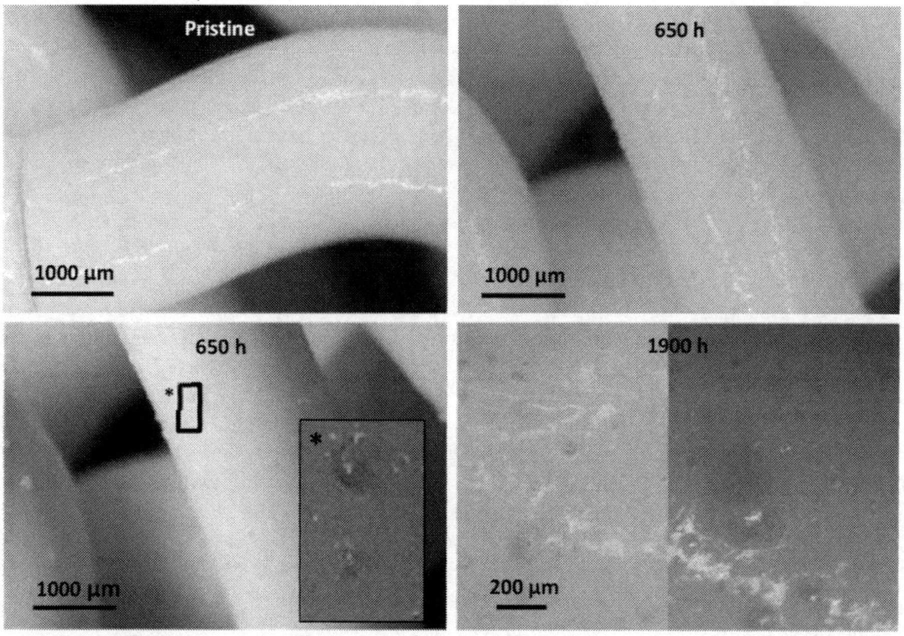

Figure 1. Micrographs of the hygrothermally aged greases for different aging times. The pictures on top are under full ring illumination, the bottom left under partial ring illumination, and the bottom left image is under full ring (left) and partial ring illumination (right). Images of aged greases are representative to all the non-pristine samples.

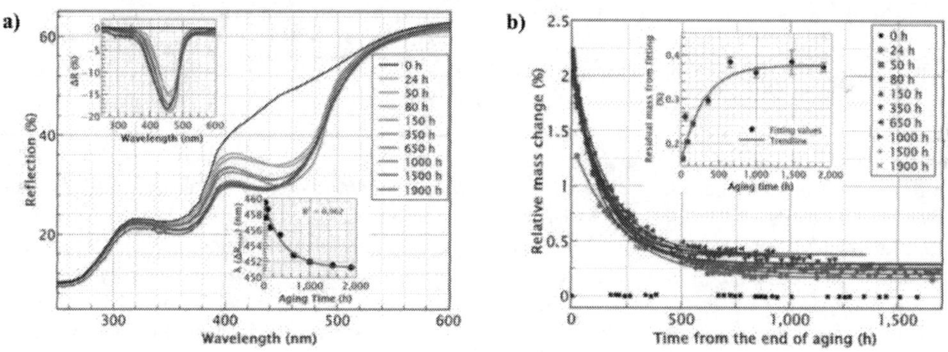

Figure 2. a) UV-VIS spectra of hygrothermally aged greases. The top left inset corresponds to the subtraction with respect to the pristine sample. The bottom right inset shows the blue-shift of the absorption center with respect to the aging time. b) Mass changes of the hygrothermally aged greases after aging for different aging times. The inset shows the evolution of the stabilized mass with respect to the aging time.

After removing the samples from the climatic chamber and in parallel to this yellowing process, a small mass increase was observed. Mass measurements were repeated along several days. The results are reported in Figure 2 b). According to the measurements, the mass intakes tend to saturate even after a short exposure around 2.20 wt% and to decrease with an exponential decay up to stabilization. The experimental data were fitted with an exponential decay in the form of $A \cdot e^{Bx} + C$ where A is an amplitude factor in weight percentages, B a decay factor and C the value of the mass stabilization in weight percentages. The resulting adjusted R^2 ranged between 0.985 and 0.998 suggesting a very good correlation between the fitting and the experimental values. These fits permit to observe with reliability the trend of the stabilization level of the mass intake depicted on the inset. These results demonstrate that the mass intakes never go back to zero and the residual intake has a saturation level of around 0.38 wt%. In practice this means the more the grease is aged, the more the stabilized mass increases up to saturation value.

To investigate the origin of mass intake two actions were carried out. Firstly, an experiment of 524 h was performed under 85 °C in an oven with no relative humidity control. Therefore, during the heat treatment, the relative humidity can be estimated at below 5 %RH. After the sample removal, no coloration and no oozing or bubble-like features were observed. Their origin is consequently attributed to the moisture effect. Besides, the sample mass decreased only a little (-0.07 wt%) and reached back its original mass value within 48 h. Secondly, TGA and DSC measurements were performed right after the sample removal from the aging chamber after 1900 h (not shown). TGA displayed a clear mass loss between typically 80 °C and 160 °C of about -1.7 wt% while it was about -0.2 wt% after two weeks of resting period. On DSC diagram, melting events are seen in the same temperature range while boiling events are present between -20 °C and 0 °C and completely disappear after two weeks of resting period. These values are coherent with the mass measured, especially when considering the stabilization plateaus. The exact nature of these volatile species remains unknown. It is supposed to be mainly water moisture (and eventually light hydrocarbons or volatiles) which might be embedded and adsorbed by the bentonite additive. Indeed, the bentonite is a clay known to display water sorption, adsorption and swelling properties [44].

4.2. Thermal gravimetric analyses and the atmosphere effect on the thermal degradation

Figure 3 a) shows the TGA measurements of Braycote® 601 EF samples under N_2 atmosphere. Three different mass losses steps can be distinguished. They are typically located from 300 °C to 450 °C (up to ≈ 75 wt%), 450 °C to 540 °C (up to ≈ 23 wt%) and 540 °C to 600 °C. The two first mass losses are attributed to the base oil degradation. The last mass loss even corresponds to the end of PTFE filler decomposition. The last change permits to evaluate the weight proportion of the PTFE filler with respect to the PFPE base oil. It is estimated at around 23 wt%. This value is coherent with the oil content reported by Morales et al. [13] and is in very good agreement with the estimation of the PTFE proportion reported by Prat et al. [15]. No consistent trend between the mass losses and the aging time was observed. Eventually, after aging, base oil degradation may occur at slightly lower temperatures. Interestingly, when the TGA was ran under air, the thermal stability of the grease was notably reduced. This can be observed in the Figure 3 b) where the grease degradation is faster and happens at lower temperatures. In addition, the results were found to be scattered without any relation to the aging time. Finally, the two firsts mass losses steps tend to be convoluted. These notable differences are attributed to the degradation paths changing owing to the oxygen present in the air favoring the grease decomposition through hydroxyl radicals and grease oxidation. The scattering of the results is attributed to the sample preparation where it is assumed that a sample having a higher effective surface (and hence a higher surface in contact with the air and oxygen) will deteriorate faster. Last but not least, the residual masses remaining in crucibles after the TGA measurements were lower when performed under air (versus N_2 atmosphere) and their color was different independent of the aging time: orange under air and black under nitrogen. A short investigation of this residue is described in the next paragraphs.

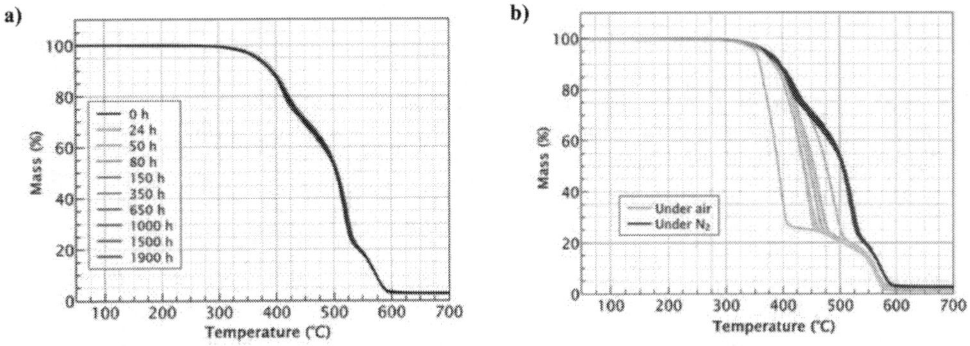

Figure 3. TGA measurements performed on hygrothermally aged greases. a) under N_2 atmosphere; b) effect of the atmosphere on the TGA measurement independent of the aging time.

The powder residues were first analyzed by Raman spectroscopy. The resulting spectra are reported in Figure 4 a). According to the database of KnowItAll software, the powder obtained under nitrogen and air are identified as amorphous carbon and iron oxide (Fe_2O_3), respectively. These findings are coherent with the powder color being black and orange, respectively.

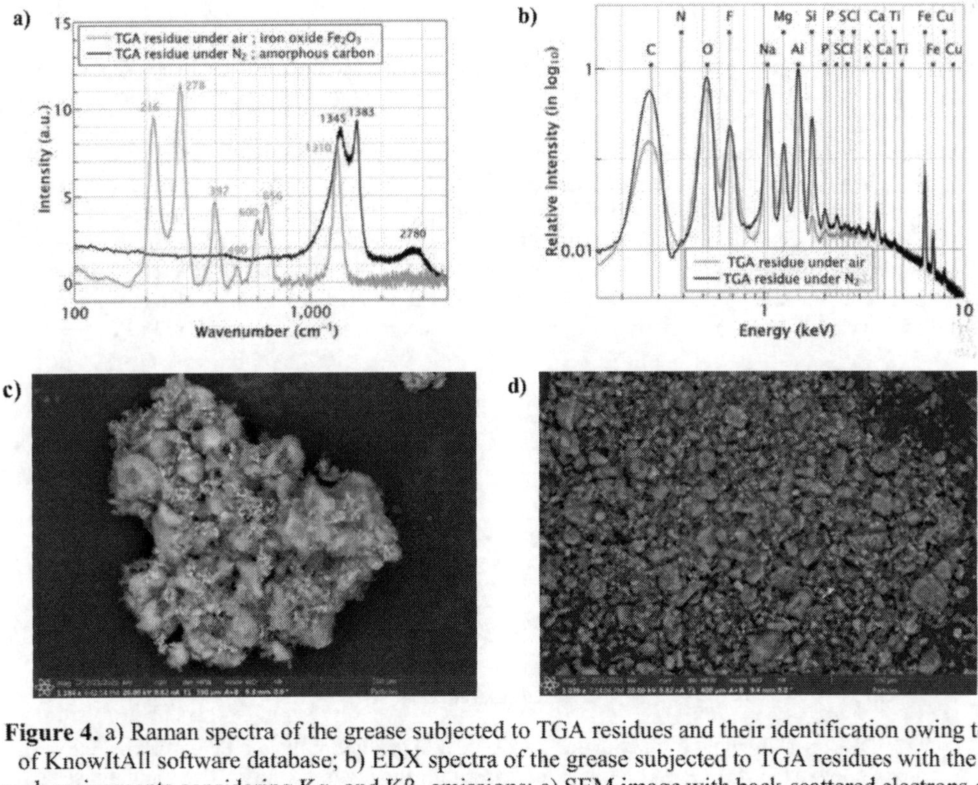

Figure 4. a) Raman spectra of the grease subjected to TGA residues and their identification owing to of KnowItAll software database; b) EDX spectra of the grease subjected to TGA residues with the peak assignments considering $K\alpha_1$ and $K\beta_1$ emissions; c) SEM image with back-scattered electrons of the grease subjected to TGA residue obtained under air, the image is representative of the powder obtained under air; d) SEM image with back-scattered electrons of the grease subjected to TGA residue obtained under N_2, the image is representative of the powder obtained under N_2.

The analyses by SEM/EDX also show evidence of different morphologies. While the powder identified as iron oxide occurs in particles shapes that coalesced, the amorphous carbon is random and appears in thinner grains (Figure 4 c) and d)). EDX analyses revealed the presence of the same elements in both cases (Figure 4 b)). They are probably ascribed to the next origins:

- A majority of C, O and F atoms are coming from the PFPE oil and PTFE thickener.
- Si, Al, Na, K, Ca, Mg are mainly coming from the bentonite structure.
- Ti, Fe and Cu are probably impurities, coming from the bentonite or additional additives.
- Traces of P and S probably come from organic additives.

In agreement with Raman analyses, the relative carbon content is notably higher for the black powder than for the orange powder. These results demonstrate that thermal degradation of Braycote® 601 EF is dependent on the atmosphere, yielding different products. The smaller residual mass obtained under air could be explained by a more complete decomposition than under N_2. Moreover, owing to these analyses, it could be hypothesized that the yellowing of the grease arises from the oxidation of the iron content by moisture.

4.3. DSC studied of Hygrothernally aged Braycote® 601 EF

The DSC analysis performed on the investigated grease allowed to confirm the composition of the grease. Figure 5 displays two cycles for the pristine and aged grease after 1900 h. This section describes first the spectrum of the grease and then the effect of the accelerated aging.

During the first cycles, two peaks can be observed around 11 °C and 21 °C, shifting to 19.5 °C and 29 °C during the second cycle. These could be explained in terms of PTFE phase transition β from triclinic to hexagonal phase and a disordered intermediate structure occurring between the two peaks [45,46]. These shifts from the first to the second cycle are attributed to the PTFE filler, which is probably being partly amorphous. Then, a very small and broad peak appears at 45 °C. It was not possible to interpret this peak, but it is supposed to be related to some unknown additives or impurities because it disappears when the sample is warmed-up between 50 to 100 °C (not shown). At 164 °C, a clear small peak is present. It is shown in inset of the Figure 5 for the pristine sample but it is also present for aged samples. It is attributed to the phase transition within the $NaNO_2$ additive. According to Ahmed and Ali [47,48], sodium nitrite undergoes (two successive) phase transitions from the disordered paraelectric to the ordered commensurate ferroelectric phase. The next peak located at 280 °C is also attributed to the $NaNO_2$ additive and, more particularly, to its melting (or decomposition) temperature [49].

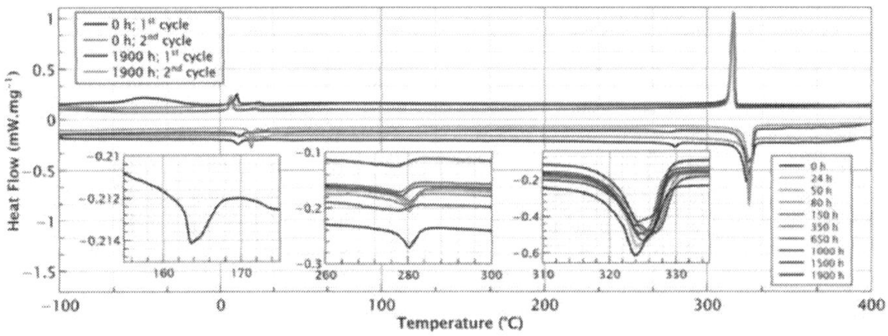

Figure 5. DSC measurement of the pristine and 1900 h hygrothermally aged grease. The left inset corresponds to the $NaNO_2$ phase transition shown for the pristine grease. The middle inset corresponds to the $NaNO_2$ melting peak for different aging times during the first heating phase. The right inset corresponds to the PTFE melting for different aging times during the first heating phase.

After the first cycle, the peaks related to $NaNO_2$ disappear probably because of its full decomposition. The last peak located at around 324 °C to 327 °C corresponds to the PTFE melting temperature [45,46,50]. Note that the base oil Brayco® 815Z was also analyzed (not shown). In this last case, no peaks were present. This confirms that the described above features correspond only to the PTFE filler or the grease additives.

All in all, hygrothermal aging is not drastically modifying the grease response measured by DSC. The appearance of a small peak at around -5 °C can be observed after the aging (not shown). It disappears after the first cycle. It is attributed to the remaining mass intake, as described above (section 4.14.1.). The peak at 280 °C tends to lose its symmetry because of the aging process (Figure 5, left inset). This suggests that the $NaNO_2$ phase is altered. This change in peak characteristics is also observed during cooling where the peak is slightly shifted (not shown). The PTFE melting peak is also flattened after aging and shifts to slightly higher temperatures (Figure 5, right inset). This suggests that the PTFE thickener is also slightly affected by the aging process with a structure locally less ordered. After the first heating cycle reaching 450 °C, all the peaks not attributable to the PTFE thickener disappear. PTFE melting temperature tends to be at higher values (typically +0.5 °C to +1 °C) after aging while the differences observed during the first cycle disappear almost completely. This suggests that the changes of the PTFE thickener properties after aging are superficial and are almost completely reset when the PTFE phase is melted and cooled down.

To sum-up, the DSC tests confirmed the presence of $NaNO_2$ additive and the nature of the PTFE filler, suggesting that this thickener is not fully crystalline, and hygrothermal aging affects slightly both the PTFE filler and the $NaNO_2$ additive.

4.4. Chemical analyses: FTIR and GC-MS

FTIR analyses were performed on the aged grease to check for evidence of any chemical changes. The normalized spectrum of the pristine and the aged sample after 1900 h in the climatic chamber is shown in Figure 6 a). Other aging times are not displayed for the sake of clarity because no changes were observed between pristine and aged samples. Indeed, no new peak was observed after aging suggesting that the grease is not or only slightly oxidized during the aging process. The following peaks were found and interpreted with the references [17,51–53]: 514 cm^{-1}, 554 cm^{-1}, 626 cm^{-1}, 640 cm^{-1}, 687 cm^{-1}, 758 cm^{-1}, 815 cm^{-1}, 832 cm^{-1}, 906 cm^{-1}, 956 cm^{-1}, 1000 cm^{-1}, 1072 cm^{-1}, 1096 cm^{-1}, 1154 cm^{-1}, 1206 cm^{-1}, 1400 cm^{-1}, 1546 cm^{-1}, 1578 cm^{-1}, 1954 cm^{-1}, 2044 cm^{-1}, 2372 cm^{-1}, 2847 cm^{-1}, 2918 cm^{-1} and 3621 cm^{-1}. The large majority of the peaks is attributable to C-F bonds vibrations including, for example, CF stretching, CF_2 bending, wagging and rocking, CF_3 deformations and CF overtones. C-O-C ether groups probably contribute to the signals between 800 cm^{-1} – 1100 cm^{-1}. Small traces of alkanes are also seen at 2847 cm^{-1} and 2918 cm^{-1}, probably arising from organic additives. The same conclusion can be drawn for the peaks at 1546 cm^{-1} and 1578 cm^{-1} which might be interpreted as N-H vibrations in additives (such as Oleanitrile, see GC-MS discussion below). A broad and tiny peak is also present in the hydroxyl region with center at 3621 cm^{-1}.

To complete the chemical characterization of the grease, GC-MS analyses were performed on the pristine sample and the sample aged for 1500 h. The outcome spectra are reported in Figure 6 b). One broad and large peak is present, along with sharp and intense peaks. The broad peak refers to the fragmentation of the fluorine species composing the PFPE grease and the PTFE filler. The sharp peaks typically refer to the additives or impurities. The shift toward shorter retention times of the main broad peak after aging means the hygrothermally aged specimen has a smaller averaged molecular weight. Therefore, the average polymer chain length is probably reduced by the aging increasing the averaged product volatility. However, the aged grease does not contain molecules having, in absolute, a higher volatility than the pristine grease since the corresponding peaks appear at the same retention time. The reduced full width at half maximum of the aged grease with respect to the pristine sample, with the loss of the peak part at higher retention time, also implies the disappearance or splitting of the largest fluorine-containing molecules present in the grease caused by the aging process. Finally, the effect of the hygrothermal aging on the GC-MS observations can be summarized as the alteration of the high molecular weight for fluorine-containing molecules increasing the average volatility of the grease without the formation of light fluorine-containing molecules that were not present in the pristine grease. In addition to the main changes discussed above, some observations shall be noted concerning the presence of sharp peaks. Among these peaks, the following ones are particularly intense and notable in the pristine grease: 1-hexadecanol (28.123 min), heptadecenal (30.388 min), dimethyl-n-hexadecylamine (31.558 min), octadecanal (32.727 min), dimethyl-n-octadecylamine (33.501 min), oleanitrile (36.879 min), n-methyl-n-benzyltetradecanamine (39.318 min) and other non-identified species. These species are ascribed to additives either to improve the grease properties, for manufacturing reason or coming from impurities or chain end-groups. The detection of dimethyl-n-hexadecylamine and dimethyl-n-octadecylamine are confirming the presence of decyldimethyloctylammonium chloride since this last is known to be an additive according to Morales et al. [13]. After the aging process, all these species remain undetectable except the oleanitrile and the n-methyl-n-benzyltetradecanamine. In consequence they are probably decomposed in the course of aging either by hydrolysis or oxidation and released in the chamber. A new peak is also present at 25.959 min and attributed to methyl 2,4-dichlorobenzoate. Even though its exact origin cannot be identified, it probably originates from the by-products of grease and additives generation during their alteration by aging. Finally, it should be noted that numerous non-ascribed sharp peaks are still present for the aged grease.

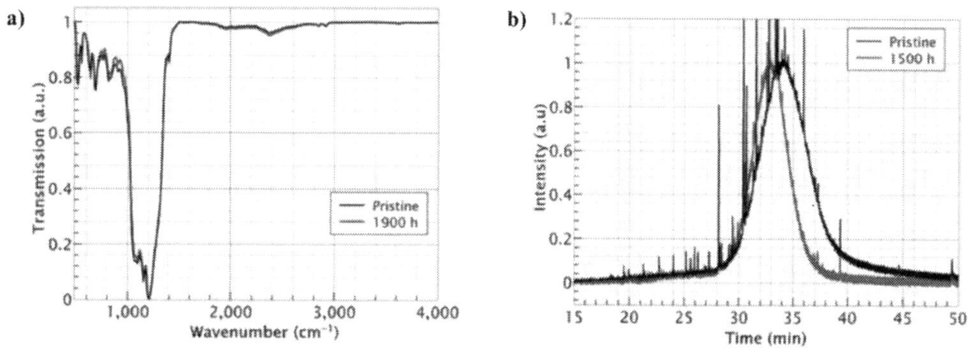

Figure 6. a) FTIR measurement of non-aged and hygrothermally aged grease after 1900 h (normalized); b) GC-MS measurement of non-aged and hygrothermally aged grease after 1500 h.

To summarize, FTIR shows no specific features when comparing the aged and pristine grease. This suggests a good chemical stability of the grease and of its base oil. Nevertheless, GC-MS measurement is useful to complete FTIR analysis which is not suitable to characterize the average molecular weight of polymeric structures. Even though the grease does not have strong chemical changes induced by the aging, a lowering of the average molecular weight is detected jointly with the partial disappearance of organic additives. In practice, this may lead to a lower thermal stability and a change in its rheological and tribological characteristics.

4.5. Rheology and tribology

The rheological properties of the grease were characterized by means of its viscosity as a function of the imposed shear rate. The obtained values are reported in Figure 7 a). The viscosity deceases with the shear rate from a magnitude of 10^4 to 10^0 Pa·s for shear rates ranging from 0.01 to 5000 s^{-1}. Hence the Braycote® 601 EF grease, which is a non-Newtonian fluid, has a shear thinning behavior. At an intermediate shear rate, a transition plateau is present. This plateau is attributed to a wall slip (or wall depletion) phenomenon [54–56]. According to Sochi [55] one of the possible mechanisms is: "*the depletion of the boundary layer where the particles in a sheared dispersed system, such as emulsions and suspensions, migrate away from the boundary region resulting in a very thin low-viscosity layer adjacent to the wall which acts as a lubricating film that facilitates fluid movement*". This wall slip is often considered as a measurement artefact (artificial lowering of the viscosity) because of the test parameters (e.g. gap between the plates) and the plates used with the rheometer. It can be partially or completely cancelled owing to the use of rough plates. However, it must be understood that in a real case such wall depletion may be present if the environmental conditions are suitable. This transition area is reproducible when the measurements are renewed by increasing the shear rate. When the shear rate is increased and then decreased, a characteristic hysteresis appears in this region (Figure 7 b)). It was found to be reproducible over cycles. This suggests that the grease probably has a time-dependent shear thinning property called thixotropy. It should be noted that the transition was found to shift at lower shear rates and higher viscosities (with smaller amplitude) when the gap is increased.

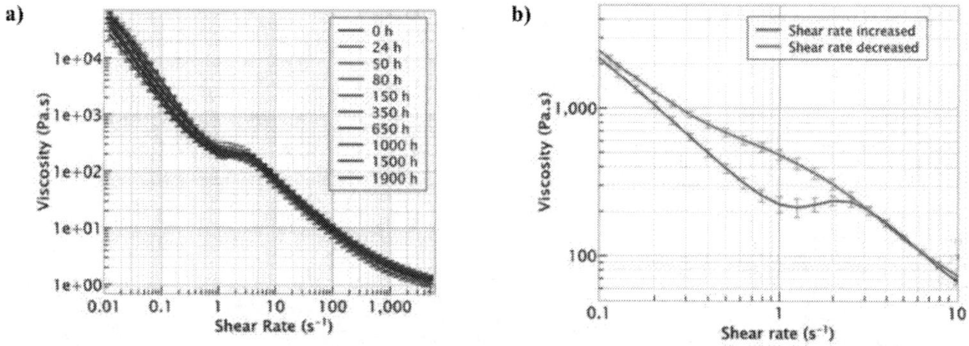

Figure 7. Rheological measurement of Braycote® 601 EF grease. a) viscosity for various hygrothermally aging times as a function of the shear rate, the error bars correspond to the standard deviation obtained over three independent measurements; b) hysteresis phenomenon in transition area corresponding to the wall slip for pristine grease, this feature is also present for aged greases, the error bars correspond to the standard deviation obtained over three cycles increasing and decreasing the shear rate.

Comparing now the grease once it has been hygrothermally aged, a trend can be supposed where the viscosity is slightly increases after aging at low shear rate and lower at high shear (Figure 7 a)). Nevertheless, these trends lie within the measurement reproducibility where error bars (corresponding to the standard deviation, being 1σ) are overlapping. In consequences this interpretation should be considered with care. It is also noted that, quantitatively, the pristine grease shows better reproducibility compared to the aged greases. The wall slip phenomenon remain unchanged after aging.

As last results, tribological properties were measured by determining the coefficient of friction (COF) in function of the applied velocity. The Figure 8 a) displays the measurements in a limited velocity range up to 5 rad.s^{-1}. In this region, measurements were found to be reproducible without any changes in the sliding characteristics due to hygrothermal aging. When this range is extended to 30 rad·s^{-1}, the lubricant starts to fail to lubricate the system around 10 rad·s^{-1} (Figure 8 b) and c)). This is manifested by an increase of the friction coefficient. In practice, an intense shrill sound is heard. During the first cycle, the lubricity is similar for the pristine and hygrothermally aged grease (Figure 8 b)). After the first cycle, the system appears to be less stable, and the COF increase is partially reversible. Differences can also be spotted between the pristine and aged grease. Indeed, the COF appears to be slightly smaller for the aged greases at low velocities once it goes above 10 rad·s^{-1} (Figure 8 c)). When the velocity is increased above 1 rad·s^{-1}, the aged grease friction coefficient increases faster, and the lubrication is deteriorated at lower velocities than for the pristine grease. In consequence, once the aged grease crosses the lubricating breaking point of the system around 10 rad·s^{-1}, the lubrication stability is decreased. Considering the error bars, these interpretations should, however, be regarded with care. Besides, although these measurements permit to draw a first trend on the grease tribological behaviors, they do not permit to assess the grease lifetime performances. For such purposes, experiments with SOT (Spiral Orbit Tribometer) should be considered [57–59].

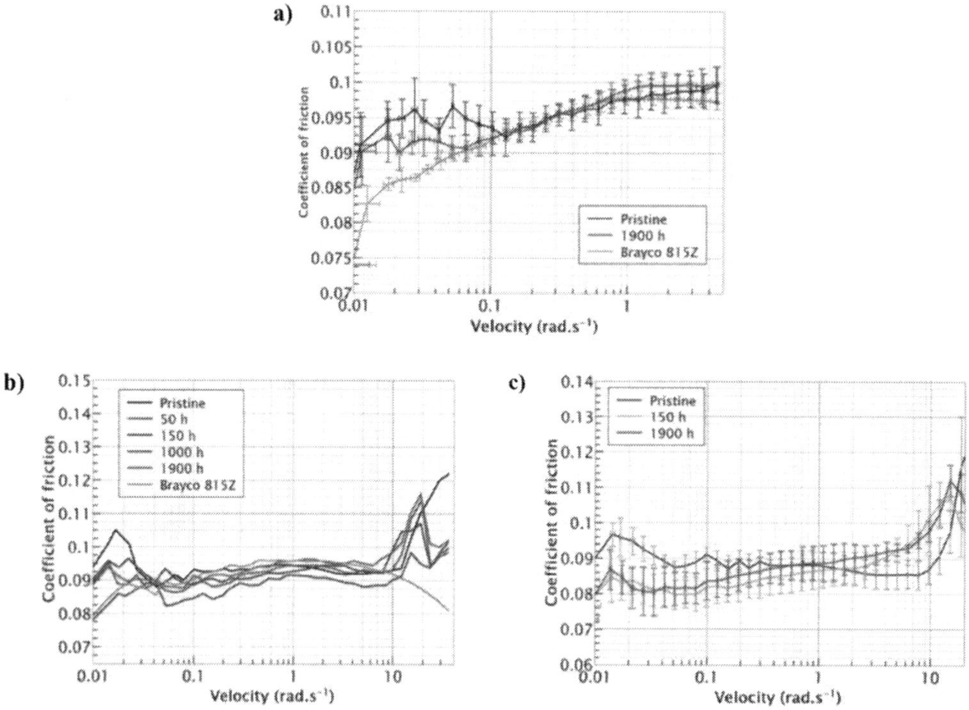

Figure 8. Tribological measurements of the pristine and hygrothermally aged greases. a) averaged COF in function of the velocity in a limited velocity range over four cycles (one cycle being increasing and decreasing of the velocity), the error bars correspond to standard deviation; b) COF in function of the velocity in an extended velocity range when the velocity is increased during the first cycle; c) averaged COF in function of the velocity in an extended velocity range over two cycles, the error bars correspond to standard deviation.

4.6. Accelerated aging of the grease summary

An accelerated aging experiment has been successfully conducted on the Brayote 601 EF grease up to 1900 h at 85 °C and under 85 %RH. After aging, the grease is gaining mass, up to a saturation level of around 2.20 wt%. This mass intake decreases with time and stabilizes up to 0.38 wt%. This stabilized intake mass is dependent on the aging time with an exponential shape: a more aged sample shows higher stabilized mass intake. This is attributed to the adsorption of moisture, and eventually light organic compounds (present inside the chamber during the aging), probably by the bentonite additive. The surface of the aged grease looks drier and contains bubble-like features of microns sizes. Their origin is not clear and might be connected with moisture adsorption or local oil bleeding and oozing. The aging process also induces a yellowing of the grease visible by naked eyes and confirmed with UV-VIS spectroscopy with an absorption center at around 455 nm. The yellowing origin is unlikely to arise from significant chemical changes of the grease since FTIR shows no specific feature. It might be related to the oxidation of the iron content present in small amounts in the grease. This last was detected by EDX in TGA residue. Thermally, the grease is stable even after aging. No evident changes are present on the TGA diagram. Similar conclusions could be drawn from DSC analyses even if slight changes of the melting properties of the $NaNO_2$ additive and the PTFE thickener are observed. According to GC-MS measurements, the hygrothermal aging induces a reduction of the averaged molecular weight and the loss of some organic traces (residues or additives). Viscosity measurements showed no significant effect of aging. The wall slip phenomenon is still present and the viscosity seems to decrease slightly at low shear rates and to increase at high shear rates. This trend is not much pronounced and lies within the reproducibility incertitude range. Finally, tribological measurements suggest a lower stability of the grease after aging once the grease fails to lubricate the system. In a limited velocity range, the COF was not affected by the aging. If the velocity range is extended above 10 rad·s^{-1}, the lubricity trends may change slightly, and the system is losing stability.

5. Real aging: a grease stored over 17 years, results and discussions

In addition to the analyses performed on the hygrothermally aged greases, experiments were performed to characterise the properties of Braycote® 601 EF manufactured in 2005. This section briefly sum-up the outcomes. From optical inspection, this old grease does not appear drier than the fresh grease and no oozing is present on its surface. However, an obvious oil separation can be observed (Figure 9 a)). UV-VIS spectroscopy allows to identify a slight yellowing of the grease with a peak center at 454 nm (Figure 9 b)). It is ascribed to a passive aging phenomenon. The reflective changes seen in the UV region and above 520 nm with respect to pristine grease are supposed to be related to intrinsic properties of the grease decoupled from the passive aging (for example due to slight formulation changes).

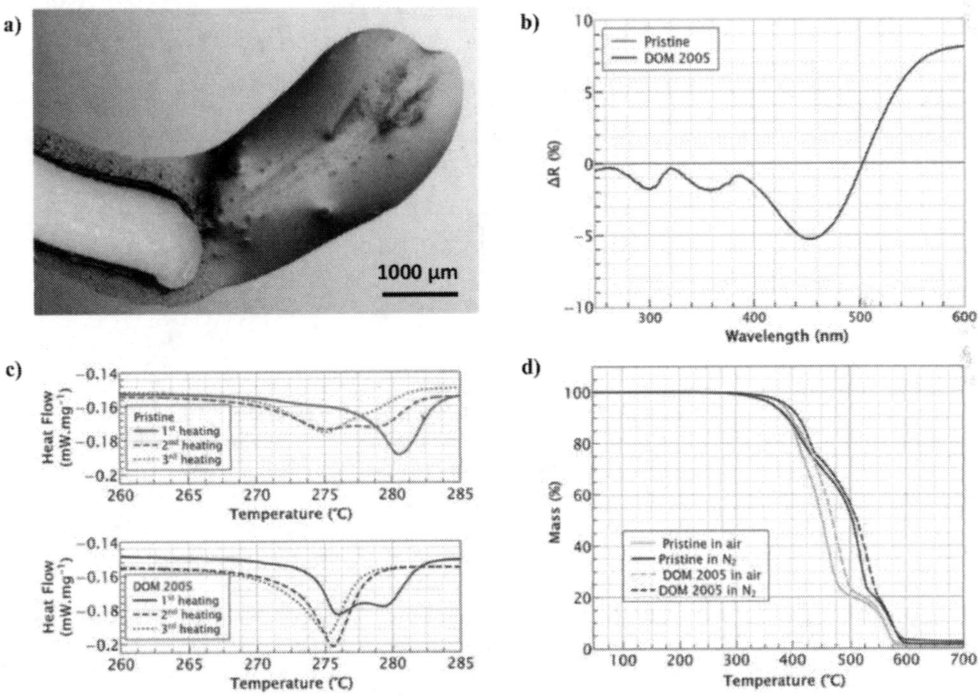

Figure 9. a) Micrograph of the grease manufactured in 2005 demonstrating an evident oil bleeding; b) reflectivity difference of the grease manufactured 17 years ago with respect to the pristine variant manufactured in 2020; c) DSC measurement of the old grease manufactured in 2005 compared with the pristine grease manufactured in 2020 focusing on the $NaNO_2$ melting event as the only difference seen; d) TGA measurements of the old grease comparing the manufactured grease in 2005 and the pristine grease manufactured in 2020.

Thermal analyses show only minor changes. In DSC analyses, the only noticeable modification is the melting peak of the $NaNO_2$ additive: it is at lower temperatures and is convoluted in two peaks for the old grease around 275 °C and 279 °C (solid blue line, Figure 9 c)) while the pristine grease has a single peak around 280 °C. This is attributed to a passive aging of the additive. Indeed, when the fresh grease is heated above the melting temperature (300 °C), cooled down and heated up again, this characteristic convoluted peak is observed (dash red lines, Figure 9 c)). If the heating step is repeated, these convoluted peaks shift and merge into a single peak centred at 275 °C for both samples (dash blue line and dot lines, Figure 9 c)).Hence the alteration (presence of convoluted peak instead of a single peak) seen for the stored grease is likely to be related with passive aging, having similar alteration path as when a sample is heated. On TGA results the only characteristic diverging form the pristine grease is a thermal degradation present at slightly higher temperatures (Figure 9 d)).

This thermal degradation at higher temperatures for the old grease is in agreement with the GC-MS results showing that the old grease has a higher average molecular weight (Figure 10). This feature is attributed to a variation in the manufacturing process. Indeed, it is unlikely that the grease would polymerise overtime to longer chains or heavier molecules. Besides, this remains in the range of the base oil products (orange curve, Figure 10). Finally, based on the manufacturer information, the product is adjusted to reach a given consistency and the raw material targets a given viscosity rather than having very precise chemical characteristics. It should be noted that, according to GC-MS analyses, the grease additives are the same as for the fresh grease without the peak of n-methyl-n-benzyltetradecanamine. This is attributed to a small change in the grease formulation between 2005 and 2020. No differences are detected with FTIR.

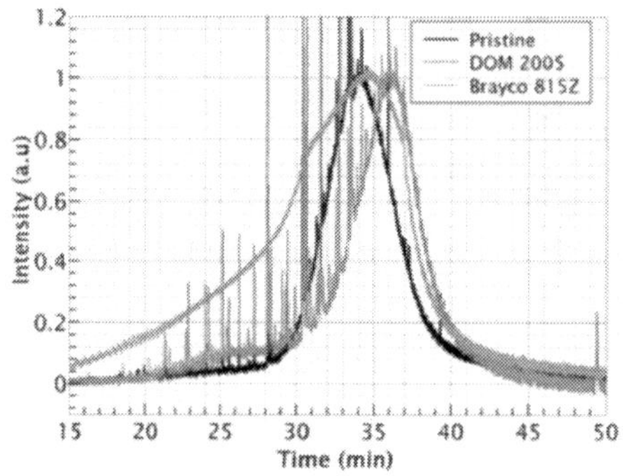

Figure 10. GC-MS comparison of the old grease manufactured in 2005 and the pristine grease manufactured in 2020, the curve of the base oil is shown as additional information.

Viscosity measurements show quantitatively similar values as for the fresh grease. Eventually, the old grease is slightly more viscous at high shear rates (Figure 11 a)). The main difference in comparison to the fresh grease is the cancellation of the slip wall phenomenon. Once again, this phenomenon is dependent on the lubricant itself and the test conditions: a slight reappearance of the wall slip was observed after the second sweep if the gap between the two plates was small enough (not shown). This is probably linked with the oil bleeding, modifying the interface with the plates. From a tribological point of view (Figure 11 b)), the old grease has a similar behavior as the other greases. No clear performance alteration is seen. Eventually, at low velocity, its behavior is closer to the hygrothermally aged grease with a slightly lower COF than for the pristine sample. However, taking into account the error bars, no strict conclusion could be drawn.

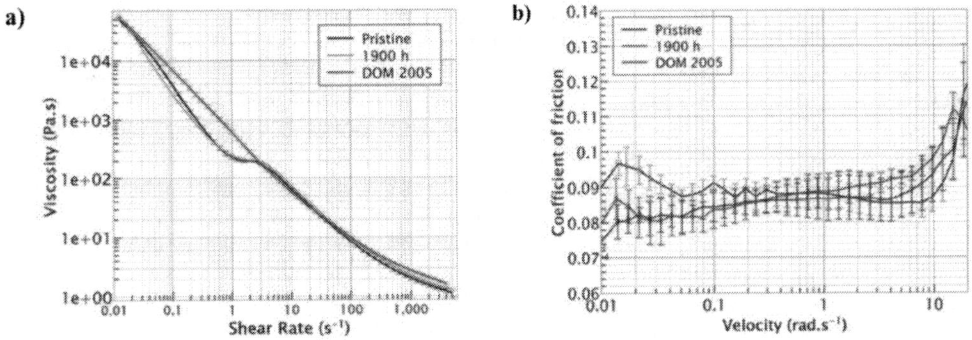

Figure 11. a) Rheological comparison of the old grease manufactured in 2005 and the pristine grease manufactured in 2020, the curve of the hygrothermally aged grease during 1900 h is shown for comparison; b) averaged COF as a function of velocity for the grease manufactured in 2005 in an extended velocity range over two cycles, the error bars correspond to standard deviation, the curve for the hygrothermally aged grease during 1900 h is shown for comparison.

6. Summary and comparison of the aged and stored greases

As already introduced previously in this paper, Braycote® 601 EF grease has been reported as a stable lubricant for space applications. However, additional insights and assessments of its behavior are desirable because of its extensive use. Besides, with the arrival of long-term storage problematic arising from constellation satellites manufacturing schedules or launch delays, finding out how the grease is supposed to behave over years in cleanroom storage environment is of interest. Considering this scope, accelerated aging was performed in this study. On the contrary to most common practices, the aging here considered not only the temperature, but also the relative humidity. In fact, the idea behind this work was to try to accelerate the aging of the grease when stored in a cleanroom environment (and not to assess its aging over mechanical work or lifetime during operational time). Because this lubricant is known to be stable, relatively highly stressful parameters were chosen, being 85 °C and 85 %RH. Although a reliable extrapolation time of the aging process is not possible due to the lack of confidence on the degradation processes, the unknown degree of the test representativeness and variability of extrapolation models that can be obtained by error propagation, attempt to estimate the acceleration factor was carried out. The corresponding real time duration of the accelerated aging process could be expected at around 17 years by the Q_{10} approach ("10-degrees rule") or longer considering other models. These values should, however, be considered as indicative and not as a quantitative assessment.

The outcomes of the results and their interpretations are coherent with the literature highlighting the high stability of the investigated grease. Only minor changes can be observed even after a very long time of aging under hygrothermal conditions. The major differences in reference to the pristine grease are the swelling and yellowing of the grease evident from optical inspection, the modification of the averaged molecular weight observed by GC-MS toward lower values and the alteration and decomposition of additives and some traces of organic compounds (additives or residues coming from the manufacturing process). The slight changes of the viscosity values should probably not affect drastically the reflow characteristics of the grease over the mechanism contact points. Deeper investigation should, however, be considered to ensure this hypothesis is correct, especially if the used condition differs from the temperature used for the viscosity measurements (20 °C). Beyond the chemical and thermal stability, the only common alteration between the old grease and the hygrothermally aged grease is the slight yellowing. The main discrepancies are the GC-MS features. The average molecular weight of the grease is probably not altered during aging real time, contrary to the grease aged in accelerated manner. Finally, no oil separation was observed for the hygrothermally aged greases while this is obvious for the old grease. This last phenomenon is supposed to modify slightly the rheological properties of the grease and be led by passive sedimentation due to the gravity effect. Based on the tribological measurements, COF values are not or only slightly affected by the aging process as a function of the velocity. These measurements do not cover grease lifetime properties.

Based on all these comments, even if some similarities exist, it is not possible to assess if the hygrothermal aging is representative of the real time aging. Beyond these differences, the greases are from different batches and in the absence of data of the old grease at its early ages, it is difficult to determine what properties changed, as well as their trends. Besides, the old grease remained in the syringe over years. Consequently, the grease has not experienced the same relative humidity level as under cleanroom conditions. Finally, one of the biggest weaknesses of the accelerated aging under hygrothermal aging is the absence of the sedimentation owing to gravity, leading to oil bleeding. This drawback can be tackled by using the setup (conical sieve) described in the ASTM D6184 standard However, the grease composition might be modified by the oil separation that may reach in some cases tens of weight percent of the initial grease mass. Besides the oil bleeding test is mainly qualitative.

7. Conclusion and perspectives

Accelerated aging of Braycote® 601 EF grease was performed under 85 °C and 85 %RH up to 1900 h. These test parameters were considered as very stressful. An old grease manufactured 17 years ago was also under investigation. The main outcomes are listed below:

- The results lead to the conclusion that the grease is very stable, even if small changes can be discerned. This conclusion is in accordance with NASA reports [13,16].
- Among the observed changes, hygrothermal aging induces a small mass intake, a yellowing of the grease, an oozing effect, a slight alteration of $NaNO_2$ additive and of the PTFE thickener (DSC), the disappearance of some organic additives or impurities, a lowering of the average molecular weight (GC-MS) and an alteration of tribological properties once the grease has failed to lubricate the system. No notable change is seen by TGA, FTIR and viscosity measurements.
- The old grease undergoes a yellowing and an oil separation inducing slight changes on the rheological measurements. The other changes are attributed to mild modifications of the manufacturing process between the old (2005) and the pristine grease (2020).
- Based on the comparison of hygrothermally aged and the old grease, the accelerated aging is only partially representative of the grease alteration. The main similarity is the grease yellowing. The main differences are the absence of oil bleeding for the hygrothermally aged grease, and the variation of the average molecular weight which is reduced by the accelerated aging process. Beyond the test representativeness, the fact that the old grease was conserved into its original tube reduces the potential interaction with the ambient moisture which may limit the grease alteration over time.
- The humidity is a factor that can notably modify the alteration of the lubricant by accelerated aging. Using milder humidity level might permit to reach a more realistic aging simulation.

Hence, these results obtained herein are considered as favorable for the investigated lubricant use in the framework of LTS. In perspective, lubrication lifetime measurement with Spiral Orbit Tribometer (SOT) might be considered to reassure the lubricating properties over time (e.g. reduction of the grease molecular weight might reduce the grease lubricating lifetime) and have insights on the functional properties of the aged greases. Additionally, in order to complete a full de-risking activity, oil bleeding over years of storage should be investigated.

Acknowledgments
The authors thank Alex Bolkhovitinov, Gerard Van Papendrecht and Sarah Rodriguez Castillo, Giorgio Parzianello for their support on GC-MS, FTIR, TGA / DSC and tribological analyses, respectively.

Conflict of Interest
The authors declare to have no conflict of interest.

References

[1] Senese S, Strube B, Lattner K, Fusco G, Deuschle T, Förster U, Thiel M, Svedevall A, Toledo A A, Sghedoni M *et al.* 2019 Meteosat Third Generation (MTG) Mechanisms: Impact on Instrument Performances 18th European Space Mechanisms and Tribology Symposium (Munich, Germany)

[2] 2016 Long Term Storage Workshop (ESA/ESTEC, The Netherlands)

[3] Locked up for 20 years – long term storage of components *https://www.esa.int/Enabling_Support/Space_Engineering_Technology/Shaping_the_Future/Locked_up_for_20_years_long_term_storage_of_components*

[4] Puig C 2018 Real prediction of ground aging of materials using thermokinetic analysis – correlation with real time aging, 14th International Symposium on Materials in the Space Environment (ISMSE), Biarritz, France

[5] Esther A C M, Dey A, Sridhara N, Yougandar B, Bera P, Anandan C, Rangappa D and Sharma A K 2015 A study on degradation of germanium coating on Kapton used for spacecraft sunshield application *Surf. Interface Anal.* **47** 1155–60

[6] Oliveira C M, Retherford K, Conard S J, Barkhouser R H and Friedman S D 1999 Aging studies of LiF-coated optics for use in the far ultraviolet SPIE's International Symposium on Optical Science, Engineering, and Instrumentation ed O H W Siegmund and K A Flanagan (Denver, CO) pp 52–60

[7] Zuppella P, Corso A J, Nicolosi P, Windt D L and Pelizzo M G 2011 Innovative multilayer coatings for space solar physics: performances and stability over time SPIE Optics + Optoelectronics ed R Hudec and L Pina (Prague, Czech Republic) p 807608

[8] Fontorbes S 2022 Study of the impact of cleanroom environment on CFRP epoxy, 15th International Symposium on Materials in the Space Environment (ISMSE), Leiden, The Netherlands

[9] Bauer R 1988 Sulfide corrosion of silver contacts during satellite storage *J. Spacecr. Rockets* **25** 439–40

[10] Oh J, Jang S and Yoh J J 2019 Towards understanding the effects of heat and humidity on ageing of a NASA standard pyrotechnic igniter *Sci Rep* **9** 10203

[11] Buttery M, Lewis S, Kent A, Bingley R and Cropper M 2020 Long-Term Storage Considerations for Spacecraft Lubricants *Lubricants* **8**

[12] Jones W R 1995 Properties of Perfluoropolyethers for Space Applications *Tribol. Trans.* **38** 557–64

[13] Morales W, Street K W and Zaretsky E V 2012 Performance and Analysis of Perfluoropolyalkyl Ether Grease Used on Space Shuttle Actuators—A Case Study *Tribol. Trans.* **55** 77–85

[14] Verhoeven D 1999 Space Mechanisms and Tribology for the Joints of the European Robot Arm (ERA) *EUROPEAN SPACE AGENCY-PUBLICATIONS-ESA SP* **438** 335–40

[15] Prat P, Vergne P, Pochard M and Sicre J 1995 Optimisation of the Rheological Behaviour of Thickened Liquid Lubricants for Spacecraft Applications *Tribology Series* vol 30 (Elsevier) pp 207–16

[16] Anon NASA Engineering & Safety Center Technical Bulletins 2007-2020

[17] Wolfberger A, Hausberger A, Schlögl S and Hołyńska M 2021 Assessment of the chemical degradation of PFPE lubricants and greases for space applications: implications for long-term on-ground storage *CEAS Space J.* **13** 377–88

[18] Helmick L S and Jones W R 1990 Determination of the Thermal Stability of Fluids by Tensimetry: Instrumentation and Procedure *Tribol. Trans.* **33** 519–28

[19] Li L, Jones P M and Hsia Y-T 2004 Effect of Chemical Structure and Molecular Weight on High-Temperature Stability of Some Fomblin Z-Type Lubricants *Tribol. Lett.* **16** 21–7

[20] Hoshino M, Kimachi Y and Terada A 1996 Thermogravimetric behavior of perfluoropolyether *J. Appl. Polym. Sci.* **62** 207–15

[21] Gschwender L J and Snyder C E 2009 U.S. Air Force Perfluoropolyalkylether Experiences *Tribol. Trans.* **52** 165–70

[22] Buttery M, Gaillard L, Rajala S, Roberts E, Rohr T and Merstallinger A 2013 Fomblin Z25: A new method for its degradation assessment & proposal for safe operation in space 15th European Space Mechanisms & Tribology Symposium – ESMATS 2013 (Noordwijk, The Netherlands)

[23] Chun S W, Talke F E, Kang H J and Kim W K 2003 Thermal Characteristics of Multiply Alkylated Cyclopentane and Perfluoropolyether *Tribol. Trans.* **46** 70–5

[24] Lewis S, Buttery M, Poyntz-Wright O, Kent A and Vortsellas A 2018 Accelerated Testing of Tribological Components - Uncertainties and Solutions 44th Aerospace Mechanisms Symposium (NASA Glenn Research Center)

[25] Pepper S V 2011 *Effect of Test Environment on Lifetime of Two Vacuum Lubricants Determined by Spiral Orbit Tribometry* (Glenn Research Center, Cleveland, Ohio: NASA)

[26] Carfagno S P and Gibson R J 1980 *A Review of Equipment Aging Theory and Technology* (The Parkway at Twentieth Street Philadelphia, Pennsylvania 19103: FRANKLIN RESEARCH CENTER)

[27] Caruso H and Dasgupta A 1998 A fundamental overview of accelerated-testing analytic models *Annual Reliability and Maintainability Symposium. 1998 Proceedings. International Symposium on Product Quality and Integrity* Annual Reliability and Maintainability Symposium. 1998 Proceedings. International Symposium on Product Quality and Integrity (Anaheim, CA, USA: IEEE) pp 389–93

[28] Viswanadham P and Singh P 1998 *Failure Modes and Mechanisms in Electronic Packages* (Boston, MA: Springer US)

[29] Roduit B, Hartmann M, Folly P, Sarbach A and Baltensperger R 2014 Prediction of thermal stability of materials by modified kinetic and model selection approaches based on limited amount of experimental points *Thermochim. Acta* **579** 31–9

[30] Maxwell A S, Broughton W R, Dean G and Sims G D 2005 *Review of accelerated ageing methods and lifetime prediction techniques for polymeric materials* (Hampton Road, Teddington, Middlesex, TW11 0LW: National Physical Laboratory)

[31] Avenel C, Raccurt O, Gardette J-L and Therias S 2018 Review of accelerated ageing test modelling and its application to solar mirrors *Sol. Energy Mater. Sol. Cells* **186** 29–41

[32] Plota A and Masek A 2020 Lifetime Prediction Methods for Degradable Polymeric Materials—A Short Review *Materials* **13** 4507

[33] Feinberg A 2016 *Thermodynamic degradation science: physics of failure, accelerated testing, fatigue and reliability applications* (Chichester, West Sussex: Wiley)

[34] Escobar L A and Meeker W Q 2006 A Review of Accelerated Test Models *Statist. Sci.* **21**

[35] Feller R L 1994 *Accelerated aging: photochemical and thermal aspects* (Marina del Rey, Calif.: Getty Conservation Institute)

[36] Šimon P 2004 Isoconversional methods *J. Therm. Anal. Calorim.* **76** 123–32

[37] *ASTM F1980-16 Guide for Accelerated Aging of Sterile Barrier Systems for Medical Devices* (West Conshohocken, PA, United States: ASTM International)

[38] Peck D S 1986 Comprehensive Model for Humidity Testing Correlation *24th International Reliability Physics Symposium* 24th International Reliability Physics Symposium (Anaheim, CA, USA: IEEE) pp 44–50

[39] Hallberg Ö and Peck D S 1991 Recent humidity accelerations, a base for testing standards *Qual. Reliab. Engng. Int.* **7** 169–80

[40] Kececioglu D and Sun F 1995 *Environmental stress screening: its quantification, optimization and management* (Upper Saddle River, NJ: Prentice Hall)

[41] Zhao X, Liu T, Burlingame Q C, Liu T, Holley R, Cheng G, Yao N, Gao F and Loo Y-L 2022 Accelerated aging of all-inorganic, interface-stabilized perovskite solar cells *Science* **377** 307–10

[42] Avenel C, Raccurt O, Gardette J-L and Therias S 2019 Accelerated aging test modeling applied to solar mirrors *npj Mater. Degrad.* **3** 27

[43] Gillen K T, Bernstein R and Celina M 2015 Challenges of accelerated aging techniques for elastomer lifetime predictions *Rubber Chem. Technol.* **88** 1–27

[44] Uddin F 2018 Montmorillonite: An Introduction to Properties and Utilization *Current Topics in the Utilization of Clay in Industrial and Medical Applications* ed M Zoveidavianpoor (InTech)

[45] Anon 2009 *Characterization of PTFE by DMA and DSC* (CH-8603 Schwerzenbach, Switzerland: Mettler-Toledo AG, Analytical)

[46] Sperati C A and Starkweather H W 1961 Fluorine-containing polymers. II. Polytetrafluoroethylene *Fortschritte Der Hochpolymeren-Forschung* Advances in Polymer Science vol 2/4 (Berlin, Heidelberg: Springer Berlin Heidelberg) pp 465–95

[47] Ahmed S A and Ali M H 1996 Electrical and thermal studies of $NaNO_2$ *Phys. Status Solidi B* **194** 517–24

[48] Naberezhnov A, Fokin A, Kumzerov Yu, Sotnikov A, Vakhrushev S and Dorner B 2003 Structure and properties of confined sodium nitrite *Eur. Phys. J. E* **12** 21–4

[49] Kourkova L, Svoboda R, Sadovska G, Podzemna V and Kohutova A 2009 Heat capacity of $NaNO_2$ *Thermochim. Acta* **491** 80–3

[50] Hanford W E and Joyce R M 1946 Polytetrafluoroethylene *J. Am. Chem. Soc.* **68** 2082–5

[51] Kuptsov A H and Zhizhin G N 1998 *Handbook of fourier transform Raman and infrared spectra of polymers* (Amsterdam New York: Elsevier)

[52] Socrates G 1994 *Infrared characteristic group frequencies: tables and charts* (Chichester New York Brisbane [etc.]: J. Wiley & sons)

[53] Liang C Y and Krimm S 1956 Infrared Spectra of High Polymers. III. Polytetrafluoroethylene and Polychlorotrifluoroethylene *J. Chem. Phys.* **25** 563–71

[54] Barnes H A 1995 A review of the slip (wall depletion) of polymer solutions, emulsions and particle suspensions in viscometers: its cause, character, and cure *J. Non-Newtonian Fluid Mech.* **56** 221–51

[55] Sochi T 2011 Slip at Fluid-Solid Interface *Polym. Rev.* **51** 309–40

[56] Sochi T 2010 Non-Newtonian flow in porous media *Polymer* **51** 5007–23

[57] Buttery M 2010 An Evaluation of Liquid, Solid, and Grease Lubricants for Space Mechanisms Using a Spiral Orbit Tribometer 40th Aerospace Mechanisms Symposium (NASA Kennedy Space Center)

[58] Pepper S V and Kingsbury E P 2003 Spiral Orbit Tribometry — Part I: Description of the Tribometer *Tribol. Trans.* **46** 57–64

[59] Pepper S V and Kingsbury E P 2003 Spiral Orbit Tribometry — Part II: Evaluation of Three Liquid Lubricants in Vacuum *Tribol. Trans.* **46** 65–9

High temperature high voltage potting adhesive

Isabelle Liémans[1], Geoffroy Petiniot[1], Philippe Vandeplassche[1], Déborah Daclin[1], Malgorzata Holynska[2], Andreas Franke[2]

[1] Thales Alenia Space in Belgium, rue Chapelle Beaussart 101, 6032 Mont-sur-Marchienne, Belgium
[2] ESA-ESTEC, Keperlaan 1, 2201 AZ Noordwijk, The Netherlands

geoffroy.petiniot@thalesaleniaspace.com

Abstract. This study presents the results of a new resin survey for potting of high voltage electronics. The main objective consists in improving the European manufacturing capability for high voltage (>6 kV) and high temperature applications (maximum temperature: 180°C). The targeted improvements include enhanced thermal stability and insulation properties of adhesive potting for high voltage applications for a target lifetime of 15 years. In addition, a specific attention has been given to EU sourced materials for development. Class H resins were targeted. A trade-off based on technical data sheets, tests at material, sample and breadboard levels has been conducted. The complete test sequence performed lead to identify a resin suitable for the requirements and processes. TRL4 has been reached, but there is still a long way to go and use it in a space product. The next steps are the implementation of this new resin in the current equipment and the qualification of magnetics parts impregnated or moulded with this resin.

1. Introduction

The application of high temperature adhesives and potting becomes more challenging when not only a very good thermal endurance is required, but also when the application is in high voltage ranges. For Telecom missions, this applies especially for payload applications and potting used for travelling wave tubes, high voltage potting, power transfer, electrical propulsion power supplies, connectors and power lines as operating voltage tends to be generally higher.

This activity targets the dedicated development of adhesives with enhanced thermal stability and insulation properties for high voltage applications. It consists in the development of materials or processes that can bring improved performance with respect to the existing systems used. Any selection assumes a nominal lifetime equivalent to existing applications.

The main objective consists in improving the European manufacturing capability for high voltage and high temperature applications. The targeted improvements include enhanced thermal stability and insulation properties of adhesive potting for high voltage applications. In addition, a specific attention has been given to EU sourced materials for development.

The materials that have been searched and tested have to fit several criteria. The main technical issue is the high voltage insulation: the material shall demonstrate above 6 kV operation with no loss of performance. The mission life time is also a key. For Telecom satellites, a 15 year in-orbit operation is required. The temperature range is a crucial requirement. A class H resin (e.g. a resin with a range -55°C to +180°C) was targeted. A global overview of the steps for finding a suitable resin is given in table 1.

Activities performed in this project improved the Technology Readiness Level, according to ECSS-E-HB-11A, from TRL2 to TRL4.

Table 1. Project steps and candidate resins selection. N_0 and N_f: number of resins at the beginning and the end of the step, respectively

Step	Goal	Source/Tests	N_0 resins	N_f resins
State-of-the-art	Find the categories of materials that are convenient for the application	Literature review	-	>50
Material	Find materials that adjust best to product requirement, including Class H	Technical data sheet	>50	11
Process: Mix properties	Acceptable properties of the uncured material after mixing for process ability and hardness	Viscosity, hardness	11	7
Process: Curing conditions	Find optimum curing schedule and temperature (if any)	Curing, visual aspect, shrinking measurement	7	4
Performance evaluation	Process the selected resin in actual set-ups and check long term behaviour with regard to requirements	Tests at material, sample and breadboard levels.	4	1

2. Critical review of requirements and applications

First of all, the critical technologies which could be dependent on strategic materials have been identified. This analysis has been performed at satellite level and in Thales Alenia Space in Belgium (TAS-B) High Temperature/High Voltage (HT/HV) applications.

Two TAS-B applications are concerned:

- A Power Processing Unit (PPU) is an equipment which provides all the power supplies for an electric thruster and ensures their sequencing. TAS-B has an extensive heritage on PPU used to drive thrusters. The particularity of the thruster supplies is that some of them are high voltage and high power units,
- An EPC is an Electrical Power Conditioner. It is a high voltage power supply for a travelling wave tube (TWT). The particularity of the EPC output is to provide high electrical potential (>6000 V) required by the TWTs. The insulating material has to sustain high voltage at high temperature and long duration. To be qualified, such material has to be submitted to life test in representative conditions (design and high voltage at high temperature and at long duration). Another parameter is the thermo-mechanical behaviour of the compound objects.

The processes involved in those two HT/HV applications are impregnation and potting processes. Impregnation and potting are used for the electrical insulation of coils and magnetics in order to warrant the electrical insulation between the wires and between the winding and the external environment. For space magnetics, impregnation and potting play different roles:

- Mechanical role: the resin gives mechanical support to the wires. This prevents any loss of electrical insulation during the vibrations caused during the launch of the satellite,
- Thermal role: the resin helps evacuating the heat to the outside of the winding and makes the temperature distribution uniform within the various windings (thermal paths),
- Electrical role: the resin has to maintain its isolation properties at the high electrical fields caused by the voltage and the geometry of the magnetic, for the duration of a standard GEO (Geostationary Earth Orbit) mission.

IOP Conf. Series: Materials Science and Engineering 1287 (2023) 012012 doi:10.1088/1757-899X/1287/1/012012

3. State of the art of materials and processes for HT/HV potting and adhesive applications

The state of the art performed at materials and processes levels has been focused on impregnation-potting process in HT/HV applications.

3.1. Materials

As a general rule, the polymeric materials have good dielectric properties, they are light in weight and have a large range of thermal and mechanical properties. Within the polymeric materials, the thermoplastic polymers present a lower temperature resistance than the thermosetting ones (resins and enamels). Therefore, the search has been limited to this latter abundant family of materials.
There are several types of thermosetting polymers which are used in electrical insulation applications: This list includes some silicones, unsaturated polyesterimides (enamels), bismaleimide adhesives and polyimide resins (used as matrix in structural composites) that can offer even better high temperature resistance than the best epoxy.
By selecting a resin whose T_g (glass transition temperature) is above the expected service temperature, the risk of inadequate mechanical properties or thermal stresses can be reduced.
Polyurethanes and polyesters do not reach the minimum temperature resistance for this application, therefore they will not be considered for the materials trade-off.
A priori, the restrictions of viscosity, pot life, dielectric strength, thermal resistivity and outgassing are attainable by all the polymer families but considering the process requirements (casting and curing processes, shrinkage, …), this means that the two types of resins that have been considered for the materials trade-off are epoxies and silicones. As an example, figure 1 below shows the results of dynamic viscosity tests carried out for two commercial products at different temperatures.

On the one hand, silicones reach higher dielectric strengths than epoxies. Also, the exothermicity of their polymerization reaction is generally lower, which makes their curing cycles easier to handle and less dangerous for the magnetic coil. On the other hand, epoxies have much better adherence than silicones.

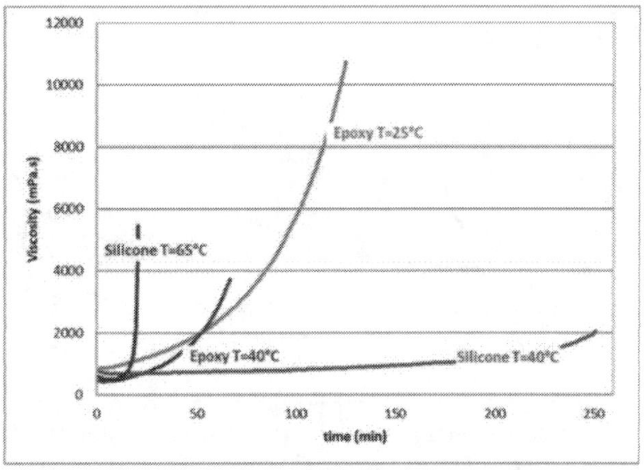

Figure 1. Shear Viscosity at different temperatures.

It has to be noted that it is also possible to add thermal fillers to the discussed resins. Metallic fillers, ceramic fillers, fibres etc. However, in TAS-B impregnation process, fillers are not often used because they may increase the viscosity of the system and they could prevent the resin from penetrating the windings by clogging the space between wires. Nanofillers can also induce the segregation phenomenon

IOP Conf. Series: Materials Science and Engineering 1287 (2023) 012012 doi:10.1088/1757-899X/1287/1/012012

of the two parts (filler and resin). Use of Nanofiller technology implies too many changes and induces contamination risks. This technology has thus been discarded for TAS-B applications.

3.2. Processes
Several process parameters have been taken into account in the study:
- Dosing and Mixing,
- Impregnation process. Two main processes used worldwide have been considered: the Vacuum Impregnation (VI) and the Vacuum and Pressure Impregnation (VPI),
- Curing process,
- Non-destructive tests: visual inspection, X-ray inspection, ultrasonic inspection, partial discharge test.

4. Materials and process trade-off
This materials trade-off includes a broad range of products, including those that have not been used in space applications yet. European and American resin manufacturers have been considered.
From the long list obtained, the materials whose technical data (from their TDS-Technical Data Sheet) adjusted best to the product requirements (see table 2) have been selected. The materials that did not reach the thermal class H (180°C) were discarded. Manufacturers that presented a risk in terms of long term supply relationship were also discarded. Eleven candidates remained (in bold in table 2).

4.1. Dynamic viscosity
The dynamic viscosity behaviour of the mixed resins is a key factor in the impregnation process, so a dynamic viscosity test has been performed on the 11 remaining resins. The results are presented in figure 2. The dynamic viscosity tests showed that the silicone systems (P1, M1) were not convenient for the application. The materials that showed a short pot life have also been discarded.

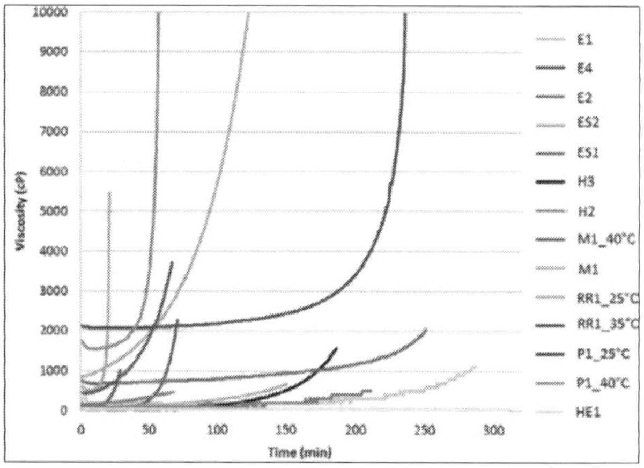

Figure 2. Viscosity as a function of time for the different resins tested.

4.2. Cure cycle
The curing cycle should be as short as possible provided that the material is completely cured and that its properties are optimal. For resins E1, E2 and E4, the cured samples presented a very heterogeneous aspect (see figure 3). E4 resin even exploded during a relatively mild cure cycle.
Given the results so far, only 4 resins remained as possible candidates, illustrated on figure 4.

Table 2. Summary of selected materials properties from their technical data sheet. Some values are presented relatively to the reference resin (REF).

Product	Pot life (rel.)	Viscosity (rel.)	Colour	Dielectric Strength (kV/mm)	T_g	Shore Hardness	Volume resistivity ($10e^{xx}$)	Temperature Class	Type
H1 (REF)	100	1	Yellow	~20	Good	85 (D)	16	F	Epoxy
H2	10	6	Yellow	~20	Very good	-	-	H	Epoxy
H3	0.6-10	10	Yellow	>20	Very good	-	16	H	Epoxy
P1	4	28	Clear	>20	Extremely low	35 (A)	14	C	Silicone
L1	10	39	Clear	>20	Extremely low	50 (A)	15	F	Silicone
E1	10	3	Clear	~20	Good	67 (D)	13	C	Epoxy
E2	3	6	Clear	~20	Low	80 (D)	12	C	Epoxy
E3	1.5	28	Amber	~20	Low	82 (D)	13	C	Epoxy
E4	2.5	8	Clear	~20	Good	87 (D)	13	C	Epoxy
R1	0.8	2.5	Clear	~20	-	86 (D)	-	F	Epoxy
R2	0.8	4	Clear	~20	-	86 (D)	-	F	Epoxy
R3	20	0.5	Brown	~20	-	89 (D)	-	C	Epoxy
M1	2.5	14	Clear	>20	-	30 (A)	15	-	Silicone
V1	70	57	Clear	-	Very Good	88 (D)	-	H	Epoxy
HE1	300	3.5	Yellow	<20	-	85 (D)	15	C	Epoxy
RR	1	0.9	Clear	>20	Very Good	90 (D)	14	H	Epoxy
ES1	0.2	0.1	Yellow	>20	Very Good	92 (D)	16	H	Epoxy
ES2	0.5	0.1	Yellow	~20	Good	88 (D)	17	H	Epoxy

Figure 3. Bad cakes appearance after cure.

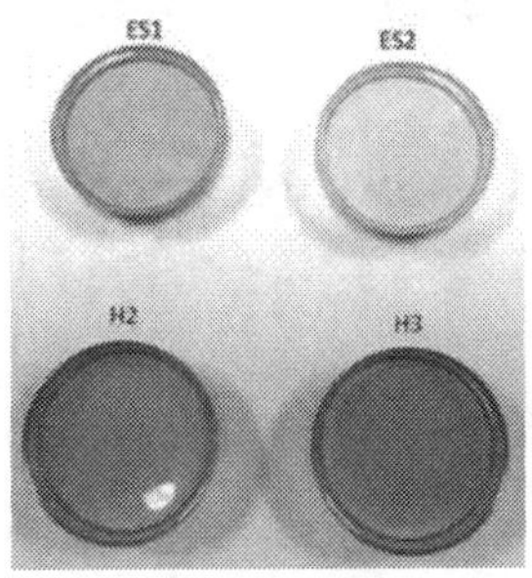

Figure 4. The 4 candidates resins cakes appearance after cure

4.3. Shrinkage

The degree of shrinkage that a resin reports during its cure is a key aspect in potting applications. The method applied here is based on ASTM D2566 which is valid for comparative measurements (which is the case of these analyses). As shown in table 3, in every case the shrinkage of the systems is below 1.5% which is quite low for a hot cure epoxy.

Table 3. Summary of shrinkage tests results. Values relative to reference REF. CTE is the coefficient of thermal expansion.

Material	CTE below T_g	Shrinkage (%)
ES2	100%	1.22
ES1	100%	0.71
H2	100%-110%	1.06
H1	110%	1.25
REF	100%	0.48

4.4. Hardness tests
The hardness of all resins, presented in table 4, are similar and also very close to the current resin considered as reference.

Table 4. Summary of hardness tests results (Shore D).

Measure #	ES1	ES2	H2	H3
1	84	85	84	78
2	85	86	84	80
3	84	84	86	83
4	85	85	84	79
5	85	85	85	82
Average	84.6	85.0	84.5	80.4

4.5. Shear Strength tests of the adhesive bond
The test has been performed according to the standard ASTM-D-1002 and the results given in table 5. The goal here was to demonstrate that new resins can sustain approximatively the same level of stress than the reference (current resin). So, as a comparison, the adherence of every resin is in most cases even better than the reference. In every case, the failure mode was adhesive.

Table 5. Summary of tensile tests results.

Maximum load	REF	ES1	ES2	H2	H3
Sample 1 (kg)	412	426	656	378	438
Sample 2 (kg)	466	586	484	354	424
Sample 3 (kg)	394	436	518	384	434
Sample 4 (kg)	462	470	426	366	416
Sample 5 (kg)	420	354	506	568	544
Average of maximum stress (kg/cm²)	133.5	140.9	160.6	122.8	139.9

IOP Conf. Series: Materials Science and Engineering 1287 (2023) 012012 doi:10.1088/1757-899X/1287/1/012012

4.6. Outgassing tests

The test was perfomed according to ECSS-Q-ST-70-02C and the results gathered in table 6. All the tested candidates comply to the outgassing test requirements. The infrared analysis indicated that no substances were detected on the condensation plate.

4.7. Water absorption

The four resins selected are epoxies, a family of materials known for its low water absortion. In every case, the water absorption data (collected from their TDS) is similar or lower to that of the current system. Therefore, their performance have been considered satisfactory for the application.

4.8. Radiation resistance

The level of radiation inside TAS-B equipments is not high enough to cause significant damage to thermoset resins. Figure 5, extracted from the database of the European Organization for Nuclear Research (CERN, [1]), shows that epoxy resins can withstand much bigger doses than those in the equipments. Specifically, an utility up to 5x10^7 Gy is often satisfactory. The test conditions presented for figure 6 chart are by far over what the equipment can see in any space mission.

Table 6. Outgassing results. TML: Total Mass Loss, RML: Recovered Mass Loss, CVCM: Collected Volatile Condensable Material

Ratio	Requirement	ES1	ES2	H2	H3
TML(%)	<1.00%	0.467	0.777	0.511	0.421
RML(%)	<1.00%	0.239	0.562	0.289	0.202
CVCM(%)	<0.10%	0.001	0.006	0.001	0.002

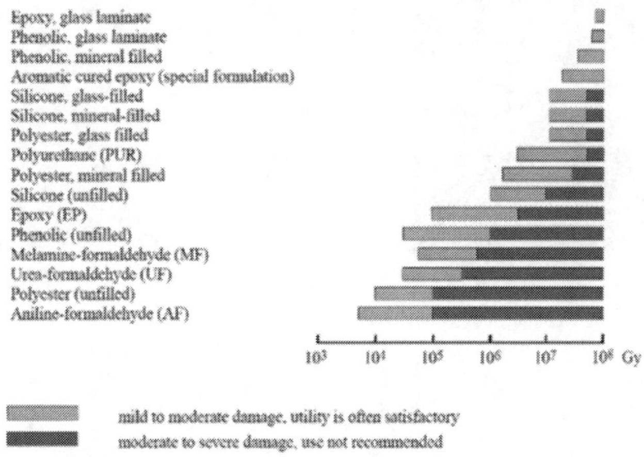

Figure 5. Classification of thermoset resin families and composites in terms of radiation resistance. Radiation effects are due to energy deposition by particles.

IOP Conf. Series: Materials Science and Engineering 1287 (2023) 012012 doi:10.1088/1757-899X/1287/1/012012

4.9. Conclusion
In this materials trade-off, a broad range of commercial thermoset resins were evaluated for their suitability for replacing the current resin set as a reference in this study. Finally, four materials resulted positive in all the different assessments.

5. Evaluation
The evaluation of the 4 candidate materials has been divided in 3 phases: tests at material level, tests at sample level and tests at breadboard level.

5.1. Tests at material level
The objective was to assess the suitability of the 4 candidate epoxy resins for the chosen application and to classify them in terms of their ability to withstand severe conditions of high temperature and high voltage.
Using a finite element multiphysics software (Comsol Multiphysics), the electrical field in the test sample was computed and a mapping is illustrated in figure 6. Then, the simplified samples have been manufactured and submitted to a life test (high electrical field at hot temperature) in order to evaluate and compare the dielectric property of the 4 resin candidates. The expected evolution of the lifetime with the temperature is an exponential increase in the lifetime ***t*** when temperature ***T*** decreases, following an Arrhenius law (*ln(t[s])=A+B/**T**[K]*), as highlighted in figure 7. Resins ES1 and ES2 showed a very large dispersion in the results. Multiple experiments have been performed, the dispersion is so high that, on the graph, only a limited set of points following a plausible trend have been kept. This could be explained by the fact that the applied electrical field was too high for them.
Based on the ageing test, the resin H3 is the best candidate.

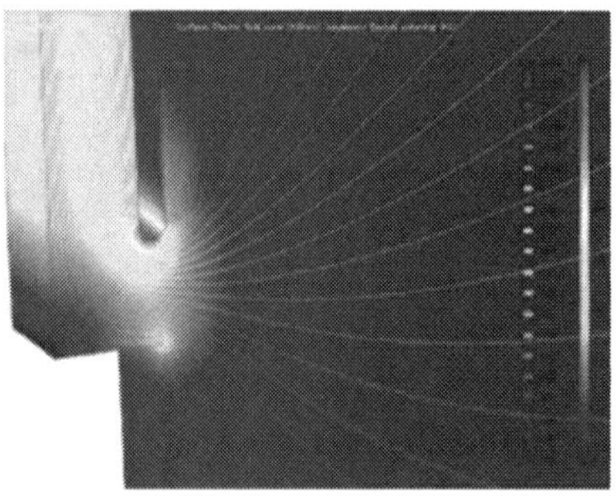

Figure 6. Electrical model created to assess the maximum voltage in the test sample.

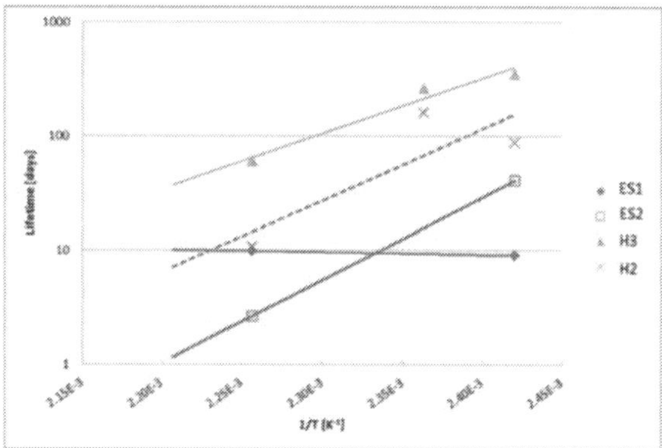

Figure 7. Life test results of the samples for the candidates resins and the reference.

5.2. Tests at sample level.
The selected resin (H3) has been implemented in representative test samples in order to be evaluated. The test sample is a magnetic part from TAS-B current applications. The impregnation, illustrated in figure 8, is in conformance with the expectations. Then, the test samples have been submitted to thermo-mechanical and electrical tests and have undergone 250 thermal cycles [-40°C; +125°C].

5.3. Tests at breadboard level.

Tests at breadboard level consisted in a technological evaluation in order to quantify the compliance with the mission duration and environment. Tests levels have been defined based on a GEO mission longer than 15 years. The breadboard evaluation focused on 4 aspects:

- Mechanical tests (vibration),
- Thermal characterization under vacuum,
- High voltage life test at high temperature,
- Damp heat test.

Four breadboards have been manufactured, each of them being able to highlight specifically the result of the test due to, for instance, geometrical or mass specificities. The breadboards have been electrically evaluated and underwent partial discharge test before proceeding to the specific test.

5.3.1. Mechanical tests (vibration - multi-platforms envelop approach). The breadboard #1 has been placed in a representative environment to be tested. Vibrations tests sequence have been applied for the 3 axes: sine survey, high level sine and random vibrations. The results were conform to TAS-B requirements ; No noticeable event or untightening was observed during the test during the tests along any of the Y, X or Z axes. The electrical measurements, performed after each axis, did not show any deviation in the performance of the transformer

5.3.2. Thermal characterization under vacuum. This test consisted in heating the breadboard #2 in order to simulate its dissipations. Evolution of the temperature has been measured at several locations and under different dissipation and environmental conditions. The predictive mathematical 3D model has been also recalibrated based on these measurements. The test showed, in different conditions, quite coherent results. Tests performed in over-dissipation and at higher interface temperature allowed to cover a wider operating temperature range. A good thermal finite element model has been built and validated that can be used all over the temperature range.

Figure 8. Sample during the impregnation process (vacuum step).

Figure 9. Life test set up.

5.3.3. High voltage life test at high temperature. The breadboard #3 has been submitted to a life test at high temperature under high voltage in order to assess the behaviour of the resin selected in a relevant environmental exposure. A picture of the set-up is given in figure 9. The test conditions have been defined thanks to the result of DC High voltage ageing tests performed on the samples previously. The life test has been pursued at 10 kV DC. No significant drift between initial measurement and the final measurement performed at the end of the life test has been observed. The Life test has been successfully performed during more than 20.000 hours (10.000 hours initially targeted).

IOP Conf. Series: Materials Science and Engineering 1287 (2023) 012012 doi:10.1088/1757-899X/1287/1/012012

5.4. Damp heat test. The damp heat test is an accelerated environmental aging test that determines the ability of a material (here, the selected resin) to withstand long-term exposure to high temperature and penetration of humidity. Two breadboards have been tested: one with the resin used in TAS-B current application (reference) and another secondary winding potted by using the new resin under evaluation (breadboard #4). No defect or anomaly has been detected by visual inspection before and after the test. No significant evolution of the dimensions and weight has been observed. The electrical measurement performed before and after the test did not show significant drift. Based on all those results, the damp heat test has been considered successful.

6. Conclusion

In the frame of this study, the materials trade-off led to the identification of a broad range of commercial resins that were evaluated for their suitability for replacing the current resin set as a reference in this study.
The complete tests sequence performed lead to identify a resin suitable for TAS-B requirements and processes.
The target to reach Technology Readiness Level (TRL) 4 has been also achieved.
Even if this project allowed to find a resin compatible with the requirements and applications, there is still a long way to go and use it in space product. The next steps are the implementation of this new resin in the current product and the qualification of magnetics parts impregnated or moulded with this resin. For future products, the good performances of this resin give the opportunity to work with higher power supply of TWT in EPC product and higher electrical field.

References

[1] Tavlet M, Fontaine A and Schönbacher H 1998 *Compilation of radiation damage test data. Part II, 2nd edition: Thermoset and themoplastic resins, composite materials* CERN 70–10

Acknowledgements

Thales Alenia Space in Belgium would like to thank ESA for its support and BELSPO for the funding of this study.

Interaction Study of Energetic Protons and Electrons with the Vacuum Ultra Violet Source used in the Complex Irradiation Facility

Erik M. Klein[1,2], Patric Seefeldt[2], Maciej Sznajder[2,3], Thomas Renger[2]

[1] University of Bremen, MAPEX Center for Materials and Processes, Robert-Hooke-Str. 7, 28359 Bremen, Germany,
[2] German Aerospace Center (DLR), Institute of Space Systems, Mechanics and Thermal Systems, Robert-Hooke-Str. 7, 28359 Bremen, Germany,
[3] PW Sznajder, P3RUN - Radiation Hardness Assurance & Data Science Expertise, Dolina Zielona 19A, 65-154 Zielona Góra, Poland

E-mail: erik.klein@uni-bremen.de

Abstract. *Context:* One important factor to be considered for space missions is the radiation environment. It is composed of energetic particles, such as protons, electrons and ions, and electromagnetic radiation, photons ranging from Vacuum Ultra Violet (VUV) to Far Infrared (FIR). Each of these radiation sources have an unique energy spectrum that further depends on the location inside the solar system, e. g. Earth orbits or interplanetary space.

For material qualification and material engineering, individual irradiation is considered and their effects are superimposed. However, effects occurring when two and more radiation types simultaneously are present have barely been investigated yet. Further complexity is added, when additional interactions take place due to laboratory hardware.

Aims: This publication reports on the efforts made to analyse and measure the interaction of a VUV and corpuscular radiation sources. The aim is to quantify the deviations due to interactions and conclude consequences for the simultaneous operation of these radiation sources.

Methods: In order to quantify the effects, the Complex Irradiation Facility (CIF), located at the DLR Bremen has been used. It connects proton and electron accelerators together with VUV and UV light sources to an ultra high vacuum chamber. The corpuscular and VUV radiation sources have been operated simultaneously. The reaction of the operation parameters, such as current measured with FARADAY Cups (FC) has been tracked and post-processed.

Results: It has been discovered, that protons considerably interact with the gas mixture used to operate the VUV source. This interaction decreases for higher beam intensities of the corpuscular irradiation. It was found, that this is likely due to protons ionizing gas atoms which are then measured by the FC as current. For electrons, this phenomena was not observed due to their smaller stopping power. The discovery restricts the acceleration factor of the CIF for proton together with VUV irradiation, but does not necessarily limit the range of application of the CIF depending on the requirements of the material and/or qualification test. Energy range of electrons stays unrestricted.

Keywords: vacuum ultra violet radiation, electrons, protons, qualification testing, material engineering

1. Introduction

The interplanetary space environment is composed of electromagnetic and corpuscular radiation originating from the Sun and from outside of the solar system. Particle radiation can be further subdivided into electrons, protons, alpha particles and further heavier ions. Each of these combinations of origin and radiation type are characterized by a spectrum of flux over particle or photon energy. For material design and qualifications testing, it is desirable to reproduce these spectra as accurate as possible. But, under terrestrial laboratory conditions, it is difficult to replicate a certain spectrum of a given particle species. This is also due to the reason, that accelerators are only capable of generating mono-energetic particles. Additionally, simultaneous operation of two or more particle sources can cause interaction of these and possibly influence radiation test outcomes. Furthermore, it is important to differentiate between interactions that also occur in space and those due to the laboratory setup.

At the German Aerospace Center (DLR) in Bremen, the Complex Irradiation Facility (CIF) is operated. It is an ultra-high vacuum system equipped with electron and proton accelerators, both with particle energies from 2.5 up to 100 keV, for sample irradiation. Additionally, light sources for the Vacuum Ultra Violet (VUV), ultra-violet (UV), visible and infrared range are available [1]. See figure 1 for a sketch of the setup. Thus, important properties of the space environment can be reproduced.

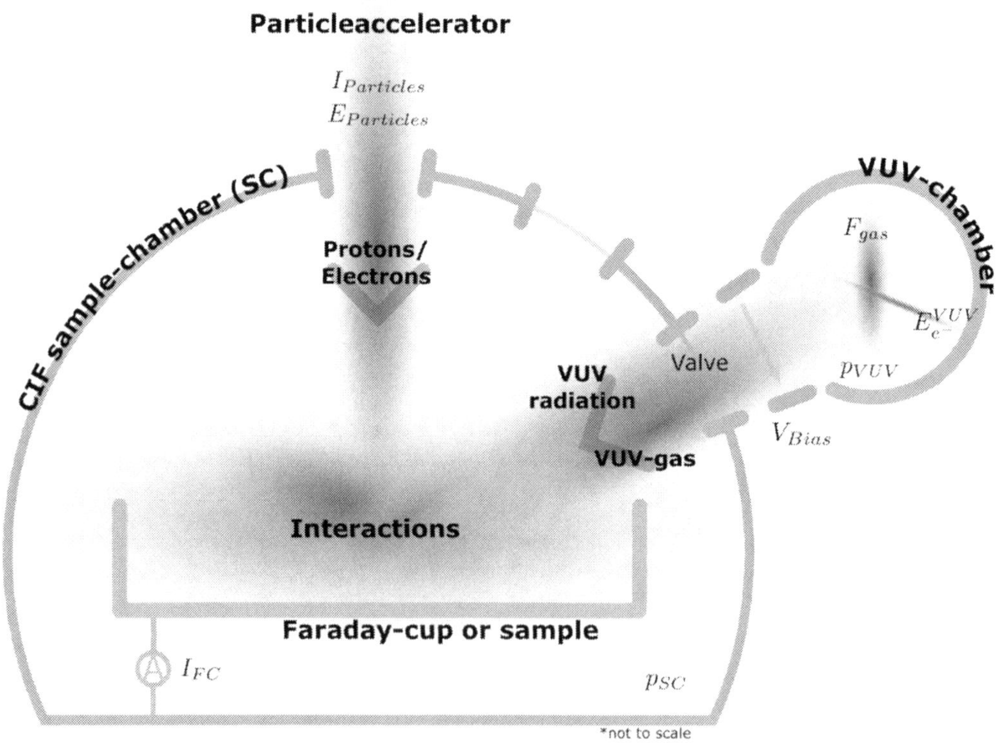

Figure 1. Sketch of the working principle of the CIF. Next to particle source and sample chamber, essential process parameters, such as current $I_{Particles}$ and energy $E_{Particles}$ of the charged particle beams, are displayed. Depiction not to scale.

IOP Conf. Series: Materials Science and Engineering 1287 (2023) 012013 doi:10.1088/1757-899X/1287/1/012013

The VUV-light source employs a gas mixture (98% argon, 1.5% krypton, 0.5% helium) irradiated with electrons to produce a calibrated light spectrum from 40 nm to 410 nm [2]. The energetic electrons for the gas irradiation are produced by a dedicated particle accelerator inside of the VUV source. The majority of the argon gas used is pumped out of the system, but a non-negligible amount enters the sample chamber, where it interacts with the materials under test, but also with incoming particles from other radiation sources. Thus, the availability of VUV radiation comes at the cost of having to deal with interactions between these traces of gas and other radiation types used in the experiment.

The following content will initially describe the employed methods and explain the procedure and post-processing to obtain the data. After that, the results will be presented and discussed. Possible explanations for seen phenomena will be elaborated. Eventually a conclusion summarizes the presented results and suggests approaches to further investigate this matter.

2. Methods

In order to quantify the influence of the VUV radiation and the gas mixture on the particle beam, both have to be operated together and key parameters have to be measured. Key parameter, that can be observed, is the current I_{FC} measured at the centre FARADAY cup (FC), see figure 1. It is assumed to be composed by following parts:

$$I_{FC} = \underbrace{I_{Particles}}_{\substack{\text{Protons/electrons} \\ \text{from accelerator}}} + \underbrace{I_{VUV}(E_{e^-}^{VUV}, F_{gas})}_{\substack{\text{VUV source} \\ \text{Photoelectric effect} \\ \text{Ions/electrons} \\ \text{from VUV chamber}}} + \underbrace{I_{Interaction}(I_{VUV}, I_{Particles}, E_{Particles})}_{\substack{\text{Ionized atom} \\ \text{secondary electrons} \\ \text{caused by interactions}}}, \quad (1)$$

where $I_{Particles}$ is the preset proton or electron beam intensity measured at the FC without VUV interaction. I_{VUV} depends on F_{gas} and $E_{e^-}^{VUV}$, gas inflow and energy of the electrons respectively, which describe process parameters of the VUV source, but have been kept constant throughout the experiments.

I_{VUV} was beforehand determined to be -1.38 ± 0.71 nA. This was conducted by irradiating the FC in the centre of the sample station, see figure 1, only by the VUV source repeatedly throughout the experiment.

Thus, I_{VUV} and $I_{Particles}$ each are the measured current when the FC is irradiated solely by either VUV or energetic particles, respectively.

$I_{Interaction}$ is here the remaining unknown quantity. The interaction current is dependent on I_{VUV}, $I_{Particles}$ and $E_{Particles}$ since these three variable describe quantitiy and properties of the interactions partners. To gain knowledge on dependencies, energy and current of the energetic particles have been swept over the range from 2.5 to 100 keV and a few nA up to 100 μA, respectively. As already mentioned, I_{VUV} has been kept constant.

For each combination of energy and current the procedure has been kept simple and was as following:

(i) The valve between Sample Chamber (SC) and VUV chamber is closed.
(ii) Particle source with either protons or electrons is running. Current $I_{Particles}$ can drift. See the dashed lines in figure 2.
(iii) VUV is running. Gas inflow is set to 1200 Standard Cubic Centimeters per Minute (SCCM). Electron energy of the e-gun is set to 1 keV. This is the default operation mode of the VUV source. See reference [2]. For this mode, the emitted spectra is available.
(iv) Pressure inside the SC is below 10^{-6} mbar.

(v) I_{FC}, prior to opening of the valve, is tracked. It is later on used to estimate I_0.
(vi) The valve between VUV chamber and SC is opened.
(vii) The response of the current is tracked.
(viii) The valve is closed again.
(ix) New energy and/or current are set for the next iteration/test.

Figure 2 shows two exemplary plots of FC current and SC pressure. The plots refer to key points (v) to (vii). The valve opens for $t = 0\,s$. It can be immediately seen, that the pressure rises, possibly by several magnitudes, to values close to $2 \cdot 10^{-6}$ mbar, which is the operating pressure of the VUV source.

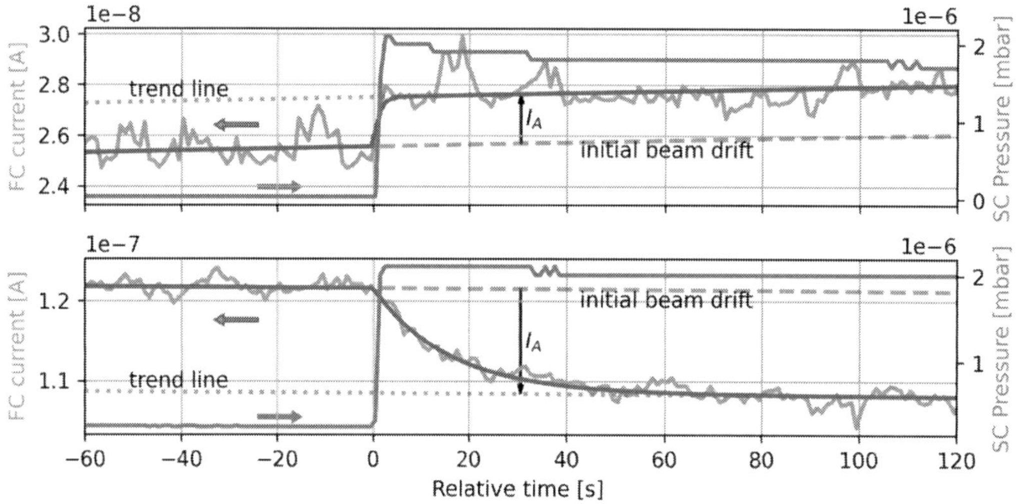

Figure 2. Exemplary plots to show the general sequence conducted for each test. Plots show test #26 and #34. Both show protons interacting with the VUV source. The dashed line depicts drift of the initial beam drift (equation 2: $d_I\,t + I_0$), the dotted line is the trend line, to which the new current tends (equation 2: $d_I\,t + I_0 + I_A$). It can be seen, that apart from similar initial states, the current decreases for one and increases for the other. Possible reasons for this will be discussed in section 3.

2.1. Post-Processing

During post-processing, the initial drift of the FC current has been estimated by a linear regression. See the dashed lines in figure 2 and equation 2.

$$I(t) = \begin{cases} d_I\,t + I_0, & \text{if } t \leq 0s \\ d_I\,t + I_0 + I_A[1 - \exp(-t/t_H)], & \text{otherwise} \end{cases} \tag{2}$$

After fitting the initial drift d_I and the current $I(t = 0s) = I_0$, the current change can be found via an least-square fit. I_A denotes the total current change or rather the vertical distance between initial drift and the trend line, t_H denotes the speed of change. The relative current change is than defined by $\Delta I = I_A/|I_0|$. $I(t)$ is plotted for two runs in figure 2. Since $I_{VUV} \ll I_{Particles}$, it is assumed that $I_A \approx I_{Interactions}$ and $I_0 \approx I_{Particles}$. ΔI therefore gives a good estimate of the influence of the interactions.

2.2. FARADAY *cup bias voltage*

FCs in the CIF can be operated with bias voltage, in order to prevent secondary electrons from escaping and thus causing false current. It is stated, that up to energies of 160 MeV, a FC can be operated without bias voltage with sufficient efficiency [3, 4]. The shape of the FC, being an elongated cylinder, closed at one end, further hinders electrons from escaping. Therefore, and in order not to influence the interactions between gas, photons and charged particles, the FC was operated without bias voltage.

3. Results & Discussion

The procedure described in section 2 has been conducted for varies energies and currents. The number is broken down in figure 3. For protons, apparently, little changes in the initial conditions could have a big influence on the final result of a test run. Due to this unrepeatable behaviour of the protons interaction with the VUV source, the number of runs with protons was higher to gather further data.

Electrons behaviour is overall predictable. ΔI fluctuated little from run to run. See figure 4.

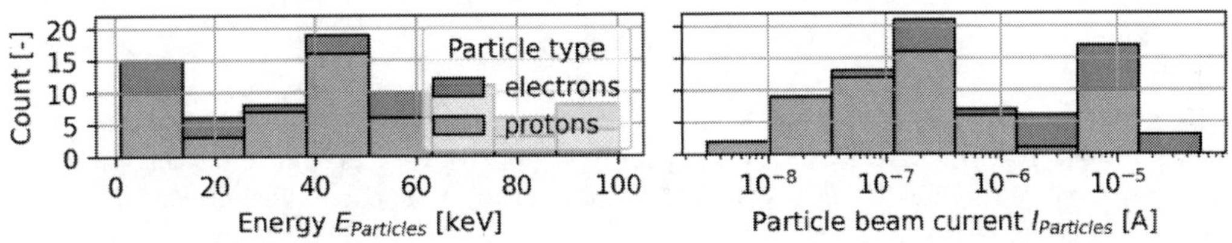

Figure 3. Stacked histograms of the tests conducted. Shown are the number of experiments for various energies and currents. Height of histogram gives number of runs conducted in given energy or current bin. A total of 83 runs have been performed.

The current on the central FC, when irradiated solely by the VUV source, was measured to be of a few nA, see above. It is unclear whether the measured current is caused by ions, electrons or due to the photoelectric effect. The current increases slightly with age of the e-gun cathode.

Furthermore, an angle dependency was investigated. As seen in figure 1 the FC is irradiated at an angle of 60°, turning the FC towards the VUV source causes an increase in current up to ~30 nA when the FC is pointed towards the VUV source.

Figure 4 shows the relative current change ΔI of the particle beam after opening of the valve, which is I_A divided by absolute of initial beam current I_0. It is immediately seen, that protons interact differently with the VUV source, than the electrons.

The data is further sampled into dependencies on particle energies $E_{Particle}$ and initial currents I_0 in figure 5. Next to it, the total current change is plotted. Here, positive values indicate increase in positive charges measured at the FC and vice versa for negative values. There is no correlation between particle energy and current change, especially *total* current change. The relative current change decreases with increasing initial current, however the total change increases.

What is further remarkable, is that the adaptation time, see t_h in equation 2 varied widely, from only sub-seconds to several seconds, indicating a dynamic process. Further, t_h appears to correlate with the absolute current change $|I_A|$ (correlation factor of 0.4), indicating that a

IOP Conf. Series: Materials Science and Engineering 1287 (2023) 012013 doi:10.1088/1757-899X/1287/1/012013

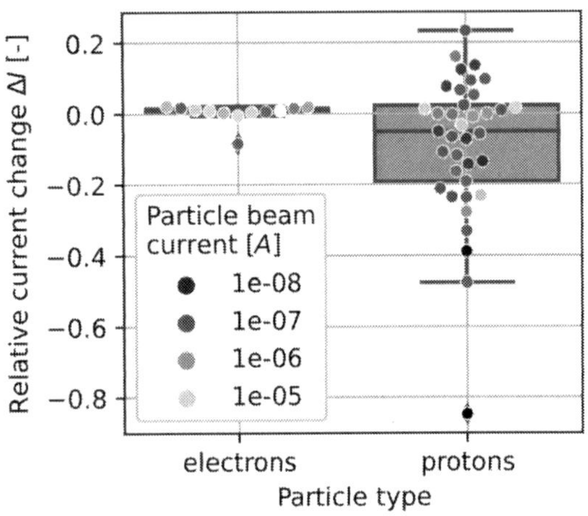

Figure 4. The figure show the distribution of the relative current change ΔI for all conducted runs divided by particle type, see figure 3. Protons reaction on the interaction with the gas and the VUV radiation is far more diverse, than that of the electrons. The measured current can halve itself or increase by 20 %. The brighter the dot, the higher the particle beam current. It can be seen, for protons, that higher currents are gathered in the centre of the distribution, but not entirely. On average, the measured FC current changed by -8.3±31.9% after interaction of the VUV source with protons and +0.5±3.5% with electrons. A further differentiation shall be done below.

charge is build up in the gas and than discharging at the FC.

Mainly relevant for the creation of charge in a penetrated medium is the stopping power of protons and electrons in argon gas for this application. Ionized atoms and their counterpart, the secondary electrons, inside the gas can than strike the FC and be there perceived as charge, just like the accelerated protons or electrons.

In figure 6 the stopping power for protons and electrons in argon gas is plotted [5]. The reader can see, that for any energy inside the investigated range, the stopping power of protons is bigger, than that of electrons. Protons are more likely to ionize atoms than electrons along their path through matter [6]. This can be attributed to the electrons property, to only after reaching energies of 10 MeV and higher, to interact with the shell electrons of the absorber [7]. Below this threshold, the interaction takes place with the COULOMB field of the nucleus without ionization. Furthermore, the penetration depth of protons is, due to the higher stopping power, by magnitudes smaller than that of electrons [5].

From this it can be concluded, that the current change of electrons is likely attributed to shadowing due to the entering gas from the VUV source. Therefore, the perceived negative charges by the FC decreases and fluctuates only little due to few ionizations taking place.

For protons it is different. Due to their higher stopping power and therefore higher energy transfer to the penetrated matter, they create a plasma along their trajectory. These charges, either ions or secondary electrons, are than measured by the FC. Due to the plasma dynamics, the perceived charge on the FC can vary widely, and even de- and increase the measured current. Conceivably as seen in figure 4 and 5. This process is independent of the particle energy, due to the stopping power, describing the energy transfer, being sufficiently high enough to ionize molecules in any run.

Figure 7 summarizes this behaviour for electrons and protons. The initial current has been binned to emphasize the dependence of the current change on the beam intensity. For electrons, the influence is negligible for all beam currents. For protons, the deviation decreases with increasing beam intensity. For protons currents, of approximately 1 μA and higher, tests with protons in combination with the VUV source can be conducted.

IOP Publishing
doi:10.1088/1757-899X/1287/1/012013

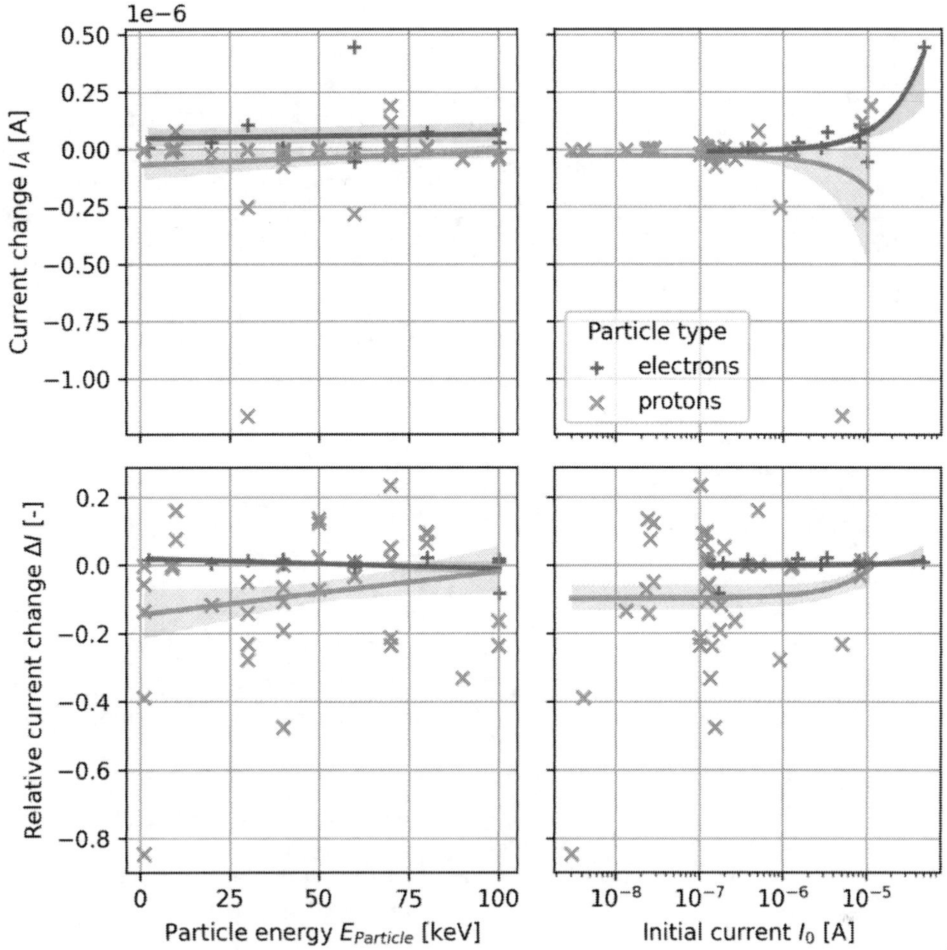

Figure 5. Initial current I_0 and energy $E_{Particle}$ plotted over relative and absolute current change, ΔI and I_A respectively. Furthermore, a regression line is plotted, accompanied by a confidence interval of 68% denoting the standard deviation. No correlation between energy and current change can be detected. For higher initial currents I_0 the relative change disappears, however the total change increases.

Looking at this phenomenon from the point of view of material qualification, the created charges can alter the outcome depending on the composition of the material under investigation. Even though, the created plasma contains ions and free electrons with small kinetic energies and charges [8], it can ionize further atoms and thus alter the analysis and outcome. For higher currents, the influence disappears compared to the actual effect of proton irradiation.

In figure 8 the acceleration factors of the CIF for a Low Earth Orbit (LEO), Geostationary Orbit (GEO) and interplanetary mission at one Astronomical Unit (AU) distance are plotted. The acceleration factor is defined by the ratio between possible flux in laboratory and that in the space environment. Since the flux in the CIF is adjustable in a certain range (See figures 3

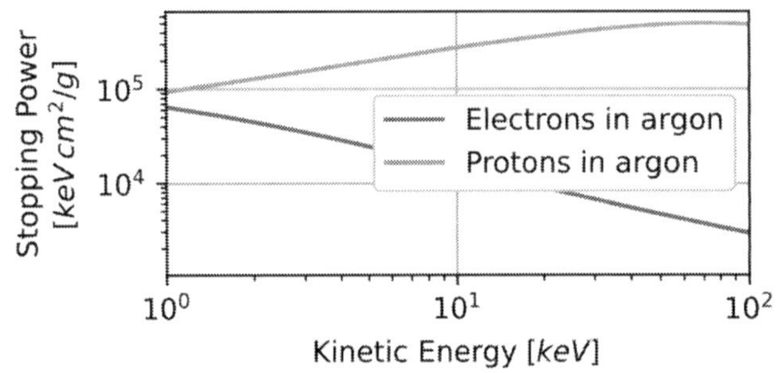

Figure 6. Comparison of stopping powers for electrons and protons in argon, denoting the gas mixture, calculated from [5]. For any energy employed in the CIF, the stopping power of protons is higher than that of electrons. This can vary for different materials.

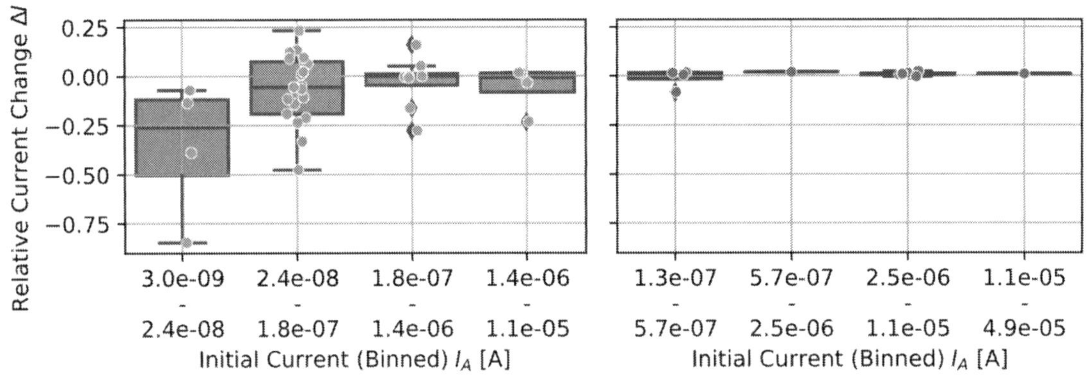

Figure 7. Box plots for binned initial current. The plot shows change in fluctuation for increasing beam intensities compared to figure 4. For protons, it is remarkable, that for increasing intensity the deviation from the original current not only decreases, but also the fluctuation decreases. As already mentioned, electrons act over all stable.

and 7), the acceleration factor also can vary inside limitations. The restricted current range of the proton source, when operated with the VUV source is located at the top of the full range.

Whether or not the restriction of the proton source can affect quality of the radiation testing depends strongly on environment and material selection. It has been shown, that e. g. aluminium oxide tends to form hydrogen blisters at higher fluxes [12, 13], which has to be weighted into the planning for material testing. This does not necessarily restrict the testing options, since the acceleration factor also depends on the environment. And whether or not a high acceleration factor can influence the test outcome depends further on the material.

Eventually it is noteworthy to mention, that it is uncertain what role the geometry of the FC, being a cylinder, plays and needs further investigation. Gas originating from the VUV source can be blown into FC body, could conceivably get caught there and be increasingly exposed to the particle irradiation. Comparing measurements with a simpler FC, consisting of a plate, will likely solve this open question.

4. Conclusion

This publication investigated the interaction between charged particles and the VUV source of the CIF at the DLR Bremen. It was observed, that the gas, used to produce the electromagnetic

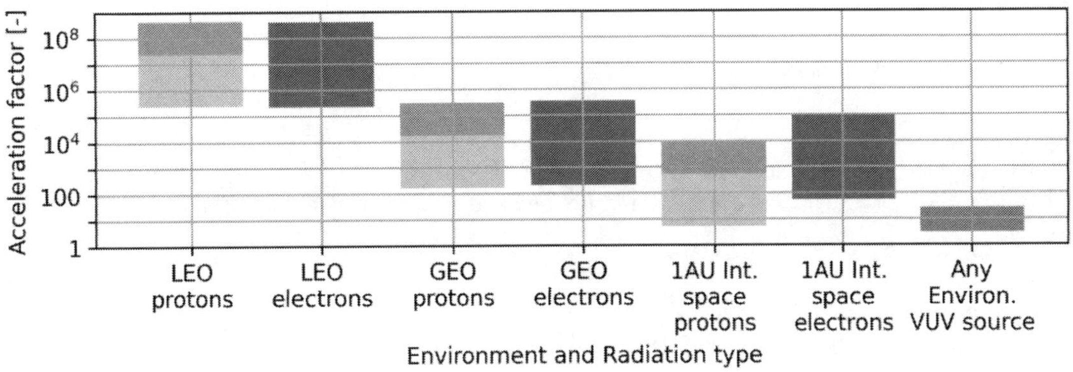

Figure 8. Exemplary acceleration factors for the proton, electron and VUV source inside the CIF for LEO, GEO and interplanetary space at 1 AU. Fluxes considered are taken from [9–11] in the range of 1 to 100 keV. Light orange denotes the full range, while the dark orange marks the restricted (1 μA and higher) share of that range when testing in combination with the VUV source. Electrons and VUV are unrestricted. The particles are swept over an area of 6 by 6 cm^2. The current varies in the range shown in figure 7. The acceleration factor of the VUV source ranges from 3 to 26.3 [2], which is valid anywhere outside the Earth atmosphere at a distance of 1 AU to the sun.

light, has significant influence on the current measured. It was found that, while for electrons the interaction is restricted to shadowing due to the gas, protons show different mechanisms as their ionization rate exceeds that of electrons. The created charges are than perceived as either positive or negative by the measuring FC. Figure 7 shows the relative current change ΔI for bins of initial current I_0 to differentiate further between low and high beam intensities. As already indicated in figure 5, for high initial beam currents, the deviation vanishes. For a minimal relative deviation from the initial current it is recommended to use currents of 1 μA and higher. For electrons, there is no restriction.

Nonetheless, it is necessary to firstly further investigate this effect, by follow on experiments and simulations. Secondly to implement the gained knowledge into the calculation of uncertainties for total fluence. And thirdly to improve the setup of the VUV source, for instance by adding another turbo molecular pump to either the VUV or sample chamber.

Follow on experiments should include other shapes of FCs, as the geometry, a cylinder, possibly traps gas and eases the plasma production by incident charged protons. A possible geometry can be a simple plate. Escaping secondary electrons will than have to be considered. Mass spectrometer measurements shall be further investigated as well. Furthermore, the gas temperature is of high interest in order to estimate the number density.

For simulation, either Geometry and Tracking (GEANT4) or Spacecraft Plasma Interaction System (SPIS) are being considered.

Conflict of Interest Statement

The authors declare that the research was conducted in the absence of any commercial or financial relationships that could be construed as a potential conflict of interest.

Author Contributions

All authors listed have made a substantial, direct, and intellectual contribution to the work and approved it for publication.

Acknowledgments

We acknowledge financial support through the APF project 'Materials on Demand' within the 'Humans on Mars' Initiative funded by the Federal State of Bremen and the University of Bremen.

Data Availability Statement

For original data, inquiries can be directed to the corresponding author.

References

[1] Renger T, Sznajder M, Witzke A and Geppert U R 2014 *Advances in Solar Sailing* (Springer) pp 541–557
[2] Sznajder M, Renger T, Witzke A, Geppert U and Thornagel R 2013 *Advances in Space Research* **52** 1993–2005 ISSN 02731177
[3] Verhey L J, Koehler A M, McDonald J C, Goitein M, Ma I C, Schneider R J and Wagner M 1979 *Radiation Research* **79** 34 ISSN 0033-7587
[4] Holmes-Siedle A G and Adams L 2007 *Handbook of radiation effects* 2nd ed (Oxford: Oxford Univ. Press) ISBN 019850733X
[5] Seltzer S Stopping-Powers and Range Tables for Electrons, Protons, and Helium Ions, NIST Standard Reference Database 124
[6] Seltzer S SHIELDOSE: a computer code for Space-Shielding Radiation Dose Calculations. Final report
[7] Krieger H 2012 *Grundlagen der Strahlungsphysik und des Strahlenschutzes* 4th ed Springer eBook Collection (Wiesbaden: Vieweg+Teubner Verlag) ISBN 9783834822383 URL `https://http://dx.doi.org/10.1007/978-3-8348-2238-3`
[8] Ozawa R 1990 *Cathodoluminescence: Theory and applications* (Tokyo: Kodansha) ISBN 9784062040334
[9] Sicard A, Boscher D, Bourdarie S, Lazaro D, Standarovski D and Ecoffet R 2018 *Annales Geophysicae* **36** 953–967
[10] Ginet G P, O'Brien T P, Huston S L, Johnston W R, Guild T B, Friedel R, Lindstrom C D, Roth C J, Whelan P, Quinn R A *et al.* 2013 *Space Science Reviews* **179** 579–615 ISSN 0038-6308
[11] Klein E M, Sznajder M and Seefeldt P 2022 *Frontiers in Space Technologies* **3** 14 ISSN 2673-5075
[12] Sznajder M, Geppert U and Dudek M R 2018 *npj Materials Degradation* **2** 1–8 ISSN 2397-2106 URL `https://www.nature.com/articles/s41529-017-0024-z.pdf`
[13] Sznajder M, Geppert U, Dudek M and Renger T 2018 *Proceedings of the 14th ISMSE & 12th ICPMSE*

Coatings Made by Atomic Layer Deposition for the Protection of Materials from Atomic Oxygen in Space

Maido Merisalu[1], Kaisa Aab[1], Väino Sammelselg[1], Kaupo Kukli[1], Harry Alles[1], Ivan Netšipailo[1], Uno Mäeorg[2], Johanna Wessing[3], Agnieszka Suliga[3], Adrian Tighe[3], Leo Nyman[4], Marko Pudas[5], Emmi Manninen[5]

[1]Institute of Physics, University of Tartu, W. Ostwaldi 1, 50411 Tartu, Estonia
[2]Institute of Chemistry, University of Tartu, Ravila 14a, 50411 Tartu, Estonia
[3]European Space Research and Technology Centre – ESTEC, Materials' Physics & Chemistry Section. TEC-QEE, Keplerlaan 1, 2201 AZ Noordwijk, The Netherlands
[4]Aalto University - School of Electrical Engineering, Department of Electronics and Nanoengineering - Space Science, 02150 Espoo, Finland
[5]Picosun Oy, Masalantie 365 02430 Masala, Finland

maido.merisalu@ut.ee

Abstract. Atomic Layer Deposition (ALD) has been investigated for the possible protection of various materials against atomic oxygen (ATOX) at ESTEC Materials and Electrical Components Laboratory facility. ALD is a conformal coating process, that can be used to apply ultra-thin films of metal oxides on various materials, that may have a sophisticated three-dimensional shape, such as the internal and external components of satellites. The challenge with metal oxides on soft and/or flexible surfaces arises from the brittle nature of these ceramic films if their thickness exceeds 30 nm. Different substrates, including silicon, Printed Circuitry Board (PCB), polyimide, and Carbon Fibre Reinforced Polymers (CFRP) were coated by ALD with 20 nm thick metal oxide films at 125 °C, then exposed to ATOX and characterized by photographing, reflectance measurement and scanning electron microscopy (SEM). The studies showed good performance of protective films prepared by ALD on polymer substrates, which suggests that the nanometer-scale coatings can improve the lifetime of these materials at low Earth orbit, where they are inevitably exposed to ATOX. In contrast, the uncoated substrates suffered near-surface damage after exposure to ATOX, which resulted in microscopic features on their surface that were visible in SEM. Damage caused by ATOX to the uncoated substrates was also visible in photographs and observable in reflectance studies. In the latter case, the changes in the reflectance spectrum were caused by the change of surface morphology and/or chemical and elemental composition due to corrosion by ATOX.

1. Introduction

The environment in low Earth orbit (LEO) is particularly corrosive for some materials due to the presence of energetic atomic oxygen [1]. Therefore, corrosion tests have been performed over the last decades at the International Space Station [2,3], as well as in laboratory conditions [4] in order to assess the performance of materials for possible use in LEO. These studies have shown that certain metals (e.g. silver) and organic materials are particularly vulnerable to corrosion caused by atomic oxygen and should not be directly exposed to the environment of space. Therefore, the list of materials for safe use in space is actually quite short and testing, although costly, is necessary to implement new ones.

A possible solution to the latter problem may be atomic layer deposition (ALD) of solid material layers, which was already investigated for its potential for corrosion protection in 1999 [5]. The greatest advantage of ALD over other coating preparation technologies is the fact that it allows one to uniformly coat substrates that may have a sophisticated three-dimensional shape [6]. A recent study has also demonstrated that some metal oxide coatings are barely affected by atomic oxygen and may thus provide excellent corrosion resistance at LEO [7]. In the latter study [7], only the top layer of the coating (~10 nm TiO_2) was slightly modified during testing with atomic oxygen, which could be regarded as equivalent to one year of exposure to atomic oxygen at LEO, i.e., at an altitude that is comparable to that of the International Space Station (~400 km). In another paper, plasma-enhanced chemical vapor deposition (PECVD) was used to prepare a functional coating, which also had a TiO_x top layer and provided excellent protection against atomic oxygen [8]. Furthermore, the ceramic coatings grown by ALD onto polymer-based substrates can reduce outgassing [9] which is problematic for sensitive instruments used for satellites. However, such coatings need to be grown to thicknesses below 30 nm [10]. Namely, in a recent study [10] it was shown thinner ceramic films grown by ALD have a higher critical tensile and compressive strains which would reduce the risk of coating failure on flexible substrates such as polymers.

In the present study, we report the possibility of using thin ceramic coatings prepared by ALD to enhance the corrosion resistance of various polymer-based materials and flexible substrates against atomic oxygen.

2. Experimental

2.1. Sample preparation

The experiments in this work were done with uncoated and coated 20 × 20 mm² substrates: silicon (Si, manually cut from prime P-type silicon wafer), polyimide (Kapton HN, DuPont), carbon fiber reinforce polymer (CFRP, provided by the German Aerospace Center, DLR), a printed circuit board (PCB, manufactured from ISOLA Group's 370HR, 180°C glass transition temperature FR-4 by Aspocomp Oy) and polymer film infrared Fresnel window (IRW, Edmund Optics). The substrates were, prior to the film growth, examined to detect the presence of dust particles and evaluate the smoothness of surface for uniform coating. The samples were cleaned with pressurized air to remove loose dust particles. Additional cleaning was performed with isopropanol, followed by drying in a laminar flow cabinet. The ceramic coating of 20 nm thickness was applied by ALD in a commercial reactor P300 (Picosun Oy, an Applied Materials Company), using the coating recipe which was provided with the instrument.

2.2. Testing with atomic oxygen

Atomic oxygen testing was performed at the low Earth orbit simulation facility (LEOX) based on the Laser Pulse Induced Breakdown (LPIB) principle in ESA's Materials & Electrical Components Laboratory [4]. The experimental atomic oxygen fluence was comparable to approximately 1 year of exposure of a direct flux of atomic oxygen at the altitude of the International Space Station at an average of 400 km. The precise values of fluence for each sample slot are given in Table 1. Prior to the main test, pre-tests were performed to measure the average flux and the spread in atomic oxygen flux across the sample plate calculated from the mass loss of Kapton witnesses. To control the adsorbed and absorbed moisture, all samples were placed in a conditioning cabinet for at least 24h before the mass measurements using a Sartorius ME5 micro balance. For the main test, the samples were weighed and photographed before (BOT) and after (EOT) the exposure, as well as characterized thoroughly with various techniques that will be described in the Section 2.3 below.

Table 1. Atomic oxygen fluence (atoms/cm^2) for tested substrates.

Substrate	Coating	Fluence (atoms/cm^2)
Polyimide (Kapton HN)	-	2.28E+21
Polyimide (Kapton HN)	ALD	2.35E+21
Carbon fiber reinforced polymer	-	2.86E+21
Carbon fiber reinforced polymer	ALD	2.86E+21
Printed circuit board	-	1.53E+21
Printed circuit board	ALD	1.20E+21
Polyethylene (infrared window)	-	8.82E+20
Polyethylene (infrared window)	ALD	1.04E+21
Silicon (1)	ALD	2.06E+21
Silicon (2)	ALD	2.23E+21

2.3. Surface characterization
The surfaces of the samples were inspected visually for discoloration using digital photography (Canon EOS 550D/Canon EOS 700D) with a stand and a macro lens. High resolution scanning electron microscopy (HR-SEM) studies and localized elemental analysis on most of the substrates were carried out with Helios Nanolab 600 (FEI Company) microscope that was equipped with an energy-dispersive X-ray spectrometry (EDX) analyzer INCA Energy 350 (Oxford Instruments) for X-ray microanalysis. Additional elemental analysis was done with a wavelength-dispersive X-ray fluorescence spectrometer (WD-XRF) ZSX 400 (Rigaku). The roughness of coatings on Si substrates was studied with an atomic force microscope AutoProbe CP II (Park SI/Veeco) in non-contact mode. The thickness of the ALD coatings was measured with ellipsometer SE-2000 (Semilab). The solar absorptance, α, and total optical reflectance in the spectral range of 250 – 2500 nm were analyzed, using a Cary 5000A (Agilent) spectrometer. The fourier transformation infrared (FT-IR) spectroscopy studies were performed with a Nicolet 6700 FT-IR spectrometer (Thermo Scientific) that was equipped with a Smart Orbit diamond micro-ATR accessory (refractive index of 2.4, active sample area diameter of 1.5 mm and angle of incidence of 45°). The spectrometer has DLaTGS detector, Vectra Aluminum Interferometer, a sealed and desiccated optical bench with CsI optics. The FT-IR spectrometer was protected from atmospheric moisture by constant purging with dry air.

3. Results
3.1. The effect of energetic atomic oxygen on the uncoated materials
The digital photos and SEM images of uncoated polyimide, CFRP, PCB and an IR-window material samples are depicted in Fig. 1. Visual inspection and photographs of the samples (Fig. 1a–d) showed that exposure to atomic oxygen had an impact on the surface of these materials. The circular unmasked area, which was directly exposed to the flux of atomic oxygen, could be clearly distinguished on all samples. A more detailed analysis with SEM provided insights into the damaging surface effects caused by ATOX. (Fig. 1e–l). SEM studies of polyimide (Fig. 1e,i) showed that the tested area was eroded rather uniformly and resulted in a rough topography with cavities that had a diameter of a few hundred nanometers. SEM studies of the CFRP substrate revealed that atomic oxygen had caused damage to both the carbon fibers (Fig. 1f,j) as well as the binding polymer (not shown). In this case, the fibers (Fig. 1j) appeared to be smoother than polyimide (Fig. 1i) after exposure to atomic oxygen. On the PCB substrate, SEM studies demonstrated minor damage to the metal area (Fig. 1g,k) and moderate damage, similar to polyimide and CFRP, on the green support material (not shown). SEM studies of the IR-window material on the other hand showed that the exposure to atomic oxygen resulted in the formation of

micrometric cavities which were filled with semi-loose nanoparticles (Fig. 1h,l). The carpet-like structure caused by ATOX as observed in Fig. 1i,j,l had been described in an earlier study on polymer materials as a typical ATOX effect [11].

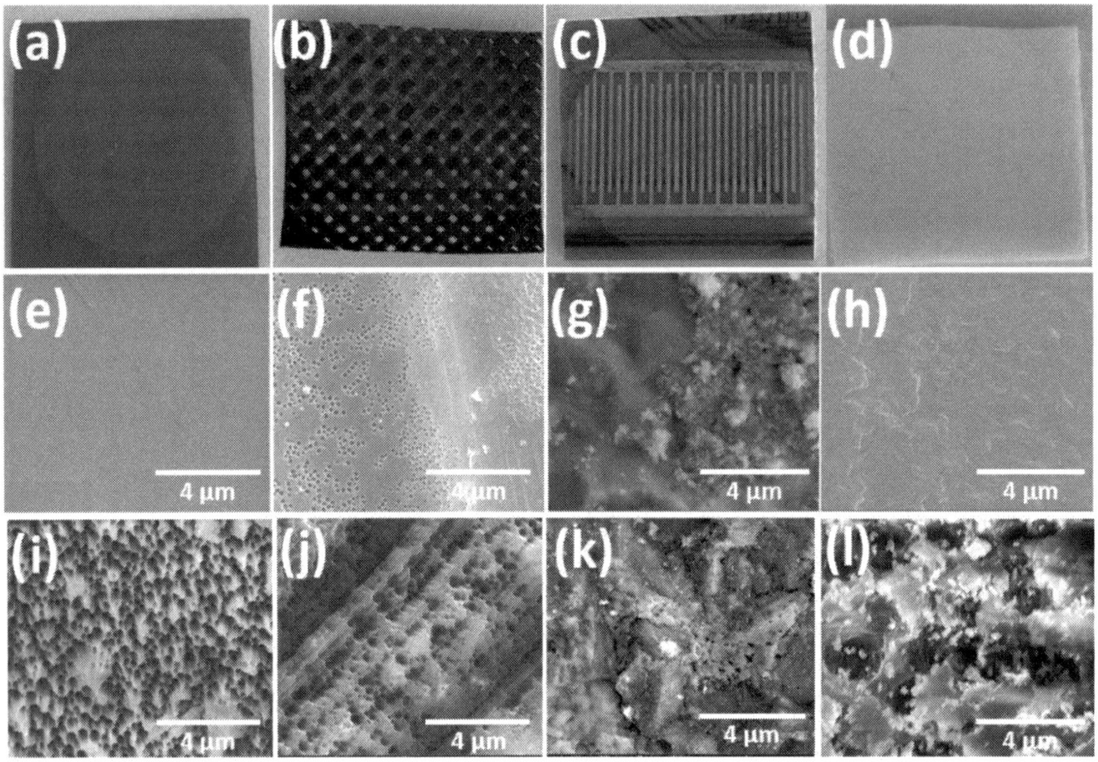

Figure 1. Photos (a-d) and HR-SEM images (e-l) of polyimide (a,e,i), CFRP (b,f,j), PCB (c,g,k) and IRW (d,h,l) before (e-h) and after (i-l)) atomic oxygen test. On PCB, the HR-SEM images (g,k) were made of the metal part.

The mass measurements of polyimide, CFRP, PCB and the IR-window film (Table 2) before and after exposure to atomic oxygen are in good agreement with the photos and HR-SEM studies depicted in Fig. 1. The highest mass erosion was observed for polyimide, which also exhibits a visual change of appearance in the photo (Fig. 1a). Another substrate with high mass erosion was CFRP (Fig. 1b, Table 2), which had suffered damage both to the polymer and carbon fibers (Fig. 1j). The PCB exhibited the lowest mass erosion value, which coincides with the HR-SEM studies. Namely, the PCB is mostly covered with metal, which suffered only minor damage (Fig. 1k) in comparison with the green support. The IR-window material was eroded less than polyimide and CFRP (Table 2), which is also in agreement with the visual observation and photos in Fig. 1a,b,d.

IOP Conf. Series: Materials Science and Engineering 1287 (2023) 012014 doi:10.1088/1757-899X/1287/1/012014

Table 2. Mass measurements of uncoated substrates before and after exposure to atomic oxygen.

Measured parameters	**Materials**			
	Polyimide	**CFRP**	**PCB**	**IRW**
Mass before test (mg)	88.712	58.596	1080.473	138.283
Mass after test (mg)	60.478	42.555	1077.040	132.500
Erosion (mg; %)	28.234 (31.8%)[b]	16.041 (27.4%)	3.433 (0.3%)	5.783 (4.2%)
Fluence (Atoms/cm^2; %)	2.28E+21 (259%)	2.86E+21 (324%)	1.53E+21 (173%)	8.82E+20 (100%)[a]

[a] Normalization of the fluence level: the fluence for IRW is taken as 100 %.
[b] Normalized erosion.

The corrosion mechanisms of polyimide and polyethylene were studied further by infrared spectroscopy (Fig. 2) and elemental analysis by XRF and SEM-EDX (Table 3). The IR spectrums measured before and after ATOX exposure were nearly identical in shape but differed in overall transmittance. In particular, the exposure to ATOX had increased the transmittance, which can be explained by the increase of surface roughness as observed in Fig. 1e,h. Such an increase of surface roughness effectively reduces the density of the material in the surface layer, which also reduces the optical path and therefore contributes to the increase in transmittance. Furthermore, IR spectroscopy studies showed that exposure to ATOX did not introduce characteristic changes in the individual absorption peaks, which would indicate the formation of C-O bonds. However, minor changes in the proportions of peaks intensities refer to the degradation of the polymers.

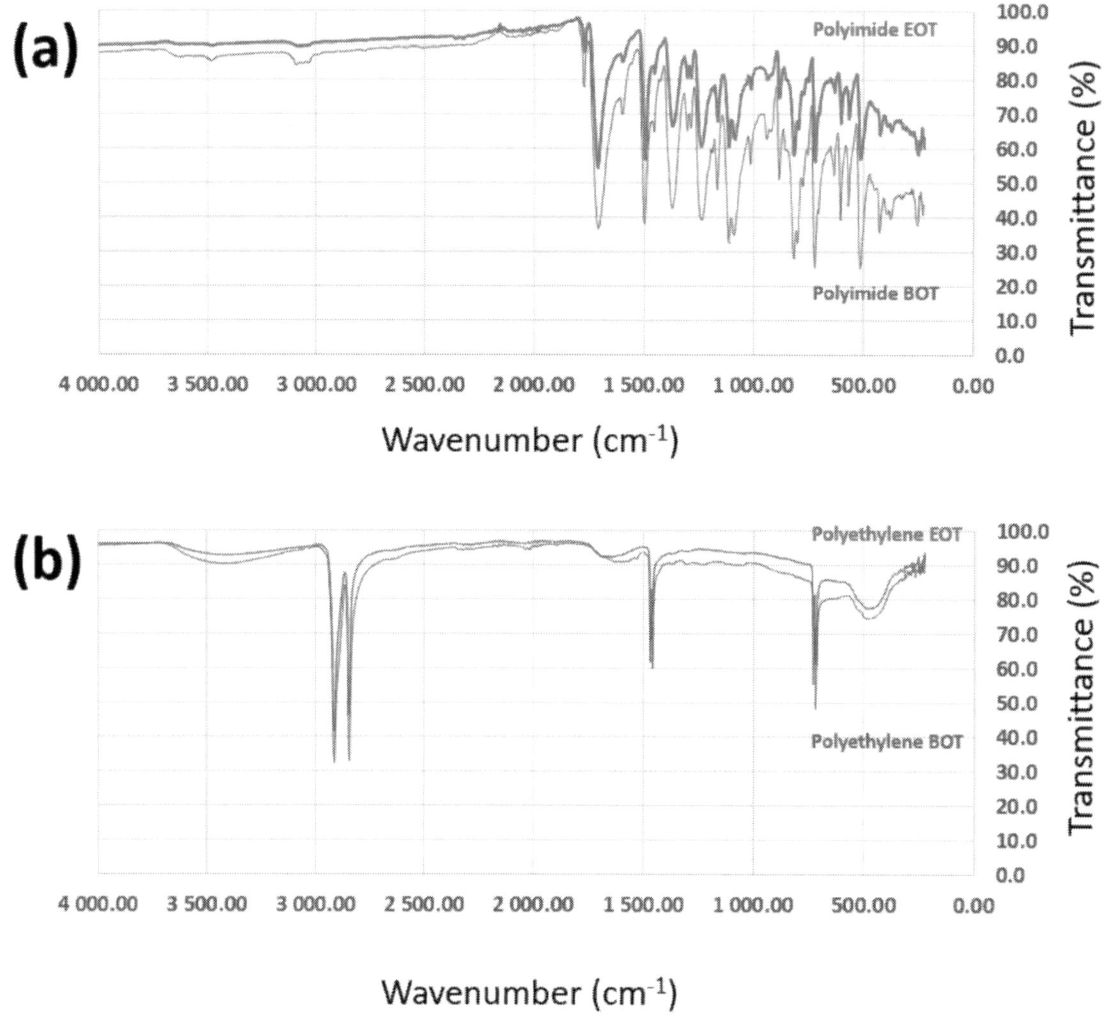

Figure. 2. Infrared spectroscopy studies of polyimide (a) and polyethylene (b) before (BOT) and after (EOT) exposure to ATOX.

The effect of energetic atomic oxygen on the elemental composition of uncoated substrates was investigated by WD-XRF and SEM-EDX, having the results summarized in Table 3. The analysis of polyimide shows that the overall elemental composition remains the same after the exposure which suggests that a layer of corrosion products does not form on the surface. This finding is in good agreement with IR spectroscopy studies (Fig. 2a). In the case of CFRP, the content of oxygen decreased, which is likely caused by the erosion of the adhesive and increased exposure of the carbon fibers. The exposure of the IR window to atomic oxygen increased the content of oxygen and zinc in its top layer. In contrast, the content of carbon had decreased.

Table 3. The elemental analysis of substrates before (BOT) and after (EOT) exposure to atomic oxygen. Less significant low content elements are not shown in the table.

Samples	Elements					
	C	O	N	Zn	Cu	Ag
Polyimide BOT	62.9	26.6	10.5			
polyimide EOT	63.0	28.7	8.3			
CFRP BOT	81.4	10.2	8.4			
CFRP EOT	86.9	5.61	7.5			
PCB BOT[a]	7.5	3.2			69.1	20.2
PCB EOT[a]	5.3	1.2			81.7	11.9
IRW BOT	96.8	2.7		0.5		
IRW EOT	82.6	16.7		0.8		

[a] The elemental composition of the metal part of the PCB was measured by SEM-EDX may be inaccurate due to the inhomogeneous nature of the alloy surface. Other substrates were studied by WD-XRF.

In order to gain a deeper insight into the corrosion effects on the IR window, additional SEM-EDX studies were done at the edge of the region exposed to ATOX (Fig. 3). The secondary electron image depicted in Fig. 3a clearly shows which region had been exposed and which part had been masked during the ATOX test. In the border region, EDX was used to map the distribution of C, O and Zn (Fig. 3b–d). These maps show that exposure to atomic oxygen had increased the content of O (Fig. 3c) and Zn (Fig. 3d), which is in good agreement with WD-XRF studies (Table 3).

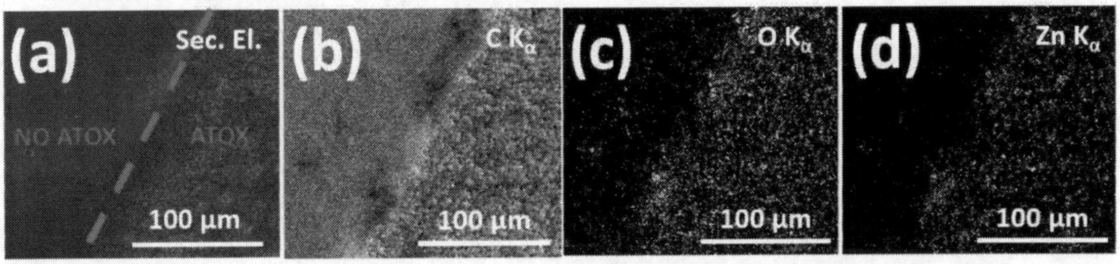

Figure 3. SEM-EDX studies of IR window at the edge of the region that was tested with atomic oxygen depicting the secondary electron image (a) and the distribution of C (b), O (c) and Zn (d).

SEM-EDX was also used to study the complex corrosion behavior of the PCB by mapping the distribution of Cu, Ag, O, C, Al, Si and Mg at the edge of the region that was tested with atomic oxygen (Fig. 4). The damage to the metal part seemed to be negligible (Fig. 1g, k) and the measurements were likely inaccurate due to the inhomogeneous nature of the surface (Table 3). However, a minor change was observed in the maps of Cu (Fig. 4b) and Ag (Fig. 4c). After the ATOX exposure the content of Cu had decreased and the content of Ag increased. A more distinct change was observed in the supporting polymer, where the exposure to atomic oxygen reduced the content of C (Fig. 4e) and increased the content of O (Fig. 4d), Si (Fig. 4g) and Mg (Fig. 4h).

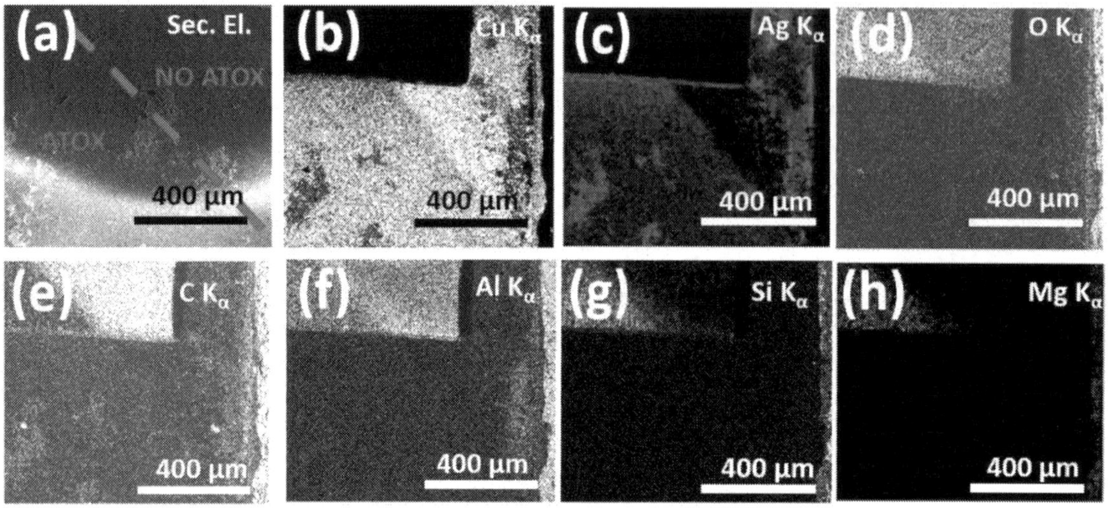

Figure 4. SEM-EDX studies of PCB at the edge of the region that was tested with atomic oxygen depicting the secondary electron image (a) and distribution of Cu (b), Ag (c), O (d), C (e), Al (f), Si (g) and Mg (h).

A more detailed study was performed to investigate the effect of atomic oxygen on the ceramic coatings grown by ALD on Si substrates as no change in visual appearance was observed. For that purpose, HR-SEM was used first, which revealed that exposure to atomic oxygen had slightly modified the top layer of the coating by introducing small ~10 nm in diameter cavities (Fig. 5). Overall, the coating appeared to be intact. Ellipsometry studies showed that the thicknesses of the ALD coatings of 22 nm did not change within measurable tolerances of +/- 1 nm. (The original coating run consisted of 5 samples, which of 2 were ATOX treated and 3 were kept as refences, all showing thickness of 22 ±2 nm). Exact measurement error from ellipsometer is hard to define, analytical error in ellipsometer model was <0.1nm. This suggests that the coating had suffered only negligible damage, similarly to the nanolaminate in an earlier study [7]. Therefore, the only notable change was in the roughness of the surface, which was in AFM studies 0.41 nm before and 0.62 nm after exposure to atomic oxygen.

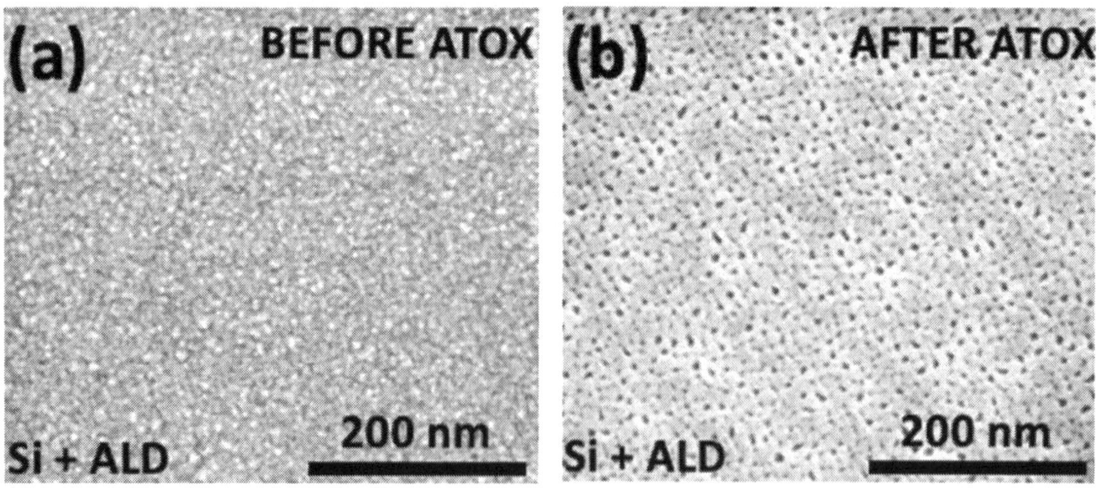

Figure 5. HR-SEM images of metal oxide coating made by ALD on Si substrate before (a) and after (b) testing with atomic oxygen.

The reflectance spectra measured for Si substrates with metal oxide coatings show that exposure to atomic oxygen slightly reduced the reflectance (Fig. 6). At shorter wavelengths, this effect could be attributed to the increase of surface roughness, which was confirmed by the AFM and HR-SEM studies (Fig. 5). However, the most dominant decrease in reflectance occurs at the infrared portion of the spectrum.

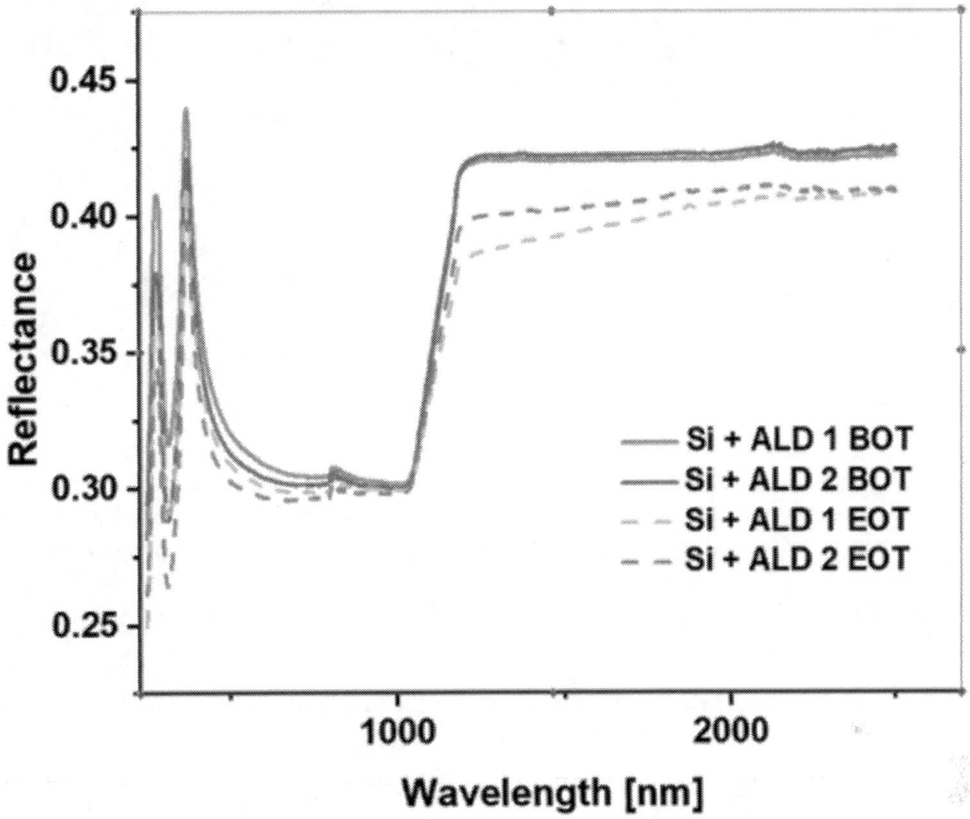

Figure 6. Reflectance spectrum of Si substrates with identical metal oxide coating grown by ALD before and after exposure to atomic oxygen. The atomic oxygen fluence for the samples is given in Table 1.

3.2. Protection of materials with ceramic coatings grown by ALD
The photos of substrates with ceramic coatings after exposure to atomic oxygen are depicted in Fig. 7. As can be seen from these photos, the ceramic coating successfully protected polyimide (Fig. 7a), CFRP (Fig. 7b), PCB (Fig. 7c) and the IR window (Fig. 7d) against energetic atomic oxygen, which can be regarded as significant improvement in comparison with the uncoated substrates (Fig. 1). This result is also in agreement with the HR-SEM (Fig. 5), ellipsometry and AFM studies, which showed that the ~20 nm coating protection is highly effective against ATOX erosion effects

Figure 7. Photos (a-d) and SEM images (e-h) of substrates with ALD-based metal oxide coating after exposure to atomic oxygen depicting polyimide (a,e), CFRP (b,f), PCB (c,g) and IR window (d,h).

The mass measurements of coated substrates show that the use of ceramic coatings grown by ALD can significantly reduce the erosion caused by atomic oxygen (Table 4). For instance, the mass erosion of polyimide with ceramic coating was only 0.137 mg, which is over two orders of magnitude lower than the mass erosion measured for the uncoated polyimide substrate (Table 2). The lowest change was observed with the PCB, where the mass erosion value of the coated sample is reduced by a third compared to the one of the uncoated samples (Table 2,4), which is still a significant improvement of performance.

Table 4. Mass measurements of coated substrates before and after exposure to atomic oxygen.

Measured parameters	Materials				
	Polyimide+ALD	CFRP+ALD	PCB+ALD	IRW+ALD	Si+ALD
Mass before test (mg)	82.662	59.705	1032.951	141.478	440.147
Mass after test (mg)	82.525	58.711	1031.818	140.870	440.144
Erosion (mg; %)	0.137 (0.2%)[b]	0.994 (1.6%)	1.133 (0.1%)	0.608 (0.4%)	0.003 (0.0%)
Fluence (Atoms/cm^2, %)	2.35E+21 (266%)	2.52E+21 (286%)	1.20E+21 (136%)	1.04E+21 (118%)[a]	2.23E+21 (253%)

[a] Normalization of the fluence level, see Table 1.
[b] Normalized erosion, see Table 1.

The reflectance spectra of bare and coated substrates before and after exposure to atomic oxygen are depicted in Fig. 8. These measurements show that the reflectance of unprotected polyimide (Fig. 8a), CFRP (Fig. 8b) and PCB (Fig. 8c) drops after exposure to atomic oxygen, which is likely caused by corrosion induced surface roughness (Fig. 1i–l). In contrast, the same substrates with ceramic coating showed no significant change in reflectance after the atomic oxygen test (Fig. 8a–c). Despite the visible corrosion effects observed in Fig. 1d,l, the reflectance of the uncoated IR window did not change much after exposure to atomic oxygen. Similarly, the use of ceramic coating on the IR window only had a negligible effect on its reflectance spectrum (Fig. 8d).

Overall, the reflectance measurements were in a good agreement with the visual inspection (Fig. 7) as well as with the mass measurements, showing that the ceramic coatings can assist in retaining the original condition of functional materials.

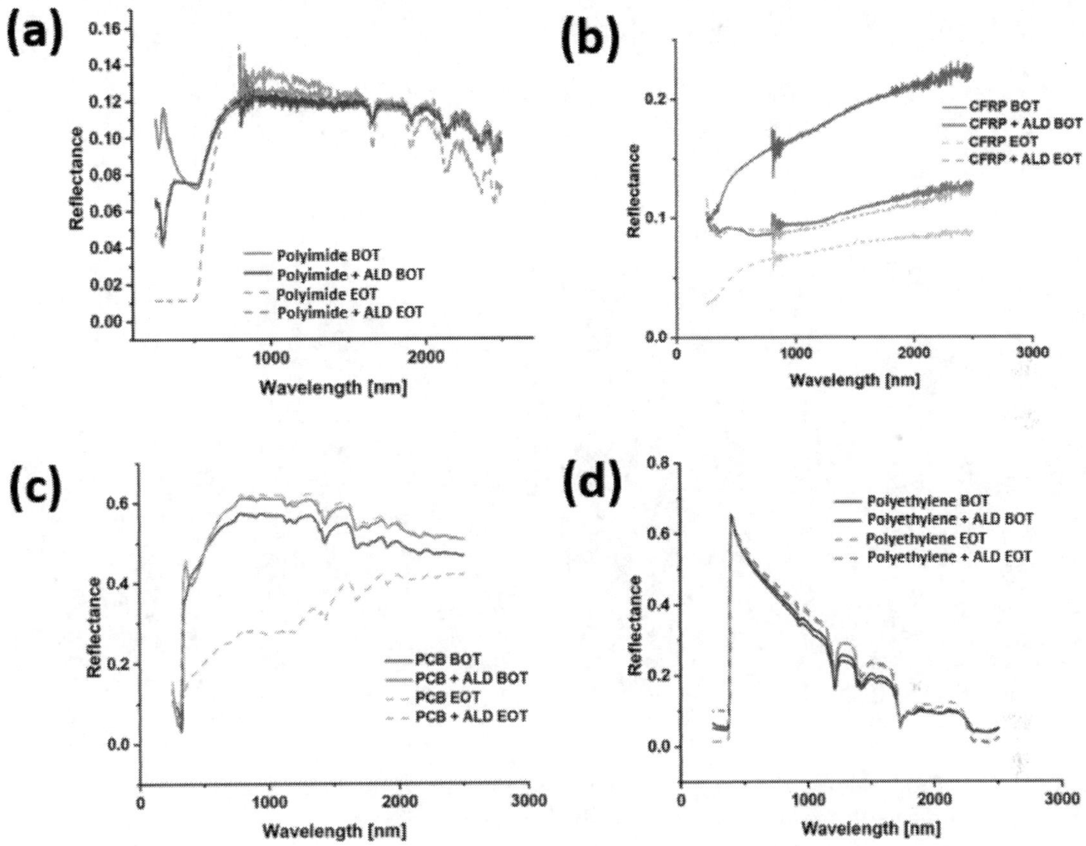

Figure 8. Reflectance spectra of bare and ALD coated polyimide (a), CFRP (b), PCB (c) and IR window (d) before (BOT) and after (EOT) exposure to atomic oxygen.

HR-SEM studies of ALD coated polyimide and IR window materials after testing with atomic oxygen revealed spatially separated sites, where the coating coverage was discontinued (Fig. 9a–d). The shape of these areas (Fig. 9a,c marked with arrows) suggested that the ALD coating cover was mechanically damaged during sample handling prior to testing with atomic oxygen. Images taken at greater magnification (Fig. 9b,d) allowed one to conclude that these coating discontinuities do not grow much when exposed to atomic oxygen. The likely explanation could be that the polymers would be damaged only after exposure to a direct flux of atomic oxygen. Undercutting likely happens but its scale is negligible, so the tests suggest that it does not have an impact in practical applications when dealing with larger parts.

Figure 9. HR-SEM studies of corrosion on polyimide (a,b) and IR window (c,d) caused by atomic oxygen in sites where the coating had been locally damaged.

4. Discussion

4.1. Corrosion mechanism

The characterization of samples before and after exposure to atomic oxygen in this study showed that the corrosion mechanism is uniquely characteristic for each material. Generally, corrosion involves the oxidation of materials, the formation of corrosion products and typical carpet-like surface structure.

4.1.1. The effect of atomic oxygen on polyimide. In the case of polyimide, the exposure to energetic atomic oxygen causes a rather uniform erosion of the material, which results in a rough pattern with cavities on its surface also observed in an earlier study [11]. The pattern exhibits a carpet-like structure typical for ATOX erosion and is likely caused by the internal structure of polyimide, making some regions more prone to erode than others. As no increase of oxygen was observed in the composition of polyimide during the test (Table 3), it is clear that the main corrosion product was gas phase CO, which immediately leaves the surface upon formation. Therefore, the gradual oxidation results in the erosion of the material. However, the exact erosion process could not be observed with infrared spectroscopy as the method collects average data from a depth of several microns of the surface while the depolymerization process likely takes place in a nanometric depth.

4.1.2. The effect of atomic oxygen on CFRP. Similarly to polyimide, the exposure of CFRP to energetic atomic oxygen caused a rather uniform erosion of the material, where the main corrosion product is likely gas phase CO which leaves the surface upon formation. However, in the case of CFRP, the material consists of two components which are apparent to corrode at a slightly different rate. According to HR-SEM studies, the fibers (Fig. 10a) are less vulnerable to atomic oxygen than the binding material (Fig. 10b).

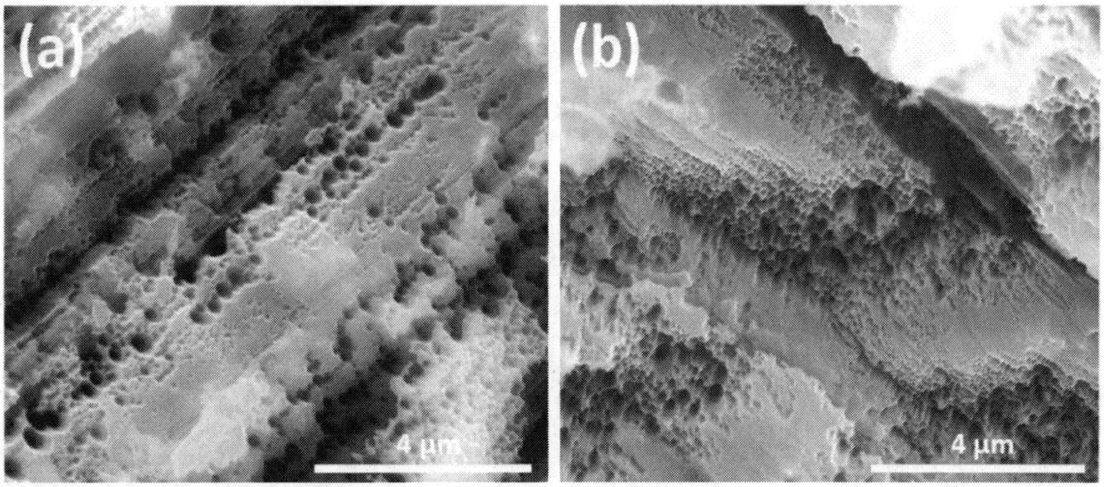

Figure 10. HR-SEM studies of CFRP after exposure to atomic oxygen depicting the fibers (a) and the binding material (b).

4.1.3. The effect of atomic oxygen on PCB. The green solder mask, as well as the metal part of the PCB both suffered damage from atomic oxygen (Fig. 1k, Fig. 4). Generally, for PCBs, the metal part is most critical for the operation of an electronic system and according to EDX studies it consists mainly of copper and silver (Table 3). SEM-EDX studies (Fig. 4) demonstrated that the metal part is less vulnerable to atomic oxygen than the support material, which also explains why the overall erosion of PCB was lower than the erosion of polyimide or CFRP (Table 2). Due to the deterioration of both the metal as well as the green solder mask, a protective coating may still be needed on a PCB when used in LEO. That's because in this study the atomic oxygen test simulated only 1 year of exposure to space in LEO, which already introduced damage to the surface of the PCB. However, in practical applications the duration of a mission can be several years and during that time the damage would accumulate.

4.1.4. The effect of atomic oxygen on IR window. The corrosion of the IR window is somewhat similar to that of polyimide and CFRP, where the dominant corrosion product seems to be gaseous CO. However, the increase in the oxygen content in the elemental composition of the IR window after testing suggested that some amount of the oxygen has either constituted solid corrosion products or was physically trapped on the surface region. Such trapping may happen if the atomic oxygen particles interact with each other inside the polymer and form O_2 and/or H_2O which are less likely to escape the material. The IR window contains a certain amount of zinc, which may easily oxidize (to a solid corrosion product) and contribute to the overall increase in the oxygen content. However, the amount of zinc was relatively low, which implied that it cannot be solely responsible for the significant/high increase of the oxygen content. Furthermore, the spectra in the IR range are nearly identical before and after testing, which suggests that oxygen has not been chemically bound to carbon in the polyethylene chains. However, X-ray photoelectron spectroscopy studies carried out and reported in an earlier paper [12] allowed one to conclude that some oxygen may become chemically bound to polyethylene, which can partly explain the increase of oxygen content seen also in the present WD-XRF studies (Table 3).

Finally, the relevant bonds would only form in the very top layer of the IR window, making it too small to be observable in infrared spectroscopy studies.

4.1.5. The effect of atomic oxygen on the ceramic coating made by ALD. The ceramic coating investigated in this study is a metal oxide by nature. Therefore, theoretically, it would be impossible for the coating material to chemically bind additional oxygen in conditions used in this work, effectively making it corrosion resistant in an oxidizing environment. However, it should be mentioned here that oxygen vacancies in the ceramic coating could still be filled with oxygen. This could have an effect on the conductivity of the coating but not the overall corrosion resistance. The effect of atomic oxygen on the conductivity of ceramic coatings may be relevant for mitigating surface charging related issues. Another component of the corrosion process would likely be the physical erosion during the interaction of the material with atomic oxygen but the kinetic energy of atomic oxygen cannot possibly be very high. This might explain why in the HR-SEM studies of the present work (Fig. 5) as well as in a paper published earlier [7], the damage caused to the ceramic coating by atomic oxygen remained negligible and affected only the top layer. This damage may have been caused by physical sputtering, where some of the coating material may also be redeposited on the surface. This would also explain why the thickness of the ultra-thin ceramic materials does not change when exposed to atomic oxygen. However, the exposure to atomic oxygen did lower the reflectance of the coating, which can possibly be explained by the changes in surface roughness, refractive index and optical absorption, somewhat speculatively also the electrical properties. In particular, the reduction of reflectance at lower wavelengths can quite likely to be attributed to the increase in surface roughness and the decrease in average refractive index. At higher wavelengths, the reduction of reflectance may be caused by the increase of electrical resistance according to Drude model [13] and/or due to higher absorption in the infrared region due to vibrational excitations.

4.2. The protection efficiency of ceramic coatings grown by ALD

The good protection efficiency of ceramic coatings on different substrates investigated in this study can be explained by the chemical inertness of the coatings towards atomic oxygen, the ability to coat uniformly the whole surface by ALD, and a good adhesion of the coating to the underlying substrate. Even so, some erosion was observed in the mass measurements (Table 4), which could be attributed to localized sites where the coating coverage was disrupted or damaged during the sample handling (Fig. 9). One can, possibly, overcome this in practical applications, where the coating preparation methodology has to be optimized for specific materials, using appropriate pre-treatments and coating synthesis parameters for the chosen ALD process. Furthermore, the handling procedures of substrates in practical applications can be developed, which should improve the overall quality of the coating. Additionally, even if the coatings were damaged, it is unlikely that delamination due to the corrosion of underlying atomic oxygen would take place and, therefore, the majority of the materials surface would still be sufficiently protected. Nonetheless, what has not been covered in this study are the effects of cyclical of temperature fluctuations and exposure to ionizing radiation in combination with energetic atomic oxygen. These synergistic effects may lead to an enhanced delamination of the ceramic coatings on polymers but this needs to be studied further.

5. Conclusions

In this study we demonstrated that ultra-thin ceramic coatings grown by ALD can enhance the corrosion resistance of polyimide, CFRP, PCB and the IR window substrates against energetic atomic oxygen.

Our studies with the bare substrates not covered by ALD coatings, at first, provided deeper insight into the corrosion mechanism of different materials when exposed to atomic oxygen. The studies indicate that in the case of polyimide, CFRP and IR window, the dominant corrosion product is most likely volatile CO in gaseous form. As a result, polymer materials gradually eroded under the flux of atomic oxygen and passivating films on such materials were not formed. However, on the microscopic scale, such erosion occurred unevenly, which resulted in a carpet-like, rough surface structure and, consequently in a decreased of the reflectivity. The uneven erosion was most notable in the case of CFRP, PCB and IR window which consisted of two different (raw) materials. In the case of CFRP, the binding material eroded faster than the carbon fibers. With regards to the PCB, the metal parts suffered little damage while the green support material was depleted of carbon and exhibited an increase of oxygen. In the case of the IR window, the erosion of carbon-based polyethylene exposed Zn particles which were originally in the composite material.

The use of ceramic coatings on polyimide, CFRP, PCB and the IR window reduced the rate of erosion by a factor of over 100, which means a significant improvement in the respective material's corrosion resistance. Furthermore, the coating's corrosion prevented the material's degradation with regards to the reflectance spectra over a broad spectral range.

The useful performance of the ceramic coating mainly lies in its chemical inertness against atomic oxygen, assisted by a good adhesion to the substrates and the natural advantage of ALD to uniformly coat the whole substrate surface compared to other coating techniques. As a result, the ceramic coating has, in the present study, hindered the oxidation of carbon by energetic atomic oxygen on the surface of the studied materials and therefore effectively prevents corrosion. HR-SEM studies showed that the very surface of the coating was slightly damaged after the atomic oxygen exposure, however the overall thickness of the coating remained unchanged as confirmed by the ellipsometry measurements. Furthermore, ceramic coatings did not delaminate even in areas with defects when exposed to atomic oxygen flux. This means that the majority of the substrate surface remained protected even if the coating suffered some mechanical damage during handling.

The findings of this study suggest that the use of ceramic coatings made by ALD may become highly useful and suitable for enhancing the performance of various materials in low Earth orbit applications. In addition, the results may lead to broadening of the list of materials that can be used in space.

6. References

[1] de Rooij A 2010 Corrosion in space *Encyclopedia of Aerospace Engineerings* ed Blockley R and Shyy (Wiley) W
https://onlinelibrary.wiley.com/
https://onlinelibrary.wiley.com/doi/abs/10.1002/9780470686652.eae242
http://esmat.esa.int/publications/published_papers/corrosion_in_space.pdf;
Last visited in August 2022.

[2] de Groh K K, Banks A B, Dever J A, Jaworske D A, Miller S K, Sechkar E A and Panko S R 2008 NASA Glenn Research Center's Materials International Space Station Experiments (MISSE 1-7). *NASA/TM-2008-215482*
https://ntrs.nasa.gov/api/citations/20090005995/downloads/20090005995.pdf

[3] de Groh K K and Banks B A 2019 Atomic oxygen erosion data from the MISSE 2-8 missions. *NASA/TM-2019-219982*
https://ntrs.nasa.gov/api/citations/20190025445/downloads/20190025445.pdf

[4] ESA-ESTEC LEOX Facility
https://esamultimedia.esa.int/multimedia/industry/ESA_TECQE_AO_013375.pdf

[5] Matero R, Ritala M, Leskelä M, Salo T, Aromaa J and Forsen O 1999 Atomic layer deposited thin films for corrosion protection. J. Phys. IV France 9 pp 493–499. https://doi.org/10.1051/jp4:1999862

[6] Ritala M and Leskelä M 2002 Atomic layer deposition. *Handbook of Thin Films* **1** pp 103–159. https://pubs.acs.org/doi/10.1016/B978-012512908-4/50005-9

[7] Merisalu M, Aarik L, Piirsoo H M, Kozlova J, Tarre A, Zabels R, Wessing J, Brieva A and Sammelselg V 2022 Nanostructured coating for aluminum alloys used in aerospace applications.

Journal of the Electrochemical Society **169** https://iopscience.iop.org/article/10.1149/1945-7111/ac7bb2

[8] Smith C T G, Delkowki M, Anguita J V, Cox D C, Haas C and S Ravi P Silva 2021 Complete atomic oxygen and UV protection for polymer and composite materials in a low Earth orbit ACS *Applied Materials and Interfaces* **13** pp 6670-6677 https://pubs.acs.org/doi/10.1021/acsami.0c21552

[9] Kestilä A, Nordling K, Miikkulainen V, Kaipio M, Tikka T, Salmi M, Auer A, Leskelä M and Ritala M 2018 Towards space-grade 3D printed, ALD-coated small satellite propulsion components for fluidics *Additive Manufacturing* **22** pp 31-37 https://doi.org/10.1016/j.addma.2018.04.023

[10] Jen S H, Bertrand J A and George S M 2011 Critical tensile and compressive strains for cracking Al_2O_3 films grown by atomic layer deposition *AIP Journal of Applied Physics* **109** 084305 http://dx.doi.org/10.1063/1.3567912

[11] Banks B A, Rutledge S K, Hunt J D, Drobotij E, Cales M R and Cantrell G 1995 Atomic oxygen textured polymers *NASA Technical Memorandum* **106769**

[12] Zafar R, Park S Y and Kim C G 2021 Surface modification of polyethylene microplastic particles during the aqueous-phase ozonation process *Environmental Engineering Research* **26(5)** 200412 https://doi.org/10.4491/eer.2020.412

[13] Drude model: https://en.wikipedia.org/wiki/Drude_model

Acknowledgments

The authors thank dr. Aarne Kasikov for fruitful consultations in the interpretation of optical measurements and Peeter Ritslaid for the assistance in X-ray fluorescence spectroscopy studies. The studies were partly funded by the EU through the European Regional Development Fund under project TK141 (2014-2020.4.01.15-00). The atomic oxygen testing was performed in the framework of the "Announcement of opportunity for atomic oxygen in the ESTEC Materials and Electrical Components Laboratory/ESA-TECQE-AO-013375)," through a collaboration with Picosun Oy.

IOP Conf. Series: Materials Science and Engineering 1287 (2023) 012015 doi:10.1088/1757-899X/1287/1/012015

Design of Innovative High-Performance Polymer for Passive Lunar Dust Mitigation

Guido Saccone[1], Nunzia Favaloro[1], Claudio De Rosa[2], Rocco Di Girolamo[2], Giuseppina Di Stefano[2] and Fabrice Cipriani[3]

[1] CIRA – Italian Aerospace Research Centre, Capua (CE), 81043, Italy
[2] University of Naples Federico II, Via Cintia, 21, Naples (NA), 80126, Italy
[3] ESA – European Space Agency, Noordwijk, The Netherlands

g.saccone@cira.it

Abstract. Dust in space environments is now recognized as a major concern for successful manned and robotic exploration and colonization missions. Indeed, a thin layer of dust covers the lunar surface made of loose granular material, consisting of a broad range of shapes, sizes, and types of sediments, which mainly consists of silicate minerals in the form of micrometric, sharp, abrasive, porous, chemically reactive dust particles. The aim of this paper is to present preliminary results related to the design and development of an innovative, lightweight, high-performance polymer with an elevated strength-to-weight ratio able to mitigate particle contamination. In detail, dust mitigation or the minimization of the surface energy and consequently the adhesion forces among the external layers and the granular dust micrometric particles can be achieved through modifications of the surface properties by means of both chemical and physical methodologies. The proposed approach, to realize an innovative material and not a hydrophobic coating, is potentially applicable to a wide range of technological conditions and it relies on reproducible and controllable chemical modifications of the material's surface properties through the design and synthesis of suitable base materials i.e., aromatic polyimides and copolyimides and incorporation of special low molecular weight additives, i.e., surface migrating agents, loaded directly within the reacting mixture during the intermediate phases of the copolymerization. The materials investigated in this work exhibit mechanical properties able to withstand the extreme space environment conditions and an elevated non-sticking behaviour of its surface layers in contact with granular, micrometric dust particles of lunar regolith i.e., abhesion capacity.

1. Introduction

Dust in space environments, especially Moon and Mars has been recognized by several Space Agencies, including NASA and ESA, as a major concern for the successful robotic and manned exploration and colonization of extra-terrestrial bodies for both short and long terms missions.

Regarding the Moon environment, the micrometric lunar dust particles were an unforeseen problem and they caused several issues, as verified during the Apollo missions 11 and 17, from 1969 to 1972, and during the more recent exploration and observation activities [1].

Dust issues may include vision obscuration during landing [2-3], abrasion of visors, gloves and boots of astronaut's suits [4], degradation of radiators and thermal control surfaces [5], deterioration of solar panels efficiency [6], deposition on optical surfaces [7], creation of wrong responses from measuring devices and false instrument reading [8], interference with the operation of Extra-Vehicular Activity (EVA) systems including suits, airlocks, suit ports tools, etc. [3].

The Lunar surface is covered by regolith i.e., a loose granular material, consisting of a broad range of shapes, sizes, and types of sediments [6]. It mainly consists of silicate minerals [9] crushed by meteoric bombardment causing the fracturing, scattering and agglutination of lunar regolith in the form of micrometric (approximately 50% of the lunar regolith consists of granular dust particles with diameters less than 60 μm [10]), sharp, abrasive, porous, chemically reactive dust particles. The particulate with a radius ≤ 25 μm comprises a significant portion and it is the most hazardous component of the lunar regolith [6]. Indeed, these granulometric particles are electrostatically charged by the interactions with the local plasma environment and the solar ultraviolet (UV) radiations, as well as the solar wind due to the photoelectric effect i.e., the UV and X-ray induced photoemission of electrons of the grain-materials [11-12]. Figure 1 displays a microscopic image of a lunar dust particle.

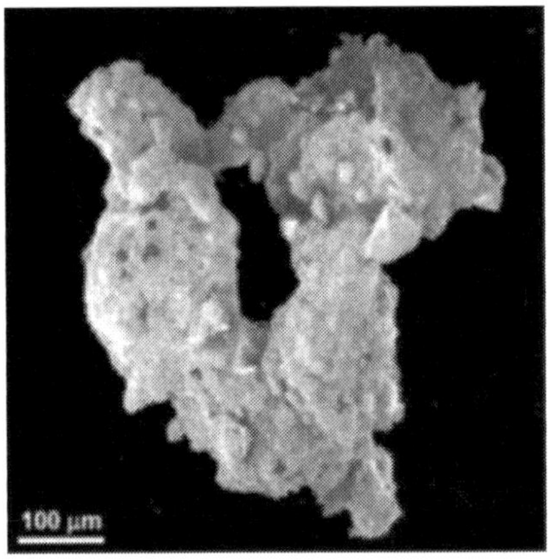

Figure 1. Microscopic image of a lunar dust particle [Credit: NASA].

Along with the jagged nature of the lunar dust, the moon's electrostatic environment is extremely complex and lends itself to dust adhesion. The day side of the lunar surface is charged slightly positively (10 V) due to the photoelectric effect, while the night side of the moon charges negatively on the order of kilovolts because of plasma electron interactions [13]. The cohesive properties of the lunar dust, due to the electrostatic charge on the dust particles, tend to make the lunar dust adhere to anything it contacts [14].

To efficiently overcome the serious issues represented by the challenging lunar dust environments, innovative technologies must be investigated and employed. They can be generally subdivided into passive and active approaches.

Passive dust mitigation technologies consist in preventing the sticking and adherence of the loose granular fraction of lunar regolith i.e., micrometric dust, without the application of any kind of external source of energy. Otherwise, passive dust mitigation technologies require the suitable design and development of special materials with the lowest possible surface energy. Indeed, this condition corresponds to the establishment of the minimum possible adhesion forces considering both electrostatic and chemical e.g., van der Waals interactions among the materials' outermost layers and the micrometric dust particles This ambitious goal can be achieved through a tailored modification of the chemical structure further improved by an appropriate alteration of the topographical/morphological features of the material's external layers e.g., the formation of micrometric conical pillars surface features associated to a decrease of the contact area between the external surfaces and solid, granular particulates. The last result can be obtained by applying several possible chemical and physical methods.

IOP Conf. Series: Materials Science and Engineering 1287 (2023) 012015 doi:10.1088/1757-899X/1287/1/012015

Instead, active dust mitigation technologies foresee the direct cleaning of the examined surfaces or their protection from dust deposition and contamination. Active technologies utilize external forces: electrodynamic, electrostatic, fluidal, mechanical, etc.

The active and passive methods can be considered to belong to the wider family of dust mitigation strategies comprising also the approaches aimed to avoid the generation of dust and the design criteria specifically conceived in the conceptual design phase in order to bypass or at least mitigate the problems caused by the extremely complex and hostile lunar dust environments [1, 15]. Figure 2 summarizes schematically this proposed taxonomy [15].

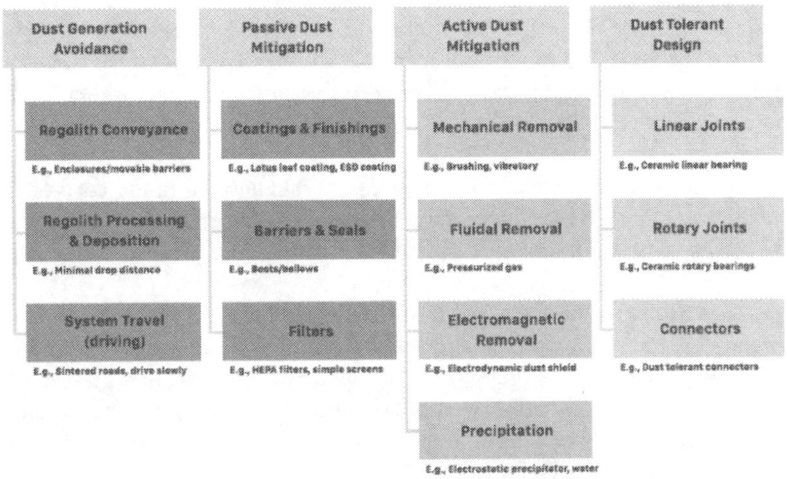

Figure 2. Taxonomy comparing the main dust mitigation technologies [15].

The present paper deals with the topic of passive dust mitigation technology and presents the design and experimental results related to the application of the surface chemistry approach. The proposed method consists essentially of the selection of suitable monomers and solvents and the incorporation of a small amount (< 5 wt. %) of special additives i.e., the Surface Migrating Agents (SMA). The SMA can spontaneously migrate toward the external layers of the material drawn by thermodynamic forces, activated by a more favourable balance between enthalpy and entropy of the intermolecular interactions along the interface with the surrounding air in comparison to the intermolecular forces in the bulk of the polymer matrix, as schematically illustrated in Figure 3.

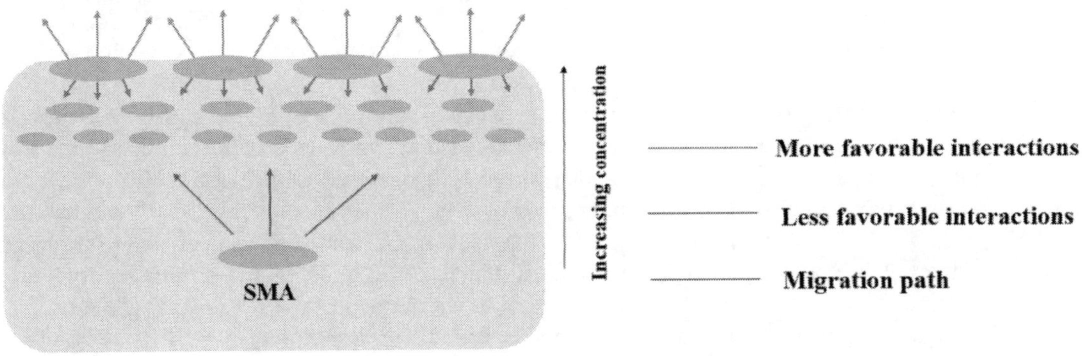

Figure 3. Scheme illustrating the generation of the SMA concentration gradient within the polymer matrix.

SMAs are light molecular weight oligomers, which can be easily loaded in the reacting mixture during the first step of polyimide synthesis i.e., formation of the Poly Amic Acid (PAA) when it is in an uncured or solvated liquid state.

The incorporation of SMAs offers several advantages compared with other surface methods and the more conventional approach of applying hydrophobic coatings. The main ones are:

- Reduced manufacturing complexity since the substrate generation and the surface modification occur in the same phase.
- Possibility to more easily retain the excellent thermo-mechanical properties of polyimides because a minimum loading of SMA is sufficient to improve the surface properties. Although the greatest concentration of SMA is along the external layers at the interface between the material and the surrounding environment, differently from coatings, SMAs are placed also within the first underlying layers according to the concentration gradient within the polymer thickness as sketched in Figure 3.

Therefore, the hydrophobic surface can be more easily regenerated in the case of undesirable wear and abrasion phenomena. Thus, the SMA method allows a longer endurance of the material's service time compared to the application of coatings and consequently better protection against ageing-induced damages.

This work was carried out in the framework of the LUMI project, which is pursuing by a research consortium between the Italian Aerospace Research Centre (CIRA) and the University of Naples Federico II under the European Space Agency (ESA), Open Space Innovation Platform (OSIP) financial support. OSIP is part of a wider ESA effort to support the future competitiveness of the European space industry with early technology development, implemented through the Space Technology Strategy and the technology program "Discovery & Preparation".

2. Reference Material

To start with lightweight materials endowed with an outstanding strength-to-weight ratio, the selection was oriented towards the class of high-performance polymers.

In this group of materials, the polyimides (PI) were chosen in agreement with the main international space agencies including NASA [16] as the most promising initial candidates for two main reasons:

1) They exhibit intrinsic very good mechanical, thermal, chemical and electrical insulating properties [16]. In detail, regarding the electrical properties, PIs possess an Ohmic resistivity in the order of 10^{17} Ω m [17] and an electrical breakdown strength higher than 109 V/m [18].
2) They benefit from a consolidated space heritage already used in several missions as flexible solar arrays and thermal blankets, i.e., the external layer in a Multilayer Insulator (MLI) structure of Geostationary Earth Orbit satellites protecting them against solar radiation [19].

PIs are advanced, high-performance and heat-resistance engineering polymers, originally developed by Bogert and Renshaw, who successfully completed the first synthesis of a polyimide sample from 4-aminophtalic acid in 1908 [20]. However, due to the lack of processability via melt polymerization, significant advances in PIs synthesis and processes were achieved only in 1955 by Edwards and Robinson at DuPont™ Company with the development of the first high-molecular weight aromatic polyimide i.e., Poly(4,4'-oxydiphenylene pyromellitimide – PMDA-ODA PI), commercially known as Kapton®, which can remain stable in a wide range of temperatures from 4 K to 673 K [21-22].

Additionally, PIs were also evaluated by the Jet Propulsion Laboratory of NASA as good plastic support for solar sails, due to their long duration in the harsh space environment [21].

Regarding PIs properties, their dimensional stability is mainly attributed to the strength of the chemical bonds, which join different groups in the polymer backbone, and the presence of highly stiff and rigid heterocyclic systems into the polymer chains i.e., the heterocyclic, inert imide ring exhibiting high interchain interactions and elevated cohesion forces among the polymer backbones. Moreover, their thermal stability is further significantly improved by incorporating aromatic i.e., -C_5H_6 rings on the backbone and/or side groups, as shown in Figure 4.

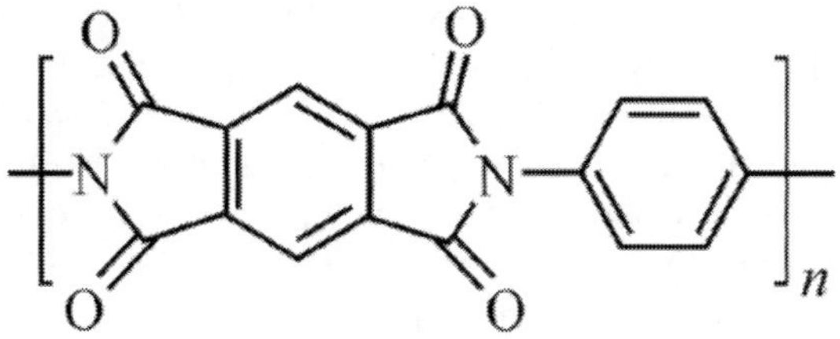

Figure 4. Representative structure of aromatic polyimide.

Aromatic PIs gain such incredible mechanical and thermal properties that they are used in place of metals and glass in many high-performance applications, not only in electronics but also in the automotive and even the aerospace sector.

In addition to such high thermal stability, the nature of the chemical structure, consisting of rigid imide and aromatic rings, always provides excellent mechanical and dielectric properties as well as elevated chemical resistance. It is worth highlighting that a variety of specific functionalities can be easily added to the backbone and/or side moieties of PIs to meet specific demands in applications [23-24].

Moreover, the properties of PIs can be dramatically altered by minor modifications in the chemical structure because of the specific selection of the starting monomers i.e., diamines and dianhydrides. Indeed, also subtle variations in the structure of the monomers produce a tremendous effect on the overall properties of the final PI and several features are of paramount importance towards understanding the PI structure-property relationships e.g., the effects of changing the dianhydrides and diamines or the overall flexibility of the polymer chains on the basic parameters i.e., glass transition and the melting temperatures.

Therefore, the exceptional versatility of PIs is to be used for designing and tailoring the final properties of this class of high-performance polymers to meet the diverse specific needs in different fields of activity with a suitable selection. Furthermore, an accurate experimental combination of the starting monomers is another very appealing merit of PIs, which is unparalleled by most other classes of macromolecules [25].

3. Selection Criteria

PIs are usually polymerized through the two-step process of mixing the starting monomers in a suitable aprotic polar solvent e.g., dimethyl acetamide, n-methyl pyrrolidone, tetrahydrofuran, etc. yielding the production of polyamic acid (PAA) intermediate precursor, which is a polyamide.

In this work, the starting monomers for PI synthesis were selected according to three fundamental criteria, detailed below:

a) Chemical compatibility.
b) Chain length of diamines.
c) Dipole-dipole repulsion moieties.

a) The chemical compatibility between the monomers i.e., dianhydrides and diamines, is necessary to shift the thermodynamic equilibrium of the first step toward the formation of PAA and yield a final PI with a very elevated degree of polymerization. This leads to the achievement of high molecular weight, which is an essential feature for the complete exploitation of the excellent thermo-mechanical properties of this class of advanced heat-resistant polymers. Elevated chemical compatibility can be ensured by selecting dianhydrides owing to very high electron affinity i.e., electrophilicity of the

IOP Conf. Series: Materials Science and Engineering 1287 (2023) 012015 doi:10.1088/1757-899X/1287/1/012015

carbonyl groups and diamines showing elevated ionization potential and basicity i.e., the nucleophilicity of the amine groups.

b) Related to the chain length of diamines, another characteristic, associated with the attainment of outstanding thermo-mechanical properties, is the correlation between the number of benzene rings i.e., the chain length of the aromatic diamines and the position of the amino groups along the $-C_6H_5$ moieties and the glass transition temperature T_g of the final PI. Indeed, the introduction of too many benzene rings in the diamine monomer involves the incorporation of additional flexible linkages, requiring less thermal energy to activate the vibrational motion, decreasing the T_g.

c) Probably the most important chemical feature, which should be satisfied to produce materials with minimum surface energy is the presence of pendant trifluoromethyl $-CF_3$ groups within the molecular backbone of both monomers. This is ascribed to the most elevated electronegativity of the fluorine atom and the associated highest polarity of the C-F covalent bond in organic chemistry leading to the formation of electric dipoles able to exert the most elevated possible chemical and electrostatic repulsion forces against micrometric dust particle i.e., *abhesion* capacity.

In the intermediate phase, the incorporation of SMAs within the reacting mixture can be efficiently accomplished since the system is in a liquid state. However, for the successful migration of the SMAs toward the outermost layers, the rheological properties i.e., density and the inherent viscosity of the PAA polyamide intermediate solution play an important role.

4. Experimental Methods

Attenuated Total Reflectance Infrared (ATR-IR) spectra were recorded with a Thermo Nicolet 5700 spectrometer using Zinc Selenide (ZnSe) as an ATR crystal. ATR uses a property of total internal reflection resulting in an evanescent wave.

Wide-angle X-ray scattering (WAXS) profiles were obtained with Ni-filtered Cu K-α radiation with a PANalytical Empyrean powder diffractometer performing continuous scans of the diffraction angle 2θ, in the range of 5-40°.

Calorimetric measurements were performed with a Differential Scanning Calorimeter (DSC) METTLER TOLEDO 822, in a flowing N_2 atmosphere, at a heating rate of 10 °C min^{-1}.

In all experiments, a small amount of the sample was heated from room temperature up to 400 °C, then cooled to room temperature and finally heated again up to 400 °C. The second heating scan was used to determine the glass transition temperature T_g.

5. Results and Discussion

Several PAA samples were designed and synthesized as shown in Figure 5, belonging to two alternative classes: samples realized using more conventional monomers and intermediate precursors obtained by mixing innovative diamines and dianhydrides including pendant perfluoro groups.

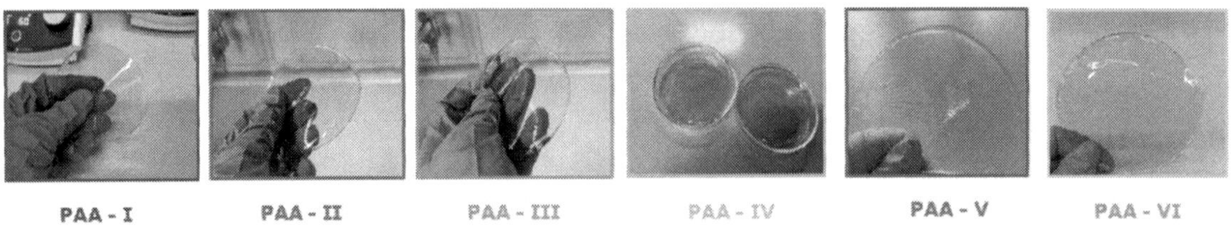

Figure 5. Photographs of six films of polyamic acid obtained with different combinations of monomers.

PAA–IV is a representative sample of the first class and it was polymerized through a polycondensation reaction between 4,4'--oxydiphthalic anhydride (ODPA) and 4,4' -diaminodiphenyloxydianiline (4,4' ODA) in N-methyl pyrrolidone at room temperature in a flowing nitrogen atmosphere as schematically shown in Figure 6.

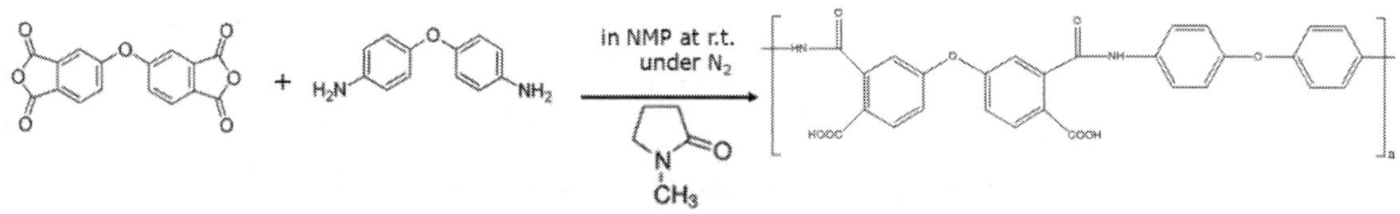

Figure 6. The synthesis procedure of PAA–IV.

Then the reacting mixture was stirred overnight and dried at 60 °C for 24 hours leading to the production of the intermediate polyimide precursor in the form of polyamic acid.

Instead, PAA–V is a demonstrative coupon of the second class, in which both the monomers i.e., the dianhydride and the diamine possess trifluoromethyl pendant side groups endowed with the highest possible dipole-dipole repulsion forces due to the most elevated electron-withdrawing effect in the C-F covalent bond determined by the highest electronegativity of the fluorine atom. The presence of these lateral moieties enhances the chemical and electrostatic repulsion forces exerted by the complete imidized material against micrometric granular particles i.e., passive dust mitigation. Indeed, the trifluoromethyl groups as well as other fluorocarbon moieties, incorporated within the PIs backbone along the main or the lateral chains, decrease the material's outermost layers' surface energy and consequently enhance its hydrophobicity [26]. Moreover, Wohl and coworkers of the NASA Langly Research Center theoretically and experimentally demonstrated the existence of a significant correlation between the hydrophobicity of a polymer and the actual particle adhesion for simulants of the lunar regolith dust [27].

The successful accomplishment of the first step of PI synthesis i.e., the formation of the polyamic acid for both samples was verified through ATR spectroscopic measurements reported in Figure 7 and Figure 8.

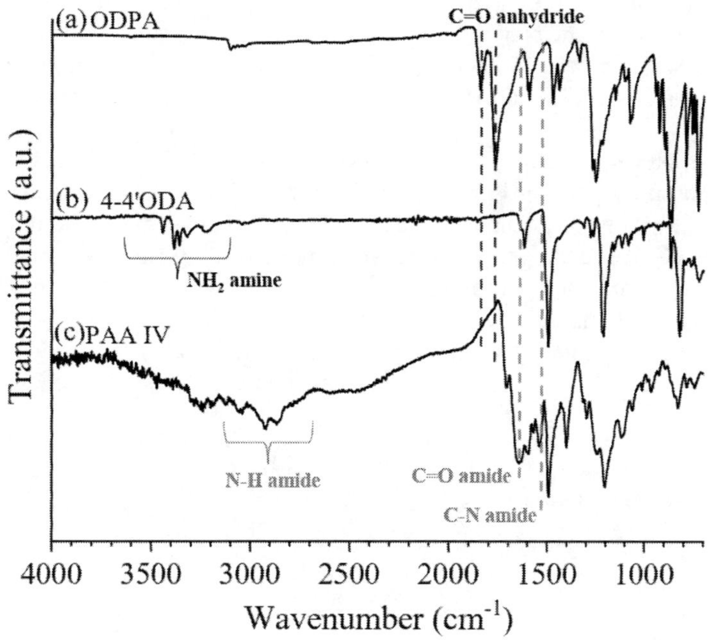

Figure 7. The ATR spectra of dianhydride (a), diamine (b) and the corresponding polyamic acid PAA–IV.

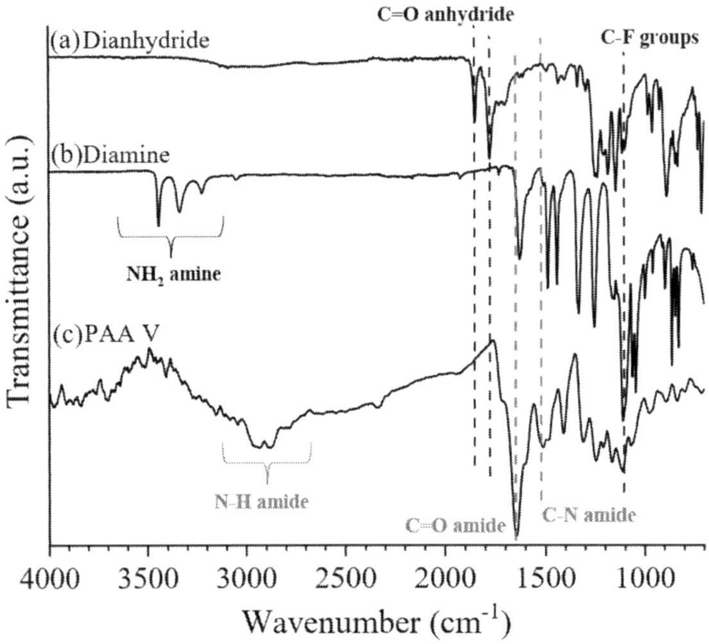

Figure 8. The ATR spectra of the dianhydride (a), the diamine (b) and the corresponding PAA–V.

It can be clearly observed that the asymmetric stretching vibration of carbonyl groups in the dianhydrides observed at wavenumber 1860 cm^{-1} (Figure 7a – Figure 8a), but also the characteristic absorptions of amino groups NH_2 of the diamines observed at 3300 cm^{-1} (Figure 7b and Figure 8b), disappeared in the PAA–IV and PAA–V spectra, respectively. The absence of characteristic absorbance peaks related to the monomers' functional groups implies the complete conversion of the dianhydrides/diamines during the reaction step. In fact, the characteristic amide bands at 1545 cm^{-1} (C-N stretching) and at 1720 cm^{-1} and 1660 cm^{-1} (C=O asymmetric and symmetric stretching) were present in both PAA's spectra (Figure 7c and Figure 8c), suggesting the successful formation of the poly(amic acids).

Moreover, the structures of the coupons were analysed using WAXS techniques, as shown in Figure 9, revealing an amorphous pattern for both samples with a broad halo centred around 18-20°. The amorphous nature of the polyamic acids could be attributed to their asymmetric structural units and aryl ether linkage, which decreased the intra- and inter-polymer chain interactions resulting in loose polymer chain packing and aggregates. It is worth noting that the PAA–V coupon exhibits a broad main halo at a lower value of the scattering angle and an additional weak halo near 25° probably due to the introduction of strong electronegative fluorine atoms exerting elevated repulsive forces between molecules leading to a significant increase of the distance among the polymer chains [28].

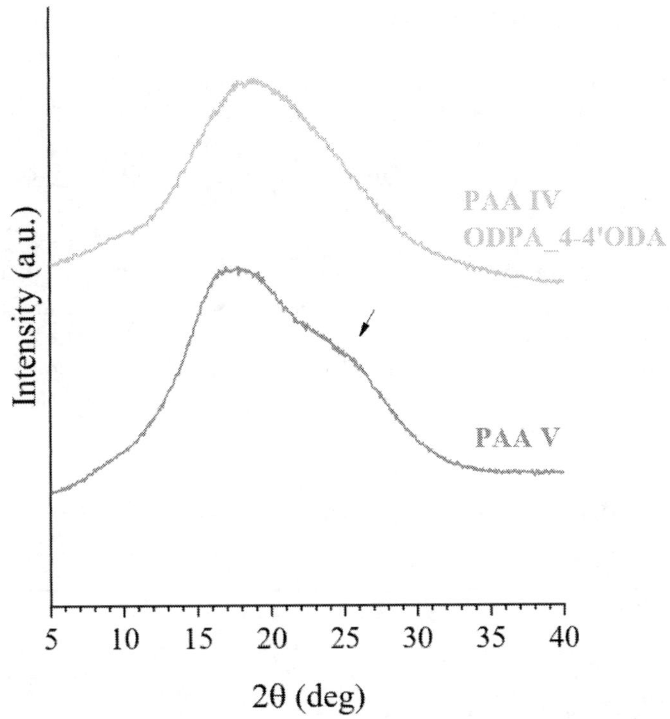

Figure 9. WAXS profiles of PAA–IV and PAA–V samples, representative of the two investigated classes.

Furthermore, both the classes of polyamic acid solutions were characterized in terms of thermal properties i.e., imidization temperature (T_i), glass transition temperature (T_g) and rheological properties i.e., density and inherent viscosity. This was carried out for assessing the ability of both the intermediate polymer matrices to incorporate surface migrating agents and to promote their motion toward the outermost layers without hindering this process due to the excessive agglomeration of the reacting mixture during the solvation and curing stages. The DSC scans are reported in Figure 10.

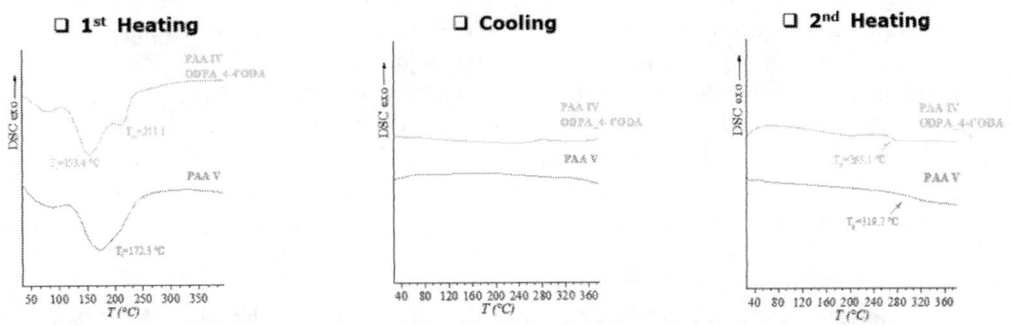

Figure 10. DSC profiles measured for both the investigated samples during the first (on the left) and the second (on the right) heating and during the intermediate cooling (in the centre).

The first heating scans show broad endothermic peaks associated with the imidization reactions with the removal of the water as a byproduct and the residual solvent. The cooling scans confirm the absence of crystallization peaks, while the second heating scans reveal the glass transition temperatures T_g for both samples.

PAA–V coupon exhibits the highest T_g equal to ~319 °C. This can be justified because the inclusion of perfluoro $-CF_3$ pendant groups along the polyamic acid backbone decreases the intra-segmental and inter-molecular mobility leading to polyetherimide chains with more elevated conformational rigidity.

The density and inherent viscosity of either the intermediate polymer coupons were summarized in Table 1 and the comparison between the samples representative of the two classes indicates that the incorporation of trifluoromethyl side groups along the monomer's backbone provides two fundamental advantages:

i. Significant decrease of the polyamic acid solution viscosity from 0.60 dL g^{-1} for the PAA–IV achieved through polycondensation of conventional monomers to 0.25 dL g^{-1} for the sample containing $-CF_3$ groups.
ii. Remarkable growth of the glass transition temperature from ~265 °C to ~320 °C corresponding to a parallel improvement of the thermo-mechanical properties of the final material completely imidized.

Table 1. Rheological and thermal properties of both the investigated samples.

Samples	η_{inh} [dL g^{-1}]	d [g cm^{-3}]	T_i [°C]	T_g [°C]
PAA–IV	0.60	1.29	153.4	265.1
PAA–V	0.25	1.42	172.3	319.6

6. Conclusions

In the present paper, passive dust mitigation technologies for the lunar environment were analysed and the chemical strategy, based on the use of innovative high-performance polymers combining at the same time minimum surface energy and superior thermo-mechanical resistance, was discussed.

A chemical approach based on the selection of suitable monomers and incorporation of special additives, such as surface migrating agents, was described and motivated.

Preliminary experimental results, regarding the synthesis and both the rheological and thermal characterization of polyimide intermediate precursors of polyamic acid solutions, were presented and commented on.

They show promising achievements justifying the effort pursued on this chemical strategy.

Furthermore, they confirm the inclusion of trifluoromethyl pendant side groups along the main polymer chains of the starting monomers as a valuable method to provide crucial advantages compared to the more conventional precursors, synthesized with monomers lacking moieties with elevated electronegativity. These useful benefits consist of more favourable rheological properties of the polymer matrix in the intermediate solvation and curing phase at the beginning of the imidization process, and in higher glass transition temperature i.e., improved thermo-mechanical characteristics of the final completely polymerized material, which is a suitable candidate for applications in harsh environment as the Moon.

It is important to highlight that all the substances used in the LUMI project are compliant with the REACH regulations. Indeed, the only chemical species, included in the REACH-Restricted Substance List, i.e., N-methyl-pyrrolidone was used in a concentration lower than 0.3% w/w according to the upgraded regulation (9 May 2020). Moreover, other solvents, to be used in replacement of NMP, are under investigation.

7. Acknowledgements

The authors thank the ESA/OSIP funding contract No. 4000137002/21/NL/GLC/ov.

References

[1] Afshar-Mohajer N Wu C-Y, Curtis J, Gaier J R 2015 *Advances in Space Research* vol 56 p 1222
[2] Calle C I Buhler C R McFall J L Snyder S J 2009 *Journal of Electrostatics* vol 67 p 89
[3] Gaier J R 2005 *NASA/TM 213610*
[4] Goodwin R 2002 *Apollo 17: The NASA Mission Reports* (Apogee Books, Ontario, Canada)
[5] Gaier J R Waters D L Misconin R M Banks B A Crowder M 2011 *Proc. 41st Int. Conf. on Environmental Systems Portland, Oregon*
[6] Wohl C J Atkins B M Belcher M A Connell J W 2010 *Proc. International SAMPE Symposium and Exhibition*
[7] Murphy Jr T Adelberger E Battat J Hoyle C McMillan R Michelson E Samad R Stubbs C Swanson H 2010 *Icarus* vol 208 p 31
[8] Wagner S A 2006 *NASA, TP-2006-213726*
[9] Fateri M Gebhardt A Gabrielli R A Herdrich G Fasoulas S Groβmann A Schnauffer P Middendorf P 2015 Additive Manufacturing of Lunar Regolith for Extra-terrestrial Industry Plant *30th ISTS Conference* Kobe, Japan
[10] Taylor L A Schmitt H H Carrier W D Nakagawa M 2005 *1st Space Exploration Conference: Continuing the Voyage of Discovery* Orlando, Florida (American Institute of Aeronautics and Astronautics)
[11] Abbas M Tankosic D Craven P D Spann J F LeClair A West E A 2007 *Planetary and Space Science* vol 55 p 953
[12] Stubbs T J Vondrak R R Farrell V M 2006 *Advanced in Space Research* vol 37 p 59
[13] Calle C I 2011 *Journal of Physics: Conference Series* vol 301 p 012006
[14] McDivitt J A et al. 1970 *Apollo 12 mission report MSC-01855*
[15] Cannon K M Dreyer C B Sowers G F Schmit J Nguyen T Sanny K Schertz J 2022 *Acta Astronautica* vol 196 p 259
[16] Mathews A S Kim I Ha C-S 2007 *Macromolecular Research* vol 15 p 114
[17] Griseri V 2020 Polyimide Used in Space Applications *IntechOpen publication*
[18] Liaw D-J Wang K-L Huang Y-C Lee K-R Lai J-Y Ha C-S 2012 *Progress in Polymer Science* vol 37 p 907
[19] Yokota R J 1999 *Photopolym. Sci. Technol.* vol 12 p 209
[20] Bessonov M I Koton M M Kudryavtsev V V Laius L A 1987 *Polyimides Thermally Stable Polymers* (Consultants Bureau, New York)
[21] Ha C S Mathews A S 2011 Polyimides and high-performance organic polymers. *Advanced Functional Materials* (Springer, Berlin, Heidelberg) pp 1-36
[22] Edwards W M Robinson I M 1959 *U.S. Patent 2867609*
[23] Kim Y Goh W H Chang T Ha C S Ree M 2004 *Adv. Eng. Mater* vol 6 p 39
[24] Ha C S Park H D Frank C W 2000 *Chem. Mater* vol 12 p 839
[25] Gouzman I Grossman E Verker R Atar N Bolker A Eliaz N 2019 *Advanced Materials* vol 31
[26] Wu X Csi J Cheng Y 2022 *Applied Polymer Science* vol 139 p 51972
[27] Wohl C J Connell J W Lin Y Belcher M A Palmieri F L 2011 *3rd AIAA Atmospheric Space Environments Conference* Honolulu, Hawaii
[28] Liu H Zhaia L Bai L He M Wang C Mo S Fan L 2019 *Polymer* vol 163 p 106

IOP Conf. Series: Materials Science and Engineering 1287 (2023) 012016 doi:10.1088/1757-899X/1287/1/012016

Smart heaters based on flexible electronics for future spacecrafts

B Pozo [1], L Machón [1], D Cantero[1] and I Muñoz [1]

[1] Tekniker, Iñaki Goneaga Road, 5, 20600, Eibar, SP

E-mail: borja.pozo@tekniker.es

Abstract. Thermal control applications deployed in space environments require innovative, lightweight, and flexible systems. This smart heater is designed to adapt the implementation of electronics, heaters, and sensors by using Kapton material as a replacement for the typical rigid substrate (FR4) used in past planetary and scientific missions. To realise the benefits of these new flexible technologies using state-of-the-art materials (Kapton, Copper, and Inconel) for space thermal electronic applications, the smart heater has been designed for vacuum operation from -55 ºC to 100 ºC and non-operational from -150 ºC to 150 ºC. In addition, the systems must be able to reach the set point temperature consecutively with an error margin of ±1 ºC and read the temperature value with an accuracy of 0.5 ºC for the operating temperature range.

1. Introduction

Thermal control is one of a spacecraft's most critical systems since it ensures the proper operation of the mechanical and electrical components. The environmental issue that could adversely impact spacecraft surface missions includes wide ranges of temperature varying from -225 ºC to +464 ºC [1] depending on external environmental heating factors, such as direct sunlight, sunlight reflected off a planet or moon, infrared energy emitted from Earth and free molecular heating effect experienced during low orbit flights [2].

Due to the operational requirements of every spacecraft, an accurate and stable control despite environmental changes needs to be guaranteed. ESA's future large-scale science missions for 2035-2050 [3], will focus on moons of the giant planets of the Solar System, temperate exoplanets or the galactic ecosystem, and new physical probes of the early Universe, continuing the scientific and technological exploration of space with increasingly ambitious and powerful space missions. Latest heating system requirements and technological road-map developments are focused on the design of smart systems, reducing the weight and the costs, and increasing the integration and adaptability on spacecraft.

As a result of using flexible materials as a substrate in several printed circuit boards for control electronics applications, e.g., temperature and relative humidity sensors [4], the required specifications and a definition of a "smart" heater have been defined in this research for the electronic systems to be taken into consideration in future applications.

For this reason, the main objective of this work has been to define and develop a smart heater concept [5], focusing on configurable of temperature set-points, re-distribution capability of heating power, self-control capability, robustness and redundancy, simplicity, communications with

IOP Conf. Series: Materials Science and Engineering 1287 (2023) 012016 doi:10.1088/1757-899X/1287/1/012016

Engine Control Unit (ECU) of the spacecraft, easy mechanical integration, reduction of mass and weight, minimisation of cost, and compatibility with the present and future space developments.

The requirements have been searched, analysed and set the possible limitations to the development of the smart heater concept about the space environment and the requirements of the work. On the one hand, some limitations could be the EEE components availability [6], the compatibility with vacuum, the radiation tolerance of all the components and materials [7], the operational temperature, the cost limitations, the size and weight, and the implementation in a single flexible substrate [8].

On the other hand, some critical requirements for the system are communications, such as the use of CAN Bus [9] with the main spacecraft control; configurability, to obtain a self-controlled system; power redistribution capability, within the heating system; mass and weight reduction, focusing on higher payloads in the rockets; robustness and redundancy; easy mechanical integration, through different sizes and shapes; and cost minimisation.

2. Block diagram of the system

The general architecture based of conventional thermal systems [[10], [11]] has been analysed as the first step in the design process. Furthermore, a solution based on a smart thermal control circuit composed of control, power, heater, sensing, and communications (Fig. 1), and on Kapton, which was used as a flexible substrate for the PCB, has been proposed. In order to manufacture a flexible electronic circuit, several requirements described in ECSS-Q-ST-70-12C [8] have been considered, including the necessity of two layers of flexible circuit board, the packaging of the integrated circuits, the thickness of the board, the precise manufacturing process, and the use of connection fingers type as connectors.

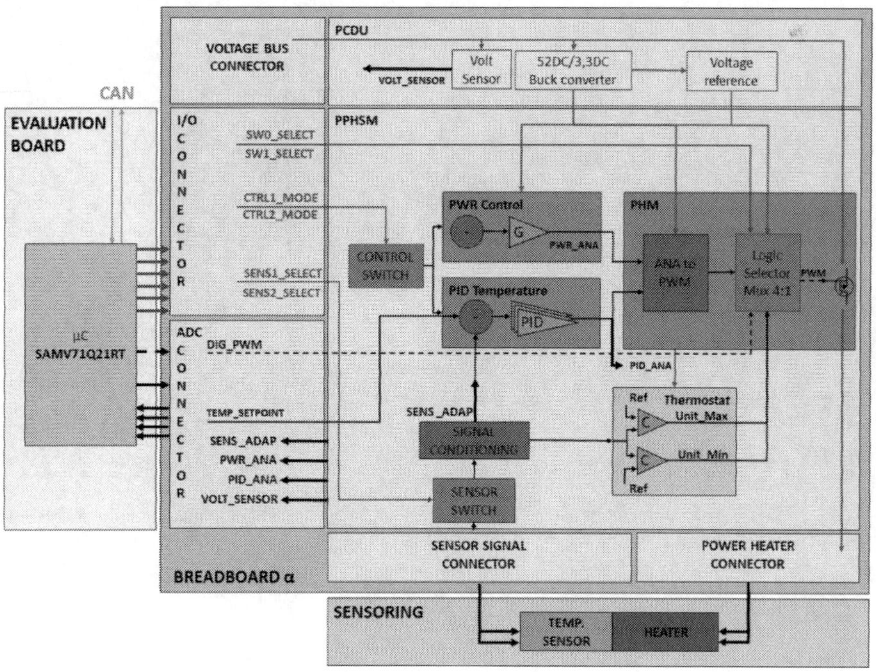

Figure 1. Smart heater block diagram.

In the initial design phase, the overall architecture based on previously used thermal systems has been analysed. A solution based on an intelligent thermal control circuit - control, power, heater, sensors and communications - has accordingly been proposed. A two-step concept, α Breadboard and β Breadboard, has been carried out. The first has been allowed the verification of the principles on a rigid breadboard before incorporation of the modules into the design of the flexible β Breadboard.

3. α Breadboard

3.1. Thermal simulations and design of heater

Among the elements necessary to constitute an electric heater, the selection of the conductive element is one of the most important and critical phases. This element must have three main properties: high-strength, non-magnetic materials, and high temperature coefficient to allow the self-regulation. For the α Breadboard design, thermal simulations have been carried out to analyse the required power density, dimensions, designs, and possible conductive materials with the objective of obtaining $1W/cm^2$ in a $100cm^2$ board for the power heat characteristics with different thermally conductive materials, e.g., Copper, Silver, and Inconel.

Electrical heaters are based on Ohm's and Joule's laws. This has been resulted in two heater designs printed on Kapton material, as shown in Fig. 2, with 0.15 mm of thickness. On the left is a Copper circuit 0.5 mm wide and 9 m long with a total of 90 lines while on the right is an Inconel circuit 2.5 mm wide and 1.8 m long with 18 lines. After the manufacturing process, it has been verified that the heater complies with the theoretical resistance and that there are no short circuits between its lines. With a measuring instrument, the resistance of the Copper heater has been measured to be 11.8 Ω compared to the theoretical resistance of 12.9 Ω. At this first phase, the Cooper heater has been connected directly to the PCB connector between the power PADs.

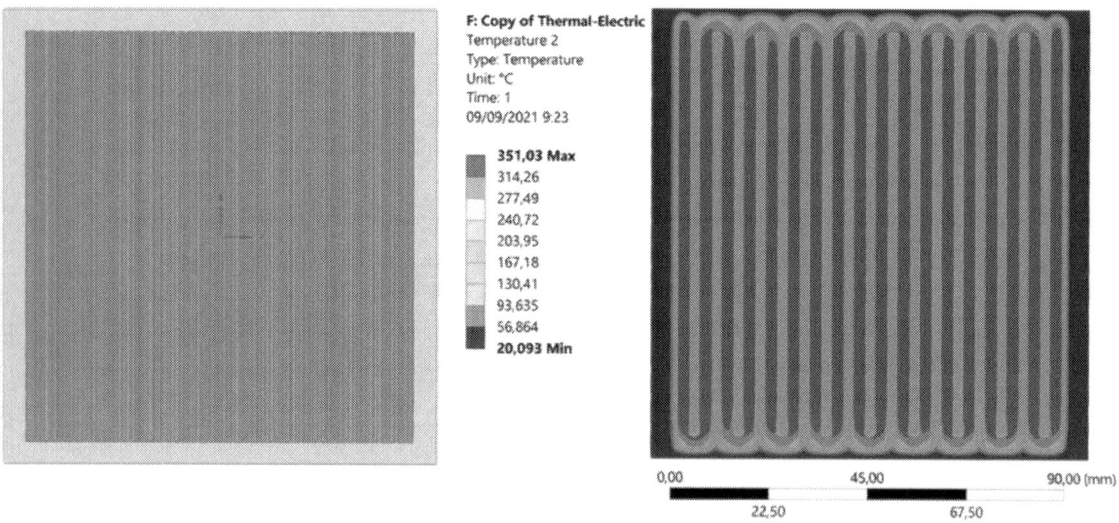

Figure 2. Heater design simulations based on Copper/Silver (left) and Inconel simulations (right).

3.2. Flexible electronics

The main control operation of α Breadboard has been based on operational amplifiers, thermostats, analogue PIDs, and digital PIDs. With most promising conductive materials (Copper and Inconel), thermal simulations of the flexible heater have also been carried out at the required bus power (52V), as it is described in section 3.1.

The choice of heater materials with good capabilities against oxidation, vacuum conditions, high temperatures, thermal resistance, etc. has been one of the main focuses of this work. Copper has often been utilized extensively in rockets to transform electrical energy into thermal energy [12]. As a result, our work proposes using Inconel, a material with excellent heat transfer performance. For demanding space applications, the nickel-chromium alloy Inconel has been used in a variety of printing processes [13] as well as fabrication methods [14]. It has been suggested in this work as a result.

Furthermore, smart heater system is based in two blocks with a hybrid electronic system analog-digital: temperature control and system communications. The thermal control system is based on operational amplifiers, thermostat, analog and digital PID, implemented in SAMV71Q21RT microcontroller based on the high-performance 32-bit ARM Cortex-M7 processor. The power management system is set to work with standard spacecraft bus power of 52 V and convert to lower typical voltages, such as 5V or 3.3V, required for all the control elements. In addition, thermal simulations have been carried out to analyse the required power density, dimensions, designs, and conductive materials (Copper and Inconel) with the objective to obtain $1W/cm^2$ in a $100cm^2$ breadboard. Furthermore, the system communication is based on standard CAN ECSS-E-ST-50-15C [8] protocol with the objective to maintain management capability from the spacecraft main control system (ECU) to each smart heater system.

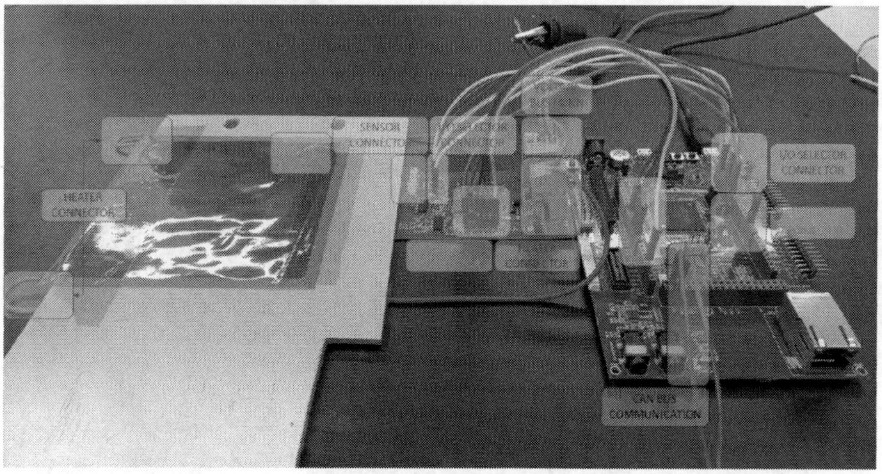

Figure 3. Setup for heating and cooling first cycles.

3.3. Climatic chamber tests

The early prototype tests shall be performed in a climatic chamber first and in a thermal vacuum chamber later, fulfilling the defined thermal operating range. The first tests will consider the requirements about temperature uniformity, measurement accuracy, control hysteresis, control accuracy, output accuracy and voltage variability order to fulfil the test campaign with the pre-established criteria. See Figure 4. The total preliminary set of tests performed were baking

IOP Conf. Series: Materials Science and Engineering 1287 (2023) 012016 doi:10.1088/1757-899X/1287/1/012016

tests between -20°C and 100°C non-operational temperature range.

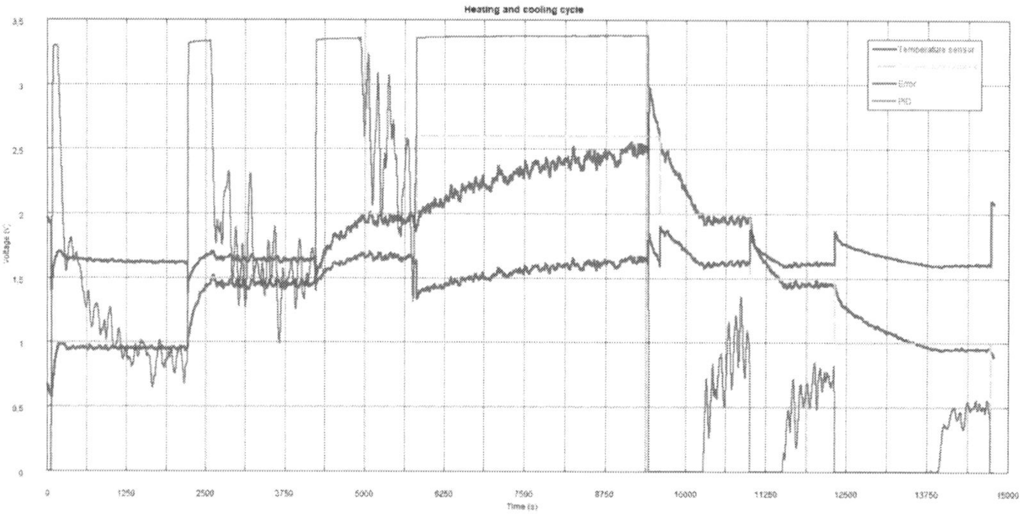

Figure 4. PID control from 25 ºC to 100ºC. Heating and cooling cycles.

Additionally, during the climatic chamber tests, screenshots have been taken from the results obtained by the thermographic camera at the different desired setpoint temperatures. Thus, Figure 5 shows the properly temperature variation on the surface of the Copper heater with a range of 25 °C to 100 °C colour scale. The system is capable of heating from 25 ºC to 100 ºC and, subsequently, cooling to ambient temperature of 20 ±2 ºC within a climatic chamber with a temperature reading error of ±0.5 ºC.

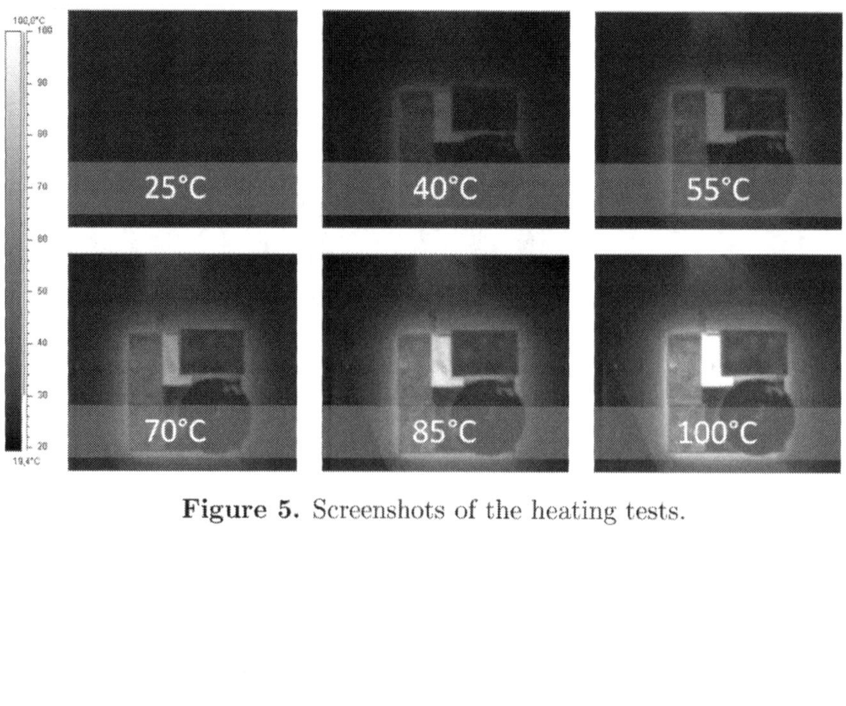

Figure 5. Screenshots of the heating tests.

3.4. Vacuum chamber tests

As shown in Figure 6, by changing the system environment to a vacuum environment, with a 1.9 E^{-5} mbar in the chamber, a variation in the PID control and temperature sensor parameters has been observed for different desired temperature values from 25 °C to 80 °C. Vacuum tests have been performed on the heating system with satisfactory results, but with an increase in the temperature reading error of ±1.5 °C. The configuration of the isolated heater, without air gaps, for heat transfer tests in thermal vacuum environments has been shown in Figure 7.

According to the results of these tests, the system is suitable for advanced electronics for upcoming space missions and is capable of heating in the flexible Copper over Kapton design.

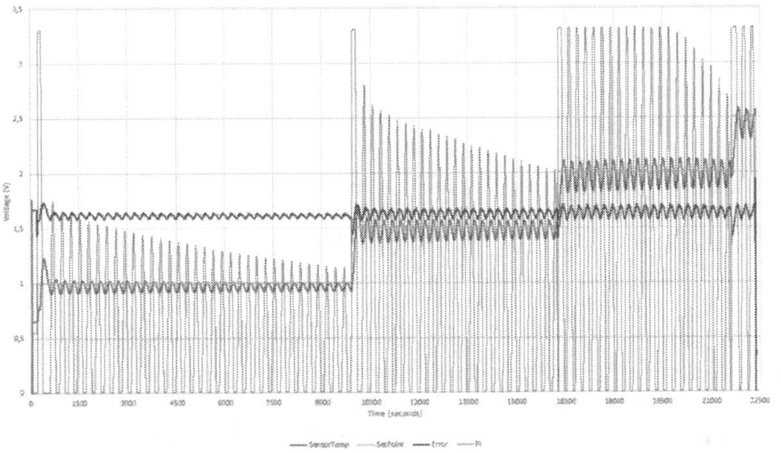

Figure 6. Heating vacuum cycle from 25 ºC to 85ºC.

Figure 7. Vacuum chamber heater setup.

4. β Breadboard

After the α Breadboard, the definition and establishment of the roadmap for the development of the future smart heater, including its qualification, improvements and limitations has been analysed. The design of the complete system including the characteristics of the space systems, including EEE qualification [6], operational improvements and limitations of the parameters have been implemented in the β Breadboard, integrating the digital part in the analogue and power electronic design.

Therefore, the objective of this final system has been to develop a design in which the heater, temperature sensor, microcontroller and the full analogue and digital system are included in a single substrate of Kapton material as it is shown in Figure 8. For the β Breadboard, it has been necessary to examine certain tolerances and distances between tracks, as well as the conductive materials to be used in the manufacturing process of the Kapton substrate material for the printing of the electronics, the type of manufacturing for outgassing problems in space environments, solder pastes for the electronic components to withstand space radiation and the manufacturing process, etc.

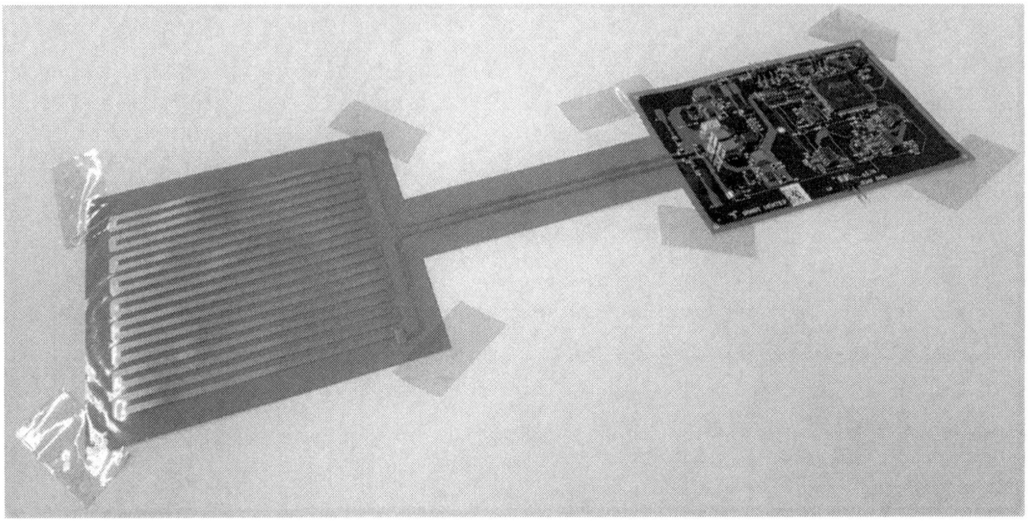

Figure 8. Flexible Kapton printed circuit board.

5. Conclusions

The main purpose of this work based on the definition, development, and implementation of a smart heater concept for future spacecrafts has been verified and validated with two preliminary breadboards (α and β) tested in a controlled environment, demonstrating the viability as an alternative option for future space exploration missions based on the operative, redundancy, safety, and in the reduction of the weight and costs.

Considering the requirements established, the definition and the main concept of the smart heater, the theoretical hardware design, the microcontroller selection, the manufacture of the printed circuit board, and the heater have been accomplished. A general setup has been presented for a real and comprehensive implementation. Moreover, integrating the communication protocol, three types of possible controls, the standard power supply voltage, the sensing acquisition circuit, and the heater definition. Moreover, a shortlist of possible risks about the initial concept design have been accomplished focusing on reconfigurability of wide range temperature set-points, redistribution capability of heating power, self-control capability, system robustness, redundancy, easy mechanical integration, reduction of mass, minimisation of cost, and compatibility to present and future space developments.

The thermal control system has been verified and validated in different operation conditions (thermal and vacuum) and different wide range temperatures established for space environments. The validation process consisted of checking that the system complies with the established requirements through a series of tests with different heating situations from -20 °C to 100 °C. The operating results of the thermal system have been influenced due to the testing environment where the system is located. The ambient temperature has been varied under controlled conditions and has been eliminated the possible air gaps which have caused differences in the temperature measured by the temperature sensor and in the oscillation of the PI controller.

To ensure that the β Breadboard is fully accurate, the system must now perform some tests (such as bake-in tests, thermal shock tests, thermal vacuum cycle tests, etc.), as well as start up and function properly in terms of both operation and communications.

References

[1] Jones B, Discovering the Solar System (2nd ed.). Wiley. ISBN: 9780470510797. 2007. Retrieved from: https://www.perlego.com/book/2757878/discovering-the-solar-system-pdf
[2] Gilmore D, Spacecraft Thermal Control Handbook Volume I: Fundamental Technologies, American Institute of Aeronautics and Astronautics, Inc. ISBN 978-1-884989-11-7. December 2002.
[3] Tacconi L J, Arrdge C S, et al, Voyage 2050 senior commitee. Voyage 2050. Final recommendations from the Voyage 2050 Senior Commitee, May 2021. Retrieved from: https://www.cosmos.esa.int/documents/1866264/1866292/Voyage2050-Senior-Committee-report-public.pdf
[4] Barmpakos D and Kaltsas G, A review on humidity, temperature and strain printed sensors—Current trends and future perspectives. Sensors 2021,21, 739. DOI: DOI: 10.3390/s21030739
[5] Bodie M, Russell G, Mccarthy K, Lucas E, Zumberge J and Wolff M, Thermal analysis of an integrated aircraft model. Proc. 8th AIAA Aerosp. Sci. Meeting Including New Horizons Forum Aerosp. Expo., 2010, p. 288.
[6] European Cooperation for Space Standarization, ECSS-Q-ST-30-11C Rev.2: Derating – EEE components. ESA-ESTEC Requirements and Standards Division. June 2021. Retrieved from: https://ecss.nl/standard/ecss-q-st-30-11c-rev-2-derating-eee-components-23-june-2021/
[7] European Cooperation for Space Standarization, ECSS-Q-ST-70-06C – Particle and UV radiation testing for space materials. ESA-ESTEC Requirements and Standards Division. July 2008. Retrieved from: https://ecss.nl/standard/ecss-q-st-70-06c-particle-and-uv-radiation-testing-for-space-materials/
[8] European Cooperation for Space Standarization, ECSS-Q-ST-70-12C - Design rules for printed circuit boards. July 2014, p. 173. Retrieved from: https://ecss.nl/standard/ecss-q-st-70-12c-design-rules-for-printed-circuit-boards-14-july-2014/
[9] European Cooperation for Space Standarization, ECSS-E-ST-50-15C, Space engineering: CANbus extension protocol. ESA-ESTEC Requirements and Standards Division. May 2015. Retrieved from: https://ecss.nl/standard/ecss-e-st-50-15c-space-engineering-canbus-extension-protocol-1-may-2015/
[10] Carvey A, Ferguson P, Thibault J F, Aero E, Haddad E SSC19-WKII-06 An Experimental Study of a Smart Radiator Device for Enhanced Passive Thermal Control of Small Spacecraft. February. 2021.
[11] Osiander R, Firebaugh S L, Champion J L, Farrar D and Darrin M A G, Microelectromechanical devices for satellite thermal control, in IEEE Sensors Journal, vol. 4, no. 4, pp. 525-531, Aug. 2004. DOI: 10.1109/JSEN.2004.83029
[12] Oshman C et al. Flat flexible polymer heat pipes. Journal of Micromechanics and Microengineering, 2012, vol. 23, no 1, p. 015001. DOI:10.1088/0960-1317/23/1/015001
[13] Brown A, Jones Z, Tilson W Classification, Effects, and Prevention of Build Defects in Powder-bed Fusion Printed Inconel 718. En TMS 2017 Annual Meeting and Exhibition. 2017. NASA ID: 20170003400
[14] Godec M et al. Hybrid additive manufacturing of Inconel 718 for future space applications. Materials Characterization, 2021, vol. 172, p. 110842. DOI: 10.1016/j.matchar.2020.110842

IOP Conf. Series: Materials Science and Engineering 1287 (2023) 012017 doi:10.1088/1757-899X/1287/1/012017

A Simple Method for Determining Shallow Charge Distributions in Dielectrics via Pulsed Electroacoustic Measurements

Zachary Gibson and J R Dennison

Physics Department, Utah State University, Logan, UT, 84322, USA

zack.gibson@usu.edu

Abstract. The understanding of charge dynamics in dielectric materials is paramount in mitigating electrostatic discharge events for spacecraft. The most critical spacecraft charging events are found to result from incident electrons in the energy range of 10 keV to 50 keV. The charge embedded in dielectric materials in this energy range are deposited a distance into the material on the order of a few to tens of microns. One way to measure and understand the deposited charge is via pulsed electroacoustic measurements (PEA). However, the typical PEA spatial resolution of ~10 μm is not sufficient to resolve or discern charge deposited at the lower end of this incident electron energy range, where deposited charge distributions are obscured by the superposition of the signal originating from induced mirror charge on the electrode of the pulsed electroacoustic system. A simple method is proposed and demonstrated in which reference measurements from a pristine sample are used to separate the effect of the induced mirror charge from the measured embedded charge to obtain a more accurate determination of the deposited charge distribution.

1. Introduction

Understanding and predicting internal charge distributions in dielectrics are paramount for myriad of applications from high voltage DC power transmission to spacecraft charging [1-3]. Arguably, one of the best means to measure embedded charge is via the pulsed electroacoustic (PEA) method [4-6]. This allows for a nondestructive observation of the internal charge distribution in dielectric materials. Of course, there are limitations such as the spatial resolution [7], defined as the full-width-at-half-maximum (FWHM) of the leading interfacial peak, which is typically ~10 μm. This limits the ability to resolve multiple charge distributions on the order of the spatial resolution. In a typical ambient PEA system, there is an electrode at either surface of the measured sample. This means that if there is embedded charge, the measurement includes a signal due to mirror charge induced on the electrodes by the embedded charge distributions. Thus, charge distributions deposited ≲10 μm into the sample cannot be clearly and accurately resolved [8, 9].

For spacecraft charging, this becomes an issue as charge is often deposited on these length scales. The critical charging regimes for spacecraft are found to be incident electron energies from approximately 10 – 50 keV [10], which result in deposition depths in typical spacecraft materials on the order of a few to tens of microns [8, 9]. This means that classic PEA methods cannot accurately resolve the charge distributions resulting from the lower end of the critical energy range of incident electron energies for spacecraft charging, particularly ≲30 keV.

IOP Conf. Series: Materials Science and Engineering 1287 (2023) 012017 doi:10.1088/1757-899X/1287/1/012017

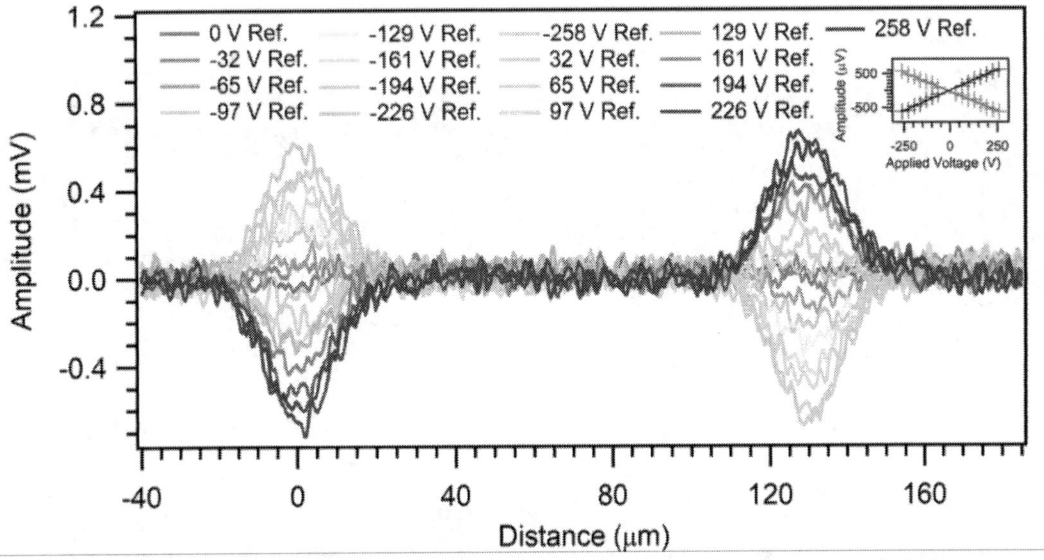

Figure 1. Measurements of pristine PEEK with varying DC bias applied. Inset shows amplitude as a function of applied voltage (markers) with a linear fit for the left (red) and right (black) interfacial peaks.

This paper presents a proof of concept study for a simple method to remove the effect of the interfacial charge from the PEA measurement to accurately observe shallow charge distributions. The method proposed is to simply remove contribution to the the signal from the induced mirror charge on the electrode near the deposited charge distribution. This is accomplished by using a pristine reference sample measured with varying applied voltages. This reference can then be directly subtracted from the raw data of the original PEA measurement of the irradiated sample with the shallow charge distribution.

This paper begins by outlining the experimental details and is followed by an explanation of the proposed method. It ends with a discussion of the results, conclusions drawn from the study, and suggested future work.

2. Experimental Details

This section describes the material properties, irradiation details, and an overview of the PEA method. The measurements presented in this paper were obtained with an ambient PEA system at the Université of Paul Sabatier in the DSF group at LAPLACE. Similar results made with the PEA system at Utah State University are presented elsewhere [8, 9].

2.1. Samples

The samples used in this experiment are polyether-etherketone (PEEK) from APTIV Victrex PEEK Film Technology and are nominally 125 μm thick (1000-125G) [11]. The samples were baked out at 100 °C for ≥72 hrs under <10^{-4} hPa vacuum to remove volatile compound content and to minimize water content. The PEEK samples were shielded from ambient light which could potentially result in photo-induced charging or discharging because the optical bandgap is ~3.1 eV [12].

2.2. Irradiation Details

One sample was irradiated with a monoenergetic electron beam at Utah State University with a high energy electron gun (Staib EH-80) at a vacuum level of ≤10^{-6} hPa at room temperature. A carbon scattering foil was used to produce a broader, more uniform beam spot. The scattering foil decreased the nominal 30 keV incident beam to an average incident energy of 24 keV. The flux was ~65 pA/cm^2 with a duration of 4 minutes. Further details are presented in [8].

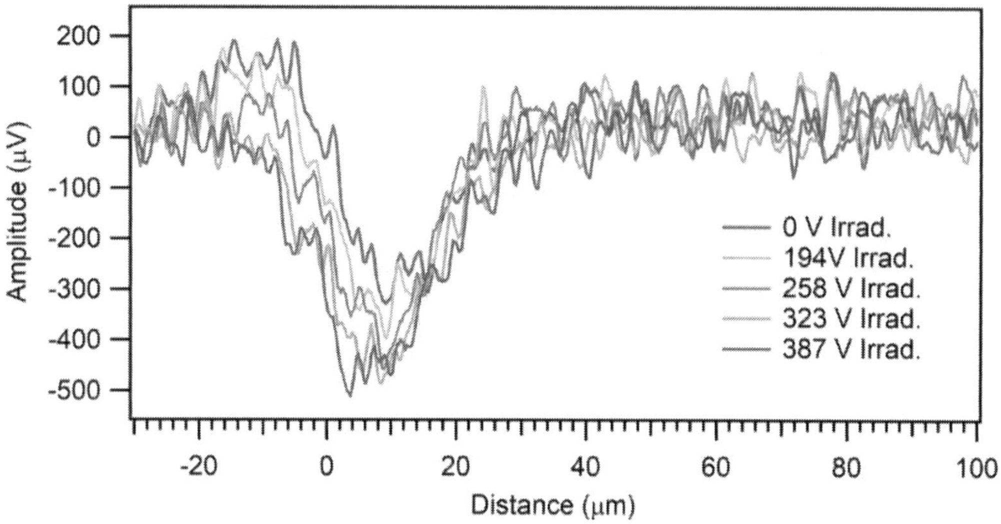

Figure 2. Measurements of 24 keV irradiated PEEK with varying DC bias applied.

2.3. Pulsed Electroacoustic Method

The PEA method works in a parallel plate capacitor configuration where a pulsed voltage induces an acoustic pressure wave that is measured by a piezoelectric sensor. Using simple time-of-flight and the relevant speeds of sound, the internal charge distribution can be obtained [6]. Note that the data presented in this paper are not calibrated for charge magnitude as the aim of the paper is to demonstrate a proof of concept for the proposed method. The raw data from the PEA system used has minimal artifacts and is quite noisy due to a low charge magnitude deposited in the sample.

The samples were measured with an ambient PEA system at Université Paul Sabatier in Toulouse, France. The PEA system employed a ~250 V pulse amplitude, ~3 ns pulse width, aluminium electrodes, polyvinyl chloride coupling layer at the high voltage/excitation electrode (no coupling film on ground electrode), silicone oil at sample interfaces and the sample was clamped in the typical parallel plate capacitor configuration. Note that all samples were measured with the apparent incident electron direction coming from the left (near the ground electrode).

3. Analysis

The proposed analysis method involves measuring two samples. The first sample is pristine and has no embedded charge. This sample was measured with varying applied DC bias from -258 V to +258 V to obtain a set of reference measurements to use for removing the signal originating from the electrode sample interface; refer to figure 1. The second sample contains embedded charge near the surface. Typically a DC bias is not applied during measurement of the irradiated sample. However, this was done here to provide multiple test measurements to attempt the proposed method with varying amounts of charge on the electrode sample interface. The applied DC bias to the irradiated sample was 0 V to +387 V; refer to figure 2.

An example outlining this novel method is presented in figure 3. The measurement of the irradiated sample with no DC bias, the reference measurement used to remove the signal from the charge at the electrode sample interface, and the resulting determination of the deposited charge distribution are displayed in red, blue, and green respectively. The analysis can be accomplished by processing the data in such a way to remove the signal originating from the interfacial charge. This can be done in two ways, either by adding a reference obtained through applying a negative DC bias to the pristine sample or subtracting a reference obtained through applying a positive DC bias to the pristine sample. The response of the PEA system should be linear and it should not matter which reference is used. However, it is suggested that the same sign of charge is removed to avoid any potential nonlinearities in the measured

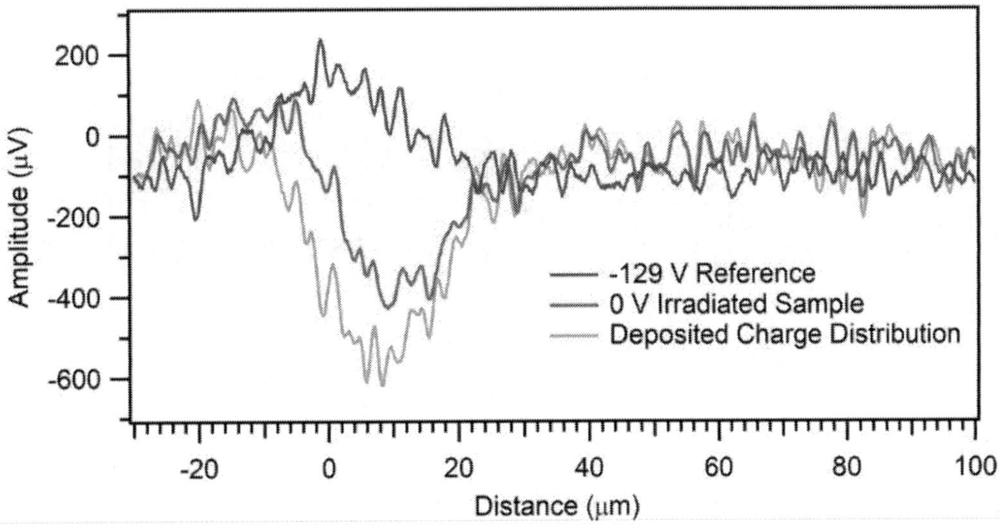

Figure 3. Example of measurements used to find the actual deposited charge distribution without effects of surface charge.

PEA signal. That is, for the positive induced charge it is best to remove a negative DC bias reference signal, as this induces positive charge at the electrode sample interface near the embedded charge distribution.

For analysis, a specific reference measurement must be chosen to subtract from the irradiated sample measurement that has a charge signature close to the actual amount of charge that should be removed by subtraction from the irradiated sample measurement. An iterative process involving choosing the closest reference measurement and an appropriate linear scaling factor to best account for the actual charge to be removed is used to find the optimal combination of reference signal and scaling factor. In this study the scaling factor was adjusted by hand until a best fit was determined.

The best fit can be determined in the following way. Starting with the reference signal closest to the optimal amount of charge to be removed from the irradiated sample measurement and a unity scaling factor, a difference curve is calculated. Considering deposited negative charge near the surface, if there is a positive charge peak in the difference curve, the scaling factor is increased to remove more charge. Iteration continues until the apparent embedded charge distribution appears to increase in magnitude (rising edge moves closer to the interface). Plotting a horizontal line at zero amplitude can be helpful when adjusting the scaling factor until there is no positive charge peak and the embedded charge distribution does not appear to increase. Ideally, a specific protocol or optimization algorithm to obtain a best fit is needed to remove the potential bias from the current subjective method.

This scaling of the reference signal is justified in the following ways. The amplitude of the measured PEA signal scales linearly, as shown by the inset in figure 1. The FWHM of the reference signals are 21.5 ± 0.8 μm (error as determined by the standard deviation), with no correlation with the measured amplitude or applied voltage. The interfacial peak position was determined to have an error of ±0.7 μm, similar to the error of the FWHM. Better uncertainties can be obtained with less noisy data [13]. Typical PEA measurements use applied voltages in the kilovolt range, where the applied voltage is only 100's of volts. Due to the low amplitude signals, the signal to noise ratio is lower and there is higher uncertainty in the peak position.

4. Results

It is clear in the measurements of the irradiated sample in figure 2 that it is difficult to accurately distinguish the charge distributions from the induced mirror charge and the embedded charge. However,

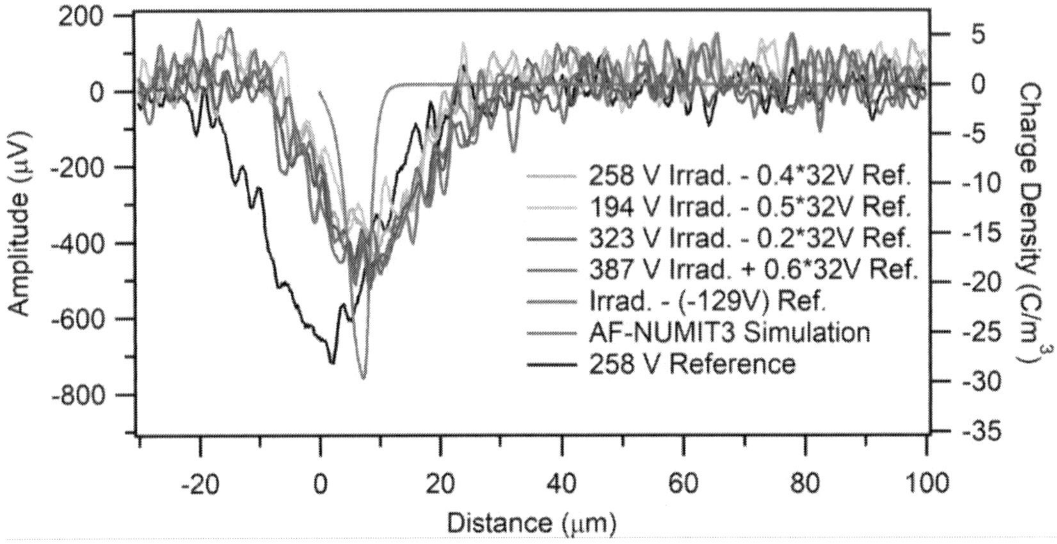

Figure 4. Results from all measurements of irradiated sample with corresponding interfacial charge removed. AF-NUMIT3 [14] simulation is superimposed and references the right axis. A reference signal is plotted in black to indicate the position of the sample surface. Legend indicates the irradiated sample measurement analyzed, whether the reference was added or subtracted, the scaling factor, and the DC bias used to obtain the reference signal (e.g. +32 V).

once the signal from the sample electrode interface is removed, the embedded charge distribution becomes clear, as is apparent in figure 4. All measurements of the irradiated sample with varying DC bias with the induced charge removed agree quite well, as depicted in figure 4 where the resulting charge distributions are superimposed. Note that no fitting was attempted to force the results to agree. The FWHM of the determined shallow charge distributions were 19.6±1.7 μm. This is lower than the FWHM for the reference signals, but is within error. The effective amount of charge removed by subtracting the reference signal was determined for each measurement as a test of the validity of the analysis method and it was found to scale linearly with the applied DC bias. This is another indication that the method is valid.

As a further test of the measurements and method, a simulation of the deposited charge distribution was determined with AF-NUMIT3, a deep dielectric charging code [14]. This can be seen as the narrower green distribution superimposed on the plot in figure 4. The simulation is narrower as it does not take into account the broadening due to the spatial resolution of the PEA system. The peak deposition depth as determined by the measurements and the simulation are 8.2±1.3 μm and 7.3 μm, respectively. This slight discrepancy of ~1 μm is already excellent agreement and within error. The results could perhaps be further in agreement if effects from delayed radiation induced conductivity were included in the simulation (AF-NUMIT3 has the capabilities but was not used to obtain these results). When a surface of a sample is grounded during irradiation (the back surface in this case), delayed radiation induced conductivity will cause a slight shift in the peak position of the deposited charge distribution towards the grounded surface driven by the electric field due to the induced mirror charge. Note that in this case the samples are irradiated with no surface electrode, but are mounted to grounded aluminium on the back surface. The samples are measured in the PEA system in a parallel plate capacitor configuration after radiation induced conductivity effects have decayed and the charge is essentially stationary.

5. Conclusions

The results indicate that this is a valid method for obtaining measurements of charge distributions near the sample surface using an ambient PEA system. It should be noted that measurements can be obtained using "open PEA" methods where the electrode near the irradiated surface is not actually in contact with the sample and there is no signal measured from that interface. This allows for measurements of less resistive materials where this is not possible when using an ambient PEA system, as charge may migrate during transport from an irradiation chamber to the measurement system if the sample is not sufficiently resistive. However, it then becomes more difficult to determine the exact position of the surface of the sample. The peak position of the signal shows the position of the charge and this does not change with magnitude, as demonstrated in figure 1. This means that the reference signal can be used to accurately determine the location of the surface of the sample to an accuracy of fractions of a micron in a well characterized PEA system [13]. The black curve in figure 4 is a reference measurement indicating the position of the surface of the sample. This allows for determination of the peak deposition depth of the charge distribution to be determined with a high degree of accuracy.

There are improvements that should be made for this method as well. As mentioned, an optimization algorithm should be implemented to remove the "human error" of this method. More work is needed.

The sample measured contained a relatively low magnitude of charge and resulted in relatively noisy measurements, as the charge magnitude was approaching the sensitivity limits of the PEA system. Further work should be done with a higher magnitude of charge embedded and studies should also be extended to lower energies. The demonstrated measurement of a charge distribution with a deposition depth of ~8 μm is impressive, but is near the spatial resolution of the system. This method should be pushed to explore the limits in resolution of deposition depth down to 10 keV incident energy, and perhaps lower.

It should also be noted that the spatial resolution limits the amount of useful information that can be determined about the shallow charge distributions. The width of these charge distributions are clearly narrower than the spatial resolution of the PEA measurements. However, other parameters such as asymmetry and the charge magnitude can be obtained in addition to the peak deposition depth with further analysis and processing of the data.

Acknowledgments

This material is based upon research supported by the Chateaubriand Fellowship of the Office for Science & Technology of the Embassy of France in the United States. Gibson acknowledges support from USU Presidential Doctoral Research Fellowship. The authors would like to acknowledge the help of Virginie Griseri and for allowing access to their PEA system. The authors would also like to acknowledge the help of Brian Beecken with use of AF-NUMIT3.

References

[1] Dennison J R and Pearson L H 2013 Pulsed electro-acoustic (PEA) measurements of embedded charge distributions *Nanophotonics and Macrophotonics for Space Environments VII*, 887612

[2] Hastings D and Garrett H 2004 *Spacecraft-Environment Interactions* Cambridge University Press

[3] Liu R, Takada T and Takasu N 1993 Pulsed electro-acoustic method for measurement of space charge distribution in power cables under both DC and AC electric fields *Journal of Physics D: Applied Physics* **26** 986

[4] Imburgia A, Miceli R, Sanseverino E R, Romano P and Viola F 2016 Review of space charge measurement systems: Acoustic, thermal and optical methods *IEEE Transactions on Dielectrics and Electrical Insulation* **23** 3126-3142

[5] Hussaini H, Adam A A and Susan A A 2015 Review of space-charge measurement using Pulsed Electro-Acoustic Method: Advantages and Limitations *J. Eng. Research Applications,* **5** 90-95

[6] Maeno T, Futami T, Kushibe H, Takada T and Cooke C 1988 Measurement of spatial charge distribution in thick dielectrics using the pulsed electroacoustic method," *IEEE transactions on Electrical Insulation* **23** 433-439
[7] Hole S 2008 Resolution of direct space charge distribution measurement methods *IEEE transactions on Dielectrics and Electrical Insulation* **15** 861-871
[8] Gibson Z and Dennison J R 2023 Relevancy of Pulsed Electroacoustic Measurements for Investigating Spacecraft Charging *IEEE Transactions on Plasma Science* doi: 10.1109/TPS.2023.3244058
[9] Gibson Z and Dennison J R 2022 The Relevance of Pulsed Electroacoustic Measurements for Spacecraft Charging *Spacecraft Charging Technology Conference Proceedings*
[10] Olsen R C 1983 A threshold effect for spacecraft charging *Journal of Geophysical Research: Space Physics* **88** 493-499
[11] Aptiv Vixtrex PEEK Film Technology, "APTIV 1000 Series Films for Electrical Insulation"
[12] Mackova A, Malinsky P, Miksova R, Hnatowicz V, Khaibullin R, Slepicka P and Svorcik V 2014 Characterisation of PEEK, PET and PI implanted with 80 keV Fe+ ions to high fluencies *Nuclear Instruments and Methods in Physics Research Section B: Beam Interactions with Materials and Atoms* **331** 176-181
[13] Gibson Z and Dennison J R 2022 Uncertainties of the Pulsed Electroacoustic Method: Peak Positions of Embedded Charge Distributions *2022 IEEE 4th International Conference on Dielectrics (ICD)* 539-543, doi: 10.1109/ICD53806.2022.9863513
[14] Beecken B P, Englund J T, Lake J J and Wallin B M Application of AF-NUMIT2 to the modeling of deep-dielectric spacecraft charging in the space environment *IEEE Transactions on Plasma Science* **43** 2817-2827

IOP Conf. Series: Materials Science and Engineering 1287 (2023) 012018 doi:10.1088/1757-899X/1287/1/012018

Euclid cleanliness and contamination control

Mauricio Portaluppi[1], Sandra Fontorbes[2], Antonio Saverino[3], Riccardo Rampini[4]

[1] European Space Agency ESTEC, TECQEE, Keplerlaan 1, 2201AZ Noordwijk, mauricio.portaluppi@ext.esa.int

[2] Airbus Defence and Space, Avenue des Cosmonautes 31, 31402 Toulouse, sandra.fontorbes2@airbus.com

[3] Thales Alenia Space Italy, Strada Antica di Collegno 253, 10146 Turin, antonio.saverino@thalesaleniaspace.com

[4] European Space Agency ESTEC, TECQEE, Keplerlaan 1, 2201AZ Noordwijk, riccardo.rampini@esa.int

Abstract. Euclid is a space-based optical/near-infrared survey mission of the European Space Agency (ESA) to investigate the nature of dark energy, dark matter and gravity by observing the geometry of the Universe and the formation of structures over cosmological timescales. It will operate at L2 and it will survey about 36% of the sky on a 6 year mission, all with high image quality and stability. To allow such scientific achievements, an integrated cleanliness and contamination control (CCC) have been put in place covering all phases of Euclid development until end of operations. This document describes the design challenges, the engineering and AIT solutions used to control and to monitor the entire mission life time contamination from organic and particulate sources as well as the in-orbit water ice contamination.

1. Euclid spacecraft description

Euclid has the scope to study the dark energy and dark matter of the universe, by using galaxy clustering and weak lensing over a vast portion of the sky for 6 years. The satellite will orbit at L2 Sun-Earth Lagrange point. The spacecraft is designed in three main sections, the instruments, the Payload module (PLM) and the Service Module (SVM). The PLM contains the scientific elements of the mission, the telescope, optical cavity and the two instruments, near infrared spectro-photometer NISP and visible imager VIS. Science objectives, mission and Euclid spacecraft have been described by Racca et al[1].

IOP Conf. Series: Materials Science and Engineering 1287 (2023) 012018 doi:10.1088/1757-899X/1287/1/012018

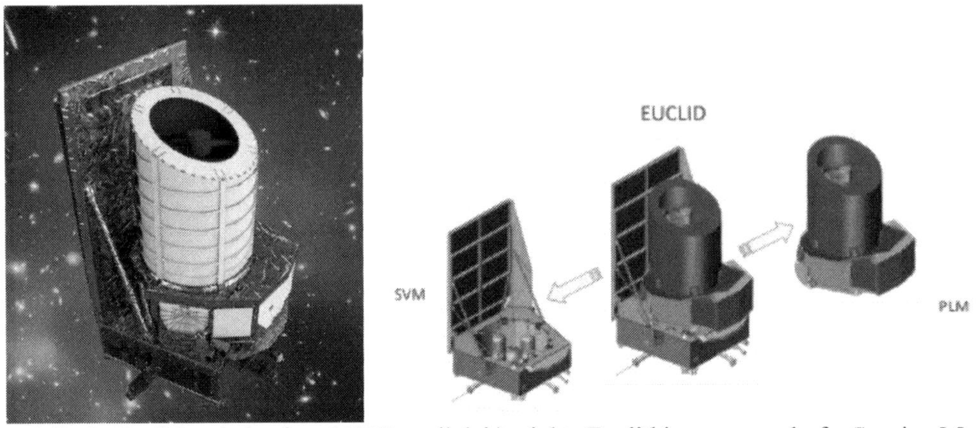

Figure 1. Left: Euclid artistic view (ATG medialab); right: Euclid is composed of a Service Module (SVM) and a Payload module (PLM).

1.1. Payload module

The Payload module contains the scientific core elements of the mission. It is composed of two cavities: the front cavity in which resides a telescope with M1 and M2 mirrors and the instrument cavity with VIS, NISP and FGS. A common SiC baseplate supports mechanically the PLM front and rear cavities and the interfaces with the SVM. The PLM schematic as well as a sketch of the instrument cavity and the front cavity can be seen on Figure 2. All contamination sensitive items can be clearly identified on these pictures. Instruments are described in a dedicated section.

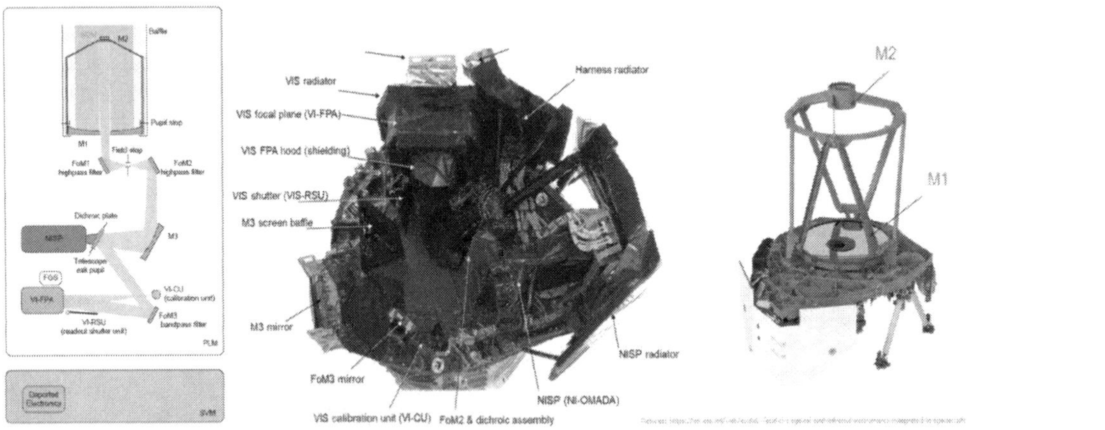

Figure 2. Left: PLM optical schematic; center: PLM rear cavity showing instruments and optics integrated and; right: PLM front cavity highlighting M1 and M2, baffle and MLIs removed.

1.2. Service Module

The SVM is mainly composed by two large sub-assemblies: the platform and the sunshield. The platform is an irregular hexagonal-base prism built around a central cone hosting the propellant tanks, providing the mechanical interfaces with the PLM and the launcher. The platform lateral panels accommodate all subsystems equipment, grouped by function, all radiators and insulators for the internal equipment thermal control and host the scientific instruments warm electronics, which functionally belong to the

PLM. The sunshield, is a wide structure protecting the PLM telescope and instruments from solar radiation and hosting the Photovoltaic Assembly.

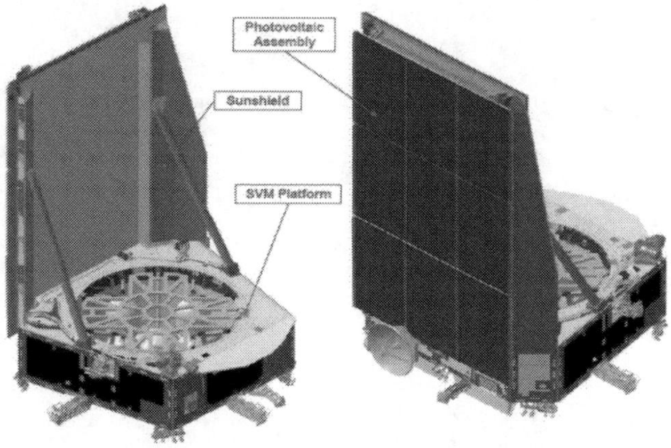

Figure 3. Service Module (SVM)

1.3. Scientific Instruments

Euclid carries two instruments, a Visible imager VIS and near infrared spectro-photometer NISP. Regarding VIS characteristics, VIS has a large focal plane, enabling weak lensing measurements. It consists of 36 CCDs, 604Mpix, pixel size 12µm - 0.1arsec pixel sampling size, spectral band 550-900nm over a field view 0.5deg^2 and survey area 15000deg^2 [2]. VIS units are sparse on the instrument cavity volume thus from a cleanliness perspective they share the PLM volume and cleanliness constraints.

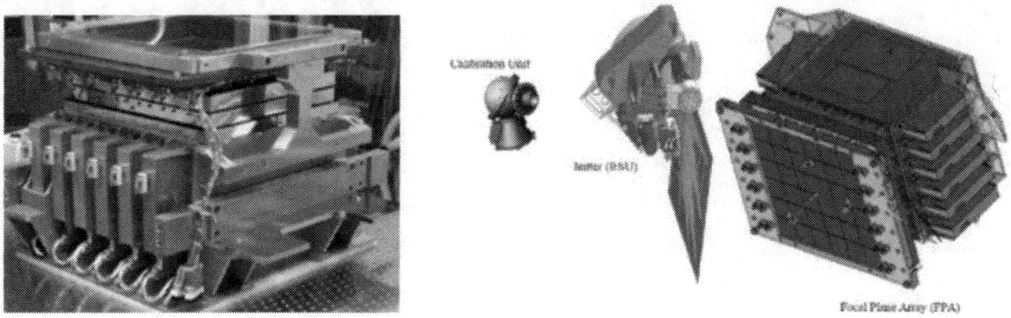

Figure 4. Left: VIS focal plane and; right: VIS units calibration unit, shutter and focal plane array

The NISP instrument operates in the 0.9-2.0µm range at a temperature lower than 140K, and the detectors at ~95K. The photometric mode is for the acquisition of images with broad band filters, and the spectroscopic mode is for the acquisition of slitless dispersed images on the detectors [3].

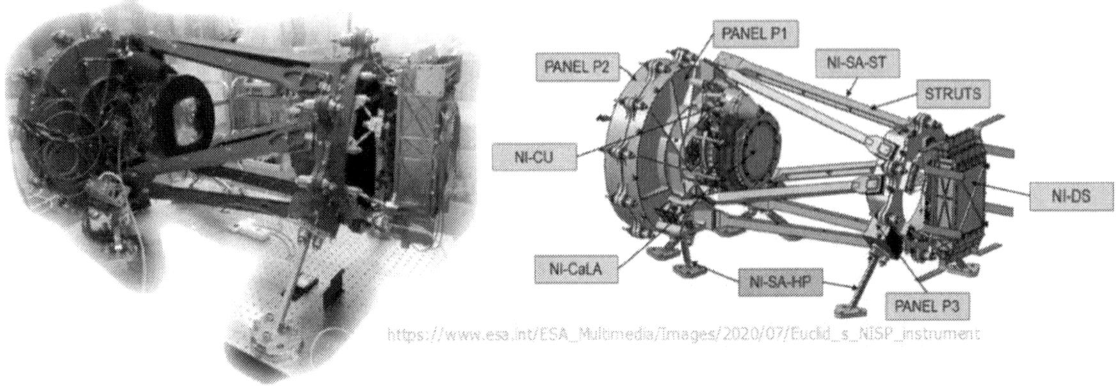

Figure 5. Left: NISP picture during integration, and right: NISP as designed with units described.

2. Euclid cleanliness and contamination control

2.1. Cleanliness requirements

From a contamination point of view, the most sensitive parts of the S/C are clearly those related to the scientific investigations and measurements, thus all items composing the optical chain (mirrors and dichroic) and the VIS, FGS and NISP detectors. Fully based on performances calculations, the maximum allowable PAC and MOC levels were split on the major mission phases: at delivery to PLM integration, at the end of PLM AIT, thus at delivery to the Spacecraft, at Beginning of Life and finally at End of Life. PAC is here most related to straylight issues, while MOC to signal intensity reductions.

Table 1. Euclid contamination specification of sensitive items on the optical path

	At delivery to PLM	At delivery to SC	At BOL	At EOL
VIS Focal plane	260ppm - 0.5µg/cm2	600ppm – 2.2µg/cm2	1200ppm – 2.7µg/cm2	1200ppm – 4.0µg/cm2
NISP ext.optics	700ppm - 1.0µg/cm2	1200ppm – 1.6µg/cm2	1600ppm – 1.7µg/cm2	1600ppm - 2.0µg/cm2
FGS	200ppm – 0.5µg/cm2	500ppm - 2.2µg/cm2	1200ppm – 2.7µg/cm2	1200-ppm - 4.0µg/cm2
M3,FM2,FM3, Dichroic	150ppm - 0.05µg/cm2	1050ppm – 0.5µg/cm2	1200ppm – 0.55µg/cm2	1200ppm - 0.55µg/cm2
M1,M2,FM1	150ppm – 0.05µg/cm2	1050ppm – 0.5µg/cm2	3000ppm – 0.85µg/cm2	3000ppm - 1.0µg/cm2

2.2. Contamination design approach

Being a big telescope with a wide aperture on the top, the Euclid S/C, and in particular the PLM, presents a design intrinsically prone to contamination. In fact the PLM internal side is organised in many cavities, the baffle front cavity, the optical instruments cavity and the NISP cavity it is evident how they represent enclosures where outgassed molecules during flight can more easily stick on colder surfaces rather than find the way to escape towards space through the baffle. Furthermore, working at cryogenic temperatures, ice formation represents one of the most critical design aspects and drivers.

In addition, considering the PLM external configuration, it is easy to understand how particulate contamination can enter the front cavity from its large aperture during on-ground phases, if not properly protected or where protecting it is not an option, like for instance during the PLM thermal vacuum test, during optical testing or verification needed both at PLM and at S/C level or during the S/C launch phase, from encapsulation to separation. Aside these considerations, because of its peculiar design from the moment of PLM integration and all subsequent phases, no cleaning is possible as well as direct monitoring of what happens inside.

For these reasons, an intense iterative process focused at matching all design needs with the many contamination control constraints took place early on the project, which has been refined continuously until the critical design review and beyond, progressively following all the manufacturing and assembly phase implementation. At this stage the design and the cleanliness provisions have been defined. In this view, contamination engineers role was of primary importance in all design aspects, other than in realizing how to implement it in a way to maintain cleanliness levels within the limits required, both on-ground and in-orbit.

Figure 6. End of PLM AIT. Picture from https://twitter.com/ESA_Euclid

As usual within missions with high sensitivity to contamination, the cleanliness process has been initialised identifying the technical performance losses associated to accumulation of contamination on scientific and non-scientific equipment. Starting from this crucial set of requirements, an end of life budget has been split throughout the main mission phases and budgets for ground and in-orbit contamination have been predicted. Ground contamination predictions tools have been used to determine how to meet each phase specification, by calculating how particulate and molecular contamination deposition rates on each surfaces and locations on each step of the AIT.

Based on first results, contamination engineers started iteratively refining all design aspects and on-ground activities possibly constrained by contamination risks, progressively optimizing a contamination control baseline, fully harmonizing needs and efforts to be exploited at all levels, being instruments, PLM, SVM or integrated S/C. All identified boundary conditions have been taken into account, which main are: cleanroom classes, geometries, time of exposure, use of mitigations, specific protections, stringent materials selection, bake-out plans, exploiting to the maximum extent all previous and relevant experiences. This process has been tuned until sufficient margin was identified for each project phase. In this frame, as said common to applications highly sensitive to contamination, the big challenge for the Euclid project was represented by its complicated design and by the strong interconnections of the three main S/C branches, the Instruments Set, the PLM and the System Level. This implied an harmonized and shared work of the entire team of contamination engineers of the three different parties, collaborating to build and achieve the most contamination control design driven as possible. Whenever not meeting the specifications, the Euclid team has increased the facilities cleanliness classes, created one portable ISO 5 Euclid dedicated cleanroom, designed and implemented barriers and customized protections or directly changed physical/material design finding acceptable levels for the ground phases.

For the cruise and operational phases, contamination modelling analysis of outgassing and thruster plume have been carried-out to evaluate effects from self (PLM-instruments) and cross contamination (SVM-PLM), bringing to the establishment of a common full scale bake-out plan, requiring that all materials of the PLM, of the instruments, and of the majority of the SVM were submitted to TQCM monitored bake-outs. In this view, a common and agreed bake-out protocol has been developed by the whole contamination engineering integrated team, to optimize objectives and effort.

A sketch of the overall strategy is shown in the following Figure 7. Each choice in terms of design have been validated until EOL by modelling or by using ground prediction tools, once feasible dedicated contamination control would be implemented and receive a feedback loop about their technical feasibility or trigger a new change to compensate for lack of margins.

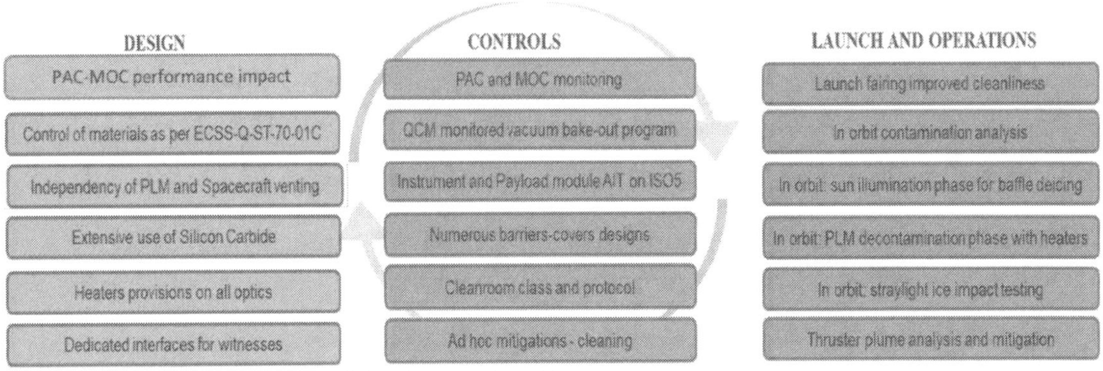

Figure 7. Contamination control implementation flow

Worth highlighting the differences between PLM and SVM. The PLM is designed to work as an independent contamination "system" than the spacecraft itself during operations, limiting connections. This separation allowed to broadly implement two different schemes of contamination control, one more stringent at PLM level where all instruments and PLM AIT required extensive use of ISO 5 facilities and another at SVM level which has been developed and integrated on ISO8. Once PLM and SVM have been mated, a mixed approach was necessary, implementing a more stringent control when the optics are exposed to cleanrooms or in vacuum to reduce the cross contamination during thermal vacuum tests.

In-orbit, the PLM cleanliness control is managed separately without any impact from outgassing or venting from SVM, in practice simplifying the outgassing and decontamination management in-orbit. Ground vacuum tests however need to have a mixed approach where cross contamination is considered, not only from SVM to PLM but also from the facility towards Euclid. Spacecraft contamination control is explained on a specific section.

2.3. Payload module contamination challenges and strategy

The main PLM challenges related to contamination control are connected to the high sensitivities of mirrors and scientific instrument detectors, to be preserved both during on-ground activities and during mission, in the frame of a wide open volume design, critically to be protected.

The PLM contamination control strategy has been based initially on control of materials selection process based on low outgassing properties assisted by an extensive bake-out campaign. Materials selected are highly inorganic, Silicon Carbide (SiC) has been used on a number of structural and optical elements, such as the M1 and the baseplate (Figure 8). All the PLM parts containing non-metallic materials were submitted to a TQCM monitored bake-out, implementing a common and agreed Euclid bake-out protocol and procedure, summarised as 72 hours of minimal duration, pressure lower than 1E-5 mbar, and stoppage criteria deviation from linearity of outgassing rate less than 1%. The success of the bake-out has been each time followed and agreed by the entire project contamination engineering teams. It is important to note that all bake-outs had a very advantageous temperature (between 333K and 353K) versus the operational temperature (between 125K and 150K), aspect found on cryogenic missions that comes handy to support cleanliness engineering.

M1 Flight Model (left picture) and baseplate rear face (right) – ADS courtesy.

Figure 8. Depicted the M1 flight model (left) and the tilted baseplate (right), two of many parts in silicon carbide

The entire design of the PLM and of its interfaces with the SVM has been studied in such a way to completely segregate the two modules with the aim to avoid any contamination impact of the SVM towards the PLM. In fact, when mated, no part of the PLM internal volume is in direct communication with the SVM. In this view and with this objective, a specific design has been implemented on the main PLM struts in order to orient the outgassing coming from the glass fibre composite away from the PLM optical cavity.

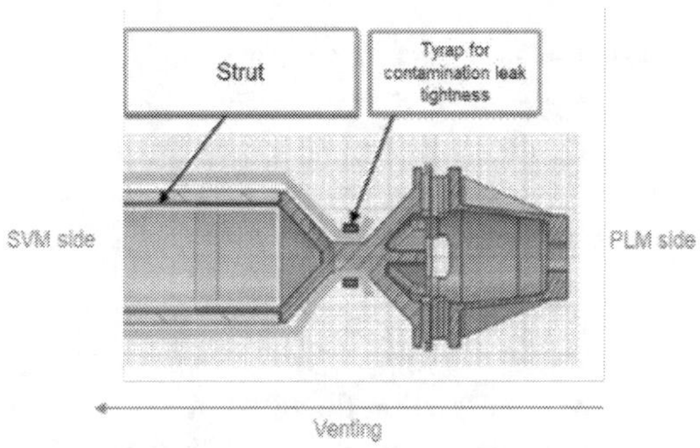

Figure 9. PLM-SVM struts showing a dedicated venting designed for Euclid

Finally, to further preserve the PLM from possible contamination coming from the external environment, all its external envelope is completely closed, so that the only aperture is represented by the baffle opening, protected on ground by suitable covers.

The PLM AIT has been performed in ISO 5 environment. In situ monitoring of particulate and molecular contamination has been implemented all along the integration and during the thermal vacuum test. The Figure 10 (red circle) highlights the presence of the witness samples representative of the front cavity exposed during the integration phases and during the thermal vacuum test thanks to the implementation of a dedicated bracket compatible with molecular contamination witness samples and cryogenic particulate witness plates. The same approach has been used for the instrument cavity- not visible in the picture. Due to the close vicinity to the optics the brackets have been designed to be dismountable from the external side, using a simple but efficient bracket. Furthermore, to avoid risk of damage to the PLM, such brackets with the witness samples have been dismounted only during the mechanical test campaign and during the transport. The frequency of the monitoring has been at each main step of the integration, e.g. at each integration of an optical surface, and before and after each test sequence. On the right, as protection versus on-ground contamination Euclid utilised extensively hard covers with more than 99% protection efficiency and soft covers, made of SCC1000 supported by an aluminium frame, which are mechanical testing compatible. The rear cavity of the instrument is closed and fully protected by the MLIs, designed with dedicated venting paths and by the instruments radiators. The efficiency of the covers has been verified using a specifically built mock-up of the external baffle.

IOP Conf. Series: Materials Science and Engineering 1287 (2023) 012018 doi:10.1088/1757-899X/1287/1/012018

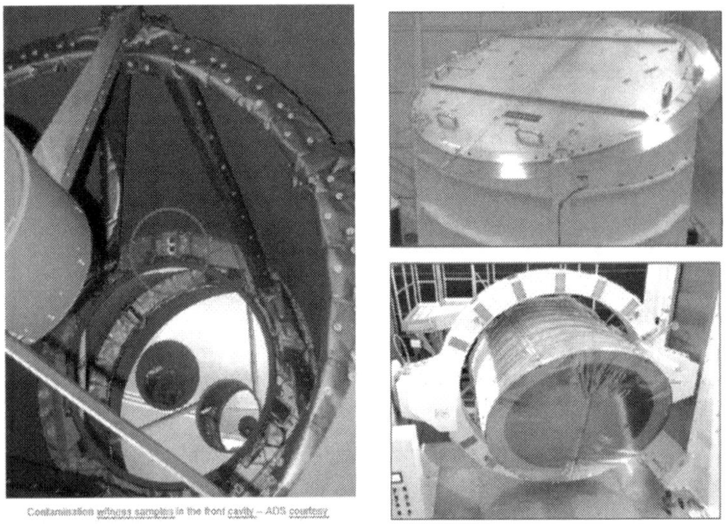

Figure 10. Left: Internal side of baffle, showing M1 and the witness samples installed; right: hard cover installed on PLM baffle and soft cover for mechanical testing.

Finally, heaters are implemented in the PLM, especially on the back side of the mirrors, and instruments have in-flight decontamination capabilities. Even though being negligible in term of mass, the heaters have been submitted to bake-out before their installation on the structure of the mirrors to limit their outgassing during decontaminations phases. In addition, the heaters' location, materials, and outgassing behaviour have been modelled with in-orbit analysis, the impact has been very limited. Water ice growth kinetics and risks affecting the optical performance have been studied [4][5][6] and decontamination strategies are implemented to remove ice by heating.

Mounted on the SiC baseplate, VIS units have been treated as integrated part of the system PLM with no physical separation. Knowing that the VIS CCDs cannot be cleaned, a PLM challenge has been to design each AIT step to avoid impact on VIS cleanliness budgets principally. To depict this aspect following PLM thermal test a potential cleanliness anomaly has been declassed as non-impacting thanks to the PLM mitigations and monitoring managed at PLM level rather than VIS level.

On the other hand, NISP instrument can be considered until a certain extent standalone. NISP contains its own volume which is linked to the optical cavity through ventings and the externally directed perforated MLI. Decisions at NISP level not necessarily have an effect on the PLM (and VIS) and vice versa. The highest risk however on this set-up is in case water ice keeps bouncing inside the NISP cavity or within the filter wheel.

Figure 11. PLM instrument cavity picture being integrated at ADS Toulouse, in front middle the VIS focal plane, right side the NISP instrument.

The cleanliness status of the Payload Module at delivery to Spacecraft has been outstanding, in particular regarding molecular contamination, which confirms the efficiency of all implemented design choices or mitigations during AIT. The only exception was the particulate contamination level of the M1, due to its location in the bottom of the baffle and its dimension, the cleaning was not easily feasible. Nevertheless, it has been demonstrated by updated Scatter Sensitivity Analysis, that this particulate contamination level has no detrimental effect on the performances. Below the summary status at PLM delivery to spacecraft, concluding one of the major milestones of Euclid. At Euclid spacecraft level, no cleanliness deviations have been identified at this date, in fact considering the remaining AIT until BOL and EOL. Future publications will include the actuals monitored values by the Euclid spacecraft including the launch campaign

Table 2. PLM contamination levels at delivery to S/C

	M1	M2	M3	FoM1	FoM2	FoM3	Dichro	VIS	NISP	FGS
MOC ($x10^{-7}$ g/cm²)	2.50	2.03	1.58	2.20	2.51	2.07	2.16	3.53	6.49	5.95
PAC (ppm)	1599	636	770	970	970	870	770	874	620	620

2.4. In-orbit contamination analysis – organic and water-ice contamination
The PLM in-flight outgassing has been studied using Systema Outgassing software. On this tool, Dynamic outgassing data of materials, in accordance with VBQC ECSS-Q-TM-70-52A, are injected in a 3D model which considers surface temperatures at all phases, during cruise and during operations. Emission and re-emission of molecules from different nodes of the model are then computed using mass transfer factors. A Monte-Carlo Ray-tracing approach has been used.
The focus has been the potential outgassing at cold and cryogenic temperature following an early decontamination phase. Two scenarios have been studied:

- Extrapolation of the outgassing data from standard dynamic outgassing tests, where no outgassing below 150K occurs, both in terms of emission and re-emission, the result is positive and no decontamination will be needed during the entire operation phase, Figure 12b.
- Extrapolation of the outgassing data from standard dynamic outgassing tests, the result is less optimist and identifies that one decontamination per year will be necessary, Figure 12c.

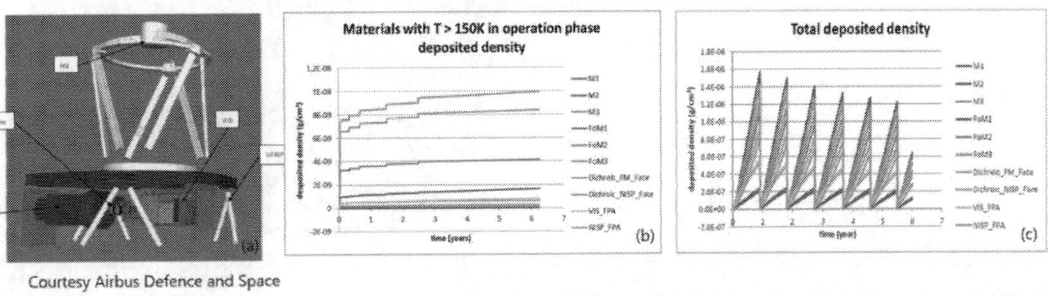

Figure 12. left PLM 3D with applied materials, the baffle and MLI has been removed to facilitate visualisation, middle: total deposited best case results; right: worst case results requiring one decontamination a year.

2.5. In-orbit contamination analysis - decontamination phase

The decontamination phase profile is planned during early commissioning phase. It consists of an initial transient warming phase based on electrical heaters, located on the rear cavity sensitive elements and by a sun illumination phase, the scope is to outgas light organic molecules and remove water ice. The profile has been defined using the in-flight contamination analysis. supported also by an in-situ ice kinetic testing and scattering effects on flight representative materials and conditions, the work has been presented by B. Bras on [5] and [6] (see 1.8.1). The early decontamination phase is planned to be as long as six days, four under isothermal conditions, where temperatures of sensitive items, such as optics and detectors, will be maintained between 200K to 273K.

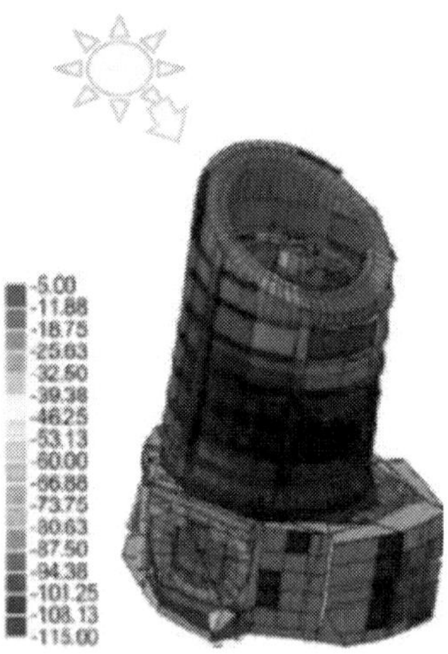

To support the required heating power, Euclid will be tilted slightly towards the Sun (Figure 13). This consequence of the is an increase temperature, on the external baffle where the lower surfaces will still be cryogenic, for such reason the risk of having ice has been studied. The dynamics of water release from the cryogenic parts of the external baffle have been characterised by testing different thicknesses of water-ice deposited on a MAP black PNC paint.

Figure 13. Sun illumination phase thermal map, the achieved minimum peak temperatures at the bottom external baffle surfaces, increased the risk of water ice condensation, which has been studied in details with in-situ water-ice measurements.

The ice time-dependency re-emission behaviour has been derived both with quartz crystal microbalances (QCMs) and by in-situ reflectivity testing [5]. The tests have been carried out at ESA-ESTEC and the ice kinetics have validated for the duration of the ice decontamination phase, see Figure 14.

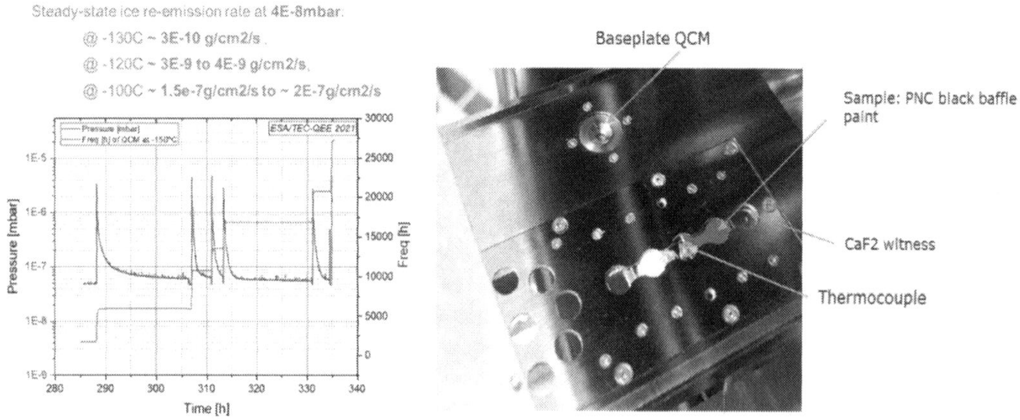

Figure 14. left: Ice re-emission rates at different temperatures; right: picture of set-up, for more details see presented work [5].

2.6. In-orbit contamination analysis – ice straylight impact on performance

No specific ground phase water absorption mitigation has been possible on Euclid, although an extensive vacuum bake-out campaign has been carried out. Since water is reabsorbed by materials when re-exposed to ambient conditions, the strategy to control water on Euclid relies on in flight decontamination phases. In order to predict ice performance impact on flight optics and use this information as a potential trigger for decontamination and for data model calibration purposes, a series of in-situ ice straylight characterisations on flight representative mirrors and optics have been tested at ESTEC, the results are presented by B.Bras et al. [6].

2.7. Service module and Euclid spacecraft

The big challenges relevant to the Euclid S/C design related to the contamination control engineering were mainly aimed at avoiding any contamination of the PLM sensitive surfaces coming from the Service Module (SVM), to assure all the SVM support to the decontamination phases and to maintain a very low contamination budget, especially for particles, during all phases from the mating of the two modules until launch.

In order to avoid any contamination coming from the SVM towards the PLM, a stringent and constraining materials selection process based on low outgassing properties was implemented. Furthermore, an extensive TQCM monitored bake-out campaign, involving the majority of items containing non-metallic materials has been performed, using the common and agreed Euclid bake-out protocol and procedure applicable to the PLM parts. It is interesting and important to highlight that, differently from the PLM, the mass and volume of SVM parts containing non-metallic materials is substantial. All the SVM harness, the entire sunshield, all SVM panels and MLIs, were submitted to bake-out. In order to further protect the S/C sensitive surfaces from outgassing, the same approach was followed during the Euclid S/C TVAC test planning, where all GSE items containing non-metallic materials were requested to be submitted to bake-out.

In parallel to an overall design aimed at avoiding any possible contamination coming from the SVM towards the PLM, the biggest challenge at S/C level was creating a contamination control program able to maintain a very limited contamination budget from the PLM and SVM mating to launch on sensitive surfaces. The main critical aspect was that all AIT phases from that moment on, including the whole environmental testing campaign, were planned to occur under ISO 8 conditions and without the support from purging. For this reason, the PLM had to be protected with a series of specifically designed baffle covers as mission development matured. The initially designed cover provided a 98% of efficiency and later it has been dramatically improved to values well above 99% in efficiency. Nevertheless such cover, which was fundamental during the nominal AIT phases and made by a metallic honeycomb structure, wasn't compatible with the three most critical tests under a particulate contamination point of view: the acoustic test, the vibration test and the thermal vacuum test.

Figure 15. Euclid spacecraft mated and ready for environmental and mechanical tests

Thus, two covers were further designed: a "soft cover", consisting of an aluminium circular ring holding a plastic foil, fully compatible with acoustic and vibration loads, and a "thermal cover", to be used during the S/C thermal vacuum test, made by a hexagonal metallic ring, on which 6 metallic black anodized extremely low outgassing petals are mounted. These two covers have a 98% of efficiency against particulate contamination and on a case it was validated during the S/C STM itself. Following this philosophy, to avoid performing the vibration and acoustic tests with star trackers lens and baffles internal surfaces unprotected, as normally done, a dedicated soft cover has been designed to be mounted during mechanical testing.

In order to allow all covers swaps and installations, implying optics exposure to the external environment, a brand new dedicated fully portable ISO 5 has been procured by TAS-I. Other than the portability, this facility can create an ISO 5 environment of specific parts of the satellite. In this view, it has the peculiarity to be movable with the Euclid S/C horizontal axis, supporting localised implementation of extremely clean conditions. The Euclid ISO 5 portable tent will be utilised also during the launch campaign.

The detailed contamination control programme defined at design phase, has been later validated during the Euclid S/C STM AIT campaign. The STM has been used to prove the CCC programme efficacy, to validate hypotheses and contamination rates used in the contamination predictions, to prepare procedures and fine tune them for the PFM campaign, to optimize the cleanliness inspections plan and monitoring plan and to train personnel to constraints related to contamination control. Such validation phase has been of the outmost importance prior to PFM and highly recommended for any scientific satellite.

Concerning the orbit phase, a contamination analysis has been carried out for both outgassing and for thruster plume. The favourable geometry of the SVM has been of benefit to avoid cross contamination with the PLM. In a matter of fact, while common spacecraft architectures present major contamination sources in field of view of radiators or instruments, on Euclid the PLM is fully shielded from the SVM outgassing, being the SVM overall view factor with the PLM baffle aperture reduced to zero. About potential self-contamination impacting the SVM sensitive items, both extremely limited view factors, and the bake-out campaign effect brings molecular depositions in-orbit to negligible values. About thruster plume possible contamination, covers have been installed to avoid cross contaminating thermal control paints and OSRs, as shown on Figure 16.

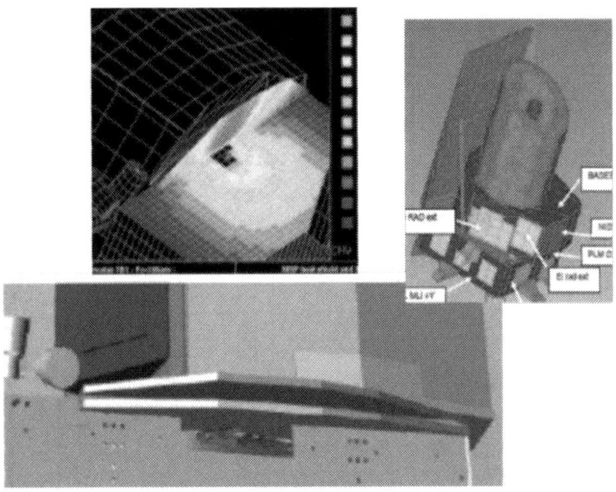

Figure 16. Upper left: thruster plume in-orbit analysis; upper right: PLM to SVM in-orbit outgassing analysis, and bottom: implemented plume shield used to protect radiative surfaces.

2.8. Pre-launch and launch phase

As already mentioned in the above, one of the most critical part of the Euclid design is that the PLM presents a very large aperture towards the external environment, bringing it to be prone to contamination, mostly particulate. While this disadvantage can be managed during phases until encapsulation by means of high efficiency covers, once the S/C is under the fairing its configuration becomes very critical. First, during the days before lift-off, the PLM mirrors and instruments sensitive surfaces are directly exposed to the internal fairing air, with its natural particulate fall-out. Then, from the launcher engines switching-on, vibrations and acoustic loads start causing particles detaching from the fairing internal surfaces, which a high portion is directly facing the PLM aperture. Then, from lift-off, air starts to move, hazardously transporting those particles possibly in the sensitive surfaces direction. In this view, even if a large portion of the particulate contamination budget was allocated to this very limited phase, the launch phase imposes important constraints to Euclid contamination engineering. To mitigate cross contamination during launch, the entire Euclid external surfaces will be cleaned to a VC-HS + UV level, and at the very last accessible moment before fairing encapsulation. Considering the pressure and constraints of the launch campaign, this activity is expected to be a challenge. The launch provider will see its share of critical activities. Two main improvements are required when compared to typical missions. Firstly, an extensive cleaning of all items facing Euclid when encapsulated, the fairing internal surfaces are required to achieve a level of particulate of only a few hundreds parts per million. Secondly, an improved control of the fairing air quality to reduce ingress of contamination comparable to an ISO 5.

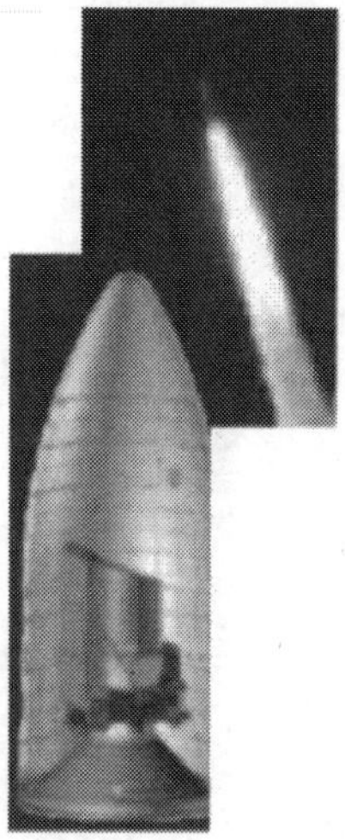

Figure 17. CAD of Euclid in launch configuration.

In terms of release mechanism contamination, the second stage adaptor clamps aren't exposed to sensitive surfaces, thus no specific effects are planned at early launch phase.

All in all, despite these precautions, the particulate contamination predictions are relatively high. Analysis shows that from combined operations to separation, a minimum of 1300 ppm will enter the PLM aperture. Although this contamination represents a noticeable amount with respect to the overall Euclid budget allocations, such level is considered sufficient to comply with the Euclid needs.

3. Conclusions

The Euclid observatory is a demonstration of a successfully integrated contamination control management at Scientific Instruments set, Payload Module and Spacecraft level. The challenges and efforts in contamination control engineering have been described as well as their implementation. Once in-orbit, molecular and water-ice condensation effects have been evaluated, in particular water ice has been characterized with unique testing capabilities and facilities at ESTEC [4][5][6]. End of 2022 – early 2023 will cover final spacecraft acceptance testing followed by kick-off of launch preparations, currently planned for early 2023. All these collected efforts throughout the Euclid development until flight operations will contribute to produce unique dark energy and matter scientific discoveries.

References

[1] Racca G *et al* 2016 The Euclid Mission Design. *Proc. of SPIE* Vol. 9904 99040O-1
[2] Cropper *et al* 2018 VIS: the visible imager for Euclid. *Proc. SPIE* 10698: doi:10.1117/12.2315372
[3] Maciaszek T *et al* 2016, Euclid Near Infrared Spectro Photometer instrument concept and first test results at the end of phase C, *Proc. SPIE* 9904-22 doi:10.1117/12.2056702
[4] Bras B *et al* 2015 In-situ ice formation investigation. *Proc. ISMSE13*, Pau, France, 22-26 2015.
[5] Portaluppi, Bras, Fontorbes, Saverino, Rampini 2021 Euclid Cleanliness and Contamination Control and Challenges Due to Ice contamination - *CCMPP2021* (in preparation)
[6] Bras B *et al* 2022 Euclid ice contamination testing. *Proc. ISMSE15*, Leiden, The Netherlands 2022 (accepted)

Acknowledgments

The entire Euclid team and consortium for their outstanding work.
The instrument teams, VIS lead by MSSL-UK and NISP lead by LAM-FR.
Daniele Stramaccioni (ESA) for the support regarding the ice kinetic re-emission testing.

Evaluation of advanced cleaning processes for sensitive surfaces in optical instrumentation

D Cheung[1], **D Faye**[2]
[1] ECP, 395 rue Louis Lepine, 34000 Montpellier, France
[2] CNES, 18 avenue Edouard Belin, 31401 Toulouse Cedex 9, France

[1] david.cheung@ecp-cleaning.com, [2] delphine.faye@cnes.fr

Abstract. Contamination has become a major issue in all high-tech industries, with increasingly stringent cleanliness requirements. It is the case for sensitive parts of space flight hardware, such as optics or other embedded equipment for which surface cleanliness is a critical topic, since even a thin layer of contaminant can adversely affect performance. Sometimes, cleaning may be needed to reach a specified cleanliness surface level at different integration phases of instrument or satellite, or as a corrective action in case of anomaly. Thus it is recommended to consider cleaning techniques without direct contact with the sensitive or fragile surface to prevent any damage after treatment. Five processes or techniques among those described for instance in the ECSS standard dedicated to ultracleaning of flight hardware (laser, vacuum bakeout, carbon dioxide (CO_2) snow, supercritical CO_2 and plasma) have been chosen for a test campaign to assess their efficacy on representative samples of sensitive optical parts (e.g.: glasses, mirrors, treatments, coatings, etc.). The study has been focused on the removal of molecular residues from substrates intentionally contaminated by a mixture of typical chemical contaminants. After a description of the selected cleaning techniques and their pros and cons, this paper will present the methodology adopted for the process validation using analytical chemistry and the comparative results of the performance of each cleaning technique. The impact on materials and the ease of implementation have been evaluated as well. In fine, the objective of this study is to provide some guidance to choose one or several non-contact processes suitable for cleaning sensitive space hardware, in order to guarantee contamination-free surfaces before flight and limit any risk of failure.

1. Selection of test surfaces

For this study, the surfaces were selected to be representative of the most sensitive surfaces of satellites and of the critical surfaces in a close environment. The objective here was to evaluate some cleaning processes on test specimens or samples provided by the project partners (CNES, Thales Alenia Space, Airbus Defence & Space, Sodern), such as:

- sensitive surfaces (lenses, mirrors, etc.)
- surfaces that can be sources of contamination for sensitive surfaces (ceramics, composites, thermal braids or strands, paints, coatings, specific treatments, etc.)

Therefore, decontamination tests were carried out on a total of 10 specimens representative of the materials concerned (see Figure 1).

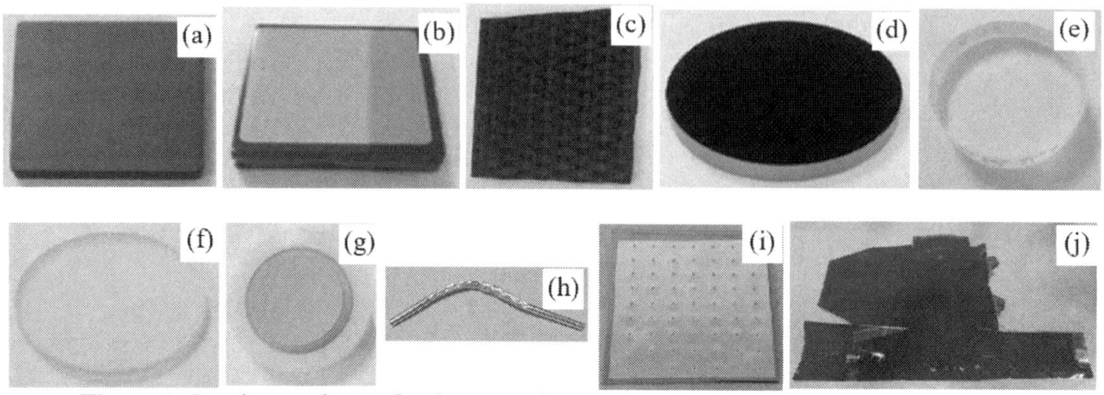

Figure 1. Specimens chosen for the tests: a) ceramic (SiC) ; b) mirror (Ag-coated SiC) ; c) composite (epoxy with carbon fibers) ; d) aluminium substrate with Acktar® coating ; e) lens (SiO_2) ; f) anti-reflective witness (SiO_2 + Hydra treatment) ; g) mirror (Ag-coated Zerodur®) ; h) thermal braid (Silver Copper) ; i) SSM Ag-Sheldahl® ; j) painted MLI (with MAP® PNC black paint)

2. Selection of five cleaning processes for the study

2.1. Laser cleaning

Laser cleaning is based on a photomechanical reaction induced by the interaction between an instantaneous high-power laser and a polluted layer (grease, oil, oxide, paint, varnish, cooking residues…). The light pulse transforms the first microns of the layer to be eliminated into highly compressed plasma. The expansion of this plasma generates a shock wave which ejects the pollutants in the form of fine particles (see Figure 2).

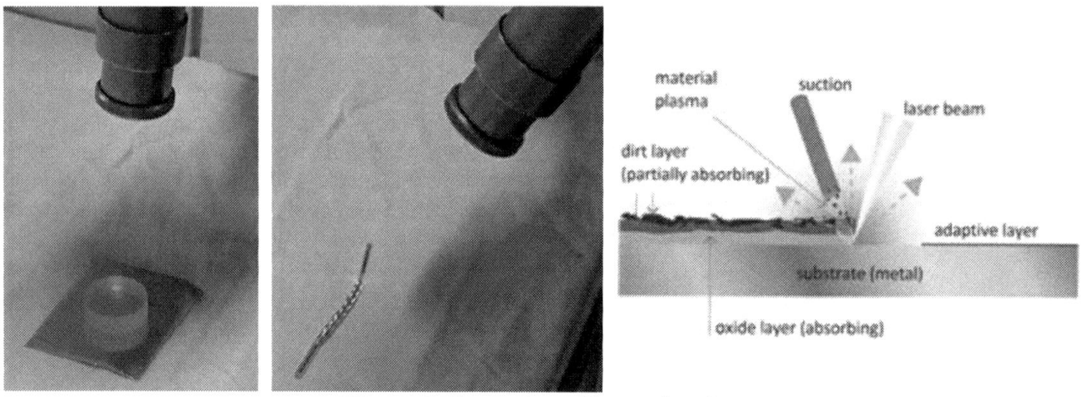

Figure 2. Examples of laser cleaning

In terms of benefits, it is a dry process; mobile use is possible (depending on configurations) and it can be automated by a robot arm. However, the settings remain very complex (in particular the adjustment of the focus on the surface to be treated); it is often poorly compatible with polymer materials (with a risk of denaturation or degradation). In addition, the strictest precautions are to be taken into consideration, in terms of protection of operators (adequate glasses, gloves…). Finally, it is recommended to assist the process with a suction device in order to evacuate the gaseous and particulate residues potentially generated.

Operating conditions:

- wavelength = 1064 nm
- average power = 8W
- max energy = 160 mJ

- distance ≈ 100 mm

2.2. Vacuum desorption process

It is a steaming process in a vacuum chamber, allowing the removal of volatile contaminants present on the surface of the materials as much as possible. In the present case, a secondary vacuum level is reached thanks to a turbo-molecular pump and the heating is ensured by a heater ribbon outside the chamber (see Figure 3).

The process is applicable to most materials and does not require any consumables (fluid or solid). However, the process duration may be more or less long, in order to maximise the outgassing process of the materials. In addition, the size of the parts is limited due to the dimensions of the available desorption chamber (diameter: 100 mm; height: 140 mm).

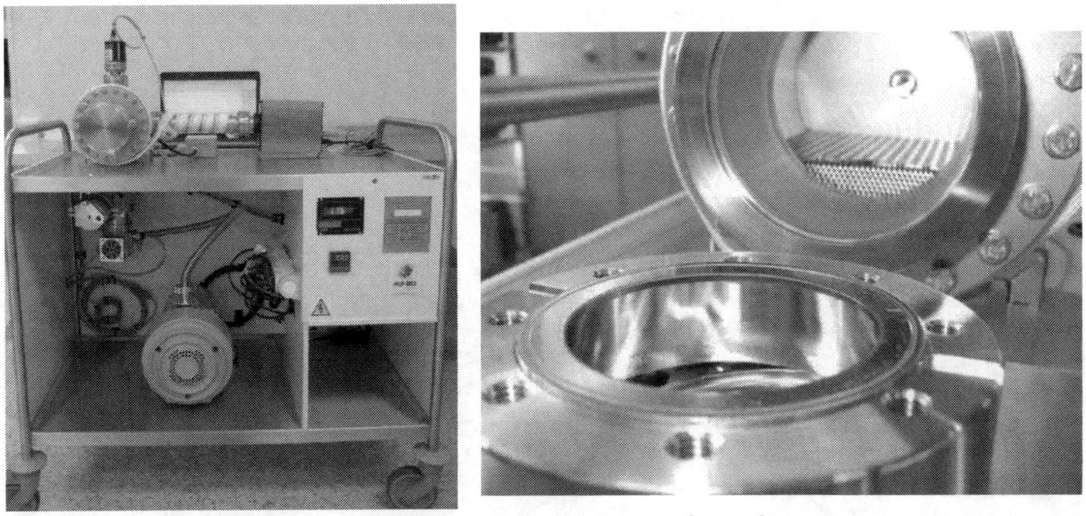

Figure 3. Equipment used for the vacuum desorption tests

Operating conditions:
- vacuum pressure = 10^{-8} mbar
- temperature = 100°C
- duration = 24h

2.3. Supercritical CO_2 cleaning

Supercritical fluid cleaning is based on the principle of the fourth state of matter. In addition to the solid, liquid and gaseous states, there may be, in some cases, a so-called supercritical state which takes place at very high pressure and temperature (see Figure 4).

This supercritical state is produced in autoclave where the part to be decontaminated is installed. Typical supercritical fluids are CO_2 (in this study) and water. This technique is particularly used to remove any kind of molecular residues, especially organic contamination such as cutting oils, hydrocarbons, undesired compounds…with thus a high degreasing and extraction power (as an example, the removal of trichloroanisole present in cork and responsible for cork taste is performed via supercritical CO_2). Nevertheless, this technique is poorly compatible with elastomeric materials [6]. Due to the use in a closed vessel, this limits the size of the parts to be cleaned.

IOP Conf. Series: Materials Science and Engineering 1287 (2023) 012019 doi:10.1088/1757-899X/1287/1/012019

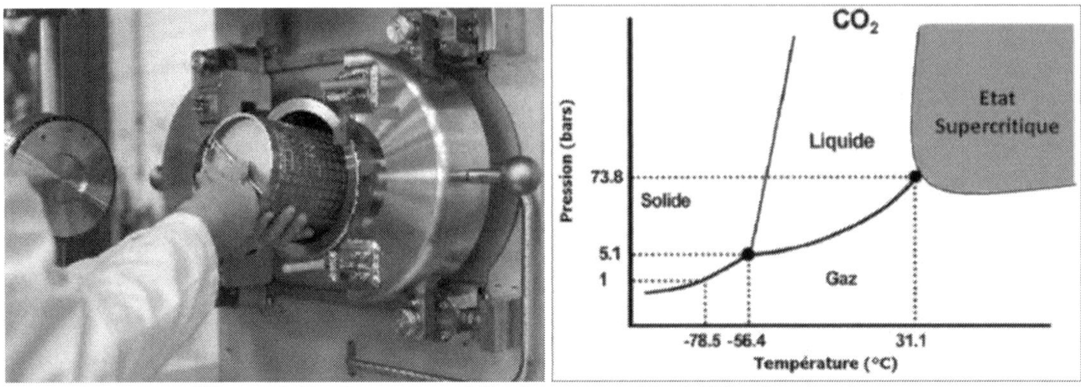

Figure 4. Equipment used for the supercritical CO_2 cleaning tests

Operating conditions:
- vessel pressure = 150 bars
- temperature = 35°C
- duration = 30 min

2.4. Atmospheric plasma cleaning

This type of treatment is based on the use of a plasma gas, generated at atmospheric pressure. In contrast to low-pressure plasma, which requires the use of a vacuum chamber, atmospheric plasma can be used in ambient conditions, using a nozzle which can be directly oriented to the surface to be decontaminated. One often speaks of a plasma torch, because of the visible "flame" which is produced at the nozzle outlet (with different geometries of plumes).
The process is dry but requires the use of specific gases (e.g.: O_2, N_2, Ar), with generation of gaseous residues. Furthermore, depending on the chosen parameters and the type of polymer material for instance to be treated, it may present a risk of surface degradation (see Figure 5).

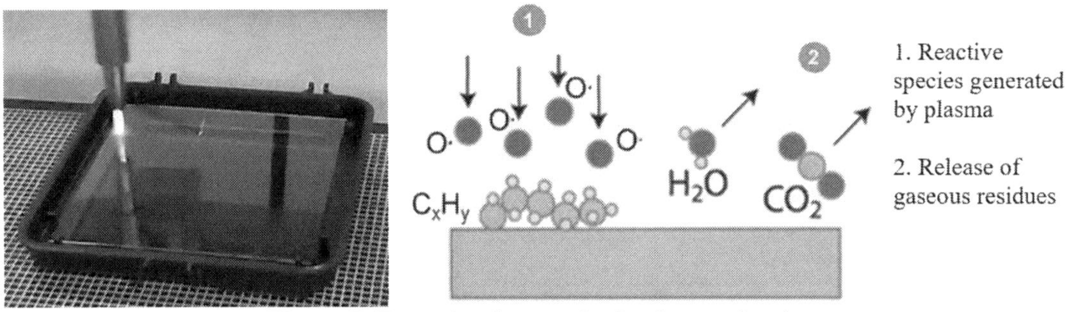

Figure 5. Example of atmospheric plasma cleaning

Operating conditions:
- plasma gas = air
- power = 800W
- distance ≈ 20 mm
- treatment speed = 200 mm/s

2.5. CO_2 spray cleaning

CO_2 spray cleaning, also called cryogenic cleaning, is characterized by the projection of solid CO_2 pellets at a temperature of approximately -80°C under high pressure on the surface to be cleaned (see Figure 6). Thanks to the sublimation effect, the pellets are transformed directly into gas upon impact on the surface. This technique is particularly efficient on organic residues, particles and for deburring. The

operation is generally performed under a fume hood to immediately evacuate the contaminant residues after the sublimation effect.
A major advantage of this method is that it is a dry process. However, the generation of gaseous CO_2 requires very special monitoring of the ambient air in the room where the equipment is located, for safety reasons with respect to operators. Finally, from lessons learnt, this technique is less suitable to flexible materials and sensitive coatings, due to high risk of damage.

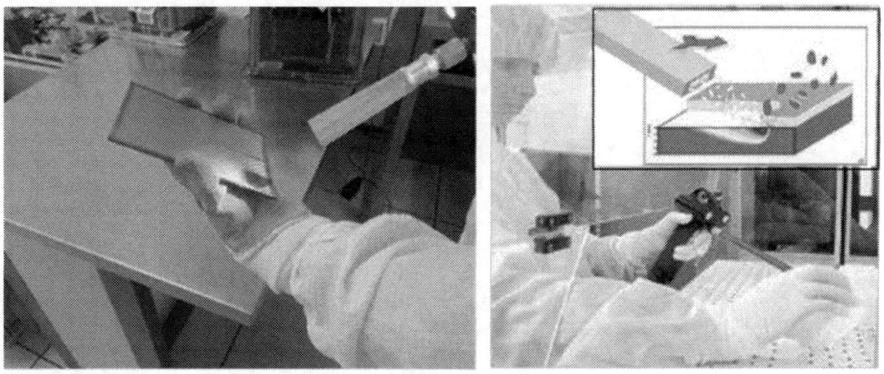

Figure 6. Examples of CO_2 spray cleaning

Operating conditions:
- pressure ≈ 2-3 bars
- distance ≈ 50 mm

3. Methodology and associated molecular contamination control

3.1. Methodology
In order to assess the decontamination efficacy of the various tested treatments, the test specimens were pre-contaminated before applying the cleaning test with the selected techniques. By comparing the residual contamination level after treatment to the initial contamination of the specimen, each process performance could be evaluated [4,5].

3.2. Protocol for pre-contamination of the samples
Three organic compounds, from families of chemical contaminants that are systematically monitored for space activities, were chosen for the study with reference to ECSS-Q-ST-70-05C standard [1-3]:
- Decamethylcyclopentasiloxane (from silicones family)
- Tributylphosphate (from hydrocarbons family)
- Dibutylphthalate (DBP) (from esters family)

A multi-contaminant stock solution containing the 3 chemical compounds mentioned above was prepared in methanol, at a defined concentration. A microdrop of this stock solution was placed on the sample to be contaminated. Once the dry residue has been obtained, the samples were wrapped in aluminium foil, then in double heat-sealed bags. For all the tests carried out, there was systematically one witness sample (kept in its original packaging, and serving as a contamination reference) and one test sample (having undergone the assessed cleaning process).

3.3. Molecular contamination control
As the contaminants to be detected are organic species, the GCMS (Gas Chromatography coupled with Mass Spectrometry) technique was chosen for the identification and quantification of the various molecular compounds. Two distinct extraction protocols were applied, depending on the targeted molecular contamination:

- for organic surface contamination: by leaching or swabbing the part, using an appropriate solvent. The solvent (methanol) is then directly analysed via GCMS;
- for volatile molecular contamination: through the use of an outgassing microchamber, in which the sample can be introduced, then heated to the required temperature. The chamber is flushed by a gas flow (carrier gas) which transfers all the outgassed volatile compounds to a tube containing an adsorbent (generally Tenax®), in a cold trap. A thermal desorption of this tube is then performed at high temperature and all the trapped compounds are analysed via GCMS.

As part of the study, the following conditions were used: outgassing at 60°C under nitrogen for 48 hours.

4. Results

4.1. Cleaning efficacy of the tested processes: synthesis

Table 1 summarizes the average values of treatment efficacy on molecular contamination, obtained at the end of the various tests carried out on the selected samples, according to the type of process, excluding laser cleaning since the results, in that case, were not conclusive for a proper assessment. The measurements show more or less consistent results, except for a few samples, which may be related to uncertainties or possible handling errors. Those errors are estimated to be less than 5%.

Table 1. Synthesis of cleaning efficacy results (in %) of the tested processes on selected samples, pre-contaminated in a same way

	Vacuum desorption	Supercritical CO_2	Atmospheric plasma	CO_2 spray
Treatment duration	24h	30 min	< 1 min	< 1 min
Ceramic (SiC)	100%	99%	60%	99%
Mirror (SiC + Ag)	Non tested	97%	91%	93%
Composite (epoxy)	99%	97%	86%	94%
Al + Acktar® coating	100%	100% (siloxane)#	97% (siloxane)#	69%*
Lens (SiO2)	50%	98%	73% (DBP)#	95%
Anti-reflective lens (SiO2)	99%	99%	97%	99%
Mirror (Zerodur + Ag)	83%	98%	99%	Qualitative test*
Thermal braid (Ag-Cu)	100%	99%	NA	77%*
SSM Sheldahl®	100%	100%	100%	100%*
MLI (Kapton® + PNC)	Non tested	Non tested	Qualitative test	Qualitative test*
Average	91%	98%	89%	91%

*samples showing marks of surface alteration after treatment
#efficacy measured only for that compound

Note: The case of the Acktar® coating is quite specific due to the high porosity of the coating, thus possessing adsorbing properties towards molecular contamination. In many cases, higher values of some organic compounds (especially the Tributylphosphate and Dibutylphthalate) were observed after treatment. One of the hypotheses would be that the molecular contamination protocol applied is not suitable for porous coatings (e.g. insufficient temperature or outgassing duration), leading to a poor detection of the intentional contaminants on the witness samples, still adsorbed in the materials. Some of the treatments applied on the samples could potentially start the desorption process of those contaminants, and thus facilitate their detection during analysis. This would explain the increase in the contamination values after treatment, whereas it is expected to have a decrease in the concentrations compared to the witness values.

Globally, supercritical CO_2 treatment shows the best cleaning performance results for most samples, leading to the highest efficacy average compared to the other evaluated processes. Then the performance ranking of the tested processes, in terms of cleaning efficacy, is as follows:

Supercritical CO_2 > Vacuum desorption > CO_2 spray > Atmospheric plasma

The laser cleaning process, considered as too invasive, was not included in this ranking due to the inconsistent obtained results.

4.2. *Materials compatibility assessment according to the tested processes*

Among the cleaning techniques tested in this study, only two processes revealed, visually, some surface degradation after treatment. These are laser cleaning and CO_2 spray cleaning.
On painted surfaces (MAP® PNC black paint on MLI and Acktar® coating on aluminum substrate), the laser treatment left marks due to the impact of the pulsed beam. The type of degradation would be more like a "discoloration" of the surface. On mirror coating (Zerodur® + silver) and on SSM Sheldahl® sample, laser treatment tends to ablate the material, then generating particles.
Similarly, the CO_2 spray treatment generated different types of degradation depending on the various surfaces evaluated:

- "discoloration" on Acktar® coating
- aspect modification on SSM Sheldahl® sample and thermal braid (silver copper)
- material removal on painted MLI (PNC) and mirror coating (Zerodur® + silver)

5. Discussions

Laser cleaning did not lead to exploitable results. Apart from two samples, the obtained values from the treatment tests are not very usable and do not allow a proper evaluation. Furthermore, the process has shown some alterations on painted surfaces, or even a degradation of the functional surface. Consequently, the laser cleaning process is not a technique to be recommended for sensitive space materials, at least with the actual operating conditions.
The vacuum desorption process gave globally very good results on most samples, in terms of cleaning efficacy (average of 91%). This process is similar to the "bake-out" technique, commonly used to outgas polymer materials. The major disadvantages of the technique are the treatment duration that could be long (typically between 1 and several days), and limitations of the size of the chamber and thus the size of the parts to be treated.
Supercritical CO_2 cleaning is the process that showed the best treatment efficacy results on all the samples ($\geq$ 97%). As for the vacuum desorption process, the main drawback of this technique, to date, is the limitation of the size of the parts to be treated due to the size of the treatment chamber on the market. Indeed, the partner specialized in supercritical CO_2 cleaning with whom the tests were carried out, has an autoclave with the following dimensions: diameter 300 mm x depth 440 mm. This equipment would be suitable for parts that can fit in this volume (mechanical parts, individual elements), but not for larger structures or other big parts of satellites.
The atmospheric plasma cleaning process also gave good results on some samples, in terms of cleaning efficacy (average of 89%) with no damage of the treated samples. However, for safety reasons, it is not recommended to use the plasma torch in manual mode but by coupling the plasma nozzle with an xyz table or a robotic arm. In addition, as gaseous residues (e.g.: ozone) are generated during the treatment,

it is highly recommended to apply the process in a ventilated room or in a specific cabin equipped with an extraction system.
Finally, the CO_2 spray cleaning process gave good results on most samples, in terms of cleaning efficacy (average of 91%, similarly to vacuum desorption process). But this process generated the most defects in terms of surface alteration, from simple discoloration to the removal of material, including aspect modification. As for the atmospheric plasma cleaning, the gaseous residues induced during the treatment have to be considered and removed by ventilation or extraction. Consequently, the CO_2 spray cleaning process cannot be generalized to any type of material, in particular flexible materials or sensitive coatings which may be subjected to the impacts of microscopic CO_2 crystals and damaged. Thus, it is strongly recommended to carry out compatibility tests before large-scale treatment.

6. Conclusion

This study allowed to evaluate the cleaning ability of various advanced processes implying no direct contact with the sensitive surfaces to clean. Among the five techniques tested, it appeared that the supercritical CO_2 treatment presented the best average removal performance, for most of all samples, and according to the molecular compounds selected. Contrary to other techniques, no visual damage has been observed on the samples treated by supercritical CO_2. The main limitation could be the size of the part, meaning that a large volume chamber would be necessary to treat big parts of satellites.

Acknowledgements

CNES and ECP would like to thank Thales Alenia Space, Airbus Defence & Space and Sodern for providing the samples for this study, funded by CNES.

References

[1] NF EN ISO 14644-13, Cleanrooms and associated controlled environments - Part 13: Cleaning of surfaces to achieve defined levels of cleanliness in terms of particle and chemical classifications (2017)
[2] ECSS-Q-ST-70-05C, Space product assurance - Detection of organic contamination of surfaces by infrared spectroscopy (2019)
[3] ECSS-Q-ST-70-54C, Space product assurance - Ultracleaning of flight hardware (2017)
[4] Cheung, D.; Faye, D., Evaluation of decontamination processes adapted to large optical components, *14th International Symposium on Materials in Space Environment* (2018)
[5] Cheung, D.; Faye, D., Assessment of advanced cleaning processes for sensitive optical surfaces, *17th ContaminExpert Symposium* (2021)
[6] Menon, N.; Nissen, A.; Horton, R.; Fitzjames, R.; Anderson, M.; Elbakhshwan, M.; Colgan, N., Compatibility of polymers in super-critical carbon dioxide for power generation systems: High level findings for low temperatures and pressure conditions, *7th International Super-critical Carbon dioxide Power Cycles Symposium* (2022)

Improvement of Bake-Out Prediction thanks to Realistic Species Separation

Jean-François Roussel[1,5], David Lansade[1], Delphine Faye[2], Guillaume Rioland[2], Véronique Perrin[3], Sylvie Brosse[3], David Nguyen Van Sang[3], Christophe Théroude[4], Christopher Laurent[4]

[1]ONERA–The French Aerospace Lab, 31055 Toulouse cedex (France)
[2]CNES–Centre National d'Études Spatiales, 31401 Toulouse (France)
[3]Thales Alenia Space, 06150 Cannes (France)
[4]Airbus Defence and Space, 31400 Toulouse (France)
[5]Author to whom any correspondence should be addressed at Jean-Francois.Roussel@onera.fr

Abstract. A new approach to material outgassing modelling based on chemical species separation was applied to Bake-Out modelling. The outgassing of the Scotchweld EC-9323-2 epoxy glue was first characterized experimentally, both before and after Bake-Out. Based on the former, the latter was modelled, following the classical approach on the one hand and the new approach on the other hand. The MS-based species separation of this new approach allowed measuring the emitted flux of each species individually, even during the outgassing phase. The evolution of these fluxes all along the outgassing tests, during the five classical 24 h plateaus was successfully modelled by diffusion-limited outgassing laws, while it proved inconsistent with a desorption-limited outgassing. Their modelling by diffusion laws allowed a very consistent modelling of the total mass measurements in TML and CVCMs, with a small number of chemical species of different volatilities. The effect of a preliminary Bake-Out on this outgassing was finally modelled and compared to the experimental characterization of the baked material. The modelling accuracy still remains comparable to that of the traditional modelling based on mathematical species. Yet, the realistic species separation of the new approach allows simpler assumptions on water regain (its first species) than the traditional one, which needs to assess the amount of water contained in each mathematical species. Progress in the new method should come from an improved species separation and a direct characterisation of the water flux.

1. Introduction

Baking-out materials before integrating them on spacecraft is a widespread technique to reduce their condensable contaminant content. Predicting this contaminant reduction before actually performing the Bake-Out (BO), in order to specify its duration and temperature to fulfil contamination requirements, is another interesting and common practice. In Europe, this assessment is usually done following the residence time approach defined in the 1980s [1], including thermal activation of species desorption. However, this approach is now thought to be closer to a mathematical fit by "mathematical pseudo-species" than an actual separation of different chemical species. This may reduce the accuracy of the model predictions, and make the assessment of water regain after BO a bit more complex to model.

This is one of the reasons why a new approach to contamination assessment, aiming at a realistic species separation, was started in the late 2000s years by ONERA and other teams [2][3]. The principle consists in performing Thermogravimetric Analyses (TGA) by ramping up deposits temperatures to (partially) separate condensed contaminant species, and identify them by mass spectrometry (MS). This coupled

TGA-MS method, somewhat similar to a gas chromatography-mass spectrometry (GC-MS) that would be performed in situ (with poorer separating power yet), was already presented with increasingly accurate results (see e.g. [4], [5] or [6]). Once chemical species and their mass spectra are identified this way, MS data acquired during the initial outgassing phase can be processed to also track their outgassed flux. The simplest way is through mass fragments than only show in the spectrum of one single species, usually heavy ones (with mass-to-charge ratio *m/z* close to, or larger than 100 unified atomic mass unit (a.m.u.)). Otherwise, combinations of species spectra during outgassing must be pursued, which may induce more uncertainties, depending also on data accuracy.
This experimental and theoretical protocol were implemented on several space-used materials. In another communication made during this 15th ISMSE [5] by some of these authors (from ONERA and CNES, in collaboration with other ESA authors), species separation achievements were reviewed, with a special focus on recent results on the epoxy glue EC9323-2, presented quite in details.
TGA/MS-based species separation was then applied to the analysis of other physical processes, such as outgassing. The principle simply consists in tracking *m/z* fragments specific of a species during outgassing, to determine its outgassing law. This was also performed on a few other materials, and reviewed in [7].
In this communication, we go one step further. On the basis of mass spectrometry measurements acquired at ONERA during the outgassing, we first analyse and model the outgassing of the seven species that were separated and identified in reference [5]. Next, based on this physical modelling, we add a Bake-Out step to the outgassing history, to see how the forthcoming outgassing was reduced by BO. This is reported in section 3.
Before that, in section 2, the classical approach to BO modelling, as performed by industry and CNES, is reported as a basis for comparison. All of these modelling results are compared to experimental outgassing data obtained at CNES on baked materials.

2. State of the art approach

2.1. *Introduction*

In order to model the material outgassing and predict the induced molecular contamination on a spacecraft, dynamic outgassing test has been performed. This test consists in applying temperature steps on the tested material, usually four from 25°C to 125°C, and to measure the outgassing from the deposited mass on at least one cold Quartz Crystal Microbalance (QCM), typically -170°C. For desorption modelling, fitting of deposited mass is performed using several mathematical species coupled to Arrhenius law, linearized or not. Finally, each mathematical species is characterized by the mass that can emit, also called emittable mass, an emission time constant and a reemission time constant.
Up to now, simulated bake-out is obtained by recalculating the emittable mass of each species from the mass emitted during bake-out. To take reabsorption of water vapour into account after the bake-out, the emittable mass of the first species is reintroduced i.e. the emittable mass before and after bake-out is considered equal.

2.2. *Reintroduction of the first species*

As explained previously, the current method consists of recalculating the emittable mass of each species, commonly known as W_0, from the emitted mass during the bake-out except for the first specie:

$$W_0' = W_0 \exp[-t_{BO}/\tau_e(T_{BO})] \tag{1}$$

$$\tau_e(T) = A_d \exp[-E_d/RT] \tag{2}$$

$$\tau_e(T) = \tau_{e0}(T_0) \exp[k_e(T - T_0)] \tag{3}$$

where W_0' is the emittable mass after bake-out in %, W_0 is the emittable mass before bake-out in %, t_{BO} is the bake-out duration in h, τ_e is the emission time constant at a temperature T in h, T_{BO} is the bake-out temperature in K, A_d is the Arrhenius pre-exponential factor in h, E_d is the desorption activation energy in J.mol^{-1}, R is the gas constant in J.K^{-1}.mol^{-1}, T is the temperature in K, τ_{e0} is the emission time constant at reference temperature T_0 in hours, k_e is the residence time-temperature dependency coefficient in K^{-1} and T_0 is the reference temperature of the material in K.

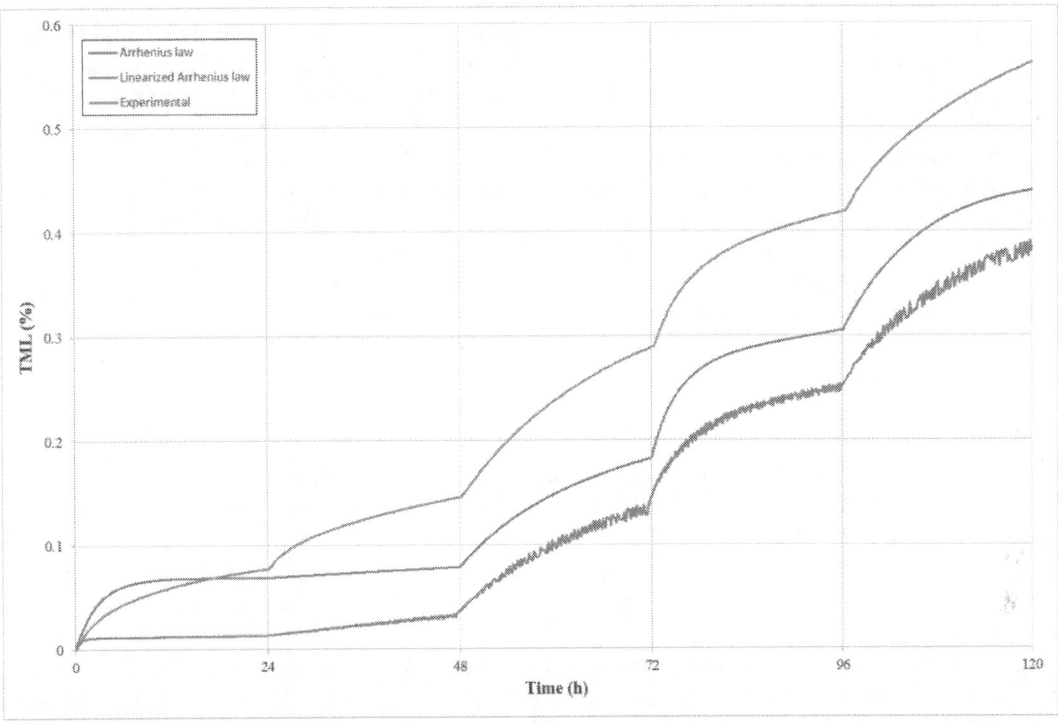

Figure 1. Comparison between simulated and experimental TML by reintroducing the first species

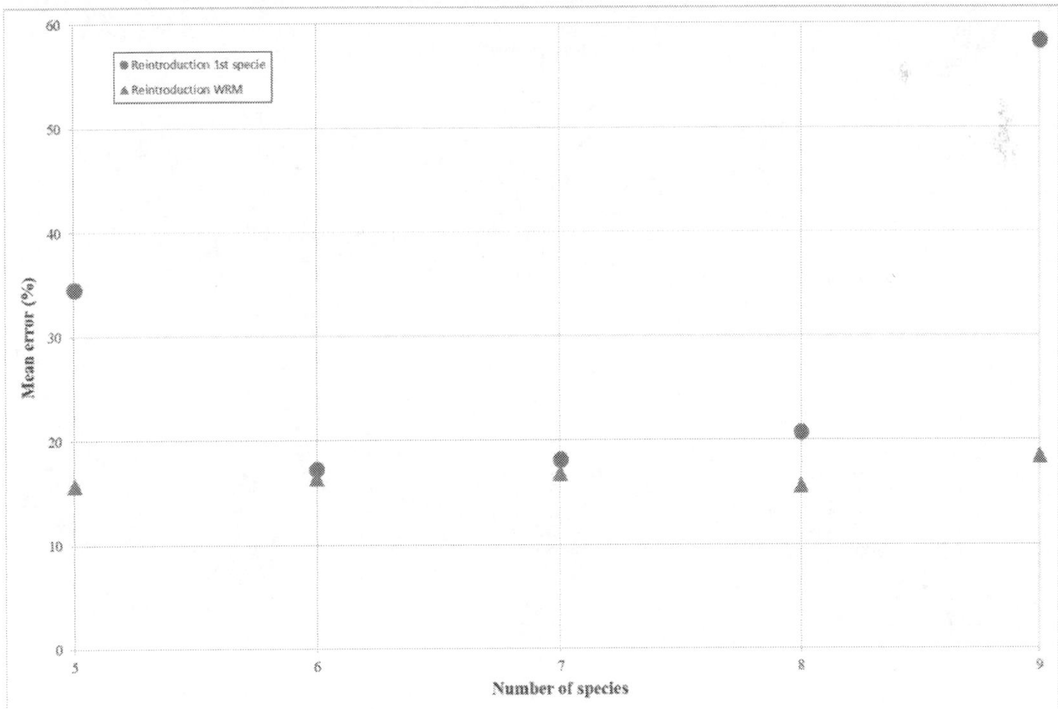

Figure 2. Evolutions of the error between simulated and experimental bake-out by reintroducing the first specie or by gradual reintroduction of WRM

IOP Conf. Series: Materials Science and Engineering 1287 (2023) 012020 doi:10.1088/1757-899X/1287/1/012020

Figure 1 shows a comparison between simulated and experimental deposited masses (TML) on the QCM at -165°C for EC9323-2 material by using this method. Both Arrhenius and linearized Arrhenius laws have been employed. Five and nine mathematical species have been used for Arrhenius and linearized Arrhenius, respectively. The experimental deposits tends to be larger than the simulated ones so the bake-out impact is overestimated. The overestimation is not systematic and depends on the number of mathematical species to perform the fitting before bake-out. Indeed, the number of mathematical species has a strong impact on the simulated bake-out and the induced error (see Figure *2*) which is another drawback of this method.

It is interesting to note that the shape of the curves is similar between 48 and 120 hours. This observation leads to think that more than the first species, usually defined as water, must be reintroduced and therefore that the water is contained in more than the first mathematical species. This explanation is corroborated by the physical species separation presented hereinafter (see §3): water, present in near surface as well as in bulk of the material, diffuses toward the material surface before its desorption. In order to fill the gap of the desorption approach, reintroduction according to Water Recovery Mass (WRM) has been studied.

2.3. *Reintroduction according to WRM*

The basic principle remains the same than those presented before i.e. recalculate the W_0 of each species according to the equation (1). However, instead of reintroducing only the first mathematical species, the reintroduction is based on the WRM. In this study, two different methods have been applied: a reintroduction from the fastest to the slowest species until the sum of reintroduced W_0 becomes equal to the WRM or the reintroduction of the first species and then distribute the remainder of WRM on all the other species.

Figure 3 shows a comparison between simulated and experimental TML for EC9323-2 material by using these two methods named gradual reintroduction and distributed reintroduction, respectively. Linearized Arrhenius law has been employed with nine mathematical species like the previous example (see Figure ***1***). More accurate predictions are obtained by using WRM instead of reintroducing only the first species. Also, less dependence on the number of species can be underlined as shown by Figure 2.

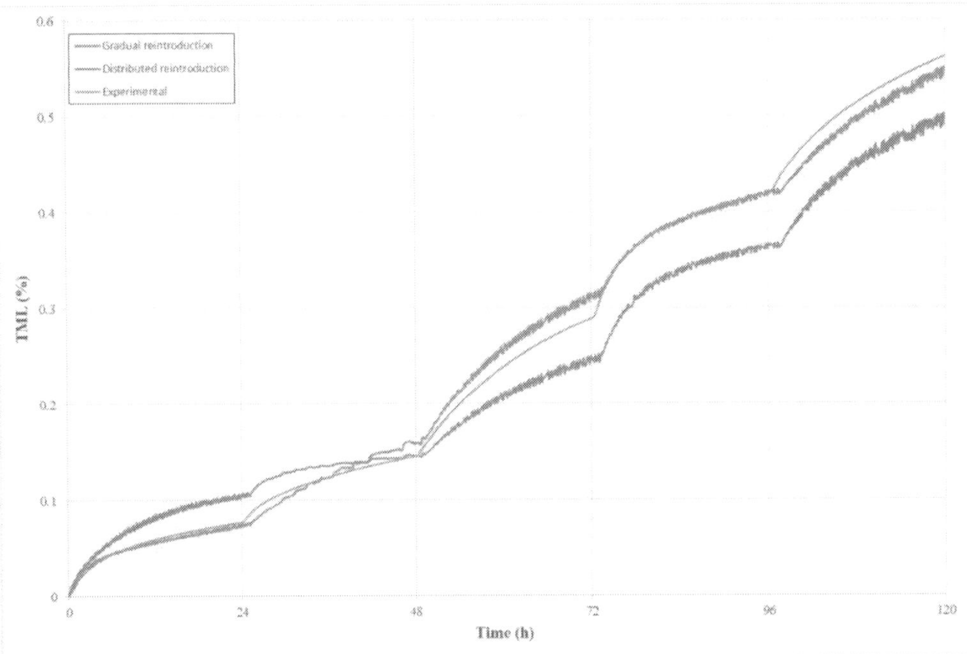

Figure 3. Comparison between simulated and experimental TML by reintroducing according to WRM

3. New approach through physical species separation

This new approach aims at characterizing outgassing at chemical species level to determine more physical and more reliable outgassing laws, and apply them to Bake-Out prediction. This goes in two steps. In a first step, the separation and often the identification of the outgassed chemical species was done through TGA/MS coupling technique as reported in the parallel communication [5], whose results are summarised next (section 3.1). In this communication, we focus on the second step, which is the analysis of the outgassing, equally at chemical species level, by following the *m/z* fragments specific of each outgassed species during the outgassing phase. This is reported in section 3.2. This allowed us determining the outgassing parameter of these species, namely their diffusion coefficients and activation energies, since their physics was found to obey Fick's law. They were next used to model the quantitative measurements made on QCMs (TML and CVCM), as reported in section 3.3, and finally how this outgassing is modified by a preliminary Bake-Out (section 3.4).

3.1. Species separation through TGA/MS method

In the parallel communication [5], the study of EC9323-2 material contaminant emission was performed by TGA/MS coupling method at ONERA. It allowed determining *m/z* fragments, as measured by mass spectrometry, that seem specific to each of the seven identified species. Indeed only six of the seven separated species were found in the set of mass spectra seen in EC9323 material GC-MS at CNES. Here we simply list these seven species and specific fragments in the
Table ***1*** below. They will be tracked during the outgassing pass as reported in the next subsection.

Table 1. m/z values specific of each of the seven separated species (from [5])

Species (TGA peaks):	Single peaks *m/z*
• Methyl-ethyl-phenyl-ethanone (MEPE) or Methyl-ethyl-benzene (MEB)	147, 162
• Octamethyl-cyclotetrasiloxane (D4)	281, 267
• Butylated Hydroxytoluene	205, 220, 145, 163, 206, 161, 177, 189, 195
• Menthol (indeed 1'-(butyn-3-one-1-yl)-menthol)	165, 180, 193
• Dodecanol	140, 168, 98, 111, 112
• Cyclohexane-dimethanol	95, 93,202, 264, 108, 67
• Unidentified molecule *M* (reemitted at 32C in 1st exper.)	58, 88

3.2. Species outgassing modelling

An important finding of [5] was that among the rich mass spectrum of these species, some fragments seemed to be specific to that species, i.e. only present in the spectrum of that species of EC9323-2 material, and not a (significant) part of the mass spectrum of any other species found in that experiment (i.e. also outgassed by EC9323-2 material or part of the chamber background). While [5] was focussed on the TGA/MS data for species separation, here we focussed on the MS data collected during the outgassing phases of these experiments.
For each chemical species, we studied the MS signal of the best of these *m/z* fragments, i.e. with the best signal level and specificity. The best two fragments were selected for the plots below. The first objective was to determine the physics ruling this outgassing. We shall see that Fick's diffusion law was always found to match MS data very convincingly. The second objective was then to determine the Fick's law parameter of each species. They are gathered in the Table 2 at the end of this section.

Figure *4* represents the MS signal of the two best specific fragments of menthol, one of the seven species found to be outgassed by EC9323-2, and their fit by the same Fick's law (with a different global factor for each fragment, accounting for their importance in menthol spectrum). The data are quite good: masses 165 and 180 a.m.u. are proportional and significantly above the noise level (around 10^{-14}). Their fit by a solution of Fick's law is also quite good.

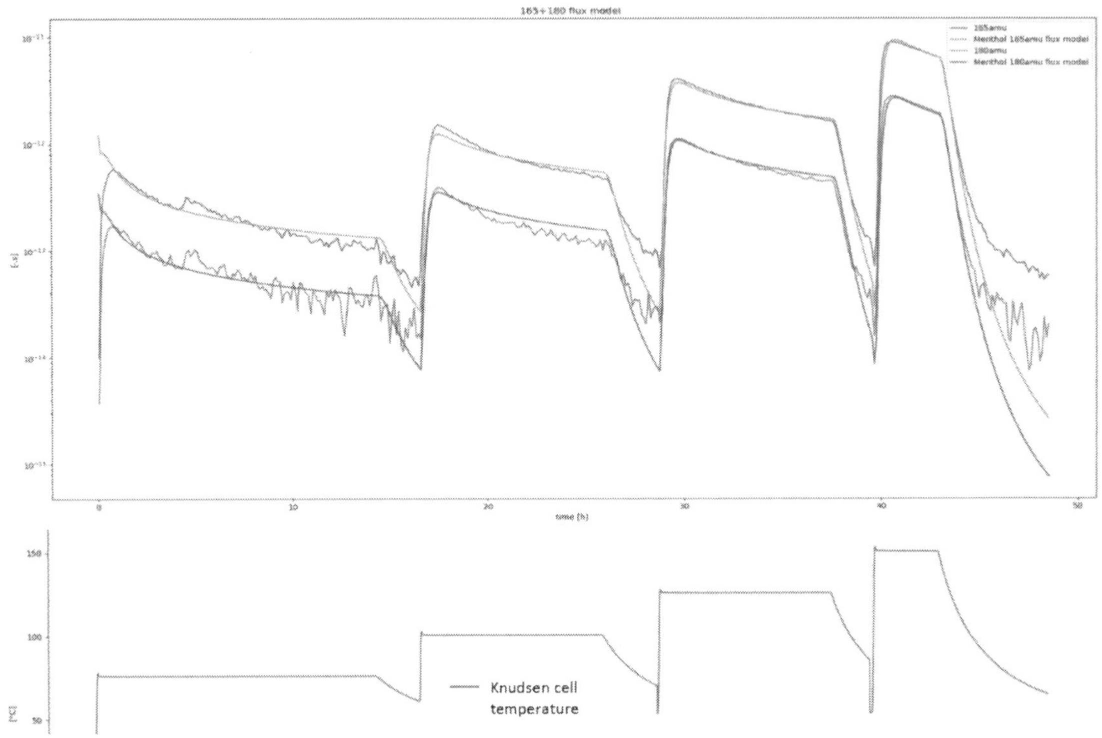

Figure 4. Two fragments representative of menthol (165 and 180 a.m.u.), compared to the same Fick's law (upper panel). The 4 clearly visible steps correspond to 4 outgassing steps at temperatures 75, 100, 125 and 150°C visible on the lower panel (each followed by a passive cooling of the Knudsen cell to stop outgassing and perform TGAs while monitoring them by mass spectrometry). Units of MS data are arbitrary.

Next figure represents the two best fragments specific of Butylated Hydroxytoluene. This time both the flux (direct MS measurement) and its fluence (partial time-integral of the MS signal) are represented. Both are quite well fitted by another solution of Fick's law again. The fit is least convincing when the signal is the smallest, closest to the noise level of the mass spec (around 10^{-14}), which probably artificially increased the measured values.

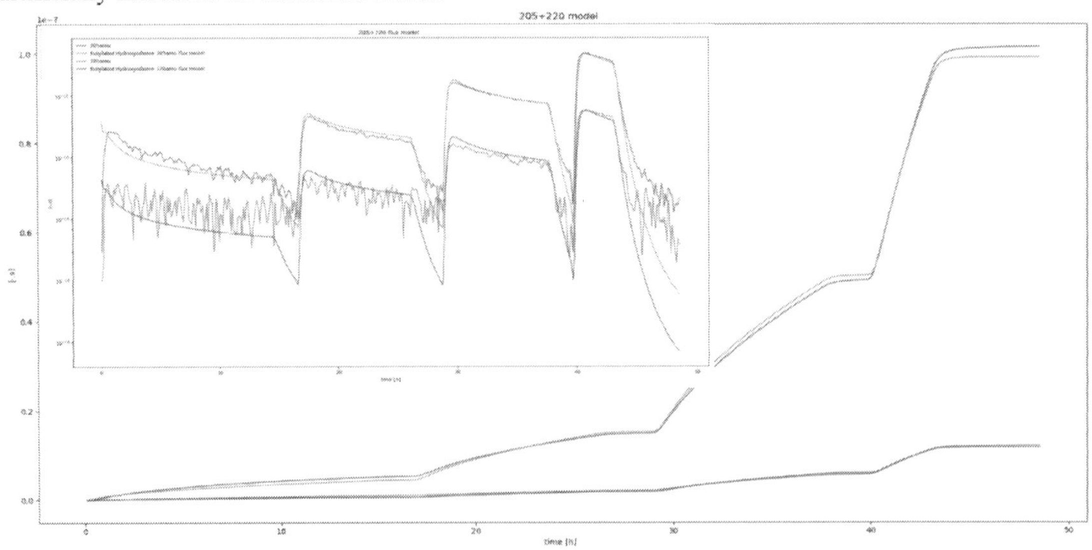

Figure 5. Two fragments representative of butylated hydroxytoluene (205 and 220 a.m.u.), compared to the same Fick's law – flux (upper left) and fluence (lower right)

Similar measurements and fits are given below for the identified molecules that could be either methyl-ethyl-phenyl-ethanone or 1,3-bis(1-methyl-ethyl)-benzene (quite similar molecules, differing only by one functional group). The fit is again quite convincing. The weakest part of the modeling is probably the start of the time interval, where experimental data do not exhibit as sharp a jump as in the model, probably because of the outgassing they experienced before actually heating the Knudsen cell at time = 0 here (kind of pre-outgassing as soon as the sample is put under vacuum).

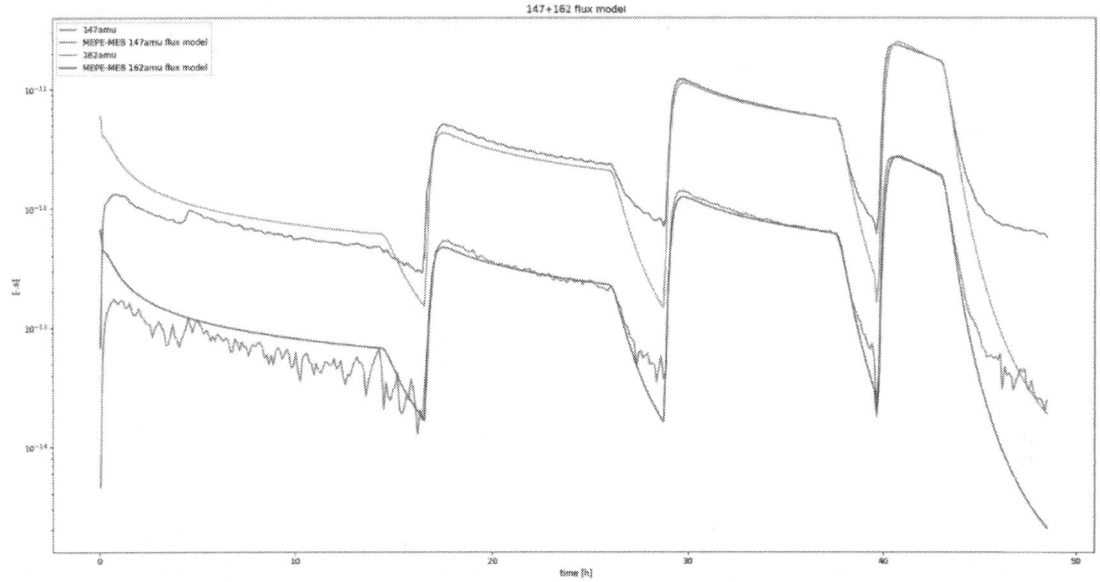

Figure 6. Two fragments representative of methyl-ethyl-phenyl-ethanone (MEPE) or methyl-ethyl-benzene (MEB), very similar molecules (147 and 162 a.m.u.), compared to the same Fick's law

Similar data for dodecanol exhibit a slightly less accurate fit by Fick's law, but still reasonable.

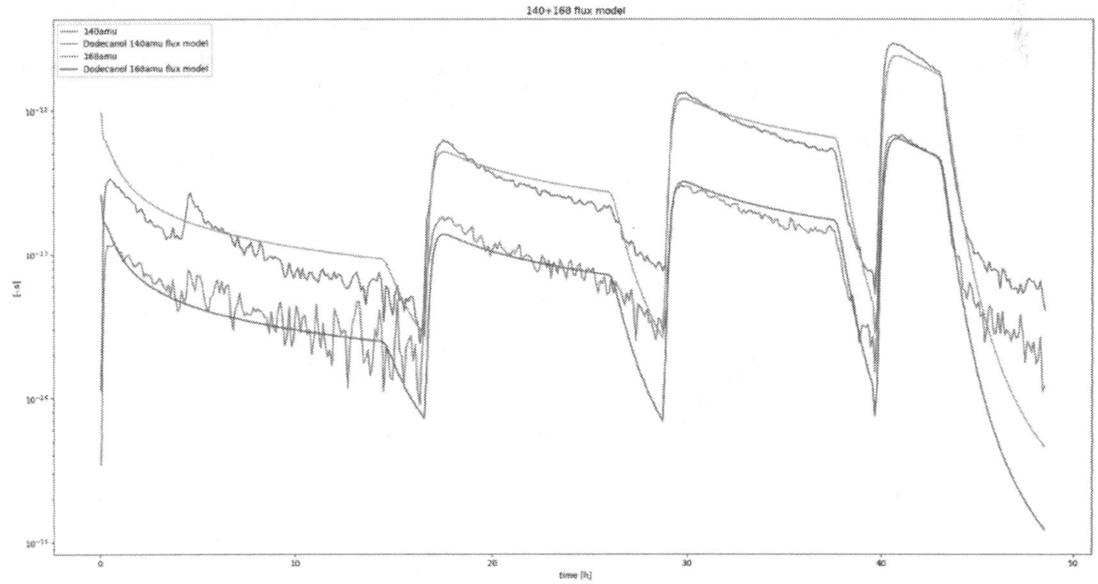

Figure 7. Two fragments representative of dodecanol (140 and 168 a.m.u.), compared to the same Fick's law

IOP Conf. Series: Materials Science and Engineering 1287 (2023) 012020 doi:10.1088/1757-899X/1287/1/012020

Next figure deals with the seventh separated molecule in [5]. We did not manage to make a link with one of the mass spectra of the species identified by GC/MS of the EC9323-2 before these experiments, so, for now we keep naming it as "the molecule whose reemission peaked at 32°C in June 2021 / 125°C experiment". Fits are still quite satisfactory. The artefact at time = 4 hours, which was already visible in some of the previous graphs, is more obvious here. It is probably due to a small uncontrolled heating of some distant part of the hot shroud or sample holder, leading to the reemission of condensed molecules. It does not seem to have large consequences.

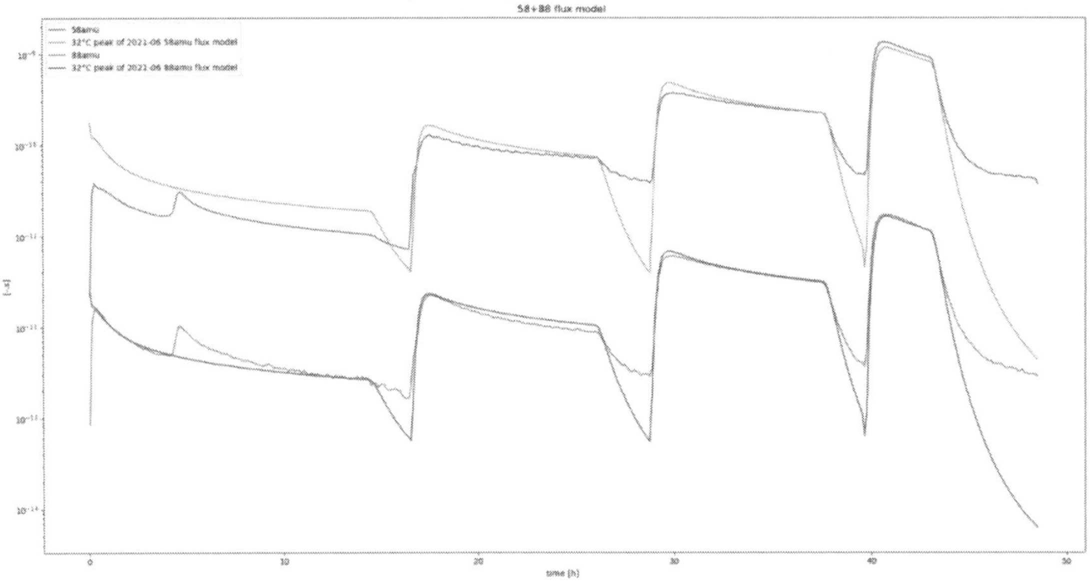

Figure 8. Two fragments representative of an unidentified molecule, the one reemitted at 32°C at 2021/06 TGA (58 and 88 a.m.u.), compared to the same Fick's law

The next graph gives similar data and model for cyclohexanedimethanol. The fit is less convincing than those of the previous separated species. The 95 and 93 a.m.u. fragments, do not have a very high atomic mass, and might be the least specific of studied fragments. It means that they might also be significantly present in the spectra of other outgassed species, contributing to these measurements by a few tens of percents. The artefact at t = 4 hours is also significantly present in these data.

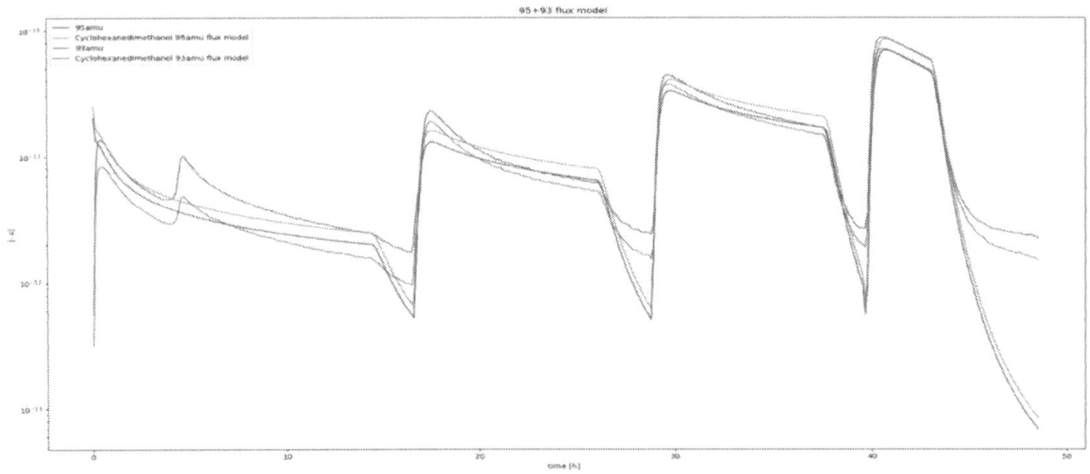

Figure 9. Two fragments representative of cyclohexanedimethanol (95 and 93 a.m.u.), compared to the same Fick's law

IOP Conf. Series: Materials Science and Engineering 1287 (2023) 012020 doi:10.1088/1757-899X/1287/1/012020

Finally, the seventh and last species represented below is octamethyl-cyclotetrasiloxane, or D4. The modeling of lower quality here again, might rather be due to the low signal of these high mass fragments (281 and 267 a.m.u.). Part of the time it is close to the MS baseline, around 10^{-14} (noise level).

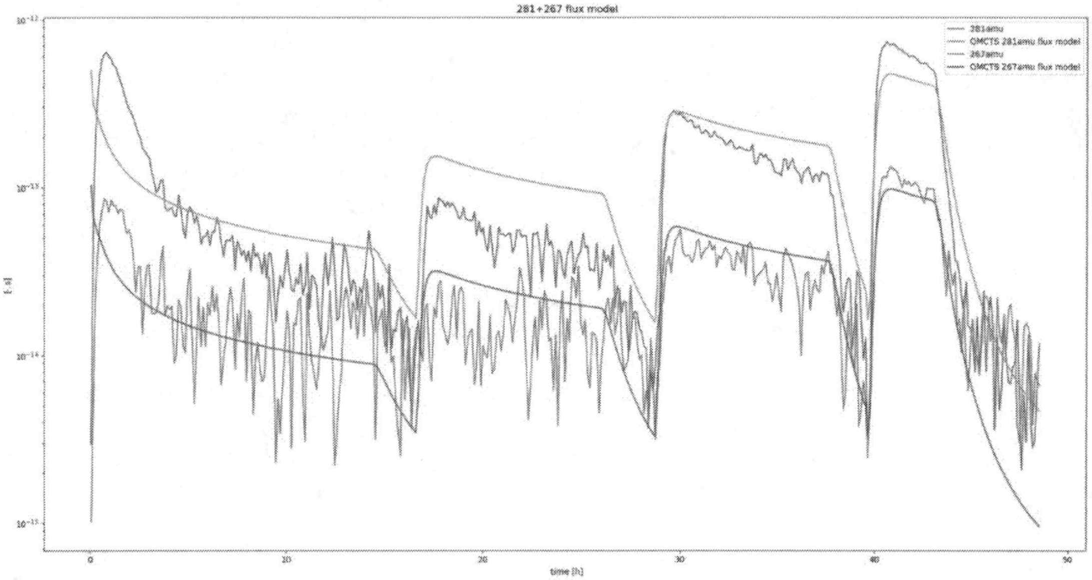

Figure 10. Two fragments representative of D4, octamethyl-cyclotetrasiloxane (281 and 267 a.m.u.), compared to the same Fick's law

The Table 2 below gives the diffusion parameters used in the modeling of the seven species displayed on the seven figures above. It was assumed to follow Fick's law

$$\frac{\partial c(x,t)}{\partial t} = D\frac{\partial^2 c(x,t)}{\partial x^2}, \tag{4}$$

which describes the evolution of each contaminant species concentration $c(x,t)$ within a material depth (space coordinate x) over time (t). The diffusion coefficient is thermally activated and should follow an Arrhenius law:

$$D = D_0 e^{-E_a/kT}, \tag{5}$$

whose coefficients D_0 and E_a are compiled in the table. In practice, the model solutions displayed above were computed by using a series development of Fick's law solutions, given e.g. in ref. [7].

Table 2. Diffusion parameters for each of the seven separated species (from the fits above)

Species (TGA peaks):	Specific MS chanels m/z	E_a kJ/mol	D_0 m^2/s	D at 125°C m^2/s
• Methyl-ethyl-phenyl-ethanone (MEPE) or 1,3-bis(1-methylethyl)-benzene (MEB)	147, 162	95	13.5	4.6×10^{-12}
• Octamethyl-cyclotetrasiloxane (D4)	281, 267	63	0.00010	6.2×10^{-13}
• Butylated Hydroxytoluene	205, 220, 145, 163, 206, 161, 177, 189, 195	122	4055	4.1×10^{-13}
• Menthol (indeed 1'-(butyn-3-one-1-yl)-menthol)	165, 180, 193	109	679	3.1×10^{-12}
• Dodecanol	140, 168, 98, 111, 112	83	0.37	4.4×10^{-12}
• Cyclohexane-dimethanol	95, 93,202, 264, 108, 67	91	4.6	5.3×10^{-12}
• Unidentified molecule *M*	58, 88	106	232	3.2×10^{-12}

IOP Conf. Series: Materials Science and Engineering 1287 (2023) 012020 doi:10.1088/1757-899X/1287/1/012020

3.3. TML and CVCM modelling

Once the outgassing laws of each of the chemical species were known individually, they were used to model the total quantities measured on the QCMs, i.e. the TML and CVCMs at -75, -50 and -25°C measured at CNES (with classical outgassing steps at 25, 50, 75, 100 and 125°C). An important question was whether each of these species was condensed at each of these three QCM temperatures, in each of the three CVCMs. Since we did not do a TGA/MS analysis of the deposits obtained at all these temperatures, we simply made assumptions based on the reemission temperatures observed for the seven species during the TGAs experiments coupled to MS performed at ONERA.

We observed in [5] that the reemission temperature of each species could vary depending on the TGA, probably due to different compositions of the deposits. We considered that the main reason of these fluctuations was the trapping of this species in a mixture with heavier species, which could delay the reemission of this species (hence to a higher temperature). So, if something like an "intrinsic reemission temperature" exists for a given species, it must be close to (or somewhat smaller than) the smallest of the reemission temperatures observed in all TGAs. Moreover, theses reemission peak temperatures correspond to a large reemission flux. For CVCM deposits cumulated over a few days, a reemission flux quite smaller than that is probably sufficient to evaporate that species (the "start of the peak" is enough for a full reemission over days, no need to go to the peak maximum as seen in TGA/MS).

If we process the reemission temperatures of all seven species during all eleven TGA/MS along this line, we get a kind of "sticking temperature" for each species. In Figure 11 the ranges of observed reemission peak temperatures is displayed on the right (organised as a function of the temperature of the outgassing that led to the reemitted deposit, to show the trend discussed above, where the presence of heavier molecules probably delays the reemission of a given molecule). For each molecule, the minimum of these temperatures, decreased of 15°C, is displayed at their left in the black rectangles. As explained above, they can be considered as approximations of the sticking temperatures of the seven species. On this basis, the presence of each of the seven species in each CVCM was assessed, and summarized in the table on the figure's left hand side.

QCM measurement	***Species considered sticking at that CVCM temperature***
CVCM -25°C	*1 species: unidentified molecule* M
CVCM at -50°C	*4 species: same 1 + CHDM + Menthol + Dodecanol*
CVCM at -75°C	*7 species: same 4 + BHT + D4 + MEPE/MEB*
TML	*8 species: all = same 7 + only water added*

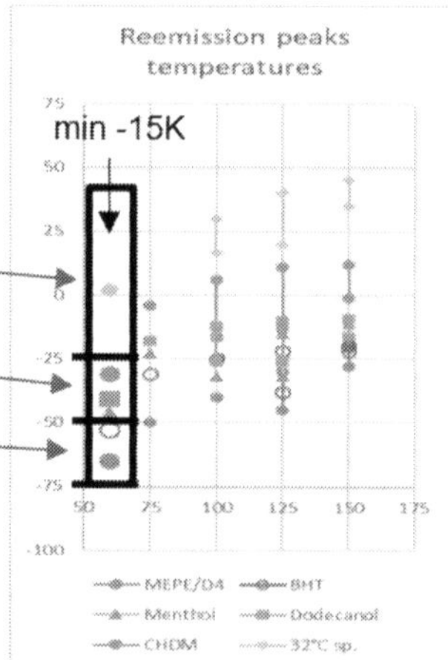

Figure 11. Ranges of observed reemission temperature for the 7 species during 10 TGA/MS from [5] organized in function of the outgassing temperatures (4 columns 75, 100, 125 and 150°C of the r.h.s graph), inferred sticking temperatures (minimum of the previous T°s decreased by 15K for each species, left column of the graph in the black bold rectangles), and assessed presence in each 3 CVCMs at -75, -50 and -25°C (l.h.s. table)

Since the TGAs of [5] were performed from surfaces at -80°C, the most volatiles species could not be detected (those not condensable at -80°C). They necessarily include water, which was thus added as an eighth species contributing to the TML. There might be some other volatile species. Yet, in the absence of actual data, we did not add another species that would add extra arbitrary fitting parameters, which we wanted to avoid. Moreover, other very volatile species would not be expected to have as big a contribution as water.

We started this process with the CVCM at -25°C, which had the advantage of involving one species only in this analysis, hence leaving very few degrees of freedom. The fit of that CVCM by that species is displayed below. The diffusion parameters of that species were simply taken from MS data (lower row of Table 1), and only the global quantity of this species was used as a fitting parameter (since MS data are not quantitative). If compared to what a desorption law would give, the fit can be considered as rather good. The model gives the right order of magnitude all over the experiment (whereas desorption exponentials give a quick exhaustion of the reemission, over a couple of temperature steps). Looking closer at the fit details, it is not so good, in particular concerning the change of deposition rate at temperature steps. The model usually underestimates the accretion rate changes at the steps. This is the sign that a species (or several) with an activation energy larger than in the model (106 kJ/mol) contributes to the CVCM (probably around 150 kJ/mol). This is not a real surprise since in the analysis of the TGA/MS data [5], when looking for specific fragments only, we were aware of probably missing a few species. For some of the observed reemission peaks, no *m/z* fragment exhibiting a "single peak" (dromedary-like peaks) was found. The reemission peaks showed in all fragments as "double-peaks" (camel-like peaks) or even multiple peaks.

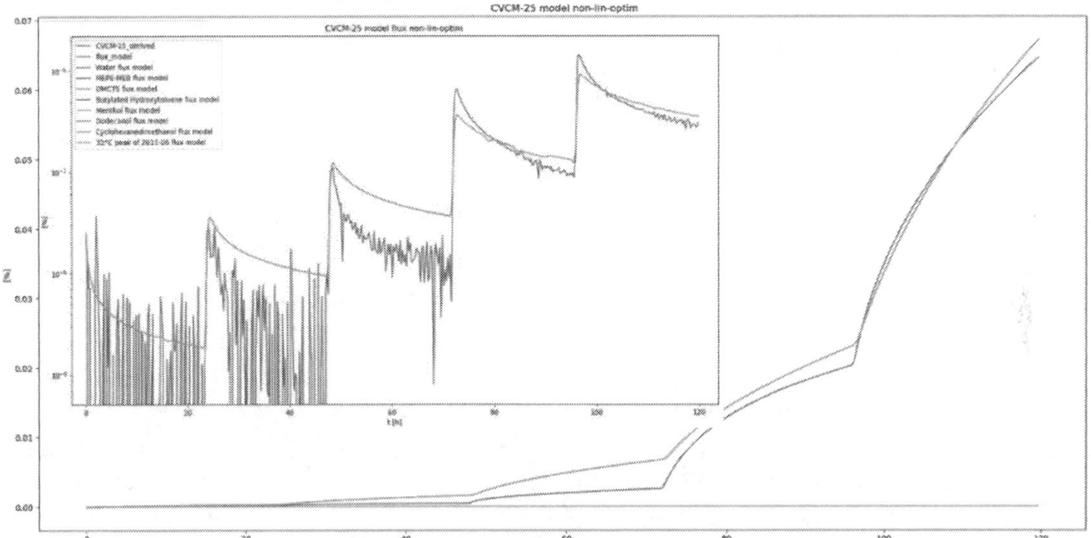

Figure 12. CVCM at -25°C : measurements and modeling (1 single species), deposit in the lower right and despoition rate in the upper left (its derivative)

The -50°C CVCM was next modeled with the four species listed in Figure 11. The fit is quite good again, with rate changes at steps that are not as accurate as one could expect yet.

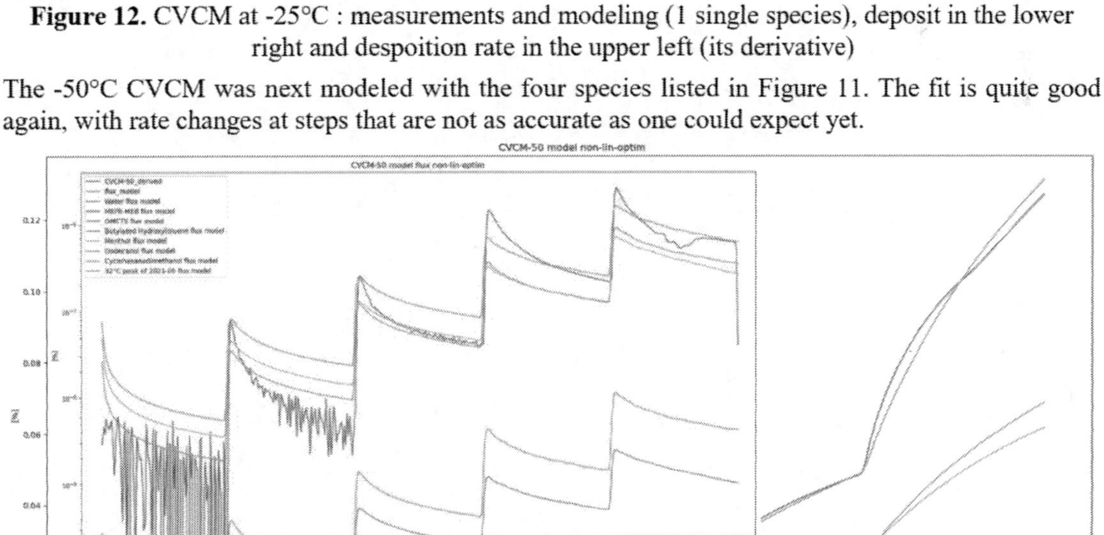

Figure 13. CVCM at -50°C : measurements and modeling (4 species), deposited mass and rate

It is good to see that adding extra species, which condensate on a colder QCM, accounts satisfactorily for the difference between a colder CVCM and a warmer QCM. This difference exists at most of the outgassing temperature steps, and is well described by a diffusion law for the outgassing of these extra species.
The same comments can be done about the -75°C CVCM displayed below.

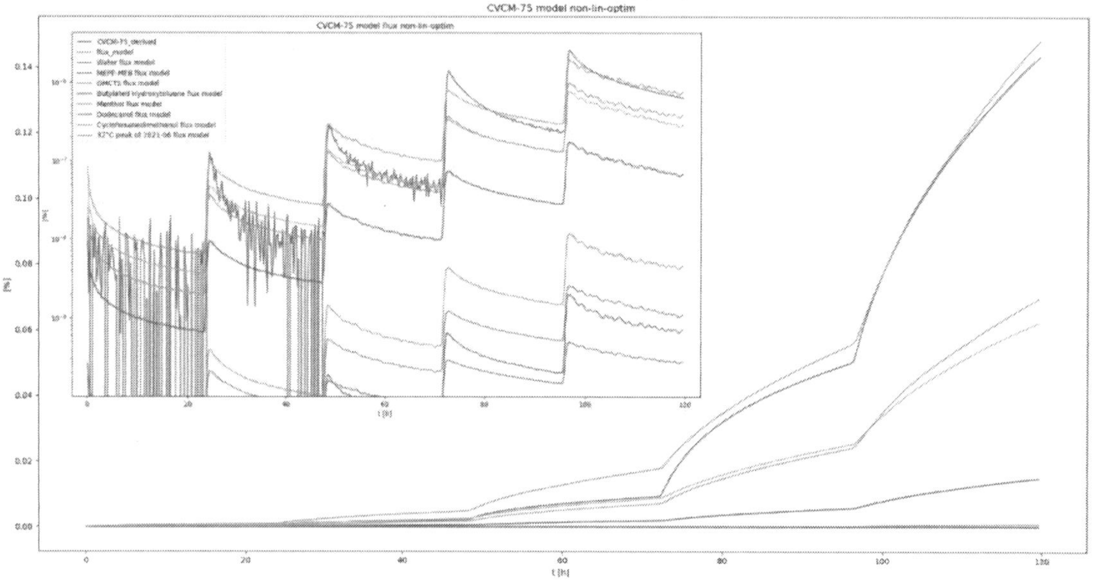

Figure 14. CVCM at -75°C: measurements and modeling (7 species), total deposited mass and rate

As explained above, we only added one extra species, water, to model the TML. So, the large difference between the TML and largest CVCM (-75°C), needed to be modeled as the contribution of one single species, water. And the figure below shows that the diffusion law used for water does a very decent job in modeling this difference, which extends over the first 4 temperatures steps and represents about 80% of the TML. Even though this was not confirmed by direct measurements of water, it is interesting to see how the model predicts the end of water outgassing before the last outgassing step (125°C). This is why the difference of deposit between the TML and the -75°C CVCM is small during that last step (almost no more water outgassed) and large difference during the first four steps (mostly due to water).

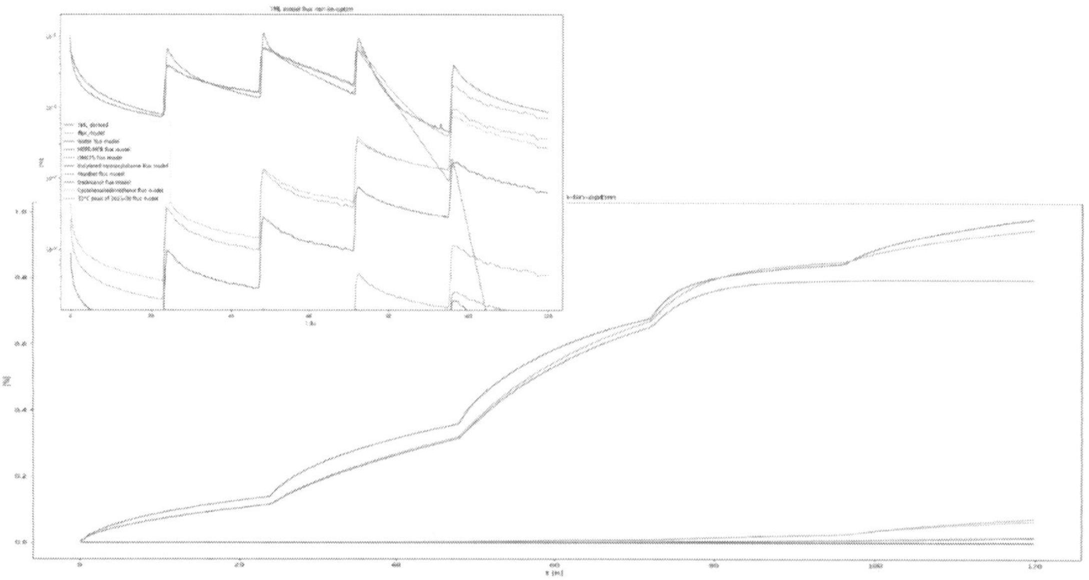

Figure 15. TML: measurements and modeling (4 species), total deposited mass and rate

3.4. *Bake-Out modelling*

With a physical model of outgassing that looks to capture its physics satisfactorily, modelling under-vacuum Bake-Out simply amounted to modelling the modified temperature history. The key points to obtain a predictive modelling of Bake-Out this way are having the right physics of outgassing, and a realistic species separation. The first point is of course needed for a good modelling with a modified temperature (+72h at 60°C before the start of the outgassing test) and the second point is needed to cope with the issue of water regain after the Bake-Out, as we shall see.

Next figure compares the post Bake-Out TML measurements with various modelling approaches. Graph (a) compares the experimental TML of a baked-out EC9323 sample to an unchanged model. Clearly, Bake-Out reduced the material TML, and the model needs to be modified to take BO into account. If the announced model change, consisting in simply modelling the BO as a pre-test temperature evolution ruled by the same Fick's law, is applied (b), the model now underestimated measurements. The cause of this discrepancy has been known for long. It is due to water regain, from the atmosphere, in the interval between the BO and the outgassing test. If we assume that water, which we now believe to be the first species, is fully regained during these few weeks (c), the model comes back to an overestimation of the measurements. It is not surprising since the Bake-Out's main effect was on water. It is also known that water regain in lab conditions in the short term is only partial. Assessing water regain is no easy. We wanted yet to keep to a simple and reasonable hypothesis, so we tried assuming that 50% of water was regained. This simple modelling approach yielded results reasonably close to data (d), within 10% of the maximum TML.

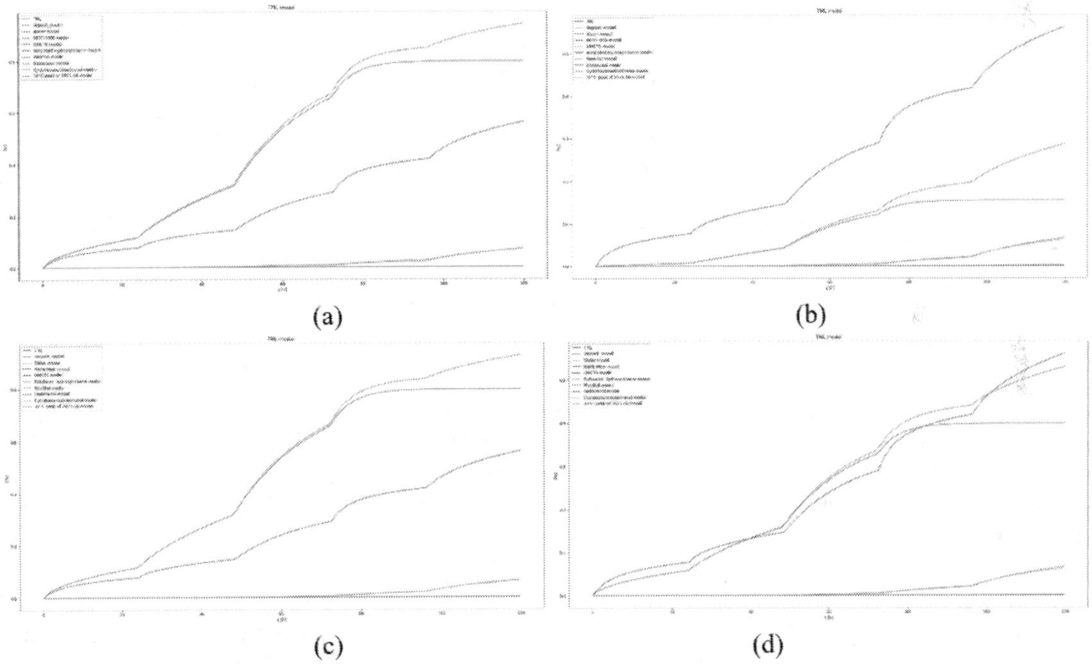

Figure 16. Comparison of post Bake-Out TML measurements (blue line) and model results (orange), decomposed in eight species (green = water, etc.): (a) unchanged model (no BO), (b) direct modeling of BO as 72 h at 60°C, (c) ditto with full water regain (1st species), (d) ditto with 50% water regain

It is interesting to compare this approach to that of section 2, the traditional one, based on mathematical species. There, since most mathematical species could in reality be partly composed of water, a partial regain needed to be assumed for all species (of variable amplitude) in order to fit the data. Here, the fact that a partial regain of only the lightest species allowed a rather good fit of the data is a very good argument in favour of the validity of our species separation.

If one (cautiously) tries to further interpret the (small) residual model-data discrepancy of graph (d), we see that the model underestimates the initial outgassing (first temperature step) and overestimates it later (third step). In the diffusion model, water regain is assumed to be uniform throughout the whole material depth. The initial water concentration in EC9323 is simply reduced by 50%, uniformly from the surface to the bottom of the material. Since water regain happens from the surface, by diffusion through the material of the molecules condensed at its surface, water concentration could eventually be larger close to the surface. This would lead to a larger outgassing at test beginning, exactly as it was seen in experimental data. It seemed to us too early to dig into this direction now, since we first wanted to verify the quality of the results of the diffusion modelling based on species separation, and no extra assumption on BO. But of course, a better modelling of water regain might be worth trying in the future.

Next figure also displays the CVCMs of EC9323-2 after BO, comparing the measurements and the models with 50% water regain. Indeed the water regain does not impact the CVCMs. These graphs can be compared to the unbaked cases of Figure 12 (-25°C CVCM), Figure 13 (-50°C) and Figure 14 (-75°C).

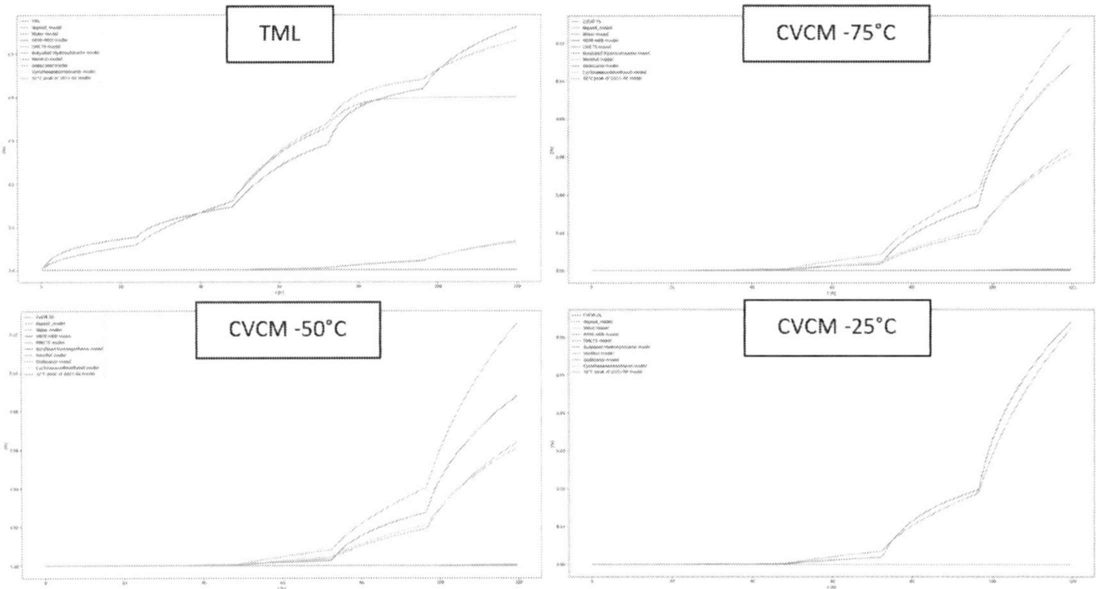

Figure 17. Comparison of post BO measurements and post BO model with 50% water regain: TML and CVCM

The species condensed at -25°C have a small volatility, and the BO has little effect on them, both in the measurements and the model. The more volatile species condensed at -50 and -75°C are more affected by the BO and the comparison is more interesting there. In both cases the model (orange lines) underestimates the data (blue lines). We think that these small discrepancies could be due again to some species missed in the species separation of [5], rather volatile species in this case. From these graphs, we can estimate these missing species as representing a total of around 0.04% of the outgassing material (versus 0.15% for the -75°C CVCM). As already suggested, improving species separation completeness is another direction of progress to pursue.

4. Conclusion and perspectives

Two different approaches were used to model post Bake-Out outgassing, and compared to experimental data. The first is the usual method which uses mathematical species and desorption laws. This study has highlighted that reintroducing only the first species after bake-out to take into consideration water reabsorption generally leads to a significant error on the simulated bake-out. Moreover, the error depends on the number of species. Methods based on the reintroduction according to WRM have been presented and have shown better results than reintroducing only the first species. The main drawback is the requirement to have the WRM value. Also, more investigations will be carried out to consolidate the reliability of these methods and their ability to predict CVCMs after a bake-out.

The second is a new method based on a realistic species separation through TGA/MS technique [5] allowed modelling outgassing of actual chemical species, through Fick's diffusion laws. A reasonable reproduction of the post-BO data was obtained by the simple assumption of a 50% water regain, water simply being the first species. To our knowledge, this is the first time this program, separating species at molecular level and modelling outgassing and BO on this basis, was implemented.

In the new approach, even though the models of the outgassing of the seven individually-separated species are quite good (comparison to MS data), the modelling of the global data (total masses and deposits, TML and CVCMs) is somewhat less accurate. It could be due to the fact that a couple of significant chemical species might still need to be found, and that water modelling might be somewhat less accurate since it could not be directly measured, due to chamber water background. In the future, these points will need to be improved. Other progress directions will be the testing of other materials, to gain more experience and understanding of these limitations, and the improvement of data processing automation.

Acknowledgements
This research activity was funded by a CNES R&T contract. The ONERA authors also want to acknowledge several funding sources over the long term, from CNES, ESA and ONERA internal resources, which supported the global development of these innovative experimental techniques and modelling capabilities.

References
[1] Van Eesbeek M and Zwaal A 1985 Outgassing and contamination model based on residence time *Proceedings of the 3rd ISMSE*, ESTEC, Noordwijk, The Netherlands, pp 25-34
[2] Roussel J-F, Faye D, Van Eesbeek M, Tondu T, Migliore R, Rampini R and Paulmier T 2009 A New Frontier for Contamination: Reaching the Molecules *Proceedings of the 11th ISMSE*, Aix-en-Provence, France
[3] Roussel J-F, Tondu T, Paulmier T, Faye D, van Eesbeek M and Rampini R 2011 Progress on the Physical Approach to Molecular Contamination Modeling, *J. of Spacecraft and Rockets*, 48-2, pp 246–255
[4] Grosjean E, Vanhove E, Roussel J. F, Rioland G and Faye D 2018 Molecular Contamination from Space Materials: Species Characterization by TGA/MS. *Proceedings of the 14th ISMSE*, Biarritz, France
[5] Lansade D, Roussel J-F, Faye D, Rioland G, Suliga A and Ergincan O 2022 Contamination Level Prediction: Progress in Species Separation by TGA/MS, submitted to *ISMSE* 2022
[6] Lansade D, Roussel J-F, Alred J M, Wong A T, Anderson J R, Soares C E, Faye D and Rioland G 2022 Separation of contaminant species by TGA/MS in European and U.S. approaches to outgassing *Proc. SPIE 12224, Space Systems Contamination: Prediction, Control, and Performance 2022*
[7] Roussel J-F, Lansade D, Alred J, Wong A T, Anderson J R, Soares C E, Faye D and Rioland G 2022 Application of contaminant species separation by TGA/MS to unraveling outgassing physics and laws, *Proc. SPIE 12224, Space Systems Contamination: Prediction, Control, and Performance 2022*

IOP Conf. Series: Materials Science and Engineering 1287 (2023) 012021 doi:10.1088/1757-899X/1287/1/012021

Measurement of Mass Gain Behavior due to Hydration and Reduction of Uncertainty Induced by Time Lag of Weighing on Pristine Kapton Film after Heating in Vacuum

Eijo Fujiwara[1, 5], Minoru Iwata[2, 6], Masahito Tagawa[3] and Sumitaka Tachikawa[4]

[1] Graduate School of Engineering, Kyushu Institute of Technology, 1-1 Sensui, Tobata, 804-8550 Kitakyushu, Japan
[2] Faculty of Engineering, Kyushu Institute of Technology, 1-1 Sensui, Tobata, 804-8550 Kitakyushu, Japan
[3] Kobe University, 1-1 Rokkodai, Nada, 657-8501 Kobe, Japan
[4] Japan Aerospace Exploration Agency, 3-1-1 Yoshinodai, Chuo, 252-5210 Sagamihara, Japan
[5] Currently, Mitsubishi Heavy Industries Compressor Corporation, Hiroshima Japan
[6] Author to whom any correspondence should be addressed.

E-mail: iwata@ele.kyutech.ac.jp

Abstract. To simulate material degradations in space environments, materials are exposed to various space environment factors in vacuum. In vacuum, low-molecular-weight component, moisture, and etc. desorb from the materials, and it follows the mass loss. The vacuum-desorbed materials, however, adsorb the moisture due to retrieval from vacuum chamber into the atmosphere condition. The mass gain due to hydration disturb the accuracy of an atomic oxygen fluence measurement and an outgassing test. In this study, we have evaluated the mass gain behaviour of vacuum-desorbed material due to hydration. To desorb the moisture, the test specimens were heated in vacuum. After dehydration in vacuum, the mass of test specimen was measured by using an ultra-micro balance inside the low-humidity controlled globe box to evaluate the mass equivalent to that in vacuum condition and then the mass was measured under controlled condition of 50% relative humidity to evaluate the mass gain in atmosphere. From the experimental results, we have evaluated the uncertainties due to hydration occurring during the time lag between sample retrieval from a vacuum chamber to sample weighing, quantitatively. The back extrapolation using the regression curve based on the solution of diffusion equation is effective on the reduction of uncertainties.

1. Introduction

In space environments, there are various environment factors, i.e. geomagnetically trapped charged particle radiations, solar wind, solar flare, atomic oxygen, thermal cycling, outgass and contamination, ultra-violet rays, X rays, gamma rays, and etc. Degradation of space-use materials are induced more or less by exposing to harsh space environments. On practical issues, the degradation of space-use materials are evaluated quantitatively by measuring with the variation of required physical and/or chemical properties depending on its usage. For example, the durability of materials are assessed by using thermo-optical properties for thermal control materials, a modulus and/or strength for structural materials, and etc. In many cases, these physical or chemical properties are measured by using the test and measurement method based on standards of the International Organization for Standardization (ISO), ASTM (American Society for Testing and Materials) International, European Cooperation for Space Standardization (ECSS), and etc.

Atomic oxygen (AO) is main component of neutral atmosphere less than 1,000-km altitude and relatively collide with a spacecraft surface having a velocity of 8 km/s at an energy of 5 eV. Atomic oxygen would erode many kinds of organic materials, and would remove matters from the substrate. Although the durability for atomic oxygen on thermal control materials would be evaluated by using of thermo-optical properties, the durability is usually assessed by using the total amount of erosion measured with mass loss because the removed volume calculated with material density from its mass loss invoke life time of bulk material and make it possible to compare between different materials, easily.

If materials are exposed to vacuum in space, volatile gas composed of low-molecular-weight components, additives, adsorbed matters, and etc., so-called outgas, are emitted from the materials. Outgas could induce the contamination on spacecraft surface and the induced contamination would provoke the performance deterioration of lens, mirror, camera, solar cell, and etc. Outgassing test is required for a material used for spacecraft, especially for the spacecraft having optically sensitive components. Outgassing test is a screening test to reject the materials having high-risk of contamination. Outgassing properties on candidate materials mainly consist of three parameters that is total mass loss (TML), collected volatile condensable material (CVCM), and water vapor regained (WVR). TML is calculated from the mass change of materials before and after vacuum heating at 125°C for 24 hours. CVCM is calculated from mass of contaminants. The contaminant are induced on the collector plate controlled at the temperature of 25°C by the outgas resulting from vacuum heating of materials at 125°C for 24 hours. WVR is calculated from mass change of vacuum-heated materials before and after exposure to 50% relative humidity at 23°C for 24 hours. Therefore, the outgassing properties are evaluated with mass change of materials.

The AO erosion and the outgassing properties are evaluated with the mass change of materials. Since AO irradiation test and outgassing test are performed in vacuum, specimen materials are exposed to vacuum, release adsorbed moisture and other gases, and become dehydration condition. Therefore, if the specimen materials are retrieved from a vacuum chamber to air atmosphere condition because the specimen would be usually weighed in air atmosphere condition, the mass would increase with time due to moisture adsorption. The mass gain due to moisture adsorption in air would disturb the results of amount of AO erosion and outgassing properties on test specimens because vacuum-stored specimen mass after finishing the tests would be required on AO durability test and outgassing test. In ASTM standards for AO interaction evaluation [1] and outgassing test [2], the sample weighing criteria are described for specimens retrieved from vacuum chamber and the effects of moisture adsorption after exposure to air could be minimized by using back extrapolation of time variation on weighing results. However, it is not enough to discuss the error caused by unavoidable moisture adsorption on mass weighing criteria. In this study, we make clear the error due to an acceptable time lag on weighing criteria by measuring the mass gain due to moisture adsorption after retrieving from vacuum to air.

2. Weighing criteria in test standards

2.1. Weighing on AO interaction evaluation

Atomic Oxygen (AO) is main component of neutral atmosphere around the Earth at an altitude of less than 1,000 km. The materials attached on spacecraft surface orbiting on low earth orbit receive the AO erosion effect and mass loss would occur. Especially, organic materials would receive significant erosion effect and total amount of thickness loss might reach more than several tens of micro meters [3]. Erosion yield E_S calculated by using

$$E_S = \frac{\Delta M_S}{A_S \rho_S F_k} \tag{1}$$

where

E_S	:	erosion yield of the sample, cm^3/atom
ΔM_S	:	mass loss of the sample, g
A_S	:	surface area of the sample exposed AO, cm^2
ρ_S	:	density of the sample, g/cm^3
F_k	:	AO fluence, atoms/cm^2

is widely used as an indicator of the resistance of material for AO erosion [1]. As can be seen from equation (1), the erosion yield means removal volume of sample material due to an atomic oxygen collision. To measure AO fluence on ground simulation test, a witness material known in-space erosion yield is used. In many cases, polyimide Kapton-H would be selected as the witness material. We can determine the AO fluence calculated with the mass loss of witness material by using equation (2) [1]. E_k is 3.0×10^{-24} cm^3/atom for Kapton-H [1].

$$F_k = \frac{\Delta M_k}{A_k \rho_k E_k} \tag{2}$$

where

ΔM_k	:	mass loss of the witness material, g
A_k	:	surface area of the witness material exposed AO, cm^2
ρ_k	:	density of the witness material, g/cm^3
E_k	:	known in-space erosion yield of the witness material, cm^3/atom

The weighing accuracy is important whether determination of the erosion yield of sample using equation (1) or measurement of the AO fluence with witness material using equation (2). During AO irradiation on ground simulation test, samples or witness materials are placed in a vacuum chamber and then would dehydrate in the chamber. After finishing the test, the samples or witness materials are retrieved from vacuum to air because an ordinary balance is unsuitable for vacuum to weigh. By retrieving to air, hydration occur on the samples or witness materials and induce the increase of mass with air-exposure time. Therefore, in ASTM standard [1], weighing criteria are described as follows:

Weighing—Because hydration occurs quickly after removal of samples from vacuum, weighing the samples should occur within five minutes of removal from vacuum dehydration chambers. Reduction of uncertainty associated with moisture uptake can be minimized by weighing the samples at measured intervals following removal from vacuum and back extrapolating to the mass at time of removal from vacuum. (p 2)

2.2. Weighing on outgassing test

Outgas emitted from materials in space could induce contaminations on spacecraft surface. The method of screening test, called outgassing test, on which the outgassing properties of materials are evaluated are described in ASTM E-595. In the outgassing test, materials are heated at 125°C for 24 hours below the pressure of 7.0×10^{-3} Pa and are weighed before and after vacuum heating. The outgassing from vacuum-heated material induce condensed matter on a collector plate kept at 25°C during the test. The mass of condensed matter is calculated with the mass change of collector plate before and after test. Additionally, the mass of vacuum-heated materials are evaluated before and after exposure to 50% relative humidity and 23°C for 24 hours. The mass loss of material due to vacuum heating, the mass of condensed matter on collector plate, and mass gain of vacuum heated material due to moisture adsorption are converted to outgassing properties, i.e. *TML*, *CVCM*, and *WVR*, respectively, by using the following equations [2] with initial mass of materials.

$$TML = \frac{S_i - S_f}{S_i} \times 100 \quad (3)$$

$$CVCM = \frac{C_f - C_i}{S_i} \times 100 \quad (4)$$

$$WVR = \frac{S_f' - S_f}{S_i} \times 100 \quad (5)$$

where

S_i : initial mass of specimen on outgassing test
S_f : mass of specimen after finishing vacuum heating
C_i : initial mass of collector palte
C_f : mass of collector plate after after finishing vacuum heating
S_f' : mass of vacuum-heated specimen after exposure to 50% relative humidity and 23°C for 24 hours

To select acceptance or rejection from outgassing test results, one of the threshold value is that "*Historically, TML of 1.00% and CVCM of 0.10% have been used as screening levels for rejection of spacecraft materials.*" (reference [2], p 1) To determine the outgassing properties, the weighing accuracy is also important on the outgassing test. During the outgassing test, materials are heated in a vacuum chamber and would dehydrate. By retrieving from vacuum to air after vacuum heating, hydration occur on the materials and induce the increase of mass with air-exposure time. The mass gain due to hydration would disturb the reliabilities of outgassing properties. Therefore, in ASTM standard [2], weighing criteria are described as follows:

Store aluminum boats with specimens and respective collector plates in desiccators (using active silica gel desiccant) immediately. After specimens have cooled to approximately room temperature, but no longer than 1/2 h, remove and weigh each specimen within 2 min of its removal from the desiccators. (pp 5–6)

In ECSS document [4], weighing criteria are described as follows:

5.3.2.10 Unloading of system

a. The supplier shall unload the system as soon as possible.

b. The supplier shall keep the specimen cups in a desiccator for not more than 30 minutes.

c. The supplier shall keep the collector plates in a desiccator for one hour.

d. The supplier shall weigh the specimen cups and collector plates.

NOTE This is done by taking them from the desiccator one by one and returned immediately thereafter.

(p 23)

3. Experimental methods and equipments

The error of sample mass is directly reflected to AO interaction evaluation (erosion yield and AO fluence) and outgassing properties (*TML*, *CVCM*, and *WVR*). Main cause of the error on mass measurement is hydration. Since moisture adsorption would induce mass gain, samples should be weighed within 5 min [1] of retrieval from vacuum dehydration chambers or 2 min [2] of retrieval from desiccators according to ASTM standard, although there is no suggestion of time restriction between removal from desiccators and weighing in ECSS document [4]. During a delay of 5 min or 2 min, the error due to hydration would occur. First, we minimized moisture adsorption effect between retrieval from vacuum chamber and weighing, and then evaluated mass gain behavior due to hydration with exposure time in air. From the results, the weighing error due to delay would be determined quantitatively. Next, we have applied a regression model to the experimental results of mass gain behavior due to hydration. From the regression analysis, we would determine a back extrapolation

IOP Conf. Series: Materials Science and Engineering 1287 (2023) 012021 doi:10.1088/1757-899X/1287/1/012021

function to minimize the error due to hydration and to predict the mass immediately before retrieval from vacuum chamber.

3.1. *Specimen and weighing*

Polyimide Kapton-H, a standard witness material for AO fluence determination, was used for a test specimen. To adjust to requirement of sample mass on outgassing test, the test specimen size was set to 20-mm wide by 70-mm long by 128-μm thick. The specimen mass was about 250 mg. The weighing is performed with a microbalance (MX5, Mettle Toledo) having 1-μg readability. The data logger (B-LOG-01, As-one) is used for data acquisition of mass change behavior at an interval of 1 sec.

3.2. *Weighing for vacuum-dehydrated specimen*

To evaluate AO interaction and outgassing properties, specimen mass immediately before retrieval from vacuum chamber is ideally required. For the in-situ weighing, it is necessary to use a balance working in vacuum. But it is not so easy to prepare the vacuum-compatible balance and to integrate it into AO irradiation test facility and outgassing test facility. In this study, the specimen mass was weighed under vacuum-equivalent atmosphere inside of a controlled atmosphere globe box (UL-1300A, UNICO) in which atmosphere is controlled to low-humidity and low-oxygen concentration environments to minimize the mass gain due to moisture adsorption. Figure 1 shows the experimental apparatus. Atmosphere in the globe box was controlled with high-purity nitrogen gas at a dew point of −76°C or less and at an oxygen concentration of 1 ppm or less. A dew point temperature of −76°C is equivalent to a moisture concentration of 1 ppm or 0.01% relative humidity. The microbalance was placed in the controlled atmosphere globe box. The vacuum chamber is attached on the left side of the globe box and an attached gate valve separate a different pressure and atmosphere condition between the vacuum chamber and the globe box. The vacuum chamber is evacuated with rotary pump (GVD-100A, Ulvac) and turbo molecular pumb (HiCube 80 Eco, Pfeiffer Vacuum) at a pressure of 4.0×10^{-2} Pa or less. To achieve outgassing test requirements, we should evacuate to a pressure of 7.0×10^{-3} Pa or less. This is an issue in the future. To desorb moisture from specimen due to heating in vacuum, a tubular furnace (TMF-500N, As-one) and a fused quartz tube are integrated in the vacuum chamber. The specimen was moved to the part of fused quartz tube in the chamber and was heated with the tubular furnace in vacuum at a temperature of 150°C for 24 hours. After finishing vacuum heating, the chamber was backfilled with high-purity nitrogen gas controlled to low-humidity and low-oxygen concentration in globe box at a room pressure. And then, the gate valve was opened and the vacuum heated specimen was transferred to the inside of controlled atmosphere globe box by using a magnetic coupling type linear motion feedthrough manipulator attached on the chamber. The vacuum-heated specimen transferred in the globe box was weighed with microbalance by hand work.

3.3. *Weighing for mass gain due to hydration*

After weighing the vacuum-heated specimen in the globe box, we evaluated the hydration behavior. To evaluate the mass gain behavior due to hydration, we should weigh the vacuum-heated specimen in an atmosphere environment maintained at a constant temperature and humidity. Since moisture concentration in globe box was too low to evaluate hydration behavior, we weighed the vacuum-heated specimen with the microbalance in a constant temperature and humidity room controlled to a temperature of 23°C and a relative humidity of 50% based on the outgassing test conditions. The globe box is located in a building that is different from the building with a constant temperature and humidity room. Therefore, to prevent the disturbance due to unexpected moisture adsorption, the vacuum heated specimen kept in moisture-free condition should be transferred from the globe box to the constant temperature and humidity room. We enclosed the vacuum-heated specimen in a transfer capsule in the globe box, and then brought the capsule to the constant temperature and humidity room. The transfer capsule shown in figure 2 is a simple vacuum component using NW40-ICF70 conversion nipple (for example, 954-7466, Canon Anelva). We can easily pick up the specimen by using a quick-release coupling. The capsule was removed from globe box to air atmosphere through a pass box attached on the right side of the globe box. The capsule was brought to the constant temperature and humidity room. The specimen enclosed in the transfer capsule was picked up in the constant temperature and humidity room and then was quickly started to weigh the mass gain behavior due to hydration with time.

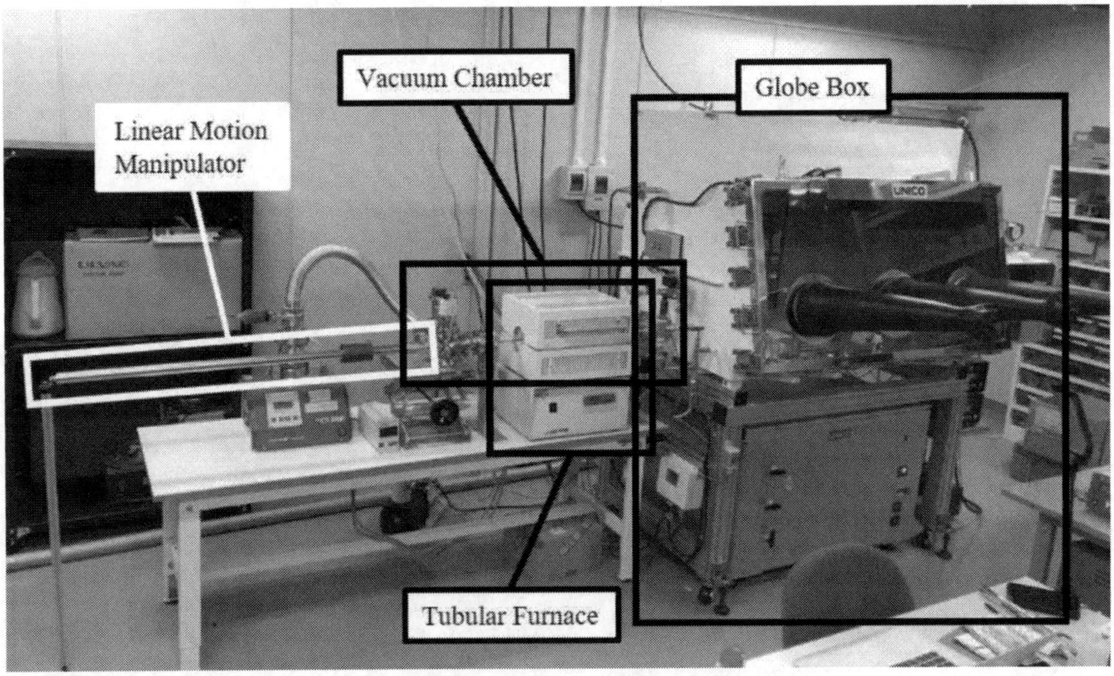

Figure 1. Experimental apparatus for sample weighing under vacuum-equivalent condition.

Figure 2. Transfer Capsule.

3.4. Regression model for back extrapolation

Time delay between at time of removal from vacuum chamber and weighing is permitted for up to 5 min on AO interaction evaluation and for up to 2 min on outgassing test. To reduce the weighing uncertainty induced by moisture adsorption during the time delay, it would be recommended that the recorded time variation of mass are back extrapolated to the mass at the time of retrieval from a vacuum chamber. However, a regression model used for the extrapolation is not specified. In this study, a mathematical model on mass gain due to moisture adsorption was constructed based on the diffusion of matter in solid. The curve fitting using the mathematical model based on diffusion equation was applied to the measurement results of mass change with time, and then we confirmed the correlation between them. On the diffusion of matter in a plane sheet material, diffusion in the direction of depth would be expressed by using one-dimensional diffusion equation.

$$\frac{\mathrm{d}C}{\mathrm{d}t} = \mathrm{D}\frac{\mathrm{d}^2C}{\mathrm{d}x^2} \tag{6}$$

where

t	:	time, sec
x	:	the direction of depth, m
C	:	concentration of diffusing matter at time t on x, g/m^3
D	:	diffusion coefficient, m^2/sec

If the concentration of diffusing matter is kept at a constant of C_1 for $t \geq 0$ on the plane sheet surface and the concentration of that inside of the plane sheet is uniform at C_0 for $t = 0$, the trigonometric series solution of non-steady state diffusion equation [5] is

$$\frac{C - C_0}{C_1 - C_0} = 1 - \frac{4}{\pi}\sum_{n=0}^{\infty}\frac{(-1)^n}{2n+1}\exp\left\{-\frac{D(2n+1)^2\pi^2 t}{d^2}\right\}\cos\frac{(2n+1)\pi x}{d} \tag{7}$$

where

d	:	thickness of plane sheet, m
n	:	number of trigonometric series

By integrating equation (7) with respect to x, total amount mass of diffusing matter inside the plane sheet at time t is calculated with the following equation [5].

$$M(t) = M_0 + M_\infty \quad 1 - \sum_{n=0}^{\infty}\frac{8}{(2n+1)^2\pi^2}\exp \quad -\frac{D(2n+1)^2\pi^2}{d^2}t \tag{8}$$

where

$M(t)$	:	total mass of plane sheet with diffusing matter after time t, g
M_0	:	initial mass of plane sheet, g
M_∞	:	total mass of diffusing matter after infinite time, g

4. Results and Discussions

4.1. Weighing dehydrated specimen

The sample before vacuum heating was stored in a constant temperature and humidity room at 23°C and 50% relative humidity for 24 hours and then the sample was weighed. The sample mass before vacuum heating was 253,211 μg. After weighing, the sample was transported into the glove box kept at low-humidity and low-oxygen concentration and was stored in it for 24 hours. The weighed sample was hanged on the front end of the rod of linear motion feedthrough manipulator delivered in the globe box and was transferred into vacuum chamber by operating the manipulator. The sample was moved to the position of quartz tube on which tubular furnace is attached. After moving the sample to the vacuum-heating position, we evacuated the vacuum chamber and heated the sample with the tubular furnace in vacuum at a temperature of 150°C for 24 hours. After finishing vacuum heating, we turned off the tubular furnace and cooled the sample by exposing quartz tube to air atmosphere resulting from the removal of upper tubular furnace unit from quartz tube. And then the chamber was backfilled with high-purity nitrogen gas controlled to low-humidity and low-oxygen concentration in globe box at an atmospheric pressure. The gate valve between the vacuum chamber and the globe box was opened and the vacuum heated specimen was transferred to the inside of controlled atmosphere globe box by using a magnetic coupling type linear motion feedthrough manipulator attached on the chamber.

To evaluate uncertainties of weighing in the globe box, the sample was weighed with elapsed time in the low-humidity and low-oxygen concentration environment. The reason of weighing inside of the globe box is to evaluate a sample mass in moisture-free condition equivalent to vacuum in-situ measurement. Therefore, the time-course change of mass in the low-humidity and low-oxygen concentration environment should be enough small compared with the mass change that we want to identify. After finishing the vacuum heating, the vacuum-heated specimen transferred in the globe box was removed from the rod of manipulator and was weighed with microbalance by hand work. Figure 3 shows the time-course change of sample mass in the low-humidity and low-oxygen concentration environment after finishing a vacuum heating. The sample mass was 250,210 μg immediately after vacuum heating in the low-humidity and low-oxygen concentration controlled globe box. As a result, *TML* was 1.185% calculated from a mass loss of 253,211 − 250,210 = 3,001 μg. The sample mass decreased in the first hour, and then gradually increased later. Although the cause of mass loss within first hour was not clear, it was deduced that the mass gain during the next five hours is due to a moisture adsorption in the low-humidity environment in globe box. The sample weighing in globe box induced about 20-μg increase during 5 hours except for mass loss within first hour. The mass gain is equal to <0.01% increase in terms of the error on TML for an initial mass of 253,211 μg. Since we could minimize the uncertainty by reducing the time for weighing in the low-humidity environment in globe box, the sample weighing in the low-humidity controlled globe box would be effective.

4.2. Weighing hydrated specimen

4.2.1. Mass change during specimen transportation. After weighing the vacuum-heated sample in globe box, mass gain behavior due to moisture adsorption was evaluated in a constant temperature and humidity room. The sample was enclosed in a transfer capsule, and then we transferred the capsule to the constant temperature and humidity room located in another building. The capsule was opened in the constant temperature and humidity room kept at 23°C and 50% relative humidity, and the enclosed sample was weighed quickly. Figure 4 shows the sample weighing results in the globe box and the constant temperature and humidity room before and after the sample transportation. Unfortunately, we spent five hours on the preparation of weighing in the constant temperature and humidity room because we have only one micro balance. During the preparation of weighing, the sample was stored in the globe box. The sample mass was 250,284 μg immediately after the removal from the transfer capsule and represented an increase of 67 μg compared with the mass before enclosing in transfer capsule. The increase of 67 μg would include the apparent mass gain less than 10 μg induced by the difference of buoyant force between a low-humidity and low-oxygen concentration environment and a constant temperature and humidity environment. The cause of mass gain during the transportation from the globe box to the constant temperature and humidity room might be the adsorption of moisture released from the transfer capsule wall and leaking through the seal on the capsule and/or the adsorption of moisture during time lag less than 11 seconds that we spent on the hand working on picking up the sample form the capsule, placing the sample on the microbalance, and getting the first weight value after a period of stability on the micro balance.

4.2.2. Mass gain due to hydration. The sample weighing was continued for 32 hours immediately after picking up the sample from the capsule. Figure 5 shows the mass gain behavior of the sample for 32 hours. The sample mass rapidly increased at the beginning of weighing, and then mass increase became moderately after 6 hours and almost saturated after 12 hours. After 24 hours, the sample mass represented an increase of 3,048 μg. If we set a dehydrated mass of 250,210 μg that is the specimen mass immediately after vacuum heating in the globe box, WVR was 1.233% calculated from the mass gain of 253,332 – 250,210 = 3,122 μg. The mass gain include the uncertainties during weighing and storage in globe box and that during specimen transportation. TML and WVR value in this paper may be larger than the outgassing data reported in JAXA, NASA, and ESA. Because we would remove moisture adsorption during storage in desiccator after picking up the sample from vacuum chamber and there is also no uncertainty induced by the time lag of 2 min between the retrieval of sample from a vacuum chamber and weighing.

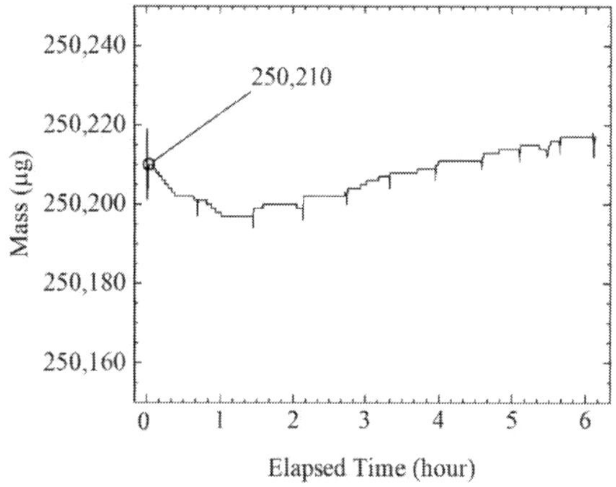

Figure 3. The time-course change of sample mass in the globe box after finishing a vacuum heating.

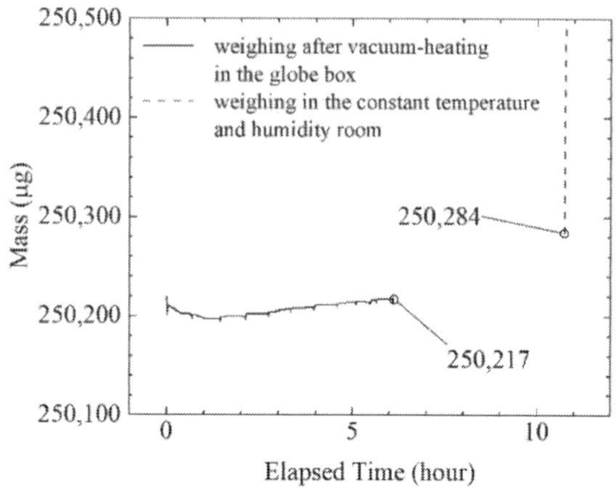

Figure 4. The measurement results of sample mass in the globe box before the sample transportation and the constant temperature and humidity room after the sample transportation.

4.2.3. Uncertainty induced by time lag on sample weighing. In ASTM E-2089-00 and E595-15, it is determined that the samples are weighed within 5 min of removal from a vacuum chamber and within 2 min of removal from a desiccator, respectively. The sample weighing uncertainty occurring during the acceptable time lag of 5 min in AO test estimation and 2 min in outgassing test were evaluated by using the measurement results of mass gain in the constant temperature and humidity room shown in figure 5. The sample mass without moisture adsorption in the constant temperature and humidity room was set to 250,284 μg that was the first-obtained mass immediately after the removal from the transfer capsule. To evaluate the mass gain due to moisture adsorption during the time lag of 5 min and 2 min at 23°C and 50% relative humidity in the constant temperature and humidity room, we excepted the uncertainty due to the mass change in globe box, the mass gain during the transportation by using transfer capsule, and the mass gain between picking up the sample form the capsule and getting the first weight value after a period of stability on the micro balance.

The sample mass at 5 min later and 2 min later after taking off the seal of transfer capsule are shown in figure 6. After 5 min, the sample mass represented an increase of 420 μg. The error of AO fluence estimation induced by hydration during the time lag of 5 min was derived from equation (2). For a exposed area having a 20-mm wide by 70-mm long that is same size as the sample, $E_k = 3.0\times10^{-24}$ cm^3/atom for Kapton-H, and $\rho_k = 1.42$ g/cm^3, an uncertainty of AO Kapton effective fluence ΔF_k is 7.0×10^{18} atom/cm^2. When the exposed area is small compared to the sample size, the uncertainty would become large.

After 2 min, the sample mass represented an increase of 243 μg. The error of TML and WVR in outgassing test induced by hydration during the time lag of 2 min was derived from equation (3) and (5). For the initial mass of 253,211 μg before vacuum heating, the uncertainty ΔTML or ΔWVR is 243 μg/253,211 μg×100 = 0.096%. Since many specimens would be tested in an outgassing test, it is tolerated that specimens are stored in a desiccator after finishing vacuum heating. The time lag of 2 min is doesn't include shipping time from vacuum chamber to desiccator and storage time in a desiccator. If we would take into account moisture adsorption during the time lag between the removal of specimen from vacuum chamber and sample weighing, the error would become larger.

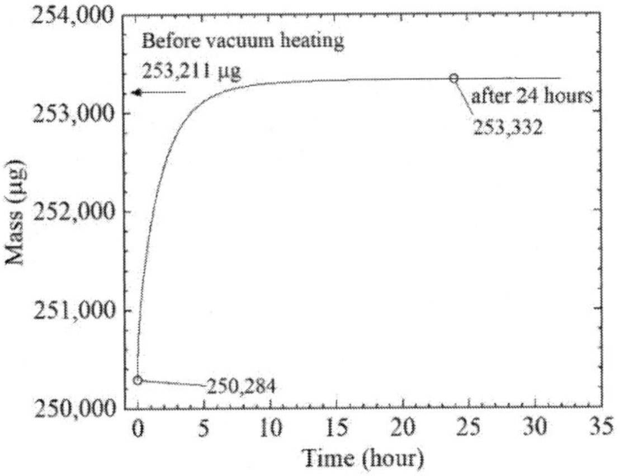

Figure 5. The mass gain behavior of the sample in the constant temperature and humidity room for 32 hours.

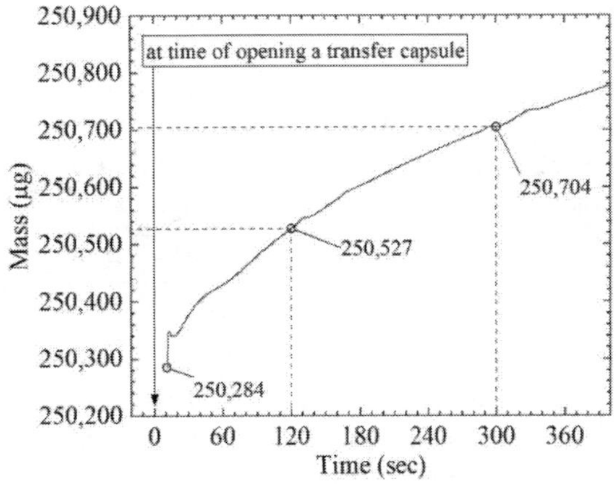

Figure 6. The sample mass at 5 min later and 2 min later after taking off the seal of transfer capsule.

4.3. Regression analysis for reduction of uncertainty due to hydration
In the ASTM standards, it is described that an uncertainty due to hydration can be minimized by back-extrapolating time variation of sample weighing after retrieval of sample from vacuum chamber or desiccator. However, it is not described what kind of model we should use for the extrapolation. In this paper, we used two kinds of function to extrapolate mass change due to hydration, and then evaluated the reduction effect of uncertainty.

4.3.1. Linear regression. The back extrapolation using linear regression is the simplest way to reduce uncertainty due to hydration if the regression model is unknown, although the mass gain due to hydration is non-linear as shown in figure 5. We assumed that the sample was weighed for 1 min from a lapse of 5 min and 2 min after taking off the seal of transfer capsule, and the recorded weight for 1 min were extrapolated to a mass at the same point in time of first-obtained actual weighing value after taking off the seal of transfer capsule. Figure 7 shows the results of linear back extrapolation in the case of a time lag of 5 min in ASTM E-2089. For a exposed area having a 20-mm wide by 70-mm long that is same size as the sample, $E_k = 3.0\times10^{-24}$ cm^3/atom for Kapton-H, and $\rho_k = 1.42$ g/cm^3, an uncertainty of AO Kapton effective fluence ΔF_k is 3.3×10^{18} atom/cm^2. Figure 8 shows the results of linear back extrapolation in the case of a time lag of 2 min in ASTM E-595. For the initial mass of 253,211 μg before vacuum heating, the uncertainty ΔTML or ΔWVR is 0.042%. In all case, the uncertainties with extrapolation was less than half of those without extrapolation.

4.3.2. Regression model based on diffusion equation. We assumed the solution of diffusion equation as a regression model of mass gain due to hydration. The solution of mass gain based on diffusion equation is obtained in the form of infinite series solution as shown in equation (8). For $t = 0$ the term inside the parentheses is non vanishing within the range of finite series. Therefore, a truncation error resulting from the finite series number n would be included in the $M(t = 0)$ if we apply the finite series to the model.

From equation (8), the truncation error due to finite term sum is expressed as following equation.

$$M(t) = M_0 + M_\infty \quad 1 - \sum_{n=0}^{m} \frac{8}{(2n+1)^2\pi^2} \exp \quad -\frac{D(2n+1)^2\pi^2}{d^2} t \quad - Error$$

$$Error \equiv M_\infty \sum_{n=m}^{\infty} \frac{8}{(2n+1)^2\pi^2} \exp \quad -\frac{D(2n+1)^2\pi^2}{d^2} t \tag{9}$$

Since the truncation error shown in equation (9) is always positive, the mass derived from finite series sum become larger than that from infinite one under same parameters. The error is maximum at $t = 0$. By using finite series sum solution, the mass back-extrapolated at the time of retrieval from a vacuum chamber include the truncation error. To determine the series number n inducing an enough small error, we evaluated the relationship between the series number n and the term inside the parentheses in the right-hand side of the equation (8) for $t = 0$ by using a following equation.

$$\frac{M(t=0) - M_0}{M_\infty} = 1 - \sum_{n=0}^{\infty} \frac{8}{(2n+1)^2\pi^2} \tag{10}$$

Figure 9 shows the error $(M(t=0) - M_0)/M_\infty$ regarding with series n. The error $(M(t=0) - M_0)/M_\infty$ decreased with series n, and became less than 1% of total mass of adsorbed matter M_∞ for $n = 20$. In this study, we constructed the regression model based on diffusion equation for $n = 20$. For $t = \infty$, equation (8) is expressed in $M(t=\infty) = M_0 + M_\infty$. From the results of mass gain as shown in figure 5, if we assign first-obtained mass immediately after the removal from the transfer capsule and rehydration mass after 24 hours at 50% relative humidity to M_0 and $M(t=\infty)$ respectively, the total mass of adsorbed matter is $M_\infty = M(t=\infty) - M_0 = 253{,}332$ µg − 250,284 µg = 3,048 µg. Therefore, on the regression analysis using the model of equation (8) for $n = 20$, $M(t=0)$ include the error of about 30 µg. By using equation (8) for $n = 20$,

$$M(t) = M_0 + M_\infty\left(1 - \sum_{n=0}^{20} \frac{8}{(2n+1)^2\pi^2} \exp\left(-\frac{D(2n+1)^2\pi^2}{d^2}t\right)\right) \tag{11}$$

we performed a regression analysis for the rehydration mass gain data for 32 hours at 50% relative humidity shown in figure 5. Figure 10 shows the result of regression analysis. The regression model would explain the mass gain behavior of Kapton-H. The largest mass difference between actual measured value and regression curve was 78.5 µg at the time of first-obtained mass.

If we can spend substantial time trying to extrapolate recorded mass data, it is also possible to perform the regression analysis by using infinite series solution. In this study, we calculated the infinite series sum by using "nsum" function of Python library "mpmath" that allow to take efficiently infinite sum based on Richardson extrapolation and shanks transformation algorithms and performed a regression analysis for the rehydration mass gain data for 32 hours at 50% relative humidity shown in figure 5. In figure 10, the regression analysis result based on infinite series solution is shown together with that on finite series one. As shown in table 1, we obtained parameters from the fitted curves. $M(t=0)$ is a mass extrapolated to $t = 0$. $M(t=0) - M_0$ means the truncation error. By taking the infinite series sum using method stated above, we confirmed that the truncation error was removed. As shown in figure 10, the difference between the regression curve based on finite series model and that on infinite one was clearly identified within first 10 seconds of hydration during which sample weighing is difficult. After 10 seconds, the regression curves are almost same. The mass gain behaviour due to hydration had a good correlation with the mathematical model based on diffusion equation. However, during the beginning of hydration, since there are large discrepancy between the measurement results and regression curve, it is necessary to improve the mathematical model on hydration in the future.

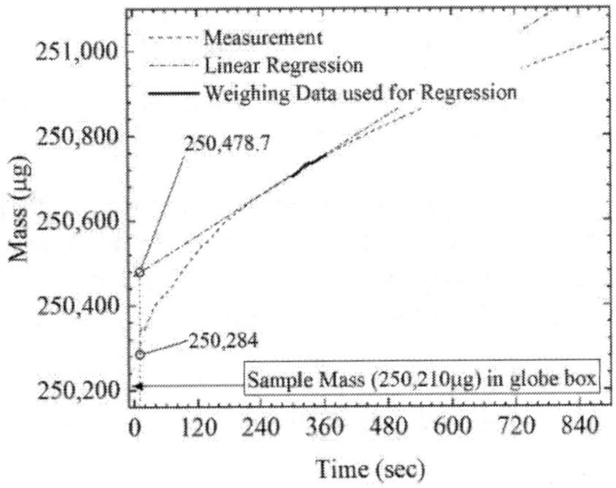

Figure 7. Results of linear back extrapolation in the case of a time lag of 5 min.

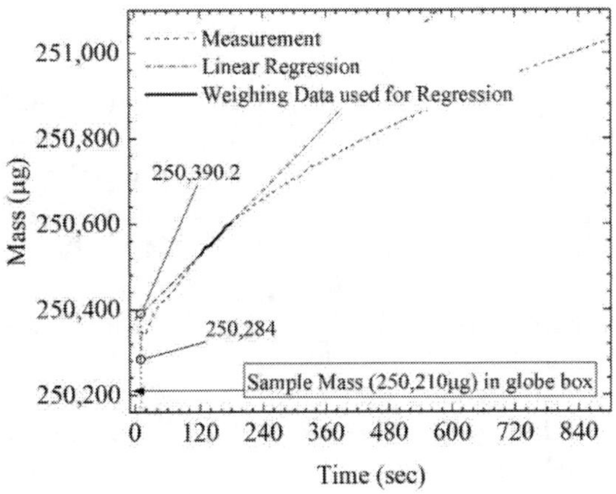

Figure 8. Results of linear back extrapolation in the case of a time lag of 2 min.

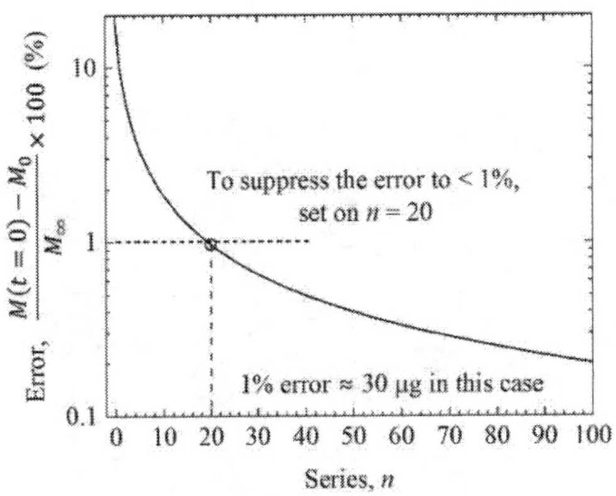

Figure 9. Truncation effect on infinite series solution of diffusion equation.

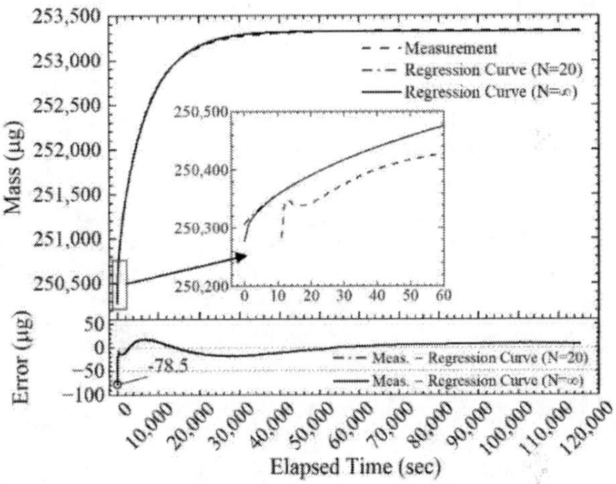

Figure 10. Regression analysis results using the solution of diffusion equation.

Tabel 1. Regression Parameters of fitted curves.

	Range of series sum on equation (8) $n = 0$ to 20	$n = 0$ to ∞
M_0	250,280.6 μg	250,280.8 μg
M_∞	3,044.3 μg	3,044.1 μg
$M(t = 0)$	250,309.9 μg	250,280.8 μg
$M(t = 0) - M_0$	29.4 μg	0.00 μg

4.3.3. Uncertainty reduction using infinite series solution of diffusion equation. Next, we evaluated the reducing effect of hydration uncertainty on the back extrapolation using equation (8) for infinite series sum. We assumed that the sample was weighed every second for five minutes from a lapse of 5 min and 2 min after taking off the seal of vacuum chamber or desiccator and that the recorded weight for 5 min was extrapolated to a sample mass at the time of removal from the vacuum chamber or desiccator. To evaluate the reducing effect of uncertainty quantitatively, the actual measured mass shown in figure 5 was compared with the regression curve of mass obtained as a result of the back extrapolation of 5-min mass data that were extracted from the mass gain behavior in figure 5. Figure

11 and 12 show the results of back extrapolation of 5-min data with a lapse of 5 min and 2 min after taking off the seal of transfer capsule. Table 2 shows the error at the time of first-obtained mass between the actual measured data and the regression curve. When we extrapolated the recorded mass for 5 min from a lapse of 5 min after taking off the seal of transfer capsule to a mass at the same point in time of first-obtained actual weighing value after takin off the seal of transfer capsule, the mass difference between the actual measured mass and the regression curve at the same point in time of that was 66.7 μg. For a exposed area having a 20-mm wide by 70-mm long that is same size as the sample, $E_k = 3.0\times10^{-24}$ cm³/atom for Kapton-H, $\rho_k = 1.42$ g/cm³, an uncertainty of AO Kapton effective fluence ΔF_k is 1.1×10^{18} atom/cm². When we extrapolated the recorded mass for 5 min from a lapse of 2 min after taking off the seal of transfer capsule to a mass at the same point in time of first-obtained actual weighing value after takin off the seal of transfer capsule, the mass difference between the actual measured mass and the regression curve at the same point in time of that was 40.9 μg. For the initial mass of 253,211 μg before vacuum heating, the uncertainty ΔTML or ΔWVR is 0.016%. As shown in table 2, the error resulting from extrapolation with regression model based on diffusion equation was six times smaller than the error without the extrapolation. The extrapolation based on the regression model of diffusion equation would be effective to reduce the hydration uncertainties for Kapton-H.

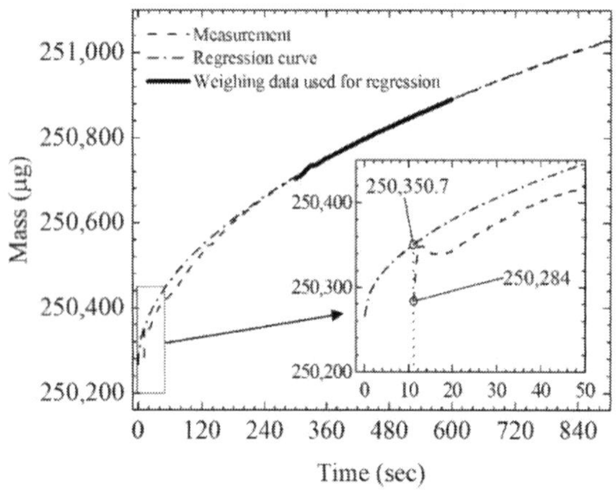

Figure 11. Results of back extrapolation using infinite series solution of diffusion equation in the case of a time lag of 5 min.

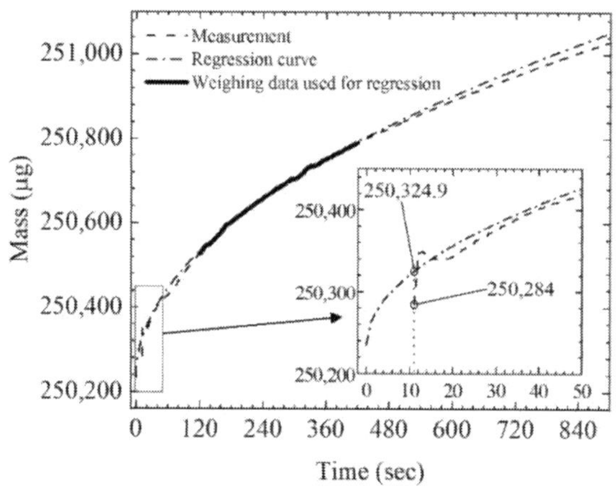

Figure 12. Results of back extrapolation using infinite series solution of diffusion equation in the case of a time lag of 2 min.

Tabel 2. Uncertainty reduction using back extrapolation with regression model.

	Uncertainty			
	5-min delay (ASTM E-2089)		2-min delay (ASTM E-595)	
	mass	AO fluence ΔF_k	mass	ΔTML or ΔWVR
Without extrapolation	420 μg	7.0×10^{18} atom/cm²	243 μg	0.096%
With extrapolation				
linear regression	194.7 μg	3.3×10^{18} atom/cm²	106.2 μg	0.042%
infinite series solution of diffusion equation	66.7 μg	1.1×10^{18} atom/cm²	40.9 μg	0.016%

5. Summary

The sample weighing inside a low-humidity controlled globe box would be useful to obtain the mass equivalent to that due to in-situ measurement. We have evaluated the uncertainties due to hydration between retrieving a sample from a vacuum chamber and sample weighing quantitatively by using of measured mass gain data during water adsorption under the constant temperature and humidity condition. We should make clear the cause of mass change during conditioning to switch weighing environment and sample transfer from the globe box to the constant temperature and humidity room and should reduce the time lag due to weighing conditioning and sample transfer in the future. The mass gain due to hydration had a good correlation with the mathematical model based on the sum of the first 20 terms in the infinite series solution of diffusion equation. In the case of extrapolating to the mass at the time of retrieval from a vacuum chamber, the regression curve using the sum of infinite series is better to remove the truncation error at the time of retrieval. If the time lapse of 5 min (ASTM E-2089) and 2 min (ASTM E-595) after retrieval from a vacuum chamber, the uncertainties due to hydration with extrapolation using linear regression was less than half of those without extrapolation and that using regression model based on diffusion equation was six times smaller than the error without the extrapolation. In the future, we will improve the mathematical model on hydration of polyimide film because there is some discrepancy between the measurement results and regression curve during the beginning of hydration. In addition to these, the hydration behaviors will be evaluated for more than one sample to confirm reproducibility and other factor (geometric shape, packing condition in a boat, heating temperature, relative humidity, and etc.) effects and will be evaluated for materials having different matter, surface morphology, degradation condition, and etc.

Acknowledgments

A part of this work was supported by the Coordination Funds for Promoting Aerospace Utilization from the Ministry of Education, Culture, Sports, Science and Technology (MEXT), Japan.

References

[1] Standard Practices for Ground Laboratory Atomic Oxygen Interaction Evaluation of Materials for Space Applications ASTM E 2089-00.

[2] Standard Test Method for Total Mass Loss and Collected Volatile Condensable Materials from Outgassing in a Vacuum Environment ASTM E 595-07.

[3] Leger L J and Visentine J T 1986 *Journal of Spacecraft and Rockets* **23** pp 505–11.

[4] Space product assurance: Thermal vacuum outgassing test for the screening of space materials ECSS-Q-ST-70-02C 15 November 2008.

[5] Crank J 1975 *The Mathematics of Diffusion* (New York: Oxford University Press) pp 44–9.

IOP Conf. Series: Materials Science and Engineering 1287 (2023) 012022 doi:10.1088/1757-899X/1287/1/012022

Contamination Level Prediction: Progress in Species Separation by TGA/MS

David Lansade[1,4], Jean-François Roussel[1], Delphine Faye[2], Guillaume Rioland[2], Agnieszka Suliga[3] and Orcun Ergincan[3]

[1]ONERA–The French Aerospace Lab, 31055 Toulouse, France
[2]CNES–Centre National d'Études Spatiales, 31401 Toulouse cedex 9, France
[3]European Space Research and Technology Centre (ESTEC). European Space Agency (ESA),NL-2200 AG Noordwijk, The Netherlands
[4]Author to whom any correspondence should be addressed at david.lansade@onera.fr

Abstract. Molecular contamination can be detrimental to spacecraft life expectancy as it may induce changes to the properties of the surfaces on which it deposits. In order to control these potential issues, manufacturers need to assess in-flight contamination levels in the quite early design phases, possibly leading to design changes or extra material bake-out to fulfil requirements. The accuracy of model assessments may thus have direct impacts on missions and costs. Prediction of outgassing and deposition at mission timescale is done by extrapolating short-term on-ground experimental results, by fitting the latter with some empirical or physical laws. Common methods are the power law interpolation (in isothermal conditions) or the residence time approach (with thermally activated time constants), which leads to exponential decays in isothermal conditions. Although they allow fitting the measured outgassed flux over few-day-long experiments, their extrapolation to ten or fifteen-year-long missions is more challenging. Validating empirical laws might involve long-term empirical comparisons, while obviously the validation of physical models should start with checking their physics. In spite of many years trying, a basic difficulty was yet found to block significant progress on physical validation. As long as measurements are limited to the total mass on QCMs it looks possible to fit data as a sum of almost any law, such as residence time or diffusion models e.g. for outgassing. Following this observation, a conclusion was drawn that the only way to discriminate between two different physics was to characterize contamination at chemical species level. Such characterizations are a prerequisite to unravel the physics at play during several phenomena: outgassing, deposit films dynamics, photochemistry, environment/ATOX interactions, etc. This innovative approach, presented herein, aims at identifying outgassed contaminants by means of coupled thermogravimetric analysis and mass spectrometry. For this purpose, this paper reports the successful species separation and identification of contaminants outgassed from a material used in space applications: the epoxy adhesive Scotchweld EC-9323-2.

1. Introduction

Upcoming space missions are becoming more and more demanding with their contamination budgets. Simulations of contamination at the satellite scale and their subsequent long-term predictions are of prime importance in order to respect contamination budgets allowed by the missions.

Contamination is the consequence of several phenomena, namely the outgassing of contaminant species from a source material, their transport towards other surfaces, their sticking to the said surfaces, their possible photopolymerization and/or reemission from the target [1]. Though transport in rarefied environment is well known, the other phenomena above-mentioned, especially outgassing and reemission, only remain partially known. So far, contamination predictions rely on several contamination outgassing and reemission models themselves relying on different laws, sometimes chosen somewhat arbitrarily: first order desorption, zero order evaporation, diffusion, bouncing or sticking on surfaces, etc. [2]–[5]

Though models currently used to fit laboratory week-long contamination experiments and allowing long-term predictions could work well in the hypothetical case where only one chemical species would outgas from a source, their application to real-case materials does not provide satisfactory results. Indeed, it is fairly easy with current methods to fit any data obtained by increasing the number of "virtual species" used to model any of the phenomena already mentioned. Long-term predictions obviously suffer from the lack of knowledge about the actual chemical species responsible for contamination and the physical laws governing their behavior.

For the past 15 years, ONERA, with the help of its partners among which ESA and CNES play an important role, has been convinced that the only way to perform accurate long-term predictions of contamination phenomena is to decipher the data gathered as a total mass in lab experiments into a sum of physical laws for each constituent of the contamination deposit characterized individually [6]. To do so, a technique known as species separation and identification of contaminant species by thermogravimetric analysis coupled to mass spectrometry (TGA/MS) is being developed [7], [8], [9]. Another interesting approach consisted in attempting a species separation based on peaks seen on QCMs during TGAs, and attributing them to different (unidentified) species [10], [11].

The purpose of this study is to strengthen the TGA/MS-technique knowledge that has started to be gathered by ONERA and its partners in the last decade, then mainly on a structural epoxy adhesive, namely Scotchweld EC-2216 [5], [9], and refine the technique leading to the characterization of contaminants. In this context, species separation and identification by TGA/MS was now applied with success to another epoxy glue, Scotchweld EC-9323-2.

2. Experimental section

2.1. Facility

Experiments dedicated to a characterization by Thermogravimetric analysis (TGA) coupled to Mass Spectrometry (MS) were carried out in ONERA's COPHOS facility (Contamination PHOton Synergy) (figure 1). Transport between the effusion cell, the quartz crystal microbalances (QCMs) and the mass spectrometer follows the "hot shroud" principle. The shroud is maintained at high temperature (< 180 °C) in such a way that the contaminants residence time is very short with regards to every other characteristic time, allowing to talk about instantaneous rebounds.

During outgassing or reemission, contaminants divide between said wells:

- Cold surfaces (QCMs, collection plates)
- Pumping orifices
- Mass spectrometer

During outgassing, both the QCMs and the collection plates are maintained at temperatures below -80 °C. During the reemission phase, QCMs and plates are brought successively to higher temperature of about 150 °C following a heating ramp of 2 °C/min, thus completing TGAs.

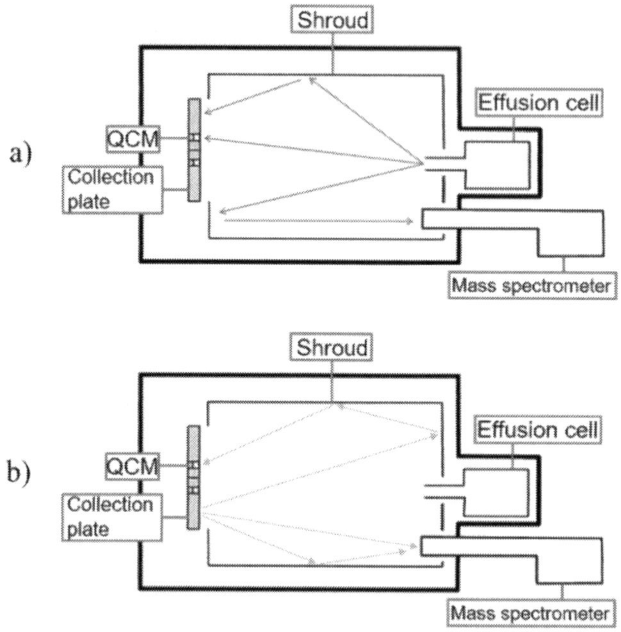

Figure 1. Scheme of COPHOS facility showing the role of the hot shroud principle a) during outgassing, b) during reemission. Contaminants are transported by successive reflections until they reach wells (cold parts, mass spectrometer or pumping orifices).

2.2. *Protocol*

A piece of Scotchweld EC9323-2 epoxy adhesive is placed in the effusion cell and heated for a few hours, until enough outgassed contaminants are condensed on the plates and the QCMs (typically equivalent to a few thousand x 10^{-8} g.cm^{-2}). Then, the outgassing flux is greatly reduced by letting the effusion cool down passively. After the outgassing flux is lowered sufficiently (undistinguishable with regards to a typical reemission flux during a TGA), series of TGAs are launched according to §2.1.

To perform species separation, 2 distinct phases are necessary:

1) Outgassing phase:
   The effusion cell is heated, thus accelerating every thermoactivated processes (diffusion, desorption). Most contaminants deposit on the plates and QCMs but a fraction is also received by the mass spectrometer.

2) Reemission phase (TGAs):
   Direct TGA
   Direct measurement of the reemission rate by QCM heating (2 °C/min). The vibrating frequency of the QCMs being sensitive to temperature, this effect must be taken into account. Note that the surface of the QCMs is quite small, inducing a weak signal in the mass spectrometer.

   Inverse TGA
   Heating of the (large) plates. Allows good signal over noise ratio (SNR) in the mass spectrometer. Deposit happens as well on the cold QCMs (direct view to be taken into account), meaning no temperature effect to be corrected.

11 TGA series were performed on four different samples. Sometimes, TGAs were carried out after each outgassing step, except for TGA series n° 7. In the latter case, the mass spectrometer used for species identification had seen its sensitivity decrease strongly. In order to counter the signal loss, a large amount of material was placed in the effusion cell and 4 steps of outgassing were carried out in order to increase the thickness of the contamination deposit on the QCMs and collection plates. Table 1 below sums up the 11 TGA series with their respective outgassing temperatures and the mass of the samples used. Samples were typically of rectangular form, with thickness and width roughly equal respectively to 2 mm and 1 cm. Their length were equal to a few centimeters. The difference in mass observed in series 7 is due to the fact that a larger number of samples were put simultaneously in the effusion cell as discussed later in section 3.5.

Table 1. Summary of the outgassing temperatures and samples masses of the 11 TGAs series performed.

TGAs series number	1	2	3	4	5	6	7	8	9	10	11
Outgassing temperature (°C)	125	150	75	100	125	150	75, 100, 125, 150	75	100	125	150
Sample mass (g)	1.38		2.28				12.72	1.85			

The amount of data gathered during these 11 TGA series being sizable, only TGA series 1&2 will be used in the following paragraphs. A quick view at series 7 at the end of this paper will also provide interesting data.

In figure 2 can be seen the temperature profile of the elements of interest for these TGAs, namely the effusion cell (or Knudsen cell), the QCMs, the collection plates and the shroud. First, the effusion cell containing the sample was heated at 125 °C for about 2.5 hours. It was then left to cool down passively as mentioned before. After that period, the 4 elements of the sample holder were heated successively from -80 (their temperature during outgassing) to 150 °C at a rate of 2 °C.min^{-1}. Each element underwent a TGA even though one QCM and one collection plate would have been enough: this was done for redundancy purpose. After all the elements that went through a TGA were cooled down again, another outgassing step at higher temperature (150 °C) was performed. Then again, the four TGAs were carried out again.

IOP Conf. Series: Materials Science and Engineering 1287 (2023) 012022 doi:10.1088/1757-899X/1287/1/012022

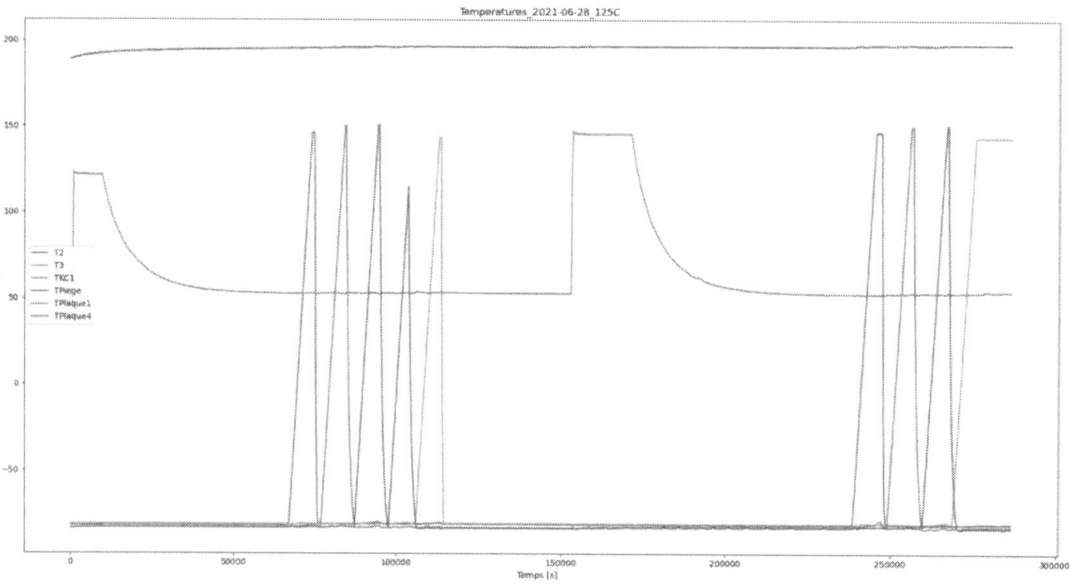

Figure 2. Temperature profiles of the Knudsen cell (TKC1), the QCMs (T2, T3), the Plates (TPlaque1, TPlaque2) and the shroud (TPiege) during tests carried out in June 2021: two outgassing steps, four TGAs after each step. The second TGA of the Plate 4 after the first outgassing step is a mistake.

During the TGAs, the mass spectrometer behaves like a well and is able to detect the reemitted molecules. For this purpose, the TGAs of the plates are very useful as their surface is large and allow good signal over noise ratio in the recorded spectra. For this reason, the MS data presented in the rest of this paper were recorded during the TGAs of one of the plates. The MS ionizes incoming molecules with an electron beam. Following ionization, these molecules are fragmented and the subsequent fragments are separated in function of their mass over charge ratio, noted m/z ratio in the rest of this document. Each molecule follows a pattern of fragmentation where some fragments are more probable than others. A graph bar consisting in the relative abundance of each fragment as a function of their m/z ratio is known as a mass spectrum.

The separation and identification of reemitted contaminants during TGAs are performed following 4 successive steps [12]:

1) Finding high m/z fragments that seem to exhibit a single peak during TGA. The combination of these two criteria allows thinking that these fragments are likely to originate from a single molecule.
2) Knowing the previous fragments' reemission temperatures, one must find other fragments peaking at the same temperatures and verify that the shape of their peaks in log scale is in accordance with a common molecular origin, that is to say that they must look parallel and thus proportional.
3) A stack of parallel fragments peaking at the same temperature is the closest one gets from a mass spectrum at a given temperature. It is then possible to reconstruct a partial spectrum of the outgassed molecule.
4) Comparison between the partial mass spectrum and a database and/or previous experiments (typically gas chromatography coupled to mass spectrometry (GC-MS) carried out during the qualification phase of the material) allows relatively reliable identification.

In the following section, species separation and identification of the contaminants outgassed by Scotchweld EC-9323-2 will be presented.

3. Species separation and identification : results

During these series of experiments, seven species were properly separated systematically and among which six were identified confidently. Though it may feel anachronous, the following sections will be named after the identified molecules even if definitive identification only happens at the very end of the process.

3.1. *Butylated hydroxytoluene (BHT)*

The four steps mentioned in the previous section were applied. For the sake of concision, steps 1 and 2 are presented in a single figure. Peaks with similar shapes and reemission temperatures during a TGA are visible in figure 3. For this first TGA, one can find many fragments exhibiting a single peak, all of them being reemitted with the same shape at around -28 °C. The shape criteria means that their signal is proportional in logarithmic scale. One deducts from this first level analysis that the isolated fragments probably originate from a single chemical species. The relative contribution of each of these fragments corresponds to their contribution in the mass spectrum of the species. It is hoped, once confirmed with several different TGAs, that part of the species' mass spectrum can be determined. That part should be in principle the one constituted of original fragments, that is to say fragments not encountered in spectra of other species.

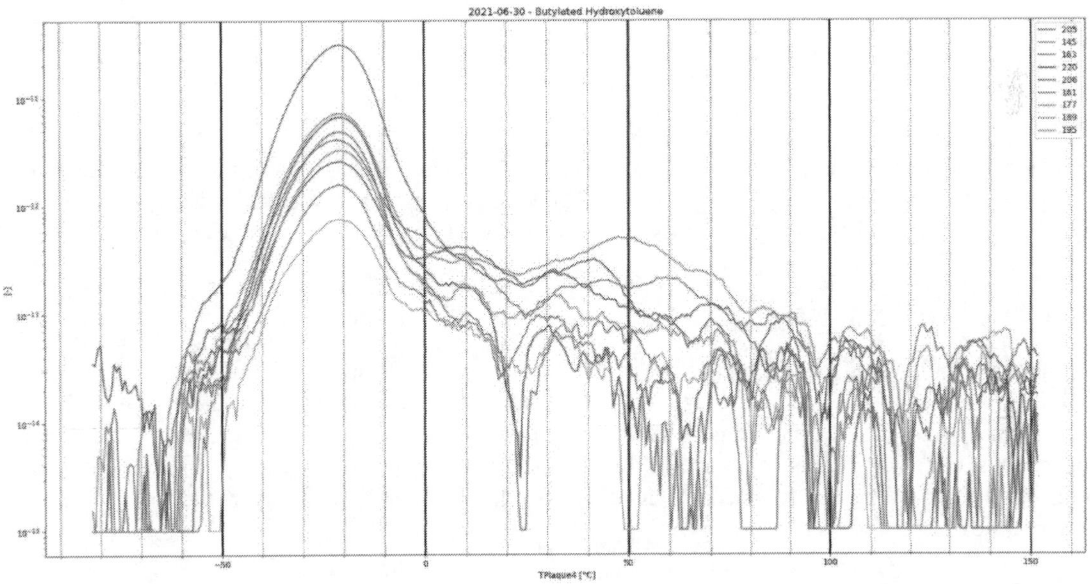

Figure 3. Similar peaks observed during the second TGA series #1 (outgassing step carried out at 125 °C) on fragments m/z = 205, 145, 163, 220 a.m.u., etc. of the mass spectrometer.

This stack of parallel fragments allows the reconstruction of a mass spectrum, visible in figure 4. In this figure, the blue dots represent the background of our chamber, that is to say masses that are detected in the mass spectrometer when no molecule is being reemitted (here, at 100 °C). This way, the orange squares are the masses that are peaking around -28 °C, that is to say the temperature at which appears the peak in TGA shown in figure 3.

IOP Conf. Series: Materials Science and Engineering 1287 (2023) 012022 doi:10.1088/1757-899X/1287/1/012022

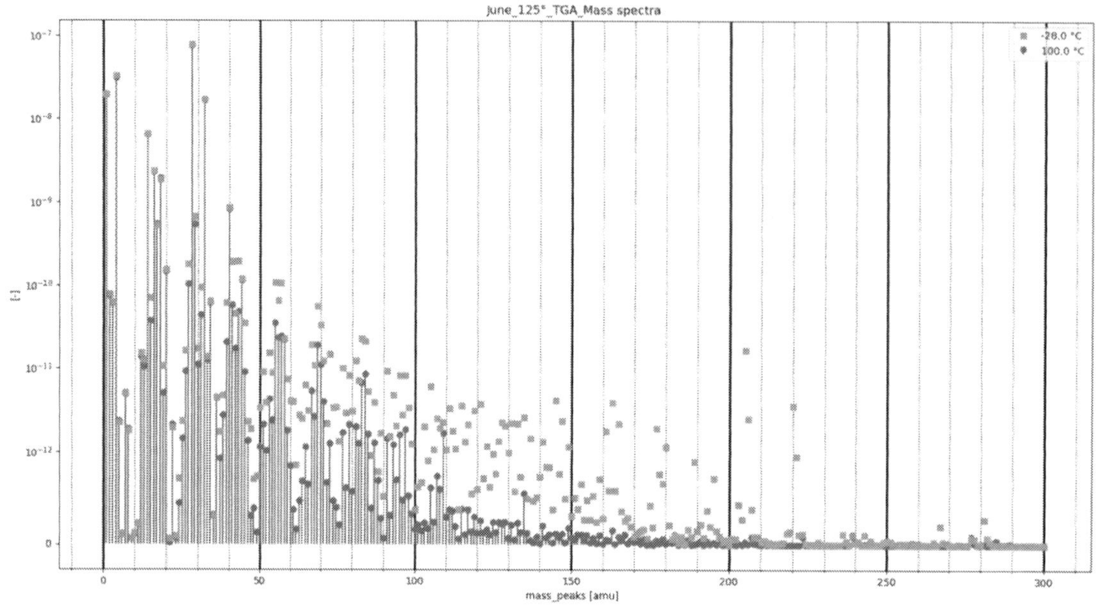

Figure 4. Mass spectrum measured during the TGA peak of the studied (-28 °C, orange squares) and the chamber's background (blue circles).

The figure 4 shown above represents the mass spectrum measured during reemission peak, compared to the facility's background (end of the TGA + 100 °C). As it was seen on the previous figure's temporal signals, this spectrum is very similar, with signal raising well above background level for:

- Fragments already tracked (205, 145, 163, 220, 206, 161, 177, 189 a.m.u.)
- But as well 221, 165, 179, 180 and sometimes 267, 281 a.m.u.

Comparison with NIST spectra [13] of species detected by GC-MS on EC-9323-2 by CNES suggested that the molecule we separated was in fact butylated hydroxytoluene (figure 5). All the tracked fragments exhibiting a single peak with common shape are present, especially the very characteristic 205 and 220 a.m.u. ones. Other fragments appearing in the reconstructed spectra do not appear in BHT's spectrum, probably because they are fragments belonging to other molecules partially reemitted at the same temperature.

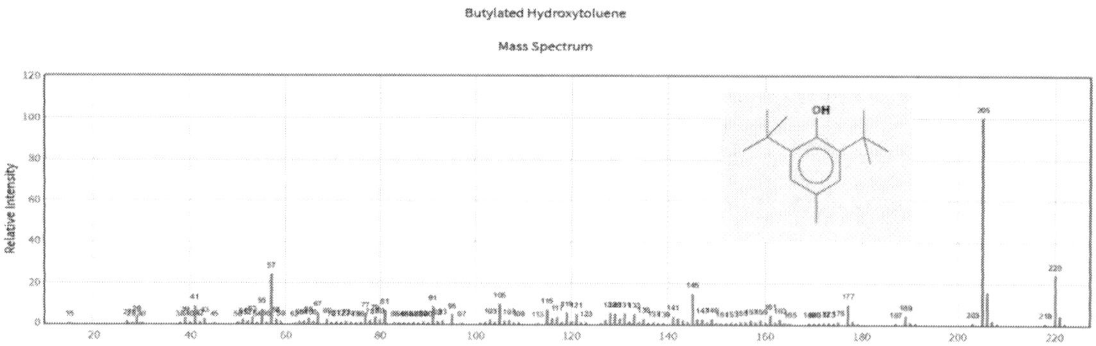

Figure 5. NIST mass spectrum and skeletal formula of butylated hydroxytoluene.

The same work was repeated 10 more times during our study, leading every time to the same conclusion. Table 2 at the end of this section sums up the reemission temperature of each separated molecule for the 11 TGA series carried out.

3.2. *1,4-Cyclohexanedimethanol (CHDM)*

The second species separated is especially identifiable thanks to canal m/z = 95 a.m.u. The signals for fragment m/z = 95, 202 and 264 a.m.u., which are single peaks, as well as a few others, less "pure", but presenting similar peaks at the same reemission temperature of 8 °C are represented on the MS data presented in figure 6.

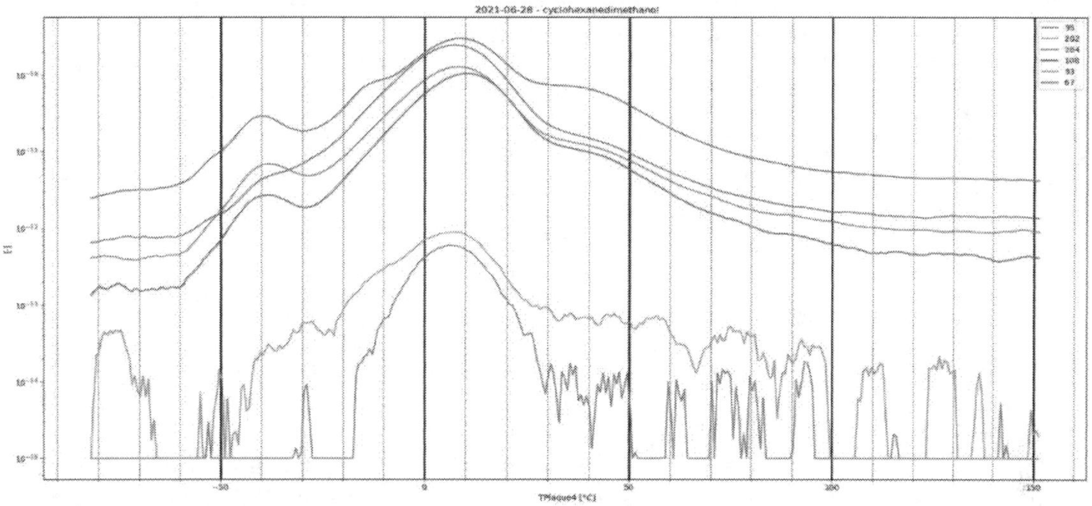

Figure 6. Fragments 95, 202 and 264 a.m.u. as well as some other important ones from the CHDM (non-single ones) reemitted during TGA series #1 (outgassing step carried out at 125 °C).

From these signals, the reconstruction of a mass spectrum was made possible. The result is shown in figure 7 below.

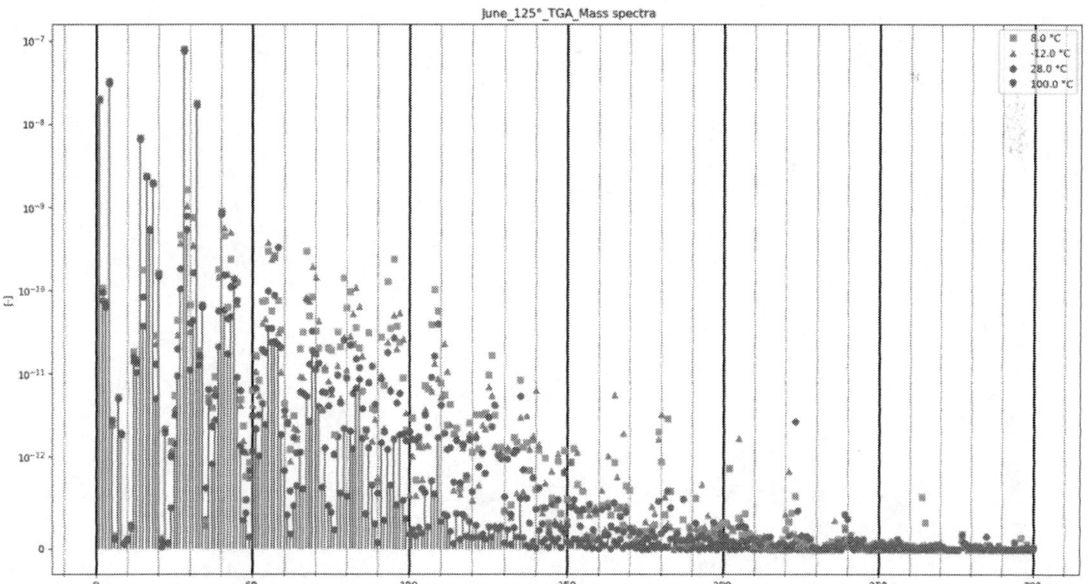

Figure 7. Mass spectrum reconstructed during the discussed peak in TGA (8 °C, orange squares), compared to the chamber's background (blue squares) and some other peaks

Similarly to what was done for BHT, the background of the chamber is represented by blue circles, while the signals detected at 8 °C is represented with orange squares. This mass spectrum exhibits on the one hand the fragments already discussed (264, 202, 108, 95, 93, 67 a.m.u.) but as well other masses present in the CHDM mass spectrum [14] (figure 8): 126, 107, 109, 81, 91, 55 and 41 a.m.u. It seems that the predominance of CHDM in the TGA peaks is almost certain especially since this molecule was observed by CNES in GC-MS experiments performed on the same material.

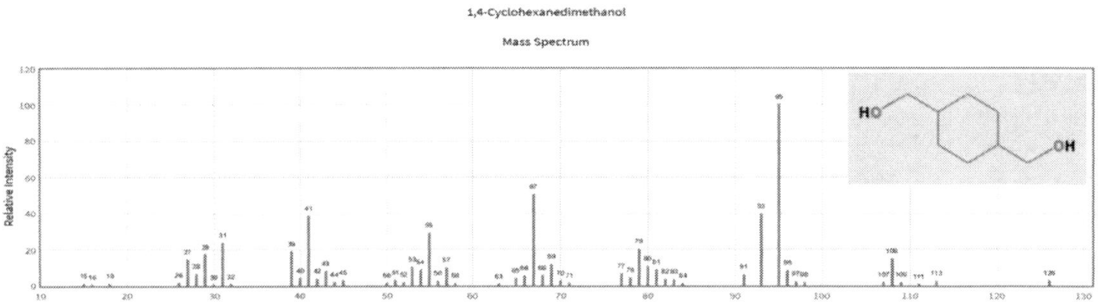

Figure 8. NIST mass spectrum and skeletal formula of 1,4-cyclohexanedimethanol.

3.3. *1-Dodecanol and Menthol derivative*

The TGA peak seen at -13 °C on the June 2021 TGA carried out after an outgassing step at 125 °C is arduous to treat. Indeed, from this first TGA (figure 9), it seems that two different shapes of reemission peak are present: some being narrow (98, 111, 112, 140, 168 a.m.u.), the others wider (165, 180, 193 a.m.u.).

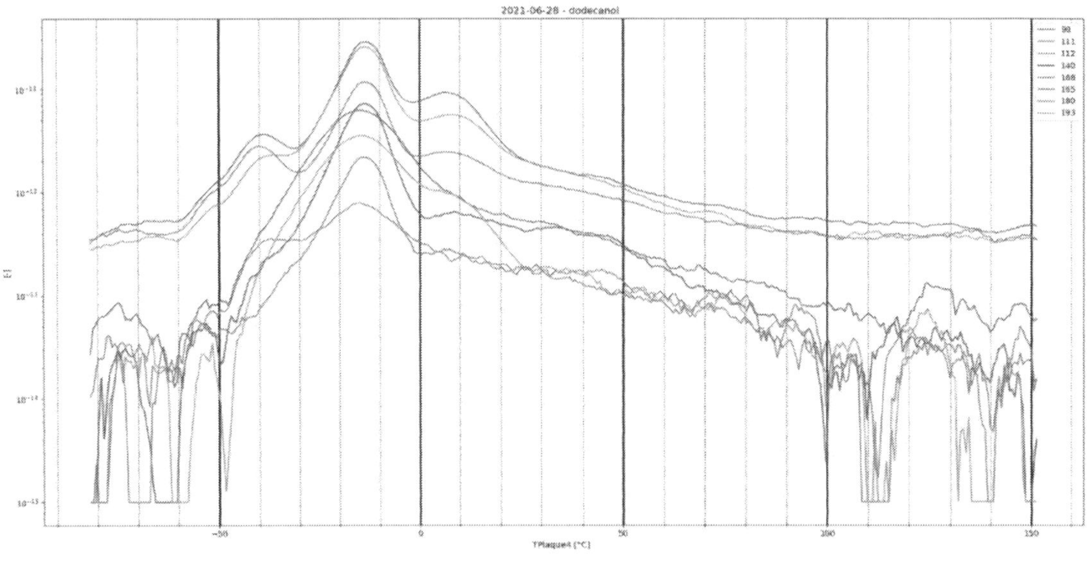

Figure 9. Several fragments reemitted around -13 °C during TGA series #1. (outgassing step carried out at 125 °C).

These results being sufficient to separate and identify two distinctive species, a typical mass spectrum as seen in MS during the reemission peak (figure 10) followed by two mass spectra from database (figures 11 and 12) are presented. As often, these species were selected among the ones seen in GC-MS by CNES. Using these data allows on the one hand to narrow the research of species to those actually outgassed by the source material EC-9323-2 (representability) and on the other hand to limit ourselves to research smaller database (practical numerical efficiency).

IOP Conf. Series: Materials Science and Engineering 1287 (2023) 012022 doi:10.1088/1757-899X/1287/1/012022

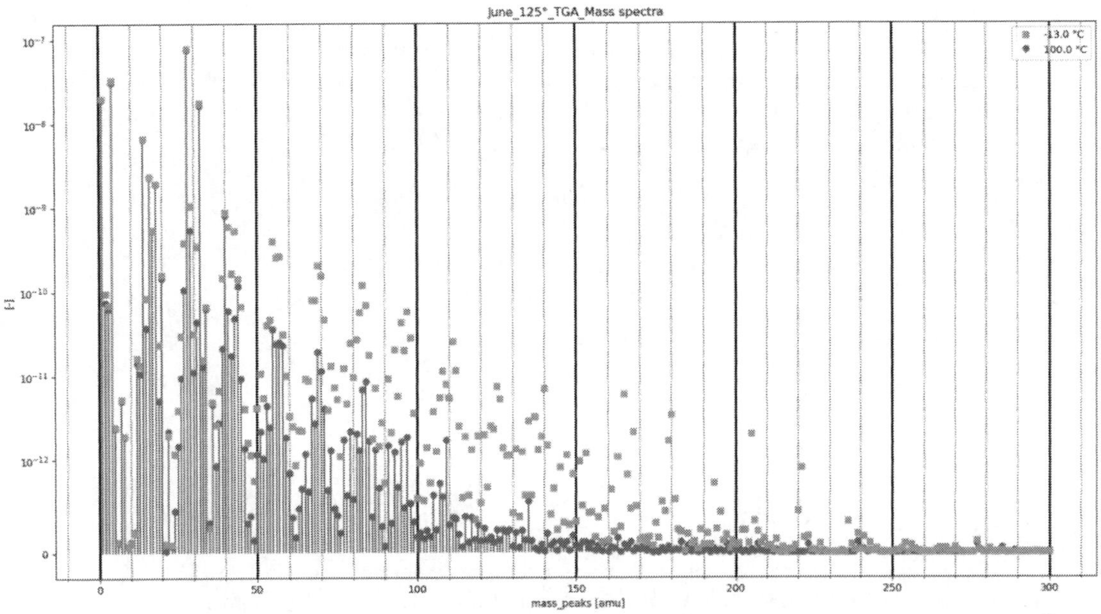

Figure 10. Typical mass spectrum measured during the reemission peaks at -13 °C during TGA series #2 after an outgassing step carried out at 125 °C (orange squares). The background of the chamber is represented with blue circles.

This slightly more complicated case is great to recap the process leading to the identification of a chemical species:

1) Among temporal measurements in TGA/MS, one can look for and find several fragments exhibiting peaks at the same temperature and being proportional with one another. Out of these canals, the most useful are the ones exhibiting single peaks (a single peak for a given m/z signal as a function of time, with a shape close to what is expected during reemission). This indicates that this fragment probably exists in the spectrum of only one molecule.
2) In order to validate this hypothesis, the selected fragments must all be found in the mass spectra of the candidate molecules from which they are thought to originate. Warning: the signals observed in mass spectrometry are often shown in log scale to demonstrate proportionality over time whereas mass spectra found in databases are often shown in linear scale.
3) Other peaks found in the mass spectra of the candidate molecules [15], [16] (figures 11 and 12) must as well be present in the measurements (figure 10), even though they may not be detected as single peaks.

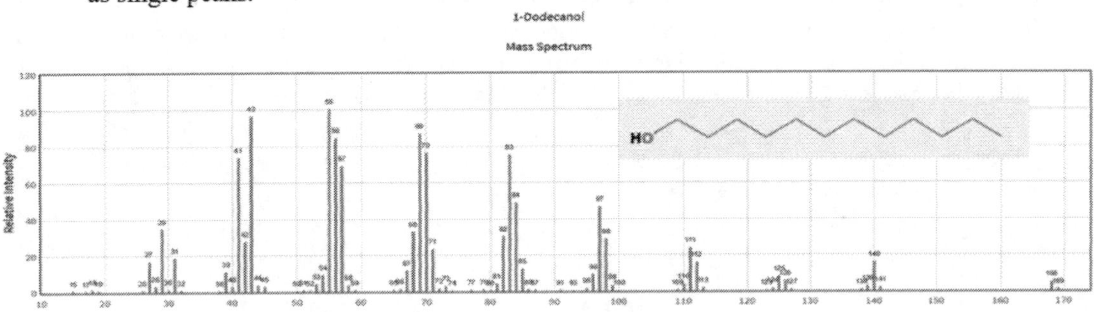

Figure 11. NIST mass spectrum and skeletal formula of 1-Dodecanol. Note that the spectrum includes fragments from the first group: 98, 111, 112, 140 and 168 a.m.u.

For 1-Dodecanol, the three steps above-mentioned give:

1) Fragments m/z = 98, 111, 112, 140, 168 all exhibit simple, proportional peaks (see TGAs above)
2) These fragments can all be found in the NIST mass spectrum of 1-Dodecanol (figure 11)
3) Other important peaks from the database spectrum can all be found in the actual measurements (figure 10) even though they were not selected as they were not detected as single peaks (they appear in other species' spectra).

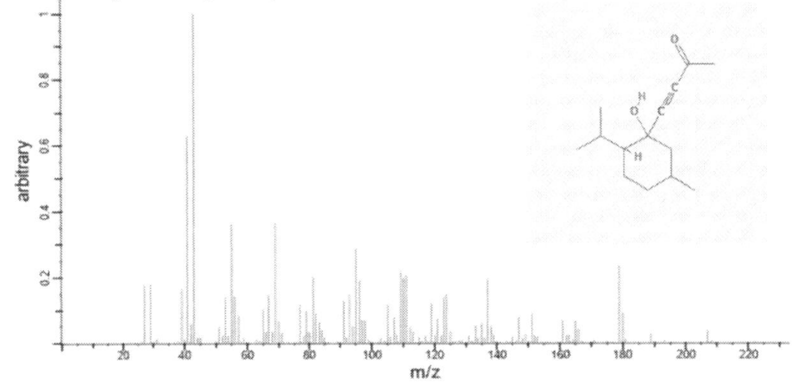

Figure 12. SpectraBase mass spectrum and skeletal formula of (1R,2S,5R),1'-(butyn-3-one-1-yl)-menthol. Note that the spectrum includes fragments from the second group: 165, 180 and 193 a.m.u.

The same thing is observed for (1R,2S,5R),1'-(butyn-3-one-1-yl)-menthol:

1) Fragments m/z = 165, 180 and 193 present proportional single peaks (see TGAs above)
2) These fragments can all be found in the database spectrum above (even m/z = 193 a.m.u., with very small signal, compatible with what was measured, represented in log-scale…)
3) Other important peaks from the database spectrum (figure 12) can all be found in the actual measurements (figure 10) even though they were not selected as they were not detected as single peaks (they appear in other species' spectra).

It must be noted that even if in the case presented here these two molecules were reemitted at the same temperature, it has not been the case in every TGA series. The actual reemission temperatures of these two molecules for each TGA series can be seen in Table 2.

The 4 species presented so far are the ones that were identified with most certainty. 3 more were detected and possibly identified and will be presented in the next section.

3.4. Other identified species: octamethyltetrasiloxane (D4), methyl-ethyl-phenyl-ethanone (MEPE)/ 1,3-bis(1-methyl-ethyl)-benzene (MEB)

Before summarizing these results, let us mention a few other species that were only characterized partially. Four new fragments seemed to show single similar reemission peaks. Measurements carried out afterwards allowed the distinction between two pairs: masses 147 and 162 a.m.u. in the one hand and masses 281 and 267 a.m.u. on the other hand. Exploitable TGAs are represented in the following figure 13.

IOP Conf. Series: Materials Science and Engineering 1287 (2023) 012022 doi:10.1088/1757-899X/1287/1/012022

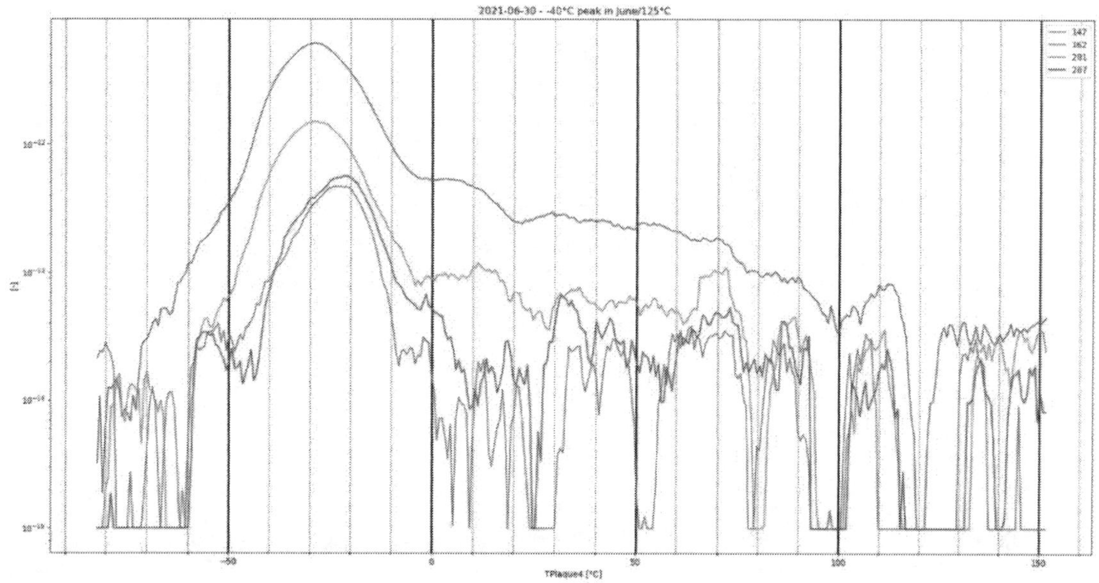

Figure 13. Fragments 147 and 162 a.m.u. as well as 281 and 267 a.m.u. detected during TGA series #3 after an outgassing step carried out at 150 °C.

In figure 13, it is clear that the two pairs of fragments mentioned earlier have their own behavior, suggesting each pair represents a single molecule. Concerning the first pair mentioned, masses 147 and 162 a.m.u., candidate species seen in GC-MS at CNES are two look-alike molecules, namely methyl-ethyl-phenyl-ethanone (MEPE) and 1,3-bis(1-methyl-ethyl)-benzene (MEB). Due to their almost identical molecular weight and structures, it was difficult to distinguish these two molecules, especially since both these molecules were found in GC-MS experiments. The measured ratio of 162 amu to 147 amu signals, is close to 0.25, quite to similar to that of MEB, while MEPE has 0.38 ratio. It is an argument in favor of the former.

The reconstructed spectra of these two molecules are represented in figure 14 below. One can see on their NIST database mass spectra [17], [18] that these two molecules share most of their main peaks (figure 15). The peaks that could have allowed better discrimination are small masses that are most of the time drowned in the facility's background.

IOP Conf. Series: Materials Science and Engineering 1287 (2023) 012022 doi:10.1088/1757-899X/1287/1/012022

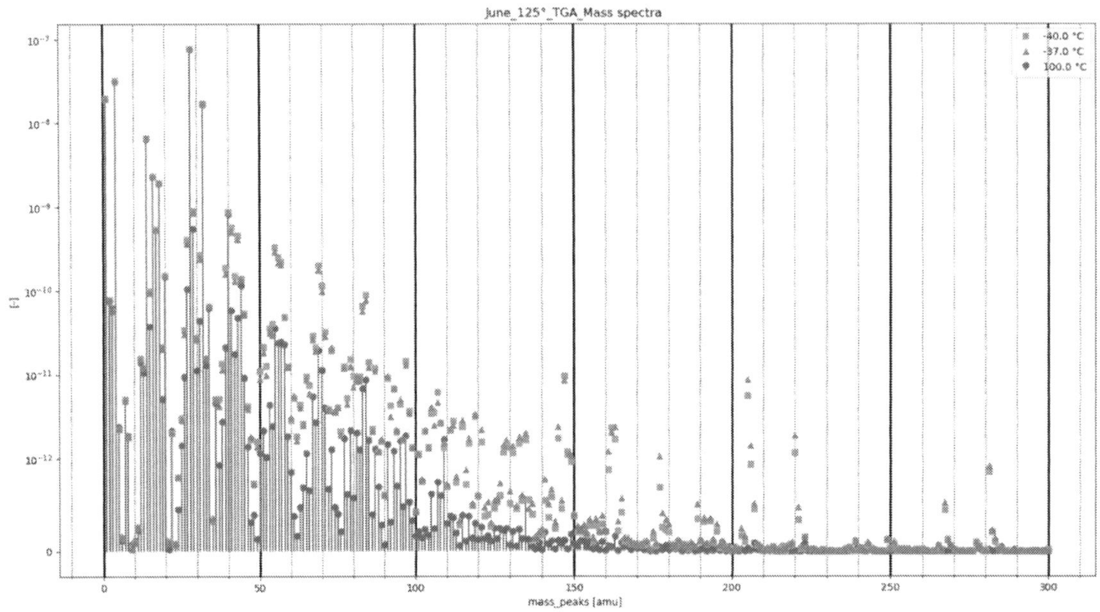

Figure 14. Spectra measured during the peaking of fragments 147, 162, 267 and 281 a.m.u. They were reemitted at two different but close temperatures of -40 °C for m/z = 147 and 162 a.m.u. and -37 °C for m/z = 267 and 281 a.m.u., represented respectively with orange squares and green triangles.

As both were seen in GC-MS, it is also quite probable that both these species are reemitted concomitantly.

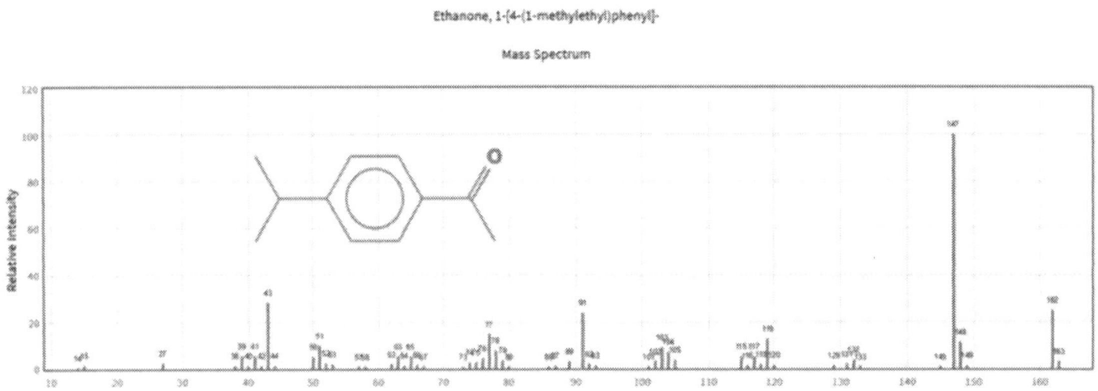

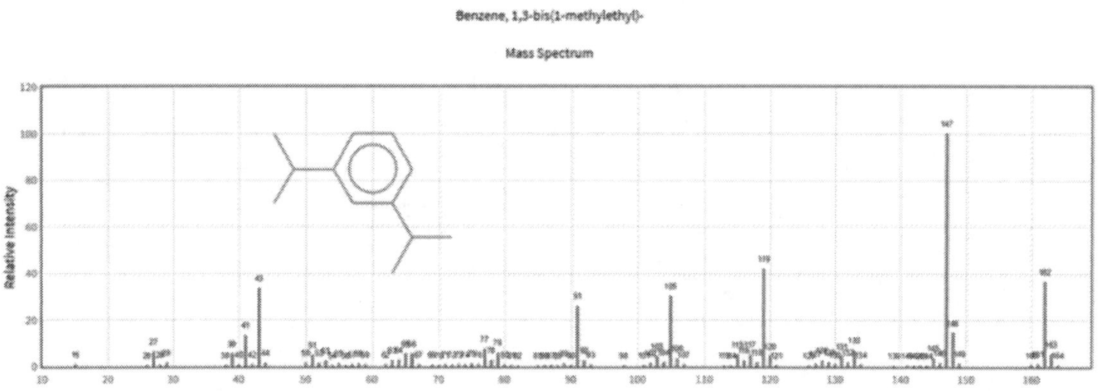

Figure 15. NIST spectra and skeletal formulas of the two candidate species seen in GC-MS having large contributions of fragments 147 and 162 a.m.u., namely methyl-ethyl-phenyl-ethanone and 1,3-bis(1-methyl-ethyl)-benzene.

Concerning fragments 267 and 281 a.m.u., the candidate species is octamethylcyclotetrasiloxane (D4), also observed in GC-MS (figure 16) [15]. This identification may not seem as convincing that the previous ones for several reasons. First of all, one would expect a little bit more signal for fragment 265 a.m.u. than for fragment 267 a.m.u. Besides, the relative amplitudes of fragments 281 and 267 (or 265, 266) a.m.u. would be expected to be in the order of 10 times, whereas they appear similar. Though one should be conscious that a mass spectrometer is not a quantitative tool, such discrepancy with databases has to be noted. Other fragments expected to be found with such molecules are present in the recorded spectra even though their signal is very weak (205, 249, 282 a.m.u.).

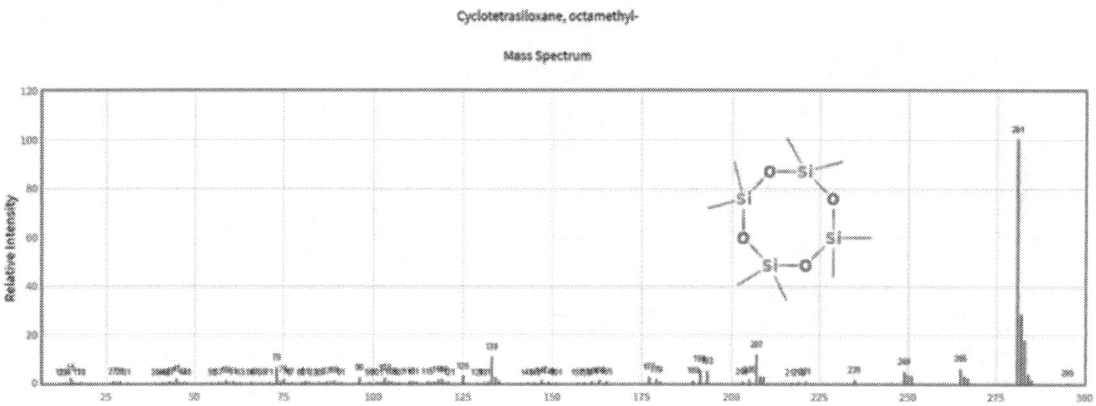

Figure 16. NIST spectrum of the candidate species seen in GC-MS showing contributions for fragments 267 and 281 a.m.u., namely octamethylcyclotetrasiloxane (D4).

Finally, one can mention fragments 58, 88, 223 and 224 a.m.u. that were judged as simple peaks in TGA series 1&2, with reemission peaks at 32 °C. No candidate molecules were identified neither for these four fragments nor for two pairs (their shapes and sizes might have suggested two different pairs: 58 and 88 a.m.u. on one side and 223 and 224 a.m.u. on another side), see figure 17 below.

IOP Conf. Series: Materials Science and Engineering 1287 (2023) 012022 doi:10.1088/1757-899X/1287/1/012022

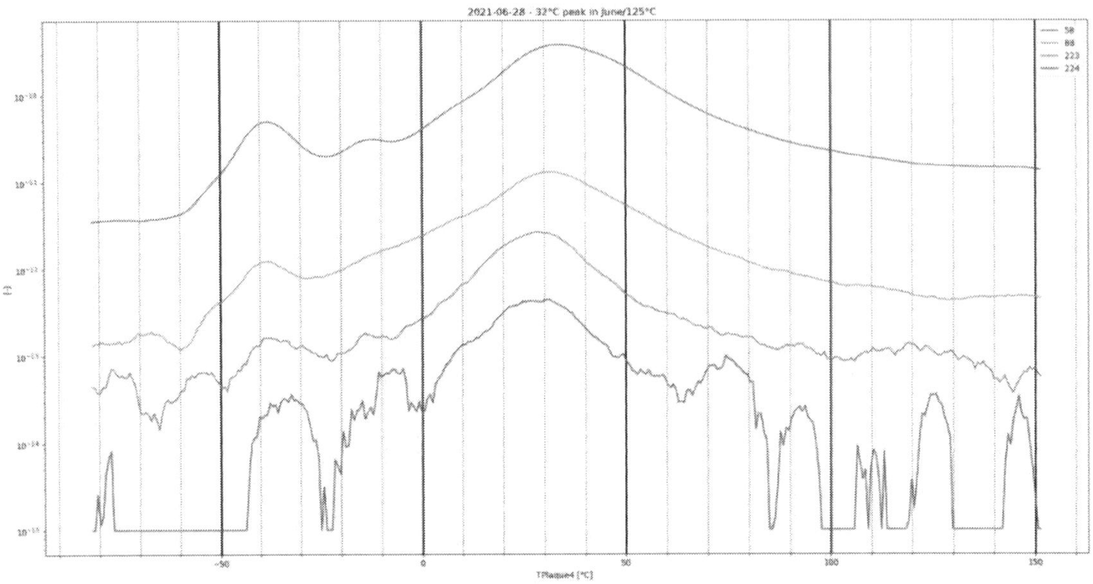

Figure 17. Fragments 58, 88, 223 and 224 a.m.u. seen as "single peaks" in TGA series 1&2.

3.5. The special case of TGAs series #7

Following a drop in sensitivity (about two decades) of the mass spectrometer used for this study, a series of TGAs were carried out trying different parameters to overcome the issue. First of all, a lot more material was put in the effusion cell: 12.72 g compared to 1.38 to 2.28 g in the other experiments presented here. Secondly, the temperature of the effusion cell containing the source material was progressively raised from 75 to 150 °C by steps of 25 °C, without any intermediary reemission (TGA) with sole purpose to increase the amount of matter deposited, thus the signal in the mass spectrometer. The outcome of these news parameters was that the deposit at the surface of the QCMs and the collection plates was huge when compared to other experiments. Whereas the deposits usually present thicknesses of about a few thousands of Angstrom, the one created in such conditions was estimated by the QCM at about 5 μm.

The consequence of such a thick deposit was that almost every contaminant previously separated was reemitted concomitantly (they shared almost the same higher reemission temperature somewhere around -6 °C) (See Table 2 and figure 18). This experiment will prove itself to be very interesting to understand reemission temperature shifts observed along the 11 TGAs series carried out for this study.

IOP Conf. Series: Materials Science and Engineering 1287 (2023) 012022 doi:10.1088/1757-899X/1287/1/012022

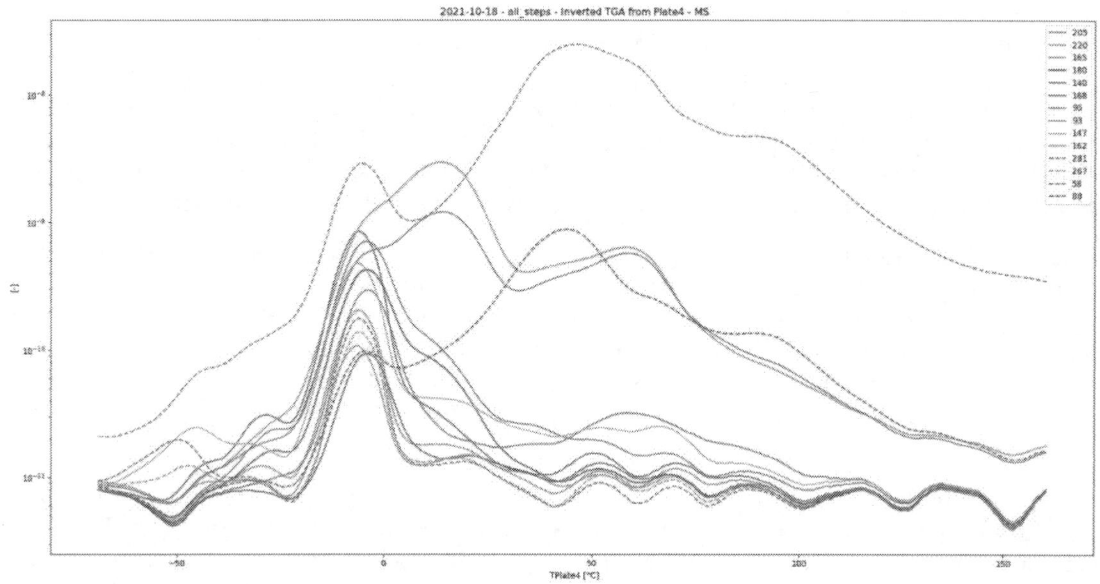

Figure 18. Several fragments representative of the seven separated species outgassed by EC-9323-2. This figure shows that during TGAs series #7, the very large deposit caused almost every species to be reemitted altogether at a higher temperature of about -6 °C.

4. Summary and discussion

The following Table 2 summarizes species identified or simply separated by MS. For each TGA, the temperature at which the reemission peak is reached is indicated. The darker the grey in the table, the better are the separation and identification of the considered species.

Table 2. Reemission temperatures of each species seen in each TGAs series

TGAs series number	1	2	3	4	5	6	7	8	9	10	11
Effusion cell outgassing temperature (°C)	125	150	75	100	125	150	75 to 150	75	100	125	150
Species (TGA peaks):											
1) MEPE/MEB **2) D4**	-40 -37	-28 -22	-30 to 0	-45 to -38	-45	-20	-7	-50 ?	-40 ?	-30	-28
3) BHT	-28	-21	-30 to -10	-3 to 5	-38	-20	-7	-31	-25	-22	-22
4) 1-Dodecanol **5) Menthol derivative**	-13	-11	-40 to -13	-26 -32	-26 -32	-16	-4	-18 -23	-12 -16	-10 -15	-9 -11
6) CHDM	8	6	-3	-16	-13	-1	-5 (et 13 et 60)	-4	6 (mult)	11	12 (mult)
7) "32 °C peak"	32	35	55 (et -46)	17	20	35	45 (et -5)	divers	30	40	45

Sets of fragments exhibiting unique and proportional reemission peaks were found. Each of these sets constitutes part of the mass spectrum of a chemical species. Seven species were separated (numbered in table 2), among which six were identified with a relatively good certainty. To our knowledge, this is the first time outgassed products from a material are characterized in that much details with such techniques.

As it will be outlined in the following section, the number of outgassed species as well as their amount do not allow qualifying this experiment as exhaustive. An important result from these series of experiment, on which emphasis was not yet put in spite of its omnipresence, is the fact that reemission peaks' temperatures vary and shift very often from one TGA to the other. Here, considering the important number of TGAs (11 for the record), the pervasiveness of the phenomenon, the metrological progresses on both the calibration and the thermodynamics of the system used, it is now believed that the shifts observed in the reemission peaks temperatures are absolutely linked to physics.

Table 2 summarizes all the reemission peaks temperatures, species by species and for every TGA. Most variations are in the order of 10 K, though they can reach more than 20 K in some cases. The very thick deposit obtained during TGA series #7 (5 µm thick) is a special one in the matter. In this case, every reemission peaks temperatures increase and pile around -6 °C, being the sign of a brutal and global reemission of a large part of the deposit. The interpretation that can be made is the one of a very thick layer of (almost entirely) mixed contaminants, among which the less volatile ones disable the migration and reemission of the more volatile ones.

If one overlooks that special case, a trend can be identified in these reemission temperatures. The higher the outgassing temperature leading a deposit, the higher the reemission temperature peak. This trend seems to be general enough, despite being sometimes a little fluctuating if one considers every data available, some of it barely standing out of the noise. If one limits itself to data gathered in TGAs series #8 to 11 (with a brand new mass spectrometer) and to the four species best identified, the trend appears very clearly (figure 19).

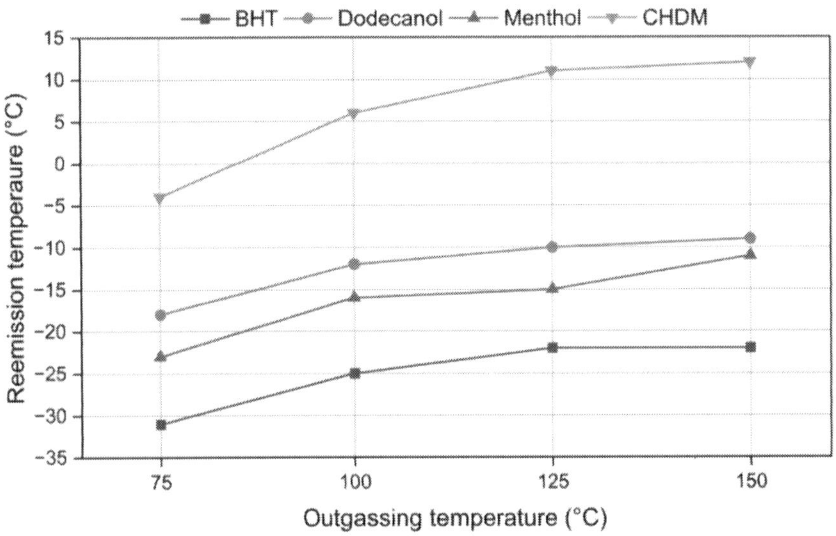

Figure 19. Recorded temperatures of the four main identified species reemission peaks as a function of the outgassing step temperature that leads to their deposition.

If one tries to determine what would a molecule's reemission temperature out of any "mixture effects", it would logically be closer from the left column (outgassing at 75 °C), and probably a little bit lower, since even at that "low" temperature of outgassing a small mixture effect can happen in the deposit made primarily of quite volatile species. That "ideal reemission temperature" that many people have in mind in an idealized or simplified vision could actually be the reemission temperature of the pure substance, unmixed. One could think that this temperature could be reached for very thin (sub-monolayer) deposits, where inter-species interactions stop being prevalent. Unfortunately, in this very case, reemission could be governed essentially by the interactions with the substrate, which could become very strong in this sub-monolayer hypothesis.

Though this stays true, in the scope of the very applicative researches presented here, one should stick to the fact that:

- Reemission peaks temperatures fluctuate and those fluctuations range from 10 to 20 K
- The presence of less volatile species in the deposit seems to drag those reemission temperatures higher
- Very thick deposits exacerbate those shifting effects that tend to pile most species' reemissions together at higher temperature
- Despite these shape and temperature fluctuations, mass spectra are very reproducible from one TGA to the other and are the species' signatures.

5. Conclusion

The work presented in this paper aimed at improving species separation and identification of contaminants outgassed from a space material, namely the epoxy glue Scotchweld EC-9323-2. The knowledge of these species, associated with a marker allowing to follow them individually (*m/z* fragments specific of each species), will also permit in other works to follow them during the outgassing phase of the source material. This is of prime importance in order to be able to predict and model the behavior of the said material once in space.

Thanks to the TGA/MS method further developed during this study, 7 species were systematically separated (that is to say they were "seen" in every TGA). Out of these 7 species, 4 were identified with a high degree of certainty, namely butylated hydroxytoluene, 1-dodecanol, (1R,2S,5R),1'-(butyn-3-one-1-yl)-menthol and 1,4-cyclohexanedimethanol. Two more were quite well identified, though only a couple of fragments were really characteristic, leaving some room to doubt concerning the actual molecules. These molecules were methyl-ethyl-phenyl-ethanone (or methyl-ethyl-benzene) and octamethylcyclotetrasiloxane. Finally, one species was always separated but could not be identified.

These separation and identification showed that the reemission temperatures of the identified species shifted from one TGA to another. This had been seen in previous studies, especially on Scotchweld EC-2216, and was then thought to be metrological artefacts. In this study, it was systematic and happened for each of the 11 TGAs carried out, allowing to think it is really a physical phenomenon. These shifts usually happened in a 10 to 20 K range. It is believed they originate from mixture effects, in which lesser volatile species trap the more volatile ones. On a very specific case, where the contaminant deposit was very large, (a few microns), the shift in temperature could be several dozens of Kelvin as almost every species were reemitted concomitantly. Despite these shifts, mass spectra reconstructed from the gathered data were very repeatable and were the signature of species. This constitutes a large progress in the separation and identification of outgassed species under vacuum and paves the way to new contamination control standards in the space industry, especially when hurdles of data processing automation will be overcome.

Acknowledgements
This research activity was funded by an ESA contract. The ONERA authors also want to acknowledge several funding sources over the long term, from CNES, ESA and ONERA internal resources, which supported the global development of these innovative experimental techniques and modelling capabilities.

References

[1] Tribble A C, "Molecular Contamination", in *Fundamentals of Contamination Control*, SPIE Press., vol. TT44, Bellingham, WA, 2000, p. 7-60.
[2] Van Eesbeek M and Zwaal A, "Outgassing and contamination model based on residence time", in *Proceedings of the 3rd ISMSE*, ESTEC, Noordwick, The Netherlands, 1985, p. 25-34.

[3] ESA-ESTEC, "ECSS-Q-ST-70-01C15 Cleanliness and contamination control". ESA Requirements and Standards Division, 2008.
[4] Garrett J W, Glassford A P M, and Steakley J M, "ASTM-E-1559 Method for Measuring Material Outgassing/Deposition Kinetics", *J. IEST*, vol. 38, n° 1, p. 19-28, 1995.
[5] Vanhove E, Grosjean E, Rioland G, and Faye D, "Comparison of molecular contamination models based on TGA/MS experiments", 14th ISMSE, Biarritz, France, Oct. 2018, p. 9.
[6] Roussel J-F, Faye D, Van Eesbeek M, Tondu T, Migliore R, Rampini R and Paulmier T, "A New Fontier for Contamination: Reaching the Molecules", 11th ISMSE, Aix-en-Provence, France, 2009, p. 11.
[7] Vanhove E, Tondu T, Roussel J-F, Faye D, and Guigue P, "In Situ Real-Time Quantitative and Qualitative Monitoring of Molecular Contamination", *J Spacecr Rockets*, vol. 53, n° 6, p. 1166-1171, nov. 2016, doi: 10.2514/1.A33505.
[8] Vanhove E, Roussel J-F, Faye D, and Guigue P, "In Situ Real Time Quantitative and Qualitative Monitoring of Molecular Contamination by Coupling TGA and MS: Recent Advances and Development of a Numerical Tool for Data Analysis", 13th ISMSE, Pau, France, June 2015, p. 12
[9] Roussel J-F, Vanhove E, Grosjean E, Faye D, Rioland G, Rampini R and Ergincan O, "Progress in a physical approach to contamination in Europe", in *Systems Contamination: Prediction, Control, and Performance 2018*, San Diego, United States, Sept. 2018, p. 8. doi: 10.1117/12.2320013.
[10] Suliga A, Ergincan O, and Rampini R, "A new dynamic outgassing procedure to predict reemission and deposition values of contaminants", Proc. SPIE 11489, Systems Contamination: Prediction, Control, and Performance 2020, 114890E (21 August 2020); https://doi.org/10.1117/12.2568022
[11] Suliga A, Ergincan 0 and Rampini R, "Modeling of Spacecraft Outgassed Contamination Levels by Thermogravimetric Analysis.", *J Spacecr Rockets*, vol. 58, n° 4, p. 1010-1016, July 2021, https://doi.org/10.2514/1.A35020
[12] Lansade D, Roussel J-F, Alred J M, Wong A T, Anderson J R, Soares C E, Faye D, Rioland G, "Separation of contaminant species by TGA/MS in European and U.S. approaches to outgassing", Space Systems Contamination: Prediction, Control, and Performance, San Diego, California, USA, 2022. doi: https://doi.org/10.1117/12.2634565.
[13] Wallace W E, NIST Mass Spectrometry Data Center, "Butylated Hydroxytoluene". https://webbook.nist.gov/cgi/cbook.cgi?ID=C128370&Mask=200 (accessed 2022/11/21).
[14] Wallace W E, NIST Mass Spectrometry Data Center, "1,4-Cyclohexanedimethanol". https://webbook.nist.gov/cgi/cbook.cgi?ID=C105088&Mask=200 (accessed 2022/11/21).
[15] Wallace W E, NIST Mass Spectrometry Data Center, "Cyclotetrasiloxane, octamethyl-". https://webbook.nist.gov/cgi/cbook.cgi?ID=C556672&Units=SI&Mask=2FF (accessed 2022/11/21).
[16] John Wiley & Sons, Inc., "Menthol, 1'-(butyn-3-one-1-yl)-, (1S,2S,5R)-", *SpectraBase*, 21 novembre 2022. https://spectrabase.com/compound/IjsWndAN1Vm (accessed 2022/11/21).
[17] Wallace W E, NIST Mass Spectrometry Data Center, "Ethanone, 1-[4-(1-methylethyl)phenyl]-". https://webbook.nist.gov/cgi/cbook.cgi?ID=C645136&Mask=200 (accessed 2022/11/21).
[18] Wallace W E, NIST Mass Spectrometry Data Center, "Benzene, 1,3-bis(1-methylethyl)-". https://webbook.nist.gov/cgi/cbook.cgi?ID=C99627&Mask=200 (accessed 2022/11/21).

A cryogenic method for testing thermal radiative properties of surfaces for space probes

J Frolec[1*], T Kralik[1], B Bras[2], M Portaluppi[2]

*corresponding author contact: frolec@isibrno.cz

[1] Institute of Scientific Instruments of the Czech Academy of Sciences, Kralovopolska 147, 612 00 Brno, Czech Republic

[2] Materials' Physics & Chemistry Section TEC-QEE at ESTEC (European Space Research and Technology Centre), Keplerlaan 1, 2200AG Noordwijk, Netherlands (contact: tec-qee.qms.team@esa.int)

Abstract. We present a cryogenic method for testing of thermal radiative properties of materials commonly used in ground-based systems as well as for space probes. It is possible to measure total hemispherical emissivity or absorptivity of metallic or dielectric surfaces (0.1% - 100%) in the range of temperatures from 10 K up to 320 K of the source of thermal radiation. Emissivity of thermal control coatings for the space probe JUICE are presented together with other materials used for space exploration or cryogenics.

1. Introduction

Heat exchange by thermal radiation between two objects at different temperatures is given by their areas, mutual orientation, absolute temperatures and radiative properties (emissivity/absorptivity) of the surfaces [1]. The control of the heat exchange by thermal radiation is crucial for design of any space probe [2, 3]. To meet the spacecraft thermal control criteria, it is necessary to know the thermal radiative properties of the used surfaces. ECSS-Q-ST-70-09C introduces several methodologies. Our method, originally developed for Cryogenics, can offer data useful also for spacecraft design.

We present here cryogenic measurement of thermal emissivity of the paint used on JUICE thermal control surfaces such as Moons and Jupiter Imaging Spectrometer (MAJIS) radiators and high gain antenna (HGA). The radiative surface is aluminium coated with a white paint Alion Z93C55 with low solar absorptivity and high thermal emissivity.

The emissivity of dielectric coatings strongly depends on its thickness because of its partial transparency [4]. For this reason, two samples with layer thickness of 110 μm ± 10 μm and 60 μm ± 10 μm were tested.

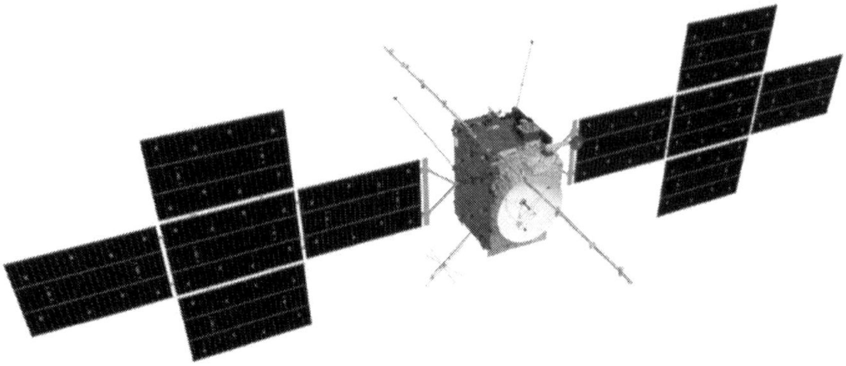

Figure 1. The Jupiter Icy Moons Explorer, JUICE. Copyright: ESA/ATG medialab.

Using the recent measurements of the Alion layers for the JUICE mission, alongside with several examples of our previous work, we introduce here our cryogenic method for testing thermal radiative properties of various materials. This calorimetric method, utilizing the principle of bolometric measurement, can cover the measurement of total hemispherical emissivity (or absorptivity) in a large range of temperatures of thermal radiation from cryogenic to room temperature values. Total hemispherical emissivity is defined as the ratio of the total emitted radiance of a specimen to that of a perfect black body radiator at the same temperature. Analogically, absorptivity is the fraction of the radiation energy incident on a surface that is absorbed. Considering the temperatures of interest, the corresponding radiation covers mid to far infra-red.

2. Materials and method

2.1. Apparatus and the measurement method

The method dedicated for measurement of total hemispherical emissivity (ε) and absorptivity (α) has been already described in detail in our former paper [5].

The method is based on measurement of radiative heat power Q_R transferred between two concentric discs 40 mm discs in diameter, separated by about 0.5 mm gap (Fig. 2). The heat flow is measured after reaching thermal equilibrium as a temperature drop on a calibrated thermal resistor linking the absorber with the LHe bath. Basic characteristics of the cryogenic apparatus are as follows:

- Upper disc – radiator at temperature T_R (maximal range 10 K – 320 K).
- Lower disc – absorber at a temperature T_A (maximal range 5 K – 120 K).
- For emissivity measurement the sample is at the radiator position (T_R), while during absorptivity measurement the sample act as absorber of thermal radiation at T_A.
- The opposite sample is a reference surface with high values of absorptivity (and emissivity) (~94% for $T_R \geq 50$ K).
- The Q_R is evaluated from temperature drop T_A-T_B on a calibrated thermal resistor (where T_B is the temperature detected at the bottom of the thermal resistor).
- The emissivity ε (or absorptivity α) of the sample is evaluated as $\varepsilon,\alpha = Q_R/(A\sigma(T_R^4 - T_A^4))$ using sample area A and Stefan–Boltzmann constant σ [1, 5].
- The apparatus is cooled in a wide neck LHe Dewar with stabilized pressure.
- Measurement range of emissivity or absorptivity: 0.1 – 100%.

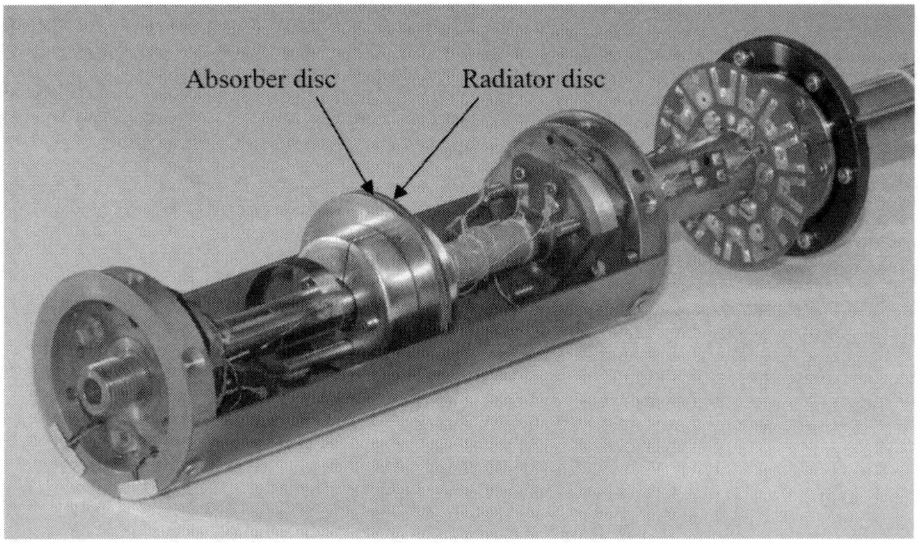

Figure 2. Opened measurement chamber of the apparatus for testing of the thermal emissivity or absorptivity. The chamber with the discs is housed in a stainless-steel casing tube (not shown here), which is further immersed in a LHe bath in a commercial LHe wide neck Dewar vessel.

2.2. Samples

Two 40 mm disc coated by thermal control Alion Z93C55 paint were tested by the presented method in order to find optimal layer for some surfaces of the space probe JUICE. First disc was coated by average thickness of about 110 μm, while the second one was covered by thinner layer (about 60 μm).

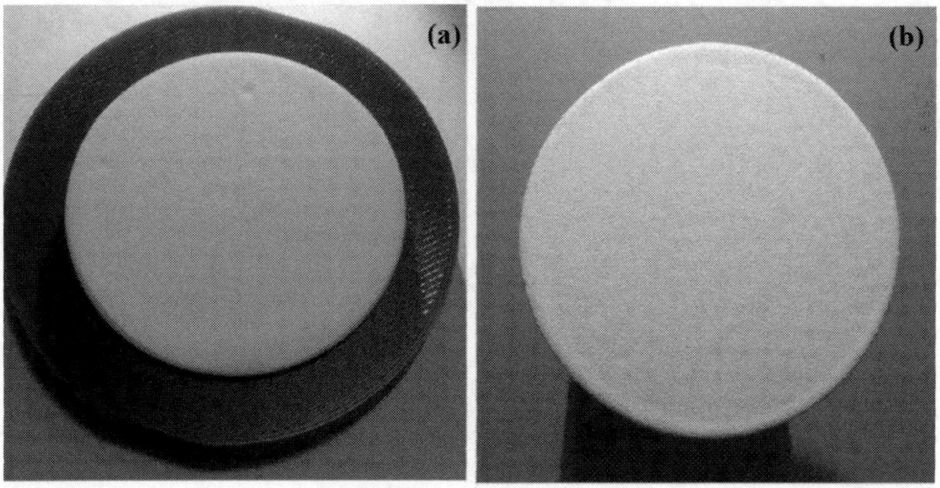

Figure 3. Tested samples of the Alion Z93C55 paint with thicknesses of 110 μm (a) and 60 μm (b) deposited on aluminium disc.

3. Results and discussion

Total hemispherical emissivities of the Alion paint deposited on aluminium discs in two different thicknesses were measured. We found only slight difference in emissivities between the tested samples. Slightly higher values were obtained in case of the 110 μm layer and especially at low temperatures below 100 K (Fig. 4). For both samples, at temperatures below 40 K emissivity values are lower, due to partial transparency of the dielectric layers for longer wavelengths of the thermal radiation, while at higher temperatures the dependence of emissivities on T_R was weak and reached very high values of about 90% (Fig. 4). Similar temperature dependences were found in case of epoxy coatings or special inorganic highly emissive layers [4].

The presented cryogenic method is suitable also for testing materials reaching lower emissivity values, including highly reflective metals with extremely low thermal emission. In one of our previous paper a unique cryogenic database, publicly available in the Mendeley repository, has been introduced [6]. It contains thermal radiative properties of various materials in a wide temperature range.

As example we present here measurements of emissivity and absorptivity (Fig. 5) of technically pure copper and Ti6Al4V alloy (Ti Grade 5), i. e. widely used materials either for excellent mechanical characteristics (Ti6Al4V) or very high thermal and electrical conductivity (Cu). Ti6Al4V represents a metallic material with a very high emissivity, while copper is known for extremely low emissivity values. Fig. 5 also shows the effect of a thin gold layer (1-2 μm), which significantly lowered the original emissivity (absorptivity) of the titanium alloy sample after the deposition. It shows the potential of gold layers for the lowering the heat exchange by thermal radiation. It also demonstrates a difference between thermal emissivity and absorptivity of pure metals. This has an origin in much lower temperature of the sample in absorptivity measurements with respect to the emissivity measurement. The response of the pure metals on the thermal radiation depends not only on the temperature of the source of the thermal radiation (spectrum) but also on the temperature of the material itself. The thermal radiative properties of metals are strongly connected with their electric conductivity which is strongly dependent on the temperature. For our temperature range this not the case for dielectrics, where the dependence of the thermal radiative properties on the material temperature is low.

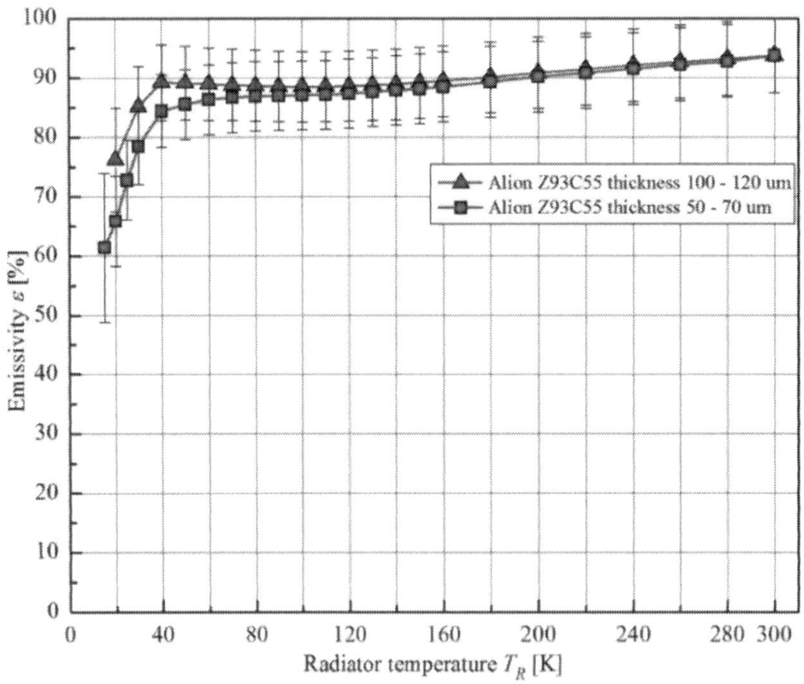

Figure 4. Temperature dependence of total hemispherical emissivity of the Alion coating with different thickness on its temperature T_R.

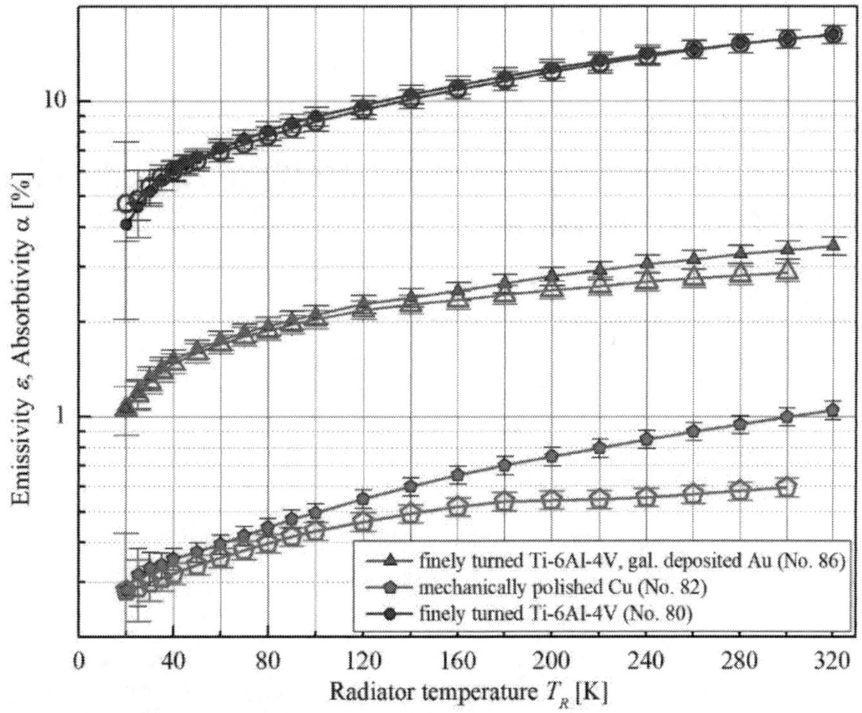

Figure 5. Measured temperature dependences of total hemispherical emissivity and absorptivity for polished Cu (internal sample No. 82), Ti-6Al-4V alloy with a bare surface (No. 80) and with galvanically deposited Au (No. 86). Full circles correspond to emissivity values (sample as radiator) and open symbols to absorptivity (sample as absorber).

4. Conclusions

Specific environment of the outer space places high demands on ground-based thermal testing of materials intended for space applications. Our cryogenic measurement method allows assessment of total hemispherical emissivity and absorptivity of various materials, including exceptionally emissive layers as well as high-purity metals with low thermal emission, in a wide temperature range (from room temperature down to 10 K). Exact values of emissivity or absorptivity in a wide temperature range are crucial for optimal thermal design of cryogenic systems as well as for spacecrafts thermal control.

References

[1] Howell J R, Mengüc M P, Daun K, Siegel R 2020 *Thermal Radiation Heat Transfer* (Boca Raton: CRC Press)

[2] Williamson M 1987 *Physics in Technology* **18** 120-127

[3] Kingston J 2016 *The Aeronautical Journal* **119** 245-246

[4] Kralik T, Katsir D 2009 *Infrared Technology and Applications XXXV* 131-139

[5] Kralik T, Musilova V, Hanzelka P, Frolec J 2016 *Metrologia* **53** 743-753

[6] Frolec J, Králík T, Musilová V, Hanzelka P, Srnka A, Jelínek J 2019 *Cryogenics* **97** 85-99

Acknowledgments
This work was supported by Czech Academy of Sciences (RVO:68081731).

Thermal absorptivity of a metallic surface contaminated by EAC-1A regolith simulant

T Kralik[1], J Frolec[1*], B Bras[2], M Portaluppi[2]

*corresponding author contact: frolec@isibrno.cz

[1] Institute of Scientific Instruments of the Czech Academy of Sciences, Kralovopolska 147, 612 00 Brno, Czech Republic

[2] Materials' Physics & Chemistry Section TEC-QEE at ESTEC (European Space Research and Technology Centre), Keplerlaan 1, 2200AG Noordwijk, Netherlands, (contact: tec-qee.qms.team@esa.int)

Abstract. The aim of this study is to estimate the impact of EAC-1A lunar regolith simulant contamination on the thermal absorption properties of a metallic surface. Masses ranging from 1.0 mg to 1016 mg of the European Astronaut Centre lunar regolith simulant 1 (EAC-1A) were placed on a chemically polished aluminium disc and the total hemispherical absorptivity (α_{TH}) was measured at temperatures T_A of the disc itself between 5 K and 100 K. Temperature T_R of the thermal radiation source ranged from 20 to 280 K. Even a small amount of EAC-1A (several μg/mm^2) increased the original α_{TH} of the bare surface up to thirty times and with increasing volumes the effect was even more prominent. However, α_{TH} significantly decreased at higher T_R in case of higher simulant volumes. The results suggest that grains of EAC-1A acted as a thermal shield, preventing the incident thermal radiation to reach the substrate surface. Our findings may have a practical relevance for the thermal management of spacecrafts for exploration missions on Moon and Mars surface.

1. Introduction

Thermal radiation, one of the basic modes of heat transfer, plays an important role in overall heat exchange especially in thermal insulations of some ground based cryogenic systems, or in outer space, where conduction and convection do not occur in vacuum of space. In order to keep the internal systems within certain temperature limits despite the harsh environment outside, the spacecraft thermal control must be considered and designed carefully [1, 2].

Heat transfer by thermal radiation between two objects is given by their areas, mutual orientation, absolute temperatures and the surface´s ability to emit/absorb thermal radiation at selected temperatures of the surfaces [3]. Absorptivity, the absorbed fraction of the thermal radiation reaching the surface, generally depends on the spectrum of the incident radiation, surface temperature and surface finish [4, 5]. However, another important factor is surface contamination that can significantly alter the original value of absorptivity and consequently affect the thermal control of a spacecraft, which is highly dependent on the thermal radiative properties of the used materials.

Moon exploration, especially mankind's effort to send humans back to the Moon, requires developing new technologies as well as extensive knowledge of the lunar soil and the characteristics of the environment on the Moon surface. Some terrestrial materials can be manufactured closely matching the properties of lunar soil (regolith) for the ground-based hardware development and spacecrafts testing. One of such lunar regolith simulants is the basanite material from Siebengebirge Volcanic Field (Königswinter, Germany) designated as European Astronaut Centre lunar regolith simulant 1 (EAC-1A) [6].

Although physical and chemical characterisation of the regolith simulant EAC-1A has been already thoroughly investigated [6], our knowledge of the effects of surface contamination by a EAC-1A or other regolith simulants on absorption of thermal radiation is very limited. Since pure metals with a clean surface generally show low absorptivity and emissivity values, any contamination can enhance these values significantly. The aim of our research was thus determining experimentally this impact and shed new light on related risks for the thermal management of spacecrafts.

2. Materials and method

2.1. Apparatus

The method dedicated for measurement of total hemispherical emissivity (ε) and absorptivity (a) has been already described in detail in our former paper [7].

Briefly, its principle is based on the measurement of radiative heat power Q_R transferred in vacuum between two parallel and interchangeable concentric 40 mm discs separated by a small gap (approximately 0.5 mm). For the presented absorptivity measurement, the examined sample is placed at the lower position of absorber (temperature T_A), whereas at the opposite position above of is placed a disc with "black" reference surface (Fig. 1) acting as the source of thermal radiation (temperature T_R). A reference made of eight circular duralumin rings coated with Fractal Black^{TM} layer, having very high absorptivity and emissivity of about 95% [7], was used for this purpose.

The absorber disc with the examined sample is connected with a thermal resistor (acting as a heat flow meter) made of a thin stainless-steel tube. At its bottom the temperature T_B is detected. The sample as well as the reference disc and the heat flow meter are placed inside a measuring chamber housed in a stainless-steel casing tube, which is immersed in a pressure-stabilized LHe bath.

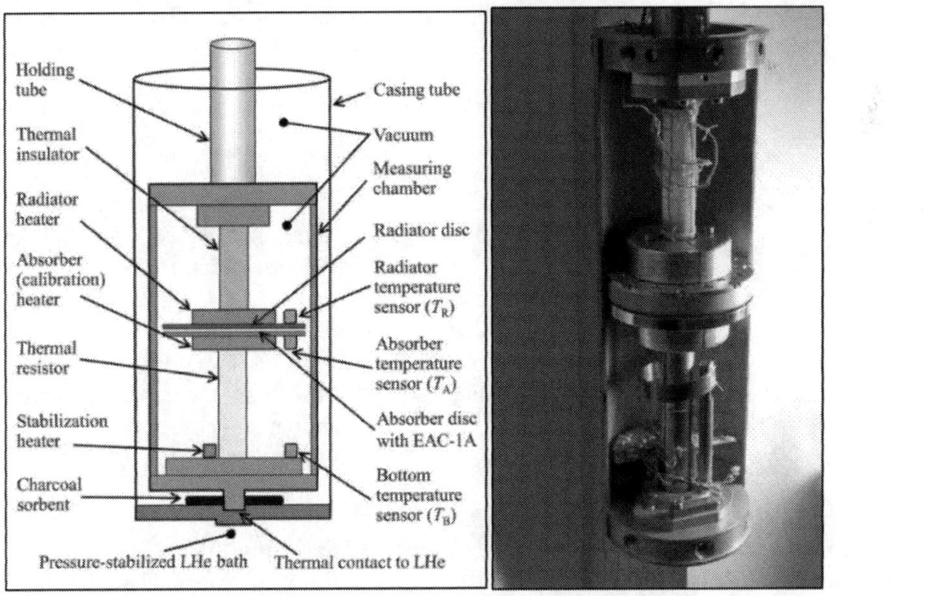

Figure 1. Left: Scheme of measuring chamber at the cold end of the instrument used for the cryogenic tests (adapted from [7]). Right: Photo of the opened measuring chamber with the regolith simulant spread on the absorber Al disc distanced by a small gap from the radiator above.

2.2. Samples preparation

Defined amount of the EAC-1A was scattered on the surface of aluminum substrate disc with elevated circumferential rim (Fig. 2). The disc was then placed into measuring chamber at the position of absorber of thermal radiation. To prevent any changes on the sample (movement of the simulant on the Al surface) the apparatus was kept in vertical position (i.e., both absorber and radiator discs stay in a horizontal position) and without shaking during the sample preparation, evacuation, cooldown and measurement. The vacuum in the apparatus is created before the cool down at room temperature by a turbomolecular pump. In order to prevent surfaces contamination by volatile deposits or parasitic heat flow conducted by emerging residual gas, all parts of the measuring chamber are kept for several hours between 310 and 320 K. From the same reason, during the subsequent cooldown process in a commercial LHe Dewar vessel the absorber disc with the regolith simulant is cooled down as the latest after the radiator and the measuring chamber. During the measurement the vacuum is passively kept by cryopumping at a level lower than 10^{-7} Pa.

Figure 2. Chemically polished aluminum disc (40 mm in diameter) that was used for the measurements in the position of thermal radiation absorber. Bare disc (left), identical disc after scattering of 10 mg of EAC-1A regolith simulant (middle) and fully covered by 1016 mg of identical material (right). Elevated rim at its circumference prevents the studied material spilling out of the substrate disc.

2.3. Measurement procedure and evaluation

After cool-down to about 5 K, the radiator is set to selected setpoints using temperature controller Lake Shore 340 and after reaching thermal equilibrium temperatures T_R, T_A and T_B were measured. The value Q_R of transferred heat power is calculated from the temperature drop (between T_A and T_B) in steady state on the previously calibrated thermal resistor. We use LakeShore sensors Cernox™ CX 1050 for measurement of T_A as well as T_B and LakeShore DT470 SD silicon diode as a sensor for T_R. The final absorptivity of the sample is calculated as a ratio of the measured Q_R and the heat power Q_B that would be measured with two perfect "black" (100% absorbing and emitting) samples $Q_B = A\sigma(T_R^4 - T_A^4)$ [W] [7]. A denotes the sample area and σ the Stefan-Boltzmann constant.

3. Results and discussion

We present our experimental data on total hemispherical absorptivity values at selected setpoints given by temperature of the thermal radiation source, ranging from 20 up to 280 K (Fig. 3). Corresponding temperatures for the substrate disc carrying the regolith ranged from 5 up to about 100 K. The actual temperature of the regolith simulant was not known, as discussed below. Eight different amounts (mass m) of the regolith simulant between 1.0 mg and 1016 mg were individually measured (Tab. 1). As the reference, also absorptivity of the identical substrate disc with bare surface (Fig. 2, left) was

IOP Conf. Series: Materials Science and Engineering 1287 (2023) 012024 doi:10.1088/1757-899X/1287/1/012024

tested. On the other hand, by application of 1 016 mg the substrate surface was completely covered by the lunar simulant (Fig. 2, right).

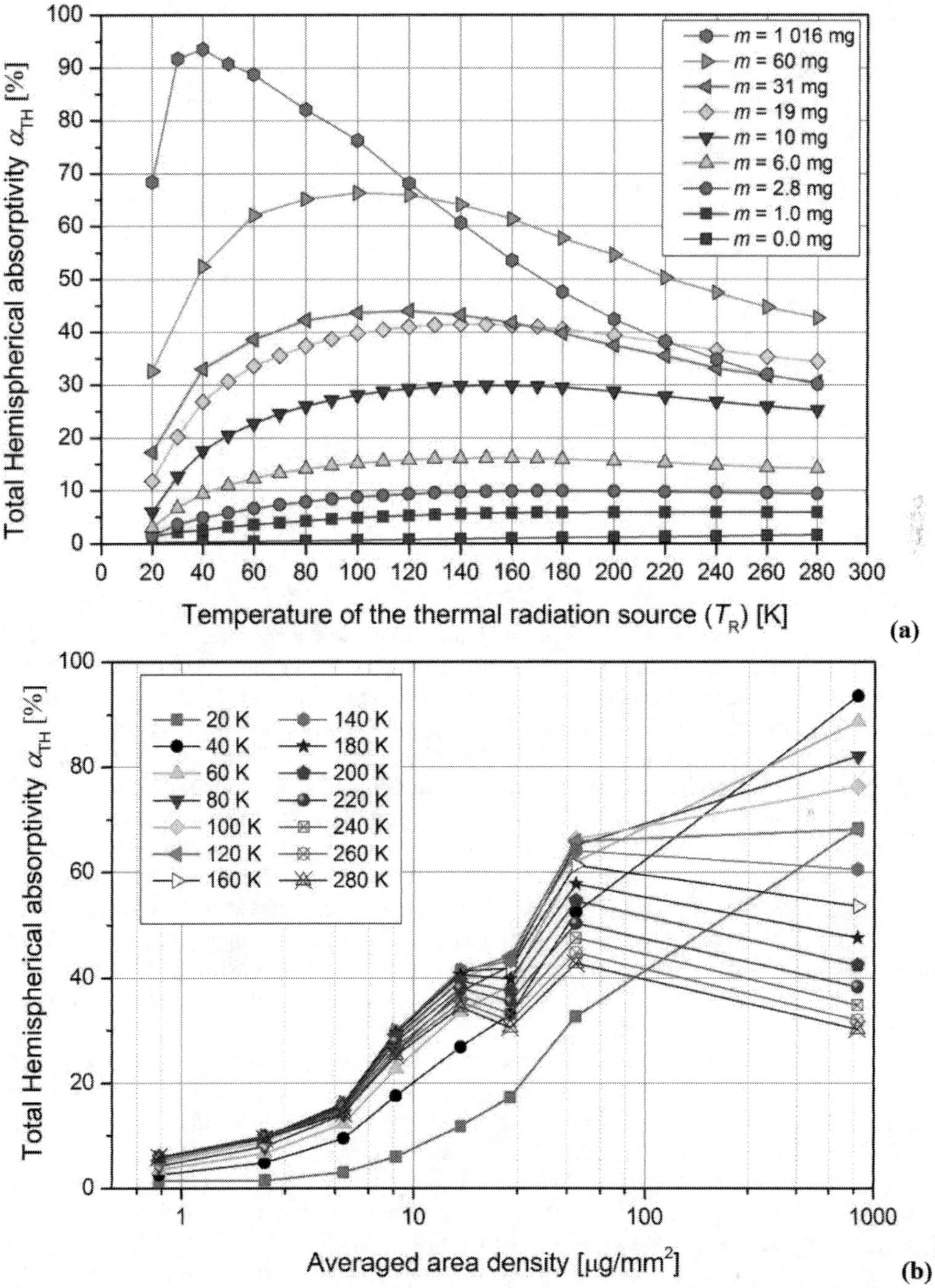

Figure 3. Dependence of the total hemispherical absorptivity on temperature of the thermal radiation source (a) and on the averaged area density of the EAC-1A (b) as specified in Tab. 1. Absorptivities were measured for bare surface of the Al disc (curve m = 0 mg) as well as for several masses of EAC-1A including the complete coverage of the surface by the EAC-1A (curve m = 1016 mg).

IOP Conf. Series: Materials Science and Engineering 1287 (2023) 012024 doi:10.1088/1757-899X/1287/1/012024

The absorptivity increases with the increasing amount of the EAC-1A from about 6% for 1.0 mg to more than 90% for 1016 mg (Fig. 3 (a), Tab. 1). We also found that after a steep increase at low temperatures the absorptivities reached maximal values and decreased at higher T_R. With increasing lunar simulant amount the maxima were more pronounced and shifted approximately from 200 K to 40 K (Tab. 1).

Table 1. List of the measured configuration of the absorber disc covered by the EAC-1A simulant.

Amount of the EAC-1A (mg)	Averaged EAC-1A area density (μg/mm^2)	Maximal absorptivity (%)	Absorptivity maximum at T_R (K)
0.0	0.0	1.8	280
1.0	0.8	5.8	200
2.8	2.3	9.9	180
6.0	5.0	16.2	150
10.0	8.4	29.9	150
19.0	15.9	41.5	140
31.0	26.0	44.1	120
60.0	50.2	66.4	100
1 016	850.0	93.6	40

The absorptivity dependence on the amount of the regolith simulant was expected and results from rising coverage of the highly reflective substrate aluminium surface by highly absorptive material. The presence of maxima in the mentioned temperature range requires more complex explanation. We assign them to the fact that the regolith simulant acts as a loosely connected thermal shield for the Al substrate surface beneath. This is possible because the regolith simulant has only weak thermal connection with the absorber surface and thus at higher radiative heat flows (i.e., higher T_R) the temperature of EAC-1A grains can be much higher than the detected temperature of the Al absorber disc. Specifically, the temperature of the radiator rose from 20 to 280 K, while temperatures ranging typically from 5 to about 100 K in the case of the absorber disc with the lunar simulant were measured. In such situation, a lower heat flow is measured as a consequence of the fact that higher fraction of thermal radiation impacting absorber is reradiated back towards the radiator (or out of the gap) by the regolith simulant and is not detected by the heat flow meter. As a consequence, the calculated absorptivity at higher temperatures is lower. For the area densities higher than 8.4 μg/mm^2 the shapes of the absorptivity curves are not directly proportional to the area density. This is evident for the curves in Figure 3 (a) after their maxima. We attribute this effect to a partial or full multilayer stratification of the dust on the dish. Two or more layers of dust lead to steeper decrease of the absorptivity with increasing radiator temperature T_R. Figure 3 (b) also demonstrates this trend: the dependence of absorptivity on increasing area density changes its character with increasing T_R (Fig. 3 b). This is consistent with the behaviour of a multilayer thermal shield.

We also observed tendency of slowing down the absorptivity decrease with warming up the sample to highest measured temperatures (about 280 K). The increase was prominent at 300 K (not shown), but the data are not shown since this was connected with impossibility of reaching thermal equilibrium for T_A, i.e., correct reading of stable temperature was not possible. It resulted from continuous releasing of volatile components from warm grains of the regolith simulant (and possibly also the radiator at T_R) inside the measuring chamber. It can enhance parasitic heat flow conducted by emerging residual gas or cause surfaces contamination by condensation of volatile components between the grains on the Al surface of cold absorber. Both options lead to increase of the measured absorptivity [8].

IOP Conf. Series: Materials Science and Engineering 1287 (2023) 012024 doi:10.1088/1757-899X/1287/1/012024

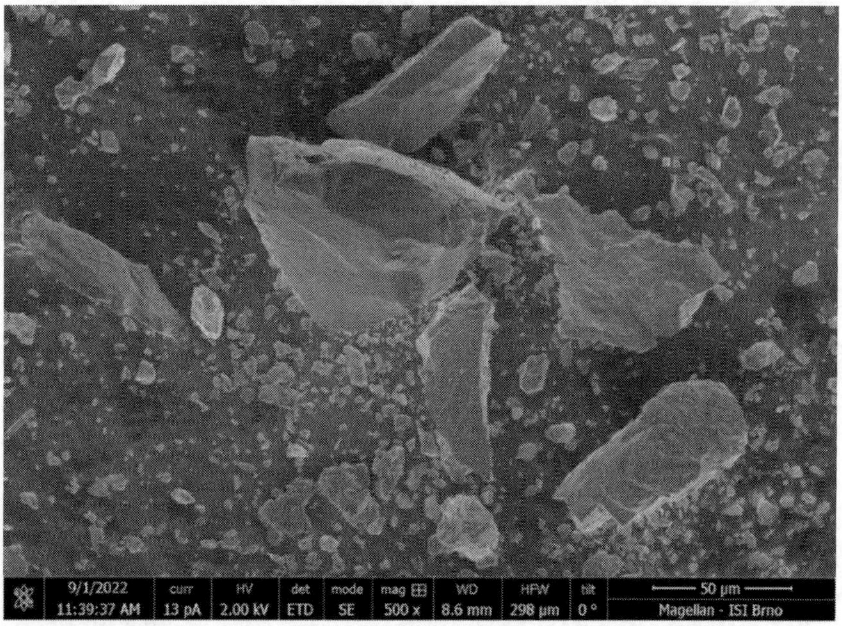

Figure 4. An image of the EAC-1A regolith simulant, from the tested batch, taken in a scanning electron microscope.

Although the microstructure of the EAC-1A is already known [6], we have used electron microscope in order to show the character of particles in the used EAC-1A material (Fig. 4). Such images are the first step for the analysis of particles sizes and distributions, which could enable us to use a mathematical model to fit the experimental data and see the nature of the interaction of the simulant particles on the surface with the thermal radiation.

4. Conclusions

This study is the first step towards enhancing the understanding of interaction of thermal radiation with metallic surface partly covered by a lunar soil analogue. Using our dedicated cryogenic apparatus, we found strong enhancement of aluminium surface´s absorptivity after its contamination by lunar regolith simulant, namely EAC-1A. Even small amount of this material (several micrograms per square millimetre of the surface) increased the original absorptivity of the bare surface up to thirty times and with increasing surface coverage the enhancement was even more prominent. Maximal α_{TH} of the EAC-1A (fully covered surface m=1016 mg) reached about 94%, close to values of 95-98% for Moon surface emissivity detected from lunar orbit [9].

All measurements were done in a wide temperature range from 20 K to 280 K of the radiation source. The absorptivity was also dependent on temperature of the thermal radiation source for each tested amount of the lunar simulant. After reaching maximal absorptivity at cryogenic temperatures, a shift to higher temperatures caused gradual decrease of absorptivity, especially in the cases of high quantities of the regolith simulant. This implies that thicker layer of the material can act as a thermal shield with low thermal connection to the disc, preventing the incident thermal radiation to reach surface of the substrate. The measurement of spectral characteristics of the lunar regolith in the far infra-red will bring important insight on the energy transfer mechanisms (e.g., regolith absorbing vs reflecting incoming radiation). The same effect could also degrade the performance of a thermal radiator by its shielding. Our research could be a useful aid for thermal design and analysing the situations where interaction of low absorbing surfaces of space probes with lunar regolith or similar materials can occur.

References

[1] Kingston J 2016 *The Aeronautical Journal* 119 245-246
[2] Williamson M 1987 *Physics in Technology* 18 120-127
[3] Howell J R, Mengüc M P, Daun K, Siegel R 2020 *Thermal Radiation Heat Transfer* (Boca Raton: CRC Press)
[4] Sabuga W, Todtenhaupt R 2001 *High Temp - High Press* 33 261-269
[5] Frolec J, Králík T, Musilová V, Hanzelka P, Srnka A, Jelínek J 2019 *Cryogenics* 97 85-99
[6] Engelschiøn V S, Eriksson S R, Cowley A, Fateri M, Meurisse A, Kueppers U, Sperl M 2020 *Scientific Reports* 10 5473
[7] Kralik T, Musilova V, Hanzelka P, Frolec J 2016 *Metrologia* 53 743-753
[8] Musilová V, Hanzelka P, Králík T 2010 *11th Cryogenics 2010 IIR International Conference* 37-42
[9] Vasavada A R, Bandfield J L, Greenhagen B T, Hayne P O, Siegler M A, Williams J-P, Paige D A 2012 *Journal of Geophysical Research: Planets* 117

Acknowledgments

This work was supported by Czech Academy of Sciences (RVO:68081731). We thank members of Microscopy and Microanalysis group from the Department of Electron Microscopy at ISI Brno for images of the regolith simulant.

Detection and characterization of contamination with fluorescence spectroscopy

E Gouisset[1], G Rioland[2], F Bourcier[2], D Faye[2], P Walter[3] and F Infante[1]

[1] Intraspec Technologies, 20 avenue Didier Daurat, Toulouse, 31400, France
[2] CNES, 18 avenue Edouard Belin, Toulouse, 31400, France
[3] LAMS, 4 place Jussieu, Paris, 75005, France

emmanuel.gouisset@intraspec.com

Abstract. In the field of failure analysis and in particular molecular and particulate contamination, being able to detect any trace of contaminants during the integration of an orbital spacecraft is crucial. In this context, fluorescence allows not only to detect but also to discriminate contaminants. We studied the fluorescence response of two epoxy adhesives, typical sources of spacecraft contamination in orbit with a portable broadband hyperspectral instrument (UV-Vis-NIR) developed in collaboration with the CNES and Intraspec Technologies, but also with a commercial spectrofluorometer. These measurements had two objectives, evaluate the performance of our hyperspectral instrument in order to identify prospect of improvement, but as well study the pertinence of fluorescence signature study in the contamination field. The first goal brings out that the hyperspectral instrument is capable of imaging the scene and allows us to extract fluorescence spectra from the image, but it still needs development, especially in term of sensitivity in UV range. The second goal shows promising results. Fluorescence studies with the spectrofluorometer emphasize that fluorescence spectra are specific to the chemical nature of the contaminant, which allows us to clearly discriminate them.

1. Introduction

Satellites are the spearheads of telecommunication and Earth and sky observations. However, the success of space missions could be compromised by molecular contamination, composed of molecules outgassed from materials (paints, varnishes, adhesives, and so on), thrusters and human activity [1]. Indeed, because molecular contamination might blind and damage observation instruments, optical communications, thermo-optical properties of coatings or/and solar cells, it is a major risk that must be controlled at every step of the life duration of satellites. Numerous space missions faced issues due to molecular contamination [2-7]. That is why during the process of satellite integration, the intervention of contamination engineer is needed to validate the integration progress. Currently, the unique solution for the operator is to trigger fluorescence of contaminants that might be deposited on sensitive surfaces during qualification or integration process. However, the fluorescence spectra is not measured, it is only detected by human eyes ; the operator can only detect contamination fluorescence spatially from the eye detection capacity, in other words from 400 nm to 700 nm.

The purpose of this paper is to study the fluorescence response of different materials used by space companies and public institutes in order to validate a hyperspectral imager (developed by the French Space Agency, CNES, and Intraspec Technologies) which is able to imaging the scene and collect

IOP Conf. Series: Materials Science and Engineering 1287 (2023) 012025 doi:10.1088/1757-899X/1287/1/012025

spectra of each pixel in a range from 330 nm to 1000 nm. This preliminary study will confirm the necessity of this development in order to give a new tool for the operator during the control process.

The particularity of molecular fluorescence spectra can be determinant for characterizing the chemical nature of the contaminant. Even if the spectrum cannot enable a straight forward chemical identification (usually, a sample can be collected for infrared spectroscopy analysis), this energetic signature of specific fluorophores can partially lead the operator to relate an unknown contamination to a potential source. In this paper, we discuss our capacity of detecting outgassed fluorescent contamination, then the development of a method which permits quantitative and spatially detection of contamination on sensitive surfaces of any types and in a non-destructive manner.

As depicted before, the method is based on the principle that the outgassed contamination may contain fluorophores, which could be excited with UV light. For the preliminary test, we used 270 nm LEDs. The use of a LED is greatly cheaper than a laser and allows us to make fast and easy measurements. The fluorescence spectra of the sample are observed with a 200-1000 nm spectral range spectrometer.

2. Material and methods

2.1. Instrumentation

2.1.1. Deposition chamber. Outgassing is carried out in CNES facility at 125 °C for 24 hours. The outgassing bench (figure 1) allows four material samples to be simultaneously outgassed with a direct view factor to the collectors. A 1.8 cm² substrate is exposed to the contaminant flow, thus making it possible to produce an equivalent deposition surface. The chamber is a stainless steel tank maintained under secondary vacuum with a Pfeiffer® TMU 520 turbopump. The chamber is equipped with four ovens (figure 2), each heated by a Minco HK6909 heater and regulated by a Eurotherm regulator.

Figure 1. Outgassing bench equipped with a primary pump and a secondary pump to ensure high vacuum. 4 ovens are facing substrates collecting the outgassing products.

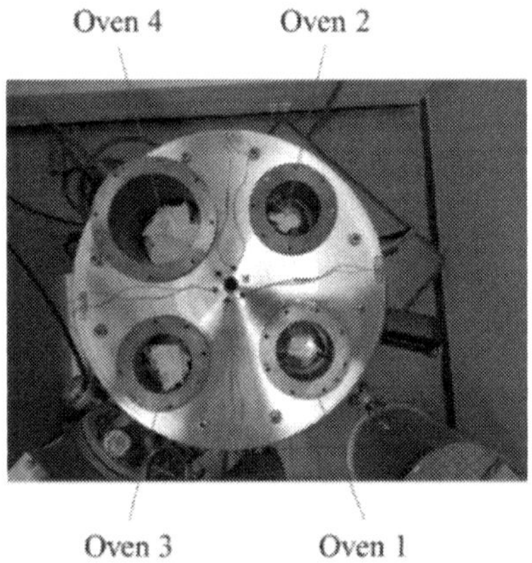

Figure 2. EC9323-2 material deposited in furnaces prior to installation of fused silica substrates. The masses deposited in ovens 1, 2, 3 and 4 are 297.5 mg, 1254.8 mg, 5087.5 mg and 19964.2 mg respectively.

2.1.2. Transmittance measurement. Transmittance measurement was performed on fused silica substrate before and after contaminants deposition. The Spectrophotometer is a Cary 5000 spectrometer from Agilent equipped with a deuterium lamp emitting in the UV range and a quartz-iodine lamp (QTH for Quartz Tungsten Halogen) emitting in the visible-infrared range. Measurements were carried out from 200 nm to 1000 nm with a resolution of 2 nm.

2.1.3. Fluorescence measurement. Fluorescence measurement was performed on fused silica substrate before deposition, then after outgassed contamination deposition. The spectrometer is a OceanOptics QEPro tuned with a 250-100 nm spectral range. An optical fiber with an aperture of 600 μm was used in order to collect special point of interest. The light source is a LaserComponents LED at 270 nm coupled to a radiator. The LED beam is collimated with a Thorlabs lens 25mm.

2.1.4. TD-GC-MS Measurement. TD-GC-MS measurement was performed on the collected contaminants after deposition on substrates. The gas chromatography (GC) measurements were carried out with a Clarus 500 chromatograph (PerkinElmer® Elite 5-MS column with dimensions of 60 m×0.25 mm×0.25 μm), coupled with a TurboxMatrix 300 TD thermal desorber (TD) and a Clarus 500 mass spectrometer (MS).

2.1.5. Hyperspectral measurement. The hyperspectral instrument was developed within a collaboration between the French Space Agency (CNES), the Molecular and Structural Archaeology Laboratory (LAMS-CNRS) and Intraspec Technologies [8]. The working interval of wavelengths is 335 to 1000 nm. The resolution is 512 pixels in a field of view of 3.5 degrees and with a spectral resolution of 1 nm in UV range, 2-3 nm in VIS range and 5 nm in NIR range, guarantying an 80% flux in each pixel of the cube image. Its development was described in a previous work [9-13], including more details about technical description and first results on real samples. The light source is a LaserComponents LED at 270 nm on a radiator. The LED beam is collimated with a Thorlabs lens 25mm to have a uniform beam in the field of view.

3. Experimental setup

3.1. Materials

We decided to analyze the outgassed contaminants on fused silica substrate from two epoxy adhesives: EC2216 and EC9323-2 materials (Scotch-Weld 3M™). Both have been widely tested for aircraft and aerospace applications. EC2216 material is at least composed of amine, kaolin, diethylene glycol, toluene, according to safety datasheet. EC9323-2 material is at least composed of fatty acid, amine, diethylene glycol, siloxane, phenol compound, toluene, according to safety datasheet.

- Samples of EC2216 were polymerized with a mass ratio of 7:5 (compounds A:B) at room temperature during 7 days as described in technical datasheet provided by suppliers. Dimensions were 30 mm x 20 mm and 1 mm thick.
- Samples of EC9323-2 were polymerized with a mass ratio of 27:100 (compounds A:B) at room temperature during 14 days as described in technical datasheet provided by suppliers. Dimensions were 30 mm x 20 mm and 1 mm thick.

3.2. Procedure

Prior to performing experiments, SiO_2 substrates were cleaned with acetone, then substrate transmittance and fluorescence were measured in order to carry out a blank before outgassed contaminant measurement. Materials were heated at 125°C during 24 hours and substrate temperature was 25 °C. The masses of condensable contaminants (CVCM) resulting from the degassing of the EC2216 and EC9323 materials are respectively 0.005% and 0.04%. These two data come from internal tests at CNES. The thickness of the deposit on the surface of the substrates is calculated according to the equations below (Eq. 1) :

$$e = \frac{mass_{deposited}}{S \times \rho} \tag{1}$$

Where e (in nm) is the thickness of the deposit, S (in cm^2) the aera of the deposit, and ρ the density of the deposit conventionally estimated at 1 g/cm^3.

IOP Conf. Series: Materials Science and Engineering 1287 (2023) 012025 doi:10.1088/1757-899X/1287/1/012025

Table 1 and table 2 present the results obtained according to the masses of material that we outgassed.

Table 1. Thickness of contaminant deposit from EC2216 material obtained as a function of the mass of outgassed material.

Sample name	**1-EC2216**	**2-EC2216**	**3-EC2216**	**4-EC2216**
Initial mass EC2216 (mg)	302.548	1266.965	5123.144	19675.456
Deposited mass EC2216 (mg)	0.015	0.063	0.256	0.984
Deposit thickness EC2216 (nm)	85	350	1420	5500

Table 2. Thickness of contaminant deposit from EC9323-2 material obtained as a function of the mass of outgassed material.

Sample name	**1-EC9323-2**	**2-EC9323-2**	**3-EC9323-2**	**4-EC9323-2**
Initial mass EC9323-2 (mg)	297.511	1254.823	5087.542	19964.215
Deposited mass EC9323-2 (mg)	0.119	0.502	2.035	7.985
Deposit thickness EC9323-2 (nm)	660	2800	11300	44300

Figure 3 and figure 4 present the pictures of each deposit on silica substrates after the outgassing of EC2216 and EC9323-2 materials.

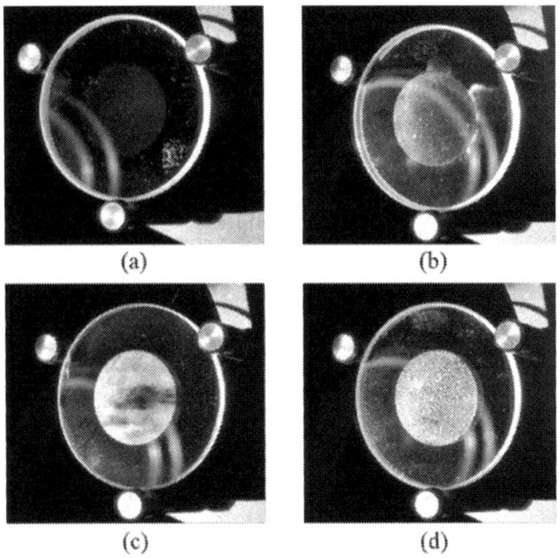

Figure 3. Deposits of contaminants resulting from the outgassing of EC2216 material on fused silica substrate. (a) 85 nm film, (b) 350 nm film, (c) 1420 nm film, (d) droplet deposition (theoretical film thickness of 5500 nm).

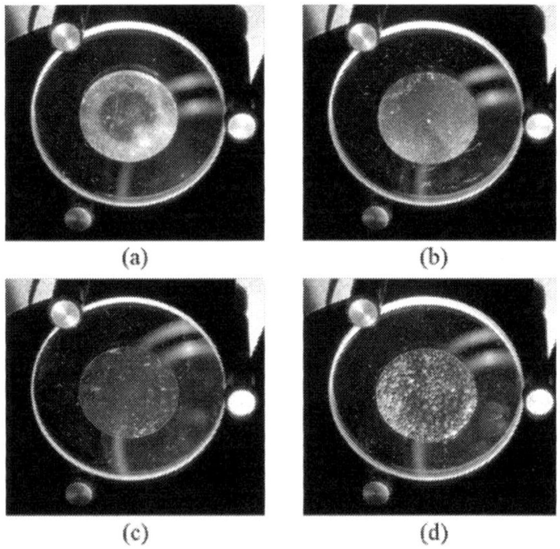

Figure 4. Deposits of contaminants resulting from the outgassing of EC9323-2 material on fused silica substrate. (a) 660 nm film, (b) 2800 nm film, (c) droplet deposition (theoretical film thickness of 11300 nm), (d) droplet deposition (theoretical film thickness of 44300 nm).

3.3. Preliminary analysis

TD-GC-MS measurement. TD-GC-MS measurements were performed on 4-EC2216 and 4-EC9323-2 samples to identify the molecules present in the deposit resulting from the outgassing of EC2216 and EC9323-2 materials. The samples were heated at 200 °C for 10 minutes in the thermal desorber at ambient pressure. Then, all the molecules were injected into the chromatograph coupled to the mass spectrometer. The chromatograph was heated from 30°C to 250 °C with a slope of 5 °C per minute. The scan was performed for molecules between 33 and 550 atomic mass units (amu).

The results of the TD-GC-MS analysis of the outgassing products of EC2216 and EC9323-2 materials are presented in table 3 and 4, respectively. The peaks at retention times of less than 4 minutes are due to internal contamination in the column.

Table 3. TD-GC-MS measurement on EC2216 material.

Retention time (min)	Molecule
1.94	Ethanol*
2.14	IPA*
2.70	methane, nitro*
3.13	n-hexane*
3.64	Trichloromethane*
7.99	**glycidol**
29.41	**1,4-Cyclohexanedimethanol**
29.95	**1,4-Cyclohexanedimethanol**
31.08	**1,4-Cyclohexanedimethanol**
31.35	**1,4-Cyclohexanedimethanol**
33.24	**Phenol, 2,4-bis(1,1-dimethylethyl)-**
35.30	**Diethyl Phthalate**

Table 4. TD-GC-MS measurement on EC9323-2 material.

Retention time (min)	Molecule
1.91	**Methylamine, N,N-dimethyl-**
2.21	**Oxirane, trimethyl-**
5.38	Acetic acid*
10.51	**Hexamethylcyclotrisiloxane**
17.32	**Phenol**
25.11	**Nonanoic acid, methyl ester**
29.70	**1,4-Cyclohexanedimethanol**
30.20	**1,4-Cyclohexanedimethanol, cis-**
31.12	**1,4-Cyclohexanedimethanol**
31.40	**1,4-Cyclohexanedimethanol, cis-**
32.37	1-Dodecanol*
33.26	**Phenol, 2,4-bis(1,1-dimethylethyl)-**

* Peaks at retention times of less than 4 minutes are due to internal contamination in the column.

3.3.1. Transmittance measurement. In order to select the most appropriate excitation wavelength for fluorescence analysis, we carried out transmission measurements for each deposit and their respective thicknesses. Transmission measurements on substrate contaminated with the outgassing of EC2216 material (figure 5) highlight an absorption range between 200 and 360 nm.

IOP Conf. Series: Materials Science and Engineering 1287 (2023) 012025 doi:10.1088/1757-899X/1287/1/012025

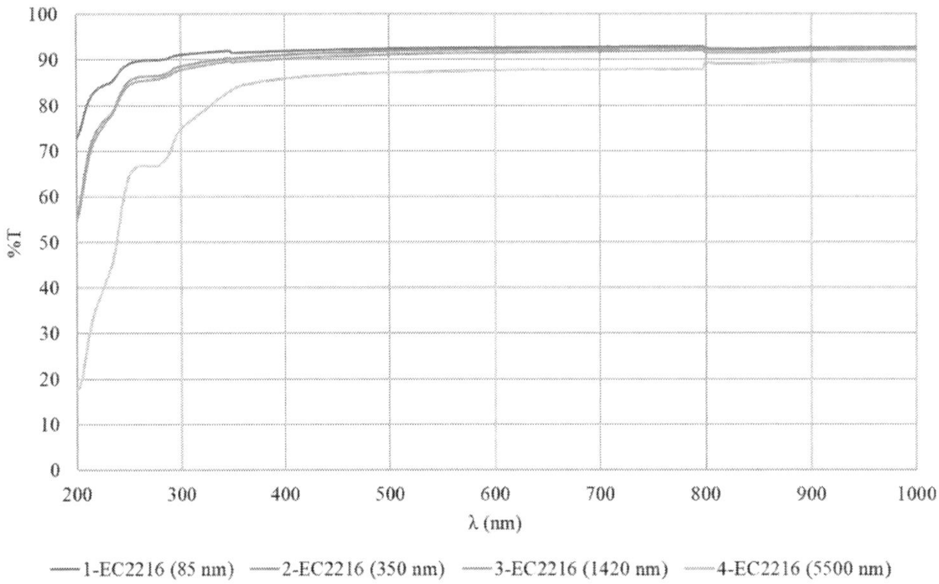

Figure 5. Transmission measurements for each deposition thickness of EC2216 contaminants material on fused silica substrates. 1-EC2216=85nm; 2-EC2216=350nm; 3-EC2216=1420nm; 4-EC2216=5500nm.

Cyclic compounds, aromatics and other derivatives such as 1,4-cyclohexanedimethanol show strong absorption peaks between 200 and 280 nm [14-16]. Phenol,2,4-bis(1,1-dimethylethyl) will absorb around 270 nm [17]. Heavier molecules like diethyl phthalate can also explain the absorption peaks in this wavelength range [18]. According to literature data, molecules of diethyl phthalate and 2,4-bis(1,1-dimethylethyl)-phenol are responsible for the absorption between 200 and 360 nm. No data were found on smaller molecules such as glycidol.

We can also observe a relatively constant absorption over the entire wavelength range for sample 4-EC2216. We assume that this must be related to the phenomenon of light scattering due to the droplets that form the deposit.

Table 5 summarize the wavelength ranges of the molecules responsible for absorption.

Table 5. Summary of outgassed molecules of EC2216 material and their absorption range.

Molecules	Absorption range (nm)
1,4-cyclohexanedimethanol	< 200 nm
2,4-bis(1,1-dimethylethyl)-phenol	~ 270 nm
Diethyl phthalate	230 et 280 nm

Transmission measurements on substrate contaminated with the outgassing of EC9323-2 material (figure 6) highlight an absorption range between 200 and 360 nm.

IOP Conf. Series: Materials Science and Engineering 1287 (2023) 012025 doi:10.1088/1757-899X/1287/1/012025

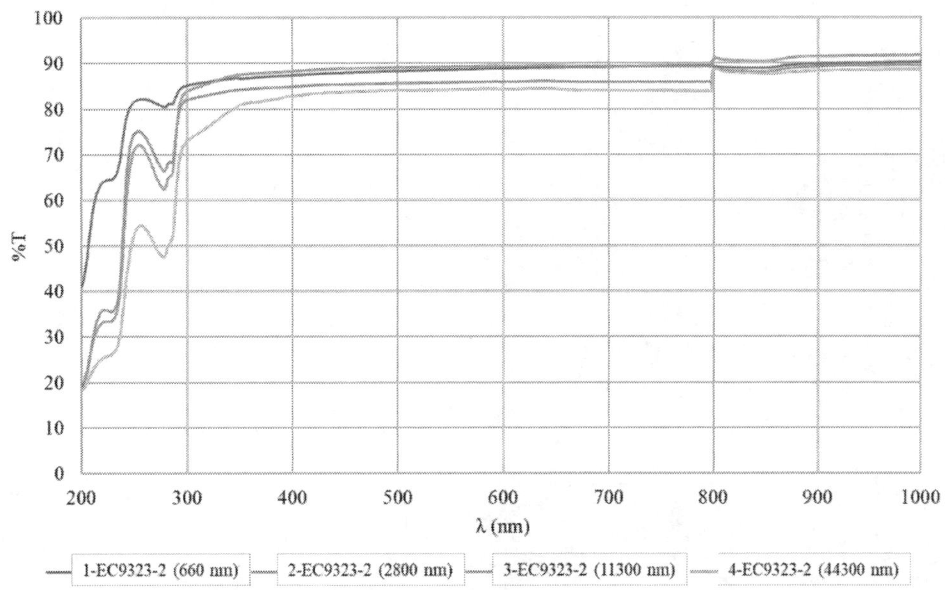

Figure 6. Transmission measurements for each deposition thickness of EC9323-2 material on fused silica substrates. 1-EC9323-2=660nm; 2-EC9323-2=2800nm; 3-EC9323-2=11300nm; 4-EC9323-2=44300nm.

We again find 1,4-cyclohexanedimethanol with absorption peaks at 200 and 280 nm as well as 2,4-bis(1,1-dimethylethyl)-phenol absorbing at 270 nm.. The silicone compounds that we found in the TD-GC-MS analysis have a very low impact on the transmission in this wavelength range compared to the other compounds [19]. According to literature data, molecules of diethyl phthalate and 2,4-bis(1,1-dimethylethyl)-phenol are responsible for the absorption between 200 and 360 nm. No data were found on smaller molecules such as methylamine, oxirane, and methyl nonanoate.

As before, we observe a relatively constant absorption over the whole wavelength range for the sample 4-EC9323-2 but also 3-EC9323-2 which must be linked to the phenomenon of light scattering due to the droplets which form the deposit.

Table 6 summarize the wavelength ranges of the molecules responsible for absorption.

Table 6. Summary of outgassed molecules of EC9323-2 material and their absorption range.

Molecules	Absorption range (nm)
Hexamethylcyclotrisiloxane	~180 nm
1,4-cyclohexanedimethanol	< 200 nm
2,4-bis(1,1-dimethylethyl)-phenol	~ 270 nm

3.3.2. Fluorescence measurement. Thanks to the measurement bench (figure 7), we were able to collect the fluorescence signal of the outgassing products of EC2216 and EC9323-2 materials

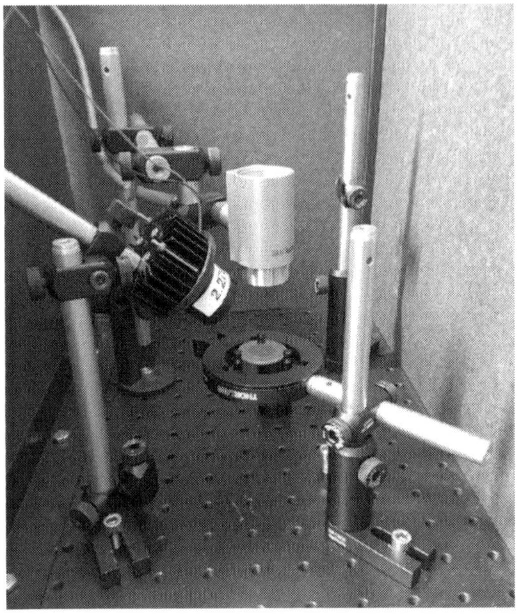

Figure 7. Measurement bench for measuring fluorescence with the OceanOptics spectrometer. It is made up of an excitation LED at 270 nm, and a light collimator injecting the flux into an optical fiber connected to the spectrometer.

The collimator and the LED are located at 5 cm and 7 cm from the sample, respectively. The parameters of the OceanOptics spectrometer are also set up for an average of 10 acquisitions of one second.

The fluorescence results of the deposits of outgassed molecules from EC2216 material are presented in figure 8 and show the fluorescence spectrum for each thickness.

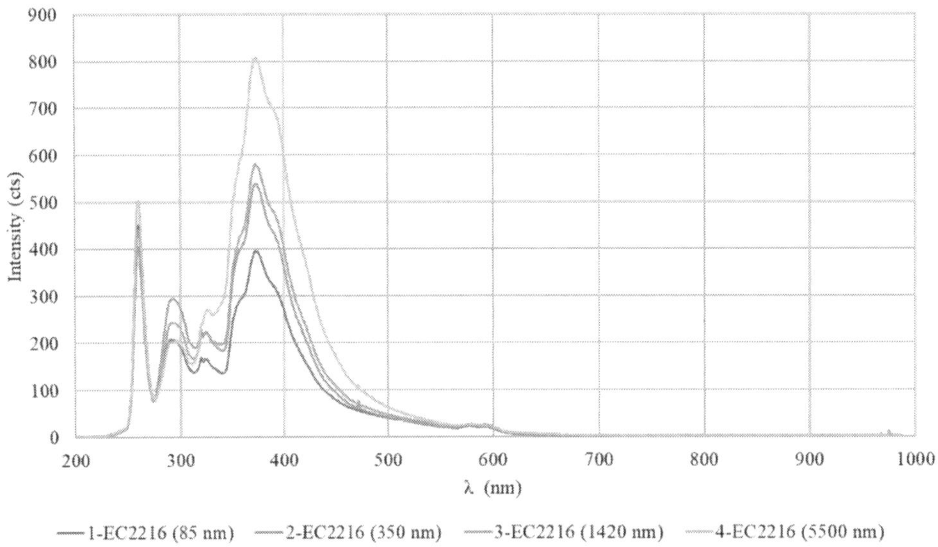

Figure 8. Fluorescence spectra of the outgassing deposits of EC2216 material acquired using the OceanOptics spectrometer.

First, we can observe an important peak at 270 nm, which is the wavelength of the incident light scattered by the sample. It is possible that the most intense peak at 290 nm is emitted by the 2,4-bis(1,1-dimethylethyl)-phenol molecule [20], as well as the second, less intense peak at 320 nm [21, 22]. The third modulation observed around 350 nm may be due to the diethyl phthalate molecule.

IOP Conf. Series: Materials Science and Engineering 1287 (2023) 012025 doi:10.1088/1757-899X/1287/1/012025

Indeed, for an excitation at 270 nm, this molecule is likely to emit at 355 nm [20]. It is likely that the emission peak having a maximum at 400 nm may result from the interaction between the 2,4-bis(1,1-dimethylethyl)-phenol moiety and that of 1,4-cyclohexanedimethanol [23, 24]. On the other hand, we do not explain the modulations around 600 nm.

The other molecules present can alter the fluorescence response of the compounds mentioned above. It is then excessively complex to isolate each of the modulations and peaks of the spectrum to ensure that each of them corresponds to such and such a molecule.

The fluorescence results of the deposits of outgassed molecules from EC9323-2 material are presented in figure 9 and show the fluorescence spectrum for each thickness.

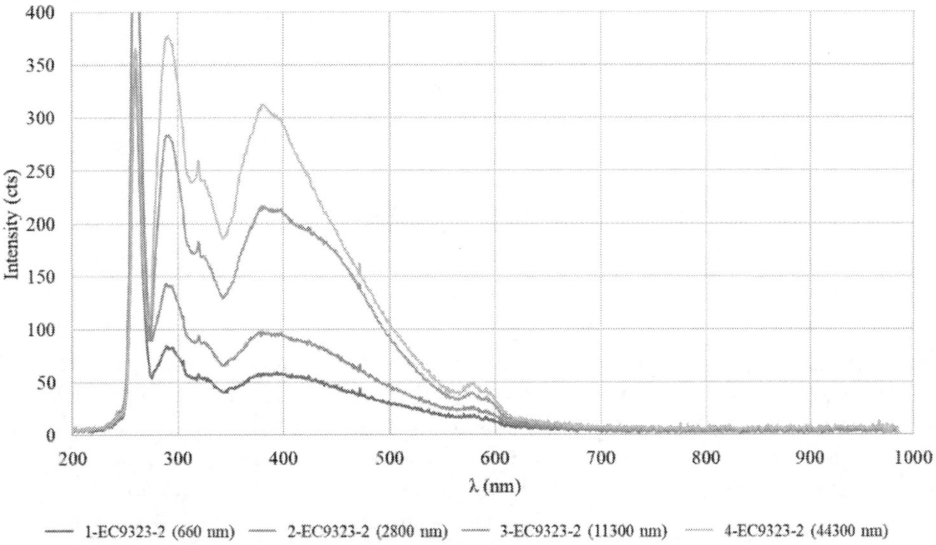

Figure 9. Fluorescence spectra of the outgassing deposits of EC9323-2 material acquired using the OceanOptics spectrometer EC9323-2.

As in the case of the deposit from EC2216 material, it is possible that the most intense peak at 290 nm is emitted by the 2,4-bis(1,1-dimethylethyl)-phenol molecule [20]. On the other hand, it is much more intense in the case of the deposit resulting from the material EC9323-2. The second peak, less intense at 320 nm, does not have the same shape as for the deposit of EC2216 material, but we can assume that it is still related to 2,4-bis(1,1-dimethylethyl)-phenol [21, 22]. Also, as for the deposit of EC2216 material, the emission peak having a maximum around 400 nm can result from the interaction between 2,4-bis(1,1-dimethylethyl)-phenol and that of 1,4-cyclohexanedimethanol [23, 24]. As in the previous case, we also do not explain the modulations around 600 nm. It is also here difficult to conclude on the roles of the different molecules with regard to their fluorescence response.

Although the analysis of the fluorescence spectra of the different samples is tedious, the comparison is relatively easy. Indeed, when we observe figure 10, the comparison of the products resulting from the outgassing of the two adhesives makes it possible to identify them. The 300 nm peak of the deposit from the EC9323-2 material leaves no doubt. Moreover, the modulations of the peaks around 400 nm are different, the latter is also wider for the deposit of EC9323-2 material.

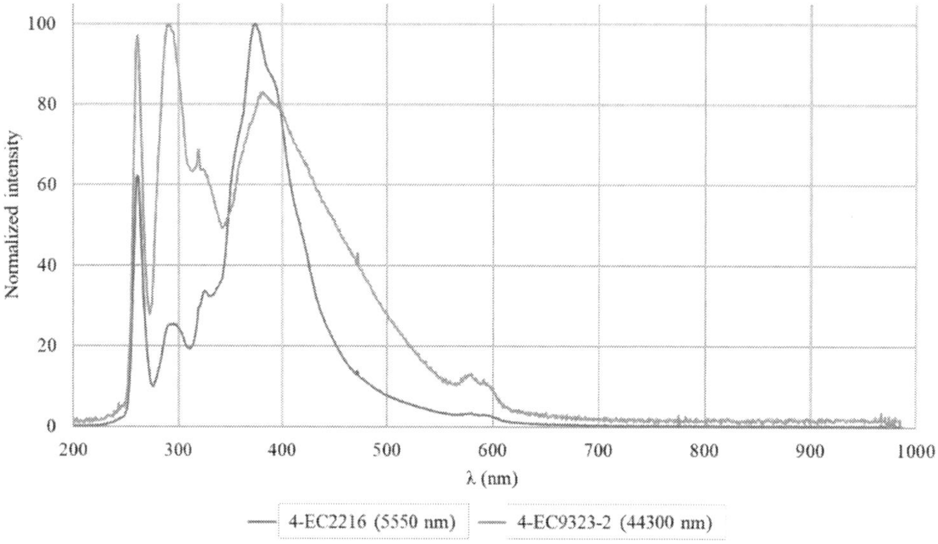

Figure 10. Comparison of fluorescence spectra of contaminant deposits from EC2216 and EC9323-2 materials.

Having identifiable spectra opens up new possibilities for contamination experts and cleanroom operators. If at the end of the development of the instrument and future experiments which would make it possible to detect deposits of contaminants of the order of a few tens of nm, the creation of a database could prove to be necessary. Indeed, if the contaminants are identifiable using their fluorescence response, it could be interesting to such a database coupled to digital processing for automatic analysis.

3.3.3. Hyperspectral measurement. The previous study allowed us to highlight a fluorescence signature of the outgassing products of EC2216 and EC9323-2 materials. We therefore carried out a series of measurements with the hyperspectral instrument in order to image the contaminants in addition to acquiring their fluorescence spectrum. The measurements of the samples with the hyperspectral imager were carried out in the optical laboratory at CNES. The instrument was mounted on its tripod, facing the sample to be measured as shown in figure 11. We set up the same measurement parameters used with the OceanOptics spectrometer, described before. That is to say, the distance between the LED and the sample is equal to that of the previous study, i.e. 7 cm.

The measurement was not convincing for samples 1, 2 and 3, in particular for the deposits related to EC2216 material. Indeed, the thicknesses and the fluorescence signal were too low for our instrument. They will therefore not be presented.

IOP Conf. Series: Materials Science and Engineering 1287 (2023) 012025 doi:10.1088/1757-899X/1287/1/012025

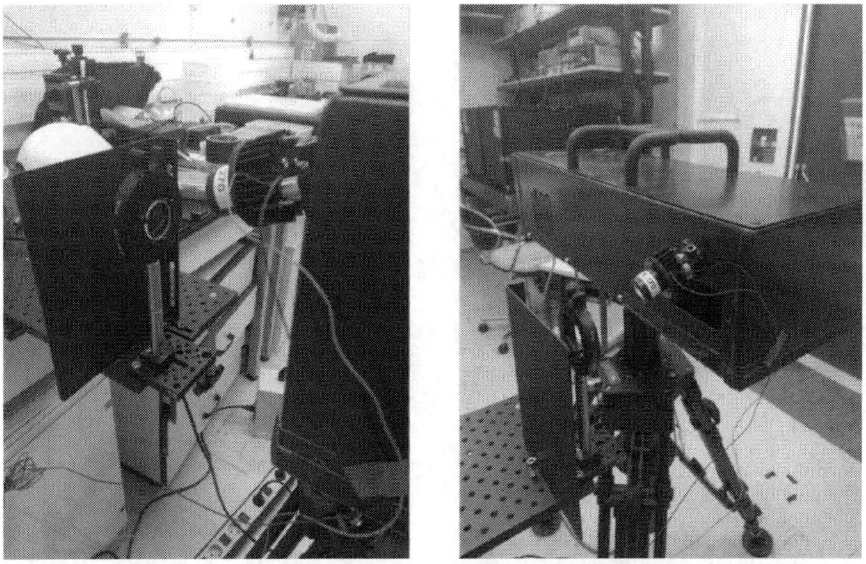

Figure 11. Analysis of contaminant samples by the hyperspectral imager. We fixed to the instrument a single LED at 270 nm, in order to reproduce the acquisition parameters used with the OceanOptics spectrometer.

We conducted preliminary tests to establish the best acquisition time that would allow us to measure a fluorescence signal, an acquisition time of 30 seconds allows us to detect a signal. On the other hand, such a long acquisition time made the measurement of the entire sample much too long. Indeed, the measurement time was 8 hours, we would have subjected our sample to significant UV damage which could modify their fluorescence response. We therefore decided to measure only a restricted area at 30 seconds of acquisition time, which would allow us to analyze the fluorescence spectrum but also to avoid too long UV exposure. And finally, a full scan to fully imaging the sample with an acquisition time of 5 seconds. The acquisition parameters are given in table 7. For each of the samples, 5 seconds acquisition time is too short to extract a spectrum on which we could work, but it allows us at least to image the scene and identify the contaminated areas.

Among other things, we present the integral measurements (with an acquisition time of 5 seconds) before the measurements of restricted areas (with an acquisition time of 30 seconds), to make the reading of these sections easier. In reality, the chronology is reversed; we first carried out the measurements of the restricted areas before the entire sample. This is to extract the fluorescence spectra in priority.

Table 7. Acquisition parameters for sample 4-EC2216 and 4-EC9323-2.

4-EC2216 / EC9323-2	Integral measurement	Restricted area measurement
Scan angle (°)	-2° à 2°	-0,2 à 0,2
Scanned area width (mm)	32,2	3,4
Acquisition time (s)	5	30
Measurement time (min)	92,8	53,4

Figure 12 shows the measurement that allowed us to image the entire 4-EC2216 sample. In the center of the substrate, we can observe the contaminant deposit, but also textile particles deposited on the sample. The integration time was set at 5 seconds, which is a correct compromise for an acceptable measurement time and sufficient signal to detect the fluorescent deposit. This measurement lasted a little over an hour and a half. However, this measurement does not allow us to extract an analyzable spectrum, the signal to noise ratio is too high. In order to have an acceptable fluorescence signal, as explained previously, we had to set the acquisition time to 30 seconds.

IOP Conf. Series: Materials Science and Engineering 1287 (2023) 012025 doi:10.1088/1757-899X/1287/1/012025

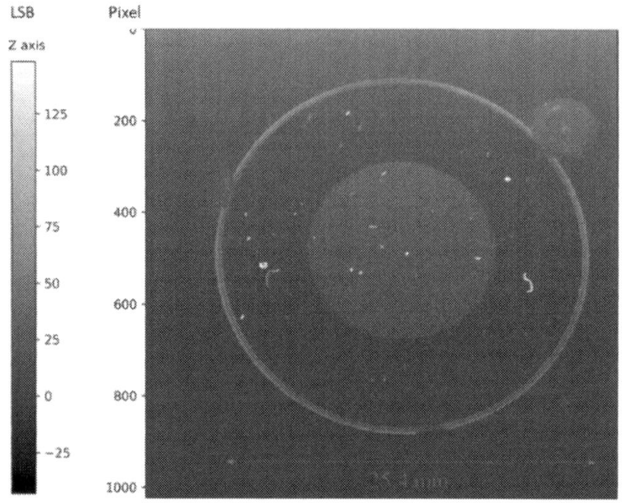

Figure 12. Measurement of sample 4-EC2216, in the center the deposit of contaminants, the very luminous particles are textile dust. The acquisition time was set for 5 seconds in order to image the entire sample. On the other hand, this setting does not allow us.

We have greatly reduced the analysis area; figure 13 highlights this more restricted area. The new measurement for 30 seconds of acquisition time took just under an hour.

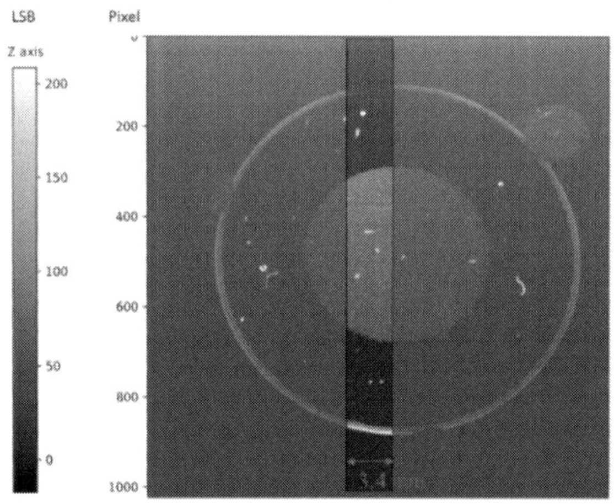

Figure 13. Highlighting of the restricted zone of sample 4-EC2216, for a measurement at 30 seconds in acquisition time. We can already appreciate the details of the deposit.

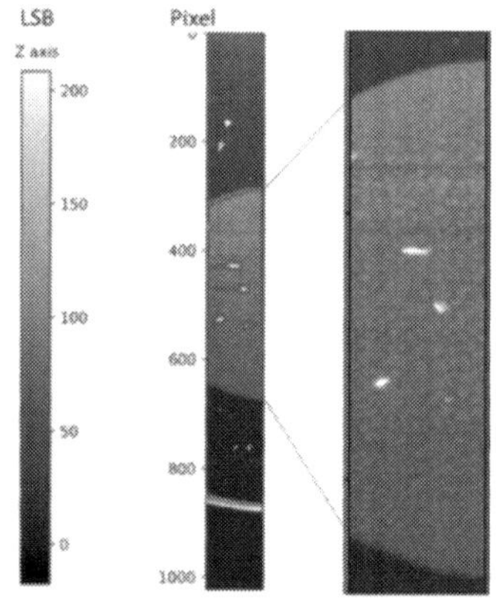

(a) (b)

Figure 14. Sample 4-EC2216: measurement of the area restricted to 30 seconds of acquisition time. (a) region of interest clipped to extract the average spectrum of each pixel. (b) Zoom on the analyzed area, we can appreciate the droplet morphology.

Thanks to the tools developed for data analysis, we can create a region of interest (ROI) which allows the user to extract the spectra of each pixel from this zone, and finally average them between them. The traced region of interest is visible in figure 14, we have outlined the entire deposit by excluding tissue particles which would distort the fluorescence spectrum.

The fluorescence spectrum thus obtained is presented in figure 15, we find the similar peaks obtained by the OceanOptics spectrometer

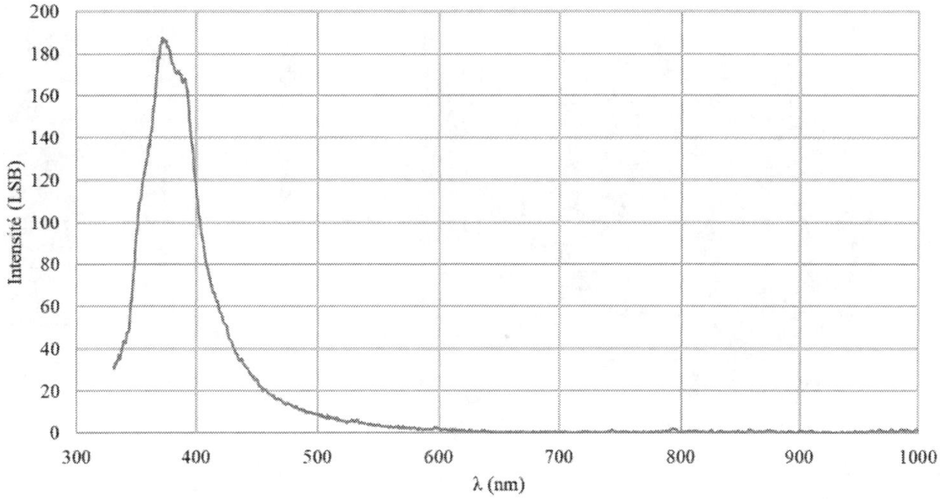

Figure 15. Sample EC2216: fluorescence spectrum averaged over the area of interest plotted in figure 14.

We encountered a similar problem with the analyses of the outgassing deposit from EC9323-2 material. Although the fluorescence signal is more important, we were not able to extract an analyzable fluorescence spectrum for an integration time of 5 seconds. We therefore produced an image of the entire sample, presented in figure 16. Always in the center of the substrate, we can observe the deposition of contaminants. The droplet morphology is more marked than the previous sample. We also note textile particles deposited on the sample.

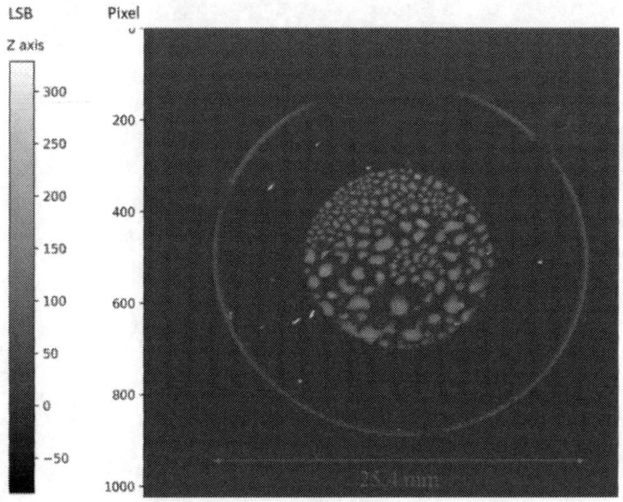

Figure 16. Measurement of sample 4-EC9323-2, in the center the deposit of contaminants, the very luminous particles are textile dust. The acquisition time was set to 5 seconds in order to image the entire sample. However, this setting does not allow us to extract spectra on the contaminated area, due to the excessive signal to noise ratio.

As for the previous sample, this measurement does not allow us to extract an analyzable spectrum, the signal to noise ratio being still too high. We therefore once again greatly reduced the analysis area and set an acquisition time of 30 seconds. The new scan took just under an hour. The smaller area is visible in figure 17.

IOP Conf. Series: Materials Science and Engineering 1287 (2023) 012025 doi:10.1088/1757-899X/1287/1/012025

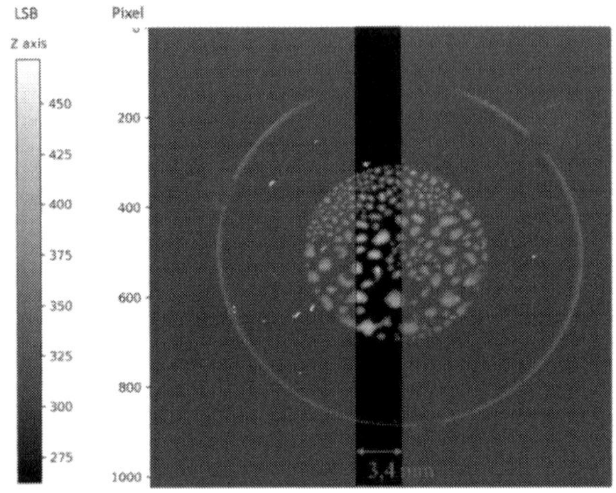

Figure 17. Highlighting of the restricted zone of sample 4-EC9323-2, for a measurement at 30 seconds in acquisition time. We can already appreciate the details of the deposit.

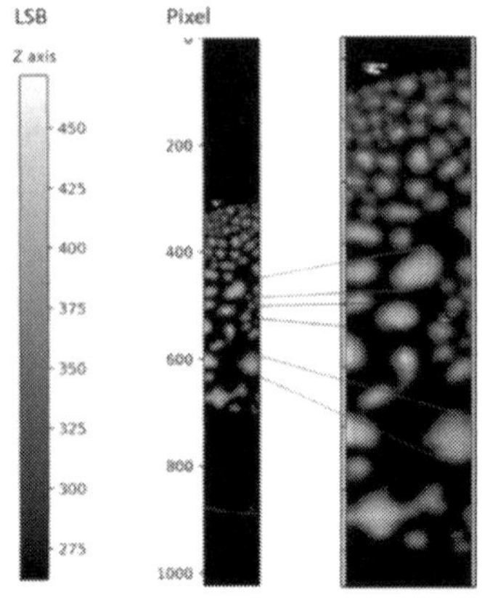

Figure 18. Sample 4-EC9323-2: measurement of the area restricted to 30 seconds of acquisition time. (a) regions of interest clipped to extract the average spectrum of each pixel. (b) Zoom on the analyzed area, we can appreciate the droplet morphology.

The droplet morphology did not allow us to achieve a single ROI. We therefore plotted three different regions of interest in order to compare the signal. These three regions show the same fluorescence spectra. figure 18 shows the three areas we selected arbitrarily.

On this same figure, we can notice that the top of the figure is not completely clear. This is due to poor optical focus of the translation stage. This bad focus has surely caused a loss of intensity of the measured fluorescence signal of the deposit.

The fluorescence spectrum for the red zone of interest is presented in figure 19, the spectrum is less easily recognizable this time compared to that obtained by the spectrometer and is also more noisy than the spectrum fluorescence of the contaminants from EC2216 material because the average region of interest contains fewer pixels.

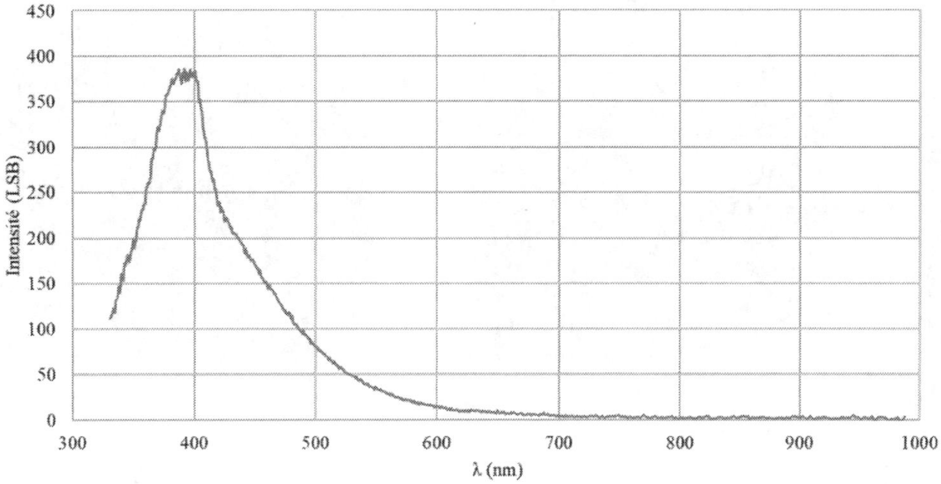

Figure 19. Sample 4-EC9323-2: fluorescence spectrum of the red region of interest plotted in figure 18.

The two samples analyzed by the hyperspectral imager allow us to detect particles present on the substrate and the deposit. These have a circular or wire-like shape and are visible to the naked eye, as we can observe in figure 20. It is most often a textile particle. We plotted an ROI around one of these particles to measure its fluorescence signal, as shown in figure 21.

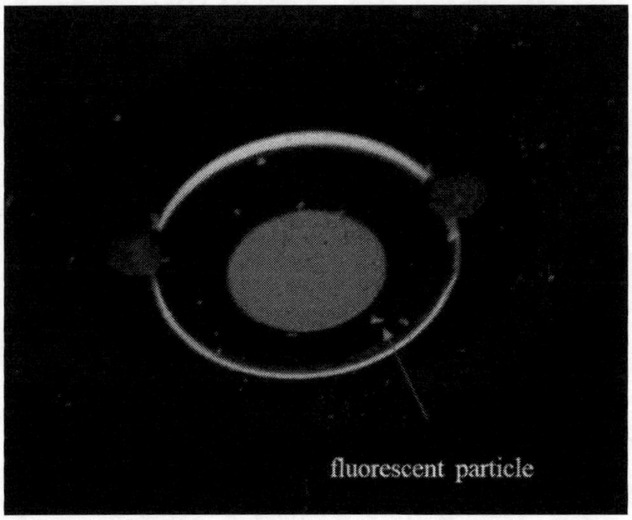

Figure 20. Fluorescent particle on a deposit from the EC2216 material.

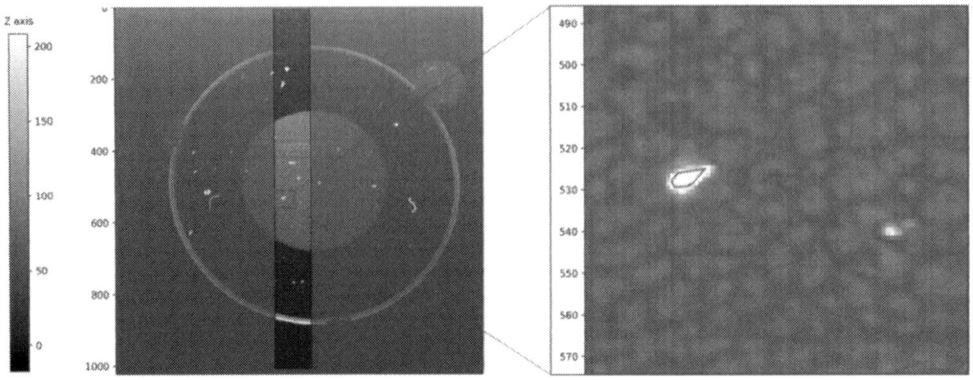

Figure 21. ROI on a particle present on sample 4-EC2216. Their signal is about three times higher than the fluorescence signal of the deposit.

The fluorescence spectrum of this particle is shown in figure 22. In order to highlight the difference in fluorescence intensity, we have also shown the fluorescence spectrum measured for the deposit of sample 4-EC2216. We can notice that this particle, probably from a white coat, has a fluorescence intensity about three times greater than the deposit of EC2216 material. This is explained by the presence of optical brightener in these textile particles, the specific signature of which we find on the measured spectrum [25]. These compounds are present in particular in detergents in order to enhance the white color of certain clothes.

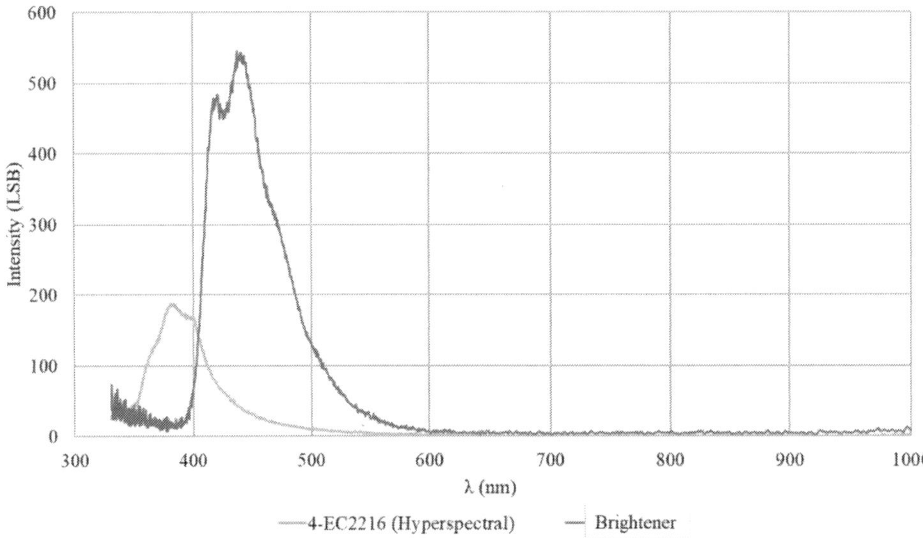

Figure 22. Fluorescence spectra of a brightening particle (in blue) and deposit of sample 4-EC2216 (in gray) measured by the hyperspectral instrument.

We can now compare the measurements made with the OceanOptics spectrometer and those with the hyperspectral imager for the two samples 4-EC2216 and 4-EC9323-2. Figure 23 represents the spectra acquired for the sample of material 4-EC2216. In order to describe the comparison between the two spectra, four zones can be defined:

- (a): the hyperspectral imager does not detect the modulation located in the UV between 330 and 350 nm.

- (b): the hyperspectral imager detects the three modulations at 350, 375 and 390 nm. We can note the good correlation between the two spectra on this range of wavelengths.
- (c): the two spectra do not show an identical slope between 400 and 550 nm. Furthermore, the imager does not detect the slight modulation visible at 425 nm.
- (d): the hyperspectral imager does not detect the modulation between 570 and 600 nm.

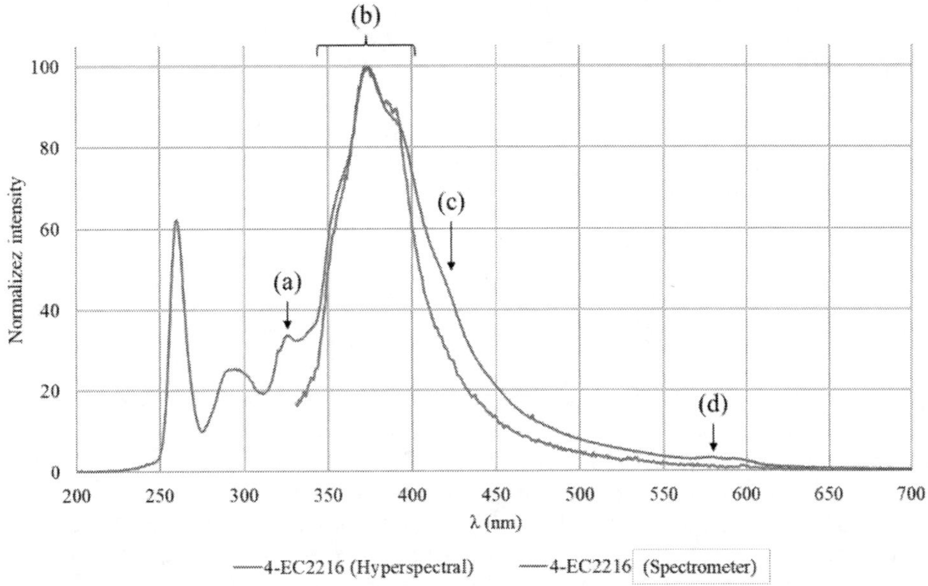

Figure 23. Comparison of normalized signals measured by the OceanOptics spectrometer and the hyperspectral imager for sample 4-EC2216. (a) The spectrometer measures a signal in this interval, while the imager does not detect it. (b): The hyperspectral imager measures the same peaks at 350, 375 and 390 nm. (c) There is a significant difference in slope between the two spectra, the hyperspectral imager does not seem to take into account the modulation at 425 nm. (d) The hyperspectral imager does not detect modulation between 570 and 600 nm.

In figure 24 we compare the spectra of the following sample (4-EC9323-2). Again, in order to describe the comparison between the two spectra:

- (a): the hyperspectral imager does not detect the modulation located in the UV between 330 and 350 nm.
- (b): the ascending slopes of the modulation at 390 nm detected by the two instruments correspond. On the other hand, the hyperspectral imager does not detect the same signal maximum as the OceanOptics spectrometer.
- (c): the two spectra do not show an identical slope between 400 and 550 nm. In addition, the hyperspectral imager detects a slope change visible at 450 nm.
- (d): the hyperspectral imager does not detect the modulation between 570 and 600 nm.

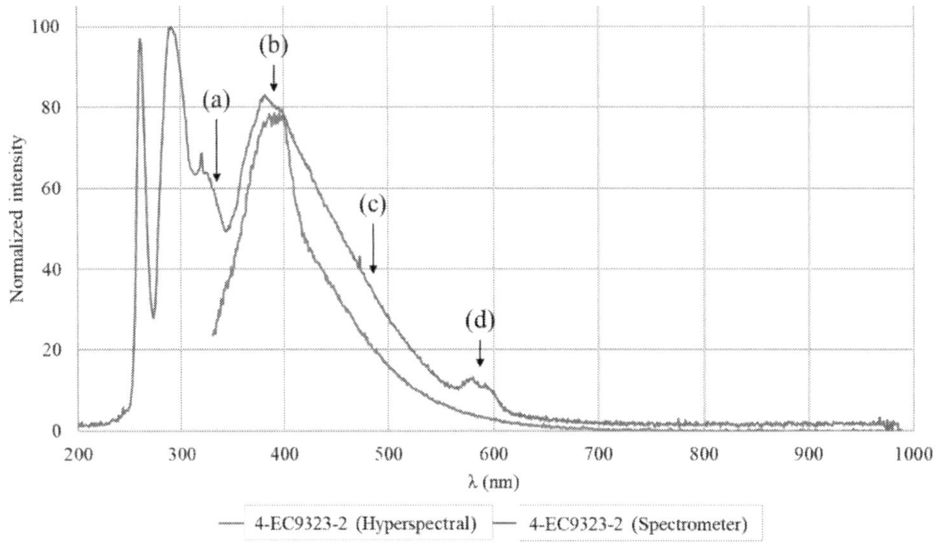

Figure 24. Comparison of normalized signals measured by the OceanOptics spectrometer and the hyperspectral imager for sample 4-EC9323-2. (a) The spectrometer measures a signal in this interval, while the imager does not detect it. (b) The hyperspectral imager does not detect the same fluorescence peak maximum. (c) There is a significant difference in slope between the two spectra, the hyperspectral imager detects a modulation at 450 nm. (d) The hyperspectral imager does not detect modulation between 570 and 600 nm.

The highlighted differences in these two comparisons demonstrate that the hyperspectral instrument is able to detect the main fluorescence peaks for both samples. On the other hand, it fails to detect the secondary peaks whose intensity is weaker. This may be due to premature aging of the aluminum treatments of the optics. Indeed, during an inspection of the optics using a white light, we noted the appearance of inhomogeneous purple reflections on certain mirrors. The reflectivity of these optics in certain wavelength ranges may have evolved and no longer correspond to the curves of reflectivity. A measurement of the reflectivity of the mirrors would be necessary in order to confirm the impact of the alteration of the aluminum treatment on the sensitivity of the instrument. Another element that could be responsible for the instrument's lack of sensitivity in detecting low intensity peaks is the presence of stray light. In figure 21 for example, we can observe a gray level gradient in the Y axis of the image. In fact, the measurement of the sample 4-EC2216 requires pushing the sensitivity of the instrument in order to detect a signal in a dynamic ranging from 0 to 125 LSB, which highlights this stray light. However, all the measurements were carried out in the dark, the only light source present during the measurement is the excitation source itself. It is possible that there is a path for this light which illuminates the CCD sensor homogeneously over its X field. In this case, we should observe a global shift (offset) on all the wavelengths of the spectrum, which it is not the case. The presence of this stray light is therefore probably not the main factor responsible for the lack of sensitivity of the instrument. In the specific case of UV, over the 330 to 350 nm wavelength range, the hyperspectral instrument does not detect any modulation, unlike the OceanOptics spectrometer.

The theoretical performance curve of the hyperspectral instrument as a function of the wavelength, shown in figure 25, clearly shows that the response of the instrument collapses in the UV for a wavelength under 400 nm.

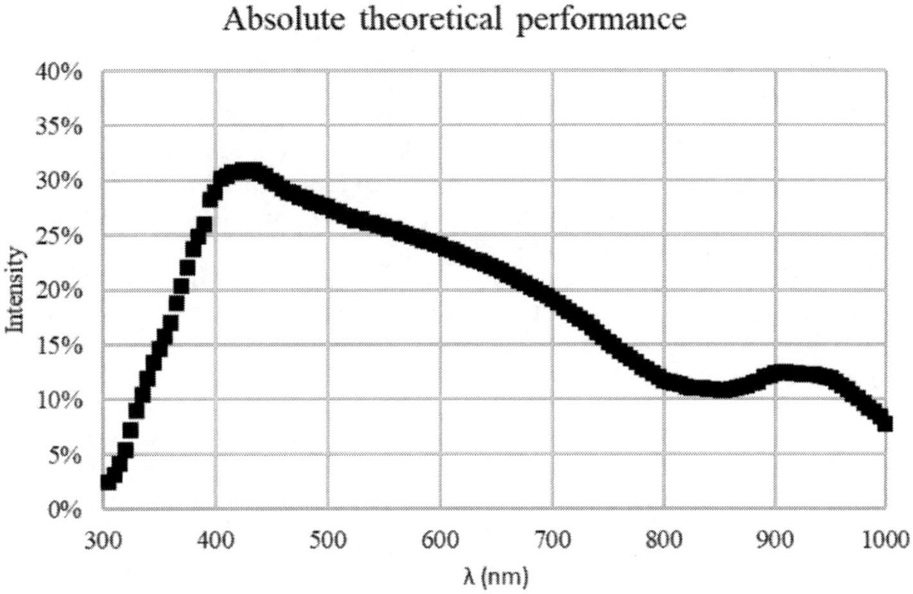

Figure 25. Absolute theoretical performance curve.

4. Conclusion

We have selected two widely used space epoxy adhesives. Using an outgassing bench developed at CNES, we outgassed each of the materials for four different masses in order to obtain deposits (in film or droplets) of various thicknesses on fused silica substrates. A first analysis using a spectrofluorimetric bench allowed us to acquire the fluorescence responses of these deposits. We have shown that each deposit has a characteristic response allowing us to easily identify them. We also noticed that the more the thickness of the deposit increases, the more the signal is intense. With this data, we knew what kind of spectrum we were supposed to measure using the hyperspectral instrument. It was demonstrated that the fluorescence spectra of contaminants from these two materials had a response below 330 nm, and therefore would not be detected by our instrument.

We therefore subsequently measured the thickest deposit from the two materials using the hyperspectral imager. It turned out that the instrument requires a significant acquisition time to detect a signal that can be used in imaging, around 30 seconds for each line scanned. Therefore, if we want to have a measurement of the entire sample, the acquisition time is relatively long (about 8 hours). This may limit the size of the scenes we want to analyze, in order to avoid too much UV exposure of the sample.

As assumed, the sensitivity in UV range is low, the instrument no longer detects a signal below 350 nm. With regard to the contaminant deposit from EC2216 material, the spectrum is almost equivalent to that measured by the spectrofluorimetric bench. On the other hand, the deposition of contaminants from EC9323-2 material is more difficult to compare. In both cases, some modulations are not detected by the hyperspectral instrument and the shape of the slopes of some peaks do not match. We assume that the radiometric correction does not make it possible to make the signal completely usable. As mentioned, we need to review our calibration protocol. Nevertheless, the results are promising and the spatial performances are correct. We are now able to trace the fluorescence spectra of the materials present in the image.

The hyperspectral instrument still requires development time, in particular on the software part allowing to control the various actuators but also the radiometric correction routines. We have to adapt the working wavelength range according to the field of application, but also improve the sensitivity of the instrument. The latter indeed imposes acquisition times which are still too long to be able to carry out measurements in a cleanroom. The need to work in the darkest environment possible, but also to block access to the spacecraft during the time of the measurement, does not allow us to consider future tests in integration room for the moment.

Spectrofluorimetry shows promising results. Indeed, this study emphasized that fluorescence spectra are specific to the chemical nature of contaminants, which makes it possible to discriminate them. The spectrometer made it possible to detect a exploitable signal, even for the smallest thicknesses (about hundred nanometers). Future test campaigns would allow us to test the detection limit of the spectrometer, in particular for different acquisition times.

Moreover, the excitation source, in our case a 270 nm LED, has a low efficiency compared to a laser. The latter, being more monochromatic and more powerful, would trigger an increased fluorescence signal from our deposits, and therefore improve the signal to noise ratio, or allow us to reduce the acquisition time. We were only interested in contamination on fused silica substrates, but other substrates such as alumina, germanium or MLI could be studied before testing on more complex assemblies such as electronic boards. Ultimately, building a fluorescence spectrum database would allow the operator to quickly and easily recognize a contaminant and take decisions in consequences. This database will have to be supplied with spectrum of contaminant alone, mixtures, and to take into account the substrate.

Acknowledgement

The author is grateful to the CNES for the funding of this study and for providing the platform and logistic for the research work. Also, the author would thanks the CNES laboratory and expertise team for their support.

References

[1] Tribble A C 2000 *Fundamentals of contamination control* vol TT44 (Bellingham: SPIE Press)
[2] Tveekrem J L, Leviton D B, Fleetwood C M and Feinberg L D 1996 Contamination-induced degradation of optics exposed to the Hubble Space Telescope interior *Optical System Contamination V, and Stray Light and System Optimization* vol 2864 (Denver: SPIE) pp 246-257
[3] Baggett S, Sparks W, Ritchie C and MacKenty J 1996 Contamination correction in synphot for wfpc-2 and WF/PC Instrument Science Report WFPC2 96-02
[4] Plucinsky P P, O'Dell S L, Tice N W, Swartz D A, Bautz M W, DePasquale J M, Edgar R J, Garmire G P, Giordano R, Grant C E *et al* 2004 An evaluation of a bake-out of the ACIS instrument on the Chandra X-Ray Observatory *UV and Gamma-Ray Space Telescope Systems* vol 5488 (Glasgow: SPIE) pp 251-263
[5] O'Dell S L, Swartz A S, Tice N W, Plucinsky P P, Marshall H L, Bogdan A, Grant C E, Tennant A F, and Dahmer M 2017 Modeling contamination migration on the Chandra X-ray Observatory IV *UV, X-Ray, and Gamma-Ray Space Instrumentation for Astronomy XX* vol 10397 (Denver: SPIE) pp 83-92
[6] Mossman D L, Bostic H D, and Carlos J R 1987 Contamination induced degradation of optical solar reflectors in geosynchronous orbit *Optical Systems Contamination: Effects, Measurement, Control* vol 777 (Orlando: SPIE) pp 12-19
[7] Skurat V E, Leipunsky I O, Volkov I O, Pshechenkov P A, Beriozkina N G, and Gatsenko L S 2011 Changes of Solar Array Surfaces on Orbital Station Mir *J. Spacecraft and Rockets* **48(1)** 53-58
[8] Gouisset E 2022 *Développement et étude de la réponse instrumentale d'un imageur hyperspectral large bande (UV-Visible-NIR) permettant la caractérisation physico-chimique de contaminants sur surfaces sensibles d'engins orbitaux* Ph. D. thesis Sorbonne Université

[9] Bourcier F, Pansu R and Faye D 2012 Fluorescence behaviour of space materials under UV illumination in cleanroom *ISMSE Symposium* (Noordwijk) pp 1-6

[10] Bourcier F, Faye D and Le Nouy P 2015 Imagerie hyperspectrale de la fluorescence des éléments organiques *Mesures et Techniques Optiques pour l'Industrie* (Pleumeur-Bodou/Lannion) pp 1-8

[11] Bourcier F, Walter P, Pedetti S, Faye D, Spezzigu P, Infante F, Le Nouy P and Zedda E 2016 Hyperspectral instrumentation to image and characterize the fluorescence of materials *Systems Contamination: Prediction, Control, and Performance 2016* vol 9952 (San Diego: SPIE) pp 160-175

[12] Bourcier F, Pansu R, Faye D, Le Nouy P, and Spezzigu P 2017 Hyperspectral characterization of fluorescent organic contaminants on optical payloads *International Conference on Space Optics - ICSO 2014* vol 10563 (Tenerife: SPIE) pp 1216-1223

[13] Bourcier F, Walter P, Faye D, Rioland G, Infante F, Gouisset E, Trecul Q, Johansson R and Lairgi Y 2022 Performance review of a UV/vis/IR fluorescence hyperspectral camera to detect contamination on spacecraft during integration *Space Systems Contamination: Prediction, Control, and Performance 2022* vol 12224 (San Diego: SPIE) p 122240P

[14] Pickett L W. Muntz M, and McPherson E M 1951 Vacuum Ultraviolet Absorption Spectra of Cyclic Compounds. I. Cyclohexane, Cyclohexene, Cyclopentane, Cyclopentene and Benzene *J. American Chemical Society* **73(10)** 4862-4865

[15] Sponer H 1942 Ultraviolet Absorption Spectra of Substituted Benzenes *Rev. Modern Physics* **14(2-3)** 224

[16] Keller-Rudek H, Moortgat G K, Sander R, and Sörensen R 2013 The MPI-Mainz UV/VIS Spectral Atlas of Gaseous Molecules of Atmospheric Interest *Earth System Science Data* **5** 365-373

[17] Dearden J C and Forbes W F 1959 Light absorption studies: part XIV. The ultraviolet absorption spectra of phenols *Canadian J. Chemistry* **37(8)** 1294-1304

[18] Xu B, Gao N Y, Sun X F, Xia S J, Rui M, Simonnot M O, Causserand C and Zhao J F 2007 Photochemical degradation of diethyl phthalate with UV/H2O2 *J. of Hazardous Materials* **139(1)** 132-139

[19] Villahermosa R M and Ostrowski A D 2008 Chemical analysis of silicone outgassing *Optical System Contamination: Effects, Measurements, and Control 2008* vol 7069 (San Diego: SPIE) pp 51-60

[20] Benmansour B, Stephan L, Cabon J Y, Deschamps L, and Giamarchi P 2011 Spectroscopic Properties and Laser Induced Fluorescence Determination of Some Endocrine Disrupting Compounds *J. Fluorescence* **21(3)** 843-850

[21] Cowgill R W 1963 Fluorescence and the structure of proteins. I. Effects of substituents on the fluorescence of indole and phenol compounds *Archives of Biochemistry and Biophysics* **100** 36-44

[22] Moon A Y, Poland D C, and Scheraga H A 1965 Thermodynamic Data from Fluorescence Spectra. I. The System Phenol-Acetatel *The J. of Physical Chemistry* **69(9)** 2960-2966

[23] Zechner J, Köhler G, Grabner G, and Getoff N 1980 The role of the molecular environment for the primary photoprocesses of phenol in solution *Canadian J. Chemistry* **58(19)** 2006-2010

[24] Barman N and Sahu K 2014 Reduced fluorescence quenching of coumarin 102 at higher phenol mole fractions in cyclohexane-phenol and anisole-phenol solvent mixtures: role of competitive hydrogen bonding *RSC Advances* **4(102)** 58299-58306

[25] Esteves M F, de Noronha A C, and Marinho R M 2004 Optical brighteners effect on white and coloured textiles *World Textile Conference - 4th AUTEX Conference* (Roubaix)

Euro Material Ageing – Ground-based testing and selection of the ESA flight candidate materials

Agnieszka Suliga, Johanna Wessing, Aurélie Hand, Adrian Tighe, Sebastien Vincent-Bonnieu, Riccardo Rampini

ESA-ESTEC, Noordwijk, The Netherlands

Agnieszka.Suliga@esa.int, Adrian.Tighe@esa.int

Abstract. One of the newest European additions to the International Space Station (ISS) is the Airbus Bartolomeo payload hosting platform, attached externally to the European Columbus Module. This will provide an unobstructed access to the external space environment offering research opportunities for a wide range of space science applications. One of the Bartolomeo missions under development is Euro Material Ageing (EMA) – a joint collaboration between the European Space Agency (ESA) and the French Space Agency (CNES). ESA is using the Bartolomeo platform as a testing opportunity to perform materials science research and calibrate its ground-based testing facilities against the real space environment. Additionally, it offers a unique opportunity to small space players of the European market for direct in-flight technology validation.

1. Introduction

Material degradation induced by the harsh space environments can be critical for mission success and astronaut safety. Spacecraft are exposed to high vacuum, thermal cycling, intense ultra-violet radiation from the sun, micrometeoroids, and debris impacts, as well as ionizing radiation, and all these effects may result in loss of key material properties. The Low-Earth orbit is a destination for most satellites, where the additional presence of highly reactive atomic oxygen causes erosion and oxidation of surfaces. Ground test facilities have been developed to simulate space environmental conditions to test and qualify materials and spacecraft components and to better understand the degradation mechanisms. Additional flight experiments provide valuable insights into synergistic effects which cannot be completely reproduced on ground. By comparing the results of ground-based and in-flight experiments, it is possible to manipulate the ground test facilities parameters to improve their accuracy in simulation of the space environment. Several experiments, including ESA's MEDET [1] and NASA's MISSE [2] projects have been launched in the past and enabled to study performance, stability, and long-term survivability of materials and components after actual in-flight exposure. As our understanding of the space environment is continuously progressing, and novel advanced materials are being developed for more and more demanding applications it is of high significance to continue these space environmental studies. In this paper, the flight test opportunity called Euro Material Ageing (EMA) campaign is described. EMA is a collaboration between the European Space Agency (ESA) and the French Space Agency (CNES) which is planned to be launched in 2024.

2. Description of the flight experiment

The EMA payload will be attached in the RAM face of the Bartolomeo, which is an external platform attached to the Columbus module of the International Space Station (ISS). The aim of Bartolomeo is to offer a scientific platform for a variety of experiments, including in-flight material science experiments for space applications, providing an unobstructed view to space and exposure of material samples to the full spectrum of synergistically acting space environment effects.

The EMA experimental set-up consists of an array of sample holders, holding up to 9 samples. Fragile and frangible samples, such as glasses, are additionally protected by a transparent plastic cover. If no damage is visible upon inspection by ISS crew, for example from launch vibrations, the cover will be removed. **Figure 1** presents the 3D model of the EMA payload (left), sample mounting (right) and a mock-up with dummy samples representing the flight configuration (right).

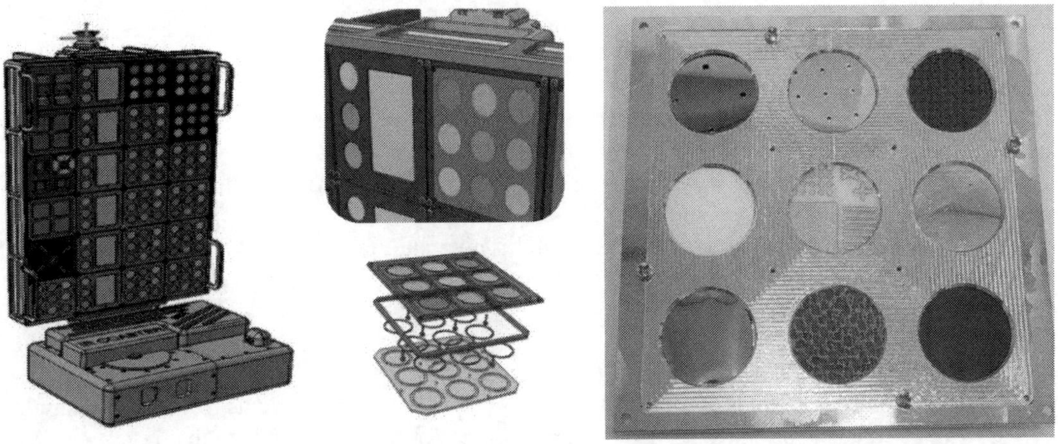

Figure 1: Left: 3D model of the EMA payload. Middle: Set-up of the sample holder, including baseplate, gaskets, spacer, and cover plate. Right: Sample holder mock-up, filled with some candidate samples.

3. Samples

Within the scope of the first EMA campaign, the samples slots will be populated by CNES, ESA and European institutions. Out of 135 sample slots which are available for passive samples, ESA is populating 18 with environmental monitors and calibration materials and the remaining 45 slots were made available for the European community of material science for research and development and in-flight qualification testing. The 15 selected experiments from the European community including their foreseen application are listed in Table 1. The remaining slots are populated by CNES, including active environmental monitors such as two types of ATOX monitors and a quartz crystal microbalance (QCM) for contamination monitoring.

Table 1: Experimental contribution of the European material science community to the EMA mission.

Sample type	Application
Polysilazane coatings on membrane	Photovoltaic arrays, drag sails and thermal components
Metallic glasses	New class of materials with exceptional mechanical and functional properties
New, MoOx based, material systems	Flexible optical solar reflectors
Carbon fibre reinforced ceramic	Low thermal expansion structural material for a reentry vehicle
Carbon fibre reinforced plastics containing toughened polybenzoxazine nanocomposites and cyanate esters	Novel material candidate for lightweight space applications
Additive manufactured liquid crystal polymers	Novel candidate material for structural components
Transparent polyimide films	Transparent films for photovoltaic arrays

Black Laser Surface Treatment	Space compatible black surface treatment to replace coatings and paints of space components
Metal/insulator patterned samples	Printable electronics
3D-Printed NiTi	Candidate material for additive manufacturing for space applications
Ge/Si and CzT samples	Sensors and Laue lenses of future gamma-ray astrophysics telescope
Diamond Like Carbon (DLC) and Plasma Electrolytic Oxidation (PEO) coatings	Tribology coatings for space applications
Additively manufactured metals	Candidate material for a high efficiency thruster
Materials for device applications in space: Silicon for sensing MEMS and NEMS and BMGs	Candidate material for storage of mechanical energy
Ultra-High Temperature Ceramic Matrix Composites	Thermal protection systems for space vehicles

Table 2 gives an overview of the samples selected by ESA to serve as reference materials or calibration samples. The aim of the ESA's internal research programme is twofold:

1. To include materials that are commonly used in spacecraft manufacturing, to serve as a reference for the new material developments.
2. To include materials that can serve as environmental monitors (ATOX fluence, contamination witnesses, UV/VUV radiation dose) and can help in calibration of ESA's ground-based testing facilities.

For instance, among the new developments, there are a few types of black and white thermal coating materials for which CeraNovis NV14 and Acktar Fractal Black will serve as reference materials since extensive space qualification testing had been performed for these materials in the past.
ATOX fluence will be monitored by measuring the erosion of thin Kapton HN layers (**Figure 2**) and transparent samples such as cover glass, calcium fluoride window and transparent FEP will serve as contamination witnesses. Several standard MLI layers will also serve as reference materials to monitor the space environment throughout the mission.

Table 2: The different ESA samples selected as reference materials

Reference	Samples
Atomic oxygen fluence calibration samples (ASTM E2089)	Kapton HN
	FEP
Contamination witness	Calcium fluoride transparent window
	FEP
UV radiation	White paint (CeraNovis NV14)
	Black paint (Acktar Fractal Black)
Standard optical samples	Cover glass
	OSR
Components of MLI	FEP/ (Ag/Inconel)
	ITO/FEP/(Ag/Inconel)
	VDA/Kapton
	SiOx/Kapton
	Beta cloth

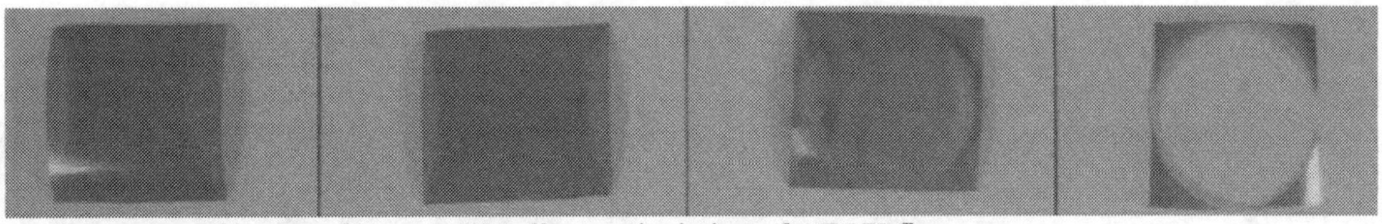

Figure 2: Erosion of layers of Kapton HN allows calculation of ATOX fluence.

4. Pre-flight qualification at ESA-ESTEC

The purpose of the ground-based pre-flight screening tests for EMA is to investigate the space environmental resistance of samples proposed for the campaign and to select the best candidates for the flight and de-risk the experiment outcome. ESA is providing access to ESTEC's Materials & Electrical Components Laboratory, as well as technical support and guidance to the European institutions which were selected to fly their experiments but are not necessarily experienced in material qualification testing for space applications.

The selected samples are undergoing characterization with regards to their physical and chemical properties. Mass measurements, visual inspection (optical and SEM) and thermo-optical properties analysis are among the standard tests conducted before and after most of the environmental tests performed in the Materials and Components Laboratory at ESTEC, such as outgassing, ATOX, UV/VUV exposure or thermal cycling. Based on the qualification results, samples are allocated into the primary or backup flight lists or are excluded from the campaign completely in case they do not meet the pass/fail criteria.

The qualification testing at ESTEC started approximately two years ahead of the flight samples integration. In some cases, test results showed that the potential flight samples would not survive the space environment exposure for the duration of the EMA flight experiment, or their key properties will be severely damaged. Consequently, in these cases ESA advised to modify the samples or to replace the respective samples by back-up samples. In the following sections, some results for the ATOX and UV/VUV exposure screening tests are presented, which visualizes the importance of material testing in a representative environment.

4.1. ATOX screening test

ATOX is very reactive, especially with organic materials, hence the objective of the screening analysis was to evaluate the extent of the erosion yield, especially for thin polymeric foils to make sure that they will survive the planned space exposure. Several other materials had been tested in the frame of the EMA pre-flight campaign, for instance to verify substrate and coating stability of selected samples which seem prone to ATOX erosion mechanism and for novel materials where the ATOX response is unknown. The typical ATOX test procedure consists of mass measurement, visual inspection of the surface with high resolution microscope and microstructure with SEM, and thermo-optical properties measurements before and after the test.

Figure 3 shows erosion due to ATOX exposure for two different materials: 1) an additively manufactured plastic (left) and 2) carbon fibre reinforced ceramic - CFRC (right) and Figure 4 shows the photographs of samples after ATOX exposure versus the plots of their thermo-optical parameters (reflectance or transmission in UV VIS NIR region).

Both materials are prone to ATOX erosion, but in terms of mass loss there is an order of magnitude difference between them. The additively manufactured polymer shows a significant erosion yield of approximately 2×10^{-24} cm^3/atoms, but no changes in optical properties were measured (Figure 4, 1). On the other hand, the CFRC sample, shows a very low erosion yield of approximately 2×10^{-25} cm^3/atoms, but under the ATOX exposure it experiences darkening and a change of solar absorptance, which will impact the thermal control properties of this structural material (Figure 4, 2).

Figure 3: SEM images of ATOX erosion: 1) Additive manufacturing polymer (left) and 2) carbon fibre reinforced ceramic composite (right) – a clear border between an ATOX masked and ATOX exposed parts of the sample is visible.

Figure 4, 3) shows an example of a normally highly transparent sample, that was developed to replace the standard cover glass for space applications. Due to the ATOX exposure, the sample lost its transparency, which is a critical feature for this material. The inner circle shows the transparency loss after the ATOX exposure, whereas the outer areas show the highly transparent unexposed material which was covered by the sample holder.

Since the test was performed in the early stages of the samples' qualification scheme, the experimenters had time to reformulate the sample and find a formulation that stayed highly transparent in the ATOX-rich environment.

1) Additive manufactured polymer

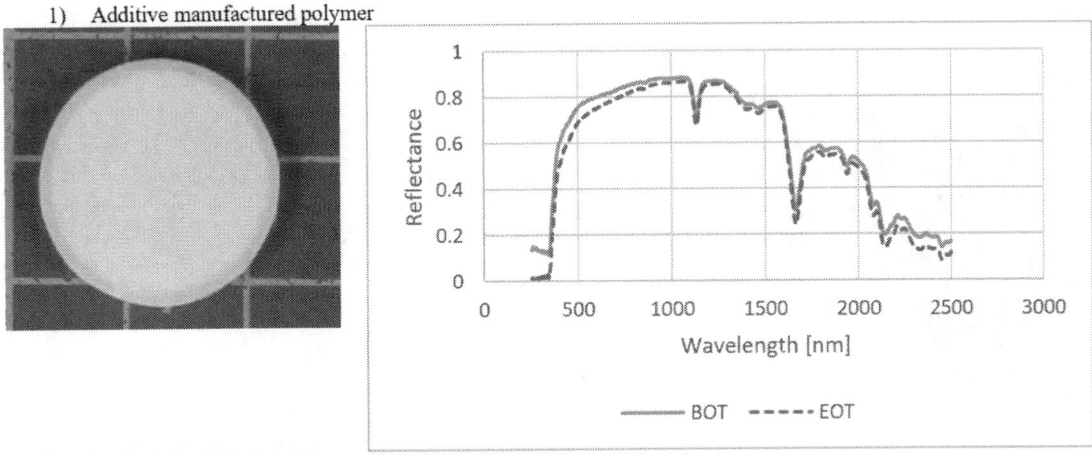

2) Carbon fibre reinforced ceramic

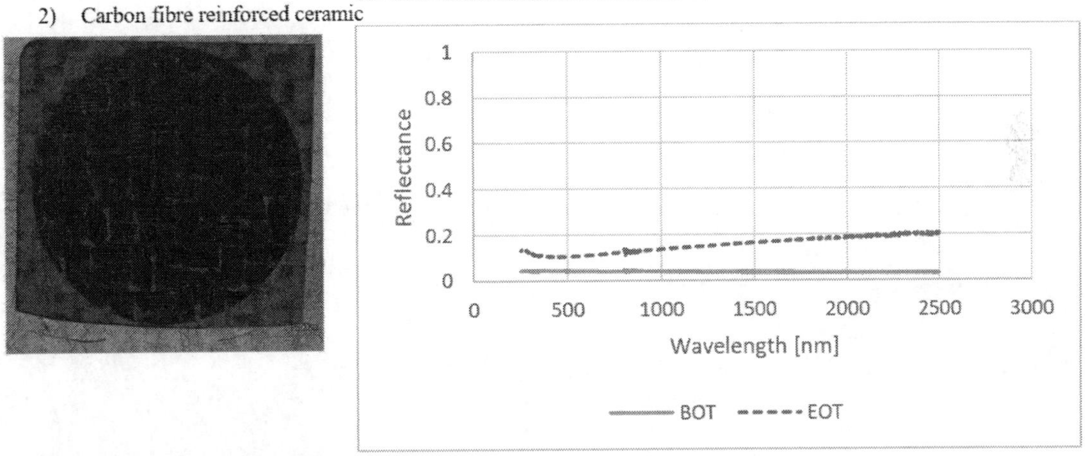

3) Thin transparent polymer film

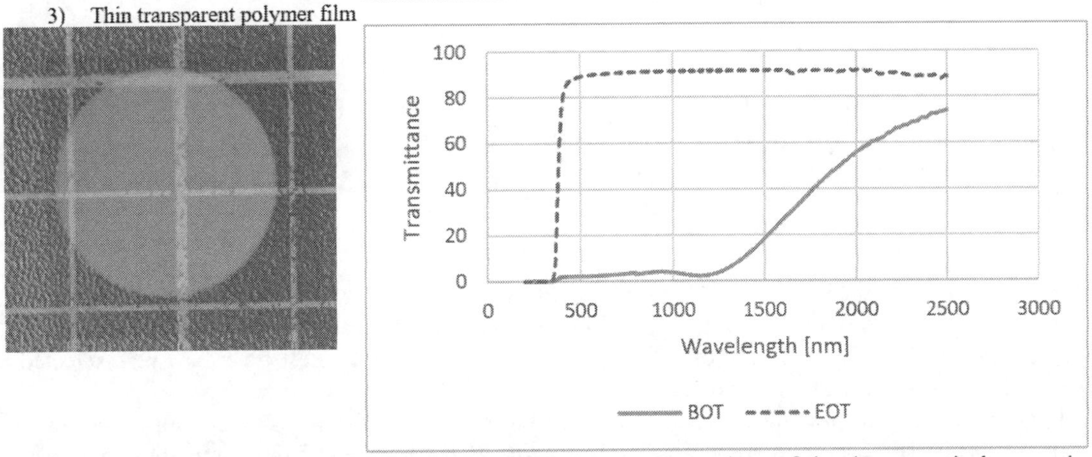

Figure 4 ATOX exposure effects on selected flight candidates. Comparison of the thermo-optical properties between BOT (beginning of test) and EOT (end of test).

4.2. UV/VUV screening test

Ultraviolet radiation (UV) and vacuum ultraviolet radiation (VUV) carry enough energy to induce significant changes of the thermo-optical properties. An example of such degradation can be observed for two polymeric samples depicted in Figure 5: 1) a transparent thin film for protection of photovoltaic arrays and 2) additively manufactured polymer. It can be clearly seen that exposure to approximately 3000 equivalent sun hours caused yellowing and significant changes in thermo-optical properties.

1) Thin transparent polymer film

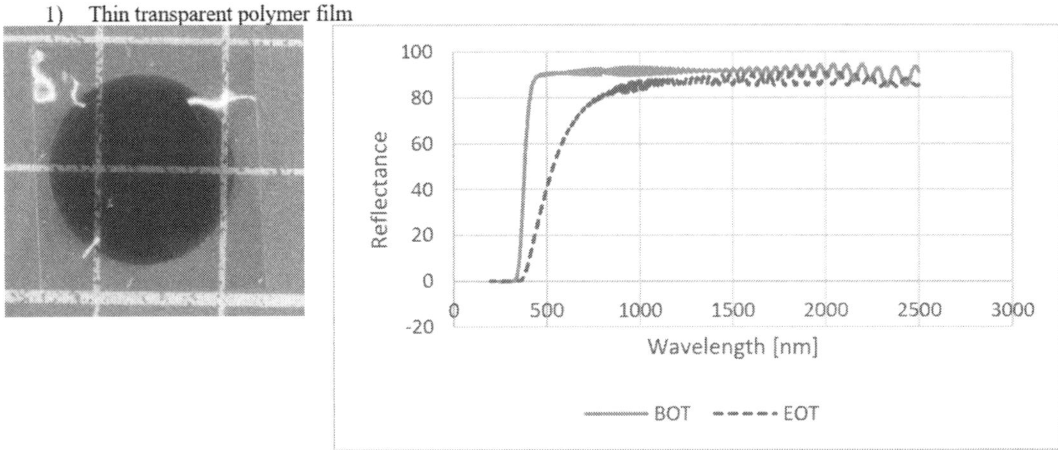

2) Additive manufactured polymer

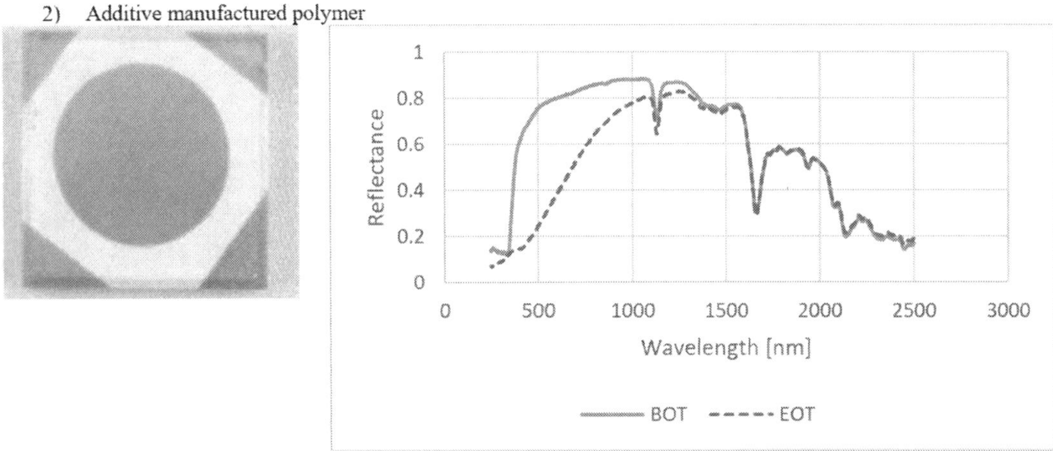

Figure 5: VUV/UV exposure effects on selected flight candidates. Comparison of the thermo-optical properties between BOT (beginning of test) and EOT (end of test).

5. Conclusions

EMA is a new scientific mission offering a platform for testing of material science research and technology directly in flight, on the exterior of the ISS. Almost 50 proposals were received by ESA after the announcement of opportunity, of which 15 proposals with the highest potential were selected to participate in the pilot EMA campaign.

Additional samples are provided by ESA and CNES to serve as environmental monitors and reference materials and to serve as calibration samples for improvement of ground-based simulation facilities.

The results of the campaign are expected to accelerate the development of European space materials and strengthen the position of the European space market. The pre-flight qualification of the external experimenter and ESA samples has demonstrated the importance of qualification testing in a reliable space environment conditions. The results originated from the qualification activities and eventually the flight exposure are planned to be openly available in a form of an database and published more extensively by the respective institutions that participate in the EMA flight experiment.

6. Outlook

The EMA qualification phase is planned to be finished by January 2023 followed by the integration of samples. The pilot flight is scheduled for 2024 with a space exposure planned for 12-18 months. The second EMA flight is likely to take place a year after. A call for new proposals will be made in due time to give European institutions the chance to participate and "fly their samples".

References

[1] Tighe A P, Iwanovsky B, van Eesbeek M et al. 2009 Overview of results from the materials exposure and degradation Experiment (MEDET) after 18 months in orbit on the ISS *11th International Symp. on Materials in the Space Environment (Aix-en-Provence)*

[2] de Groh K K, Banks B A, Mccarthy C E, Rucker R N, Roberts L M and Berger L A 2006 MISSE PEACE Polymers Atomic Oxygen Erosion Results *2006 MISSE Post-Retrieval Conf. sponsored by the U.S. Air Force Research laboratory (Orlando: Florida)*

IOP Conf. Series: Materials Science and Engineering 1287 (2023) 012027 doi:10.1088/1757-899X/1287/1/012027

High temperature atomic oxygen testing for ESA's Envision mission to Venus

Agnieszka Suliga, Abel Brieva, Gabor Milassin, Adrian Tighe*, Thomas Voirin

*adrian.tighe@esa.int

Abstract. EnVision will be ESA's next Venus orbiter, providing a holistic view of the planet from its inner core to upper atmosphere to determine how and why Venus and Earth evolved so differently. The spacecraft will be equipped with a suite of instruments including a sounder to reveal underground layering, and spectrometers to study the atmosphere and surface. The thermal fluxes in orbit around Venus are high, the solar flux at Venus distance is twice its value on Earth (around 2600 W/m2), and the reflected sunlight flux has a similar order of magnitude due to the albedo. Besides, there will also be an aerobraking phase, when a third thermal flux, the aerothermal flux, needs to be considered, with a similar order of magnitude as the two others. These high total thermal fluxes, together with the cold instruments' requirements and high-power dissipation make the thermal environment a design driver for the mission. The chemistry of the atmospheric environment during aerobraking needs also to be taken into account. Atomic oxygen exists in Venus atmosphere at the altitude range foreseen for aerobraking, with densities of up to 10^{17} oxygen atoms per cubic metres at aerobraking altitudes. Though the concentrations are small, their accumulation over up to 2000 atmospheric passes lead to total atomic oxygen fluences on the exposed spacecraft surfaces which are comparable to those encountered on typical Low Earth orbiting satellites (upwards of 10^{21} atomscm^{-2}). This drives the selection of spacecraft materials on the most exposed surfaces. Whilst significant data exists about atomic oxygen erosion effects on materials for Low Earth Orbit, there is little knowledge about high temperature atomic oxygen effects on materials under environmental conditions representative of the Venus atmosphere. To analyse the potential risks for the Envision mission, and to assist the designers in selecting the best materials, an atomic oxygen test campaign is being performed in the ESA ESTEC atomic oxygen ground-based simulator (LEOX facility). The facility has been adapted so that material samples can be simultaneously exposed to atomic oxygen and high temperatures up to at least 200°C. For the test campaign, the Envision instrument teams and designers of the prospective satellite systems are providing candidate materials to be exposed. This includes a variety of materials to be utilised on the outside of the satellite such as thermal control coatings, optical materials, coatings for appendages and antennas and multi-layer insulation. The paper describes the modifications to the facility and the first results of the testing which has been performed.

1. Introduction

Envision is an ESA mission to Venus, with a main scientific objective to investigate why Venus and Earth could have evolved so differently [1]. To achieve this, the spacecraft will carry several scientific instruments such as VenSpec spectrometers and VenSAR synthetic aperture radar, to map the atmosphere and surface morphology [2]. Such measurements require approaching the planet's surface by 220-470 km, which in the current spacecraft configuration is only possible by performing an aerobraking manoeuvre.

Envision is scheduled for launch on Ariane 6.2 in 2031. After the interplanetary transfer of 5 months, the spacecraft will start the aerobraking phase, at approximately 250 000 km and finish at a 500 km polar orbit. The aerobraking phase should last 25 months, and during this time the spacecraft will experience an aerothermal flux (a combination of high sun illumination (2650 W/m^2) and drag) and additionally with every pass the spacecraft will travel through atomic oxygen (ATOX) environment, which is known to be a highly erosive, especially for materials that show chemical affinity towards oxygen (e.g., organic materials or some metals).

This paper presents the results of the qualification campaign for various candidate materials to be used in fabrication of the Envision spacecraft and it discusses changes in material properties after exposure to the simulated aerobraking environment. This material study included two types of multi-layer insulation materials, white paint on aluminium and titanium substrates, black paint on aluminium and titanium alloy with two types of surface finish (glass blasting and black coating).

2. Experimental

The test was carried out in the Materials and Electrical Components Laboratory in ESA ESTEC in the Netherlands. The so-called LEOX facility (*Low Earth Orbit Atomic Oxygen Facility*) was modified specifically for the purpose of this test, to include a hot plate and hence simulate impinging of atomic oxygen on hot surfaces, similar to the aerobraking scenario. The LEOX facility is based on the Laser Pulse Induced Breakdown principle described by DM Buczala and Minton [3] to create a conically shaped beam of atomic oxygen. Due to the shape of the beam, the ATOX flux varies along the sample plate. The ATOX fluence was calculated based on the erosion of a standard witness sample of Kapton HN, with a known erosion yield of 2.8× 10^{-24} g/cm^3 and adjusted for individual positions using data obtained during a previous test in which Kapton HN was exposed in all locations.

Table 1 lists the samples selected for the test and the corresponding test parameters and figure 1 shows the inserted samples into the LEOX facility. 5 samples were tested in duplicates, one on a hot plate, set to the temperature of 300°C and another on a room temperature plate. Two samples were two big to be accommodated in the room temperature plate, so they were only tested in the hot conditions. The target ATOX fluence for this test was 5 x 10^{21} atoms/cm^2. The test was performed in two parts, after approximately half of the fluence was received, an inspection point was performed to measure the thermo-optical properties and visually inspect the samples.

The actual surface temperature was monitored with a Far-IR camera and varied for each sample type. To estimate the surface temperatures from the radiant heat flux measured by the camera, the emissivity (ε) of each sample was measured before and after the test and the temperature distribution on the exposed surface could be estimated using the equation: $\Phi = \varepsilon \sigma T^4$, where σ is the Stefan–Boltzmann constant.

Table 1. List of selected samples subjected to the simulated aerobraking environment.

Sample	Plate	Application	Surface temperature (°C)	Fluence (Atoms/cm^2)
Beta cloth	Hot	Reference MLI	155	4.0 x 10^{21}
	RT		21-25	3.8 x 10^{21}
Nextel	Hot	High temperature MLI	155	5.7 x 10^{21}
	RT		21-25	2.7 x 10^{21}
ITO/SiOx/Kapton	Hot	ATOX resistant MLI	160	3.5 x 10^{21}
	RT		21-25	4.0 x 10^{21}
Glassblasted Ti	Hot	Metallic alloy	260	5.4 x 10^{21}
	RT		21-25	2.1 x 10^{21}
NV14 on Al6061	Hot	White coating	275	7.4 x 10^{21}
	RT		21-25	3.2 x 10^{21}
NV14 on Ti61	Hot	White coating	258	7.5 x 10^{21}

Acktar Fractal	Hot	Black coating	265	5.6 x 10^{21}

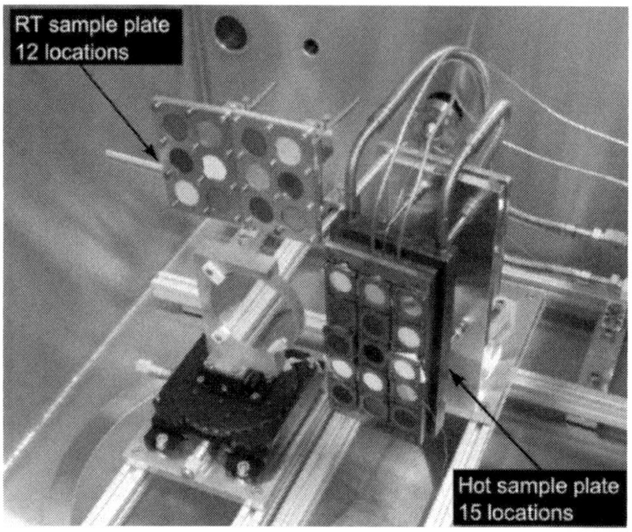

Figure 1. Hot and RT sample holder inside the LEOX test chamber (left) and close up of the samples mounted in a hot plate (right).

3. Results and discussion

The results of the performed testing show for some samples that the effects of ATOX and temperature was more detrimental combined than separately. Additionally, it was discovered that eroded materials produced significant contamination that deposited on the focusing mirror (the mirror is visible in Figure 1).

3.1. Thermo-optical properties

The solar absorptance (α) was measured with the benchtop spectrometer, Agilent Cary 5000, with an integration sphere, over the wavelength range between 2500 and 250 nm under ambient conditions. Reflectance measurements were corrected against a calibrated specular standard, 2% or 99% diffusive spectralon coupon and for the determination of the solar absorptance, total reflectance spectra are weighed against the Sun spectrum. Total hemispherical emittance was measured with the SOC ET 100 Emissometer that measures the integrated surface reflectance of a surface at two different angles of incidence and for six discreet wavelength bands in the 1.9 to 21 μm spectral range.

The thermal control of a spacecraft is determined by the ratio of solar absorptance to thermal emittance (**α_s/ε**), so this ratio was monitored to judge whether the thermal control of the spacecraft surface would be affected by effects of ATOX and temperature. Table 2 presents the summary of thermo-optical properties of samples taken before the exposure (BOT), at the inspection point and at the end of the test (EOT). The most noticeable changes in the α_s/ε ratio can be seen for glass blasted titanium on a hot plate and ITO/SiOx/Kapton on a RT plate. For other materials the ratio is stable.

Table 2. Evolution of thermo-optical properties of samples throughout the test campaign.

Sample	Plate	α_s	ε	α_s/ε	α_s	ε	α_s/ε	α_s	ε	α_s/ε
			BOT			**Inspection**			**EOT**	
Beta cloth	RT	0.27	0.90	**0.30**	0.30	0.88	**0.34**	0.27	0.89	**0.30**
	Hot	0.26	0.89	**0.29**	0.29	0.89	**0.33**	0.30	0.87	**0.34**
Nextel	RT	0.14	0.85	**0.16**	0.15	0.85	**0.18**	0.16	0.85	**0.19**

	Hot	0.14	0.85	**0.16**	0.15	0.85	**0.18**	0.16	0.85	**0.19**
ITO/SiOx/Kapton	RT	0.45	0.73	**0.62**	0.70	0.74	**0.95**	0.45	0.73	**0.62**
	Hot	0.45	0.74	**0.61**	0.49	0.76	**0.64**	0.45	0.74	**0.61**
Glass blasted Ti	RT	0.70	0.43	**1.63**	0.65	0.43	**1.51**	0.71	0.43	**1.65**
	Hot	0.70	0.44	**1.59**	0.77	0.44	**1.75**	0.83	0.43	**1.93**
NV14 on Al6061	RT	0.15	0.90	**0.17**	0.15	0.90	**0.17**	0.15	0.90	**0.17**
	Hot	0.15	0.90	**0.17**	0.16	0.89	**0.18**	0.17	0.89	**0.19**
NV14 on Ti61	Hot	0.16	0.86	**0.19**	0.17	0.87	**0.20**	0.18	0.88	**0.20**
Acktar Fractal	Hot	0.98	0.89	**1.10**	0.98	0.90	**1.09**	0.99	0.90	**1.10**

3.2. Visual and morphology changes

Visual properties were examined before the test, at the inspection point and at the end of the test with a digital camera and a macro lens. For selected samples, scanning electron microscopy (TENEO SEM) was used for morphological characterization and energy-dispersive X-ray spectroscopy (EDX) to study the elemental composition.

3.2.1. Glass blasted titanium

The most visible effect of ATOX combined with high temperature had been recorded for the glass blasted titanium. Figure 2 shows the discoloration of the sample subjected to hot ATOX conditions, a distinct transition into a rusty colour that could indicate formation of corrosion products. The second sample of glass blasted titanium was placed in the RT plate, and only experienced ATOX, and for this sample there is no discoloration visible (Figure 3). However, when the EDX spectra are compared, taken at different regions with both samples a clear trend of oxidation is visible. The lowest oxygen content was recorded for the RT sample, masked from ATOX, then RT sample, exposed to ATOX, then Hot sample, masked from ATOX and finally Hot sample exposed to ATOX (Figure 2 & Figure 3, lower images). These results indicate that synergistic exposure to ATOX and elevated temperature is accelerating the oxidation of this material, and therefore this sample is not recommended to be used in the manufacturing of the Envision spacecraft for areas exposed directly to the atomic oxygen flux.

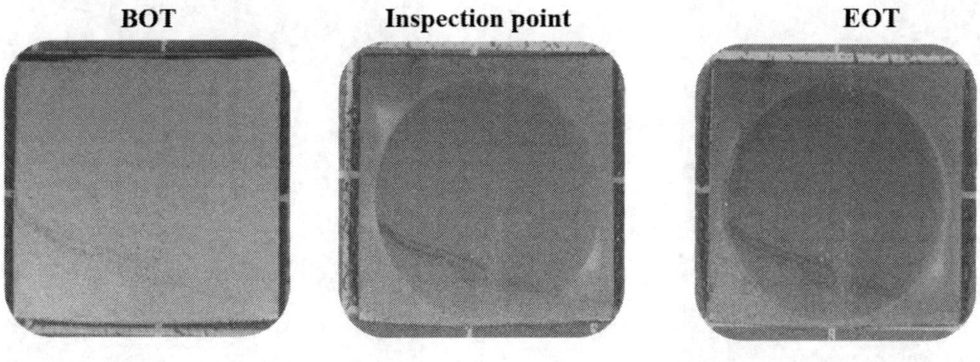

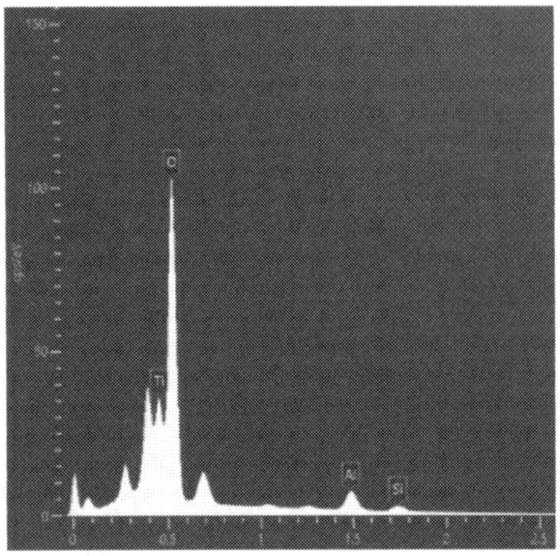

EDX spectra of non-exposed area

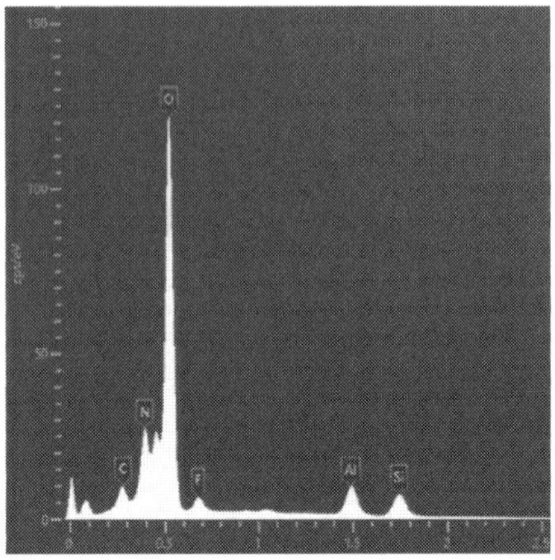

EDX spectra of an exposed area

Figure 2. Glass blasted titanium exposed to ATOX and hot temperature, visual examination and EDX spectra taken at exposed and non-exposed regions.

BOT **Inspection point** **EOT**

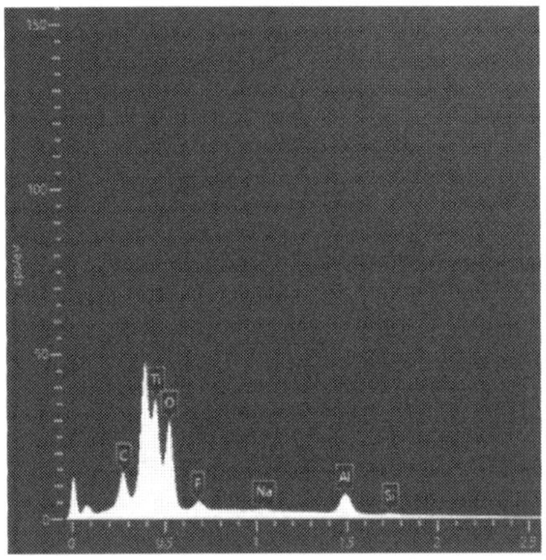

EDX spectra of non-exposed area

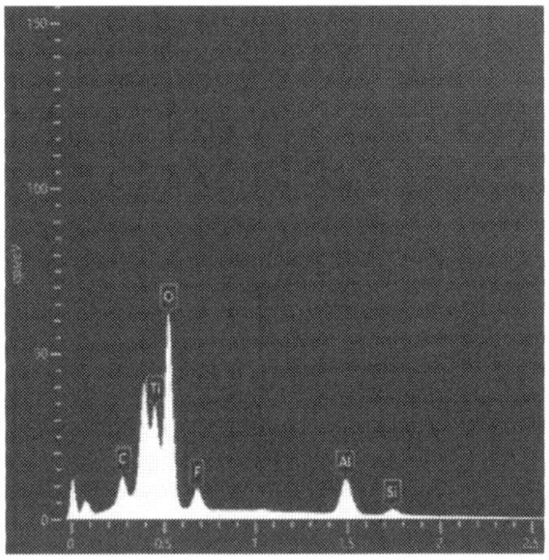

EDX spectra of an exposed area

Figure 3. Glass blasted titanium exposed to ATOX and ambient temperature, visual examination and EDX spectra taken at exposed and non-exposed regions.

IOP Conf. Series: Materials Science and Engineering 1287 (2023) 012027 doi:10.1088/1757-899X/1287/1/012027

3.2.2. ITO/SiOx/Kapton

ITO/SiOx/Kapton is often used as the outer later of multi-layer insulation material, designed to offer not only thermal insulation, but also charge dissipation (ITO) and ATOX resistance (SiOx). Figure 4 shows the visual changes in the material throughout the test campaign.

Two distinctive observations were made for this material. First the sample in the hot plate experienced localised erosion and associated mass loss of 2%. A closer look under the SEM (Figure 5) reveals two types of defects: 1) cracking of the ITO layer (Figure 5A) and 2) localised erosion, a tear in the material (Figure 5B). For this sample the thermo-optical properties are not significantly affected. On the other hand, the sample positioned in the RT plate experienced significant darkening and degradation of thermo-optical properties. This distinctive difference between the two samples suggests that perhaps the ITO/SiOx coating is not uniformly distributed over the surface of the sample.

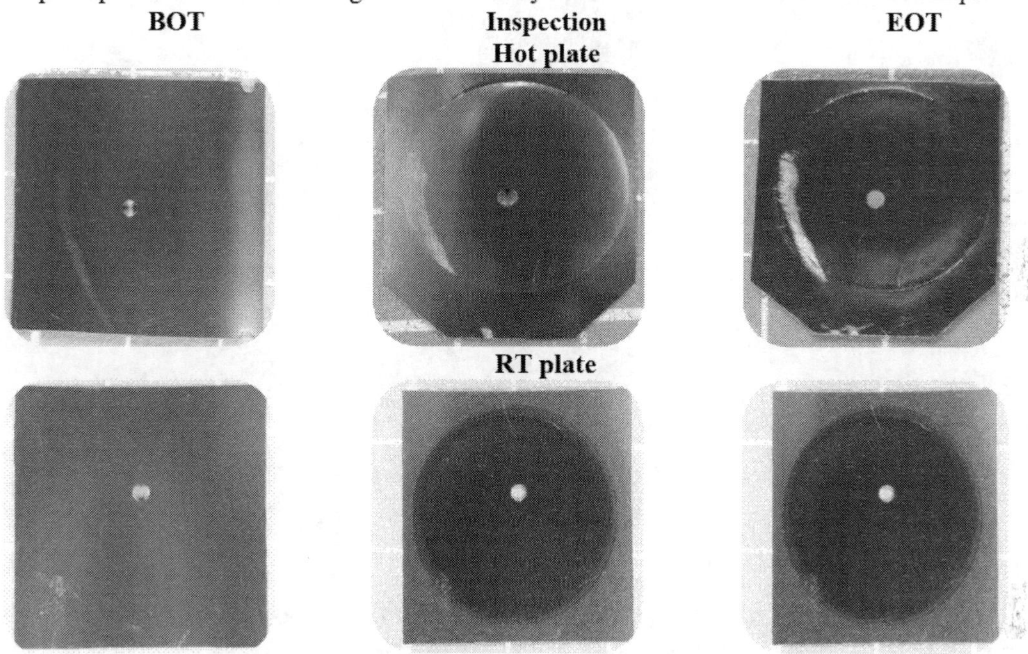

Figure 4. Visual properties of ITO/SiOx/Kapton exposed to ATOX in two configurations: on a hot plate and on a RT plate.

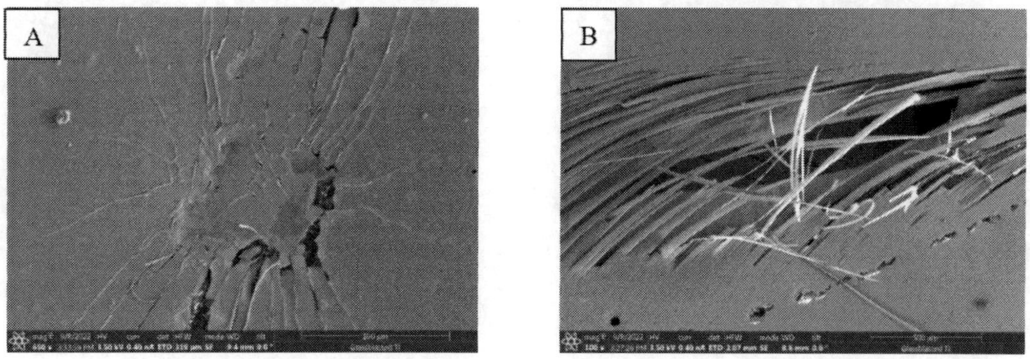

Figure 5. SEM images from the ITO/SiOx/Kapton exposed to ATOX and hot temperature showing localised damages.

IOP Conf. Series: Materials Science and Engineering 1287 (2023) 012027 doi:10.1088/1757-899X/1287/1/012027

3.2.3. Nextel and Beta cloth

Two types of fibrous multi-layer insulation materials were selected for the test, the high temperature Nextel fabric (3M) and beta cloth, a well-known MLI material with a lot of space heritage (Bron Aerotech).
Figure 6 shows the visual changes in both materials throughout the test campaign. No visual changes can be observed in the Nextel fabric, proving that the material is fully inert. In case of the beta cloth, some discoloration can be observed, indicating that the material was subjected to higher than permitted temperature levels. Additionally, comparison of the SEM images (Figure 7) shows that Nextel is a binder-free fabric and beta cloth, contains a Teflon binder that is not fully resistant to ATOX.

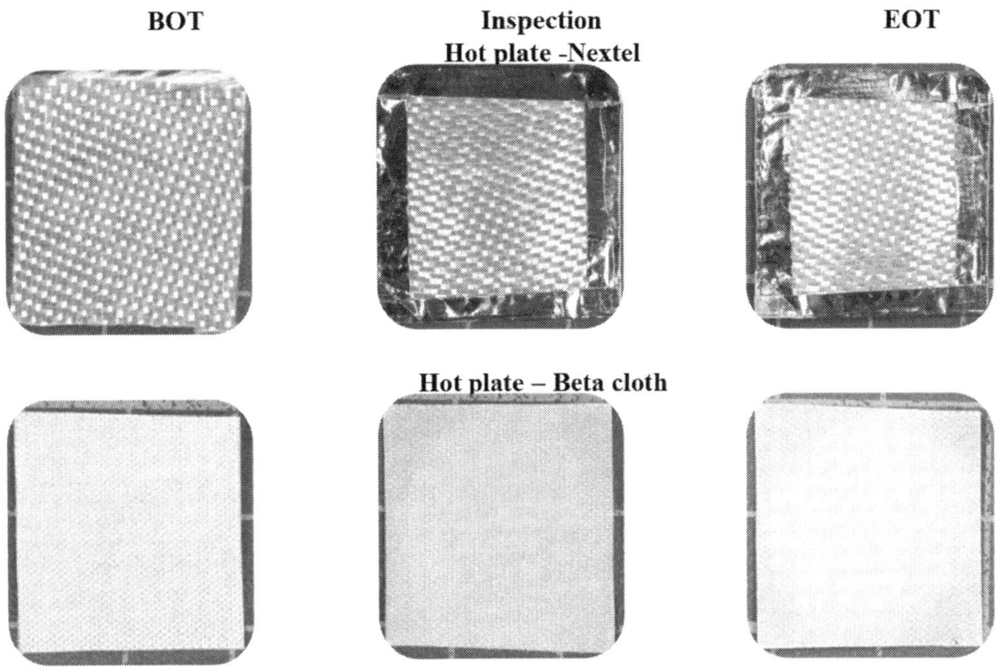

Figure 6. Visual properties of Nextel and Beta cloth exposed to ATOX and hot temperature.

Figure 7. SEM images of Nextel, high temperature fabric (A) and beta cloth (B), where erosion of the Teflon binder is visible.

3.2.4. Black and white coatings

From the family of coatings, two types were selected: 1) a black coating from Acktar (Fractal Black) deposited on a 1 mm aluminium substrate (AW-5083) and 2) white coating NV14 (Nanovation), deposited on 1 mm substate of aluminum 6061 and on 1 mm substrate of titanium 64. Acktar Fractal Black was NV14 were tested only in hot conditions, whereas NV14 on aluminum was tested in both hot and RT conditions.

Figure 8 presents the evolution of visual properties for both samples. For Acktar Fractal Black, some discoloration was observed, but this discoloration did not cause changes in the thermo-optical properties (Table 2). However, comparison of the SEM images (Figure 9) indicates changes in microstructures between ATOX masked area (A) and ATOX exposed area (B). In the case of NV14 white coating, the coating adhesion on aluminum in hot ATOX conditions is significantly worse than for the same coating on titanium. As a result, the coating was damaged at the end of the test, as shown in Figure 8 (EOT image for Hot plate- NV14 on Al), during removal of the sample cover mask.

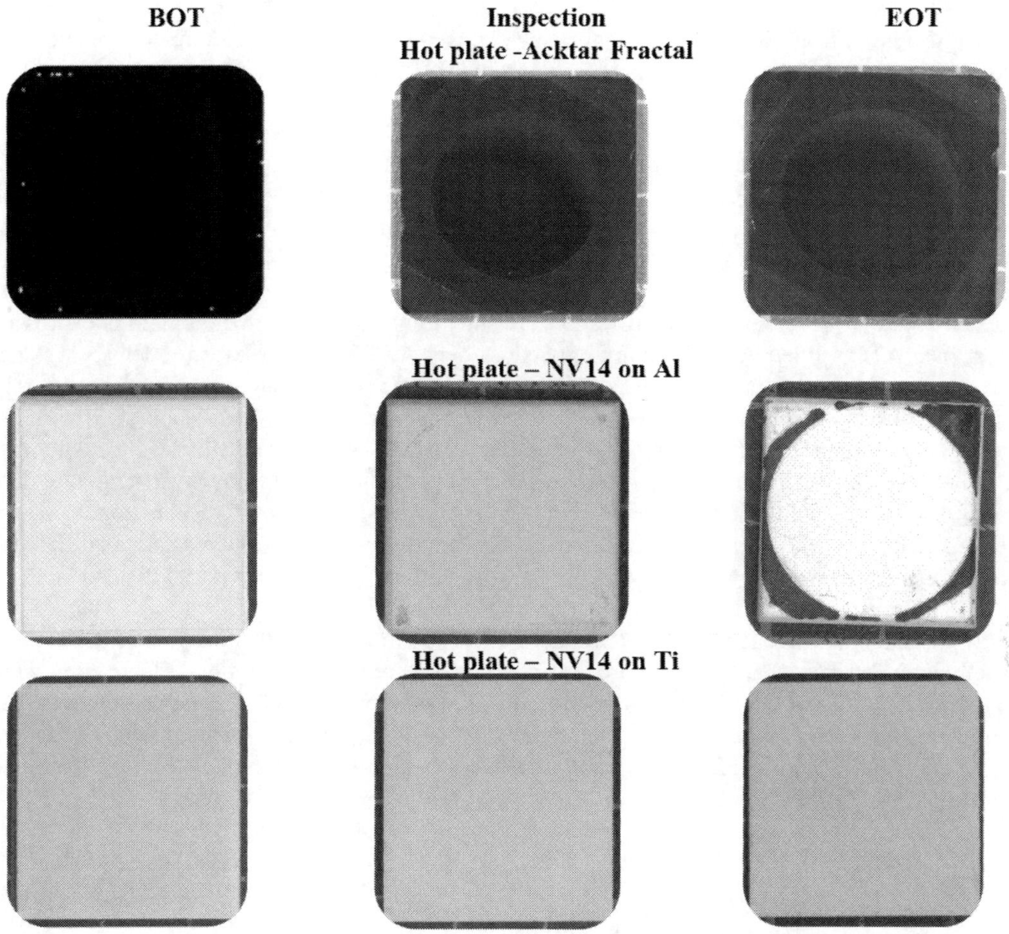

Figure 8. Visual properties of black and white coatings exposed to ATOX in two configurations: on a hot plate and on a RT plate.

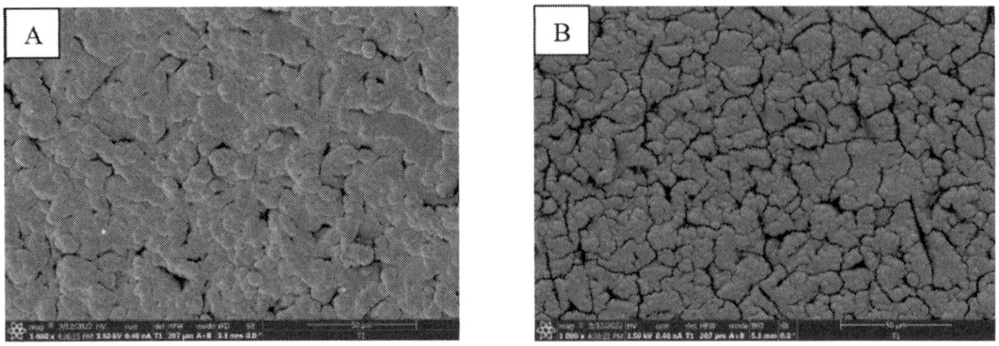

Figure 9. SEM images from the Acktar Fractal Black exposed to ATOX and hot temperature showing changes to microstructure between ATOX masked area (A) and ATOX exposed area (B).

3.3. Erosion yield

Erosion yield describes the resistance of a material to atomic oxygen. It is usually given in units of volume (or mass) loss per incident atomic oxygen atom. Erosion yield has been reported to be temperature dependent [3] and hence accurate predictions for samples exposed to ATOX and temperature simultaneously should be taking this effect into account as will be the case for the Envision spacecraft.

The results obtained for several reference materials are shown in Figure 10. Erosion yield is expressed as mass loss over ATOX fluence, rather than volume loss over ATOX fluence because density was not known for some of the materials. The exposure area was the same for all materials, therefore different materials can be directly compared. There is an almost linear dependence of the erosion yield with the temperature for both reference materials, Kapton HN and HOPG. A similar result was presented by Buczala and Minton [3], they proposed that, above 100°C, the erosion yield follows an Arrhenius-like temperature dependence while a temperature-independent process dominates the erosion at lower temperatures. Although in our study the dependence appears to be linear, it must be noticed that the temperature range in our work is too narrow to observe the Arrhenius-like temperature dependence.

In our study we note that there is an 'onset temperature' for the mechanism that dominates the erosion at higher temperatures. This onset temperature is similar for both reference materials. The erosion yield of our reference materials at temperatures close to 300 °C was approximately 2 times higher than the values below the onset temperature. The erosion below the onset temperature seems to be independent of the temperature of the sample. FEP and PE polymers were also included in the low temperature range experiments with similar results, i.e. no significant temperature dependence was observed.

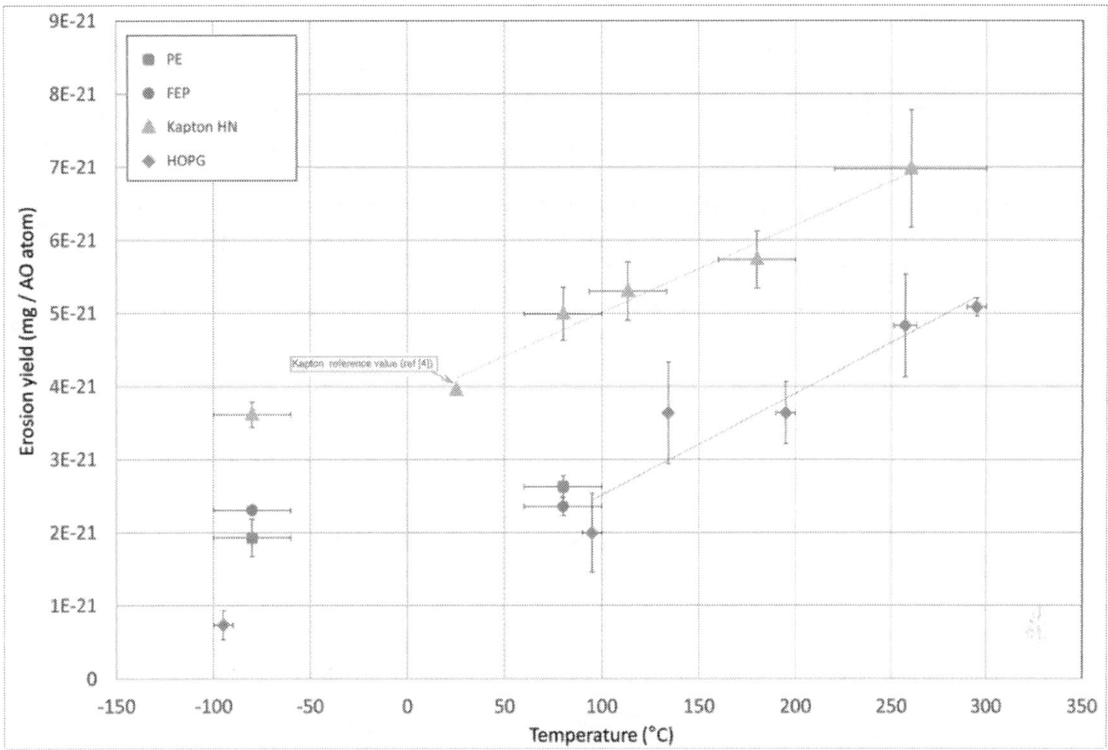

Figure 10. Erosion yield vs temperature for the different materials. The value of the erosion yield for kapton HN at room temperature has been taken from the literature [4].

4. Conclusions

To understand material interactions with the Venus aerobraking environment, a study was performed in the LEOX test facility at ESA ESTEC, specifically modified to simulate combined effects of ATOX and temperature. Hot surfaces were bombarded with ATOX, to reach a cumulative ATOX fluence of 5 x 10^{21} atoms/cm^2. Samples from different families of materials, including MLI, coatings and structural materials were tested for any detrimental visual effects and changes in thermo-optical properties.

The findings of this de-risking activity show that some materials are inert, but some show adverse reactions to the simulated aerobraking environment, therefore they should not be used in the surfaces of the Envision spacecraft that are exposed to ATOX and heat.

The most concerning effects are:

- oxidation and change of thermo-optical properties of glass blasted titanium,
- localized erosion and cracking of the ITO layer on an ITO/SiOx/Kapton MLI candidate,
- different behaviour of the same white paint NV14, depending on the substrate material (aluminium vs titanium),
- some morphology changes observed for the Acktar Fractal black coating that can lead to cracking and delamination.

The outcome of the test campaign shows that it is of great significance to test the candidate materials in adequate conditions, as the results show that even materials that are considered resistant to heat and ATOX, such as titanium, can deteriorate significantly when subjected to the same effects in combination. In the next stages of the mission preparation, more materials will be tested, including the most concerning solar panels and some ram-facing optical components.

References

[1] Widemann T and Wilson C and Ghail R 2017 EnVision: understanding why our most Earth-like neighbour is so different *42nd COSPAR Scientific Assembly* (Pasadena, California)
[2] Helbert J et al. 2019 The VenSpec suite on the ESA EnVision mission to Venus in *Infrared Remote Sensing and Instrumentation XXVII*, SPIE Optical Engineering + Applications (San Diego, California)
[3] Buczala DM and Minton T 2006 Erosion of Kapton H by hyperthermal atomic oxygen: Dependence on O-Atom fluence and surface temperature in *Protection of Materials and Structures from the Space Environment (*Amsterdam: Springer*)*
[4] De Groh K and Banks B 2018 The Erosion of Diamond and Highly Oriented Pyrolytic Graphite After 1.5 Years of Space Exposure

Acknowledgments:

The technical expertise of Tim Minton was greatly appreciated in supporting the development and commissioning of the new LEOX facility in ESTEC.

Euclid: Ice Contamination Testing

S Szmolka[1], B Bras[1,3], M Portaluppi[1], K Paterson[2], M Schirmer[2], L Venancio[1]

[1] European Space Agency, Keplerlaan 1, 2200AG Noordwijk, Netherlands

[2] Max Planck Institute for Astronomy, Königstuhl 17, 69117 Heidelberg, Germany

[3] Corresponding author, bruno.bras@esa.int

Abstract. *Euclid* is a medium-size ESA mission which will map the geometry of the Universe for a better understanding of dark matter and dark energy. Due to the mission profile and stringent performance requirements, the in-orbit contamination environment is a key aspect to control. Cryogenic operational temperatures root a particular concern over water/ice contamination on optical elements, since increased light scattering could potentially cause a significant impact on the performance of the payload. This challenging work under representative conditions closes the link between performance and ice contamination on specular surfaces (e.g. Mirror M1). The publication covers in detail the design and execution of the test campaign performed in ESA's TEC-QEE Yuta vacuum facility, resulting in a data-set of spectral information at cryogenic temperatures. The optimised translation/rotation mechanisms incorporating fiber optics allowed scanning in different angular configurations: both the specular reflection as well as scattered light at different levels of surface contamination. The test results show that the effects of ice contamination are most noticeable in specular configuration. A summary is also given about ice-kinetics measured on the black paint used on the external baffle. The work supports *Euclid*'s calibration team in predicting expected performance and tuning the frequency and duration of in-flight decontamination processes.

1. Introduction: Euclid mission

Euclid is a space-based optical/near infra-red survey mission of the European Space Agency (ESA) to investigate the nature of dark energy, dark matter, and gravity by observing the geometry of the Universe and the formation of structures over cosmological timescales from the Lagrange L2 vantage point. The mission will investigate the distance-redshift relationship and the evolution of cosmic structures by measuring shapes and redshifts of galaxies and clusters of galaxies out to redshifts 2, or equivalently to a look-back time of 10 billion years. In this way, *Euclid* will cover the entire period over which dark energy played a significant role in accelerating the expansion of the Universe. For a comprehensive mission description and objectives, the user is redirected to other sources such as [1].
For this publication, it is important to note that the spacecraft comprises a large telescope aperture (about 1.2 meters across, see Figure 1). The payload module (PLM) consists mainly of two instruments:

- VIS - Visible Imager, operating between 0.55 μm and 0.9 μm
- NISP - Near infra-red spectrometer and photometer, operating between 0.9 μm and 2 μm

IOP Conf. Series: Materials Science and Engineering 1287 (2023) 012028 doi:10.1088/1757-899X/1287/1/012028

As schematized in Figure 2 several elements are shared between the instruments, namely the baffle, M1, M2, FoM1, FoM2, and M3. The dichroic mirror makes the separation of light directly onto the NISP focal plane, or conversely towards FoM 3 (further into the focal plane array of VIS).
Cryogenic operational temperatures induce a predisposition to contamination events. The temperatures are so cold that not only organic material but also water may deposit on sensitive surfaces (see next section).

Figure 1. Artistic impression of *Euclid* spacecraft (*www.esa.int*)

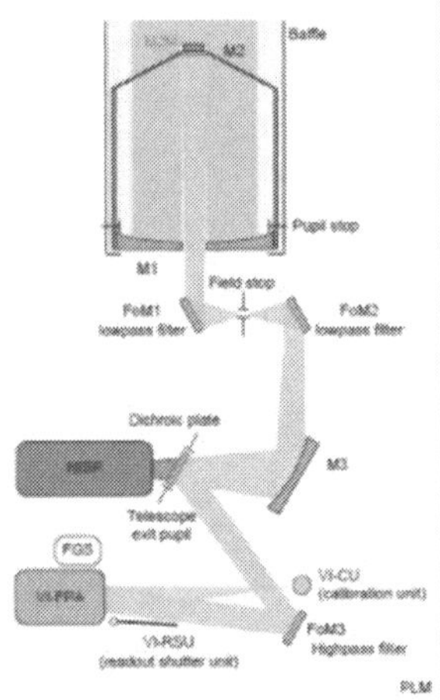

Figure 2. Schematic view of the payload module comprising NISP, VIS and several shared optical elements.

1.1 Ice contamination on the spacecraft

Ice contamination on the spacecraft becomes a significant concern when functional surfaces operate at cryogenic temperatures and there is water available from (nearby) outgassing materials. Under these circumstances, water can deposit on the sensitive surfaces and affect their performance. Possible effects to consider are light scattering, absorption, diffraction, or reflection of light. While the issue is frequently controlled and mitigated by careful design and effective decontamination measures, in-flight anomalies and exceeding performance degradation have been repeatedly documented in the literature. Examples are the cases of the MIPAS instrument [2] and AASTR and SCIAMACHY instruments onboard ENVISAT [3], with ice contamination identified as the likely root cause for the change of gain and bias on the 13.4μm channel of Meteosat/SEVERI after intercalibration against Metop/IASI [4]. Ice growth and characterizations in cryogenic conditions are the subject of much work from different sources such as [5-7], including authors of this publication [8]. In essence, it is challenging to design a test campaign that is representative of the intended environment since multiple aspects influence the properties of the ice deposit. Examples are the temperature, growth rate, angle of incidence of water molecules, roughness, and chemistry of the substrate, amongst others. There are several identified ice structures, most being metastable. Consequently, solid phase transitions can – and will – occur if thermodynamically favourable. This inherently brings an additional concern to space systems, as different phases may behave differently and cause dissimilar optical effects. Moreover, in vacuum environments ice will sublimate depending on the partial pressure of water and the surrounding temperature. The kinetics of re-emission will be presented in the results section.

Recreating representative conditions during the contamination step can be as difficult as actually measuring the performance of the contaminated surface *in-situ*. To close the gap as much as possible towards applications, in the last years the Materials' Physics and Chemistry section of the European Space Agency has performed extensive work on ice contamination and its impacts on payloads performance with emphasis on the Gaia [8] and *Euclid* [9] missions. The increasing knowledge in the field is used to predict ice effects as well as define/optimise anti-contamination measures. The current publication summarises new findings.

1.2 Ice contamination on optical surfaces: Euclid's test campaign

The *Euclid* Scientific Consortium and *Euclid*'s ESA project team joined efforts to define the most representative conditions for the test campaign. The effort was put in recreating as close as possible the conditions for ice growth as seen on *Euclid*'s optical surfaces (mirrors) and focusing directly on relevant performance measurements (spectral information of contaminated surfaces). In this way, the results can be directly used by the *Euclid* team to predict and monitor performance fluctuations during the lifetime of the mission, as well as define, and trigger decontamination events. The next chapters will describe the methods and summarise achieved results.

2. Test Approach

The test subjects were optical mirrors which were exposed to water outgassing. While at cryogenic temperature, ice condensed on the mirrors surfaces. A custom-made translation/rotation mechanism incorporating fiber optics enabled scanning reflected light in different angles, characterizing the reflectance evolution for different levels of surface contamination. Quartz Crystal Microbalances (QCMs) were used to correlate ice deposit and the spectra. For simplicity, in this work it is assumed that the ice deposit is a uniform layer with a density of 1 g cm^{-3}. With a mass sensitivity of 1.96×10^{-9} g cm^{-2} Hz^{-1}, these systems were able to monitor even the initial targets of just a few nm thick. Type K thermocouples were utilised in several locations of the baseplate and on local heaters to finely maintain the temperature conditions within the target conditions (±5 °C).

The facility adaptation required thorough validation steps to ensure test needs were achieved. Cryogenic vacuum compatible rotation mechanisms were selected to ensure an angular repositioning better than 50m°. The uncertainty of the reflection measurements is driven by the alignment between the two mechanisms and the sample surface. After careful mechanical alignment of the optical set-up, the end-to-end optical response of the system was verified against benchtop spectrometers. The specular beam was finely located by scanning the maximum signal in specular configuration. The response of the system ensured reproducibility within 1% in the specular conditions (including movement of rotation mechanisms, translation, movement of optics, etc).
The next sub-sections detail the set-up and achieved conditions.

2.1 Test conditions

ESA TEC-QEE's Yuta facility is a complex vacuum facility designed to measure *in-situ* the performance of contaminated surfaces. The facility comprises a main vacuum chamber (holding the optics assembly and the surfaces of interest) and a secondary chamber (a Knudsen cell holding the contamination source). Figure 3 provides a schematic view of the facility and of the baseplate.
Liquid nitrogen was used to cool the large cylindrical shroud, leading to a high Cryogenic surface/sample surface ratio of 341. The main chamber baseline pressure was tuned to reach less than 5×10^{-6} mbar.
The optical surfaces temperature varied according to the test specification for each test (see next sections). QCMs set at the same temperature as the optical surface monitored the contamination environment. A translation mechanism carries the optics assembly back-and-forth between a resting position and measurement locations.

IOP Conf. Series: Materials Science and Engineering 1287 (2023) 012028 doi:10.1088/1757-899X/1287/1/012028

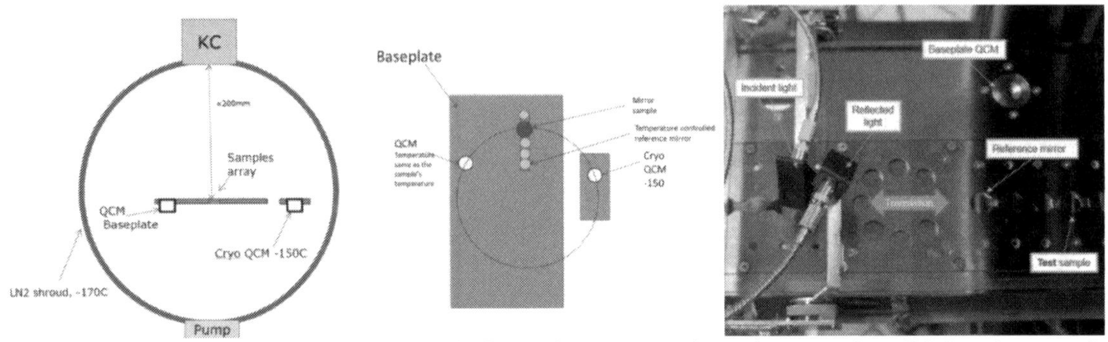

Figure 3. Cross section of the facility. Left: front view. Centre: baseplate top view. Right: photograph depicting baseplate (including optics and sample positions).

The ice deposit was accumulated over eight discrete contamination steps. Each step used a pre-determined amount of Kapton foil, the so-called "Kapton loads", ranging from about 0.24 g to 3.40 g (depending on the target amount). The equivalent thickness was computed assuming a uniform layer deposit with a density of 1 g cm^{-3}. Each test run was performed following exactly the same test approach; thus, the results are directly comparable between different mirrors.

The facility was pre-baked before the beginning of the test campaign. QCM monitoring and molecular contamination witness plates confirmed negligible background contamination levels. This was achieved by careful materials selection of the facility, increasing the surface area of cryo shrouds, etc. Once cleaned and with negligible background levels, no other tests outside water contamination have been carried for as long as 12 months.

2.2 Optical set-up: reflectance measurements

The facility was adapted to accommodate reflectance measurements in different configurations (specular and scatter angles). Both the lightsource and spectrometer are kept in ambient conditions: the set-up uses fiber optics to guide the light from the AvaLight lightsource (Avantes) into the chamber and back towards the QE Pro spectrometer (Ocean Optics). Two Avantes collimating lenses were used above the sample.
The set-up allowed us to acquire spectra between 250 nm and 1000 nm, directly covering *Euclid*'s VIS instrument.
To ensure the maximum accuracy and angular sensitivity, the reflectance collimating lens was positioned as far away from the surface of the sample as possible (L = 55 mm).
Furthermore, a 100 μm slit was mounted at the incoming light to reduce the spot size on the sample (2 mm^2). The reflected beam fell well within the detector aperture (D = 6.5 mm) when in the specular position.

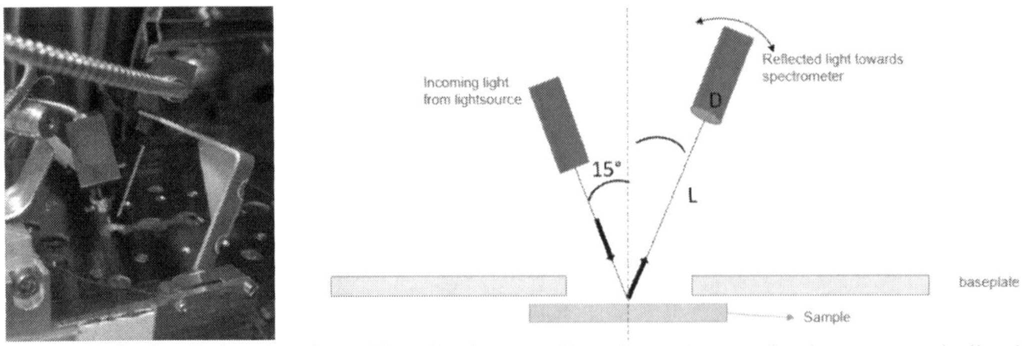

Figure 4. left: detail showing the collimating lenses, slit and rotation mechanism, arrows indicating the light path. Right: sketch of reflectance measurements for a 15° incidence.

The nature of reflectance measurements requires a "reflectance standard". In this test, the reference sample was a commercially available aluminium mirror from *Thorlabs* (Part number PF10-03-G01), thermally controlled using a local heater. Figure 3 indicates its location on the baseplate. The temperature was maintained at -80°C since it ensures no ice deposition within the test's pressure environment, which in practice means it remained clean throughout the whole campaign.

For each acquisition, the optical measurements procedure is detailed below:

1. Dark measurement (light-source off)
2. Specular measurement on reference mirror (clean background): 15° / 15° (or 45° / 45°)
3. Specular measurement on test sample: 15° / 15°, referenced to point 2 above.
4. Diffuse measurements on test sample (such as 15° / 25°, 15° / 50°, etc), referenced to point 2 above.

Note: In some cases, both 15° and 45° incidences were measured. The test specifications were defined to be closely representative of the flight configuration.

3. Materials

The surfaces to be ice contaminated were three optical mirrors (different dimensions, filmstack and substrate), each representative of specific payload components. The mirrors were tested separately (independent test).

- **DC**: Dichroic coating, partially transparent in the wavelength of interest of the test.
- **HC**: Hybrid coating, representing FOM1 and FOM2.
- **PC**: Protected silver coating, representing M1, M2, M3 and FOM3.

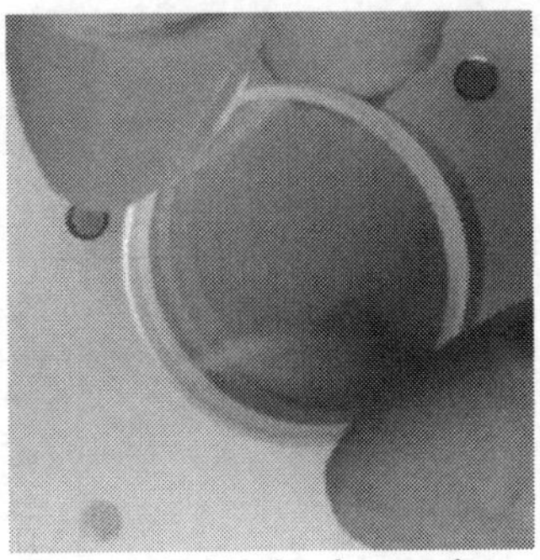

Figure 5. Dichroic coating.

Figure 6. Hybrid coating.

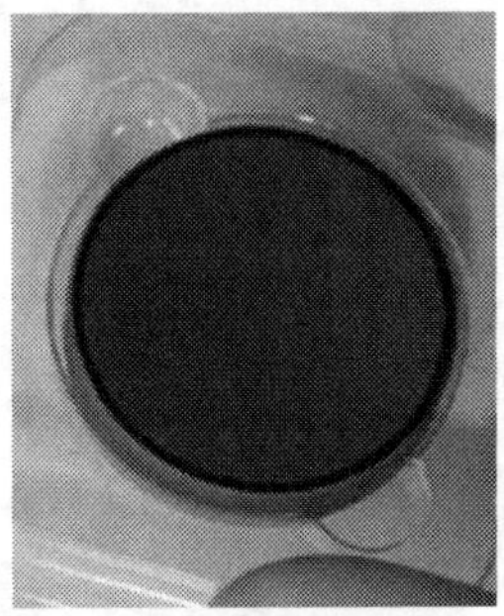

Figure 7. Protected silver coating.

The water source was pre-baked Kapton foil (Dupont clear Kapton HN RF500 48N), a known hygroscopic material that emits water in vacuum even at room temperature. Kapton foils have been vacuum baked at 120°C, at least 72hours, in high vacuum with TQCM monitoring to be depleted from any light and competing organic contamination, in particular considering a foil exposure up to 10min.

4. Optical Surfaces: Test Results

The measurement conditions achieved during the test are provided in the next table for each independent test run.

The facility log for the DC test campaign is given as an example in Figure 8. Note the slight re-emission pattern observable on the QCM curve (decrease on contamination level). This behaviour is expected

IOP Conf. Series: Materials Science and Engineering 1287 (2023) 012028 doi:10.1088/1757-899X/1287/1/012028

once the contamination source is removed from the vacuum chamber and the partial pressure decreases. The temperature (-135 °C ± 5 °C) is not cold enough to ensure prolonged residence times. This goes very much in line with the kinetics of ice re-emission that is presented in later chapters.

Table 1. Summary of the achieved test conditions.

	Test conditions		
Parameter	Test A	Test B	Test C
Mirror Material	DC	HC	PC
Mirror Temperature	-135 °C ± 5 °C	-135 °C ± 5 °C	a) -135 °C ± 5 °C b) -155 °C ± 5 °C
Angle of incidence	15°	15°	15° and 45°
Specular reflectance angle	15°	15°	15° and 45°
Scatter angles	Discrete angles between 15° and 70°		
Wavelength Range	Spectra acquired between 250 nm and 1000 nm, with about 1 nm interval.		
Target levels of contamination	The thickness information is extracted from QCM readings, considering a uniform film with a density of 1 g cm^{-3} : 1 nm, 2 nm, 5 nm, 10 nm, 20 nm, 50 nm, 100 nm, 200 nm.		

Spectral measurements were performed immediately after the contamination step is concluded, and in many cases were repeated at different moments for the same level of contamination.

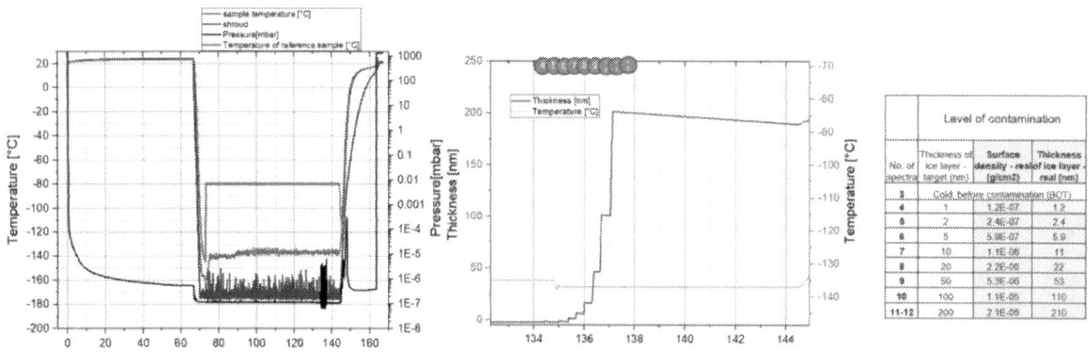

Figure 8. left: DC test run log. Centre: QCM log during the contamination step. Right: achieved contamination levels.

The QCM log during the contamination of PC sample at -155 °C ± 5 °C is given in Figure 9. As expected, very limited (if any) re-emission is observable within the test timeframe.

IOP Conf. Series: Materials Science and Engineering 1287 (2023) 012028 doi:10.1088/1757-899X/1287/1/012028

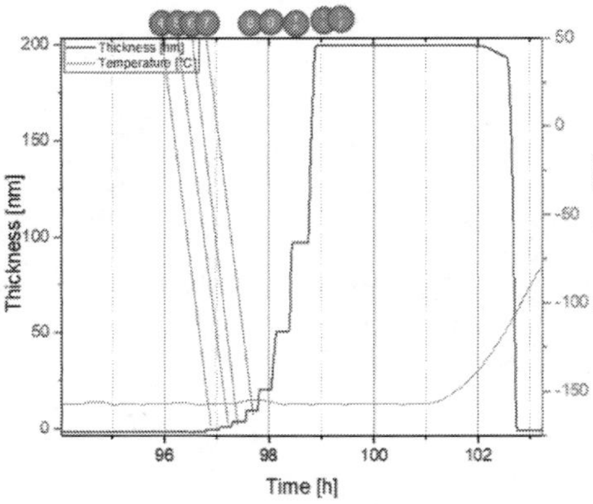

No. of spectra	Level of contamination		
	Thickness of ice layer - target (nm)	Surface density - real (g/cm2)	Thickness of ice layer - real (nm)
3	Cold, before contamination (BOT)		
4	1	1.4E-07	1.4
5	2	2.5E-07	2.5
6	5	5.6E-07	5.6
7	10	1.1E-06	11
8	20	2.2E-06	22
9	50	5.2E-06	52
10	100	9.9E-06	99
11-12	200	2.0E-05	200

Figure 9. left: QCM log for PC at -155 °C. Right: actual contamination levels.

The reflectance curves plotted in the next sections are relative to the standard mirror (uncontaminated) and therefore are not absolute reflectance values. The ratios curves are performed by dividing each contamination step by the surface's clean condition (before contamination), and therefore clearly emphasize the effects of ice contamination.

4.1 Test results: DC sample (-135 °C)

This section presents the results achieved with the Dichroic coating. Ice effects are noted almost immediately (already at 1.2 nm thick layer). Figure 10. presents the specular configuration: between 540 nm and 840 nm (the mirror effective reflectance band), all measurements fall within 1% uncertainty and thus are within the expected reproducibility of the measurements. No obvious changes are expected in the specular reflectance within the in-band reflection region due to ice growth up to 200 nm layer.
Nonetheless, the spectra clearly show optical effects in off-bands regions, with particular interest on the edges of the in-band (namely between 500 nm and 525 nm, and above 920 nm). In these regions, there is a clear effect of ice contamination which leads to increased reflected signal. It should be remarked that the highest intensity is not always registered for the highest contamination deposit (note the reflectance at 520 nm is much higher for a thickness of 110 nm ice deposit than for 210 nm).

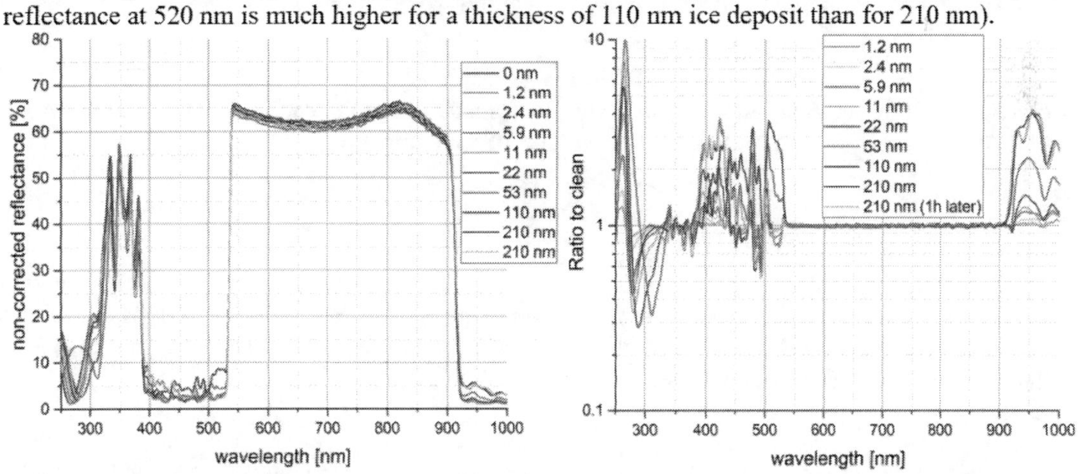

Figure 10. DC contamination test. Specular configuration. Left: non-corrected reflectance. Right: Ratios.

In scatter angles the effect is much less pronounced, as very low counts are measured throughout the test (below 0.5% relative reflectance). The ice deposit effect is detected by slight fluctuations in the lower wavelength region (below 400nm). Figure 11. provides the example of 15° / 20° configuration, while Figure 12. provides higher scatter results.

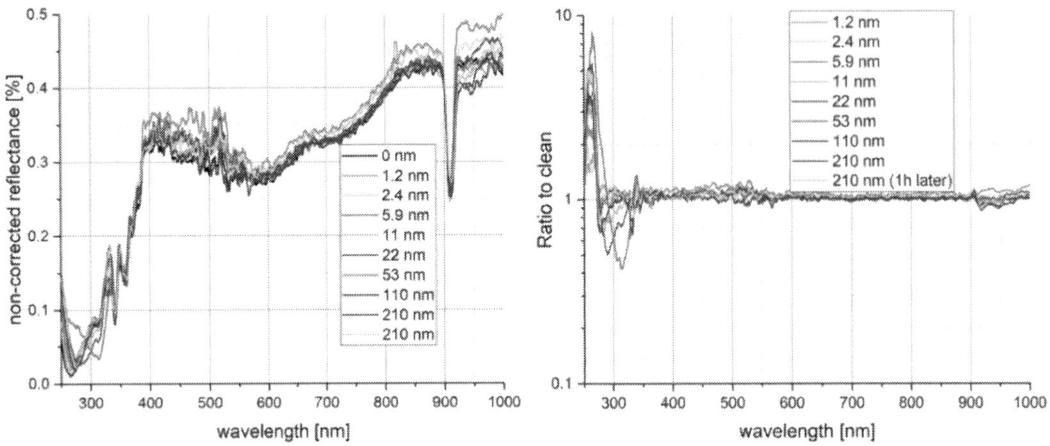

Figure 11. DC contamination test. 15° / 20° configuration. Left: non-corrected reflectance. Right: Ratios.

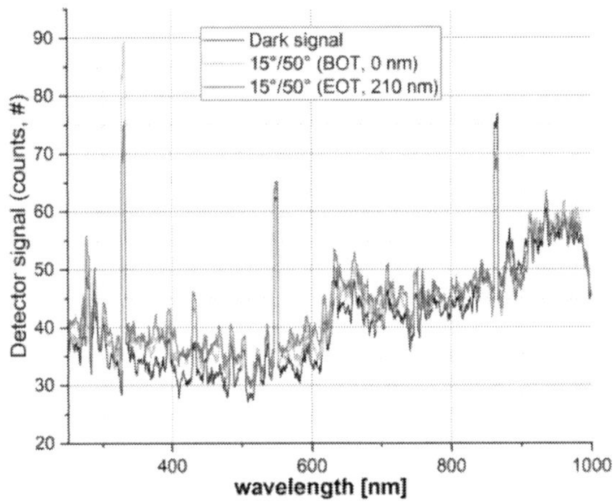

Figure 12. 15° / 50° configuration (detector counts marginally above the dark signal and do not evolve even for the highest level of contamination.

4.2 Test results: PC sample (-135 °C and -155 °C)

The protective coating (PC) mirror was subjected to four independent contamination steps for two different incidence angles (15° and 45°) and two temperatures (-135 °C and -155 °C). The levels of contamination were very similar in all tests, allowing direct comparison in all practical terms for spectra of different sample temperatures (and even incidence angles).

Figure 13 and 14 depict the 15° specular contamination phases at -135 °C and -155 °C, respectively. The performance is very similar; consequently, an important observation is that in case different ice phases are present, the optical behavior is not differently affected. The in-band region (between 500 nm and 1000 nm) does not depict any notorious effect due to ice deposit up to 200 nm. The spectral effects are observable mostly below 400 nm, within the off-band region. In these regions, there is a clear effect of ice contamination on the optical throughput. The effects are detected almost immediately (1.4 nm thick deposit of ice). As previously, the maximum thickness does not correspond to the worst-case in optical response.

IOP Conf. Series: Materials Science and Engineering 1287 (2023) 012028 doi:10.1088/1757-899X/1287/1/012028

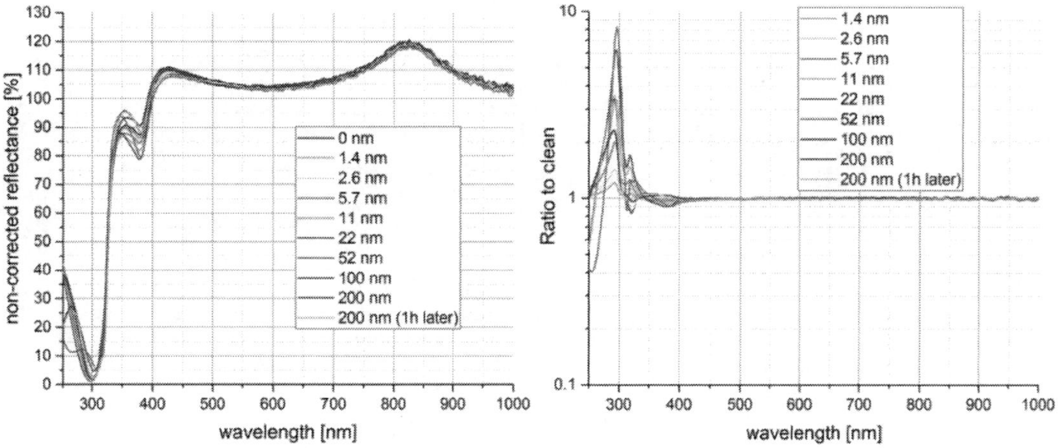

Figure 13. PC contamination test at -135 °C. 15° / 15° specular configuration. Left: non-corrected reflectance. Right: Ratio to the beginning of test (clean).

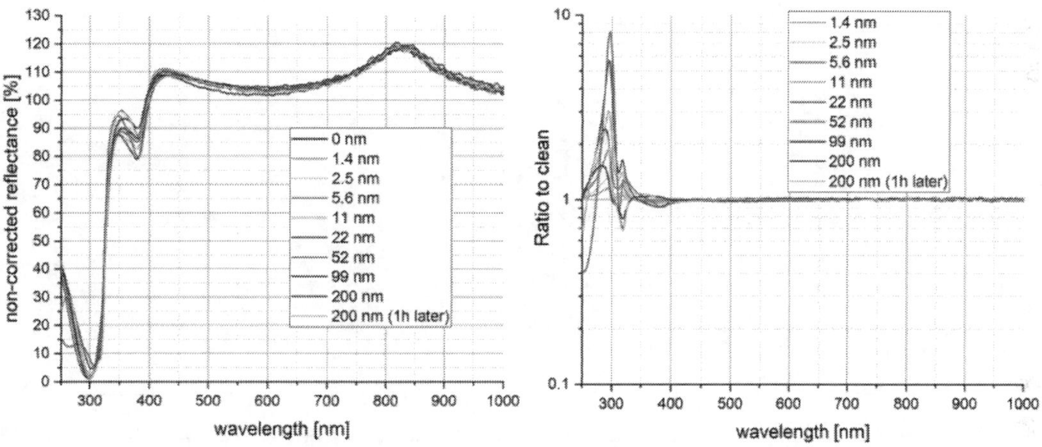

Figure 14. PC contamination test at -155 °C. 15° / 15° Specular configuration. Left: non-corrected reflectance. Right: Ratio to the beginning of test (clean).

An incidence angle of 45° alters the reflective behavior of the mirror, nonetheless the major observations remain (see Figures 15 and 16).

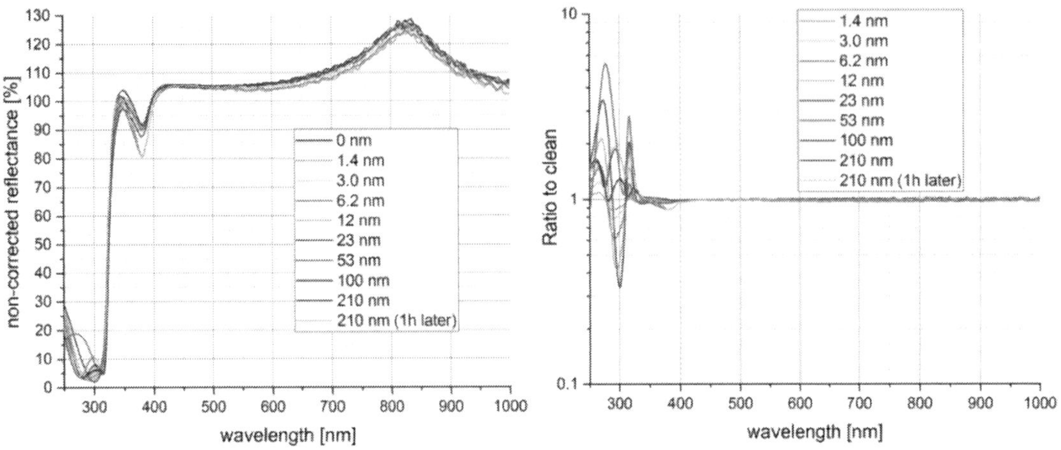

Figure 15. PC contamination test at -135 °C. 45° / 45° specular configuration. Left: non-corrected reflectance. Right: Ratio to the beginning of test (clean).

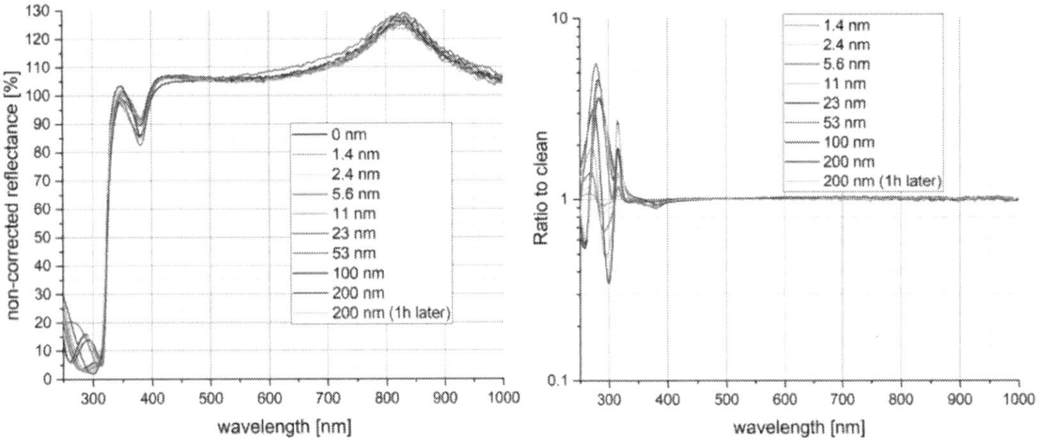

Figure 16. PC contamination test at -155 °C. 45° / 45° specular configuration. Left: non-corrected reflectance. Right: Ratio to the beginning of test (clean).

Measurements performed at higher scatter angles did not show any evidence of increased straylight/scatter in neither tested temperatures. Examples are given in Figure 17.

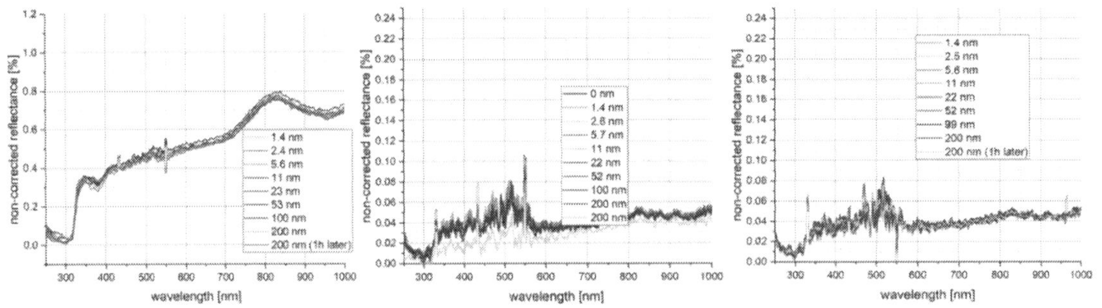

Figure 17. Scatter angles measured in PC tests. Left: -155 °C, configuration 45° / 40°. Centre: -135 °C, configuration 15° / 40°. Right: -155 °C, configuration 15°/ 40°.

4.3 Test results: HC sample (-135 °C)

The hybrid coating test was performed on two samples. The first run was performed on a sample which had undergone testing before and thus its initial surface cleanliness was not guaranteed. The second test was performed on a pristine sample.

While the previously characterized sample clearly depicts an in-band fluctuation effect (up to 10% relative oscillation), this behavior is absent from the pristine surface. The off-band (below 500 nm) shows similar effects.

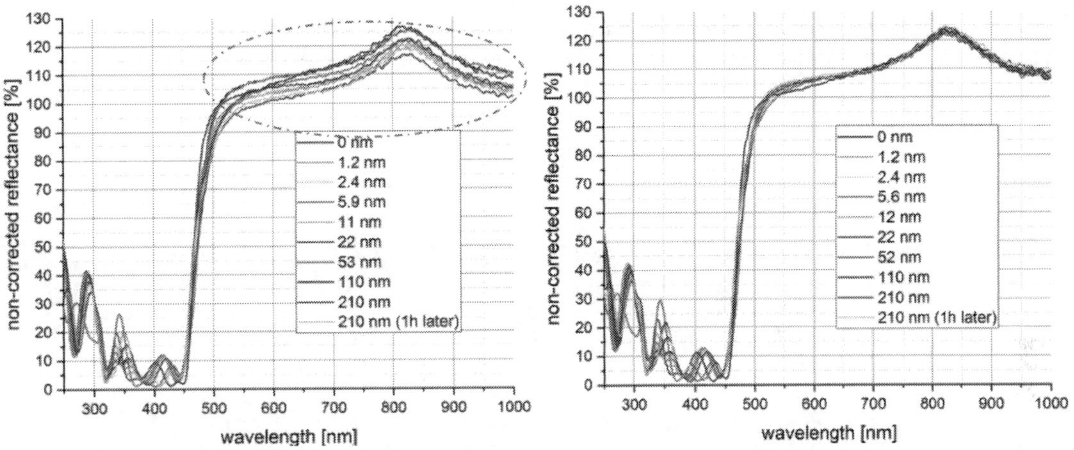

Figure 18. HC performance on 15° / 15° configuration at -135 °C. Left: sample showing features caused by previous testing. Right: pristine sample.

Scatter angles behave similar in both test campaigns, once again not depicting a remarkable evolution. Figure 19 provides few examples (attention to the low reflectance values).

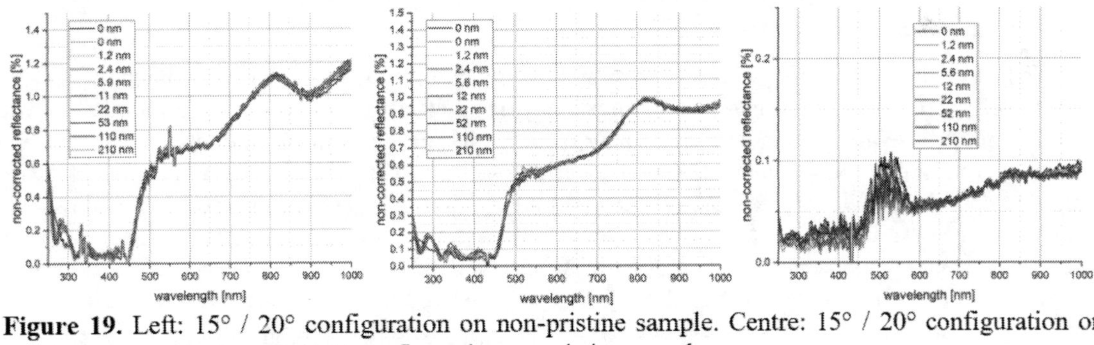

Figure 19. Left: 15° / 20° configuration on non-pristine sample. Centre: 15° / 20° configuration on pristine sample. Right: 15° / 35° configuration on pristine sample.

4.4 Dynamic observations

Throughout the vast majority of the test campaign the stability of the measurements was confirmed by repetition measurements and test operator assessments. However, in some cases a dynamic behavior was registered. This was the case on the Protected Silver coating (45° incidence, 53 nm thickness contamination step), both temporally (over the same position) and spatially (scanning over the sample surface).

i) Measurements performed according to the standard procedure (reference specular → exposed sample specular → sample scatter scan → sample specular) on the translation position 1840mm (standard location for the measurements).

IOP Conf. Series: Materials Science and Engineering 1287 (2023) 012028 doi:10.1088/1757-899X/1287/1/012028

ii) During an unintentional slight translation over the sample (about 3 mm away, bringing the optical set-up to the position 1837 mm), it was observed a sharp shift of the in-band reflectance.

iii) Moving the optics back to the nominal location (1840 mm) brought the reflectance curve exactly back to what had been measured in point i) above.

iv) It was decided to make a scan over the contaminated mirror surface and could be seen that the curve measured in point ii) represented almost all surface of the sample, however in some regions the reflectance signal was drastically higher (or lower) by up to 10 %.

v) Moving the optics back to the original location (1840 mm), the spectrum had now changed, and the curve was very similar to most areas of the sample.

Being a manual process, not all measurements were recorded. Figure 20 provides only some snapshots, unfortunately not truly representing the fast changes that were locally observed.

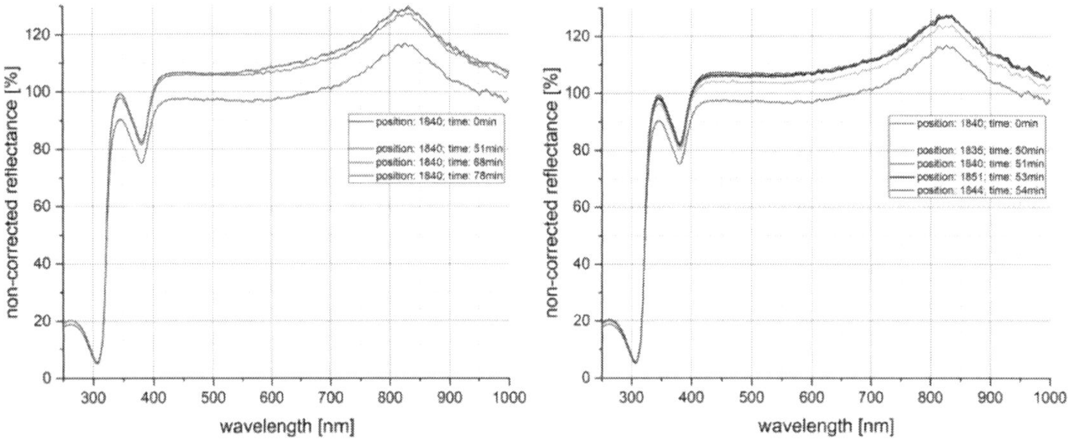

Figure 20. *Dynamic observations on PC sample. Left: temporal evolution over sample position. Right: spatial evolution.*

4.5 Ice contamination on a known substrate

The results on *Euclid*'s mirrors (maxima and minima fluctuations throughout the contamination growth) indicate reflective/refractive changes play a more important effect than scatter. To further assess this effect, a final test sequence was performed on a commercial mirror (standard aluminium mirror from Thorlabs), recreating the test conditions of Euclid's mirrors discussed in previous sections. The measured data was then compared against simulation results on this less complex substrate/filmstack. To achieve the simulations, a clean aluminium mirror was measured in a benchtop ellipsometer to determine its optical constants. Utilizing TEC-QEE's heritage [8, 10], namely the ice optical constants determined in previous test campaigns, the effects of ice contamination on the aluminium mirror were simulated/predicted within the wavelength range of the experimental test campaign.

The curves shown in Figure 21 are the ratio between contaminated and clean conditions and allow the direct comparison between modelled scenarios (blue and grey) and the measured (orange) data. To account for uncertainties such as actual ice density, deposit morphology, surface temperature, differences between the reference and exposed mirrors, and ellisopmetric model inaccuracies, two scenarios are provided for visualization against the measured data.

There is a very good correlation between the specular response (15° / 15°) and the prediction from the software tool, providing indication that the optical effects are mostly due to layer interference/reflection on the Thorlabs mirror. Within the measurement set-up capabilities, scatter effects are little evident (if present).

IOP Conf. Series: Materials Science and Engineering 1287 (2023) 012028 doi:10.1088/1757-899X/1287/1/012028

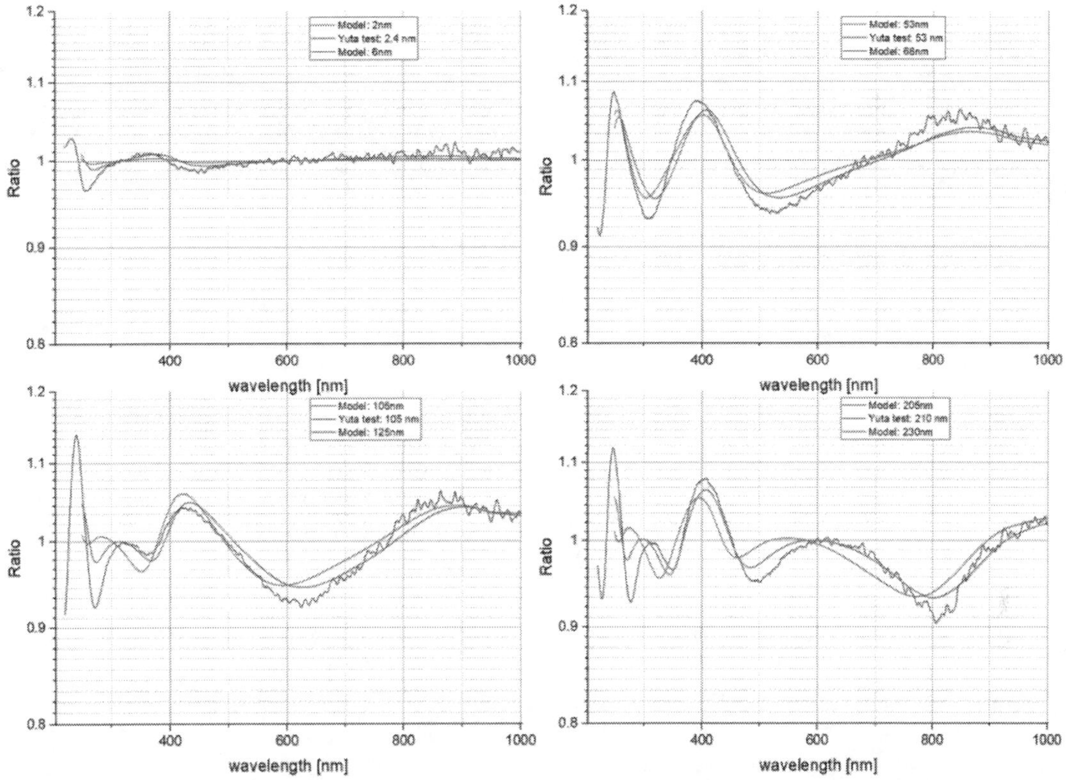

Figure 21. Comparison between achieved results and thin-film modelling simulations.

4.6 Notes about measurement reproducibility

The reproducibility within the same test is the key factor for the optical evaluations. The measurement stability and overall reproducibility were extensively evaluated before and throughout the test campaign to provide confidence about the acquired data and estimate measurement artefacts.

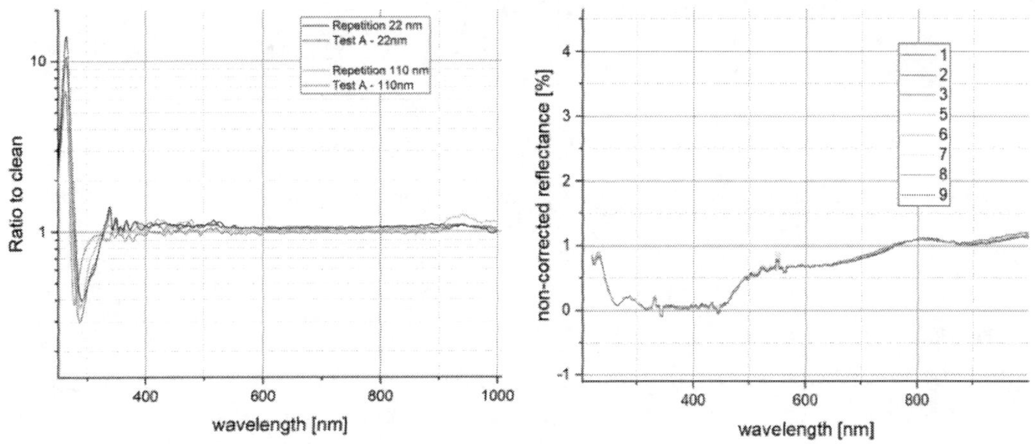

Figure 22. Left: comparison example between Test A and the Repetition Test in specular 15° / 15° configuration. Right: Measurement stability on HC sample for 15° / 20° configuration across measurements in several days.

IOP Conf. Series: Materials Science and Engineering 1287 (2023) 012028 doi:10.1088/1757-899X/1287/1/012028

5. Ice Kinetics: Test Results

An independent test campaign was performed to assess re-emission ice kinetics and size *Euclid*'s decontamination step. The full set of results were presented at the CCMPP 2021 proceedings [9], being a summary given in this section. Target surfaces (black MAP PNC baffle paint and QCM) were brought to -150 °C and exposed to a pre-established levels of ice contamination. Subsequently, both targets were brought to higher temperatures at 4×10^{-8} mbar. The monitoring was performed via QCM and optical measurements on the black paint, which allowed us to establish a direct correlation between both. Table 2 summarizes the values for steady state-remission under these conditions.

Table 2. re-emission data summary (at 4×10^{-8} mbar).

Temperature	Re-emission rate on QCM
-130°C	3×10^{-10} g cm^{-2} s^{-1}
-120°C	3×10^{-9} g cm^{-2} s^{-1} to 4×10^{-9} g cm^{-2} s^{-1}
-100°C	1.5×10^{-7} g cm^{-2} s^{-1} to 2×10^{-7} g cm^{-2} s^{-1}
-80°C	Not available (too fast to measure)

Figure 22 depicts the test results for different re-emission temperatures and different starting levels of contamination. The sublimation rates are consistent with theoretical calculations of [11] and [12] within the expected uncertainties of the measurements.

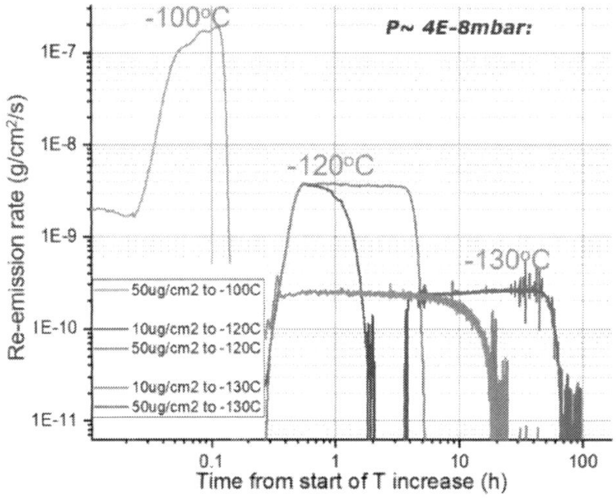

Figure 23. ice kinetics test campaign, depicting the duration it takes for each deposit to be completely removed.

6. Discussion

The campaign was designed to be as representative as possible of the environment seen by the payload during operation in space (e.g. surface temperature, outgassing source of water, ...) within a reasonable timeframe. The test samples are fully representative of the optical surfaces of *Euclid*: dichroic coating, hybrid coating, and protected silver.

The test results show that the effects of ice contamination are most noticeable in specular configuration. Contaminated surfaces up to a 200 nm thick layer of ice did not show pronounced variations in the in-

band region, nonetheless within off-band regions the effects are particularly notorious and should be evaluated with care. These are visible from very thin layers of ice contamination.
A non-pristine sample hybrid coating depicted a pronounced fluctuation during the test campaign (see Figure 18), which is attributed to a non-ideal surface status.
Several ice structures have been identified in nature, each favoured in certain environmental conditions (temperature being an important factor). It is not possible to directly determine the ice structure within Yuta facility. However possible impacts of phase changes would be measured in case they cause a non-negligible optical response fluctuation. This could be the underlying cause of the behaviour depicted in section 4.4. However, this was one of the few moments in which such behaviour was observed. Measurement stability was evident in the majority of the test campaign. Furthermore, measurements on PC mirror performed at different temperatures (-135 °C and -155 °C) did not establish a particular temperature influence: the optical performance is very similar for the same ice levels. It is speculated that a possible phase change transition may not be comparable with the time scale of the test runs.
The measurements performed on the off-the-shelf aluminium mirror from Thorlabs are in excellent agreement with simulated data using thin-film modelling software by establishing a uniform layer of ice over the substrate. This indicates that, within the measured set-up and for that surface, scatter effects are considerably less relevant than the optical interference arising from the inclusion of an extra layer of ice. In other words, the optical response could be explained and predicted by modelling techniques.

7. Conclusion

The optical performance of several *Euclid* representative mirrors were characterized for several levels of ice contamination at TEC-Q laboratories of the European Space Agency. The challenging nature of such task is evident: ice dynamics are extremely complex. The good representativeness of the space environment on ground testing facilities is essential, and effort was put to set the conditions as closely as possible to the real case.
The test campaign generated a comprehensive data-set (angular information, thickness information, temperature, different substrates, etc). The results mostly show a gradual tendency / evolution, however in some occasions uncharacteristic behaviour was observed (see section 4.4). It is speculated that time-driven effects may have occurred in some occasions and possibly transitory detected, although it was not possible to reproduce the observed effects. Large effort was undertaken to exclude set-up artefacts such as reproducibility and light-source stability concerns.

The results will be used by *Euclid* project to assess performance impacts on the VIS instrument and assist on decontamination measures.

8. References

[1] Racca G *et al.* 2016, The Euclid mission design, Space Telescopes and Instrumentation, *Proc of SPIE* **9904**
[2] Niro F, Fehr T, Kleinert A, Laur H, Lecomte P and Perron G 2009, Ice contamination on satellite IR sensors: the MIPAS case, *Geophysical Research Abstracts* **11**
[3] Smith D L 2012 ENVISAT AATSR Instrument Performance - End of Mission Report, STFC final report issued by RAL Space reference PO-RP-RAL-AT-062
[4] Hewison T J and Mueller J 2013 Ice Contamination of Meteosat/SEVIRI Implied by Intercalibration Against Metop/IASI *IEEE Transactions on Geoscience and Remote Sensing* **51** NO. 3
[5] Westley M S, Baratta G A and Baragiola R A 1998, Density and index of refraction of water ice films vapour deposited at low temperatures, *Journal of Chemical Physics* **108** NO. 8
[6] Baragiola R, 2003, Water ice on outer solar system surfaces: Basic properties and radiation effects *Planetary and Space Science* **51** Pages 953-961
[7] Labello J M 2011, Water Ice Films in Cryogenic Vacuum Chambers PhD diss. University of Tennessee

[8] Bras B, P Janik, R Martins and R Rampini 2015 In-situ Ice Formation Investigation *ISMSE 13th proceedings*
[9] Portaluppi M, Bras B, Fontorbes S, Saverino A and R Rampini 2021, Euclid Cleanliness and Contamination Control and challenges due to ice contamination, *CCMPP proceedings – in prep.*
[10] Martins R, Bras B, Milassin G and Rampini R 2015 In-situ Optical Characterization of Outgassed Compounds, *ISMSE 13th proceedings*
[11] Sack N J and Baragiola R A 1993 Sublimation of vapor-deposited water ice below 170 K, and its dependence on growth conditions *Physical Review B* **48**, No. 14
[12] Wagner W, Riethmann T, Feistel R and Harvey A 2011 New Equations for the Sublimation Pressure and Melting Pressure of H2O Ice Ih, *Journal of Physical and Chemical Reference Data*, **40**, 3103

Acknowledgements
The authors acknowledge the *Euclid* Consortium, The *Euclid* Collaboration, the European Space Agency that have supported the development of *Euclid* development. Acknowledge ESA *Euclid* Team, in particular D. Stramaccioni for supporting the definition of the baffle ice re-emission testing, and the other participants of *Euclid* Ice Working Group: TAS-I, ADS-Toulouse, and CEA.

Study of contaminants generated from space materials by impingement of atomic oxygen

R Yamanaka[1], D Faye[2], Y Kimoto[1] and N Maillard[3]

[1] Research and Development Directorate, Japan Aerospace Exploration Agency, 2-1-1 Sengen, Tsukuba, Ibaraki 305-8505, Japan
[2] CNES Laboratories and Expertise Department, Centre National d'Etudes Spatiales, 18 avenue E. Belin, 31401 Toulouse cedex 9, France
[3] ELEMCA, 18 avenue E. Belin, 31401 Toulouse cedex 9, France

yamanaka.riyo@jaxa.jp, delphine.faye@cnes.fr

Abstract. Collaborative activities are carried on between Japan Aerospace Exploration Agency (JAXA) and the Centre National d'Etudes Spatiales (CNES) for many years. These activities are dealing with contamination, space environment, electric device development etc. They are based on "Inter-Agency Agreement between JAXA and CNES concerning the cooperation in the field of space programs" which was concluded in 2010. Recently, one of our collaborative research in the field of contamination has been focused on atomic oxygen (ATOX) induced contamination. The aim of the different tests campaigns was to identify the nature of the contaminants (particulate and/or molecular) generated by ATOX impingement on polymer materials and to characterize their effects on the sensitive surfaces of optical instrument on spacecraft. In a previous work, five commonly used space materials were evaluated and the generation of particulate contamination from Beta cloth by impingement of ATOX was clarified. In the present study, several space materials were selected with different chemical natures among those used outside satellite potentially exposed to ATOX to confirm or to get new information regarding their behaviour. During 2019 and 2020 irradiation campaigns, these materials were irradiated by ATOX (at a representative fluence of 1.0E21 atoms/cm^2) and the induced contaminants were collected onto collector plates. Both of the materials and the captured induced contamination were analyzed. This paper presents the compiled results of irradiation campaigns on six materials of interest (Beta cloth, Carbon Fiber Reinforced Polymer, Ag coated FEP, UPILEX-S 125 µm, RTV-S691, MAP® SCK5N). Perspectives and future test plan between JAXA and CNES are also considered to complete this work.

1. Introduction

Atomic oxygen (ATOX) is the most abundant chemical species not only in the upper layer of the Earth atmosphere between 200 and 800 kilometers but also exists at low altitude of Venus and Mars. It is formed by the dissociation of O_2 molecules under ultraviolet radiation at short wavelengths (< 243 nm). On Earth, the density of ATOX decreases exponentially with altitude but remains significant up to about 1 000 km and varies in direct proportion to the solar activity and its 11-year cycle. For Low Earth Orbit (LEO) satellites in periods of maximal activity, the density can reach 10^8 to 10^9 atoms per cm^3. In periods of minimal activity, it may be reduced by a factor of 1 000 or even more. Thus, a satellite launched in such a period will be less exposed to this environment than one launched in a period of higher activity.

ATOX fluence also depends on many other factors such as the orbital elements (perigee, apogee, inclination), the spacecraft geometry, the mission duration, the geomagnetic activity...
With a relative velocity of about 8 km.s^{-1}, the high-speed collision of hyperthermal O-atoms with exposed materials, mainly on the front side of satellite, can generate severe damages either directly or indirectly by oxidation and/or erosion thus weakening the surface of the materials and changing their physical properties (e.g. modification of surface roughness, changes in thermo-optical properties, changes in surface conductivity…) through gradual and cumulative effects sometimes combined with UV, contamination, thermal cycling [1].
Numerous in-flight experiments showed these effects of atomic oxygen on different spacecraft sub-assemblies such as solar arrays, multilayer insulation for passive thermal control, optical baffles or antennas. In general, polymers are the most sensitive materials, but also some metals are exceptions (e.g. silver). The "reactivity" of the impinged materials is usually estimated by measuring the mass loss but induced contaminants may be produced by erosion, deposit on adjacent sensitive surfaces on satellite and then alter the performance of instruments [2], [3].
The literature is poor regarding this topic because almost all the studies up to now are focused on the determination of the erosion yield of materials. Nevertheless, there are more and more long duration and sensitive to contamination missions in LEO that means that materials are exposed to high ATOX fluence. So it is paramount to highlight the risks and quantify the level of these potential induced contaminants.

2. Experiment

2.1. Concept
Figure 1 shows the conceptual diagram of our atomic oxygen induced contamination (hereafter AIC) test. Collector plates are set on the base jig and the material sample under evaluation is fixed on the jig with two strip-shaped aluminum masks (see Figure 2) and installed diagonally. In addition, the ATOX shield jig is set over the surface of collector plates to avoid direct exposure to ATOX as shown in Figure 1. If particulate AIC is induced by ATOX, AIC can be collected by collector plates.

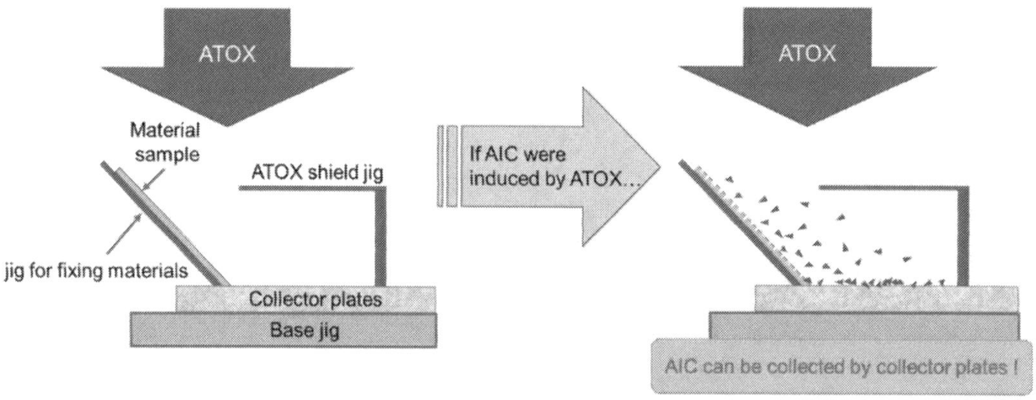

Figure 1. Conceptual diagrams of AIC test

Figure 2. Images of samples fixed on the jig with two strip-shaped aluminum masks (left: UPILEX-S 125 µm (P-1-3), middle: MAP® SCK5N (K-1) and schematic illustration of material fixed on jig (right).

2.2. *Atomic oxygen irradiation*

Combined Space Effects Test Facility at Tsukuba Space Center can irradiate materials with ATOX under high vacuum environment (10^{-3}-10^{-2} Pa) at approximately room temperature [4]. The materials which are discussed in this paper were irradiated by ATOX in this facility. The image of the facility is displayed in Figure 3. Four ATOX irradiation tests were performed. Table 1 shows the conditions of ATOX irradiation tests.

Figure 3. Combined Space Effects Test Facility @ JAXA, Tsukuba Space Center

Table 1. Conditions of ATOX irradiation tests

ATOX irradiation test number	ATOX beam average speed	ATOX fluence (atoms/cm^2)
#1	8.11 km.s^{-1}	1.01E21
#2		1.03E21
#3		1.07E21
#4		1.02E21

2.3. *Test samples*

In this paper, we focused on the evaluation of six materials (Beta cloth, CFRP, MAP® SCK5N from CNES and Ag coated FEP, UPILEX-S 125 µm, RTV-S691 from JAXA) as sources of AIC during ATOX irradiation campaigns. These six materials are widely used in spacecraft. Beta cloth and

UPILEX-S 125 μm are used for Multi-Layer Insulation blankets; Ag coated FEP is used for thermal control. CFRP is used as structural material, RTV-S691 as adhesive for bonding solar cells and MAP® SCK5N as radiofrequency transparent antistatic coating for antenna. The collector plates which consisted of aluminum plate and double-sided carbon tape were chosen to collect the particulate AIC. Table 2 shows the list of materials including the information of sample ID, collector plate ID which was installed within the special jig and ATOX irradiation test number. Reference samples which were only exposed to vacuum environment during ATOX irradiation test were prepared to be compared with the ATOX irradiated samples. The size of all materials except UPILEX-S 125 μm is 30 x 20 mm², and the ATOX irradiated area is 20 x 20 mm² (The size of UPILEX-S 125 μm is 30 x 25 mm², and the ATOX irradiated area is 20 x 25 mm²).

Table 2. List of materials

Materials	Sample ID	Collector plate ID	ATOX irradiation test number
Beta cloth 500F (PTFE-impregnated glass cloth)	B-4	Al-8	#1
	B-5	-	#1 (under vacuum only)
CFRP (Carbon fibers impregnated with mono-component epoxy)	C-4	Al-9	#1
	C-5	-	#1 (under vacuum only)
Ag coated FEP (Silver coated fluoropolymer)	F-5	Al-16	#2
	F-6	-	#2 (under vacuum only)
UPILEX-S 125 μm (Polyimide)	P-1-3	Al-23	#3
	P-1-4	-	#3 (under vacuum only)
RTV-S691 (Silicone)	R-1	Al-25	#3
	R-2	-	#3 (under vacuum only)
MAP® SCK5N (Silicone)	K-1	Al-31	#4
	K-2	-	#4 (under vacuum only)

2.4. Analysis techniques

The ATOX irradiated materials, their reference materials (not ATOX irradiated materials, only under vacuum) were all inspected and evaluated with several analysis techniques mentioned in Table 3. Collector plates were observed with optical microscopy and scanning electron microscope (SEM), and the particulate AIC were analysed by energy dispersive X-ray spectroscopy (EDS).

Table 3. Analysis techniques for materials

	Sample ID											
Analysis techniques	B-4	B-5	C-4	C-5	F-5	F-6	P-1-3	P-1-4	R-1	R-2	K-1	K-2
Cross section observation with SEM	Y	Y	Y	Y	Y	Y	Y	Y	Y	Y	Y	Y
Mass measurement	Y	Y	Y	Y	Y	Y	Y	Y	Y	Y	Y	Y
Profilometry	Y	N	Y	N	Y	N	Y	N	Y	N	Y	N
EDS	Y	N	Y	N	Y	N	Y	N	Y	N	Y	N
FTIR	Y	Y	Y	Y	Y	Y	Y	Y	Y	Y	Y	Y
XPS	Y	Y	Y	Y	Y	Y	Y	Y	Y	Y	Y	Y

3. Results and discussion

3.1. Materials

3.1.1. Observations (Visual Inspection, Cross section observation with SEM)

3.1.1.1. Visual Inspection

Figure 4 shows the photos for visual inspection of materials before and after ATOX irradiation tests. They were obtained by using a digital camera (Canon. Co., EOS kiss X70).
The visual change of surfaces can be confirmed on all of the ATOX irradiated materials compared with the reference materials.

Figure 4. Visual inspection of materials

3.1.1.2. Cross section observation with SEM

Figure 5 shows SEM images of the cross sections of materials. Shape changes can be confirmed with almost of all materials except MAP® SCK5N. In the case of Beta cloth (B-4 and B-5), it could be observed that the fluoropolymer coating was eroded by ATOX and the fiberglass became exposed. Both surfaces of Ag coated FEP (F-5) and UPILEX-S 125 μm (P-1-3) were eroded by ATOX, but the erosion depths were obviously different. The reason is that the difference of erosion yields is depending on the chemical structure of materials. The surface of RTV-S691 (R-1) seems to present more crystal like shape than the reference (R-2). On the other hand, the surface roughness of MAP® SCK5N changed slightly between the ATOX irradiated one (K-1) and its reference (K-2).

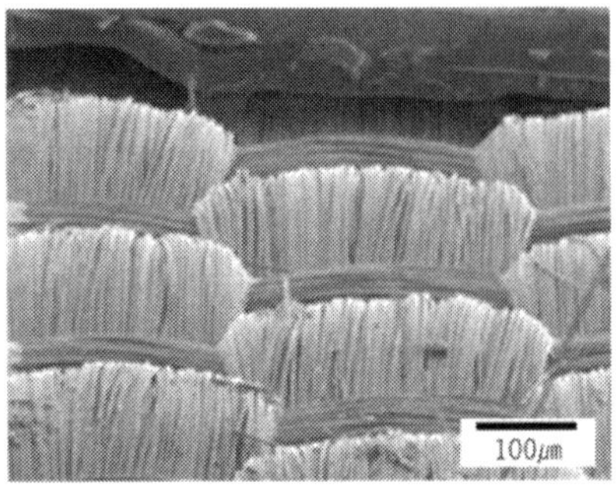

(a) B-4

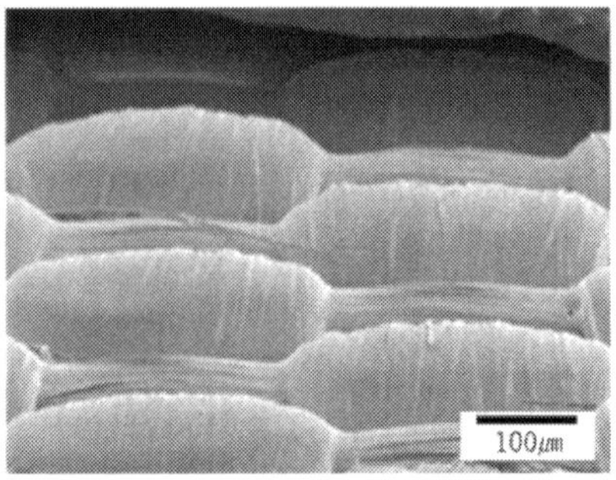

(b) B-5

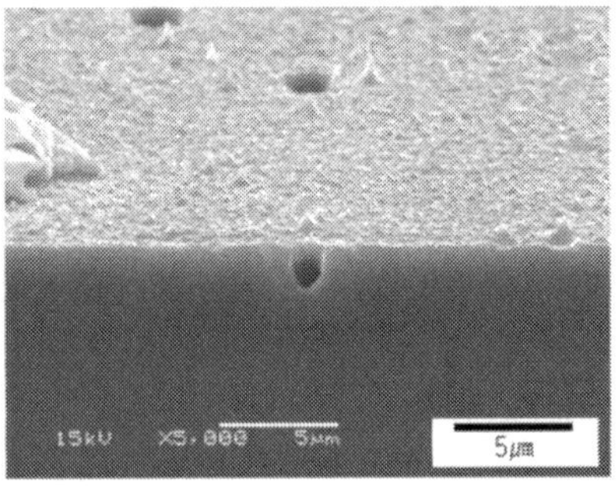

(c) F-5

(d) F-6

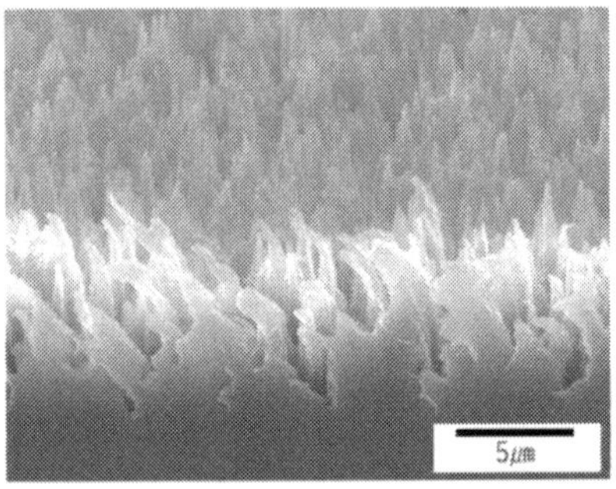

(e) P-1-3

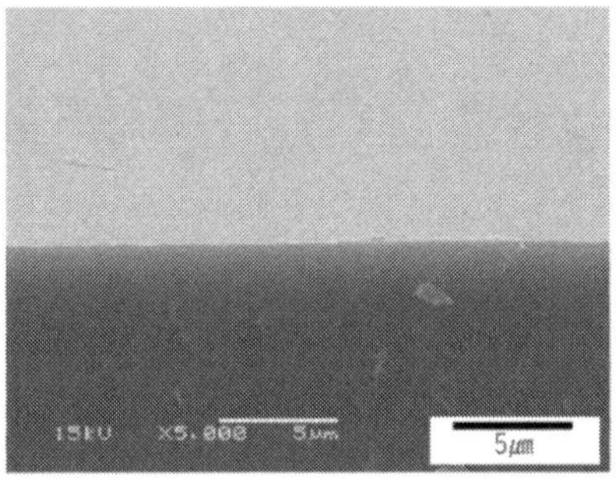

(f) P-1-4

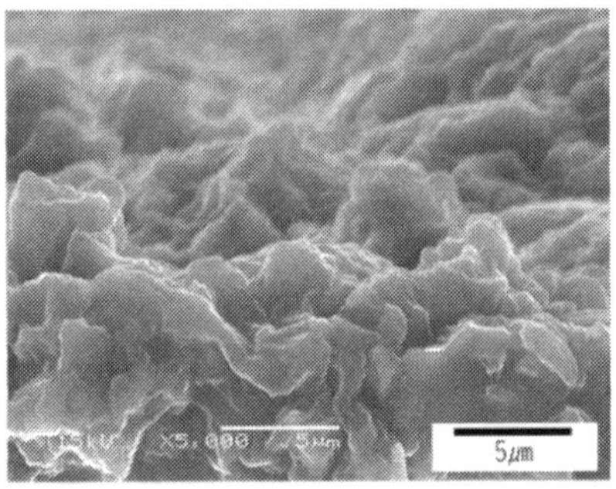

(g) R-1

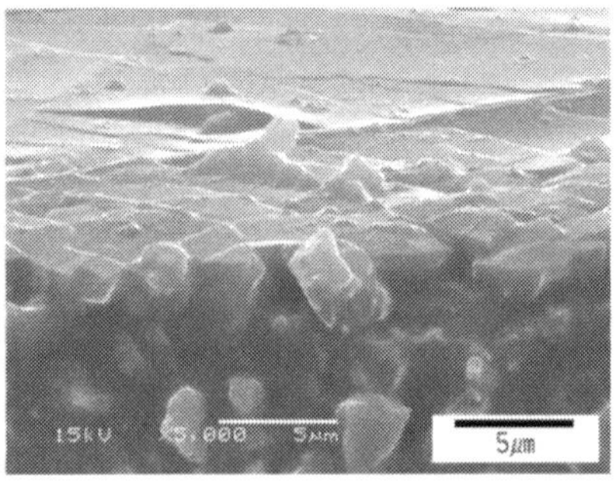

(h) R-2

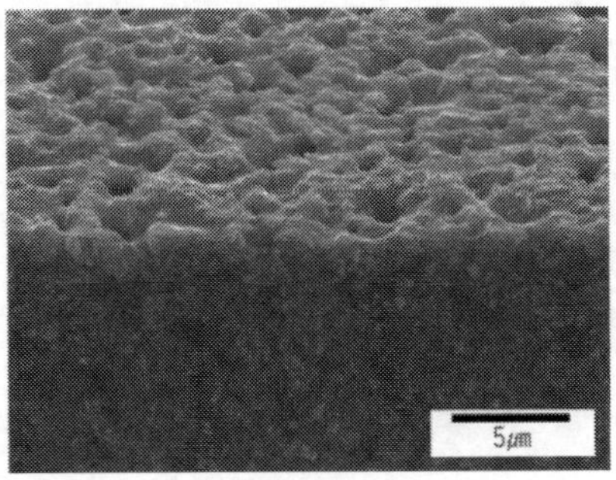

(i) K-1

(j) K-2

Figure 5. SEM images of the cross sections of the materials
(left: after irradiation, right: before irradiation)

3.1.2. Physical Measurements (mass, profilometry)

3.1.2.1. Mass measurements
All materials masses were measured before and after ATOX irradiation by using a microbalance (Mettler Toledo International. Inc., MX6). And the evaluation was made by comparing mass loss between the ATOX irradiated materials and their reference materials to subtract the outgassing effect under vacuum. Tables 4 show the mass loss of the ATOX irradiated materials and their reference materials (not ATOX irradiated materials only under vacuum), and the difference in mass loss between the ATOX irradiated materials and their references.

Table 4. Mass losses of materials

Sample ID	*Mass loss※(mg)*	*Difference in mass (mg)*
B-4	-3.400	-3.371
B-5	-0.029	
C-4	-12.216	-9.943
C-5	-2.273	
F-5	-11.744	-11.733
F-6	-0.011	
P-1-3	-12.704	-12.703
P-1-4	-0.001	
R-1	0.358	0.573
R-2	-0.215	
K-1	-0.011	0.000
K-2	-0.011	

※Mass of material after ATOX irradiation test - mass of material before ATOX irradiation test

- The mass of Beta cloth, CFRP, Ag coated FEP and UPILEX-S 125 μm decreased after ATOX irradiation (see Tables 4). Significant mass losses were measured for CFRP, Ag coated FEP and UPILEX-S 125 μm, and the lowest mass loss was measured for Beta cloth. The reason is considered that the fluoropolymer coating of Beta cloth was eroded by ATOX, on the other hand, the fiberglass of Beta cloth was difficult to be eroded. It can also be seen from the SEM images of the cross sections of Beta cloth before and after ATOX irradiation (see Figure 5 (a) and (b)).
- The mass of RTV-S691 increased after ATOX irradiation that may be a consequence of oxidation of the irradiated material because one of ATOX effect is oxidation when collision with material surfaces (see Table 4).

- No obvious mass change was confirmed on MAP® SCK5N considering the mass loss of its reference (Table 4). And there were only slight changes between the surface roughness of MAP® SCK5N before and after ATOX irradiation by observation of SEM (see Figures 5 (i) and (j)). This suggests that MAP® SCK5N has a stable chemical structure against oxidation.

3.1.2.2. Optical profilometry analysis
Surface metrology was performed directly on the irradiated samples using a non-destructive optical technique with the Sensofar S-NEOX profilometer in confocal mode. Both the hidden and exposed areas on the same sample, and especially the boundary between the two, were considered in order to compare the surface morphology and roughness (see Table 5).
When mechanical erosion occurred caused by ATOX impingement, the height differences can change more or less significantly depending on the nature of the material and steps in X-direction profile can be measured as illustrated in Figures 6 to 11.

- For the polyimide film UPILEX-S 125 μm and the fluorocopolymer Ag coated FEP, the step is well-defined with a height difference of respectively 13.44 μm and 5.10 μm between the top of the non-irradiated layer and the irradiated side, showing a stronger erosion effect for the polyimide film (see Figures 6 and 7).
- On the contrary, there is no obvious step on exposed silicone samples RTV-S691 and MAP® SCK5N (see Figures 8 and 9).
- For the composite materials, the height differences depend on the configuration of the fibers. Regarding the Beta cloth, Figure 10 shows clearly valleys and hills due to the cloth manufacturing process. The highest difference (12.45 μm) is observed when comparing the valleys of reference and exposed sides whereas the height difference between hills is only ≅ 3 μm. We can suppose that this measurement corresponds to the thickness of PTFE coating on the glass fibers that has been completely eroded on the top while a deeper erosion has occurred in the valley due to probably a thicker layer of PTFE according to the way of impregnation of the cloth.

  Regarding the CFRP, Figure 11 shows similarly valleys and hills due to the manufacturing process, under a thin layer of epoxy resin as a final stage of the process in that case. The same behaviour can also be suggested here to explain the height difference, much more important on the matrix side than on the side where the carbon fibers seem to appear (4.94 μm compared to 28.94 μm).

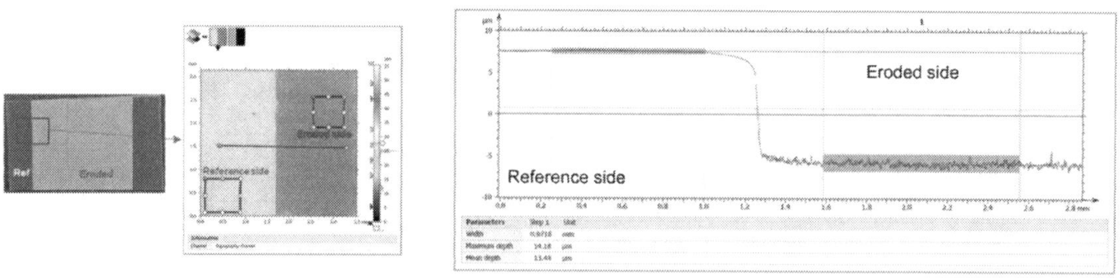

Figure 6. 2D surface texture and X-direction profile on UPILEX-S 125 μm

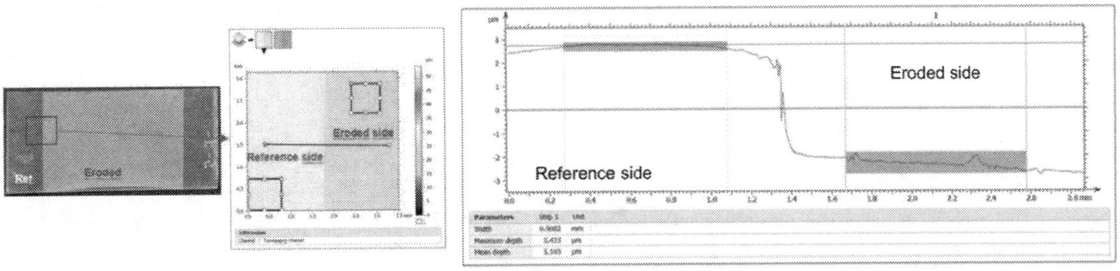

Figure 7. 2D surface texture and X-direction profile on Ag coated FEP

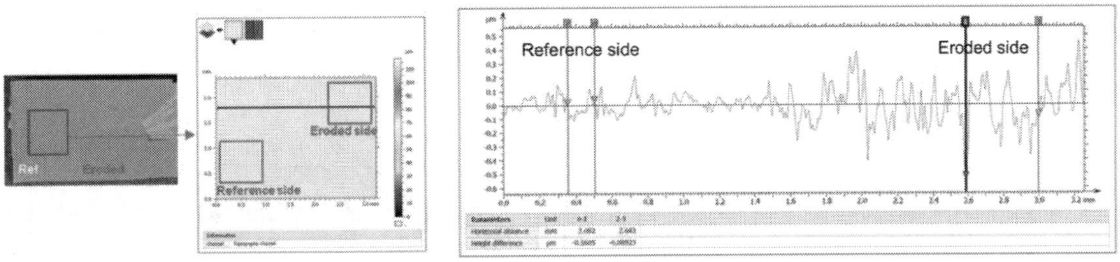

Figure 8. 2D surface texture and X-direction profile on RTV-S691

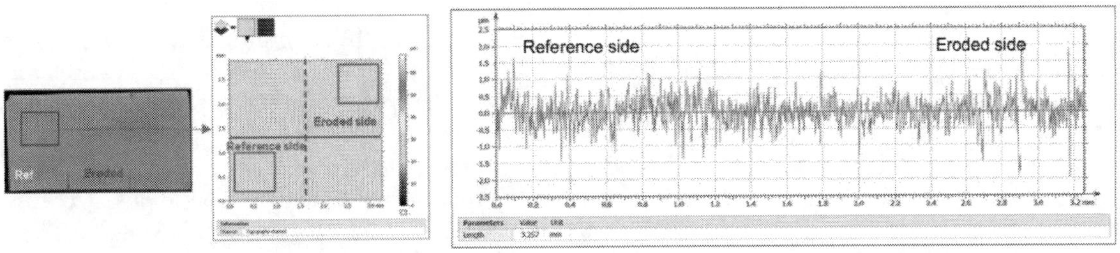

Figure 9. 2D surface texture and X-direction profile on MAP® SCK5N

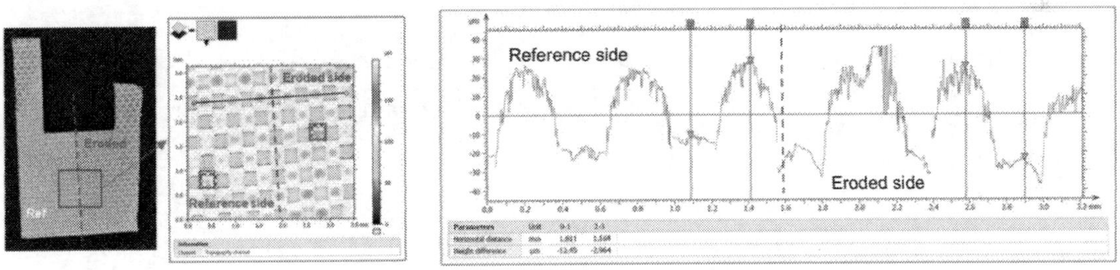

Figure 10. 2D surface texture and X-direction profile on Beta cloth

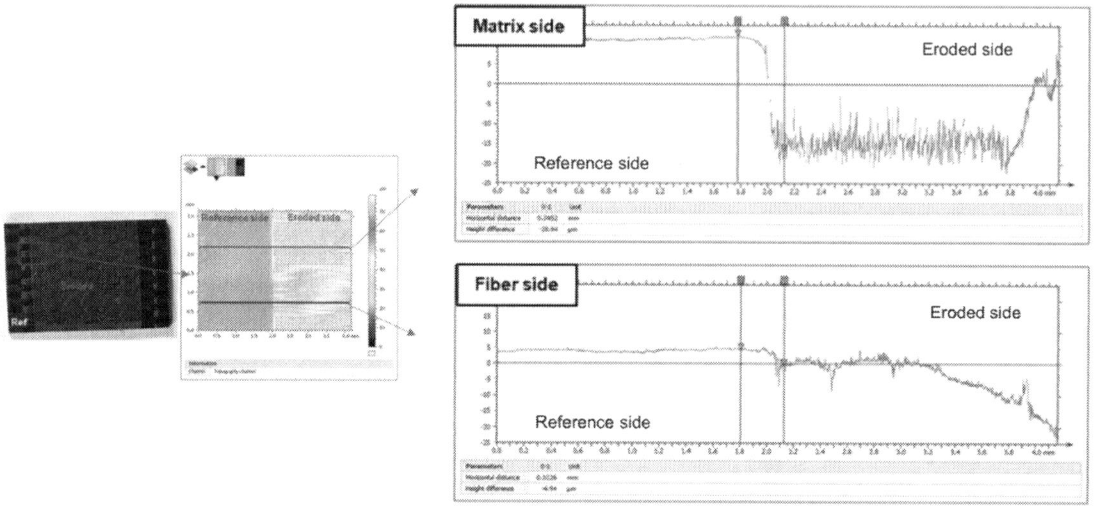

Figure 11. 2D surface texture and X-direction profile on CFRP

The roughness measurements in defined specific zones in the irradiated and non-irradiated areas showed that S_a parameter (arithmetic mean height of a line) is always higher after irradiation for all the tested materials with a much more pronounced surface roughness in the case of UPILEX-125 µm and CFRP (highlighted in red in Table 5 and illustrated in Figures 12 (a) and 12 (b)), confirming the strong erosion impact on the degradation of such polymers. Moreover, on the case of CFRP, oriented striae are clearly visible on the eroded side revealing what it will be confirmed later with the SEM analyses, i.e. the carbon fibers.

Another interesting observation can be noticed from the surface analysis at a larger scale, particularly for the silicone samples. Indeed, a network of new cracks seems to appear and be superimposed to some initial defects on the surface texture of RTV-S691 irradiated sample. Regarding the MAP® SCK5N irradiated sample, the indented lines already existing on the reference side seem to be deeper eroded, meaning that the surface local defects are accentuated.

Table 5. Height difference and roughness between reference and exposed sides

Material	Height difference (µm)	Roughness (s_a parameter) (µm)	
		Reference side	Exposed side
UPILEX-S 125 µm	13,44	0.012	0.657
Ag coated FEP	5,10	0.166	0.171
RTV-S691	non measurable	0.206	0.283
MAP® SCK5N	0,1	0.362	0.508
Beta cloth 500F	hill: 2,96 ; valley: 12,45	1.663	2.593
CFRP	hill: 4,94 ; valley: 28,94	0.195	2.152 (top of composite matrix) 1.854 (top of composite fiber)

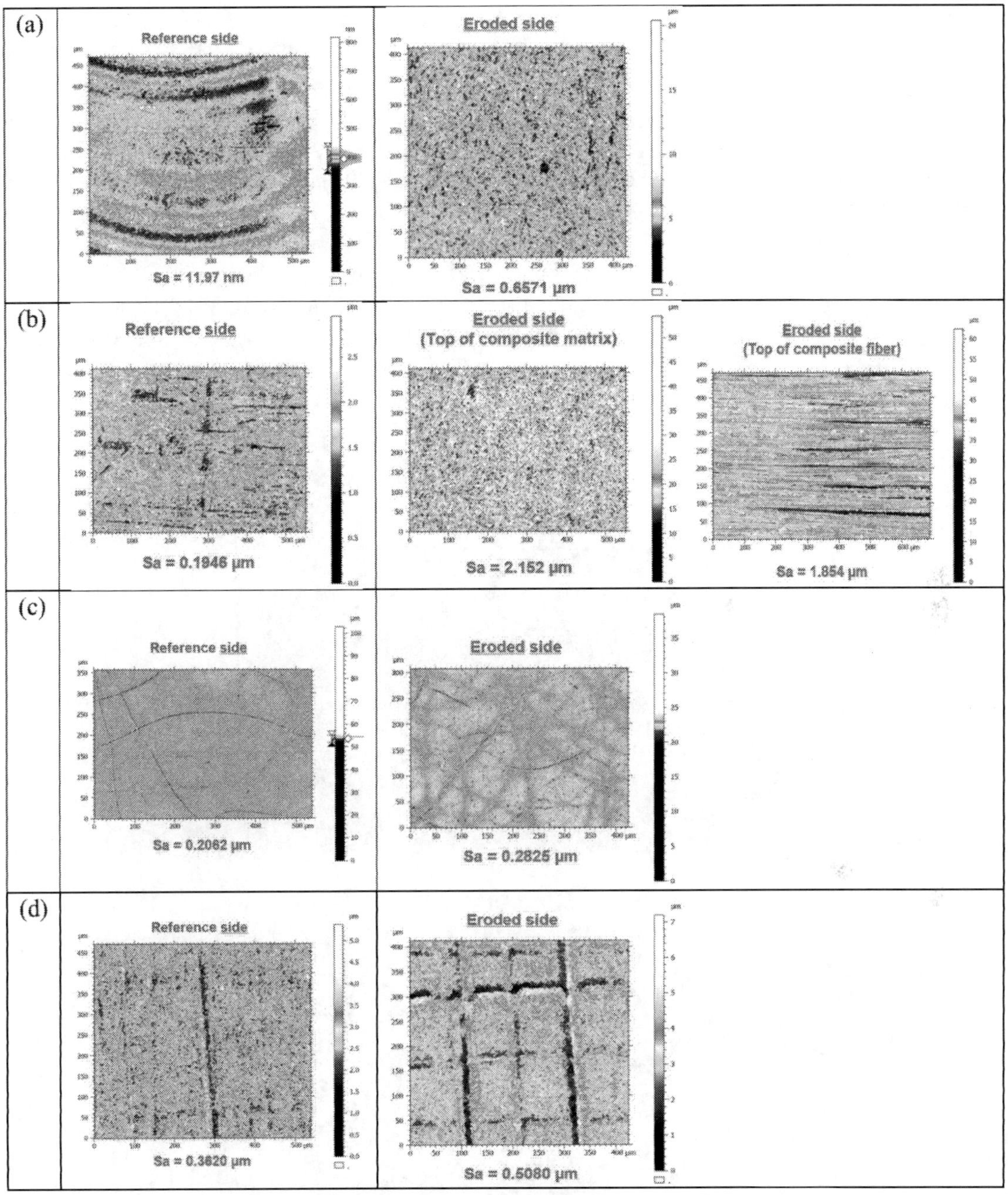

Figure 12. Comparison of roughness measurements on reference and exposed sides of the following tested materials: UPILEX-S 125 µm (a), CFRP (b), RTV-S691 (c), MAP® SCK5N (d)

3.1.3. Chemical Analysis (EDS, FTIR, XPS)

3.1.3.1. SEM observations and EDS analyses

SEM/EDS analyses of the exposed samples were performed with a Hitachi SEM and Oxford EDS spectrometer at a voltage of 15 kV and about 10 mm working distance.

The SEM images, presented below on Figures 13 to 18, were acquired at the boundary between the reference side (top) and the exposed side (down) and the main results of the EDS analysis are summarized as follows:

- The cartography of the composite materials shows clearly that the polymer of impregnation has been eroded: on one hand, for the Beta cloth, fluoride from the PTFE coating disappeared on the exposed side revealing silicon, oxygen and other chemical elements of the glass fibers (e.g. calcium). On the other hand, for the CFRP, the carbon fibers are now apparent on the magnified SEM image of the eroded side even if a very thin film of the epoxy matrix is always visible as it can be seen on Figure 14.
- The composition of the Ag coated FEP and UPILEX-S 125µm on the exposed side is almost similar to the one on the non-irradiated side (see Figures 15 and 16). For both samples, the SEM images show singularly the presence of metallic particles which protected the substrate from erosion so that we can distinguish the eroded depth all around the shaped pillars.
- The composition of the two silicone materials has slightly evolved with a weak increase of the O/Si ratio for RTV-S691 but a decrease of the same ratio for MAP® SCK5N.

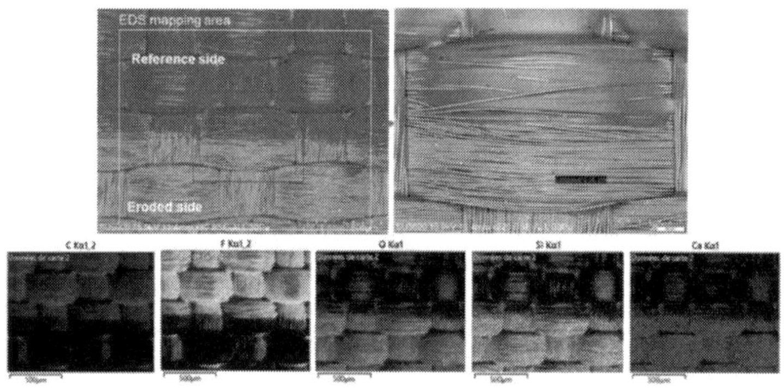

Figure 13. SEM/EDS analysis results on Beta cloth (B-4)

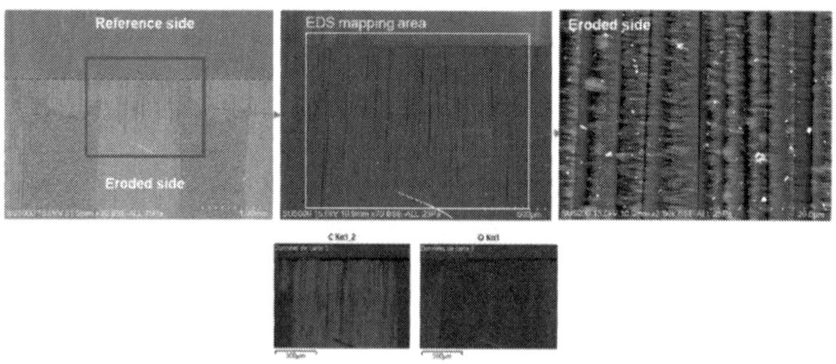

Figure 14. SEM/EDS analysis results on CFRP (C-4)

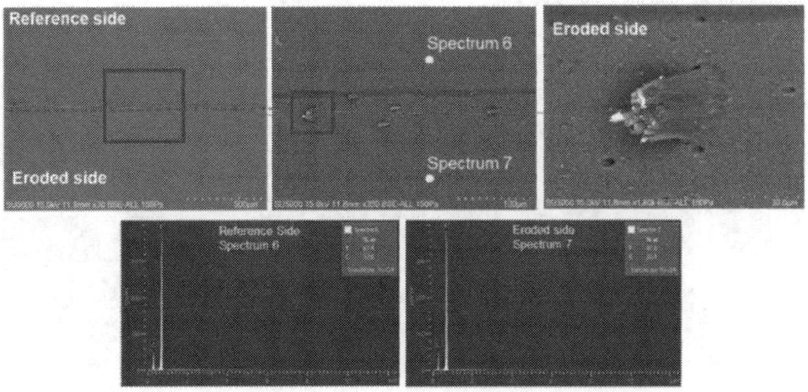

Figure 15. SEM/EDS analysis results on Ag coated FEP (F-5)

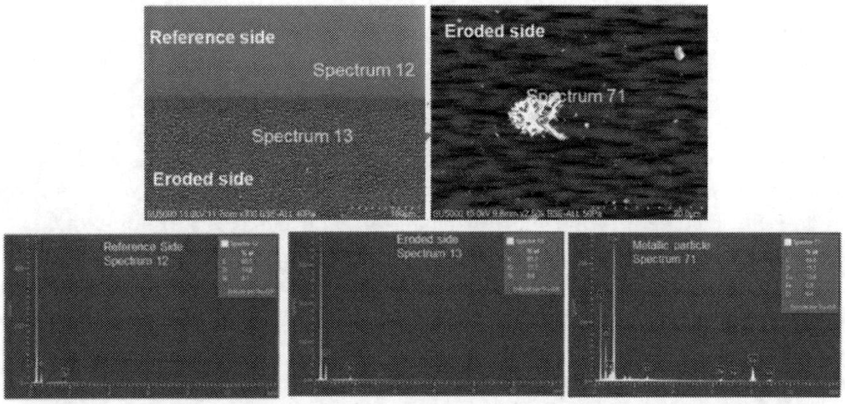

Figure 16. SEM/EDS analysis results on UPILEX-S 125 μm (P-1-3)

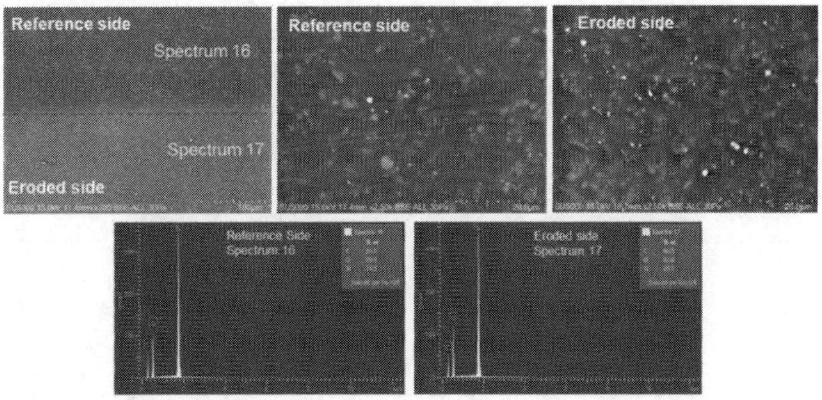

Figure 17. SEM/EDS analysis results on RTV-S691 (R-1)

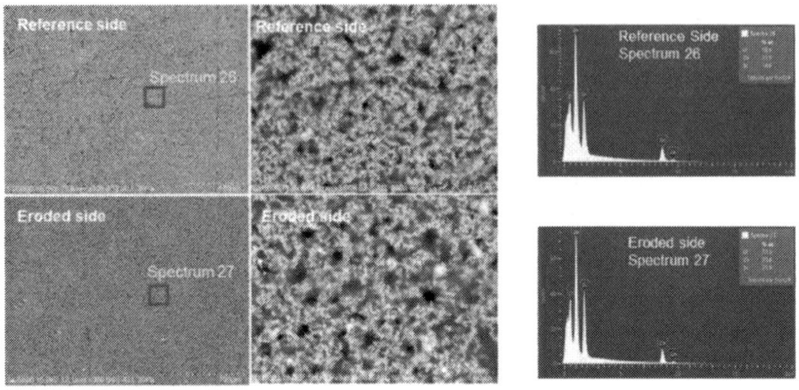

Figure 18. SEM/EDS analysis results on MAP® SCK5N (K-1)

3.1.3.2. FTIR (Fourier Transform InfraRed spectroscopy) analyses

Infrared spectra were obtained using a Nicolet IS10 spectrophotometer in ATR (attenuated total reflexion) mode on both irradiated and non-irradiated samples in the spectral range [650 – 4000 cm^{-1}]. The compared pairwise results are presented on Figures 19 to 21.

For the silicone material RTV-S691 and Ag coated FEP, the spectra can be completely superimposed that confirms there is no chemical change at the surface of the polymer. However, noticeable changes can be seen on UPILEX-S 125 µm and MAP® SCK5N spectra. The corresponding zones have been framed in dotted lines to identify the spectral range of the affected absorption bands, mostly regarding their relative intensities and shoulders e.g.:

- the carbonyl bands in the –CO–N–CO– vibrational group (around 1700 cm^{-1}) and the range of C-N bands (between 900 and 1200 cm^{-1}) for the polyimide UPILEX-S 125 µm,
- the Si-C group (1250 cm^{-1}) for the silicone MAP® SCK5N.

Besides, new absorption bands can appear, such as the O-H large absorption band in the MAP® SCK5N spectrum around 3500 cm^{-1}.

The most significant changes concern the two composite materials spectra. Indeed, for CFRP, all the absorption bands characteristic of the epoxy resin disappeared that means that the thin top layer of the material has been eroded to reveal the carbon of the fibers explaining now the absence of absorption band in the FTIR spectrum of the irradiated zone (graphite spectrum).

In the same way, for the Beta cloth, the two strong absorption bands assigned to $-CF_2-$ vibrational group in the fluoropolymer disappeared and another strong large band around 1000 cm^{-1} appeared. This new band seems to be related to Si-O-Si assymetric stretch that can be found similarly in the spectrum of SiO_2, (slight redshift to be noted) which would confirm the complete erosion of PTFE and thus the exposition of the glass fibers to external environment.

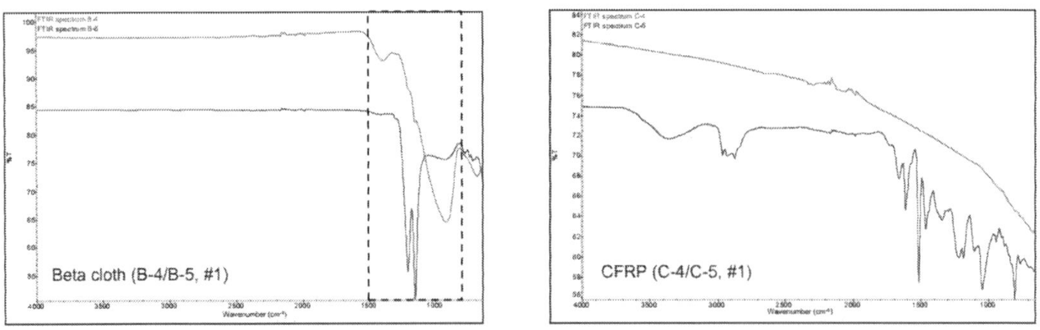

Figure 19. Comparison of FTIR spectra of the irradiated (red) and non-irradiated (blue) samples with significant changes

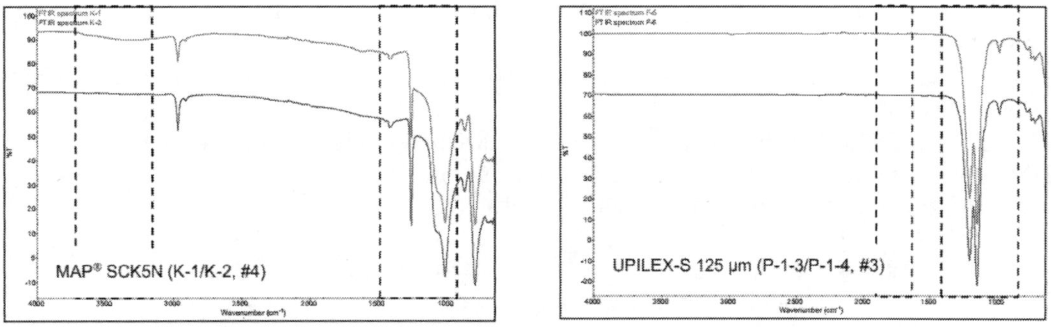

Figure 20. Comparison of FTIR spectra of the irradiated (red) and non-irradiated (blue) samples with noticeable changes

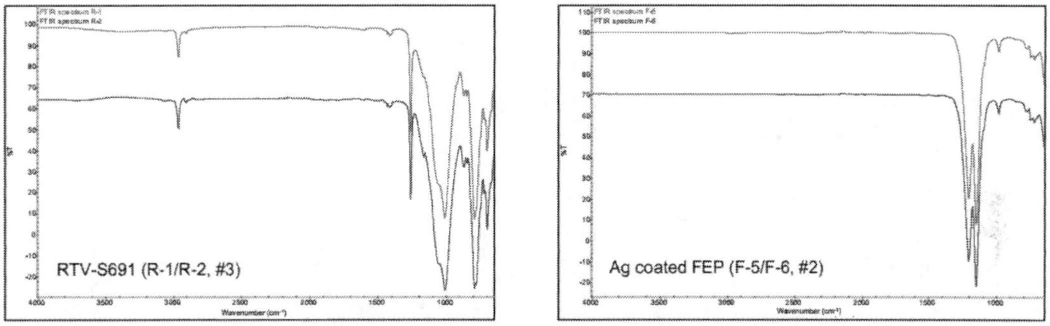

Figure 21. Comparison of FTIR spectra of the irradiated (red) and non-irradiated (blue) samples with no change

3.1.3.3. XPS (X-ray induced Photoelectron Spectroscopy) analyses

In order to determine the etching effect due to ATOX exposure, XPS analyses were performed on the substrates, before and after exposure, by using the K-alpha X-ray Photoelectron Spectrometer with Al Kα monochromator (Thermo Scientific) for high chemical state resolution.

The analytical program was defined as an elemental quantitative analysis and study of the chemical states on two areas of each irradiated and non-irradiated sample with the instrumental specifications below:

- take-off angle: 90°
- depth of analysis: lower than 10 nm (average 3 nm)
- spot diameter: 400 µm
- sensitivity limits: from 0.1 to 0.5 atomic %
- quantitative analysis: accuracy 10 to 20 % ; precision 2 to 5 % (relative) depending on the concentration

The atomic concentrations obtained from high resolution spectra are summarized in Table 6. The chemical environment of detected elements on some specific sample was also investigated to characterize new chemical forms.

Note: As there is a similar composition between not irradiated area and reference substrates, only the results on not irradiated area (hidden area) of the exposed samples are considered.

The XPS results are consistent with the conclusions of the previous analyses and the main noticeable results of the differences in the composition after irradiation are presented hereafter:

- Beta cloth: degradation of the PTFE coating (creation of fluoride form) and revelation of the glass fibers
- CFRP: detection of the graphitic chemical environment for carbon due to epoxy erosion (uncoated carbon fibers)
- UPILEX-S 125 μm: decrease of nitrogen amount, different proportions of C-O bonds
- Ag coated FEP, RTV-S691: weak oxidation of the sample surface
- MAP® SCK5N: siloxane oxidation in silica (see Table 7)

Table 6. Comparison of elemental concentration (at.%) on irradiated (irr) and non-irradiated (not irr) zones of the samples

	C	O	F	N	Si	O/Si
Beta cloth irr (B-4)	22.4	30.4	26.0	0.5	5.9	-
Beta cloth not irr (B-4)	30.0	0.5	69.2	-	0.1	-
CFRP irr (C-4)	66.8	28.4	0.8	1.4	0.8	-
CFRP not irr (C-4)	47.7	37.5	-	4.6	8.4	-
UPILEX-S irr (P-1-3)	44.3	37.8	-	5.1	8.6	-
UPILEX-S not irr (P-1-3)	68.6	22.3	-	8.9	0.2	-
Ag coated FEP irr (F-5)	30.6	0.9	68.4	-	-	-
Ag coated FEP not irr (F-5)	30.3	< 0.1	69.6	-	-	-
RTV-S691 irr (R-1)	40.1	32.2	1.0	-	25.9	1.2
RTV-S691 not irr (R-1)	45.8	28.6	0.1	-	25.4	1.1
MAP® SCK5N irr (K-1)	12.0	57.1	0.2	0.2	28.5	2.0
MAP® SCK5N not irr (K-1)	13.7	56.4	1.0	0.3	28.3	2.0

Table 7. CSi2p photopeak synthesis for MAP® SCK5N substrates

	Siloxane SiOC	Silicate $SiO_{x\ (x<2)}$	Silica SiO_2
MAP® SCK5N irr (K-1)	4	-	96
MAP® SCK5N not irr (K-1)	3	12	85

Note: Different elements as contaminants have been detected after irradiation: minerals and/or metals and probably cross contamination during the irradiation test.

3.2. Particles generated from materials by atomic oxygen

3.2.1. Observation (Optical Microscopy, SEM)

3.2.1.1. Optical Microscopy

Many particles of various morphologies (fibers and particles) and a large size range (from approximately 1 to 1000 μm) were observed on the collector plates located just below the irradiated samples (see Figure 22). The optical microscopy (KEYENCE. Co., VHX-900) was used for observation.

Figure 22. Examples of fibers on Al-8 (close to Beta cloth)

3.2.1.2. SEM /EDS analyses

The size of the particles and fibers on collector plates were measured more precisely with a Hitachi SEM and their chemical nature were identified with a Oxford EDS spectrometer at voltage 15 kV and about 10 mm working distance as for the materials themselves. These analyses showed that:

- No AIC was generated from UPILEX-S 125 µm and RTV-S691.
- Rare particles with fluorine element were collected on Al-16 coming from the fluoropolymer sample Ag coated FEP (see Figure 23).
- Rare particles of SiO_2 were collected on Al-31 coming from the irradiated silicone coating MAP® SCK5N (see Figure 24).
- Numerous particles, mainly fibers of several hundreds of µm long (up to 1 mm for some of them) were found on collector plates Al-8 and Al-9 located respectively under Beta cloth and CFRP irradiated samples as displayed on Figures 25 and 26. The chemical identification of their composition is unambiguous and confirm the origin of the fibers as AIC from the eroded and now weakened composites: SiO_2 curved and straight fibers from Beta cloth (revealed after the complete erosion of the PTFE coating) and straight carbon fibers from CFRP (revealed after the erosion of the epoxy resin coating). Their diameter (4 to 5 µm) is also consistent with the diameter of the eroded fibers remaining on the irradiated samples of the two kinds of composites, as it was easily checked on SEM images.

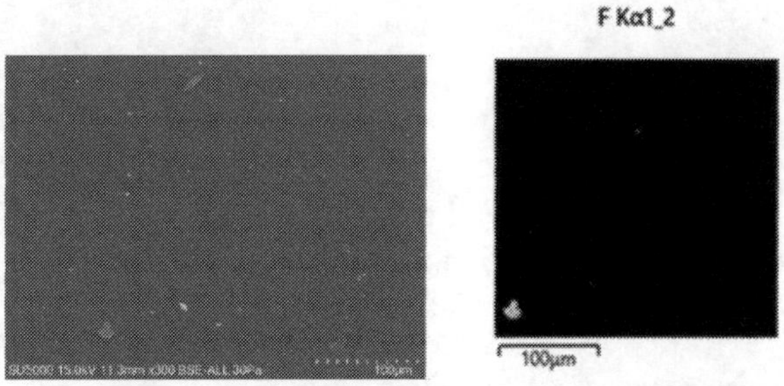

Figure 23. Particles on Al-16 (left) and corresponding chemical composition (Fluorine) (right)

IOP Conf. Series: Materials Science and Engineering 1287 (2023) 012029 doi:10.1088/1757-899X/1287/1/012029

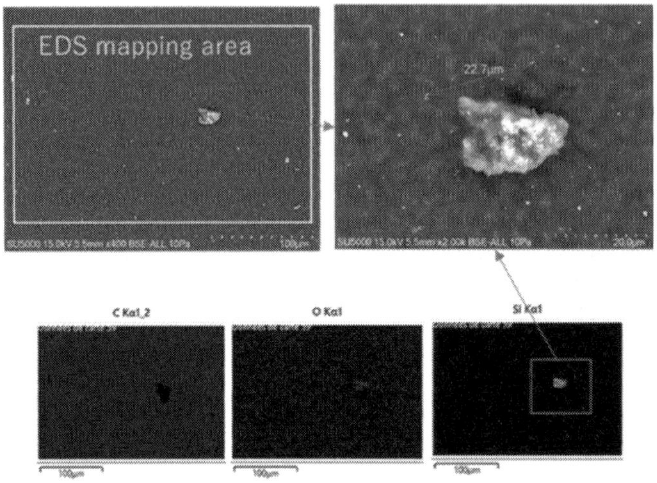

Figure 24. Particle on Al-31 (left) and chemical identification (SiO_2) (right)

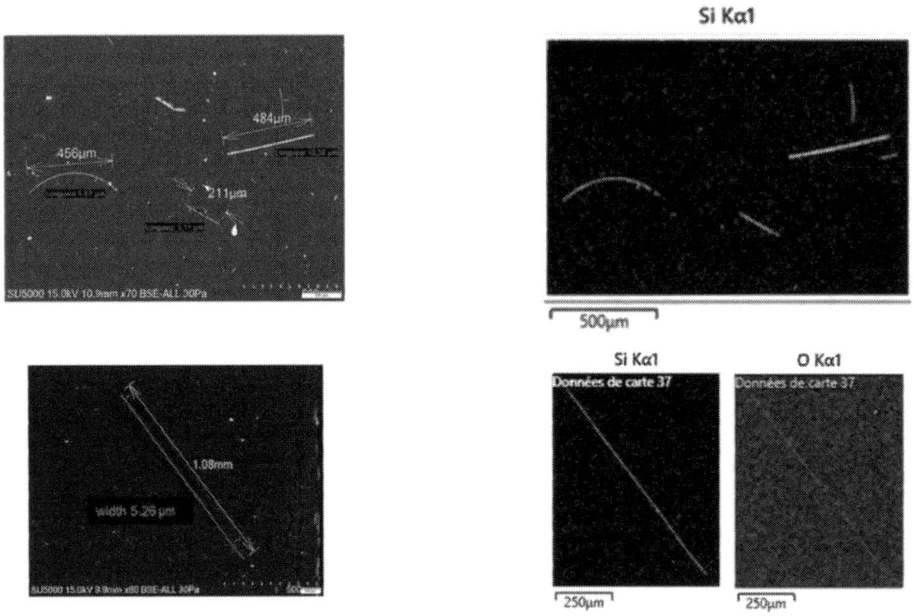

Figure 25. Curved and other straight fibers of different lengths on Al-8 (left) and corresponding chemical composition (Silicon and Oxygen) (right)

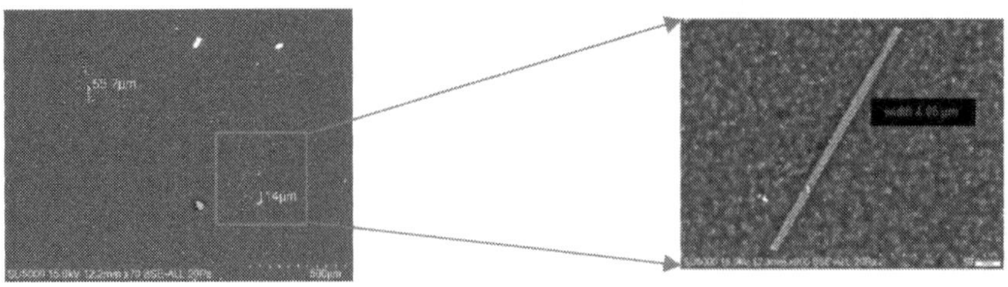

Figure 26. Straight fibers of different lengths on Al-9 (left) and corresponding chemical composition (Carbon) (right)

4. Conclusion

This study was focused on the generation of particulate contamination by ATOX erosion.

Four test campaigns in JAXA facility were performed with different polymers of interest (silicones, polyimide, fluoropolymer, composites with glass and carbon fibers). A special jig was manufactured in order to collect the contaminants on adhesive glued to aluminum plates. Both, irradiated samples and collected particles were observed and analyzed by different techniques (mass measurement, SEM/EDS, XPS, FTIR, profilometry).

Globally, the results were consistent and aimed at demonstrating the origin of the contaminants. Among the tested materials, the composites were the only impinged materials that generated particulate contamination as fibers. The PTFE coating of Beta cloth and the polymer matrix of CFRP were eroded, thus revealing the material of the fibers, respectively glass and carbon, more resistant to ATOX. For the other materials, very rare particles were found on associated collector plates and most of them were not linked to erosion.

With the same objective, another tests campaigns will be conducted to study this time the erosion of contaminated samples by outgassed products under different conditions (different concentrations of contaminants, exposure under UV).

As a perspective, a reflection will be carried out to systematize and standardize a test configuration and protocol for ATOX irradiation of materials in order to well identify and quantify the generated contaminants, not only particulate but also molecular. Besides, a database could be developed to compile all the results regarding the impact of ATOX on materials (erosion yield, AIC…) and allow their access to the space community.

Moreover, as a new opportunity of LEO experiment was offered in 2021 through the European mission called Euro Material Ageing (EMA) [5], some of the materials of this study were proposed, especially the composites, to be exposed to space environment and particularly to ATOX. Materials samples as well as collector tapes will be integrated on a dedicated structure and placed in orbit on the Bartolomeo platform of the International Space Station for a minimum of 6 months (launch date scheduled at the beginning of 2024). After retrieval, the samples and collector tapes will be characterized with identical analytical techniques to compare the degradation and AIC to on-ground results.

Acknowledgements

The authors would like to thank the manufacturers Mecano-ID, MAP Coatings and the CNES thermal Department for providing the different materials for the irradiation tests. They are grateful to ELEMCA, THALES SIX GTS FRANCE SAS and Science & Surface for respectively profilometry and EDS analyses, FTIR analyses and XPS measurements and for their helpful comments. The authors also wish to thank the staff of Advanced Engineering Services Co., Ltd., for their support in this study.

References

[1] Faye D 2005 *Space Techniques and Technology Course* vol 1 ed CEPADUES pp 337-340

[2] Yamanaka R 2017 Microparticle space debris derived from silicone contaminants affected with Ultraviolet and Atomic Oxygen *PhD thesis* Tokyo Institute of Technology

[3] Minton T K 1995 Protocol for Atomic Oxygen Testing of Materials in Ground-Based Facilities *JPL Publications* 95-17

[4] Shimamura H, Baba S and Miyazaki E 2011 Handbook of the Combined Space Effects Test Facility *JAXA-RM-10-013*

[5] Suliga A, Wessing J, Tighe A, Vincent-Bonnieu S, Hand A and Rampini R 2022 Euro Materials Ageing – Ground-based testing and selection of the ESA flight candidate materials *15th ISMSE & 13rd ICPMSE* (accepted for publication in the proceedings)

IOP Conf. Series: Materials Science and Engineering 1287 (2023) 012030 doi:10.1088/1757-899X/1287/1/012030

Commercial optical coatings in space

Johanna J. Wessing[1]*, Bruno Bras[1], Nuno Dias[1], Adrian Tighe[1], Riccardo Rampini[1]

[1] European Space Research and Technology Centre, Keplerlaan 1, 2201 AZ Noordwijk, the Netherlands, +31 (0)71 565 6565

(*) johanna.wessing@esa.int

Abstract. The use of commercial optical coatings for space applications for the aptly named "New Space" missions is a state-of-the-art trend. These products are generally less costly and time-consuming in their production and therefore, there is a much wider range of optical coatings available compared to custom-made products. However, the reliability of commercial manufacturing processes is a key concern. Additionally, the resistance of commercial products against space environmental degradation mechanism at operating temperatures needs to be evaluated. In this study, seven commercial products with multi-layer thin film optical coatings consisting of filters, mirrors and windows have been selected from various manufacturers and the results with regards to process verification and space environmental exposure testing are presented. The impact of space environmental effects on the optical performance of the coatings is evaluated, with particular emphasis on the performance at cryogenic temperature. Finally, guidelines and precautions for use of commercial optical coatings for space applications.

1. Introduction

Thin film optical coatings are used for a variety of space applications to enhance the functionality of optical components, as well as for thermal control. The development of custom-made products can be time-consuming and costly and there is generally less choice available for the instrument designer. In comparison, there is a much wider range of commercial optical coatings available, they are manufactured in larger quantities, and they can be considerably more cost efficient.

The use of such coatings is therefore becoming more widespread in the context of "New Space" missions. These missions consist of small of satellites, CubeSats based on standardized units with efficient and cheap productions [1].

However, for commercial products, the reliability of the manufacturing process is a key concern. It can be difficult to trace back the specific production batch, as well as ensure the quality and reproducibility of the relatively few samples typically required for space applications.

Different types of coatings are utilized, such as high reflectors, anti-reflection coatings, band-pass filters, polarizers. In some cases, the optical coating is exposed directly to the space environment. Consequently, coating degradation mechanisms related to exposure to the space environment including radiation, atomic oxygen (ATOX), temperature fluctuations and ultrahigh vacuum may occur over the mission duration.

ATOX is considered as one of the most serious hazards to spacecraft materials and is present at altitudes between 200 and 700 km. It is produced by the photo-dissociation of molecular oxygen in the upper atmosphere by solar radiation. Its effects include a number of processes, including chemical reactions on the materials surface and the formation of volatile and spalling oxides [2]. Some materials are known to be highly affected by ATOX, including silver which is a common material used for mirrors [3]. Ultraviolet (UV) and vacuum ultraviolet (VUV) light can also have significant degradation effects on materials used for space applications in solar exposure, e.g., orbiting spacecraft. It can lead to surface erosion and a change of the solar absorptance visible in the material's darkening [4]. Temperature fluctuations and high vacuum can enhance the material degradation additionally and synergistically. This can not only have an influence on the temperature of the component and entire spacecraft which is depending on the ratio of the solar absorptivity to its emissivity but can lead to performance loss and even failure of optical components such as windows.

Optical coatings often consisting of multiple layers ("stacks") of thin film materials, can be especially vulnerable to these mechanisms. A change or even failure of the optical performance of said materials, such as a reduced transmittance of a window might be the consequence. Additionally, the performance of a coating at operating temperature needs to be known. For infrared sensing, often operated at cryogenic temperatures, the decreasing temperatures can lead to a change in the materials properties, such as the refractive index, and thus, to a shift in the spectral response of filters [5].

For the use of commercial optical products for space applications it would therefore be important to evaluate the potential of these products with regards to their space environmental performance, as well as to verify the reliability of commercial manufacturing processes. Without endorsing specific products or manufacturers, general guidelines and precautions for the use of commercial optical coatings for space applications would be very beneficial for future missions.

2. Materials and coatings

Seven commercial optical products with multiple layers ("stacks") of thin film materials, consisting of filters, mirrors and windows were selected from various manufacturers, including Thorlabs, Spectrogon and Edmund Optics. The samples were chosen to represent various types and applications of mass-produced, off-the-shelf products, ranging from low-cost to middle-priced products all below 400 Euros per piece. An overview of the products, including their application is presented in Table 1. The last column in the table provides some information about the composition of the samples. However, the information provided by the manufacturers is normally very limited and treated without public disclosure.

Table 1. Overview over selected commercial optical products and their applications.

Sample no.	Sample type	Application	Applicable wavelength range	Composition (as available)
1	Filter	Long-pass IR filter	Block-out 200 – 3600 nm, transmittance 3.78 - 7.20 μm	Ge substrate
2	Filter	Narrow band-pass filter	Narrow band transmittance at 6415 nm, half width (HW) of 25 nm	Ge substrate
3	Filter	IR band-pass Filter	Center Wavelength (CWL) of 3.33 μm, Block-out 0.2 - 12 μm	Fused silica substrate
4	Window	IR Fresnel window	Transmittance 8000 – 14000 nm	Polymer film
5	Window	NIR II window	Transmittance 750 - 1550 nm	Fused silica with anti-reflectance coating
6	Window	Broadband Precision Window	Transmittance 3 - 5 μm	Si with anti-reflectance coating
7	Mirror	Reflecting mirror	Angle-depending reflectance of >97.5% for 450 nm - 2 μm and >96% for 2 - 20 μm	Protected silver mirror on fused silica substrate

IOP Conf. Series: Materials Science and Engineering 1287 (2023) 012030 doi:10.1088/1757-899X/1287/1/012030

Figure 1. Photographic pictures of the seven selected commercial optical coatings used for this study, consisting of (in this order from left to right) three types of filters, three types of windows and one type of mirror.

3. Testing

All samples have been tested against degradation mechanisms caused by space environmental conditions such as radiation, temperature fluctuations, atomic oxygen and vacuum. To evaluate these effects, the influence on the optical performance was a key criterion.

3.1. Optical performance

The optical performance of all samples was evaluated with regards to their reflectance, solar absorptance and thermal emittance. In addition, transmittance measurements were performed on transparent and partially transparent samples, including windows and filters. Diffuse reflectance and transmittance spectra were measured to provide insights into surface changes such as an increased surface roughness caused by degradation mechanisms and consequently potential performance loss due to straylight.
For each environmental test the optical performance parameters were evaluated before (BOT) and after (EOT) the test.

To evaluate the reliability and reproducibility of the manufacturing process, the optical performance of multiple samples of the same type and the two sides of uniformly designed samples were compared.

3.1.1. Reflectance and solar absorptance, α. The optical performance of all samples was evaluated with regards to the reflectance and solar absorptance. The total and diffuse reflectance spectra were obtained with the Agilent Cary 5000 A bench-top spectrophotometer with an integration sphere set-up between 2500 and 250 nm for the UV/VIS/NIR range under ambient conditions and corrected against calibrated specular standards. For the determination of the solar absorptance, α, the integrated total reflectance spectra weighted against the integrated sun spectrum in summation with the solar absorptance equals one. Additionally, the reflectance spectra were evaluated for the 2-25 μm NIR/IR range with the Vertex 70v FTIR (Fourier-transform infrared spectroscopy) vacuum spectrometer.

3.1.2. Thermal emittance, ε. For the determination of the thermal emittance, ε, the SOC ET 100 Emissometer was utilized. The integrating sphere captures the reflected light from the target material, integrating reflections in all directions. The data is used to calculate directional thermal emittance and derive the total hemispherical emittance.

3.1.3. Transmittance. The total and diffuse transmittance spectra were obtained with the Agilent Cary 5000 B spectrophotometer for the UV/VIS/NIR range for 2500 and 250 nm. Additionally, the transmittance spectra were evaluated for the 2-12.5 μm NIR/IR range with the Vertex 70v vacuum FTIR spectrometer calibrated with a baseline measurement.

For infrared sensing applications operated at cryogenic temperatures, the NIR/IR transmittance was measured in dependence of the temperature ranging from ambient to cryogenic temperatures down to approximately 4 K. To achieve this, a Vertex 70v vacuum FTIR spectrometer coupled with an OptistatDry cryostat was utilized. The temperature control is performed with the Mercury iTC instrument with a heater in the sample's vicinity. The silicon diode temperature sensor attached to the sample frame was calibrated to the sample and cryostat temperature before the measurement (compare Figure 2).

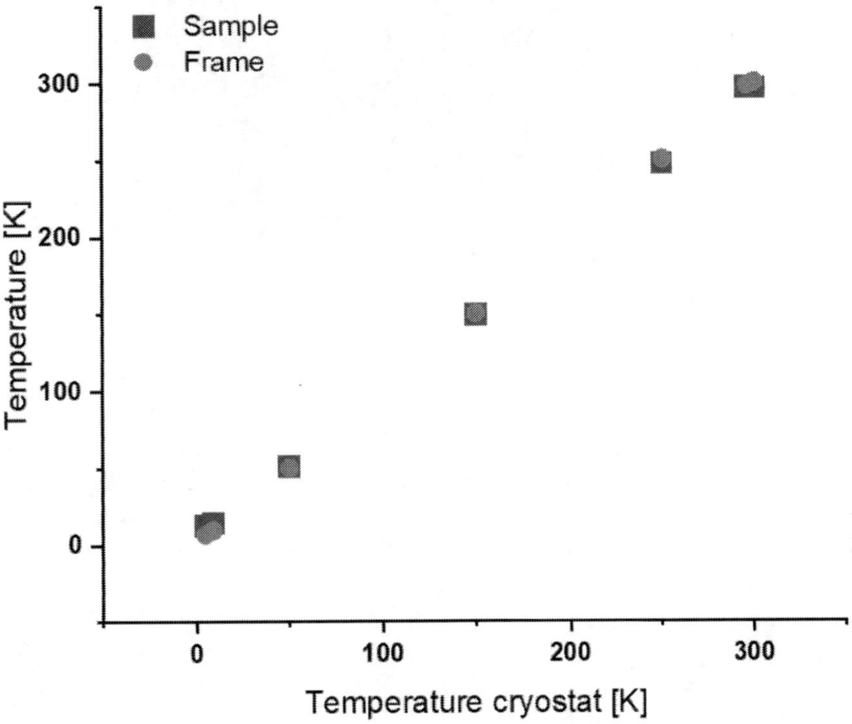

Figure 2. BOT temperature calibration for the cryostat-coupled FTIR transmittance measurements.

3.2. Atomic oxygen

The atomic oxygen testing is performed in the LEOX facility in the ESTEC's Materials & Electrical Components Laboratory. The LEOX facility simulates the near-Earth space environment by generating atomic oxygen at energy levels that are equivalent to an orbital speed of 7.8 km/s.

Four samples were included in the LEOX test, of which three can be seen in Figure 3 mounted on the sample holder. The exposure duration amounted to 90 hours total. As the exposure conditions vary slightly depending on the sample allocation, the details for each sample are given in Table 2 below. Visual inspection, mass analysis and thermo-optical measurements were conducted BOT and EOT the test as described in the previous section 3.1.

Table 2. Test conditions of the ATOX test in the LEOX facility.

Sample no.	Sample description	Fluence [Atoms/cm²]	Flux [Atoms/cm² s]
4	IR Fresnel window	9.59E+20	2.93E+15
5	NIR II window	1.08E+21	3.28E+15
6	Broadband Precision Window	1.11E+21	3.38E+15
7	Reflecting mirror	7.76E+20	2.37E+15

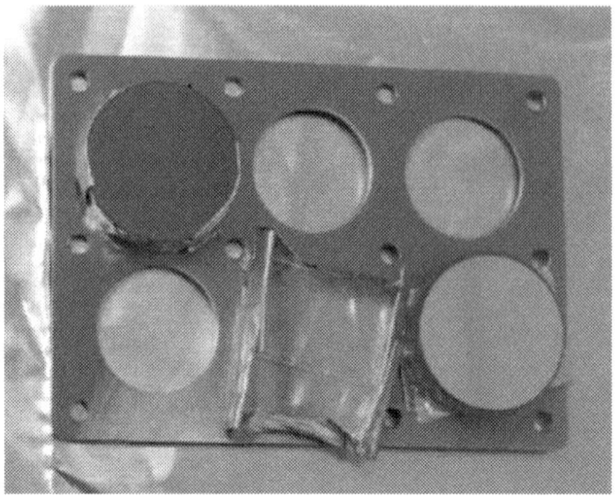

Figure 3. LEOX test samples attached on sample holder.

3.3. Thermogravimetric analysis (TGA)

TGA was performed on the IR Fresnel window in nitrogen atmosphere with a flow rate of 70ml/min. The polymer film has the lowest the temperature resistance of the selected test samples with degradation expected above approximately 100°C. An eight-step heating profile (compare Figure 4) was created to investigate its thermal stability and establish thermal boundary conditions for other tests, e.g., thermal cycling.

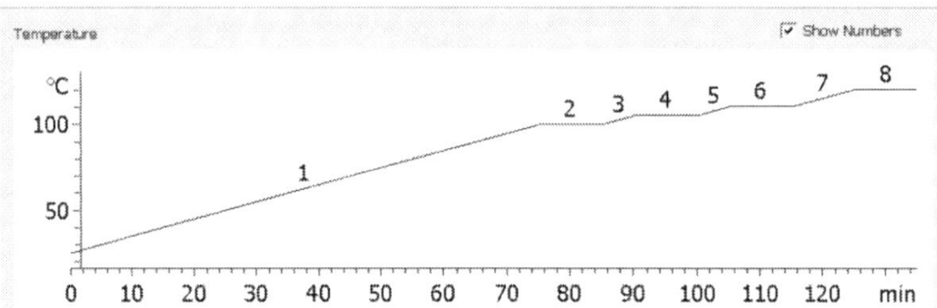

Figure 4. Temperature profile of the TGA performed on the polymer film IR Fresnel window. Isothermal steps 2-8 were included from 100 to 120°C.

3.4. UV/VUV and Thermal cycling

For the UV/VUV testing and thermal cycling, the CROSS 1 facility in the ESTEC's Materials & Electrical Components Laboratory was used. Visual inspection and thermo-optical measurements were conducted before (BOT) and after (EOT) the test, in accordance with the test procedures described in the previous section 3.1.

One sample of each type of commercial optical product was tested and mounted to the sample holder plate, as can be seen in Figure 5.

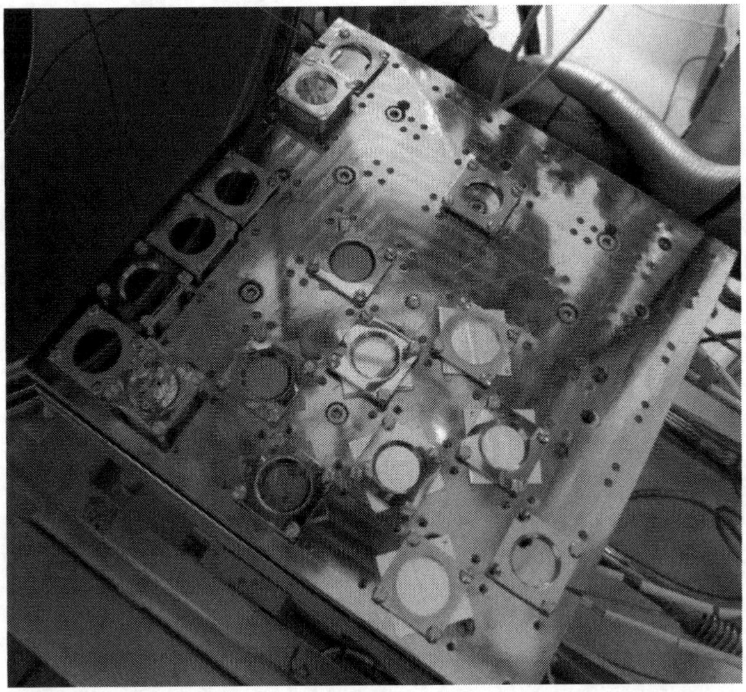

Figure 5. Sample holder plate of the Cross1 facility with mounted commercial optical samples.

For the test, the UV intensities for the UVA, UVB and UVC ranges were calibrated and compared to the solar spectrum and averaged to calculate the acceleration factor expressed in solar constants (SC). The exposure conditions can be seen in the following Table 3. The acceleration factor of acc = 3.73 solar constants for a total UV radiation test duration of 592 hours leads to 2208 equivalent sun hours (ESH) over 2 exposure cycles.

Table 3. Test conditions of the UV/VUV test in the Cross1 facility.

Total exposure [hs]	Radiation intensity [solar constant, SC]	Equivalent sun hours, EHS	Vacuum conditions [mbar]
592	3.73	2208	$<10^{-4}$

After the UV radiation test and second inspection point, thermal cycling in the same facility and vacuum conditions was performed. In Figure 6, the schematic set-up of the sample holder is shown, demonstrating how the sample plate can be heated through the thermo coax and cooled through the liquid nitrogen cooled base plate. Thermocouples in multiple locations, including the sample plate provided accurate temperature monitoring.

The thermal cycles consisted of three cooling and three heating intervals manually set. The set plateau temperatures for the cooling and heating were -60°C and 90°C for a duration of approximately 24-30 hours, respectively.

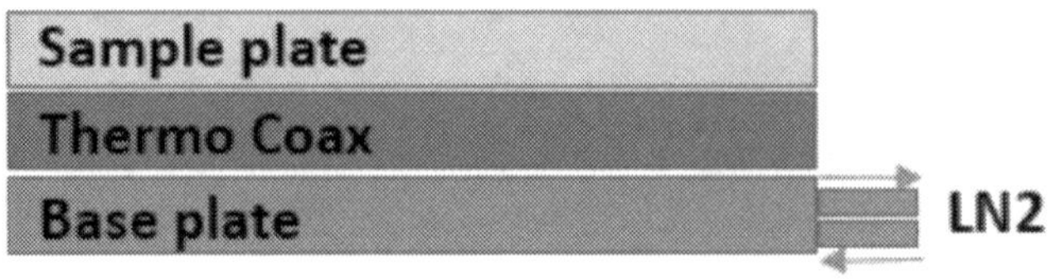

Figure 6. Schematic side view of the Cross 1 sample holder, with the sample plate on top of the thermo coax used for heating and the LN2 cooled base plate used for cooling of the samples.

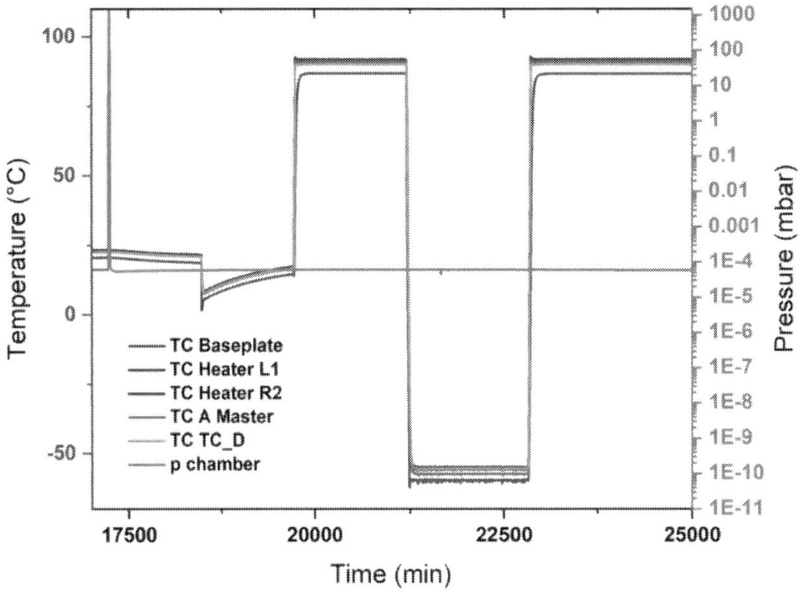

Figure 7. Temperature and pressure data of the thermal cycling test. The temperature is measured with thermocouples (TC) in different locations, including the base plate.

4. Results

4.1. Process verification

To evaluate the quality and reproducibility of the samples in the scope of this study the optical performance of multiple samples has been evaluated, as well as the two sides of samples in cases where the sides are supposed to be manufactured uniformly. For standard deviations exceeding the experimental uncertainty of ± 0.05 for the thermal emittance, ε, and ± 0.03 for the solar absorptance, respectively, the reproducibility cannot be confirmed. Overall, the results are promising as only sample 2 in Table 4 shows a standard deviation above the experimental uncertainty for the thermal emittance.

However, special attention also needs to be paid to the thermo-optical spectra. For specific applications, minor changes in the onset of peaks or intensity of the reflectance or transmittance spectra might have a crucial impact on the respective performance. These deviations might not be reflected in the solar absorptance representing integrated values over a certain wavelength range.

Table 4. Mean, x̄, and standard deviation, σ, of the thermo-optical values of multiple samples of the same type.

Sample no.	Sample description	Number of samples	Mean x̄ ± σ BOL (Beginning Of Life, air)	
			Solar absorptance, α	Thermal emittance, ε
1	Long-pass IR filter	2	0.55 ± 0.00	0.81 ± 0.00
2	Narrow band-pass filter	2	0.71 ± 0.01	0.43 ± 0.26
3	IR band-pass Filter	2	0.69 ± 0.00	0.51 ± 0.03
4	IR Fresnel window	10	0.62 ± 0.01	0.92 ± 0.04
5	NIR II window	6	0.79 ± 0.02	0.80 ± 0.02
6	Broadband Precision Window	5	0.51 ± 0.00	0.80 ± 0.00
7	Reflecting mirror	10	0.05 ± 0.00	0.02 ± 0.00

As an example, the transmittance spectra of the theoretically identical multiples of the NIR II window, sample no. 5 are shown in Figure 8. Sample 5a and 5b, as well as sample 5c-5e seem to have almost identical spectra, respectively. These seem to refer to two different batches of samples with minor deviations within one batch but distinguishably deviating spectra with up to 15% difference between the batches.

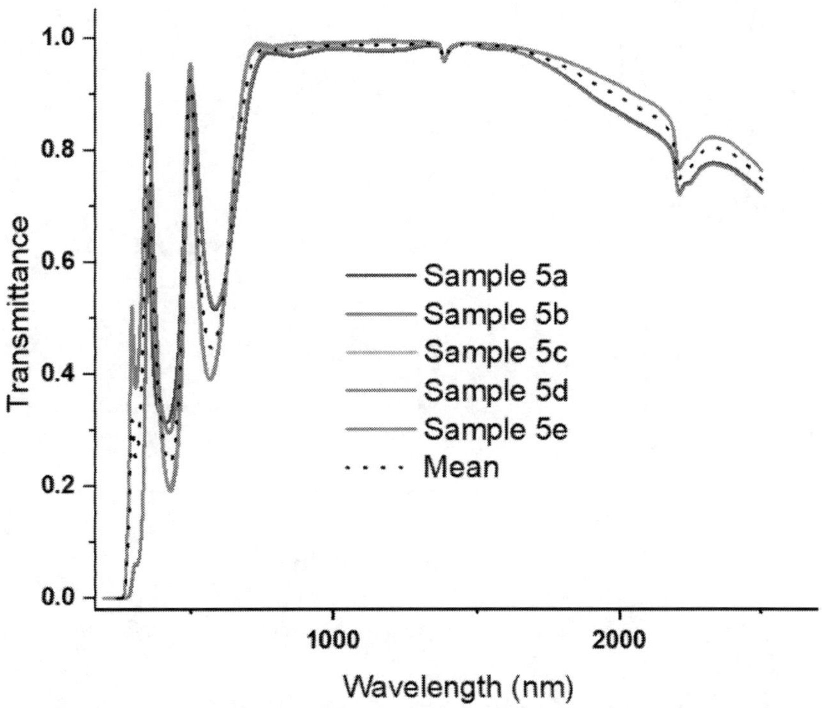

Figure 8. Transmittance spectra of multiples of the same type of NIR II window (sample no. 5).

4.2. *Environmental testing*

All samples selected for the LEOX test show a very good resistance against ATOX within the test conditions. No changes could be observed between the EOT and BOT results within the experimental error of ± 0.05 for the thermal emittance, ε, and ± 0.03 for the solar absorptance, respectively (see Table 5).

An example of the BOT and EOT total reflectance spectra is shown in Figure 9. As displayed, an overall good ATOX resistance can be observed for the protected silver mirror sample in the UV/VIS/NIR range. The protective commercial coating seems to be effective against the typical ATOX erosion of silver due to spalling oxides. Only a slight increase between the BOT and EOT reflectance can be seen in the graphic for the reflectance intensity at around 250 and 500 nm.

Table 5. Deviations between EOT and BOT LEOX for the solar absorptance, α, and thermal emittance, ε.

Sample no.	Sample description	Δα	Δε
4	IR Fresnel window	0	0.01
5	NIR II window	0	0
6	Broadband Precision Window	0.01	- 0.01
7	Reflecting mirror	0	0

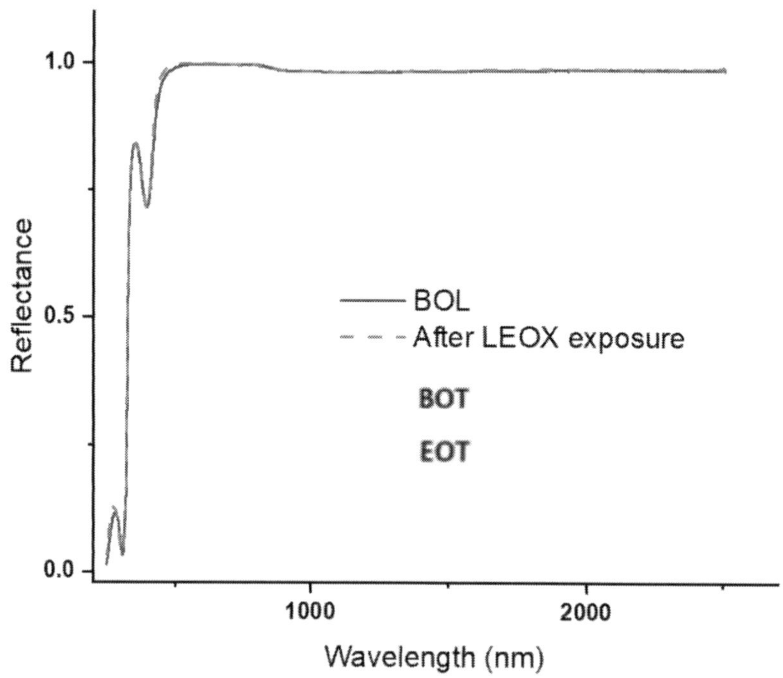

Figure 9. LEOX test BOT and EOT total reflectance spectra of the reflecting mirror sample.

In Table 6 the deviations between the solar absorptance and thermal emittance BOT and EOT the UV/VUV exposure and thermal cycling are displayed. Overall, the samples show a very good resistance against both environmental testing conditions with only minor EOT/BOT deviations. Only for two samples the deviations lay outside the experimental uncertainties of ± 0.05 for the thermal emittance, ε, and ± 0.03 for the solar absorptance, respectively. The solar absorptance values for the long-pass IR filter and the IR Fresnel window lay outside the experimental error with deviations between EOT UV/VUV exposure and BOT of - 0.08 and 0.15, respectively.

Table 6. Deviations between EOT UV/VUV exposure, thermal cycling and BOT for the solar absorptance, *α*, and thermal emittance, *ε*.

Sample no.	Sample description	Δα	Δε	Δα	Δε
		UV/VUV		thermal cycling	
1	Long-pass IR filter	- 0.08	- 0.02	0	0.03
2	Narrow band-pass filter	- 0.01	- 0.01	0.01	0.01
3	IR band-pass Filter	0	- 0.01	0	0
4	IR Fresnel window	0.15	- 001	- 0.01	0.01
5	NIR II window	- 0.01	- 0.01	0	0
6	Broadband Precision Window	-0.01	- 0.03	0	0.02
7	Reflecting mirror	0	0.01	0	0

For the long-pass IR filter, The BOT, EOT UV/VUV exposure and thermal cycling transmittance spectra are shown in Figure 10. The spectra on the left hand-side in the UV/VIS/NIR wavelength range cover the block-out region of the filter of 200 – 3600 nm. In this region, the transmittance further decreases together with the solar absorptance (Table 6) while the reflectance increases. Consequently, the environmental testing led to an enhanced block-out capacity in this region which might be acceptable for most applications. For the IR transmittance region of the filter in the range of 3.78 – 7.2 μm, as covered by the spectra on the right hand-side in Figure 10, only minor changes in the spectra due to the UV/VUV exposure and thermal cycling can be distinguished.

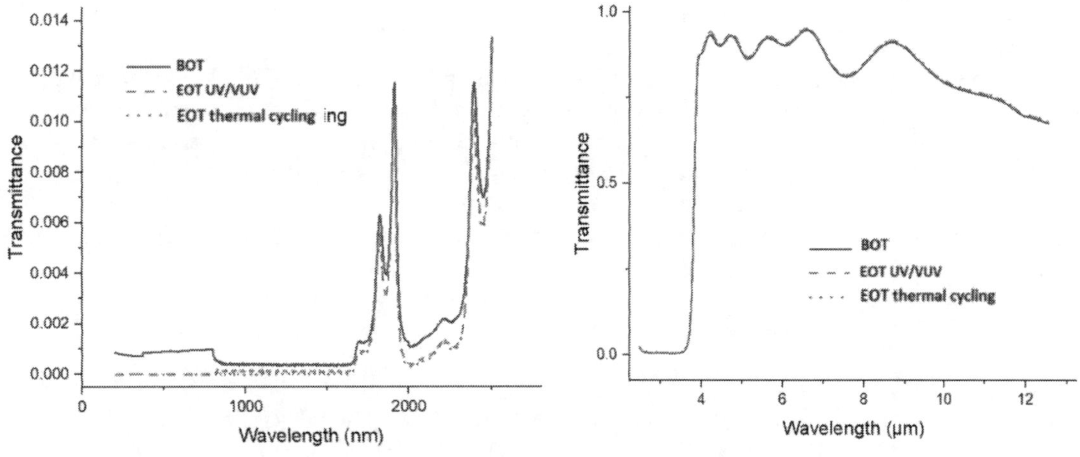

Figure 10. Transmittance spectra of the long-pass IR filter; BOT, EOT UV/VUV and thermal cycling for different wavelength ranges.

In the case of the IR Fresnel window, a major increase of the solar absorptance occurred during the UV/VUV exposure. Material degradation can also be confirmed visibly as can be seen in Figure 11. The

IOP Conf. Series: Materials Science and Engineering 1287 (2023) 012030 doi:10.1088/1757-899X/1287/1/012030

photograph demonstrates how the exposed areas darkened compared to the unexposed white areas, which were covered by the sample frame. No further major material degradation occurred for the IR Fresnel window during thermal cycling with regards to the thermal optical properties in Table 6. The thermal stability up to 90 ˚C during thermal is confirmed by the TGA results in Figure 12, which show significant mass loss only above 103 ˚C.

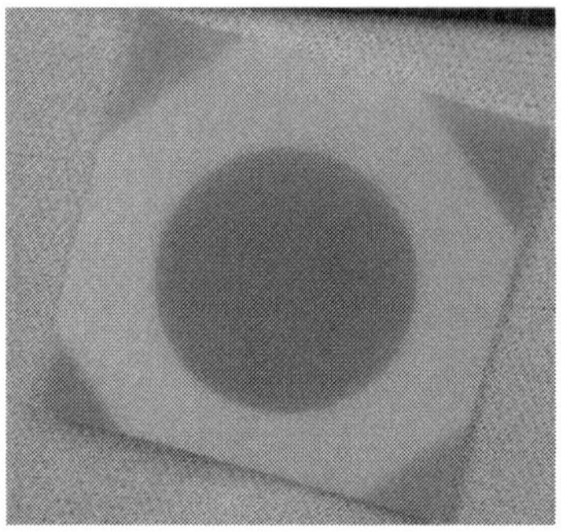

Figure 11. Photograph IR Fresnel window EOT UV/VUT exposure.

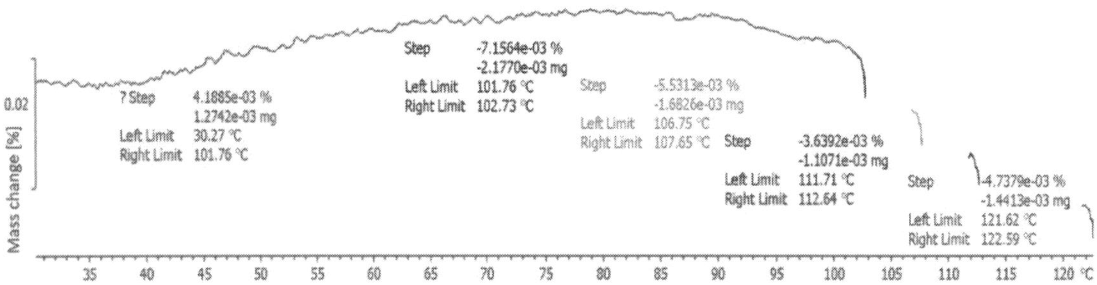

Figure 12. TGA results of mass changes [%] of the IR Fresnel window as function of temperature [°C] with significant mass loss above 103 ˚C.

In Figure 13, another example of a comparison between the BOT, EOT UV/VUV and thermal cycling reflectance spectra is shown for the narrow band-pass filter. No major deviations could be observed in this sample for the solar absorptance, and thermal emittance based on Table 6. As can be seen in Figure 13, the filter has a designed transmittance peak at 6415 nm with a half width (HW) of 25 nm. In comparing the BOT and EOT spectra a slight onset shift and intensity increase of approximately 0.2 μm and 0.1, respectively, are distinguishable. Depending on the accuracy needed for specific applications, these changes could be of significance.

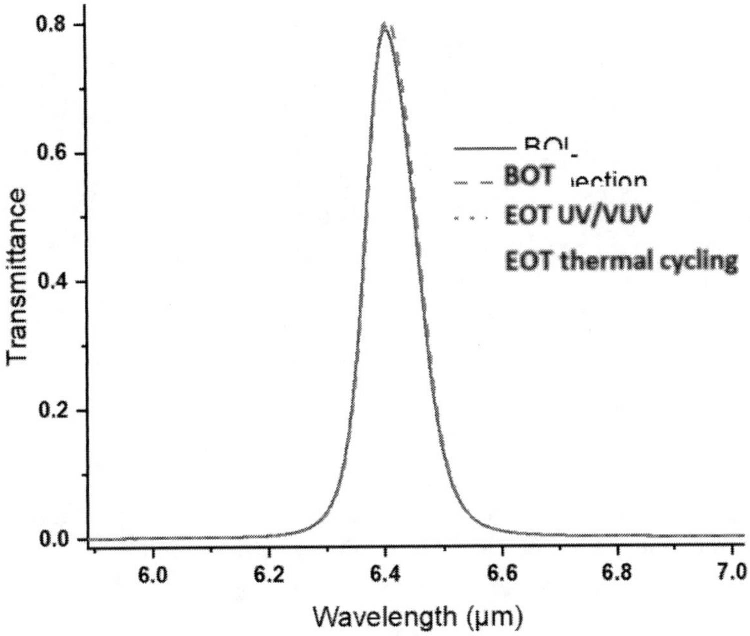

Figure 13. Narrow band-pass filter reflectance spectra, BOT, EOT UV/VUV and thermal cycling.

4.3. Influence of cold temperatures

The cryostat coupled FTIR were conducted on all transparent and semi-transparent samples. Generally, the changes in temperature over the investigated temperature range, generated a decrease or increase of the total transmittance, or lead to an onset shift of the transmittance peaks.

Two examples of the temperature dependant transmittance measurements are shown in Figure 14 (a) for the long-pass IR filter and the IR band-pass filter in Figure 14 (b). For the long-pass IR filter it is apparent from the measurements that the onset of the long-pass transmittance generally shifts to a higher wavelength with increasing temperature from approximately 3.6 to 3.7 μm. A similar trend can be observed for the long-pass IR filter where the onset of the transmittance peak shifts from about 3.1 to 3.2 μm with increasing temperatures from 4 K to 300 K. In addition, the overall intensity of the transmittance peak for the latter seems to decrease slightly with increasing temperatures.

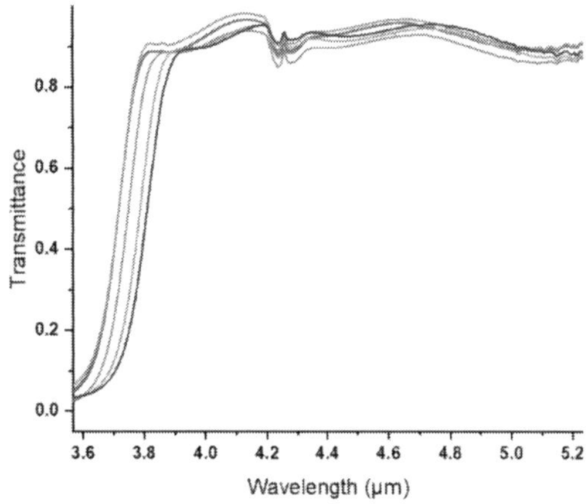

(a)

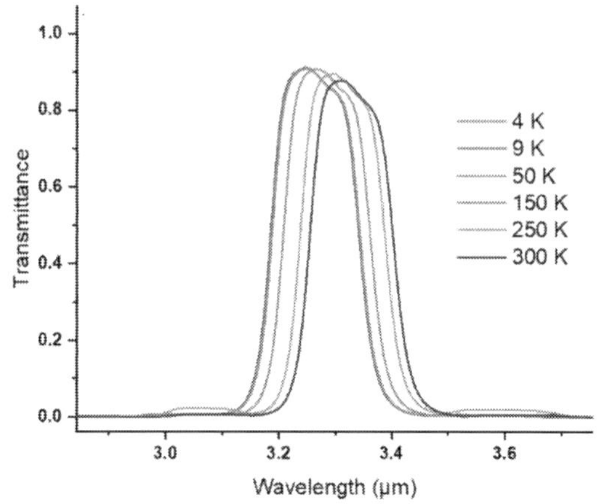

(b)

Figure 14. Temperature dependant transmittance spectra of the long-pass IR filter (figure 14 (a), left hand-side) and the IR band-pass filter (figure 14 (b), right hand-side) in the range of 4 – 300 K.

5. Conclusions

It has been demonstrated that the selected representative standard commercial optical coatings generally have good potential to be used for space applications. The reproducibility of the optical performance varies between the products and special attention needs to be paid to production batches. A general high resistance to space environmental effects was found for a high number of the tested samples, but further evaluation is recommended for specific applications. Temperature limitations and applications need to be considered, respectively. Special attention should also be paid to temperature dependant changes and shifts in the thermo-optical performance.

References

[1] ESA, 2019 ESA's 'CubeSat central' for smaller missions into space https://www.esa.int/Enabling_Support/Space_Engineering_Technology/ESA_s_CubeSat_central_for_smaller_missions_into_space
[2] de Rooij A 2010 Corrosion in space *Encyclopedia of Aerospace Engineerings* ed Blockley R and Shyy (Wiley) W https://onlinelibrary.wiley.com/doi/abs/10.1002/9780470686652.eae242 http://esmat.esa.int/publications/published_papers/corrosion_in_space.pdf
[3] de Rooij A 1989 The Oxidation of Silver by Atomic Oxygen *ESA Journal,* **13** pp 363 - 382
[4] Stiegman A 1993 Ultraviolet and Vacuum-Ultraviolet Radiation Effects on Spacecraft Thermal Control Materials *The Behavior of Systems in the Space Environment* **245** (NATO ASI Series: Springer)
[5] Pradal F, Lhuillier R and Mouricaud D 2015 Optical and mechanical properties of infrared thin film at cryogenic temperature *Proceedings of the SPIE* **9627**

IOP Conf. Series: Materials Science and Engineering 1287 (2023) 012031 doi:10.1088/1757-899X/1287/1/012031

Materials inspection of the ISS Columbus Trunnion scuff plate MLI

C Mooney[1*], N Dias[1], Y Butenko[1], O Schmeitzky[1] and G van Papendrecht[1]
[1]ESA, ESTEC, Postbus 299, Keplerlaan 1, 2200 AG Noordwijk, The Netherlands

* Author to whom the correspondence shall be addressed: cathal.mooney@ext.esa.int

Abstract. As part of the preparation for the Bartolomeo platform, various locations of the trunnion scuff plate were required. Scuff plate MLI blankets were removed and recovered from Columbus ISS in August 2019 as part of these EVA activities. These blankets having seen more than 11 years in orbit are extremely valuable from a materials-science perspective where direct analysis of the impact of the space environment can be seen and evaluated to see its effect on materials, contamination and functionality of the blankets.

1. Introduction

Materials used in the external environments of the ISS station and Columbus in low earth orbit (LEO) must possess various performance criteria which include resistance to extremes of temperature cycles, high vacuum, radiation, atomic oxygen environments (ATOX) and micrometeoroid impacts in order to be successful and fulfil their function. These trunnion MLI blankets have been exposed to over 11 years of RAM facing space environments where extreme conditions have been experienced (Figure 1). During this time, these MLI blankets must function and maintain performance under these extreme conditions.

Figure 1. ISS showing MLI locations

2. Background

The MLI used is primarily for thermal control consisting of 20 layers of β-cloth [1]. The Scuff-Plate MLI blankets have various addition elements including VELCRO HI-MEG fasteners, electrical bonding and grounding connections, Velcro hook patches and beta-cloth strips [2]. All these have been assessed along with the internal aluminised surfaces in contact with the metallic trunnion plates following exposure in order to get a greater understanding of the materials space environments interactions (Figure 2).

Of interest is the presence (or not) of micrometeoroids and foreign objects which may have been captured within the MLI during the exposed time. This paper outlines the findings following the inspection of these blankets.

Figure 2. Columbus trunnion MLI blanket

3. Results and discussion

Post flight exposure of these blanket materials included both non-destructive and destructive analysis where visual inspections, optical and electron microscopy, contamination analysis, thermo-optical measurements and other surface analysis and analytical techniques have been performed.

3.1. Visual inspections and Optical microscopy

Post flight exposure analysis of these blanket materials included both non-destructive and destructive analysis where visual inspections and microscopy was performed using white and dark light illumination. Observed under white light, the outer surface of the MLI (betacloth material which was facing the external surface of the ISS) showed some discoloration and holes. No debris was visible on the betacloth with the exception of the Velcro loop which showed some entrapped dust and fibres [3]. The presence of holes (about a few hundreds of micrometers) may be attributed to micrometeoroid impact [4]. Also shown are extensive UV discoloration and darkening of the betacloth and Velcro locations and the presence of debris, dust and fibres using UV light inspection (Figure 3).

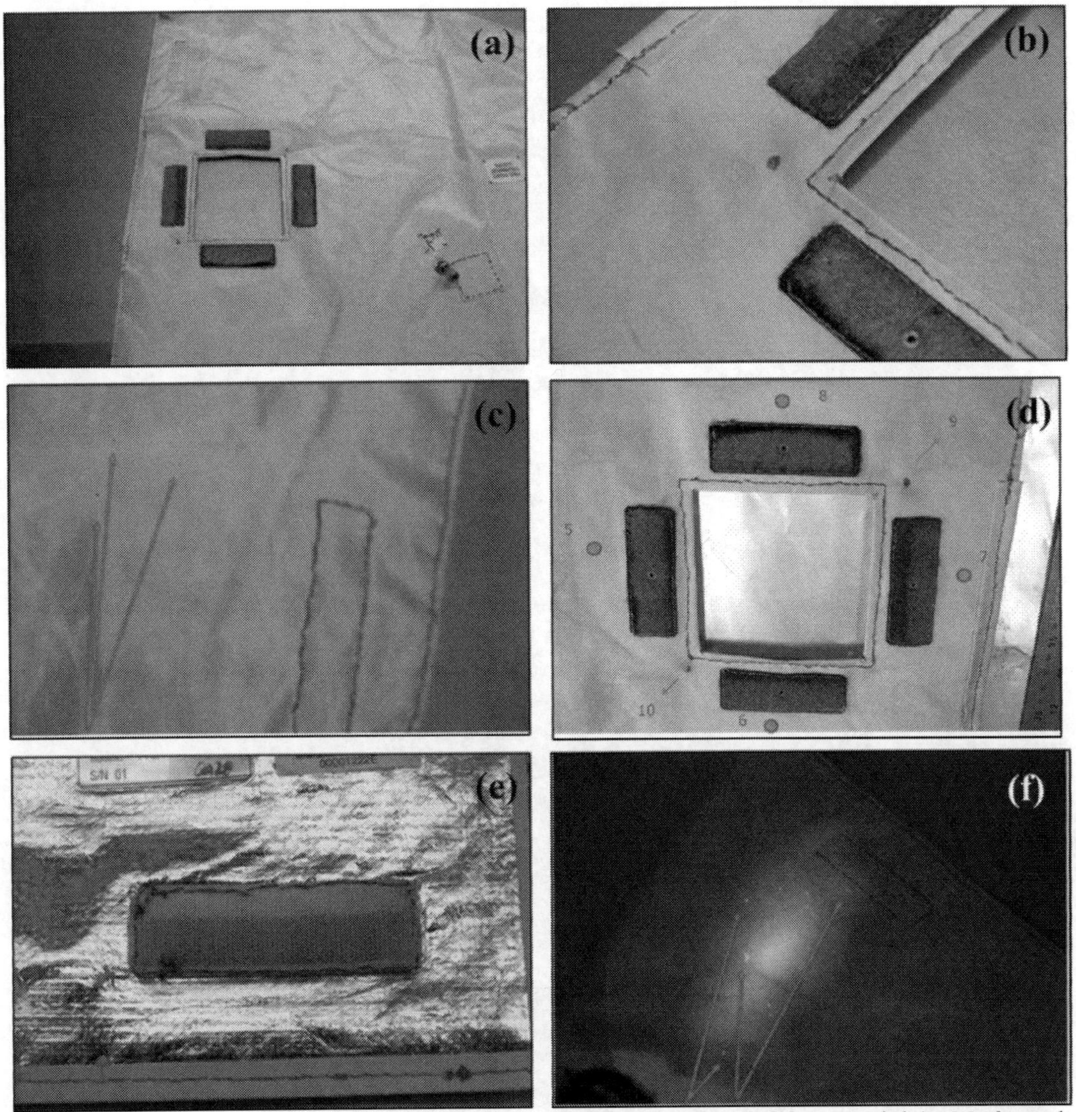

Figure 3. Columbus trunnion MLI blanket optical microscopy - (top left-bottom right) : Marks and holes/ impact spots visible on the betacloth. (a, b, c, d (blue spots indicate sample measurement locations). Discoloration is visible on Velcro hook locations (e) and UV light showing debris, dust and fibers (f).

3.2. Thermo-optical measurements

Thermo-optical measurements were performed using a Perkin Elmer UV-VIS-NIR spectrophotometer. Emissivity measurements were performed on the white coated samples using an SOC portable handheld alpha /epsilon device, according to ECSS Q 70 09C [5]. The Perkin Elmer was set to measure total solar reflectance of the samples.

Thermo-optical measurements were performed on both non-exposed Al sides (Figure 5-blue spots indicate sample measurement locations) and exposed betacloth sides (Figure 12-blue spots indicate TO and FT-IR sample measurement locations) of the MLI using SOC ET100 portable handheld alpha /epsilon devices and standalone equipment. The solar absorptance was measured in the UV-Vis-NIR wavelength range of 335-2500 nm whilst the thermal emissivity was measured in the 1.5-21 mm range.

For alpha the device measures total/diffuse reflectance spectra (UV-Vis-nIR) and the spectra are measured in bands: 335-380 nm, 400-540 nm, 480-600 nm, 590-720 nm, 700-1100 nm, 1000-1700 nm and 1700-2500 nm.

For epsilon the device measures total reflectance spectra at 20° and 60° incidence (IR) and the spectra are measured in the following bands: 1.5-2 μm, 2-3.5 μm, 3-4 μm, 4-5 μm, 5-10.5 μm and 10.5-21 μm. (Figure 4) below shows the handheld device used.

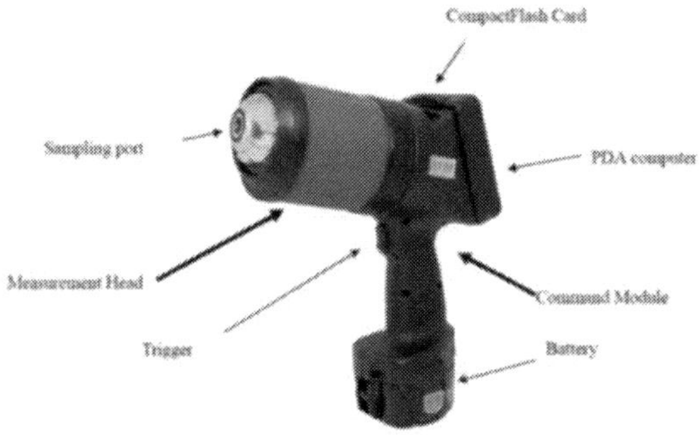

Figure 4. Alpha/epsilon ET100 handheld device used

For T-O measurements on the aluminised side point 1 is considered pristine compared to point 6, due to exposed locations. Table 1 shows the solar absorptance and emissivity measurements taken for the aluminised side of the MLI blankets. As can be seen there is some variation between pristine and darkened locations for alpha values, but emissivity does not change significantly.

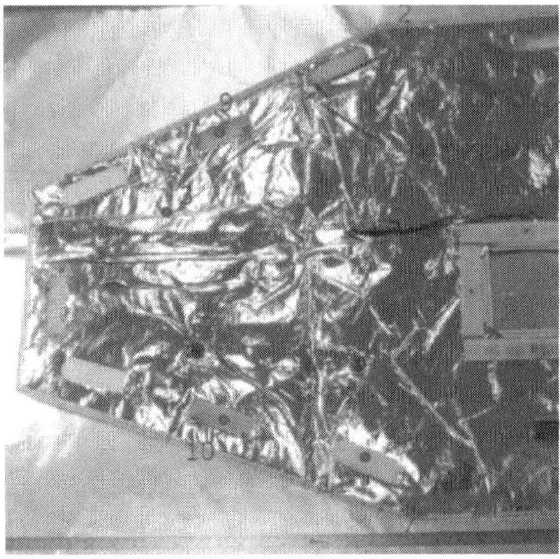

Figure 5. Alpha/epsilon measured locations on the Aluminised side (blue spots indicate sample measurement locations)

Table 1. a/e sample locations for the aluminised side

Point #	Solar Abs SOC	Solar Abs PE	Epsilon
1	0.11	0.11	0.08
2	0.38	0.41	0.36
3	0.12	0.16	0.07
4	0.10	0.30	0.03
5	0.12	0.23	0.06
6	0.39	0.45	0.50
7	0.18	0.33	0.16
8	0.36	0.32	0.91
9	0.52	0.51	0.91
10	0.49	0.49	0.91
11	0.37	0.30	0.91

The reflectance measurements are shown in the figures below. From Figure 6 and Figure 7 a clear difference was observed between locations 2 and 6 and the rest of the locations measured. The overall reflection plot in locations 2 and 6 were lower, but more so in the UV-Vis range. In Figure 8 the reflection spectra in the IR range are shown. Sample location measurement points 2 and 6 have an overall lower reflectance when compared to the other locations. This change in reflectance leads then to a higher thermal emissivity.

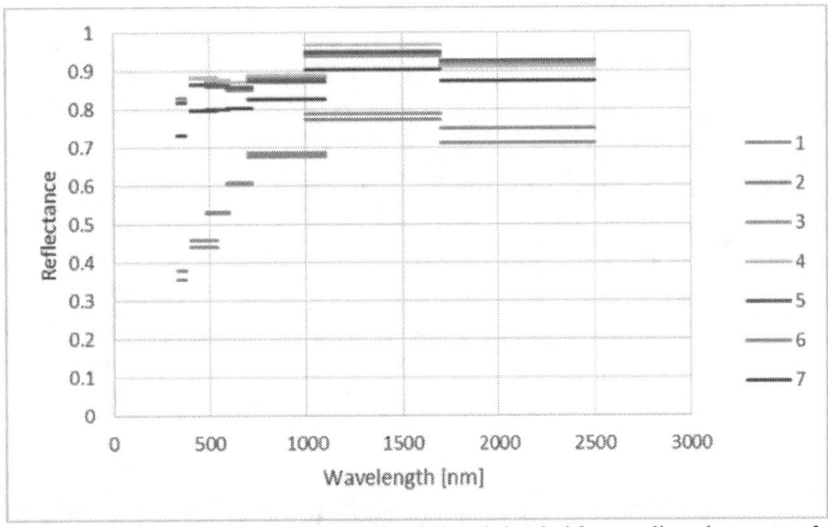

Figure 6. a/e sample locations (see **Figure 5**) of aluminised side not directly exposed- using the portable device..

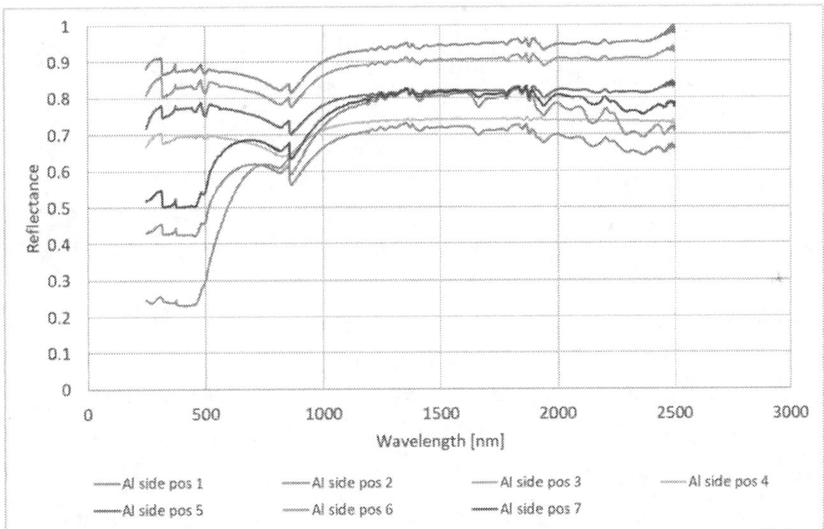

Figure 7. a/e sample locations (see **Figure 5**) of aluminised side not directly exposed using the Perkin Elmer.

Thermo-optical measurements of Velcro were also taken at positions 8, 9 10 and 11 measured as shown in Figure 9 to Figure 11 below. The pristine locations (8, 11) shown significant variation to the UC darkened locations (9, 10). For Velcro some increase in alpha and emissivity points 9 and 10 show heavy darkened degraded Velcro regions – compared to point 11 – pristine Velcro.

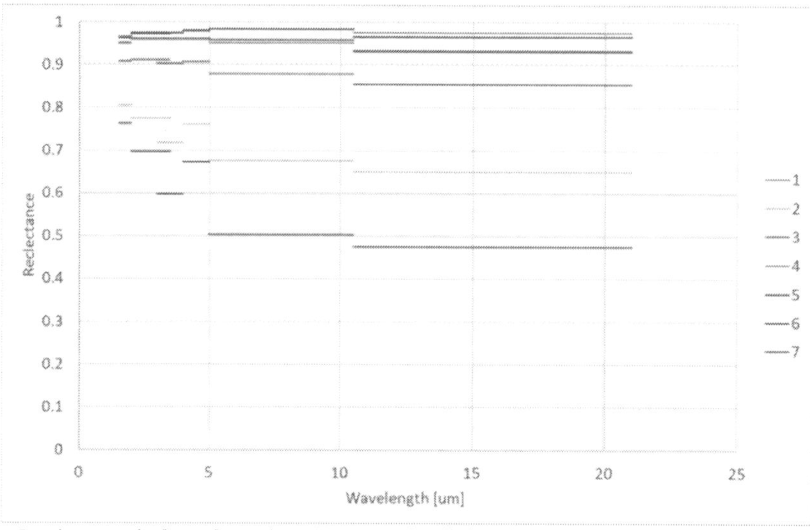

Figure 8. a/e sample locations (see **Figure 5**) of aluminised side not directly exposed IR measurements.

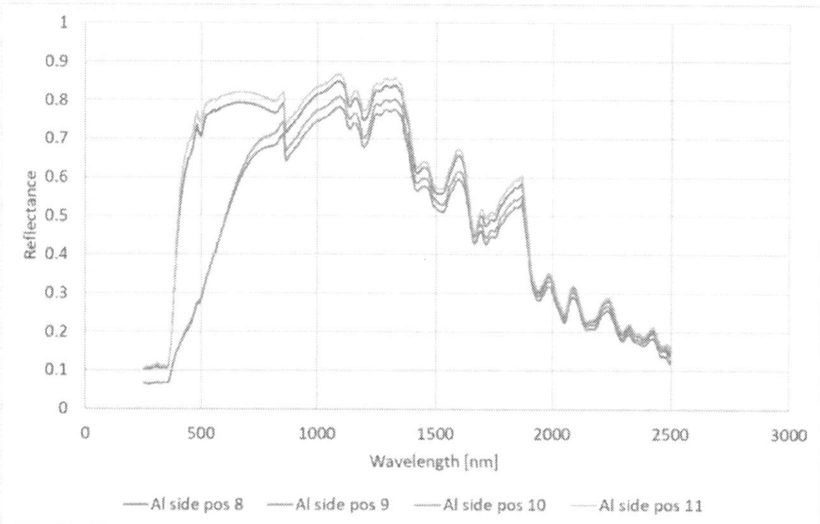

Figure 9. Reflectance measurements for Velcro locations (see **Figure 5**).

IOP Conf. Series: Materials Science and Engineering 1287 (2023) 012031 doi:10.1088/1757-899X/1287/1/012031

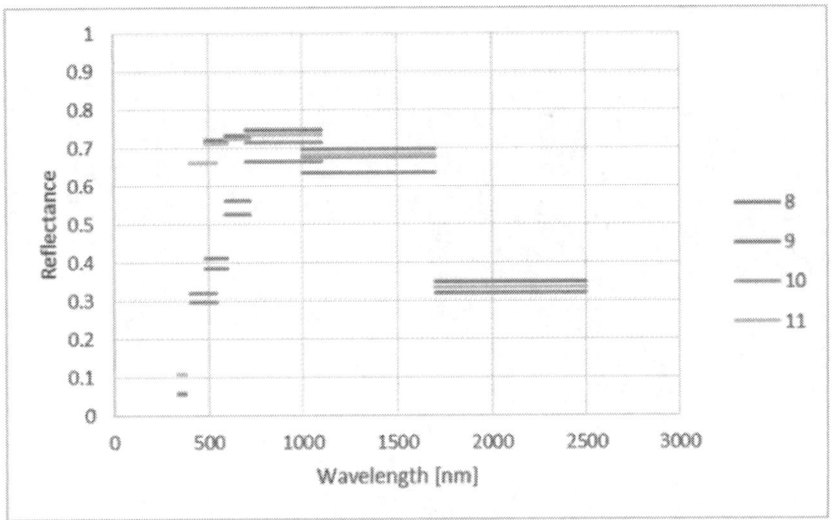

Figure 10. Reflectance measurements for Velcro locations (see **Figure 5**) not directly exposed using portable device.

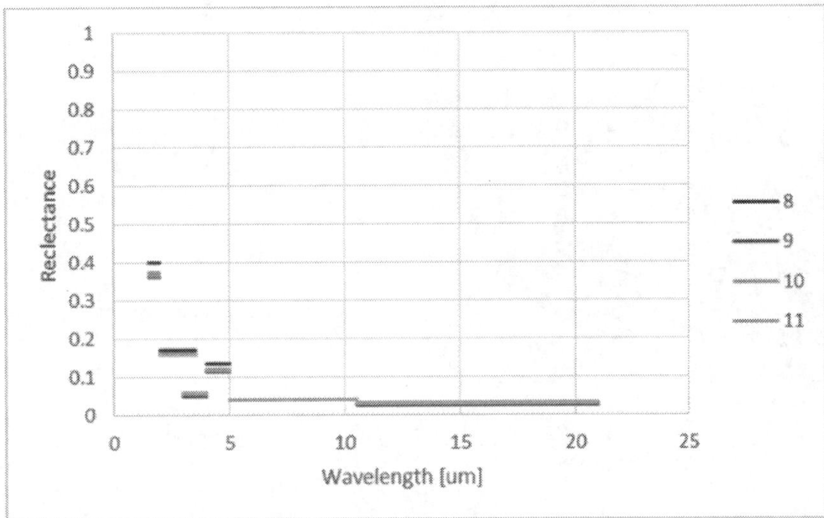

Figure 11. Reflectance measurements for Velcro locations (see **Figure 5**) not directly exposed IR measurements.

The betacloth side was also measured in various locations. Figure 12 below shows these locations with the subsequent alpha and epsilon corresponding values. Exposed betacloth areas like point 1 and 2 have 0.32 compared to the pristine betacloth point 6 with an alpha value of 0.29. – again a slight alpha increase due to browning and degradation and exposure of the betacloth [6].

IOP Conf. Series: Materials Science and Engineering 1287 (2023) 012031 doi:10.1088/1757-899X/1287/1/012031

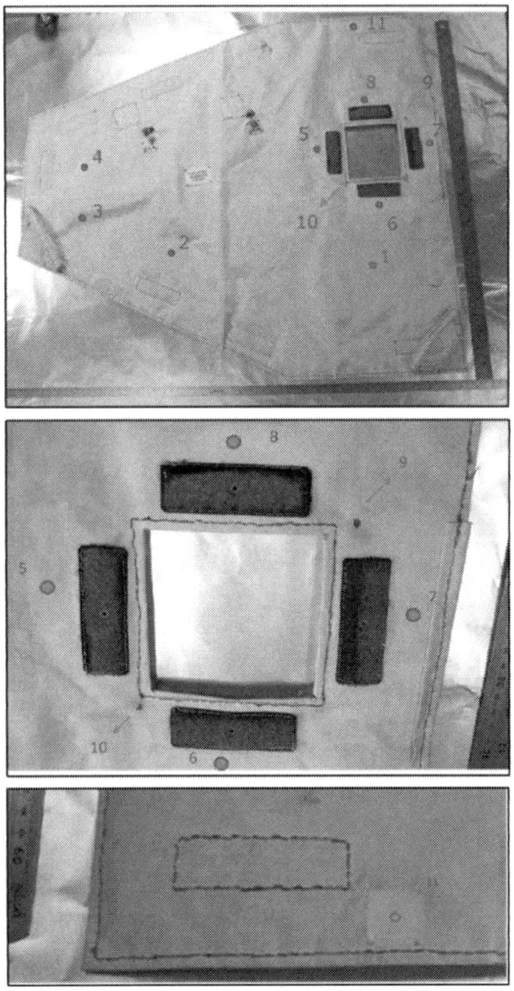

Figure 12. Alpha/epsilon measured locations on the betacloth side. (blue spots indicate sample measurement locations)

Table 2. a/e sample locations for the betacloth side

Point #	Solar Abs SOC	Solar Abs PE	Epsilon
1	0.32	0.32	0.87
2	0.32	0.35	0.87
3	0.33	0.33	0.87
4	0.33	0.35	0.87
5	0.29	0.35	0.88
6	0.29	0.30	0.88
7	0.31	0.37	0.88
8	0.31	0.43	0.88
9	0.39	0.46	0.87
10	0.36	0.55	0.87
11	0.30	0.34	0.88

Locations 9 and 10 are visibly darker than other betacloth locations. This visible change can also be detected in the UV-Vis-nIR range as seen in Figure 13 to Figure 15. In the IR range there is no clear difference in spectra between all locations

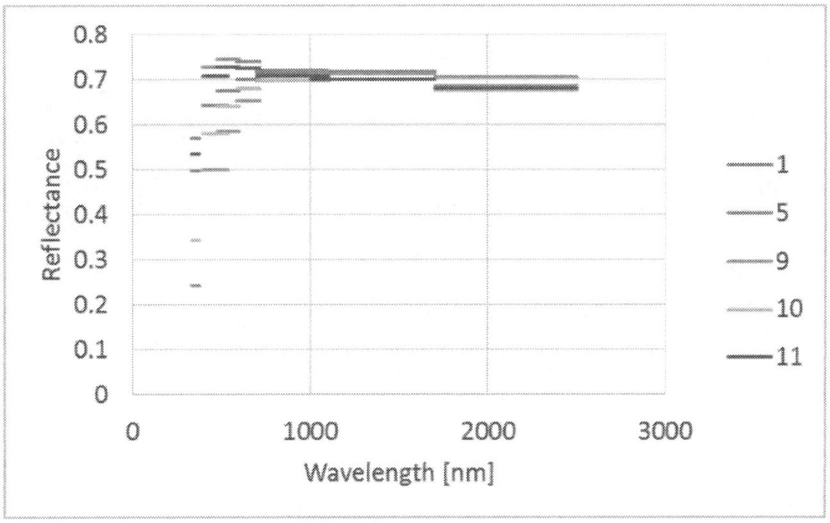

Figure 13. Reflectance measurements for betacloth locations (see **Figure 12**) for the exposed side using the portable device

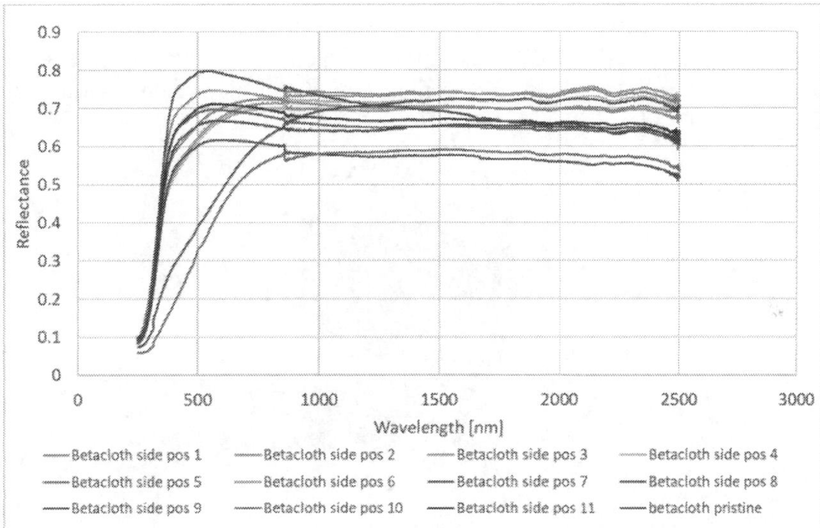

Figure 14. a/e sample locations (see **Figure 12**) of betacloth side directly exposed using the Perkin Elmer

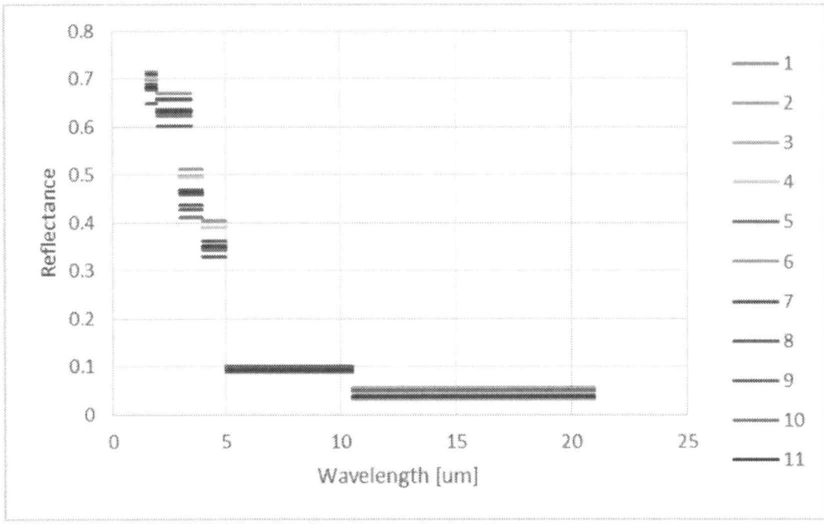

Figure 15. Reflectance measurements for betacloth locations (see **Figure 12**) directly exposed IR measurements

3.3. FT-IR measurements

FT-IR measurements were performed with the portable EXOSCAN 4100 FT-IR, with diffuse reflectance head (see Figure 16 below). The resolution was 4 cm^{-1} in the wavenumber range 4000 to 550 cm^{-1} The sample measurement locations were similar to those for alpha / epsilon measurements taken (see Figure 5 and Figure 12 above).

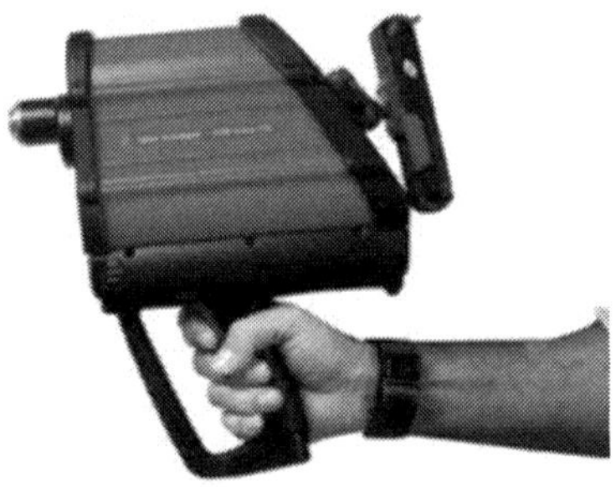

Figure 16. EXOSCAN 4100 FT-IR

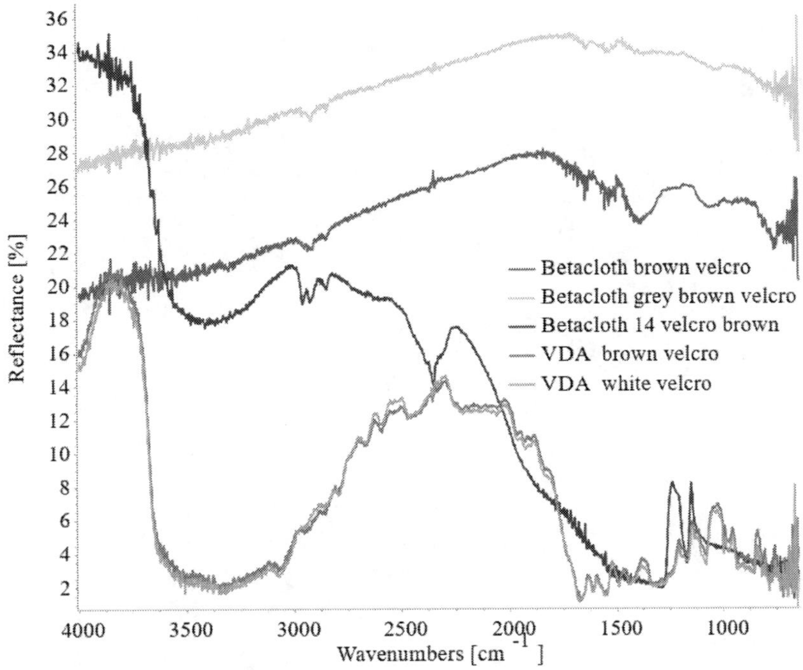

Figure 17. FT-IR measurements for betacloth locations (see **Figure 12**) using the handheld FT-IR.

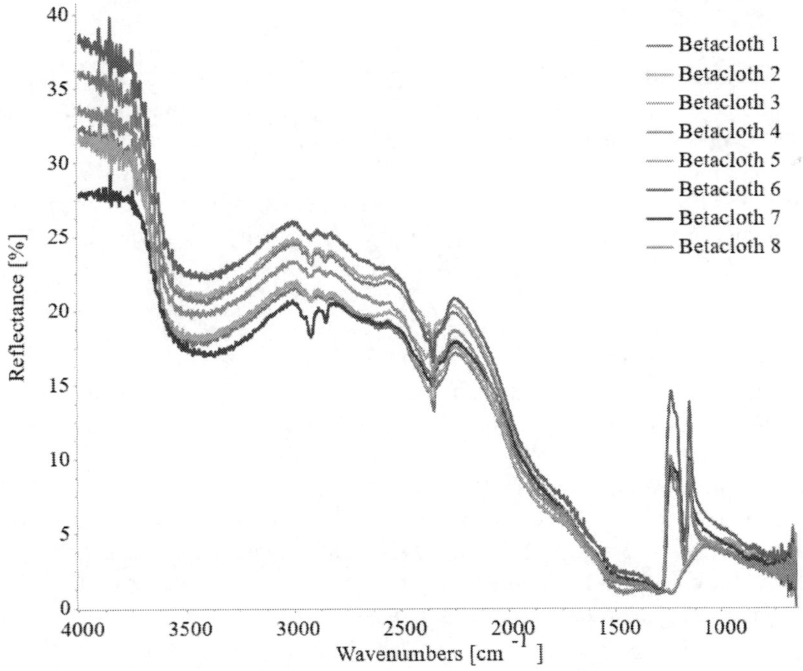

Figure 18. FT-IR measurements for betacloth locations (see **Figure 12**) using the handheld FT-IR

In Figure 19 below dd stands for diffuse head and diffuse background cap whilst the other FT-IR measurements were done with the specular reflectance head and the diffuse cap.

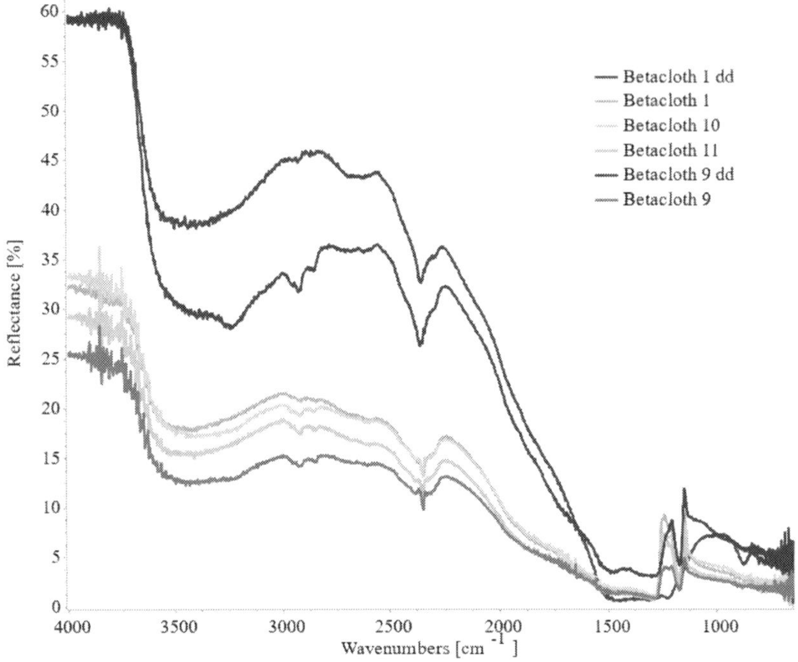

Figure 19. FT-IR measurements for betacloth locations (see **Figure 12**) using the handheld FT-IR.

FT-IR proved difficult to differentiate between exposed and non exposed locations (see Figure 17 to Figure 19) - this may have been due to the solar induced degradation of the organics, in combination with the beta cloth as reflective measurement surface. The only places with a stronger signal were organics like the Velcro locations. At those Velcro locations, the reflectance was still low. Hence it was not clear if the signal was from the base material, degraded material or (degraded) contaminants on those degraded and base material surfaces, where strong hydroxyl, some hydrocarbons on the betacloth side were present.

3.4. SEM /EDX measurements

Scanning Electron Microscopy (SEM) and Energy Dispersive X-ray (EDX) spectroscopy analysis were performed using a Teneo Scanning Electron Microscope with an electron backscattered detector (T1) at accelerating voltages of 2 and 7 kV. A low energy EDX detector was used for analysis. No special preparation or pre-treatment was applied apart from cutting from the main MLI blanket and samples were inspected as received. The estimated spot size used for EDX analysis was from 50-150nm for the accelerating voltages of 2 and 7 kV respectively. Each layer (i.e. the external outer bethacloth and 2 underlying VDA layers) were examined on both sides – namely the external -space direction facing side and the inner ISS Columbus direction facing side).

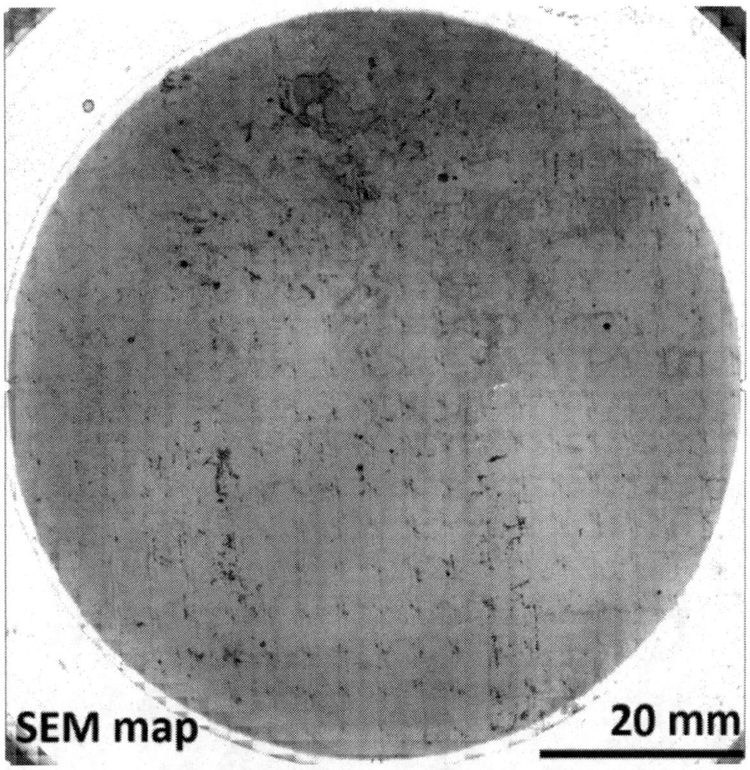

Figure 20. SEM/EDX analysis of the aluminium layer located on the inner side of MLI

SEM and EDX analysis was performed on the outer betacloth and inner VDA layer locations (Figure 20 to Figure 27). SEM/EDX analysis of the internal aluminium layers revealed the presence of contamination containing C, N, O, K, Mg, Na, Ca, F, S and P. The organic contamination appears as dark spots on the SEM map and likely originated from handling and storage. No holes were found on the underside of the blanket.

IOP Conf. Series: Materials Science and Engineering 1287 (2023) 012031 doi:10.1088/1757-899X/1287/1/012031

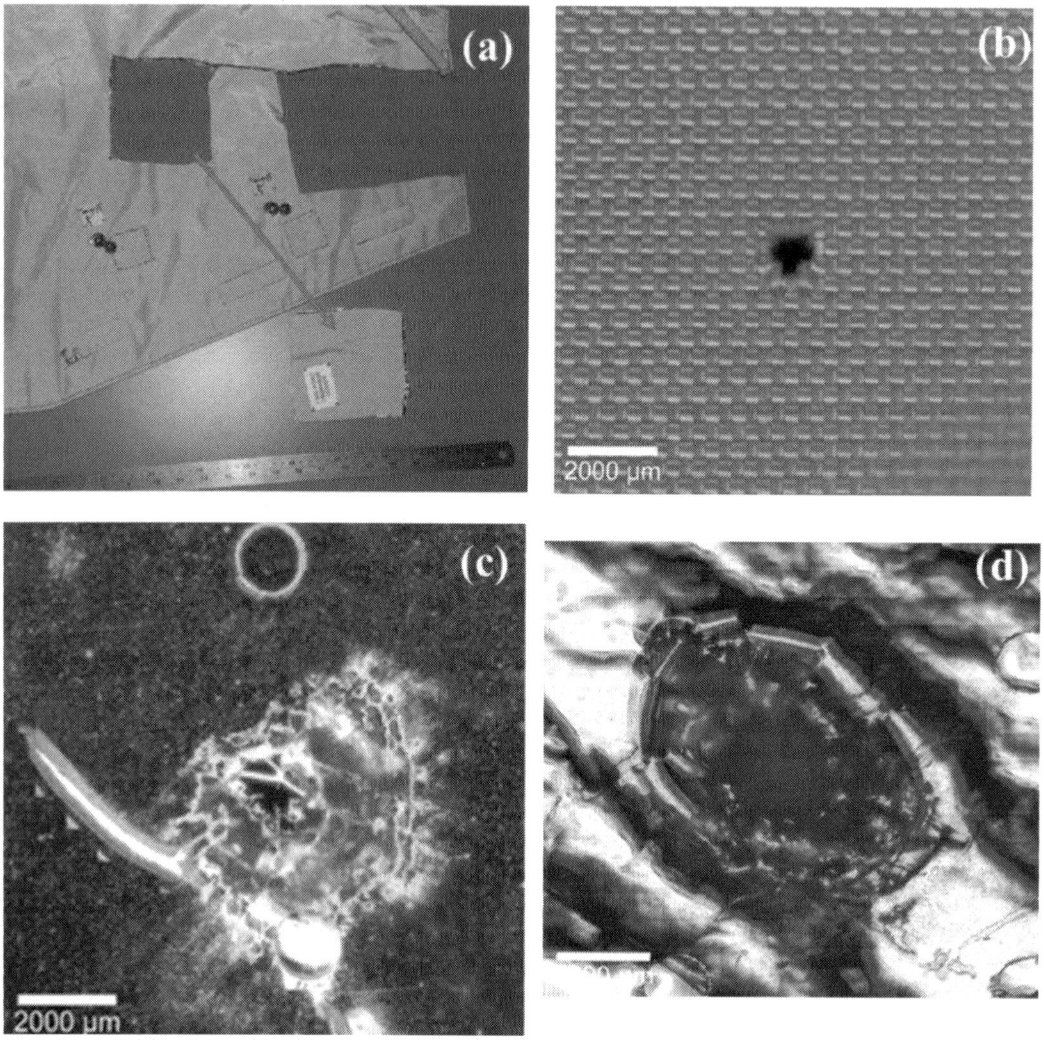

Figure 21. (a) A sample with a hole was cut for analysis. (b) An optical microscopy image of external layer 1. (c) An optical microscopy image of internal layer 2 (side facing the external layer). (d) An optical microscopy image of internal layer 3 (side facing external layers)

As can be seen from Figure 21 (d) above, the edges of the hole were bent outwards. No hole was observed on the opposite side of the blanket It can also be noted that the particle penetrated several upper layers. Of interest are the underlying yellow layers which are thought to be the exposed kapton layers underlying materials layers.

IOP Conf. Series: Materials Science and Engineering 1287 (2023) 012031 doi:10.1088/1757-899X/1287/1/012031

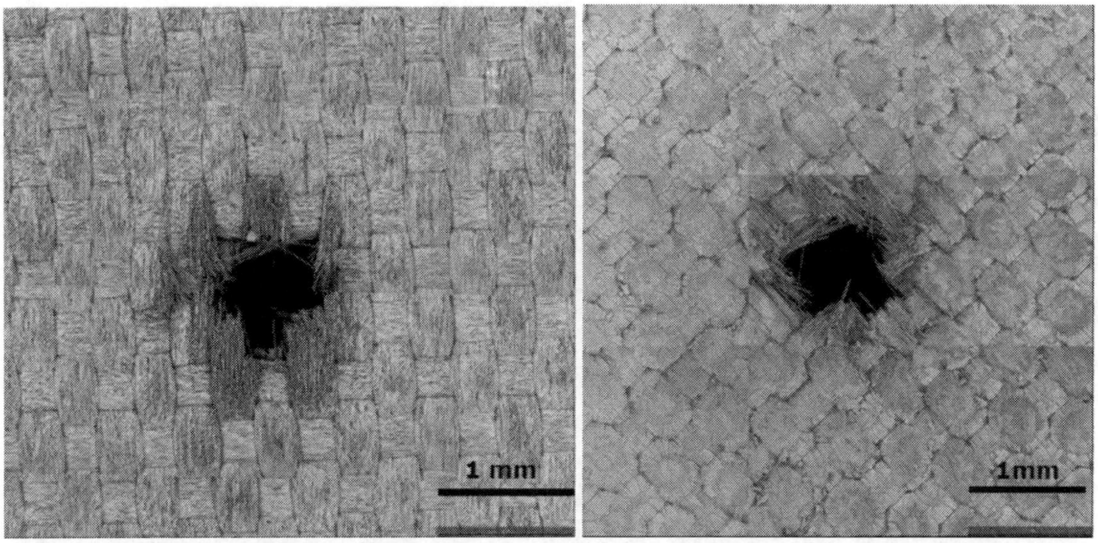

Figure 22. SEM analysis of 1st layer (betacloth). Left: the external (space facing direction) surface entrance hole. Right: Inner surface (ISS facing direction) entrance hole

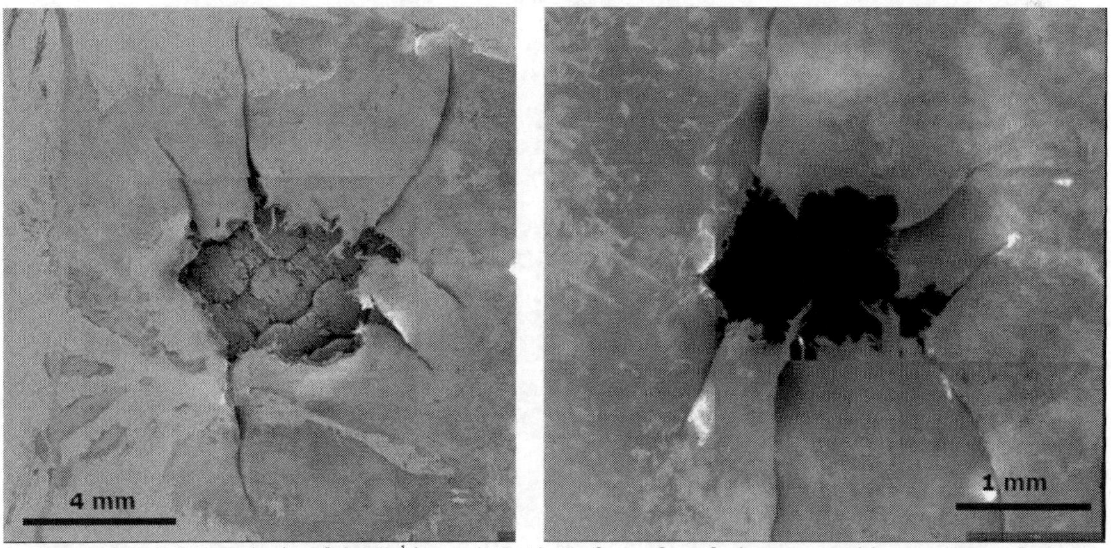

Figure 23. SEM analysis of the 2nd layer (VDA). Left: surface facing external layers (space facing direction). Right: surface facing the internal layers (ISS facing direction)

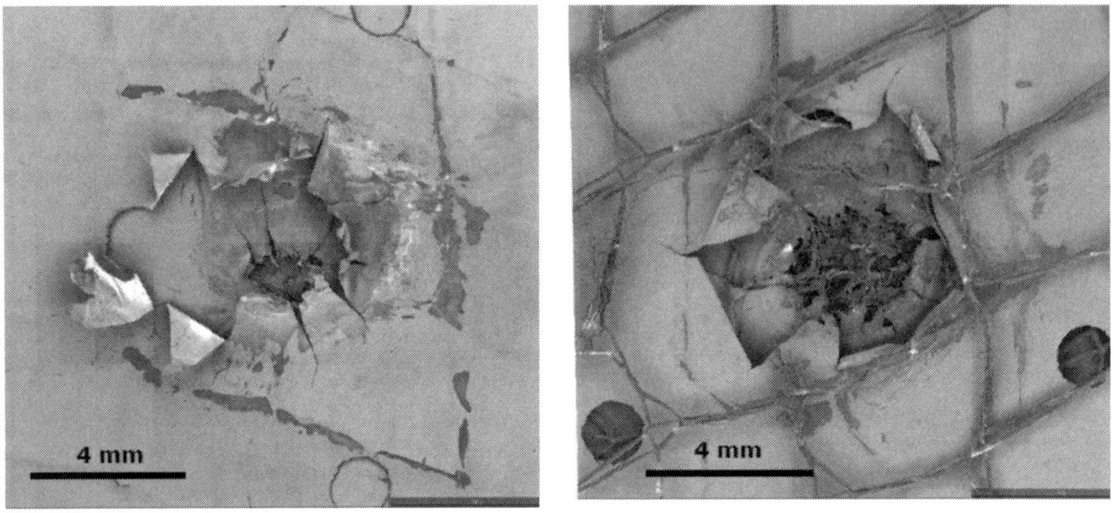

Figure 24. SEM analyses of the 3rd layer (VDA). Left: surface facing external layers (space facing direction). The bending of the edges to the direction opposite to the impact is seen in this layer. Right: surface facing internal layers (ISS facing direction). Darcon net (used as a spacer between layers) is also seen.

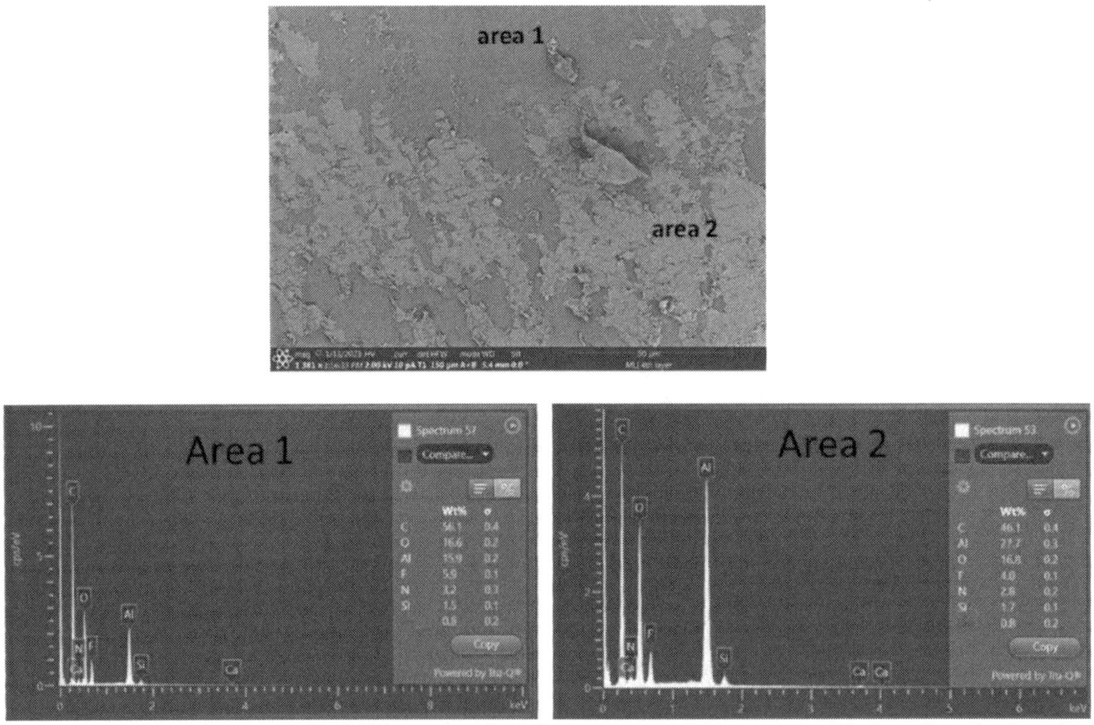

Figure 25. SEM image of the 3rd layer (surface facing external layers-space facing direction) and EDX spectra acquired in areas 1 and 2 of the analysed surface. Damaged aluminium layers (area 2) are observed on the polyimide film (area 1).

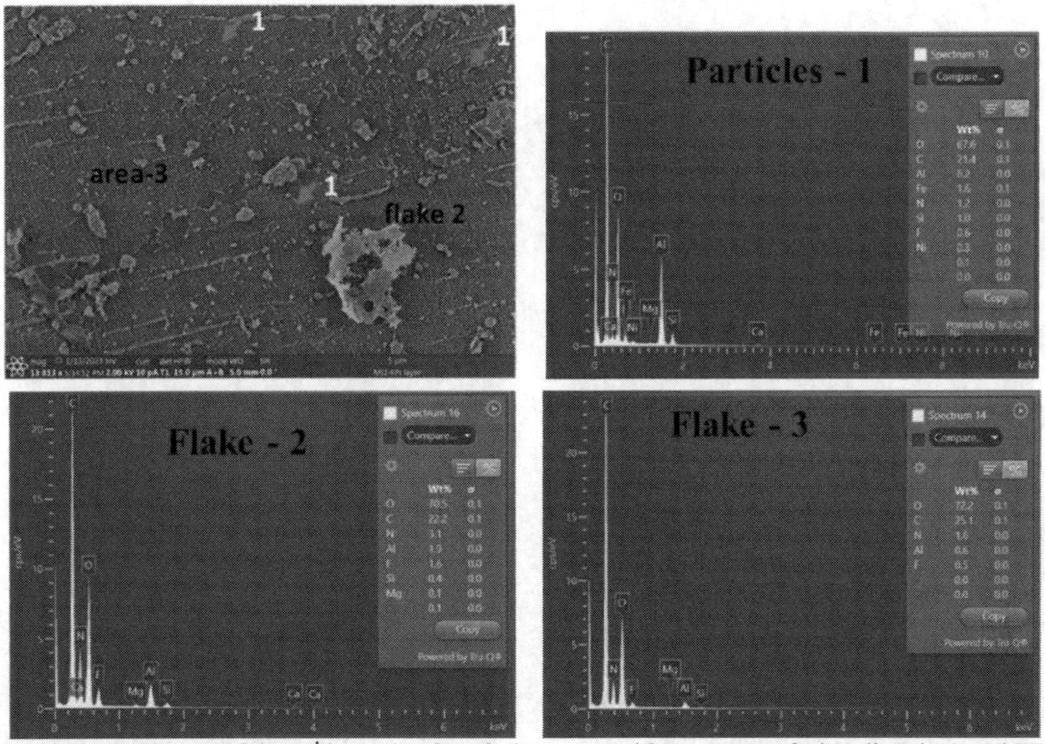

Figure 26. SEM image of the 3rd layer (surface facing external layers-space facing direction) and EDX spectra acquired in different areas: (1) a typical EDX spectrum (particles-1) acquired from sub-micrometer particles (examples are shown by the arrow in SEM image), (2) spectrum acquired from the flake-2 containing high concentrations of aluminium, (3) spectrum acquired from area 3. Particles labelled by number 1 contain iron and nickel. These particles are observed on this layer only and possibly originated from a micrometeoroid.

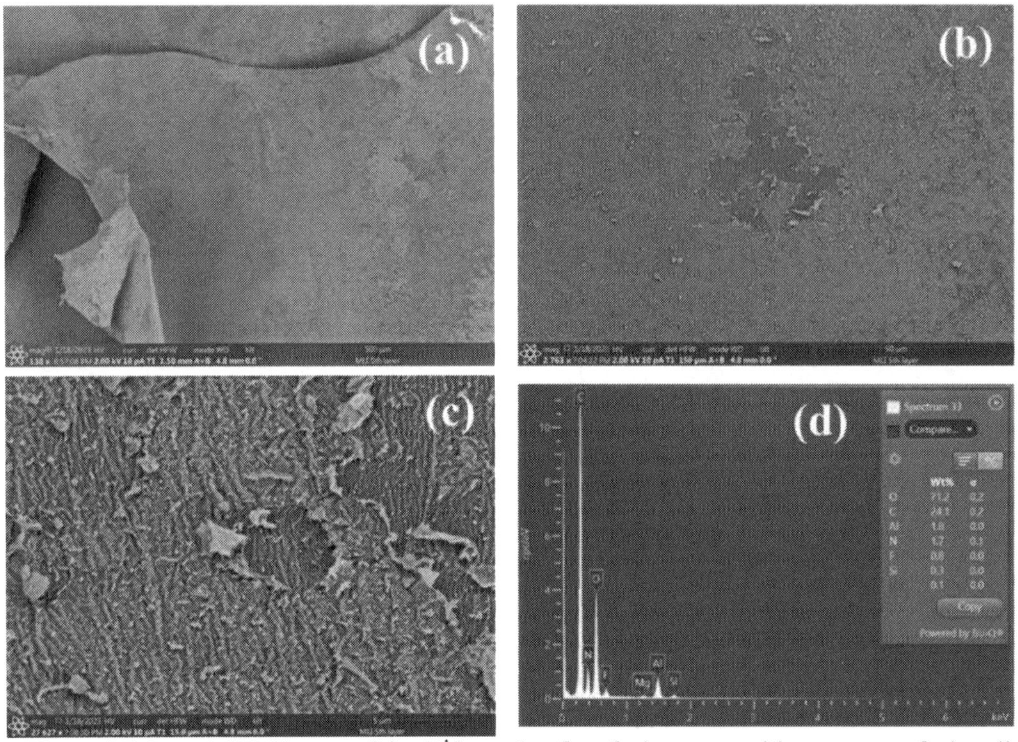
Figure 27. (a – c) SEM image of the 3rd layer (surface facing external layers-space facing direction) and an example of an EDX spectrum acquired (d). Extensive damage of the aluminium layer is observed.

4. Conclusions

Following investigation of the samples the following general conclusions can be made:

1. Visually from optical and SEM observations some locations have shown evidence of impact and what appears to be micrometeoroid impacts. These have been found in the outer betacloth layers and internally within the upper layers of the MLI blanket. It is important to note that the opposite back side layer did not show evident of penetration by debris or impact. Visually the blankets looked in good conditions after 11 years of space exposure and some localised browning or discoloration due to UV degradation was found around impact and velcro locations where protective layers may have been removed and the blankets were more exposed to UV due to geometry and configuration. Apart from some small fibres and debris the blankets were in good condition.

2. For thermo-optical measurements the pristine outer layers of the MLI blanket showed very stable thermo optical values. Velcro position 9 and 10 are visibly darker than positions 8 and 11. The UV-Vis-nIR reflection spectra showed a lower reflection in the UV-Vis range which was in agreement with the visual observations. No clear difference was observed between darkened and pristine Velcro in the IR range.

3. Chemically there was very little difference between the exposed and non-exposed locations. FR-IR proved difficult to differentiate between exposed and non-exposed locations- this may have been due to the solar induced degradation of the organics.

FT-IR spectroscopy proved to be mostly unsuccessful in differentiating between the discoloured and not visually changed / non exposed locations on the betacloth. This difficulty in acquiring a good spectral response was due to many factors. First, the betacloth itself, has, in the IR range a low mostly diffuse reflectance and secondly, the degradation of the contaminants during the solar exposure, reduced the specific absorption bands to much broader and less intense absorptions, hence reducing the spectral quality.

Hence it was not clear if the signal was from the base material, degraded material or (degraded) contaminants on that degraded and base material where strong hydroxyl and some hydrocarbons on the betacloth side where OH and HC were present. In the displayed FT-IR spectra, the response from the betacloth itself dominates the spectra, however on some locations, hydrocarbons could be detected.

4. SEM /EDX analysis showed evidence of the internal explosion of the impact of micro meteoroids with the inner layers of the MLI. This was not evident from the outer layer and shows a potential fragmentation of the impacted object with the MLI. Of importance is that this did not penetrate through the MLI blanket layers. Inside of the blanket 20 layers of polyimide foils coated by aluminium were counted. Extensive damage of the aluminium layer was observed on the first two polyimide foils coated by aluminium. The object penetrated the first 10 layers and an indent was observed on the next five layers.

No damage was visually observed in the last five layers. SEM/EDX analyses of the internal layers showed direct evidence of the presence of sub-micrometre particles containing iron and nickel indicating their likely origin was from a micrometeoroid. These particles were found only on the 3rd layer on the surface facing external layers. The entrance hole on the external layer has size of approximately one millimetre which likely corresponds to the size of the object. The bending of the edges to the direction opposite to the impact is also seen in this layer. The size of the hole in the 3rd layer is approximately four millimetres indicating the disintegrating of the object and an increased impact area due to the object's fragmentation.

References

[1] CLT-TN-AI-0051 iss 3 *TASI- MEMO*: design data for Trunion & Keel Thermal Covers Doc
[2] EADS- COL-RIBRE- DP-0009 Columbus Flight Configuration Thermal Model Description
[3] Visual Inspection of Columbus Scuff plate MLI -MC0004086 Test Report Draft OS- 2019-10-02
[4] Hyde, J L, Christiansen, E L, Lear, D M and Berger, E L 2017 Surveys of Returned ISS Hardware for MMMOD Impacts *7th European Conference on Space Debris*
[5] ECSS-70-09A Measurements of thermo optical properties of thermo control material
[6] ESA-TECQEE-LAB-HO-016225 *ESA Report:* Retrieved Columbus scuff plates MLI – TO measurements

Acknowledgments

The Authors would like to acknowledge the assistance, technical direction and advice of colleagues for their (past) help and support with this paper including Christopher Semprimoschnig, Riccardo Rampini and Marco Schena.

IOP Conf. Series: Materials Science and Engineering 1287 (2023) 012032 doi:10.1088/1757-899X/1287/1/012032

Analysis of micron size space debris impacts on materials: comparison of ground-based testing and in-flight results

Kristien Peeters (*)1, Ricardo Martins[1], Adrian Tighe[1], Riccardo Rampini[1]

[1]ESA-ESTEC, Keplerlaan 1, 2201 AZ Noordwijk (The Netherlands)

(*)Kristien.Peeters@ext.esa.int

Abstract. A new experimental facility has been built in ESA-ESTEC to simulate orbital debris impacts. The facility uses a laser driven flyer plate technique to accelerate small aluminium particles to high velocities. The area of interest is the population of small sized debris objects (<700 μm) with impact velocities up to 10 kms^{-1}. Impact damage has been observed on different types of thin foils, but fragmentation is very noticeable, both at impact and at the creation of the flyer. The size distribution seems random and velocities between 1 and 6 kms^{-1} have been established. Results from different types of targets are presented, including a new type of flyer plate target. The flyer plate method appears to be promising and the new targets show a limited fragmentation on each laser beam shot compared to the others.

1. Introduction

Space debris is an ever growing problem. With the increasing implementation of "Clean Space" policies, the real extent of the problem is slowly being revealed. Near Earth orbits are filled with all sizes of space debris, from a fraction of a micron upwards to several metres, and ground based simulation techniques are required to simulate their impact effects on materials. With 14 450 satellites sent to Earth's orbit and 9 610 satellites still in place, space is far from empty. Their launches, break-ups, explosions, collisions and other anomalous events result in debris build-up and in fragmentation. Using models (MASTER-8) the number of debris objects can be estimated. This model predicts more than 130 million space debris objects from 1 mm to 1 cm and a total mass of objects in Earth orbit of more than 10 400 tonnes. [1]

Space debris particles pose a threat to all present and future spacecraft, and their impacts can have some profound consequences. The smaller the particle, the harder it is to avoid them. For particles smaller than 10 cm, tracking from Earth is not possible, so avoidance manoeuvres become very hard. The threats and consequences can be listed as follows [2]:

- Impact velocity: high speed (up to 10 kms^{-1}), even tiny particles can pose threats.
- Population: rising population increases the potential danger to both robotic and human space missions.
- Damage:
  - Larger fragments (mm or cm size) could penetrate structural materials.
  - Smaller fragments (micron size) may cause a more gradual degradation in functional properties of materials.
- Contamination: Particle contamination arising from the vaporization of the particle itself as well as that of the impacted material.

IOP Conf. Series: Materials Science and Engineering 1287 (2023) 012032 doi:10.1088/1757-899X/1287/1/012032

- Spallation: Internal and back surface spallation of materials resulting from the shock wave produced by the hyper velocity impact.

The impact effects of larger sized space debris particles (size range upwards of a few millimeteres), are widely studied due to their immediate danger after impact, with ground simulation of these particles being well established. This is generally not the case for sub-millimeter sized particles. Whilst the impact of individual micron sized particles may be negligible, an accumalation of multiple impacts could cause a gradual deterioration in the performance of sensitive materials and subsystems exposed directly to the space environment. As the population of space debris rises drastically (including micron sized particles), and the amount of spacecraft in Earth's orbit increases, the need for experiments in this size domain becomes increasingly more apparent. In particular, this research is interesting for highly polished, brittle or coated surfaces, such as optics, solar cells, thermal control components and thin films for de-orbiting sails. This is of interest for missions which are planned to work in space for a long time or for extended life-time missions.

In this work, ground based testing of common spacecraft materials has been performed using the well-established laser driven flyer plate technique [3, 4], and the results have been compared with impacts observed on flight samples from the Long Duration Exposure Facility (LDEF). During this research, debris particles in the size range of 10-350 μm are accelerated to velocities comparable to those of space debris, corresponding to impact velocities up to 10 kms^{-1}. The technique is attractive as an acceleration technique for space debris simulation because of its relative simplicity, low cost, ease of incorporation into a vacuum facility and high shot rate capability. One limitation of the technique is the fragmentation of the flyer plate leaving the target, and alternative techniques have also been investigated to reduce this effect. This paper explains the working principle of the flyer plate generation and the experimental techniques used to assess the materials after impacts.

2. The laser driven flyer plate experiment

A laser driven flyer (LDF) plate experiment is used to accelerate small particles, sizes ranging from 10 μm up to several hundreds of μm with a thickness around a few μm, up to velocities of 10 kms^{-1}. These particles will be fired upon test samples as seen in Figure 1 and the resulting impacts can be investigated. The system uses laser ablation to produce a small particle and plasma expansion to accelerate it to high velocities. The use of a laser has multiple advantages, such as the possibility to generate very short pulses, high peak power and small, localized areas of irradiation. This makes it an interesting system for microscopic high-velocity launches, with a high controllability, reliability, and safety. [5, 6, 7] There is very little chemical contamination (compared to a light gas gun for example), and the laser can be aimed from outside the vacuum chamber.

The working principle of the experiment is based on the theory of ablation and plasma physics, visualized in Figure 2. When a laser irradiates a metal film with a high energy pulse, the laser radiation is absorbed into the metal film, which ablates and generates a plasma. Whenever the intensity of radiation increases above a certain threshold, significant interactions will start to occur within the metal lattice. The energy that is absorbed will heat up the material and cause a change in the microstructure. At high intensities, the metal target will start to sublimate, and a very dense plasma can be formed. This will be a high-temperature and high-pressure plasma, which will expand perpendicularly to the metal film. This creates a pressure gradient between the plasma pressure on one side and the vacuum zero pressure on the other side, causing the unvaporized layer of film to accelerate away and impact with a sample as seen in Figure 2. [5, 6, 7, 8]

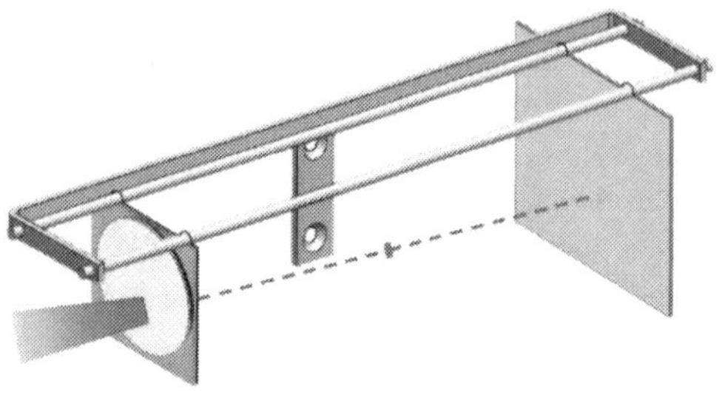

Figure 1: Overview of the laser driven flyer plate experiment: target (grey); laser beam (red); small flyer plate (black) and test sample (yellow).

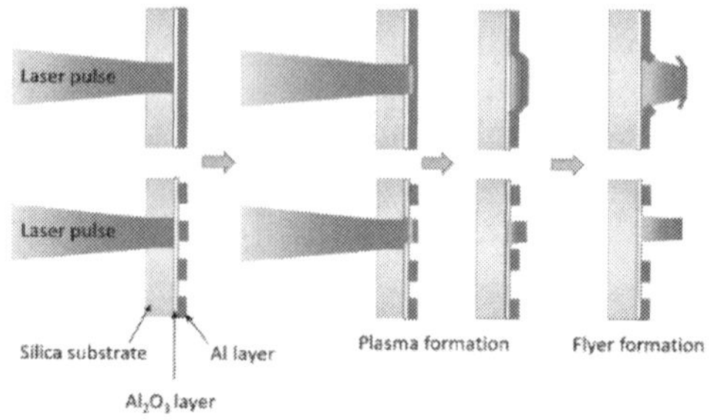

Figure 2: Flyer plate acceleration process using a fully coated target (top) and an etched target (bottom).

3. Experimental

3.1. Facility

A schematic of the experimental facility is shown in Figure 3. The laser used is a pulsed Nd:YAG Quantel Q-smart 850 with operating conditions shown in Table 3. Additionally, a He-Ne laser is used for alignment purposes. A combination of neutral density filters and a motorized attenuator are used to vary the energy of the laser beam. After this, the beam is focused onto the target using a fused silica plano-convex lens. Depending on the application, different lens focal lengths can be used to vary the spot size as well. The vacuum chamber is a stainless-steel chamber, 320 mm in diameter. The chamber houses the target and the sample on a XZ precision manipulator. The entrance window is fused silica with an anti-reflection coating and the chamber has some other ports for diagnostic and access purposes. The target and the sample are attached to the same sample holder, as seen in Figure 1, which is attached to the XZ manipulator. The manipulator translates the target and the sample across the laser beam between laser shots to make sure that each time a fresh area of the target is exposed. The rods of the sample holder are 200 mm long, so it is possible to position the target and the sample up to 200 mm from each other. Experiments were performed for different distances from 50 mm to 200 mm.

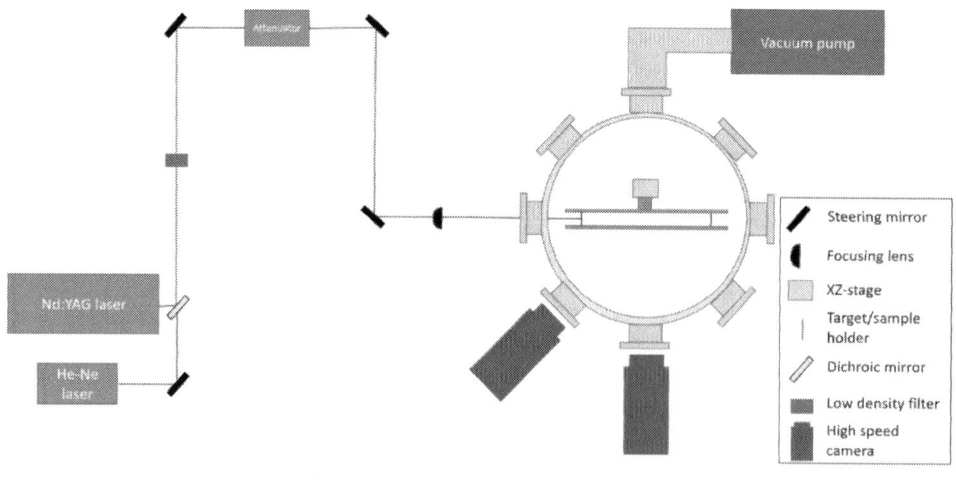

Figure 3: Experimental facility for simulation of impacts using laser driven flyer plates.

3.2. Targets

A typical LDF target is composed of a dielectric window, an adiabatic layer, and a flyer layer. The dielectric window absorbs the incident laser pulse as it generates an extremely high temperature and high-pressure plasma to drive the flyer up to high velocities. In this work the target is a transparent UV fused silica disk. The adiabatic layer, which in this case is Al_2O_3, is placed in between the window and the flyer layer to prevent the full heat being transferred to the metal film, to improve the integrity of the formed flyers. Secondly, it also plays a role as an adhesion promotor for the uppermost Al layer. [6] In the current work two types of targets have been used: Fully coated targets and etched targets seen in Figure 4. At first, the fully coated targets only consisted of in-house manufactured targets, but later also continuously sputtered targets were used. Both the continuously sputtered and etched targets were manufactured by an external coating company (Consorzio, CREO, Italy).

Figure 4: Targets used in this work (1 inch): Fully coated in-house manufactured target (left), Fully coated continuous film sputtered target (middle) and Etched target (right).

3.2.1. Fully coated in-house manufactured targets. The fully coated targets used in the first part of this work were manufactured in-house and did not have the Al_2O_3 layer. They were composed of a Ø1" x 1mm thick silica or Infrasil window and a layer of UHV aluminium foil of 38 μm thickness. Therefore, the targets were not optimal to produce the flyers. The principle of this target is that the laser beam is used to define the flyer plate area. This is one of the standard methods of flyer plate generation which has been previously reported in literature. [3, 4, 7]

3.2.2. Continuous film sputtered targets. The second type of fully coated targets are created by Consorzio CREO, and they consist of a continuously sputter-deposited aluminium layer on a Ø1" x 1 mm thick Infrasil disk, with more specs in *Table 1*. This type of target relies on the same principle as the in-house manufactured ones, namely the laser beam spot defines the flyer plate area. However, the aluminium layer is better protected due to the Al_2O_3 layer in between the aluminium and the silica disk.

Table 1: Specifics of the continuous film sputtered flyer targets created by CREO.

Substrate	
Material	Silica disk (Infrasil 302)
Diameter	25.4 mm
Thickness	1 mm
Al_2O_3 layer	
Thickness	100 nm
Al layer	
Thickness	3 μm

3.2.3. Etched targets. This is a more novel type of target, with only limited work performed previously. The principle is that the flyer plate area is pre-defined on the window, which may help to minimise the flyer plate fragmentation. The expected limitation is that the flyer plate velocity will not be as high, because the driving plasma is not confined around the edges of the flyer. The etched targets used in this work consisted of a Ø1" x 1 mm thick Infrasil disk, covered with an array of aluminium dots as seen in *Figure 5*. To achieve this pattern, part of the aluminium layer that is sputtered on the Infrasil window is etched away so that an array appears. The dots have a 50 ± 5 μm diameter and have a spacing of 1 mm apart, the specifics can be found in *Table 2*.

Table 2: Specifics of the etched flyer targets created by CREO.

Substrate	
Material	Silica disk (Infrasil 302)
Diameter	25.4 mm
Thickness	1 mm
Pattern	
Dot's diameter	50 ± 5 μm
Dot spacing	1 mm
Al_2O_3 layer	
Thickness	100 nm
Al layer	
Thickness	3 thicknesses: 2,3 and 5 μm

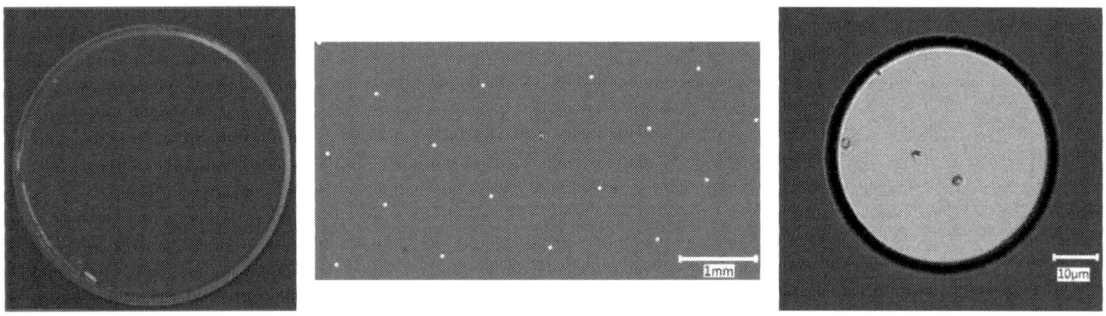

Figure 5: Images of an etched target. Full one inch target (left) zoomed image (middle) and a zoom of an individual flyer (right).

3.3. Laser

The setup requires a powerful laser capable to provide enough energy and stability to accelerate flyers to the appropriate velocities. For this a Nd:YAG laser provided by Quantel is used, specifications can be found in Table 3. The laser power was measured by a volume-absorbing thermopile sensor. It measures the average power (P_{Avg}) of the pulsed laser system and from this the power per pulse can be obtained. Knowing the repetition rate (f) of the system the individual laser pulse energy (E) can be calculated using $P_{Avg} = E \cdot f$.

Whenever a sample is used that is too thin or fragile to withstand the laser beam, the laser might cause harm to the sample on the spot that needs to be investigated. This is sometimes the case for the etched targets since excess laser radiation can pass round the edges of the dot and interact at the point of the flyer impact. For the different samples used, tests were conducted to investigate if the samples needed extra protection during the real experiments. A solution, as visualized in Figure 6, for this problem is to tilt the target and the sample with respect to the laser beam axis by ±6°. By tilting the system, the laser interaction will be at another spot as the flyer impacts.

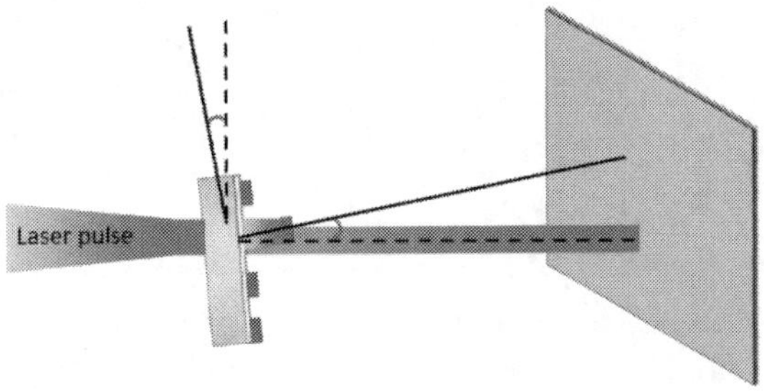

Figure 6: Overview of the tilting mechanism to avoid laser damage.

Table 3: Specifics of the Nd:YAG laser.

Laser wavelength	1064, 532, 355, 266, 213 nm
Max. laser energy (E)	0.850 J
Repetition rate (f)	10 Hz
Lamp pulse width (LPW)	170 µs
Pulse duration (FWHM) @ 1064 nm	6.4 ns
Beam width	~ 9 mm
Beam divergence	< 0.5 mrad
M^2	≤ 2

3.4. High-speed camera
High-speed cameras were used to film and detect individual particles flying by, and to film the different processes during the experiment, e.g., flyer formation, impact on sample. The cameras are provided by Photron, namely the FASTCAM SA-X2 type 480k. The system offers recording rates up to 13 500 fps at megapixel image resolution. The maximum speed 480k fps of the cameras should be high enough to be able to detect the flyer plates. However, an increasing frame rate results in a decreasing field of view and therefore a decreasing resolution. It is noted that a significant amount of light is needed inside the vacuum chamber to allow detection of the flyers via high-speed imaging. Since the experiment happens in a sealed vacuum chamber, the illumination was provided with an external light source through one of the vacuum ports. This limited the frame rate to 50k frames per second, with a resolution of 768x328. These acquisition conditions were sufficient for detecting the flyer plates in our test setup.

Image and video recordings were acquired during the different processes of the LDF experiment, with the image speed cameras. Four main parts can be distinguished: 1. Plasma formation, 2. Flyer formation, 3. Flying phase and 4. Impact on the sample and the latter three are shown in Figure 7. From the recordings it became evident that fragmentation of the fully coated target occurs directly after the flyer formation, the particle fragments into many small pieces. Through image processing of the recordings, we can calculate the velocity of the individual particles, if they fly in the correct plane of sight. By measuring the distance a particle travelled during one frame and knowing the frame rate, the velocity of each visible particle can be calculated.

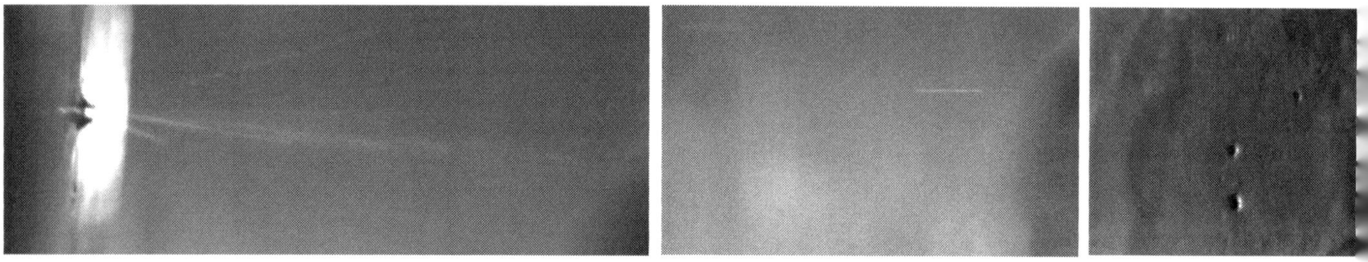

Figure 7: Different phases of the LDF experiment for a fully coated target. Flyer formation (left), Flying phase (middle) and Impact on sample (right).

3.5. Measurement of fragment size

To estimate the diameters of the particles that produced the impact craters, the following hole growth equation was used [9]:

$$\frac{d}{f} = A\left(\frac{10}{9 + e^{D/fB}}\right) + \frac{D}{f}\left(1 - e^{-D/fB}\right) \qquad 3.5.1$$

Where D is the hole diameter, d is the particle diameter, f is the foil thickness, V is particle velocity, $A = 6.97 \cdot \left(\frac{V\rho_p}{(\sigma_t \rho_t)^{\frac{1}{2}}}\right)^{-0.0723} \cdot \left({\sigma_t}/{\sigma_p}\right)^{-0.217} \cdot f^{-0.053}$ and $B = 1.85 \cdot 10^{-3}V - 0.004$.

An aluminium foil (4.5 μm) was placed in the path of the flyer plate. The impacts on this foil could be analysed in the digital optical microscope, to calculate the hole diameters. Finally, the hole growth equation, with all parameters extracted from experiments of hypervelocity impacts with aluminium particles onto aluminium foil, was used to convert these hole diameters into particle sizes.

3.6. Penetration limit equation

Numerous empirical and semi-empirical relationships have been developed to convert crater/perforation morphology to particle mass/diameter, equivalent crater size in aluminium or equivalent penetration thickness. All these methods depend on assumptions about projectile densities, velocities, and interaction with the target. The equations are developed using regression methods to approximate particle impact simulations carried out in the lab in previous work in literature. [10] Using experimental data under a wide range of different conditions, several penetration limit equations have been created over the years, as seen in Table 4. The penetration limit refers to the plate thickness at which a crater becomes a perforation, or with a known foil thickness, it can give the velocity at which an impact becomes a penetration. To calculate the penetration limit in this work, the average of all relationships in Table 4 is used, since none of the methods exactly uses the conditions applied to this work.

Table 4: Penetration limit equations, for a particle with diameter d, and velocity V, impacting foil of thickness f.

Particles used			Target	Equation	Original reference
d (μm)	V (kms^{-1})	Type			
μm size	2-20	Fe	Metallic	$f/d = 1.02\ d^{0.056}V^{0.664}$	McDonnell and Sullivan (1992) [11]
10-100	4-13	Glass	Metallic	$f/d = 1.70\ d^{0.056}V^{0.875}$	Nauman (1966) [12]
/	/	/	/	$f/d = 1.06\ d^{0.056}V^{0.67}$	Cour-Palais (1979) [13]
mm size	/	Metallic	/	$f/d = 0.70\ d^{0.056}V^{0.875}$	Fish and Summers (1965) [14]
μm size	/	Various	/	$f/d = 1.02\ d^{0.21}V^{0.880}$	Pailer and Gün (1980) [15]
/	/	Fe	/	$f/d = 0.85\ d^{0.056}V^{0.667}$	Humes (1991) [16]

4. Test Results

For the experimental ground simulation of small-sized debris impacts, different targets were used in this work. The generation of flyers leaves a mark and a hole behind on the target. The shape of the remnants that are left behind on the different type of targets differ strongly from one another as seen in Figure 8. This will also influence the shape and the fragmentation level of the flyer created.

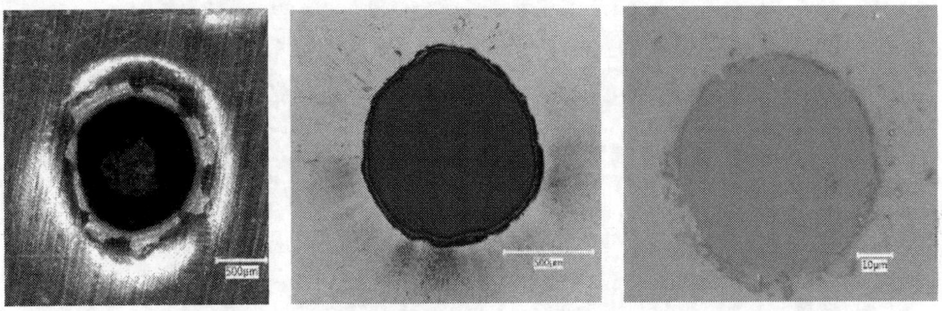

Figure 8: Remnants after laser irradiation for the different type of targets; In-house manufactured target (left), Continuously sputtered target (middle) and Etched target (right).

4.1. In-house manufactured target results

All experiments with the fully coated in-house manufactured target have been performed using a lens with focal length of 200 mm, resulting in a beam width of approximately 0.135 mm, producing flyers with a radius between 250 and 700 μm. These flyers eventually fragment in numerous particles with diameters ranging between 10 - 350 μm. Experiments have been performed on two types of samples:

- 7 μm thick copper foil
- 7 μm thick double-sided aluminized Mylar

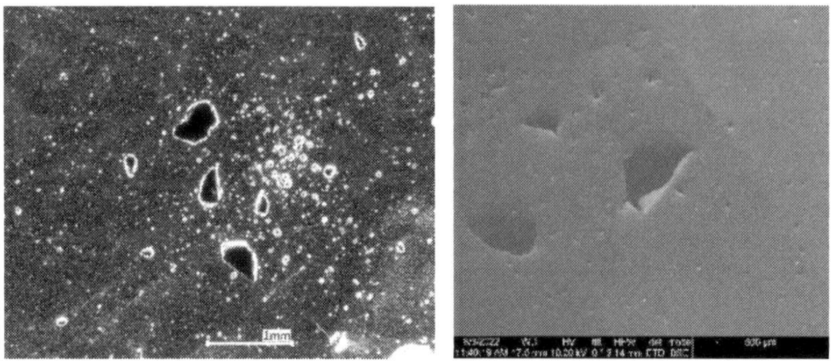

Figure 9: Impact of flyer plate fragments on double sided aluminized Mylar foil.

The samples were analysed with both a digital optical microscope and a scanning electron microscope (SEM). An optical microscope image and an SEM image of impacts from a fully coated target on a Mylar foil are shown in Figure 9. The laser beam energy for this shot was 140 mJ. In these images the fragmentation of the flyer can clearly been seen, many of the particles have made impact holes in the Mylar foil.

Figure 10: SEM map of the flayer plate impact on 7 μm thick Cu foil.

Figure 10 shows the full mapped area and Figure 11 shows zoomed in images of impacts of a fully coated target flyer plate into a 7 μm thick copper foil, shot with a laser beam energy of 128 mJ. The flyers created both fully penetrated holes and indents without full penetration. The impact craters are spread out over the target in almost a straight line covering 11.3 mm. The impacts have radii of 162 μm and 46 μm, which corresponds to approximate particle sizes of respectively 148 μm and 34 μm, as per equation 3.5.1.

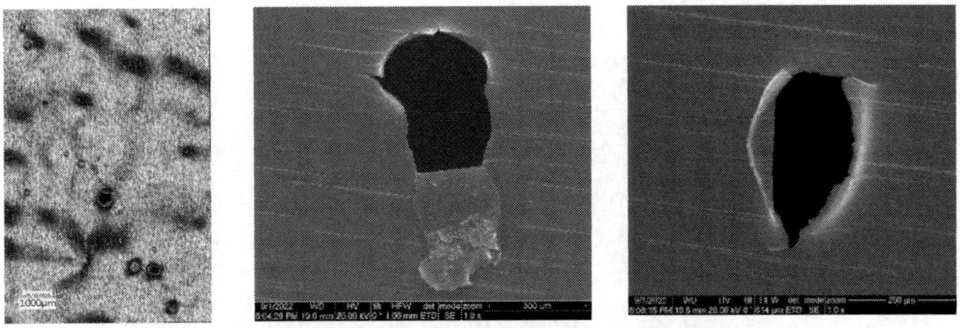

Figure 11: Impact of flyer plate fragments on 7 μm thick Cu foil.

Multiple experiments measuring the velocities of the fully coated in-house manufactured target were performed using the high-speed cameras. The particles were best visualized at a framerate of 40k or 50k, resulting respectively in a resolution of 896x368 or 758x328. Due to the fragmentation of the flyers a wide range of different sized particles with a wide range of velocities can be witnessed in the videos. Even for one laser shot the velocities can vary from 6 kms^{-1} to less than 1 kms^{-1}. With the high-speed cameras the lower limit for a clear particle detection did not necessarily depend on the velocity of the particles, however, the size of the particles is important. The bigger and thicker the particles, the easier they are seen. The velocity and fragmentation rate does impose an upper velocity limit for detection of the flyers. The particles can fly too fast to be captured nicely on camera, and when there is too much fragmentation, their trajectories will overlap with each other, making it harder to distinguish them. These upper limits are not yet exactly established.

4.2. Continuously sputtered target results

The experiments with the sputtered fully coated target were performed both using a 500 mm and a 200 mm focal length lens. The former lens created flyer sizes between 450 and 500 μm and the latter created flyer sizes of about 315 μm. These targets were still prone to fragmentation and therefore none have been caught on high-speed camera yet. The particles were too small or not in a correct state to be seen at that frame rate. It is speculated that the sputtered layer of aluminium on the target was not thick enough to create visible flyers, which will be further researched. The flyers also did not manage to penetrate or even indent the foils, even at very close distance as seen in Figure 12, supporting the previously raised idea that the flyers were not big enough. The aluminium was just deposited on to the foil, instead of going through, which could mean the flyers were not going fast enough or the aluminium was still in a heated/ soft state after the plasma formation. Using the penetration limit equations from Table 4, the upper velocity limit for the etched flyers can be calculated. Since they were not able to penetrate the 7 μm thick foils, this puts the upper limit at 513 ms^{-1}, with the actual particle velocities most probably well below this limit.

Figure 12: Impact of a continuously sputtered target on a 7µm thick Cu foil

4.3. Etched target results

A limited number of experiments have been conducted with the etched targets. The laser beam was focused with a 500 mm focal length lens and the target was deliberately moved out of the focal point. This was done to try and get a more uniform illumination over the central part of the dot. During some of the test runs, it was seen that focusing the beam onto the target resulted in damage on the target, with no flyers being created. However, other runs of the experiment with the etched targets showed that the fragmentation was significantly less. Usually there were about 2 to 6 fragments seen on the sample per shot, which were 10 to 25 µm in size. The flyers, however, were not going fast enough to really penetrate the foil, they were only deposited on the surface of the impact sample, just like with the continuously sputtered targets. Typical examples for these are seen in Figure 13.

The etched target technique looks promising to reduce the fragmentation of the flyers, since they seem to stay intact moving over long flight distances. Since the dot has a predefined size which turns into a flyer, a well-defined amount of material is accelerated with each laser shot and reaches the sample. Using the penetration limit equations from Table 4, the upper velocity limit for the etched flyers can be calculated. Since they were not able to penetrate the 7 µm thick foils, this puts the upper limit at 513 ms^{-1}, with the actual particle velocities most probably well below this limit.

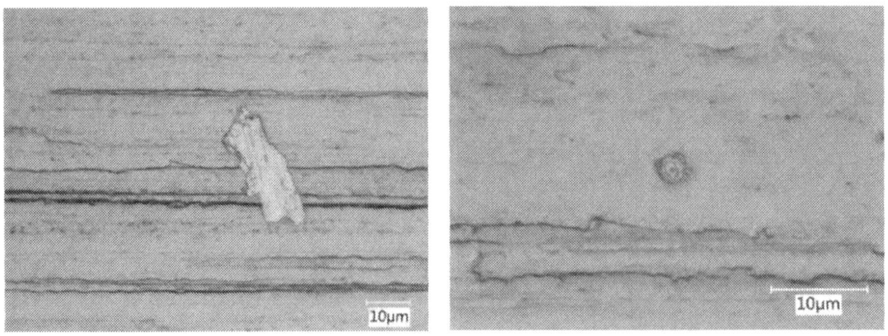

Figure 13: Impact of an etched flyer plate on a 7 µm thick Cu foil.

5. Comparison with inflight results

The ground-based results from the flyer plate experiments have been compared with craters observed on the retrieved samples of silver Teflon (Ag/FEP) blankets from the Long Duration Exposure Facility (LDEF) [17], which are available in ESTEC. The same analysis methods were used, and the morphology of the craters was compared (see Figure 14). It can be seen that many features of the craters are similar, showing that the flyer plate technique can be used to simulate space debris impacts, even when fragmentation of the flyer plate occurs.

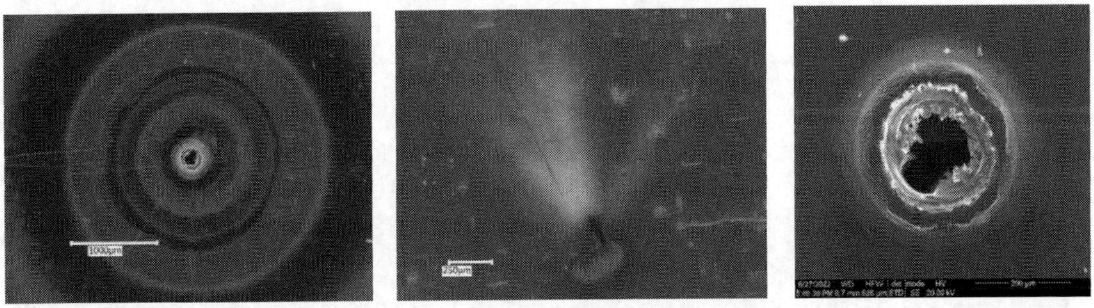

Figure 14: Digital microscope images (left and middle) and an SEM image (right) of impacts on LDEF sample.

6. Conclusions

To simulate orbital debris impacts, a new ground based experimental facility has been built, using the technique of laser driven flyer plates. While in previous works [3, 4, 8] the simulated debris flight distances were kept quite short (in the order of a few hundred microns), these distances were increased in this work up to tens of centimetres. The results of the fully coated targets show that the flyer plates tend to fragment into several hundred pieces, with a random size and velocity distribution. The fragment diameters resulting from the in-house manufactured targets were in the range of 10 – 350 µm. Using the high-speed cameras, velocities up to 6 kms^{-1} are measured. It was possible to measure the velocities of individual particles, as they fly within the cameras field of view.

The second type of targets is a novel way of creating laser driven flyer plates, with only limited references to this technique reported in the literature. The etched targets are based on an array of individual aluminium dots predefined on the substrate. From a limited set of results obtained in this work, the technique shows promise as a means of generating less fragmentation with each laser beam shot. In our work dots with a diameter of 50 µm were used, which did not fragment as much as the in-house manufactured targets, with only up to 4 flyer fragments, even over a flight distance of 200 mm. On the other hand, it is likely that the velocities of the flyer fragments were much lower, as they were not able to penetrate the 7 µm thick Cu foil. This puts an upper limit of about 500 ms^{-1} on the velocities, with the particle velocities probably well below this value. Velocity measurements, with these sputtered targets, were unsuccessful: the particles created in the sputtered targets have sizes below the detection limit of the high-speed cameras.

Further research and experimentation are required to extend the possibilities of this facility. At the moment work is being done to reduce the fragmentation, increase the velocity and improve all types of velocity measurements. Investigations into different environmental settings, such as UV, ATOX, etc. will be done to recreate the space environment more accurately and get a closer representation of real space debris impacts. The experimental work will be complemented with the modelling of the different phases of the experiment, such as the trajectory of the flyers, the level of fragmentation, the impact etc. to get a better insight on how the system behaves.

IOP Conf. Series: Materials Science and Engineering 1287 (2023) 012032 doi:10.1088/1757-899X/1287/1/012032

References

[1] ESA Space debris by the numbers, Safety & security (Accessed: 15 December 2022) Available from: https://www.esa.int/Safety_Security/Space_Debris/Space_debris_by_the_numbers

[2] Stein C, Roybai R and Tlomak P 2000 The production of contamination on spacecraft surfaces by hypervelocity debris impacts *8th ISMSE conf.*

[3] Brown K E, Shaw W L, Zheng X and Dlott D D 2012 Simplified laser-driven flyer plates for shock compression science *Rev. Sci. Instrum.* **83** 103901

[4] Curtis A D, Banishev A A, Shaw W L and Dlott D D 2014 Laser-driven flyer plates for shock compression science: Launch and target impact probed by photon Doppler velocimetry *Rev. Sci. Instrum.* **85**(4) 043908

[5] Dai F, Gong Z, Cao Y, Yang J and Zhang W 2018 Study on the technology of laser-driven flyer hypervelocity launch *IOP Conf. Ser.: Mater. Sci. and Eng.* **381** 012155

[6] Ji X, Qin W, Wu X, Wang Y, Gao F, Wang L and Hou Y 2021 Perfect metamaterial absorber improved laser-driven flyer *Nanophotonics* **10**(10) p2683-93

[7] Tighe A P 2000 Ground based simulation of orbital debris using laser driven flyer plates PhD thesis Universitiy of Southampton UK

[8] Banishev A A, Shaw W L, Bassett W P and Dlott D D 2016 High-speed laser-launched flyer impacts studied with ultrafast photography and velocitmitry *J. dynamic behavior mater.* **2** p194-206

[9] Gardner D J, McDonnell J A M and Collier I 1997 Hole growth characterisation for hypervelocity impacts in thin targets *International Journal of Impact Engineering* **19**(7) p589-602

[10] Berthoud L and Mandeville J C 1993 Empirical impact equations and marginal perforation *Proc. 1st European Conf. on Space Debris* (ESA SD-01)

[11] McDonnell J A M and Sullivan K 1992 Hypervelocity impacts on space detectors: Decoding the projectile parameters *Hypervelocitiy impacts in space* ed J A M McDonnell publ. Univ. of Kent at Canterbury GB

[12] Nauman R 1966 The near-earth meteoroid environment NASA TND 3717

[13] Cour-Palais B G 1987 Hypervelocity impacts in metals, glass and composites *International Journal of Impact Engineering* **5** p681-92

[14] Fish R H and Summers J L 1965 The effect of material properties on threshold perforation *Proc. 7th Hypervelocity Impact Symp.*

[15] Pailer N and Grun E 1980 The penetration limit of thin films *Planetary and Space Science* **28** p321-31

[16] Humes D H 1991 Large craters on the meteoroid and space debris impact experiment part 1 of LDEF-69 months in space *first post-retrieval symposium NASA Conf. Publ. 3134*

[17] Zwiener J M and Finckenor M M 1993 Micrometeorid/space debris effects on materials *LDEF Materials Results for Spacecraft Applications NASA Conf. Publ. 3257* p259-80

IOP Conf. Series: Materials Science and Engineering 1287 (2023) 012033 doi:10.1088/1757-899X/1287/1/012033

Molecular contamination control for Exomars 2022 Mission

A. Pandi[1], I. Locantore[1], M. Giuliani[1], A. Saverino[1]

[1]Thales Alenia Space, Strada Antica di Collegno 253, 10146, Torino, Italy

arianna.pandi@thalesaleniaspace.com

Abstract. The ExoMars 2022 integration facility has been built at the Thales Alenia Space Italy in Turin. The facility includes a GBT (Glove Box Train), that is an ISO3 class facility, totally aseptic and characterized by controlled airborne molecular organic contamination. This new facility provides an environment strictly controlled for what concerns cleanliness and contamination, specifically conceived to perform AIT of The Ultra Clean Zone (UCZ) of the ExoMars Rover Analytical Laboratory Drawer.
Indeed, standard methods as per ECSS-Q-ST-70-05C and ECSS-Q-ST-70-50C are not suitable for the monitoring of integration facility and Flight H/W. The environment of GBT has been required to be monitored in terms of airborne molecular contamination (AMC) on the basis of ISO 14644-8. The maximum contamination levels allowable on flight H/W after accelerated CO_2 snow cleaning and at the end of integration activities are one order of magnitude less than the detection limit of standard FT-IR techniques for molecular contamination.
For these reasons a dedicated method has been set up for molecular contamination monitoring. This paper shows the development of Thermal Desoprtion-Gas Cromatography/Mass Spectrometer (TD-GC/MS) method for the monitoring of extremely low organic contamination levels both on air samples and on witnesses representative of the Flight H/W, providing also the most recent results obtained in the frame of ExoMars 2022 UCZ QM and FM AIT Campaign.

1. Introduction

The ExoMars Spacecraft consists in a Carrier, an Entry, Descent and Landing Module hosting a 300 kg Rover. The mission scientific objectives that will be pursued by the Rover and the PPL suite are:

- To search for signs of past and present life on Mars;
- To investigate the water/geochemical environment as a function of depth in the shallow subsurface.

Scientifically, the ExoMars Rover will search for traces of past and present life, characterizing Martian geochemistry and water distribution at various locations, improving the knowledge of the Mars environment and geophysics and identifying possible hazards before landing other spacecrafts or, in the longer term, humans.

The Analytical Laboratory Drawer (ALD) is the key subsystem to accomplish the automatic sample preparation and distribution for analyses of scientific experiments. It accommodates the Pasteur Payloads Analytical Instruments and the Sample Preparation and Distribution System (SPDS), providing them structural support, thermal control and contamination and microbial controlled environment to cope with Life Detection Instruments needs in terms of cleanliness. The contamination

IOP Conf. Series: Materials Science and Engineering 1287 (2023) 012033 doi:10.1088/1757-899X/1287/1/012033

controlled volume where the Martian Samples will be processed and analysed is called Ultra Clean Zone (UCZ). UCZ remain over-pressurized on ground and until landing on Mars to prevent inflow of contaminants.

In order to minimize biological cross-contamination between Mars and the Earth and to avoid false positive and false negatives during the search for extraterrestrial life, a strict control of microbiological and organic contamination has been pursued during UCZ integration.

1.1. Ultracleanliness levels applicable to UCZ

The driving requirement applicable to all UCZ surfaces limits the maximum contamination level of organic compounds per gram of Martian sample. A bottom-up approach was used to derive the contamination of the Martian soil at the End of Mission sample as a sum of the different contributions, in order to assess the compliance with the starting above mentioned requirement. A Contamination Budget has been set for the UCZ relying on different contributors, inducing all pertaining Mission phases from on-ground integration until End of Mars operation. In particular, for the on-ground phases, two main contributions and relevant maximum allowable levels were defined:

1. Residue from UltraCleaning Treatment
2. Contamination deposition during UCZ AIT

Before the integration phase of the UCZ, it was required that all UCZ surfaces in direct contact with the Martian sample were cleaned by means of an Ultra cleaning treatment in order to fulfil the defined cleanliness level of a maximum 10 ng/cm^2 . The final qualification of the whole Ultra Clean assembly and Testing of all UCZ equipment was completed validating the entire process on the UCZ Qualification Model (QM), that allowed lowering, for UCZ Flight Model (FM), the NVR ultracleaning residue acceptability range in the order of 3-5 ng/cm^2 on the basis of QM experience.

The ultra-cleaned cleanliness levels required for the UCZ surfaces led to the definition not only of the UltraCleaning Process, but also to the design of a special UltraClean environment, a Glove Box Train (GBT) Airborne Molecular Contamination (AMC) controlled where the treatments and the AIT activities can be performed without impairing the achieved cleanliness levels. For this reason, all UCZ parts were Ultra-cleaned, assembled and tested in AMC controlled Environment (as per ISO 14644-8, AMC-7 was considered adequate after QM campaign) until the UCZ compartment is Sealed and Pressurized with Ultrapure Argon.

On the basis of the rate of molecular contamination deposition inside GBT environment, a maximum value of 12 ng/cm^2 has been allowed starting from ultracleaning process until end of integration.

2. Monitoring approach and analysis method description

The monitoring activities started with the installation, cleaning, setup and commissioning of the GBT. The quantitative method set for AMC monitoring has been optimized initially for 17 organic contaminants. Then, it was agreed to use a semiquantitative method, allowing to have a complete overview of all airborne contaminants present in Glove-Boxes (GB) or deposited on Flight H/W surfaces. This method allows a total quantification and a qualitative analysis, without a reliable quantification of each contaminant. The method, starting from sampling to the final analysis, has been tuned according to the sample to be analyzed (i.e. GB air or metallic witnesses).

2.1. Methods for GBT AMC monitoring

In order to sample the GBT air, a pump air sampler Markes ACTI-VOC for each GB was used. Each GB sampling point (1 per GB) has been equipped with one AMC sampling line/ system made of stainless steel tube with the pump downstream (see Figure 1).

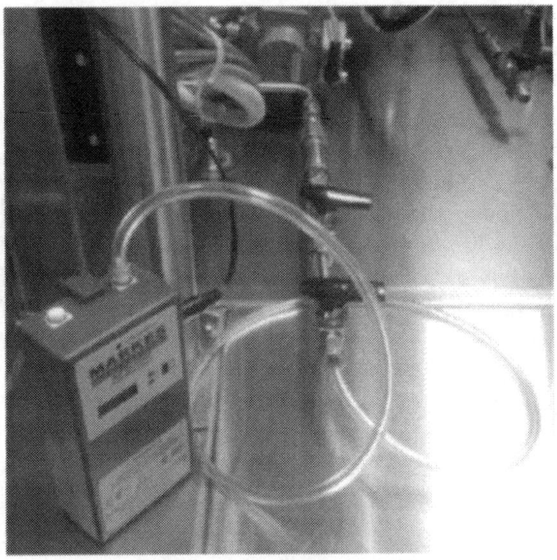

Figure 1. Glove box sampling line

The used sampling method has been verified as the most suitable (w.r.t. detection limits, sampling time, amount of sampled air required for the analyses) and is the following:

- Sampling flow of 65 ml/min in the sorbent tubes
- Sampling time of 24 hours (to sample about 100 liters of air) compatible with the sorbent matrix breakthrough volumes
- Weekly sampling and analysis of the 4 GBs in parallel
- Tenax TA 60/80 stainless steel used as sorbent tube

The tenax installation and removal can be done without any risk of cross contamination, due to the velocity of the operation and the slow diffusion through the tube. During transport, the tenax tube is locked with dedicated safe lock brasscap, packed in a double CPstat bag and transported inside a stainless steel box. Several tests have been performed to understand the possible delta contamination due to tenax transport and handling before and after sampling and no appreciable contamination was detected. The sorbent tubes are analysed with a TD-GC/MS system at TASinI Eclipse TechnoArea (see Figure 2).

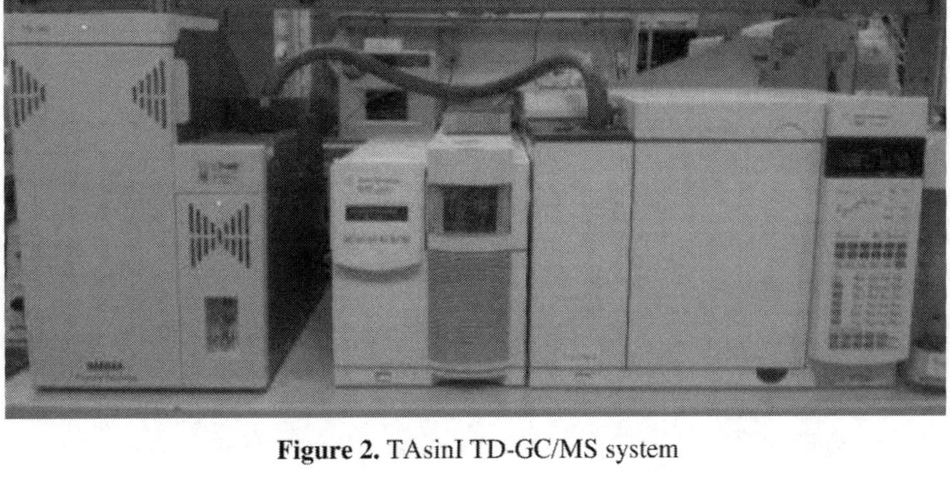

Figure 2. TAsinI TD-GC/MS system

The instrumentation includes: TD100™, an automated thermal desorption system coupled with GCMS from Agilent Technologies instrumentation with GC column: HP5-MS Ultra Inert (0,25mm x 30m x 0,25μm).

The method is semiquantitative since it is based on calibration curves of two contaminants only, dodecane (C12) and eicosane (C20), chosen respectively as representative of volatile and semi-volatile compounds. The achievable concentration limits have been evaluated on the basis of a dedicated campaign carried out during the method set-up and calibration, in conjunction with the first phase of UCZ QM activities. Semi quantitative method allowed to quantify an AMC level between AMC-7 and AMC-6 according to ISO 14644-8. According to the validated correlation between AMC values and deposited levels, the quantification performances of the method have been considered suitable to monitor GB environment and control very low molecular deposition levels. To reduce tenax background as much as possible, tubes were conditioned at high temperature (until 335°C) before using a new set of tubes. Long-term storage caps (brass caps) were tightened during each handling in order to ensure a leak tight seal. Then, tubes are checked before use by at least two consecutive runs of the sampling tube in order to check the background and reduce peaks on the blank chromatogram. The limit of detection of the method is calculated as 3σ of tube set blanks signal quantification and during UCZ QM and FM activities it varied from about 9 ng/cm^2 to 15 ng/cm^2. The analysis parameters are the following:

- Tubes desorption
  - First stage desorption: 5 min at 50 °C
  - Second stage desorption: 15 min at 300°C_No split
  - Trap temperature: from -25 to 300 °C (hold for 5 min)_Split 1:2
- GC ramp temperature: Hold @ 45°C , Ramp @ 4°C/min to 100°C for 0 min, Ramp @20°C/min to 325°C for 2 min

2.2. Witnesses use and analysis method

Metallic witnesses (see Figure 3 and 4) have been used to monitor molecular contamination deposition on UCZ surfaces during AIT. The witnesses are prismatic samples made in titanium, AISI 316L stainless steel or aluminum, fully representative of UCZ materials surfaces, also from a surface roughness point of view. They are divided in two types with different scopes:

1. Environmental witnesses (EnW), exposed inside GB environment to monitor the deposition rates during integration activities;
2. In-Process witnesses (InpW) to analyse the ultracleaning residue.

In this second case, the witnesses were used to have an indirect verification of the cleanliness level achieved on UCZ parts after the last step of cleaning with accelerated snow CO_2, since a direct verification on H/W is not possible (both due to the complex geometry of parts and since the cleanliness verification itself might re-contaminate the ultra-cleaned H/W).

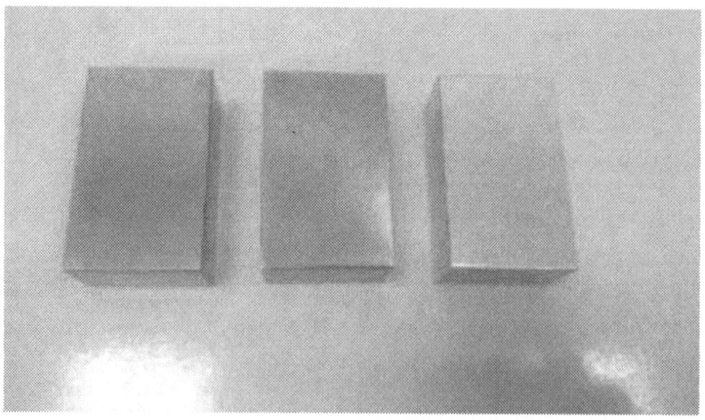

Figure 3. In-Process Witnesses (SS, Titanium, Aluminum from left to right)

IOP Conf. Series: Materials Science and Engineering 1287 (2023) 012033 doi:10.1088/1757-899X/1287/1/012033

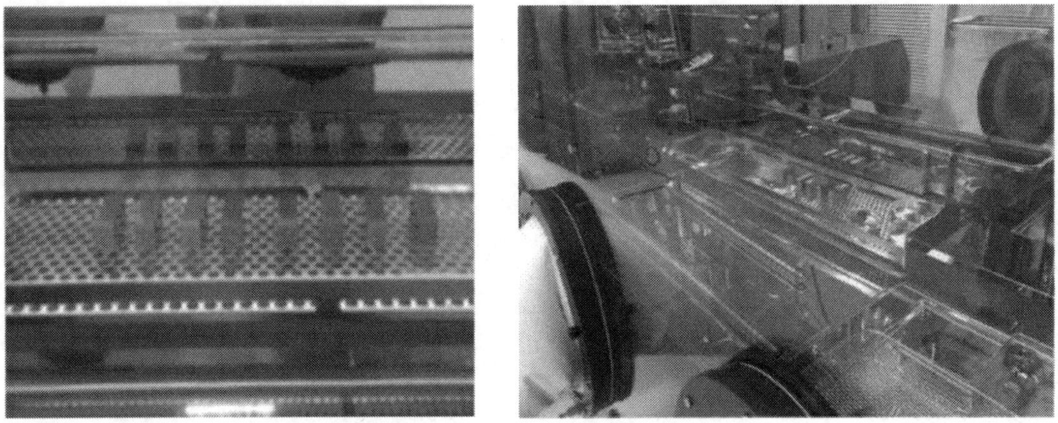

Figure 4. ENW and InPW inside GBT

Different EnWs were cleaned in the same way of the Flight H/W and one of them (one per batch of exposed witnesses) was analyzed after CO_2 accelerated snow cleaning step to quantify the value at T_0. Then, they were exposed inside GBT, placed on dedicated support to avoid as much as possible the handling by the operators and retrieved at identified timeframes to follow the organic contamination deposition rate during time.

InPW instead, were processed together with the Integration KITs. UCZ parts, after receiving from supplier, were dismounted to perform ultracleaning process and then reassembled in clean conditions inside GBT. All Flight H/W have been divided in different Integration KITs on the basis of parts dismounted, materials and AIT flow. One InPW was associated to mechanisms KITs considered critical for large surface, view/proximity of the sample and contact with the sample. From UCZ FM AIT phase start, at least two InPWs have been associated to each identified KIT to test the repeatability of cleaning process and/ to have a spare in case of high values. InPW followed from the beginning the ultracleaning treatment of the Flight H/W, starting from ultrasonic baths in multiple solvents, Dry Heat Microbial Reduction (DHMR) treatment and accelerated CO_2 snow cleaning. Then, one witness was retrieved from GBT for TD-GC/MS analyses. The witnesses were placed inside ultracleaned Alu boxes and double packed in Alu foil for Ultra High vacuum already ultracleaned. Only after this operation they were ready to leave the GBT environment, and transported outside GBT and ISO7HC cleanroom inside Cpstat or Kapton bag in order to avoid any cross-contamination during subsequent handling and transport to TASinI Eclipse Area.

The witnesses were thermally desorbed directly in Micro-Chamber/Thermal Extractor at 220°C for 10 minutes. Tenax tubes are directly connected to the chamber to collect the desorbed compounds. Then, tubes were thermal desorbed with the same method used for airborne monitoring.
The quantification followed also in this case the semi quantitative method but considering eicosane (C20) only, since the focus of the analysis is to monitor the semi-volatile compounds that can be deposited on the surfaces. The LOD is calculated considering the 3σ of chamber background and during UCZ QM and FM activities was about 2 ng/cm^2.

3. Monitoring results during UCZ FM AIT

3.1. GBT airborne environment monitoring results

Records of GBT monitoring with semi quantitative method started from February 2017 until April 2019, when UCZ FM has been delivered to RM. GBT have been sampled every week, with the exclusion of holidays and some periods used for TD-GC/MS check and calibration when analyses could not be performed. On the basis of first trails and UCZ QM campaign, the alert was set at 1000 ng/cm^3, corresponding to AMC-6 level.

Figure 5 shows all the results to understand the AMC behavior in GBT.

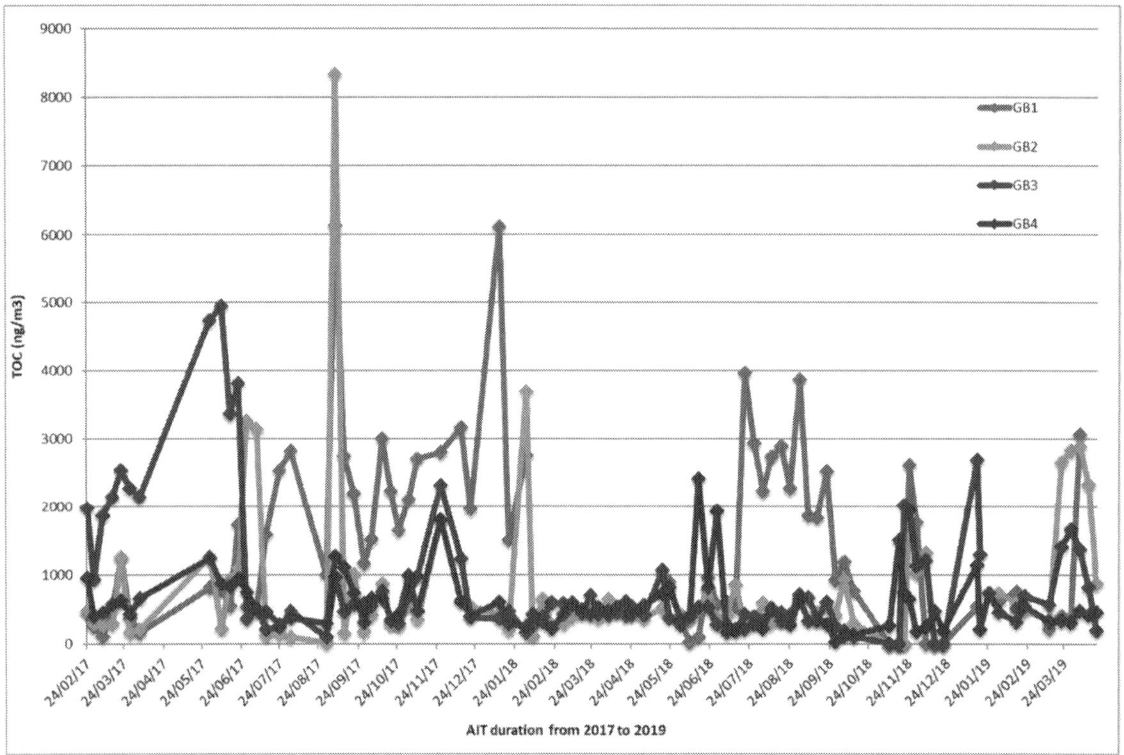

Figure 5. GBT monitoring results during UCZ FM AIT Campaign

GBT maintained about constant values throughout years. When a value above the alert limit was measured, environmental witnesses exposed inside the GB were analysed to understand the impact on contamination deposited on the Flight H/W surfaces. In all case, witnesses measurements provided very low results, around the detection limit or in any case below the maximum acceptable deposited contamination value. GBT air resulted over the alert in some cases. Higher levels were due in the most part of cases to sampling lines contamination. Indeed, they are in ISO7 HC (Highly Controlled) environment, and, even if all entrance and valves are closed, an accidental contamination can occur, as well as contamination of the Teflon ferrules, then replace, with graphites ones. On the contrary, for GB1 main cause of GBT contamination was the frequent exposure to the external environment, since it is the only glove box that was open for H/W entry. Higher level in the others GBs were mainly due to sampling lines contamination. In any case, deposition on Environmental Witnesses EnW was always compliant to the expectations, thus keeping GBT AMC level around AMC-6 or AMC-5, having so guaranteed no impact on Flight H/W.

3.2. In-process and Environmental analysis approach description

3.2.1. In-process witnesses. For what concerns the InpW, two InPW were associated to the AIT KITs considered critical from a C&CC point of view: one was analysed after ultracleaning treatment to monitor the UC residue, the other one was analysed only if the detected values were too high after recleaning with CO_2. For UCZ FM the following approach has been followed:

- a level below 5 ng/cm^2 is considered acceptable

- if for the first InpW the measured value was between 5 ng/cm^2 and 10 ng/cm^2 , the second InPW would have been analysed. If the value remained below 10 ng/cm^2 , the average value of the two analysed InPW has been considered as ultracleaning residue
- if for the first InpW the measured value was above 10 ng/cm^2 , the second InPW had to be analysed. If the second value was below 10 ng/cm^2, the average value of the two analysed InPW has been considered as ultracleaning residue if it was lower than 10 ng/cm^2. If also the second InpW was above 10 ng/cm^2, or the average of the two InpWs was above 10 ng/cm^2, the relevant KIT was retrieved form GBT and the ultracleaning process was re-started together with new associated InpWs.

Results for UCZ FM Ultra Cleaning verification are reported in Figure 6. As can be seen, some KITs were cleaner than other associated to different mechanisms and only one analysis for each KIT was sufficient. In the case of KITs associated to a particular UCZ mechanisms, at least a second analysis, and in some case a complete recleaning of several KITs was necessary. This was probably due to the fact that those KITs were cleaned with ultrasonic bath and then left in storage for about 2 months before final DHMR cycle and CO_2 snow cleaning steps. During storage they were put inside kapton bag filled with ultrapure nitrogen inside ISO5 laminar flow tent, but probably the bags were not completely leak tight and some additional contamination was inflowed/deposited on witnesses.

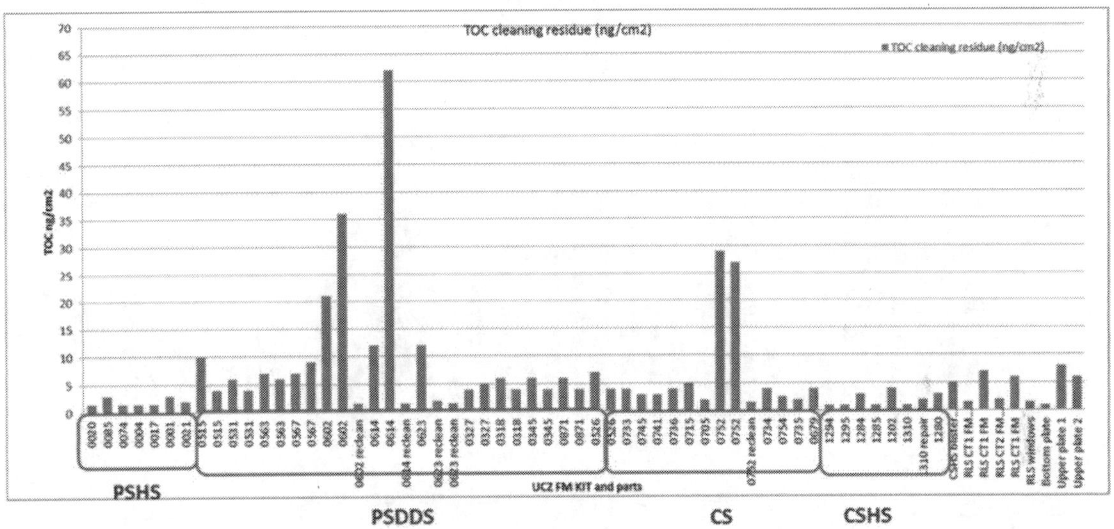

Figure 6. Ultracleaning residue expressed as TOC in ng/cm^2 for UCZ KITs in contact with Martian sample

3.2.2. Environmental witnesses. Wrt environmental witnesses for UCZ FM, a new concept have been exposed to monitor the deposition rate inside GBT:

- 2 EnW exposed in GB1 analyzed at the end of FM activities or in case of AMC NCRs
- 2 EnW exposed in GB2; one was analyzed each month and substituted with a new one
- 5 EnW exposed in GB3; one of them was analyzed every week. Every month a new set of EnW was exposed inside GB3. The fifth EnW was analysed at the end of ALD FM integration.
- 5 EnW exposed in GB4 one of them was analyzed every week. Every month a new set of EnW was exposed inside GB4. The fifth EnW was analysed at the end of ALD FM integration.
- 1 EnW to be cleaned with all the others of the batch in order to check the starting ultracleaning value

For each analysed EnW, a deposition rate has been calculated dividing the total measured amount in ng/cm^2 for the exposure days. In case the total value was below the detection limit (1.5 or 2 ng/cm^2

depending on microchamber performances during the analysis period), the detection limit itself was taken as value and divided for the exposure days.

There is not a requirement for deposition rate/day, but the reference to 12 ng/cm^2 of total deposition during UCZ AIT according to UCZ C&CC model (see Figure 6). When detected value is <LOD ng/cm^2, LOD value was considered for deposition rate calculation.

3.3. UCZ FM deposition during AIT

According to the results of ultracleaning residue and deposition during AIT, the total contamination level for each KIT associated to the UCZ FM is calculated as follows:

- Ultracleaning residue in ng/cm^2
- Identification of time period in which the KIT was exposed in GB3 and the period in which was exposed in GB4
- Calculation of a worst case deposition using values of weekly EnWs in GB3 and in GB4 in ng/cm^2/day and multiplying by the exposure days of the H/W

Sum of all above contributors using monthly cumulative values is reported in Figure 7.

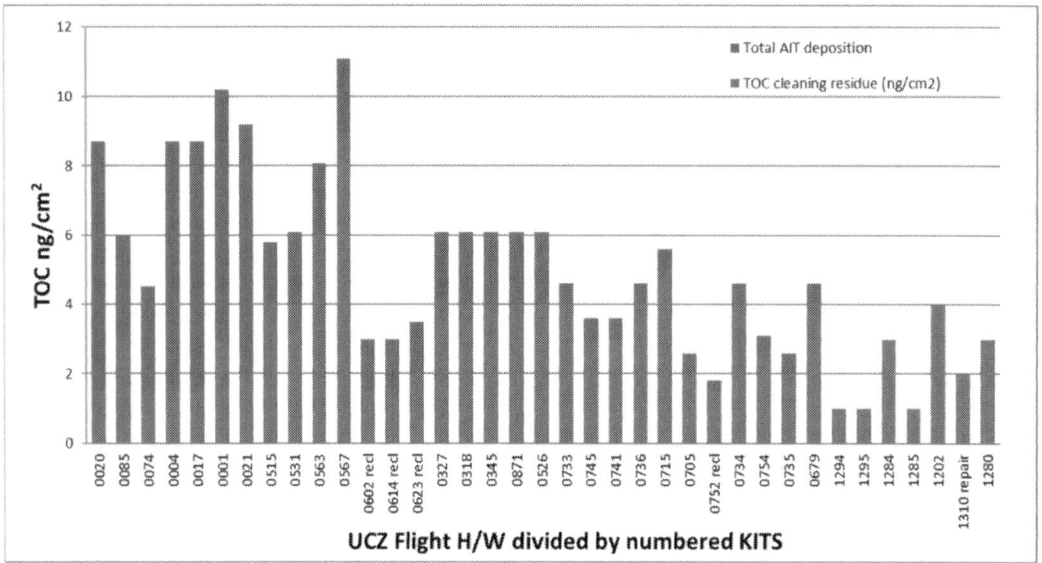

Figure 7. Total deposition calculated for different UCZ KITs in contact with Martian sample

4. UCZ volume airborne analyses during AIT

The UCZ compartment is equipped with proper gas inlet and outlet ports used for UCZ pressurization and depressurization throughout the AIT phase. Indeed, in order to protect UCZ internal volume from external contamination, a minimum overpressure from 10 to maximum 100 mbar on UCZ volume has been applied once integrated. Pressurizing and purging gases (i.e. helium, nitrogen or argon) were procured at 6.0 purity grade and FGSE have been tested through the AIT campaign before use to assure cleanliness levels, i.e the same purity level of the pressurizing gases.

During UCZ FM AIT, several checks were performed to verify the quality of the gas inside the UCZ volume, in terms of organic traces concentration. This allowed to check periodically the evolution of the contamination inside the UCZ, especially after critical tests such as Thermal Vacuum Cycling or Mars Chamber simulation. A clean UCZ volume was considered in the order of 1-5 ng/l.

The UCZ FM airborne contamination levels were more constant than for the QM (see Table 1), confirming that optimized cleaning and integration flow allows a close control of the contamination sources and possible deposition. In addition, all seals, even the ones used for testing phase only, were bake-out before use. Higher contamination levels were detected after Tvac, probably due to the contribution of non-clean environment of the TVac facility. Subsequent flushings and cleanings reset the UCZ compartment at acceptable values of the contamination levels, as can be seen from the final sampling.

Table 1. UCZ FM volume samplings results

Tested item	Sampling date	TOC [ng/l]
UCZ FM first sampling after leak check	22/11/2018	0,69
UCZ FM before mating	08/01/2019	1,63
UCZ FM after mating	21/01/2019	0,80
UCZ FM aft Pressure Relief valve integration	23/02/2019	0,54
UCZ after Tvac (re-entry in ISO7HC)	28/03/2019	2,10
UCZ resampling aft valve installation	16/04/2019	1,76
UCZ FINAL SAMPLING (after final flushing)	03/05/2019	0,63

After ALD FM delivery for integration in RM and following bathtub (the closed Rover Module compartment hosting the UCZ) closure, higher levels (between 50-100 ng/l) were detected inside UCZ. The contamination detected presented the following characteristics:

- main contribution to contamination is due to benzene, o-xylene, cyclosiloxanes, methyl substituted alcohols (i.e. penthanol, epthanol), low and high boiling siloxanes
- the contamination decreased during time, suggesting that the system cleanliness benefits from continued samplings and flushings:

After further several checks to exclude that contamination was coming from cross-contamination during tenax tubes transport and handling or internal to UCZ, Rover Module Bathtub was identified as the primary source of the UCZ contamination. Indeed, bathtub samplings confirmed the presence of same contaminants. The first sampling of the bathtub also clarified that the order of magnitude between the bathtub contaminants concentration and the UCZ contaminants concentration is about a factor of 60 (i.e. bathtub is currently 60 times more contaminated than the UCZ). Pathway for the transfer of contamination is identified via several UCZ seals, even if PRV polymeric silicone seal was assumed as the main contributor considering silicone higher permeability than other polymers, such as Viton or PTFE.

During AIT activities at Rover Module and then at Spacecraft level, a constant monitoring of bathtub and UCZ, with flushing and cleaning performed as possible, was applied. The monitoring has been quite standardized, in order to have comparable results during time. In particular:

- UCZ monitoring uses about 10 litres sample volume (fraction of 15 litres UCZ volume) with a flow rate of about 200 ml/minute.
- BT sampling is performed using Markes pump with a flow set at 65 ml/min for a maximum of 30 minutes
- UCZ flushing at a minimum of 2 shifts in order to guarantee a maximum cleaning
- TOC calculation is performed excluding some siloxanes peaks that can be due to tenax degradation by-products and not to the real sampled contamination.

4.1. UCZ airborne contamination level before and after Rover Module Thermal Vacuum Bake-out

Even if the overall requirements would have not been exceeded with measured contamination rate, the main goal was to reduce as much as possible the UCZ initial contamination level in order to avoid possible additional deposition inside UCZ surfaces during on-ground phases. For this reason, in addition to constant monitoring, Thermal Vacuum Bake-out (TVBO) of the entire Rover Module was identified as the most effective action to reduce the source of contamination.

The TVBO was performed at max non-op temperature of RM (35°C) with open Bathtub and with realtime monitoring with Quartz Crystal Microbalance (QCM) set at -40°C (coldest UCZ temperature as per thermal analysis profile). The stopping criteria was based on deviation from linearity of the QCM frequency, being monitored as a function of time, according to per TEC-QT-2014-344: f''/f' < 1% for at least 5 continuous hours with minimum 48 hours of data acquisition. The TVBO has been divided in two different phases, in order to carry on the warm purging in the UCZ volume after outgassing rate stabilization. The considered phases were the following:

- PHASE 1: started from pressure and temperature stabilization at 1E-5 mbar and at 35°C
  Total duration: 168 hours
  TQCM success criteria reached after 8 hours and maintained until the end
- PHASE 2: Outgassing rate certification with UCZ warm purging
  Total duration: 80 hours
  4 cycles of warm purging applied for a total duration of about 36 hours

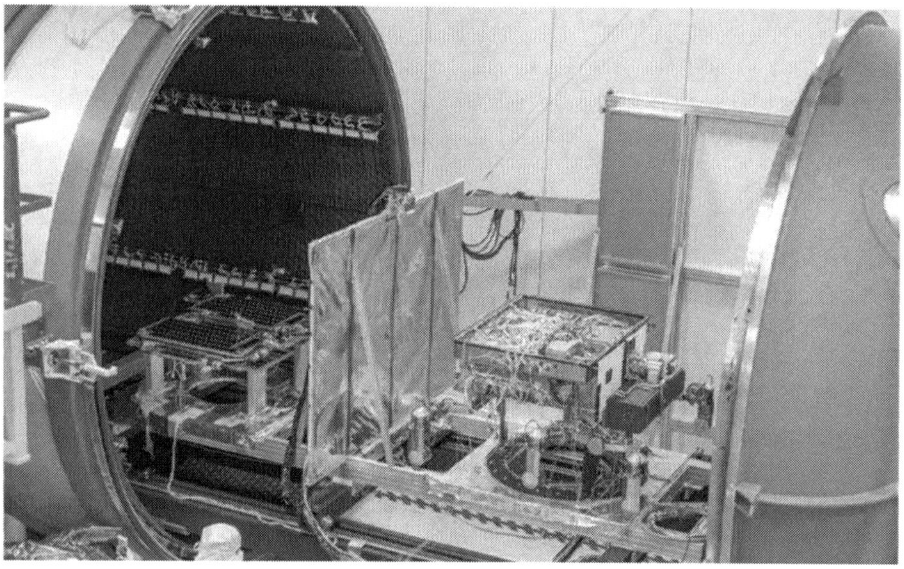

Figure 8. RM PFM TVBO configuration

The UCZ airborne contamination monitoring continued after the bake-out, with the same criteria established to have a clear comparison. From the results (see Table 2), the contamination rate was reduced of about 50%, or even better.

Table 2. Examples of UCZ FM and BT samplings results before and after RM TVBO

Tested item	Sampling date	TOC [ng/l]	TOC rate ng/L/day
UCZ sampling 1_RM PFM Stand alone	**28/07/2021**	**506**	
BT sampling_RM PFM Stand alone	**28/07/2021**	**1823**	
UCZ sampling Before BAKE-OUT - aft 1 week of BT closure	28/09/2021	68	8,2
BT Before BAKE-OUT - aft 1 week of BT closure	27/09/2021	803	106
UCZ sampling After BAKE-OUT - 10 days of closure	29/10/2021	48	4,4
BT After BAKE-OUT - 10 days of closure	29/10/2021	669	45
Other activities with Bathtub open during that months (results not comparable for contamination rate)			

Tested item	Sampling date	TOC [ng/l]	TOC rate ng/L/day
UCZ sampling after BT closure on 27.01	31/01/2022	72	
BT_ after closure on 27.01	31/01/2022	616	
UCZ sampling after BT closure on 27.01 - 4 weeks of closure	03/03/2022	95	1,4
BT after closure on 27.01- 4 weeks of closure	03/03/2022	685	4,3

5. Conclusion

Molecular contamination analysis through Thermal Desorption coupled with GCMS is a very promising method considering lower detection limit and the possibility of identification of all organics compound detected. The quantification is semi-quantitative, but considering low concentration and distribution can be considered a reliable approximation. In addition, the technique allows the determination of airborne and deposited contaminants, providing a very useful way to explain and correlate the deposition rate and possible reemission and the study of deposition plateau on different surfaces. Thus, it can become an essential instrument for the definition of adequate environment controlled also from a molecular point of view.

TD-GC/MS method was crucial to identify the contamination source of the UCZ airborne, the root cause and to check the bake-out efficiency.

6. References

[1] ISO 14644-8, and associated controlled environments -- Part 8: Classification of airborne molecular contamination
[2] SEMI MF1982 - Test method for analyzing organic contaminants on silicon wafer surfaces by Thermal Desorption Gas Chromatography
[3] VDA 278 - Thermal Desorption Analysis of Organic Emissions for the Characterization of Non-Metallic Materials for Automobiles
[4] Fisackerly R, Reimers C, Pradier A, Exploration system technology aspects in the exploration programme of the European Space Agency 2006 *Acta Astronautica* **59** 3–12

Acknowledgments

We are thankful to ESA and MOMA teams for their outstanding support and fruitful cooperation. Authors are also in particular grateful for the contributions provided by Fred Goesmann and Olaf Roders from MPS, who carried out permeation tests.

IOP Conf. Series: Materials Science and Engineering 1287 (2023) 012034 doi:10.1088/1757-899X/1287/1/012034

Using a portable IR spectrometer for materials characterization and identification

S Szmolka[1,2], **B Bras**[1], **O Ergincan**[1]

[1] European Space Agency, Keplerlaan 1, 2200AG Noordwijk, Netherlands

[2] Corresponding author, szilvia.szmolka@ext.esa.int

Abstract. This work is an assessment of contactless techniques for space hardware ground testing and characterization. It is based on Fourier Transform Infrared (FT-IR) spectroscopy, namely a commercially available portable FT-IR instrument operating in reflection and transmission measurements at a considerable distance from the measured surfaces. The two methodologies are evaluated namely related to 1) molecular contamination monitoring, and 2) materials identification. These techniques can provide important information in support of Manufacturing, Assembly, Integration and Testing activities as well as non-conformance investigations. Non-contact characterization techniques are particularly relevant when dealing with sensitive surfaces and/or surfaces with limited access such as optical components already integrated in the optical bench.

1. Introduction

Optical hardware requires specific cleanliness and contamination control techniques to ensure that the performance of the system remains acceptable throughout the mission lifetime. The ground phases Manufacturing, Assembly, Integration and Testing may expose these surfaces to contaminating environments. In case of a non-conformance, it is important to identify (and quantify) the contamination to apply adequate risk mitigation techniques or define appropriate cleaning procedures. The problem is amplified when the optical hardware is already assembled and with limited access.

Different activities were carried out to support materials remote characterization and surface contamination levels, allowing the characterization of materials and surfaces without contact. This publication describes the development, the evaluation, and the first results having in mind two objectives requiring contactless methods:

1. Characterization (identification) of contaminants on optical hardware: The set-up was adapted to support an ongoing ESA mission. The method was used to characterize silicone deposits on a gold mirror. After evaluation and characterizations in ESA's materials laboratory, the technique was utilized on flight hardware.

2. Identification of bulk materials. The sensitivity of the portable FT-IR was proven to be adequate to determine the composition of optical materials made of various types of glasses through the presence of characteristic hydroxyl (–OH) functional groups and integrated in space hardware.

IOP Conf. Series: Materials Science and Engineering 1287 (2023) 012034 doi:10.1088/1757-899X/1287/1/012034

The techniques were validated in ESTEC laboratories via systematic transmission and reflectance measurements performed from different distances and/or incidence angles – depending on the application objective.

2. Test approach

The tasks described below required the development of different contactless measurement procedures related to the two objectives mentioned in the previous section.

For the purpose of the method/procedure development, the utilized instrumentation was portable (<2kg, footprint of 180mm x 160mm x 80mm), commercially available ARCoptix FT-IR with integrated light-source, interferometer and detector. Depending on the particular wavelength of interest, two different detectors (measuring in the 2-6 µm or 2-12 µm range) were ready for operation in the Materials' Physics and Chemistry Section of ESA. Additional components, as IR optical guides and collimating lenses were also implemented in some cases to optimize the set-up.

2.1. Objective 1: Molecular contamination evaluation

The set-up was validated to support the cleanliness investigation of an optical module of space hardware -a gold coated mirror-, subject of a non-conformance review board due to the risk of having been exposed to silicone contamination. .

For this activity, the required detection level for silicone contaminants was 1E-6g/cm^2. Furthermore, the hardware was not allowed to be in contact with any tool or instrument. The measurements had to be performed remotely on the infrared mirror (in single and double reflection mode) from a given distance. Although the final objective was to measure reflection on the hardware, the technique was initially validated in transmission mode to assess sensitivity and limit of detection from the pre-determined distance of 17 cm and 30 cm – limited by the accessibility of the object. CaF_2 windows and IR reflective first surface gold mirrors were deliberately contaminated, either by outgassing products during high vacuum testing, or by direct distribution of a poly(dimethylsiloxane) solution on their surface, following the ECSS-Q-ST-70-05C[1] standard method.

Figure *1* shows the characteristic absorption peaks related to the presence of (poly/)siloxanes on the IR spectrum. Bending vibration of CH_3 in the Si—CH_3 group can be recognized by a strong, sharp band at ~1260 cm^{-1}. Another typical absorption peak referring to polysiloxanes is the two (or more) overlapping bands in the 1024-1074cm^{-1} region, caused by the Si-O-Si stretching vibrations. Measuring this very characteristic fingerprint with the portable instrumentation from a considerable distance is the challenging objective for the success of the investigation.

The instrument with the detector of the 2-12 µm wavelength range (allowing the silicone detection) was utilized in this set of activities. The drawback of the set-up is the lower intensity of the lightsource, which practically dropped to almost no signal when optical fibers were used. Consequently, the final built had a free-path between the *lightsource(+interferometer) unit – sample* and the *sample – detector,* both in transmission and reflection modes. Without optical fibers, perfect alignment is a crucial key aspect of the successful measurement and can be challenging to achieve specially in larger distances.

Different gain and apodization corrections, background measurements and subtraction are available in the software of the instrument. The parameters were optimized to reach the highest sensitivity and reproducibility.

Since the technique is an adaptation of the well-established FT-IR measurement procedures used to quantify MOC levels and standardized in ECSS-Q-ST-70-05C (e.g. hydrocarbons, esters, silicones), to a contactless form, results measured by the two techniques shall be comparable.

IOP Conf. Series: Materials Science and Engineering 1287 (2023) 012034 doi:10.1088/1757-899X/1287/1/012034

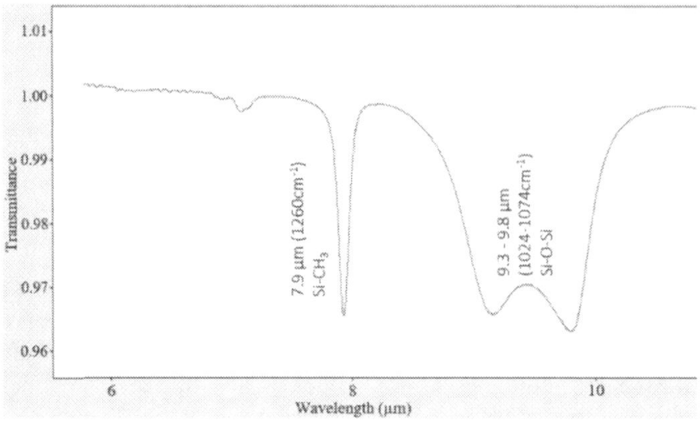

Figure 1. IR Spectrum acquired with benchtop Bruker spectrometer – fingerprint of the standard compound

2.2. Objective 2: Materials identification

The methodology was validated to support contactless identification of optical materials made of various types of glasses. Based on the precursor material, the technique and conditions of manufacturing, and the final thickness, glass products' properties vary and show different IR absorption features [2].

In this specific case, the evaluation and identification of several purities of quartz glass and borosilicate glass was done by IR spectra analysis, including -OH group quantification, by the absorption band of stretching vibrations of the hydroxyl group at ~3700 cm^{-1} (2.7 µm).

Without the need of measuring above 6 µm, the set-up with a detector optimised for 2-6 µm wavelength range was implemented in the set of the characterisations in which the lightsource had higher output intensity and the detector higher sensitivity.

Gain and apodization corrections were investigated for the best sensitivity, background subtraction was applied, optical fibers and collimating lenses were used in transmission and reflection modes. Custom-made mechanical holders were designed to measure reflection with different incidence angles.

The technique was developed to be employed on materials already integrating in the flight hardware with limited access. The main objective of the activity was to distinguish the different materials by transmission and reflection measurements of flat and curved samples.

Figure 2. Test set-up for contactless IR transmission (on the left) and reflection (on the right) measurement of glasses

3. Results

3.1. Molecular contamination evaluation

For the sensitivity and limit of detection assessment of the contactless molecular contamination evaluation, several IR transparent CaF_2 windows were prepared for transmission measurements by contaminating them with different levels of siloxane either by direct deposition via syringe or by vacuum outgassing contamination. Some examples are given in Table 1, showing the concentrations measured by a benchtop FT-IR.

Table 1. Pre-contaminated CaF_2 windows for transmission measurements

Sample	Silicone contamination (g/cm^2)
#203	1E-05
#217	1E-05
#322	5E-07
#328	1E-06
#715	7E-08

Figure 3. CaF_2 window

Transmission measurements were performed with a distance of 17 cm between the detector and the lightsource. The windows indicated in Table 1 were placed in the path of the light and the measured spectra are depictable on the right side of Figure 4. The expected silicone fingerprint is clearly observed. Other areas of the spectra depict noise due to atmospheric conditions. The set-up was deployed without utilizing inert atmosphere to be as close as possible as a representative AIT environment.

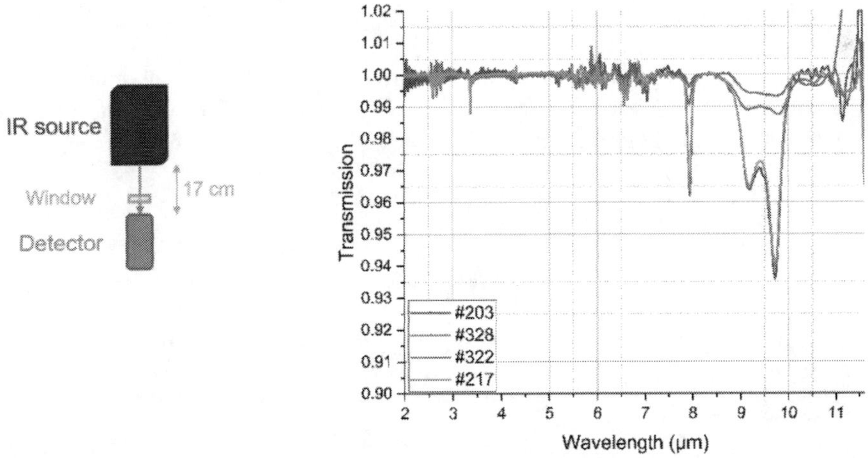

Figure 4. Configuration (on the left) and results (on the right) of the transmission measurements

In single reflection measurement mode, an unprotected first surface gold mirror (clean) was used behind the window to increase the total optical path to approximately 60cm. Results of these measurement are shown on **Figure 5**.

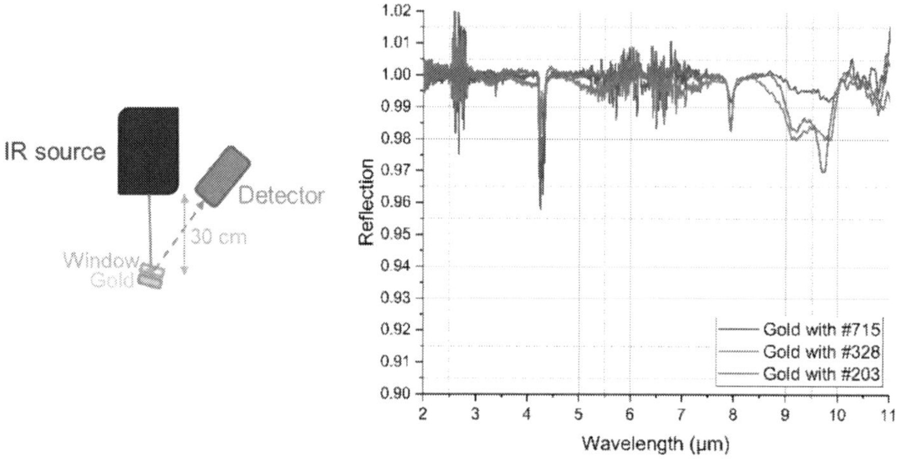

Figure 5. Configuration (on the left) and results (on the right) of the single reflection measurements

The ultimate aim of the feasibility assessment was to prove whether the required contamination level can be detected in representative configurations. For the mission at hand, this meant a double beam reflection on two gold surfaces, before reaching the detector. For this purpose, a two-phase approach was applied:

- Phase 1: Gold 1 (contaminated) + clean gold
- Phase 2: Gold 1 (contaminated) + Gold 2 (contaminated)

The silicone contamination was deposited on the surface of "Gold 1" via direct deposit, to a molecular organic contamination level of ~ 1E-06 g/cm^2. The contamination level of Gold 2 was not quantified, but it was a substantially higher amount.

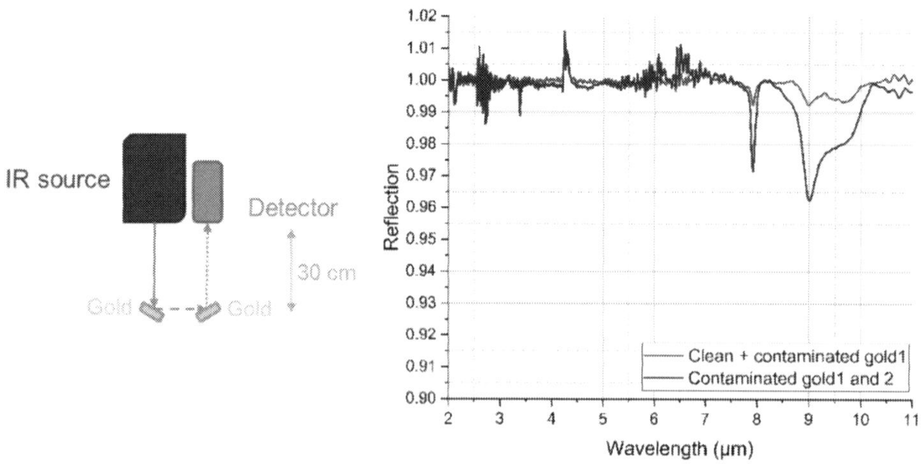

Figure 6. Configuration (on the left) and results (on the right) of the double reflection measurement

The results from Phase 1 – grey curve on Figure 6 – confirm the feasibility of the technique to detect approximately ~1E-6g/cm^2 silicone contamination in the representative configuration (double reflection, from 30 cm+30cm distance) on a reflective surface. Phase 2 – blue curve on Figure 6 – demonstrates that higher contamination in the optical path produces a sharper and even more obvious fingerprint. In case of future need, several calibrated compounds shall be measured to achieve a calibration curve (following Beer-Lambert law approach).

The best estimates for the current configuration of the test setup in an environment with limited control indicate that there may be uncertainties related to MOC measurements. A thorough evaluation to determine the uncertainty budget for the test setup has yet to be carried out. The aim of this experiment was to demonstrate that the detection limit meets the target level of contamination to be detected, which is 1E-6g/cm^2. This allows for an assessment of the detected levels of contamination on a specular gold mirror in terms of pass/fail criteria.

After characterizations performed in ESA's Materials and Electrical Components Laboratory, the technique was utilized on flight hardware.

3.2. Materials identification

Three different quartz glasses (different purities)- and a borosilicate glass sample with different dimensions were procured. These are representative of the actual hardware's materials, being the objective to clearly distinguish them.

The samples were first measured in transmittance mode with the calibrated benchtop FTIR lab-equipment (Bruker Vertex 70v) then they were subjected to transmission measurements by the portable instrument. The quantification of the hydroxyl content was accomplished through absorbance curves, following [3].

Table 2. Theoretical and measured hydroxyl concentration of quartz samples

Sample	c(OH) literature (ppm)	c(OH) benchtop FTIR (ppm)	c(OH) portable FTIR (ppm)
Quartz#1	<5	1	4
Quartz#2	<30	10	9
Quartz#3	170	132	133

By measuring IR transmission, as seen on Figure 7, the tested quartz samples can be distinguished from each other – based on their different -OH content. Furthermore, the borosilicate has a very different fingerprint and thus can be clearly identified. The -OH concentration measured in transmission mode shows exceedingly good correspondence with the standard measurement of the benchtop FT-IR results.

The samples were also measured in reflectance mode. In this case, difference between very high and low –OH contents is depictable, however the low signal from the reflection measurements leads to lower sensitivity. Consequently, it is not possible to undoubtedly distinguish between Quartz#1 and #2 materials. Most likely, a better performing detector or higher throughput lightsource would be needed.

IOP Conf. Series: Materials Science and Engineering 1287 (2023) 012034 doi:10.1088/1757-899X/1287/1/012034

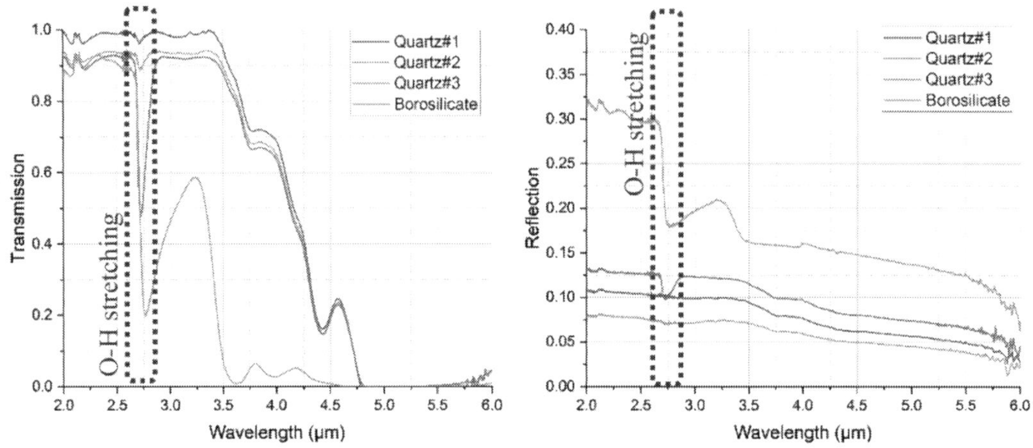

Figure 7. Contactless transmission (left) and reflection (right) measurement results of various glasses

4. Conclusion

The study confirms the applicability of a commercially available portable FT-IR instrument for two non-routine operations that can bring crucial information about materials properties and contamination status of surfaces. With the technique wide range of surfaces can be measured (e.g IR mirrors, transparent optics), however, the technique has limitations. In case of a larger contaminated area, several probed zones must be chosen with care to ensure reliability of the method, also to assess the (in/)homogeneity. As the measurements are performed through air, environmental effects should be considered. Depending on the wavelength range of interest, compensation for the presence of CO_2, H_2O might be required or the whole set-up may be placed in a closed, purged environment.

Applicability depends on the sensitivity, limit of detection and wavelength of interest. Optical guides (e.g. optical fibers) might be needed, but may affect the system throughput. Apodization and gain shall be adjusted to achieve the best signal to noise ratio. A wide range of infra-red detectors and lightsources are nowadays commercially available, therefore the chosen instrumentation for a particular task shall consider the specific objective as well as the measurement configuration.

Optimized contactless techniques can be used throughout ground life, for example to verify requirements at acceptance of hardware, post T_{vac} testing, key inspection points, amongst others. It can also be used as an inspection tool to support failure investigations and non-conformances, allowing a more complete assessment of the risk, and supporting the way forward definition. The described methods were successfully implemented and utilized in investigation campaigns of ESA projects.

References

[1] ESA-ESTEC ECSS-Q-ST-70-05C 2009 Detection of organic contamination surfaces by IR spectroscopy

[2] Nürnberg F, Kuehn B, Langner A, Altwein M, Schötz G, Takke R, Thomas S and Vydra J 2015 Bulk damage and absorption in fused silica due to high-power laser applications *Proc. of SPIE* **9632**

[3] Sloots B 2008 Measuring the low OH content in quartz glass, *Vibrational Spectroscopy* **48**

Acknowledgements

The authors gratefully thank Riccardo Rampini for his support and for providing invaluable guidance throughout the activity.

Comparison between purged and vacuum bake-out efficiency

S Szmolka[1,2], B Bras[1], R Rampini[1]

[1] European Space Agency, Keplerlaan 1, 2200AG Noordwijk, Netherlands

[2] Corresponding author, szilvia.szmolka@ext.esa.int

Abstract. Purged bake-outs can be an appealing alternative to vacuum bake-outs to mitigate molecular outgassing of space hardware. However, their suitability is still debated since there is a lack of knowledge regarding their efficiency. This publication presents a method to assess the efficiency of purged bake-outs, comparing them to vacuum bake-outs, and provides first results of a parametric study on a few materials, with particular emphasis in the effects that pressure, nitrogen flow, and gas inlet temperature have on the efficiency of the process. The study was done without a specific mission in mind, pre-determined temperature and duration profiles were applied. It is however stressed that the actual bake-out conditions will have to be related to the particular mission of interest.

1. Introduction

Outgassing from spacecraft materials is a known phenomenon that can lead to molecular contamination, compromising the performance of various space subsystems. The effect of contamination induced degradation of space hardware was reported multiple times in the past, such as the cases of Rosetta, MAP, SOHO missions [1] and Hubble Space Telescope servicing mission [2], and it remains a worrying aspect of new space missions [3],[4].

To mitigate possible performance degradation of contamination sensitive space flight hardware caused by irreversible deposition of outgassed species, strict contamination control programs, starting from design (most sensitive surfaces isolated from contamination sources), and carried on during manufacturing, assembly, integration, test, and even on operation (e.g., on-orbit bake-out) are generally implemented in space missions. The cleanliness strategies are based on careful design, materials selection, precision cleaning, thermal history, space conditioning, amongst others.

Outgassing rates are affected by several parameters, including temperature, surface area, vapor pressure and the kinetics of outgassing of each specific material. For outgassing, diffusion is the primary mechanism which is responsible for volatile chemicals' migration to the boundary layer on the material's surface and their emission to the local environment. Different standards (e.g ECSS-Q-70-02C[5], ASTM E-595 [6]) are available guaranteeing representative, comparable screening of materials' outgassing by measuring collected volatile condensable material and total mass loss in vacuum.

A general materials selection requirement is: recovered mass loss (RML) < 1.0%, and collected volatile condensable material (CVCM) < 0.1%. Nonetheless, more sensitive systems frequently have even more stringent requirements such as Total Mass Loss (TML) below a certain value (e.g. <1%).

In order to decrease the outgassing potential of materials to the least possible, the most common approach is to perform extensive bake-outs on ground - raising the temperature of the item to facilitate the outgassing of species in controlled conditions. The standard method is to perform the bake-outs in large vacuum facilities, with high budgetary and time constraints to space missions.

Purged bake-out, i.e. performing the bake-out not in high vacuum but with a certain partial pressure of inert atmosphere, was predicted or proven to be less effective than the high vacuum bake several times in the past [7],[8] yet much debate has been on-going about its efficiency and is often being proposed to control the costs and increase availability of bake-out facilities.

The Materials' Physics and Chemistry Section of ESA set-up a test campaign with an experimental approach to address the comparison between purged and vacuum bake-out.

2. Test approach

For this activity, the chosen 85°C ±1°C bake-out temperature and duration of 72 hours serve as the baseline comparative conditions. Naturally, these conditions should be adapted case-by-case for different materials and other mission requirements.

The most effective bake-out stopping criterion relates data from a Quartz Crystal Microbalance (QCM), such as reaching a certain deviation from linearity on the frequency rate. For simplicity, in the initial phase of the project, the only measurand was the sample mass loss (i.e. TML, measured immediately after stopping the bake-process). The main conditions (duration, temperature, cavity's geometry, etc) were maintained for the comparison between vacuum and the purged bake-outs.

The test was performed on silicone adhesives. After curing the samples, they were placed in a conditioning cabinet (22 ± 3°C, 55 ± 10 % humidity) for 24h before the bake-out. Ex-situ mass measurements were performed to assess total mass loss (TML) and RML. In a later stage of the study, quartz crystal microbalance (QCM) readings are also provided as in-situ efficiency monitoring. After the test, the mass was measured over several days to allow a better assessment of water uptake.

Screening outgassing tests (according to ECSS-Q-70-02C [5]) and QCM supported adapted outgassing tests were also performed to quantify the residual outgassing potential after samples being subjected to different bake-out conditions.

The two different methods (high vacuum and purged setup) required different facilities which are described in the following chapters.

2.1. *Standard method – high vacuum*

A modular, high vacuum (< 10^{-4} Pa / 10^{-6} mbar) bake-out facility (Figure 1) was adapted with temperature control capability. A gaseous nitrogen inlet was connected through a leak valve for pressure regulation. In later tests, the system was extended with a QCM for monitoring of the process. For thermal stabilization of the QCM, a cooling bath was connected and maintained at -10°C.

The samples were located in a stainless-steel vessel (= "bake-out chamber"), connected to a pumping station through KF40 connection. A heating jacket, surrounding the vessel was responsible for the temperature-control of the sample environment. The achieved temperature at the sample location was monitored via a type K thermocouple (uncertainty better than 1.5°C). After measuring the mass beginning of test (BOT) and placing the samples in the bake-out chamber, the baseline approach was:

- System pumping for 30-60 min to reach <10^{-5} mbar,
- The heating was switched on and the sample reached 85°C in 30 min with ~2°C/min heating rate

Conditions were maintained for 72 hours. In the end of the bake-out, the mass was measured again (end of test – EOT). The mass loss was monitored for several days after the test and was calculated according to ECSS-Q-ST-70-02C [5].

2.2. *Purged bake-out methods*

The task prompted the design of three ambient pressure bake-out set-ups.

The samples were placed in identical stainless-steel vessels identical to the high vacuum set-up, to maintain the same dimensions of cavity. With one end being connected to a high-purity nitrogen gas (GN2) supply and the other end assembled with a pipe, nitrogen was flushed through the bake-out vessel at different flows.

Three different heating methods were compared:

1- External heating of the metallic vessel via a Nabertherm furnace (Figure 2);
2- Use of a heating jacked directed attached to the metallic vessel;
3- In-line heater to warm-up the gas temperature at the inlet of the stainless-steel vessel

The sample's environment warming up rate was controlled to ~2°C/min to 85°C, similarly to the tests performed in high vacuum.

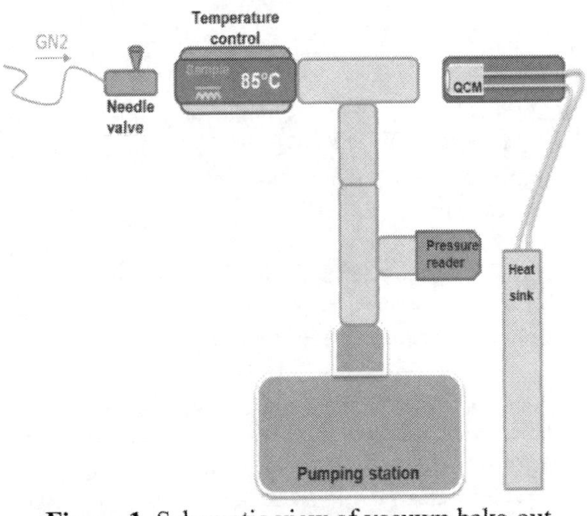

Figure 1. Schematic view of vacuum bake-out apparatus

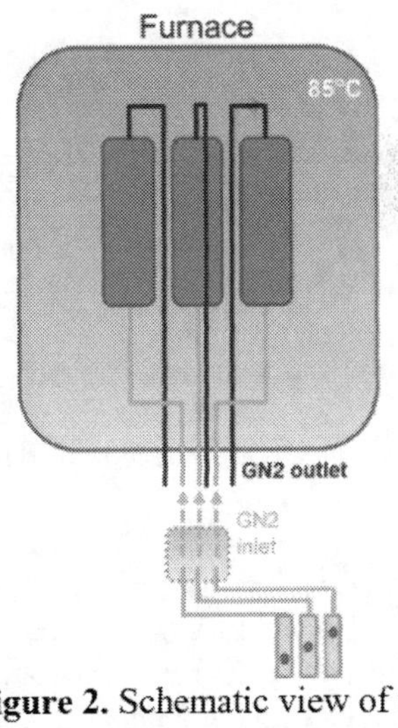

Figure 2. Schematic view of the purged bake-out apparatus

2.3. *Parameters of the study*

Pressure:

Different levels of pressure were compared, from atmospheric pressure, through 10mbar to 3E-07mbar.

Surface area/mass ratio:

Samples with various dimensions were cured to assess if the expansiveness of the surface area plays a perceptible role in out/offgassing of materials and to study whether this parameter may be a limiting factor for the different bake-out methods.

Flow rate:

The flow rate effect of the inert gas was studied between 1-15 L/min (0.6 - 10 s residence time in the sample vessel). Bake-out performed at ambient oven environments (closed vessel, no GN2 purge) was also part of the study. As a baseline, it is expected that the mass of the purge gas should exceed the mass of the outgassed species

Heating method:

To confirm representativeness of the purged method, the inlet gas temperature and the heating method was changed in some cases. The gas was either introduced at room temperature (RT) or after being heated up to 85°C.

Duration:

Besides the essentially utilized 72hours duration, elongated duration was also applied in some cases of the purged bake-out, with the time being evaluated as a trade-off to compensate for a potentially lower effectiveness of the method.

2.4. *Materials*

Silicone based adhesive I. (Si-I.): thermally curable, addition-curing silicone adhesive with low viscosity, low hardness, and low absorption of humidity. Probability of water re-adsorption is minimal after bake-out, TML, RML (in standard screening outgassing test) > 1% which facilitates the evaluation and comparability of different methods with higher confidence.

Silicone based adhesive II. (Si-II.): ASTM-E595 certified silicone based, flexible elastomer. With a lower TML and RML (< 1%) evaluation of effectivity based on mass loss differences is more complicated, but outcome of bake-outs more representative of space-grade materials.

Other materials (e.g. CFRP, epoxy) with low TML and RML are the subject of ongoing tests. Initial findings do not allow a comparative evaluation of efficiencies, due to the fact that the results are mostly driven by water outgassed/uptaken after the completion of the test. Therefore, future testing will be performed at higher temperatures (and/or with other materials) to ensure a high contribution of molecular organic contamination outgassing.

IOP Conf. Series: Materials Science and Engineering 1287 (2023) 012035 doi:10.1088/1757-899X/1287/1/012035

3. Results

In the following sections the results are presented in the form of residual (or, in some cases total) mass loss of samples after being baked out in different conditions. The presented RML results refer to those mass measurements that were done 24 hours after the bake-out, with the sample being held in a conditioning cabinet during this period.

3.1. High vacuum bake-out

3.1.1. Effect of pressure

When subjected to bake-out at ambient pressure and static environment (i.e., no flow of inert gas), the achievable RML of Si-I. is ~ 0.6% at 85°C, with the duration of 72h. Figure 3 shows that lowering the pressure even to a milder level of 10mbar improves the efficiency of the bake-out by reaching 1.9% as recovered mass loss, which can be further increased, up to 2.4% - 2.6%, by performing the same bake-out under higher vacuum.
The best efficiencies are achieved with a pressure environment of 1E-5 mbar or better in the case of the Si-I samples.

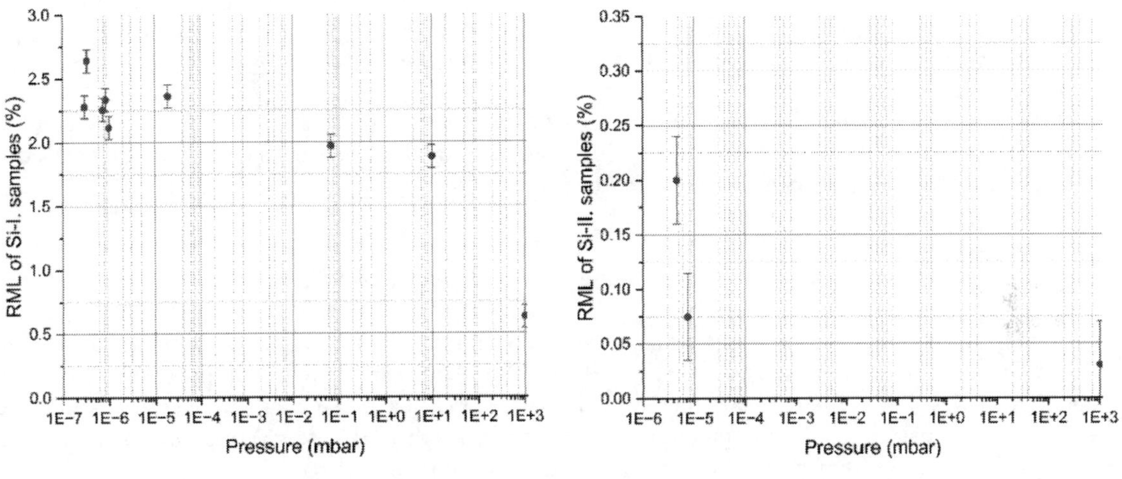

Figure 3. RML of Si-I. samples (on the left) and Si-II. samples (on the right) after bake-out at various pressures (at 85°C for 72hours)

The bake-out of the Si-II. samples in high vacuum showed a higher uncertainty proportionally, compared to the measured total mass losses and to the ratio that are seen at Si-I. samples. The results are attributed to the water uptake after the bake-out. It is possible that the conditions did not facilitate the molecular contamination outgassing and the results are driven by water outgassing and re-adsorption and/or modest differences in the curing of the material. This effect was also seen in the screening outgassing test (results are shown in Table 2).

3.2. Purged bake-out

3.2.1. Effect of flow rate

In the case of Si-I. samples introducing and increasing the level of inert gas flow improves the efficiency of the method, up to a plateau of 1.9% RML, according to Figure 4, while the RML in a screening outgassing test executed in high vacuum, at 85°C was 2.2%.

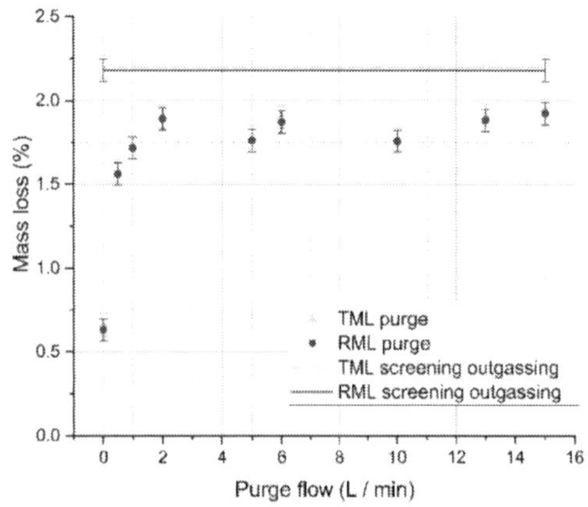

Figure 4. TML and RML of Si-I. samples baked in nitrogen purge (vs results of screening outgassing according to [5])

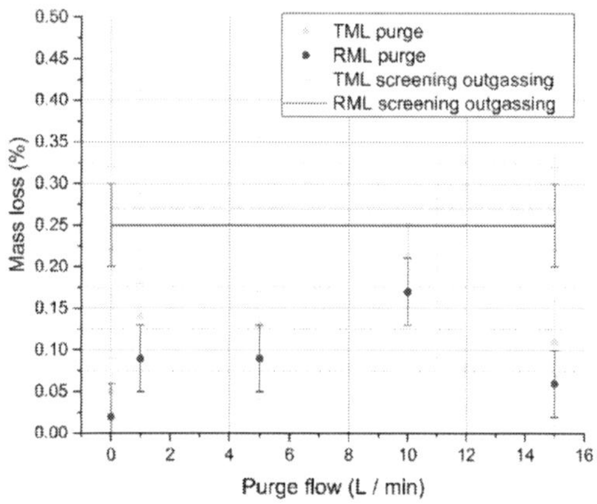

Figure 5. TML and RML of Si-II. samples baked in nitrogen purge (vs results of screening outgassing test in high vacuum, at 85°C)

Figure 5 shows that the effect of increased flow rate is not so direct for Si-II. samples and the measured mass losses are weighed with a higher uncertainty, but the presence of flow is beneficial compared to the bake in ambient oven environment.

3.2.2. Recovery of baked Si-I. samples and offgassing at ambient conditions

The mass of samples baked in purge or high vacuum stabilized, while offgassing occurred on sample baked in ambient conditions without purge. Bake-out in static air or nitrogen atmosphere, without continuous flow should not be considered. This is an important remark, as in case purge-bake out solutions are considered in future studies, it must be confirmed that there is an active flow of gas passing through the materials. Any pockets without flow likely will not be efficient.

After curing the Si-I. samples, offgassing occurs within a timeframe of weeks while being stored in a conditioning cabinet (22 ± 3°C, 55 ± 10 % RH). While the high vacuum bake-out showed ~2.2% mass loss, only by storing the material at room temperature for 60 days, a mass loss of >1.0% was depictable.

The results indicate that whereas unbaked samples progressively show a mass loss (offgassing in ambient conditions), samples baked at 85°C depict stability (no further mass loss). From these, samples baked in high vacuum show the highest mass loss; samples that underwent purged bake-out depict a lower mass loss.

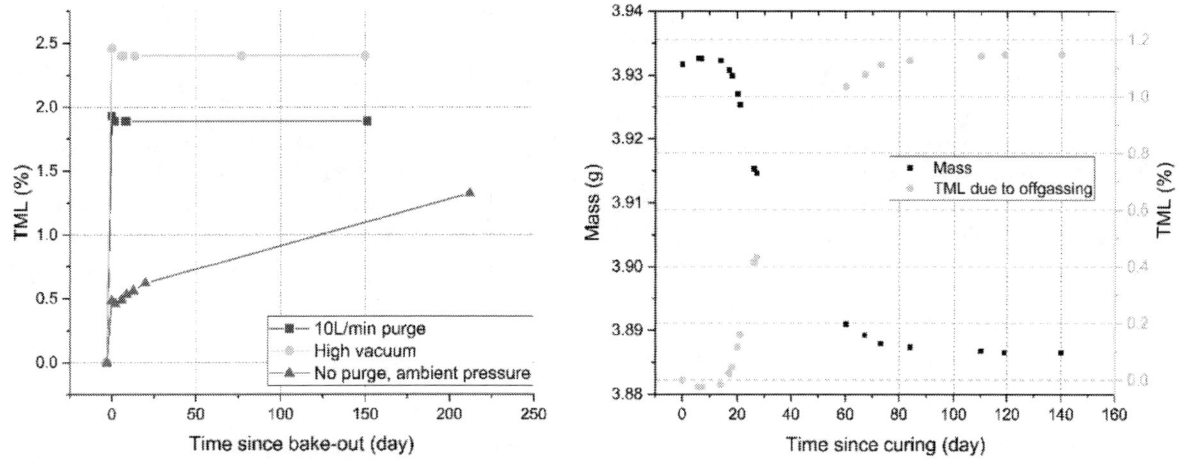

Figure 6. Offgassing effects on Si-I. samples. On the left side residual compounds' offgassing can be observed when bake-out is not efficient, on the right side the pristine material's offgassing during storage can be followed

3.3. *Effect of surface area*

The surface area/mass ratio of each sample was calculated by a simple division. The aim was to assess if samples with larger surface/mass ratio can be baked more efficiently in the same conditions. The plot on the left side of Figure 7 shows that on the studied interval, the surface area seemed to have no pronounced impact when Si-I samples were exposed to bake-out in high vacuum.

The same effect can be observed on the right plot on Figure 7, which shows the purged bake-out efficiency of different sized Si-I. samples, when being exposed to bake-out at atmospheric pressure and higher flow rates of purging nitrogen. However, in absolute values, the RML is lower than the one achieved in high vacuum conditions.

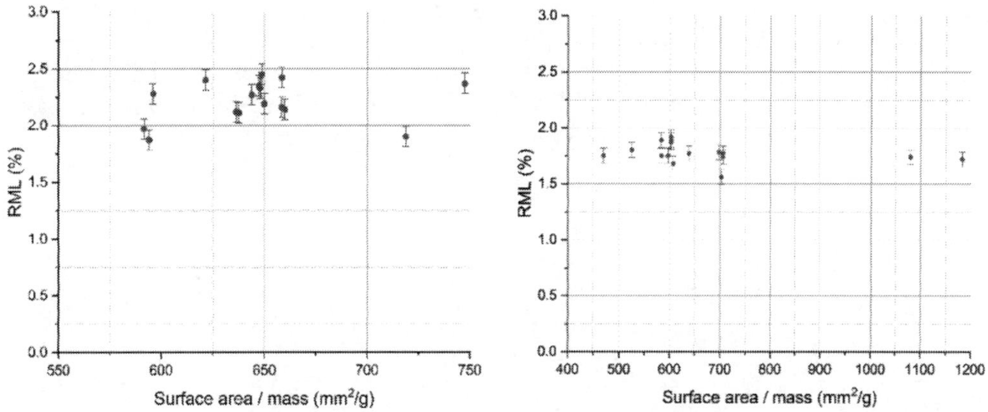

Figure 7. RML of Si-I. samples with various surface area/mass ratio after bake-out in high vacuum (on the left) and in nitrogen purge with residence time ≤2s (on the right)

3.4. Comparison of methods

After being subjected to different bake-out conditions as described in the previous chapters, several samples went through a standard screening outgassing test (according to [5]) to evaluate the residual outgassing potential and consequently the efficiency of the different methods.

Si-I samples baked out at 85°C, ambient pressure and no inert gas flow had an RML of 1.3% in the screening outgassing test at 125°C, after the bake. When baked with continuous GN_2 flow, the material still shows an RML of 1%, while the vacuum baked (at 85°C) samples only suffer a residual mass loss of 0.2 % in the standard outgassing test. This indicates that the vacuum bake-out (for the same temperature and duration) is the most effective in removing outgassing species from the test specimens.

Table 1. TML, RML results of bake-outs and following screening tests of Si-I

	TML bake-out (%)	RML bake-out (%)	TML screening outgassing		CVCM screening outgassing		RML screening outgassing	
			125°C (%)	85°C (%)	125°C (%)	85°C (%)	125°C (%)	85°C (%)
High vacuum	2.26	2.20	0.22	0.05	0.10	0.00	0.22	0.04
15L/min purge	1.94	1.92	1.06	0.49	0.37	0.40	1.05	0.48
No purge, atm pressure	0.65	0.63	1.30	N/A	0.33	N/A	1.29	N/A
11 mbar, pre-vacuum	1.91	1.90	0.75	N/A	0.29	N/A	0.74	N/A
Not baked	0.00	0.00	N/A	2.12	N/A	0.04	N/A	2.11

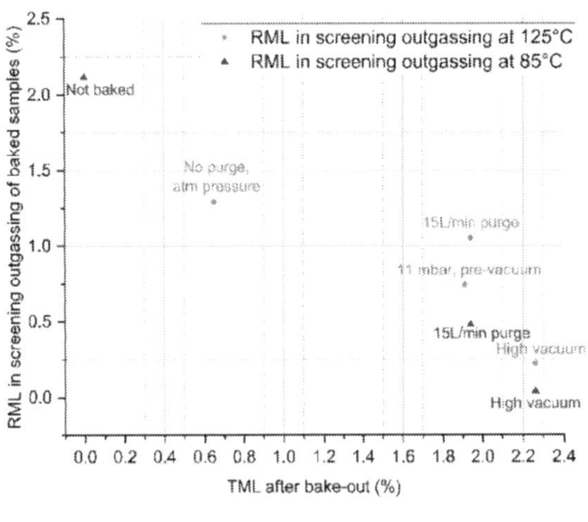

Figure 8. Screening outgassing test of Si-I. after being baked in different conditions

When baked with purge, Si-II shows a very low RML (0.02%), which matches the RML of the samples that were baked in high vacuum. This can be due to the temperature not being high enough to remove organic outgassing species (process driven by water outgassing and re-adsorption). The screening outgassing test of a pristine (non-baked) at 85°C sample results in 0.25% RML.

Table 2. TML, RML results of bake-outs and following screening tests of Si-II

	TML bake-out (%)	RML bake-out (%)	TML screening outgassing, 85°C (%)	CVCM screening outgassing, 85°C (%)	RML screening outgassing, 85°C (%)
High vacuum	0.24	0.14	0.05	0.00	0.02
High vacuum#2	0.11	0.06	0.04	0.00	0.02
15L/min purge	0.11	0.06	0.05	0.00	0.02
No purge, atm pressure	0.05	0.02	0.05	0.00	0.03
1.9E-05 mbar	0.22	0.12	0.05	0.00	0.02
Not baked	0.00	0.00	0.27	0.00	0.25

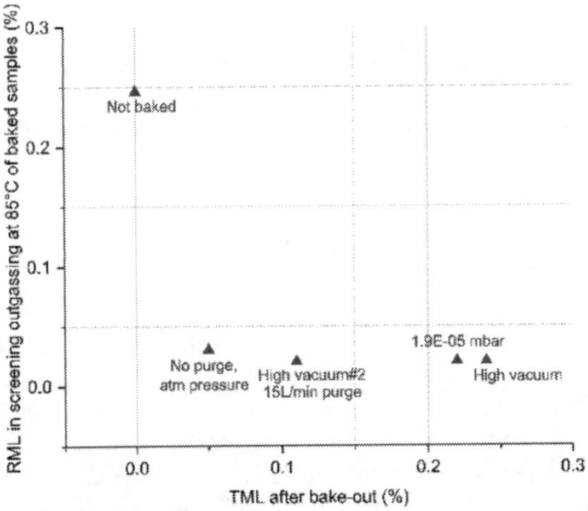

Figure 9. Screening outgassing test of Si-II. after being baked in different conditions

A QCM monitored second bake-out (85°C, 24 hours, <1E-06mbar, QCM maintained at -10°C) of previously baked Si-I samples shows the presence of residual outgassing molecules being higher after a purged bake, compared to one performed in high vacuum. Deposition of molecules can be observed on Figure 10 by the frequency change of the QCM. The area below the curve correlates to the quantity of residual molecules with outgassing potential.

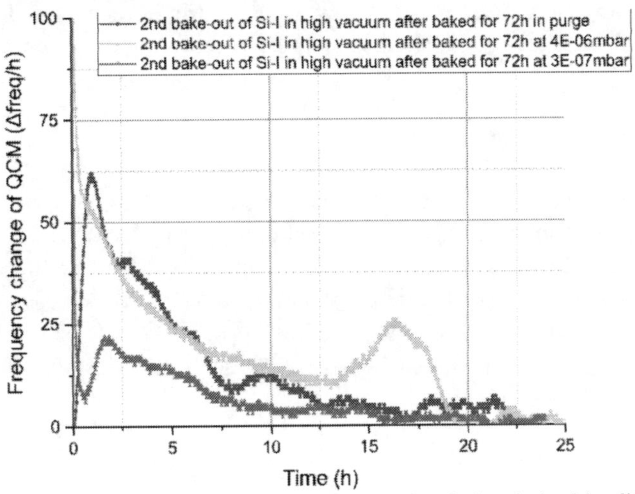

Figure 10. QCM monitored second bake of Si-I. samples after being baked in different conditions

Results of the second bake (85°C, 24 hours, <1E-06mbar) of different samples (CFRP and epoxy) are provided on Figure 11 below. Samples with a high vacuum bake-out did not depict less potential to contaminate the QCM maintained at -10°C, than the samples baked with purge. The effect was addressed to the previously mentioned water re-adsorption, as the bake-out conditions probably only facilitated the outgassing or water.

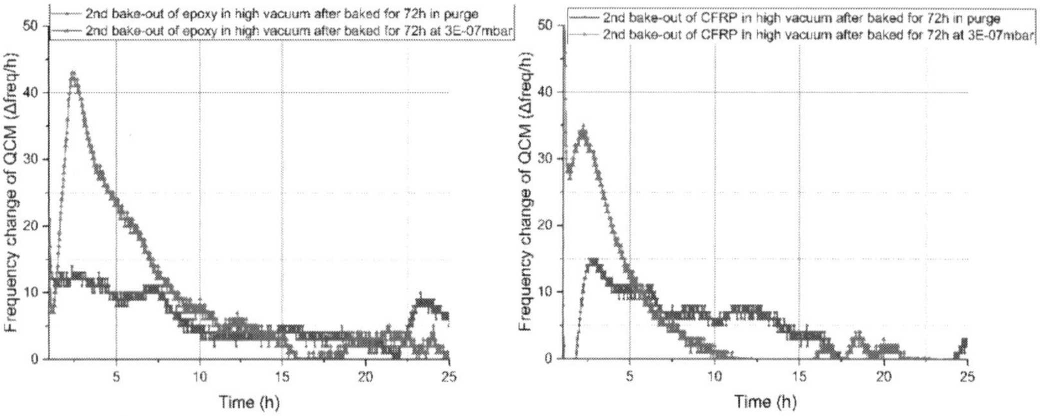

Figure 11. QCM monitored second bake of epoxy and CFRP. samples after being baked in different conditions

4. Screening method:

In summary, results confirm that vacuum bake-out achieve the highest efficiency. Nonetheless, a purge bake-out in continuous purge of inert gas may be an alternative under some situations (i.e., less stringent requirements) if care is taken to correctly deploy the purge to ensure the efficiency of the method. To allow a systematic evaluation, the following method is proposed for a material level approach:

A) Identification and ranking of critical materials, based on frequent utilization in space missions and outgassing potential according to the information in screening outgassing databases.

B) Definition of the bake-out approach for each material. The conditions and evaluation methods described in the flow chart below allow a parametric comparison. In future applications, the approach and pass/fail criteria shall be adapted considering the mission specific requirements.

C) Perform an experimental step, divided in 3 phases, using a minimum of 12 representative samples, according to the flow-chart.

Phase 1: adapted outgassing test, in high vacuum (pressure <1E-5 mbar)
Phase 2: purged bake-out with at least two conditions: gas residence time of 1 s and 10 s
As a minimum, the evaluation criterion is expected to be RML monitoring after 1, 3 and 10 days upon completion of bake-out to address water re-gain.
Phase 3: Evaluate the residual outgassing potential after Phase 1 and 2 by subjecting the samples to the same high-vacuum process, e.g: screening outgassing test or vacuum bake-out with QCM monitoring

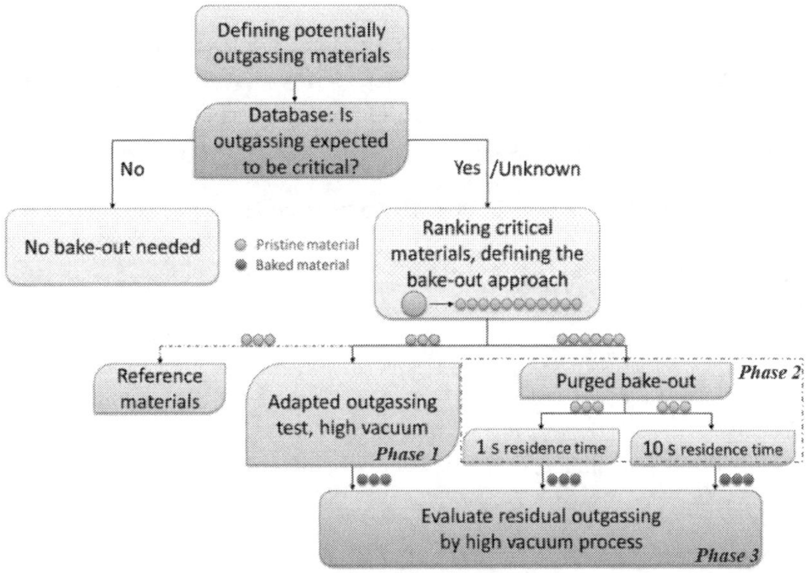

Figure 12. Scheme of proposed methodology to compare the efficiency of bake-outs

5. Conclusion

The publication, together with the poster presented at ISMSE 15 provides a methodology and the first results of a parametric study to assess the efficiency of different bake-out methods. This material level testing approach can be employed to measure the efficiency of a purged bake-out comparing to vacuum bake-out for space hardware. The evaluation criteria and test conditions should be determined on a case-by-case basis and based on the specific mission profile and requirements.

Results of silicone adhesives provide a first data-pack about the comparison between different types of bake-out. While different behaviour was observed, the vacuum bake-out is considered to be the reference for comparable temperature/duration processes, since it shows the highest efficiency in all tested cases. Furthermore, any purging system must be carefully designed since pockets with lower (or no) flow may affect the efficiency of the purged bake-out. The results indicate that static environment (i.e., no flow) is not a viable solution.

The surface area / mass ratio of samples did not impact the efficiency of both the purged and the high vacuum method in the studied range.

Water outgassing and re-gain can play a big role in ex-situ mass measurements, specifically when the conditions do not facilitate the molecular contamination outgassing. Therefore, monitoring RML for several days after the bake-out brings essential insight on the process' efficiency.

References

[1] Hässig M *et al.* 2011 Spacecraft outgassing, a largely underestimated phenomenon *2nd International Conference on Space Technology*, Athens, Greece, pp. 1-4

[2] Tveekrem J L, Leviton D B, Fleetwood C M and Feinberg L D 1996 Contamination-induced degradation of optics exposed to the Hubble Space Telescope interior – *Proc. of SPIE Opt.*

Sys. Cont. V, and S. Light and S. Optim. **2864**
[3] Suliga A, Ergincan O and Rampini R 2021 Modeling of Spacecraft Outgassed Contamination Levels by Thermogravimetric *Analysis Journal of Spacecraft and Rockets* **58**
[4] Ramanath S and Schühle U H 2020 The role of purged bakeout method in achieving the required cleanliness for the polarimetric and helioseismic imager on the Solar Orbiter mission: *Proc. of SPIE, Sys. Cont.: Pred., Cont. and Perf.* **11489**, 1148907
[5] ESA-ESTEC ECSS-Q-ST-70-02C 2008 Thermal vacuum outgassing test for the screening of space materials
[6] American Society for Testing and Materials 1984 ANSI /ASTM E-595 Standard Test Method for Total Mass Loss and Collected Volatile Condensable Materials from Outgassing in a Vacuum Environment
[7] Scialdone J J 2003 PRESSURE and PURGING EFFECTS on MATERIAL OUTGASSING and EVAPORATION *Proc. of the 9th International Symposium on Materials in a Space Environment*
[8] Chang C W 2004 Examination of material outgassing in the air and purging flow environments *Proc. of SPIE, Opt. Sys. Degr., Cont, and S.Light: Eff., Meast, and Cont.* **5526**

Characterisation of lunar and Martian dust simulants and impacts on thermo-optical properties of space materials

A Suarez Kahan[a], B Delacourt, N Dias, M Hołyńska, A Tighe

European Space Agency (ESA), European Space Research and Technology Centre (ESTEC), Materials' Physics and Chemistry Section (TEC-QEE), Keplerlaan 1, Postbus 299, 2200 AG Noordwijk, The Netherlands

[a]alice.suarez-kahan@esa.int

Characterisation analyses were carried out on several available lunar and Martian simulants in order to provide standardised comparative data to guide testing. Thermal conductivity measurements were performed on regolith powder under ambient conditions Reflectance spectra were obtained for several simulants including separated size-fractions in order to calculate solar absorptance (α_S) values. The change in thermo-optical properties of optical solar reflectors (OSRs) due to lunar simulant EAC-1A <100 µm fraction dust deposition was investigated, showing a linear increase in α_S and infrared emittance (ε_{IR}) values. Facilities are in development for future work involving dusty testing under simulated lunar environments.

1. Introduction

Planetary and lunar dust contamination must be investigated to allow for future long-term exploration and potential settlement of the Moon and Mars. Dust can obscure thermally important surfaces, abrade coatings, and deteriorate space mechanisms, amongst other problems.[1]

Apollo missions 15, 16, and 17 demonstrated a critical example of dust obscuration impacting thermal control surfaces. The fine dust deposition on the battery radiators for cooling on the lunar roving vehicle (LRV) caused the batteries to overheat far beyond their expected operation maximum that they were rated for.[2] Some work has been done using simulants to replicate these effects.[3] The complexity of lunar dust adhesion mechanisms[4] is closely interlinked with this problem, as modelling for the levels of dust expected on a surface is critical to preparing for adverse thermal effects.

Lunar and Martian regolith simulants are also of use for the development of in-situ resource utilisation (ISRU) techniques. One important case is the processing of regolith, often through sintering in some way, in order to form building blocks. Thermal conductivity is a key property of simulants being used in these sorts of tests. Measurements of regolith simulant materials might also provide data that contributes to the further development of models for the thermal conductivity of planetary regoliths.[5,6]

A large amount of testing will need to be carried out to enable the safe return to the moon whilst also allowing for more complex missions and the use of more advanced technology. Several simulants are available to researchers aiming to test contamination effects, with different intended purposes and varying levels of fidelity. These simulants however are difficult to compare between them due to the lack of standardisation of testing across laboratories. Whilst a wealth of data exists, until standardised measurements are carried out, information is lost as researchers are unable to position the results of complex tests in the context of research carried out using other simulants, or even other preparations of the same simulant.

Various different planetary simulants were used in this work, mostly lunar. EAC-1A, LMS-1, LMS-1D, LHS-1, LHS-1D, TUBS-M, TUBS-T, and Chenobi were the lunar simulants, and samples of JSC-Mars-1A and MGS-1 were the available Martian simulants.

EAC-1A is the simulant produced by the European Astronaut Centre. It was developed by the European Space Agency (ESA) using rock from the Eiffel basalt fields, which were created as a result of eruptions 45 million years ago. This was then mechanically crushed into different size range particles, of which EAC-1A is the finest.[7]

LMS-1, LMS-1D, LHS-1, LHS-1D are commercially available simulants produced by Exolith labs. LHS-1 is their lunar highland simulant, produced from anorthite mined from the White Mountain anorthosite source in Greenland, and a glass-rich basalt, and LMS-1 is their lunar mare simulant, produced mostly from pyroxene, glass-rich basalt and anorthosite. The simulants are produced by mixing separately sourced minerals in ratios to match lunar soil. The 1D type of each is the fine dust simulant of the same mineral compositions as the standard. [8,9] The Exolith simulants in this study were obtained in 2021 unless otherwise mentioned.

TUBS-T and TUBS-M are the lunar highland (terrae) and mare simulants respectively, produced by TU Braunschweig. A Scandinavian anorthositic source of a metamorphosed gabbroic complex was used as the source material for TUBS-T, and for the basaltic simulant the basalt is part of an alkali-olivine basalt occurrence in Germany, with bulk chemistry described as aluminous low-Ti basalt.[10]

Chenobi (CHemically ENhanced Olivine Bytownite-1) is a lunar highland simulant. Chenobi and OB-1 were produced to match the size distribution of Apollo 16 regolith sample 64500. Chenobi is identical to OB-1 simulant except for the glassy component being produced from plasma-melted Shawmere Anorthosite as opposed to from olivine slag. These are geotechnical simulants, of which Chenobi's more representative glass content gives it a higher fidelity.[11]

JSC Mars-1 was developed to match the reflectance spectrum of Martian soil and has since been used as a general-purpose Martian simulant. JSC-Mars-1A was produced from the same Hawaiian volcanic source after the original stock of the first batch was depleted, with a maximum particle size of 1 mm.[12] MGS-1 is the Mars global simulant also from Exolith labs. The simulant is designed to be a mineralogical simulant, produced by combining different ratios of minerals to best match Rocknest windblown soil as analysed by the Curiosity rover.[13]

In this paper the results of thermal conductivity testing for some available lunar simulants are reported. X-ray diffraction patterns are also shown for several lunar and Martian simulants. The solar absorptance was obtained from reflectance spectra of 15 samples including both as-received simulants and separated simulant size fractions. The thermo-optical impact of varying extents of lunar dust simulant deposition on OSRs is presented.

2. Experimental

2.1 Simulant preparation: Size fractions of simulant were prepared for various tests. For preparation of sample fractions 100 μm and below, wet sieving was used, otherwise dry sieving was considered adequate for fractions of less than 200 μm.[14] Moisture contents of simulants were not critical for any of the tests (as determined through testing comparisons) and therefore bake-out was not a necessary step.

2.2 Thermal conductivity: Sample: Best measurements were obtained at high compressions of the material, i.e. the maximum packing density that could be achieved without redistributing the powder by size. Pressure on the grains was controlled by applying weights to the top of the sample, which were varied with each sample holder to maintain constant pressure of 2.8-3.0 kPa. *Measurement*: The Transient Plane Source (TPS) system by Hot disk AB instrument was used. Kapton-insulated sensors between 3 mm and 10 mm diameter were used. A reference measurement of stainless steel standard sample was used to ensure validity of measurements. Powers ranging 10-40 mW were used with measurement durations of 80-160 s. Idle times between measurements were (as recommended by the manual) more than 36 times the duration of the previous measurement. *Analysis*: The Hot Disk software uses an iterative algorithm to extract thermal diffusivity and then thermal conductivity values from the values of resistivity of the nickel coil with time.

2.3 Reflectance spectra: Sample: A powder cell holder (quartz window Agilent powder accessory) was densely packed with powder cell. A number of repeat measurements were carried out depending on the homogeneity of reflectance of the grains within the sample, with fresh sample preparation in the cell each time. Repeats were carried out so that the mean value in standard deviation in reflectance value at each wavelength reflectance was kept below 0.4, after a minimum of 2 measurements. Reference measurements of solid samples were carried out by placing a sample into the cell as well as measuring it outside of the cell. *Measurement*: Ultraviolet-visible (UV-Vis) reflectance spectra were obtained using the Cary 5000 UV-Spectrophotometer in total reflectance mode with the integrating sphere attachment. The integrating sphere accessory was able to be attached at a different angle so that the incident light beam was pointed directly at the sample (incident angle 0°) which was used to also measure the diffuse component of reflection for reference samples. *Analysis*: The obtained reflectance spectrum was corrected first using baseline correction and using a diffuse calibration reference coupon (which is routinely calibrated using a traceable reference standard), and subsequently corrected for the presence of the quartz window blocking the sample from the beam. The details for this last correction are provided at the start of the appendix. To measure solar absorptance, the obtained (corrected) signal was then integrated over air mass zero solar spectrum ASTM E-490 for the spectral range from 250 nm to 2500 nm. The attributed error in values of solar absorptance is ±0.03 (when using the instrument in a standard way i.e., with a sample directly placed on the window of the integrating sphere). A further error is introduced by the subsequent correction for the glass cell which was approximated using empirical data.

Extra correction: The nature of the available integrating sphere (opening on the side) means that getting spectral data for the dust is difficult and hence a holder is required. Reflectance of as-received, dried, and sieved simulant samples was measured using a quartz powder cell (Figure 1) to hold the sample in the UV-Vis-IR spectrophotometer integrating sphere. The difficulty is that there are no calibrated standards currently in our possession that could fit into the sample holder and therefore be measured through the window. Also, the impact of the window on the sample in the holder is not a constant.

Figure 1. a) three quarter and b) top views of the powder cell accessory with lunar dust simulants inside, c) the empty cell

Reflectance of as-received, dried, and sieved simulant samples was measured using a quartz powder cell to hold the sample in the integrating sphere. An equation was used to correct for the impact of the quartz window on the reflection, before correcting the instrument with a calibration coupon:

$$R = Rg + Tg^2 * Rd \quad (1)$$

Where R is the total measured reflectance, Rg is the reflectance of the quartz glass, Tg the transmittance of the glass window, and Rd the reflectance from the dust component. This correction relies on various assumptions which can introduce error. The transmission data for the quartz window is the specular reflection obtained from light passed through the window at the normal angle. The angle of the incoming light in the reflection measurement is actually of 6°. As the dust is a diffuse sample, the light that is reflected from the dust is passed through the quartz window at many small angles, which will introduce a larger error. Figure 2 shows the difference between these two cases in a basic diagram.

The actual equation should read:

$$R = Rg + Tg_{(6°)} * Rd * Tg_{(\text{diffuse})} \quad (2)$$

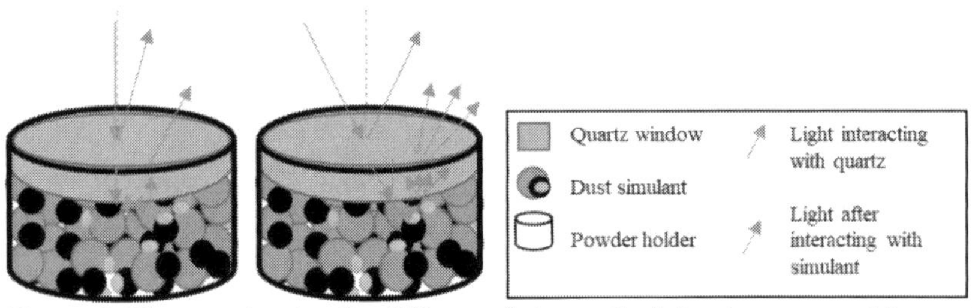

Figure 2. Diagram showing the difference between the behaviour assumed in equation 1 (left) and the more realistic behaviour of the system as in equation 2 (right)

There were no available certified calibration coupons that could fit within the quartz window so several intermediate diffuse reference samples were used to test the equation empirically before applying it to the dust samples. Several coupons were used as diffuse samples that could be measured both inside and outside of the cell. The error was seen empirically to depend on the reflectance of the sample, i.e. was less accurate with lower α_S values and more accurate for darker samples. Data for the measured reference samples can be found in the appendix. The instrumental error of the basic technique is of ±0.03 in the value of α_S, on top of which the error in correction should be considered. Using two brackets for higher (>0.5) and lower (<0.5) α_S values this maximum empirically determined correction error was combined with the existing error. We therefore tentatively attribute an error value of ±0.06 to <0.5 α_S samples and ±0.04 to >0.5 α_S samples.

2.4 Solar absorptance (α_S) / Infrared emittance (ε_{IR}) EAC-1A on OSRs: Sample preparation: EAC-1A lunar dust simulant was wet sieved to separate the <100 μm fraction for use, and thoroughly dried. The OSR surface was fully cleaned using acetone and isopropanol (IPA) solvents. The OSR substrate was enclosed by a mask with 3 circular windows of 15 mm diameter, Figure 3, (and which was demarcated with ridges to allow the thermo-optical measurement instrument head to be positioned exactly for each measurement). Dust was deposited from above and allowed to fall onto the OSR surface by repeated controlled impacts on a sieve of mesh size 200 μm holding the prepared dust fraction, then cleaned off the mask area.

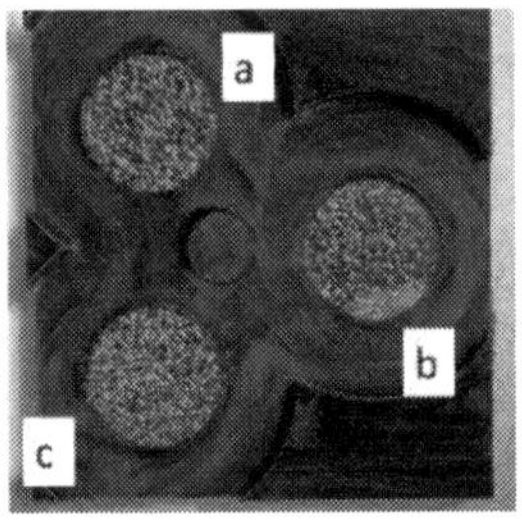

Figure 3. OSR in mask with deposited LDS. Exposed areas are labelled a, b, and c, clockwise from top.

Measurement: Mass measurements were carried out using a Sartorius RC 210 S balance after cleaning of the mask, such that the total particles on all three areas of the mask contributed to an averaged value for mass deposited per PAC (percentage area coverage). A second mass measurement was obtained after other characterisations, prior to the next dust deposition. An error value was approximated using the range in the two measurements. Particle counting was carried out using a Keyence VHX-6000 at 100× magnification using full coaxial illumination on a 1 cm^2 square area within each mask hole. Pixel resolution was 4.6 μm^2. A threshold was chosen so that the edges of the particles were fully considered by the software to be part of its area. Imaging was carried out before and after thermo-optical measurements to check for disturbances to the sample. Thermo-optical measurements were carried out on each window of the OSR sample, with 3 measurements per value. Handheld instruments from Surface Optics Corporation (SOC) were used for thermo-optical measurements. The software applies a baseline correction and uses the measurement of a NIST-traceable calibration puck (diffuse coupon and mirror for absorptance, and gold for emittance) to further correct the data. The instrument was placed on the mask, which had a thickness of 0.5 mm between window and sample. The 410 Solar SOC measures reflectance for 6 different wavelength bands between 335 nm and 2500 nm. The ET 100 SOC measures the integrated surface reflectance of a surface at two different angles of incidence (20° and 60°) and for six wavelength bands within 1.9 μm and 21 μm. The instrument was rotated between each measurement to average out any non-centring of the beam in the window area. The error in the absorptance measurements is ±0.03, and in emittance measurements is ±0.05. *Analysis*: Particles smaller than 10 μm^2 were discounted from the total area coverage to exclude noise. To measure solar absorptance, the obtained signal is integrated over the air mass zero solar spectrum for the spectral range from 335 nm to 2500 nm. The near-normal directional thermal emittance values (20°) measurements are used by the internal software to derive the total hemispherical emittance, using correlations from literature.[15]

2.5 X-ray diffraction (XRD): Measurement: The instrument set-up was optimised for the analysis of a powder composed of many constituents. The X-ray source used was a Cu source, 1.5418 Å, 40 kV, 40 mV. A Göbel mirror was added, which produces a parallel beam of X-rays on the sample surface. Although this limits the counts of X-rays, this is beneficial in producing a focused beam across a larger area of sample, especially helpful in the case of uneven surface morphology (as is the case for the simulant powder). As using the Göbel mirror produces a ghost peak in the spectrum, a transverse slit was used on the detector. The specific detector used was the Lynxeye-XE-T detector (in 1D mode). A beam blocker was used to limit any out-of-plane X-rays from reaching the detector. Lastly, the stage was set to spin so that a greater area of the sample was exposed to X-rays during the sampling interval, thus ensuring a more representative measurement of the inhomogeneous sample. Spin rate was chosen to not overly disturb the sample, 12-15 rpm. Data collection was carried out with 0.01° increments (except where noted). *Analysis*: No background subtraction was necessary, a $K\alpha_2$ subtraction was carried out. The International Centre for Diffraction Data (ICDD) database was used to source quality reference scans which matched well.

3. Results and discussion

3.1 Thermal conductivity

Thermal conductivity of powder is highly dependent on the mechanism of heat transport between the grains, as such these measurements carried out in ambient environment do not represent the lunar environment and should not be interpreted as such. These preliminary values aim to be useful to some extent in direct comparison between similar simulants, mainly for anybody running low-fidelity tests in an ambient environment. Future work is planned that would obtain thermal conductivity tests in vacuum conditions.

A transient plane source (TPS) instrument was used with the sensor lodged in the centre of a powder holder packed with regolith simulant. As the presence of voids in the material caused experimental error, packing densities had to be relatively high for each simulant. Due to the variation between the simulants (particularly particle size distribution), this high packing density was not a variable that could be controlled. Pressure was applied on the top of the sample through weights, to a value of 3 kPa. (Enough that the weight of powder on the lower grains in the sample was negligible). Compressional stress can have a very important impact on the thermal conductivity of regolith materials, due to the enhanced intergranular thermal contact from elastic deformation of the grains.[16] Volume occupancy was calculated using mass and the specific gravity for each simulant reported by the manufacturer.

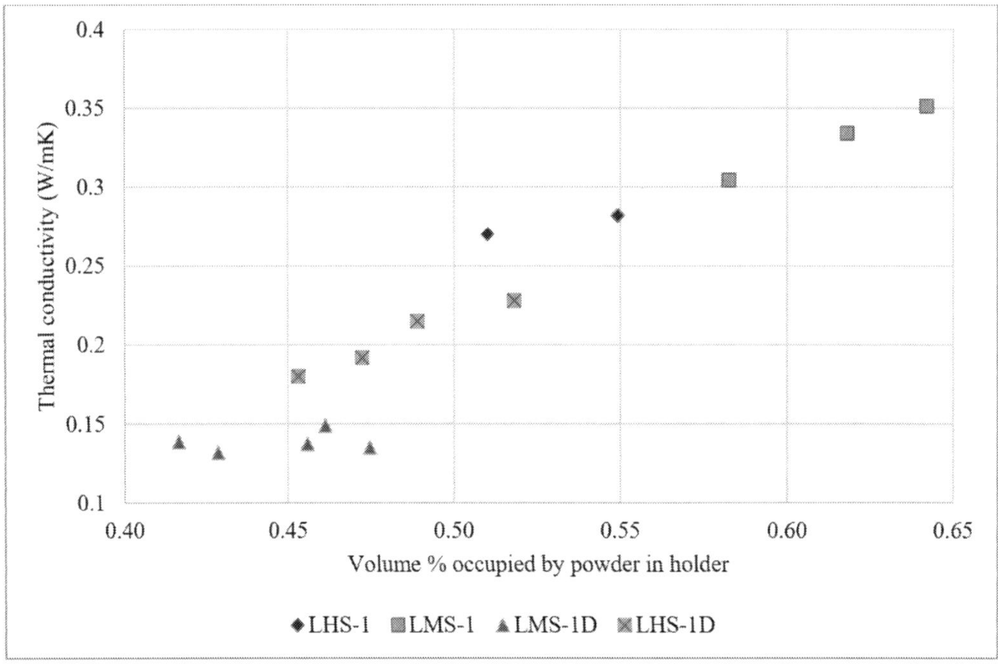

Figure 4. Graph showing the thermal conductivity in W/mK measured using TPS for varying densities of several lunar simulants.

From these data points it can be observed that the LHS-1 and LMS-1 simulants showed similar relationships of thermal conductivity with volume fraction. These simulants differ both in mineralogy and particle size distribution (PSD). Between the finer fractions, more can be inferred. These simulants differ in mineralogy (matching their large-grain counterparts) but have a near-identical size distribution and therefore reach similar volume fractions when compressed. The mineralogy of LMS-1D might therefore be interpreted as the factor responsible for resulting in a lower thermal conductivity, but other factors are still at play. The shape of the particles, bulk density of the simulant, and particle size, as well as any size sorting present, will affect the average heat transfer distance between particles.[17] The dependence of the conductivity on chemical composition of the grains might be dominated by the more important physical characteristics (e.g. as shown for angular quartz grains giving similar conductivity to silica bead grains despite their differences in solid thermal conductivity).[6]

The low thermal conductivity of regolith simulant powder presented several challenges for measurement. For example, despite having the same maximum particle size LMS-1D could not be measured using the same parameters as LHS-1D, due to its lower thermal conductivity. Also, the contribution to the measurement of conduction by gas transfer in the void spaces between particles was huge. Different packing densities were used to try and get comparative values for different samples, but packing density is not the only variable that defines how grains are arranged in the holder. Either measurements under vacuum must be achieved, or at least the grain particle size distribution parameter must be excluded from the results. Some limited existing literature data is available for thermal conductivity measurements of un-sintered regolith simulant under vacuum, i.e. JSC-1A was measured around 0.002-0.003 $Wm^{-1}K^{-1}$ [12], which is much lower than that recorded here. An important decrease of measured thermal conductivity values is expected when regolith samples are measured under vacuum as opposed to ambient [18].

Other instruments are available in the lab that allow for the possibility of measurement of powder samples, possibly under vacuum or in different temperatures. These will be investigated as another step to determine more relevant values.

3.2 Reflectance and solar absorptance

Whereas the need for more preparation and testing of thermal control surfaces with regards to dust contamination is well known, there is no database comparing the thermo-optical properties of simulants that might be used to such an end. These measurements of reflectance were conducted for various simulants in order to produce a table of α_S.

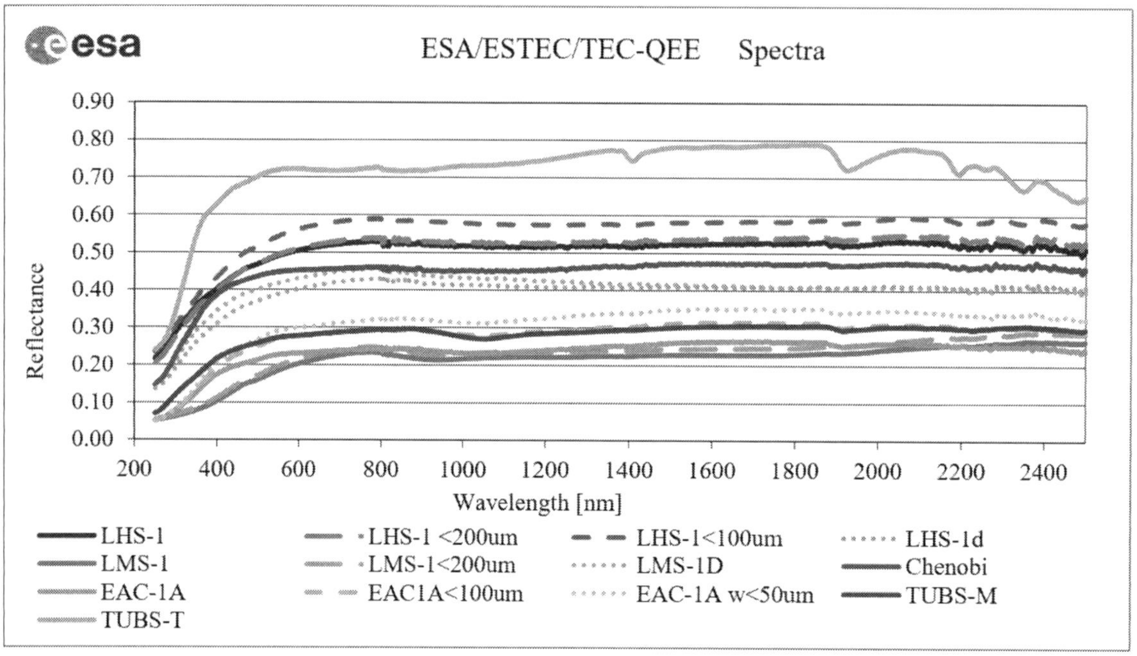

Figure 5. Reflectance spectra of lunar regolith simulants after correcting for quartz window

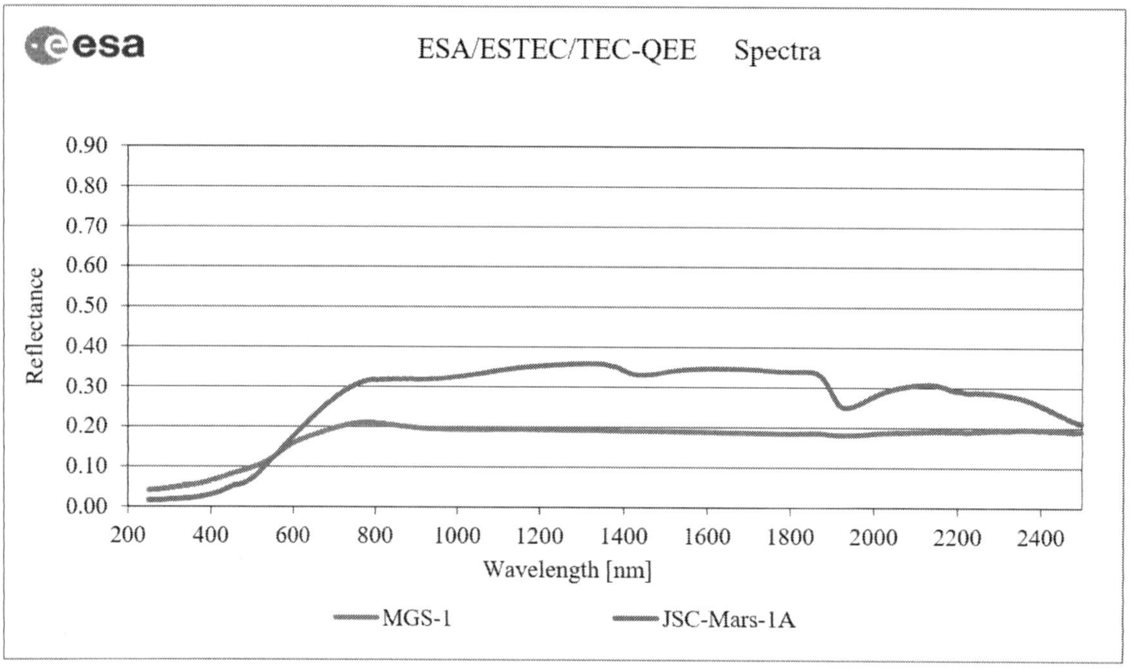

Figure 6. Reflectance spectra of Martian regolith simulants after correcting for quartz window

Table 1. α_S values for each simulant, with sample preparation description and indicative particle size ranges given. ("Bulk" here refers to the simulant as-received from manufacturer, with no size separation.) Generally, the maximum of the particle size ranges given does not make up a large fraction of the particle size distribution in the bulk simulant, but is more relevant for sieved samples.

Type	Simulant	Description	Particle size range (μm)	αS	Error in αS
Lunar Mare	EAC-1A	Bulk	20–1000 [7]	0.78	±0.04
		Wet sieved	<100	0.73	±0.04
		Wet sieved	<50	0.71	±0.04
	LMS-1	Bulk	<0.04-400 [8,19]	0.81	±0.04
		Dry sieved	<200	0.79	±0.04
	LMS-1D	Bulk	<30 [20]	0.59	±0.04
	TUBS-M	Bulk	0-2000 [21]	0.73	±0.04
Lunar Highland	Chenobi	Bulk	0-9500 [11]	0.56	±0.04
	LHS-1	Bulk	<0.04-400 [8,22]	0.51	±0.04
		Dry sieved	<200	0.50	±0.06
		Wet sieved	<100	0.46	±0.06
	LHS-1D	Bulk	<30 [23]	0.61	±0.04
	TUBS-T	Bulk	0-2000 [21]	0.29	±0.06
Mars	JSC-Mars-1A	Sieved	90-1000	0.78	±0.04
	MGS-1	Bulk	>0.04-600 [13]	0.84	±0.04

Figure 5 and Figure 6 show all of the (corrected) reflectance spectra obtained. Table 1 gives the calculated α_S for each. Generally, the mare simulants are more absorbing than the highland simulants, depending on the size separation used. It is observable that the ground samples are more highly reflective than their bulk counterparts, and have a lower alpha. The 1D simulants are composed of the same ratios of minerals as the larger fractions, but their finer grain size impacts the optical properties (e.g. scattering). These 1D simulants demonstrate how particle size is critical for determining α_S values. Any testing involving the use of simulant that has been ground must also necessarily involve a characterisation of its thermo-optical properties if these are relevant.

It is also found that for sieved samples of lower grain size maxima, the α_S values are always lower than those of the bulk simulant. In the case of the LHS-1 type simulant the ground sample, LHS-1D with a max particle size of 30 μm has a higher absorptance than the sieved <100 μm sample prepared from the bulk. The mineralogy of these two samples is very different. Upon sieving, the darker glassy basaltic particles are removed from the sample, leaving largely anorthosite grains which are much lighter. The 1D fraction still contains these basalt grains, albeit in a smaller grain size. Due to the variability in PSD and mineralogy this change in reflectance with size separation will vary with each sample.

The reflectance of both Martian simulants drops in the visible spectrum, as expected. The JSC-Mars-1A sample was more reflective.

These results highlight the need to characterise the actual fraction in use for a test. As previously remarked, (Gaier, [24]), the formulation of a lunar simulant as intended for use as a mare or highland simulant does not always serve as a guide to the absorptance values of the simulants. For example, in the case of LHS-1D and LMS-1D, the reflectance values are very similar, with the LMS-1D fraction actually higher at certain wavelengths (despite it being developed from a mare simulant). LHS-1D has a slightly higher alpha.

To the best of our knowledge, the absorptance values for most of these simulants has not been previously reported. Chenobi simulant (ground and sieved to <20 μm) was measured by Gaier et al.[25] with an absorptance value of 0.42, which as expected is lower than the obtained value in this work. The <25 um fraction of ground Chenobi was previously found to have an absorptance value of 0.39 [24]. Goulas et al.[26] report the absorbance of JSC-Mars-1A <125 μm fraction, and obtain a slightly higher reflectance also, expected considering the size separation.

Grain size was not set aside as a factor (e.g. by milling) rather, the powders as received were measured. This is to give relevant values for unprocessed simulant powder. Some fractions of simulants were separated to show how size separation affected reflectance values. Milling the samples so that the grain size of the simulants did not impact measurements would be another approach in order to compare purely the sample composition effect on α_S. This was not done in this case as milling the powders is not necessarily done when simulant is used for a thermo-optical test. Again, if a processed simulant is used for the test, then those properties will need characterisation. Dust as the most difficult fraction to remove from surfaces is the most important to characterise for most tests. Therefore the important sample to characterise here would either be the ground simulant, or the fine fraction of the simulant if this had a more relevant α_S value.

3.3 X-Ray Diffraction

XRD data for such complex samples is difficult to interpret due to the many overlapping peaks. Still though, data can be obtained from samples prepared in accordance with a standard test procedure in order to produce a reference pattern that can be used to monitor batch consistency. The variability that can be expected in simulant batches is not a well-understood parameter. [27] Another motivator for obtaining XRD patterns is the relative ease of use and macroscopic overview of the sample. Patterns were obtained for different size fractions of the same simulant for an overview of the change in mineralogy (crystalline fraction).

XRD patterns were obtained without any prior sample preparation. Grinding can impact the positions of peaks as well as the broadening of some peaks which can obscure others. Impurities can be introduced through the milling process. [8] Another benefit of using samples "as-is" is that this renders the technique completely non-destructive, which can be necessary in the case of small sample size, as well as avoiding any changes to the sample. Disadvantages include that the heterogeneous sample may be sampled from in ways that are not representative of the actual make-up of the sample such that quantification is then not possible.

Multiple minerals were identified as likely matches for each of the patterns. Due to the complexity of the mineralogical make-up of the samples, these were not able to be individually attributed. For display purposes, patterns are shown here with mineral reference data determined to be a close match and represent one of the major components. This is not meant to be a full analysis of all present minerals. A selection of the data is shown here, the rest are to be found in the appendix.

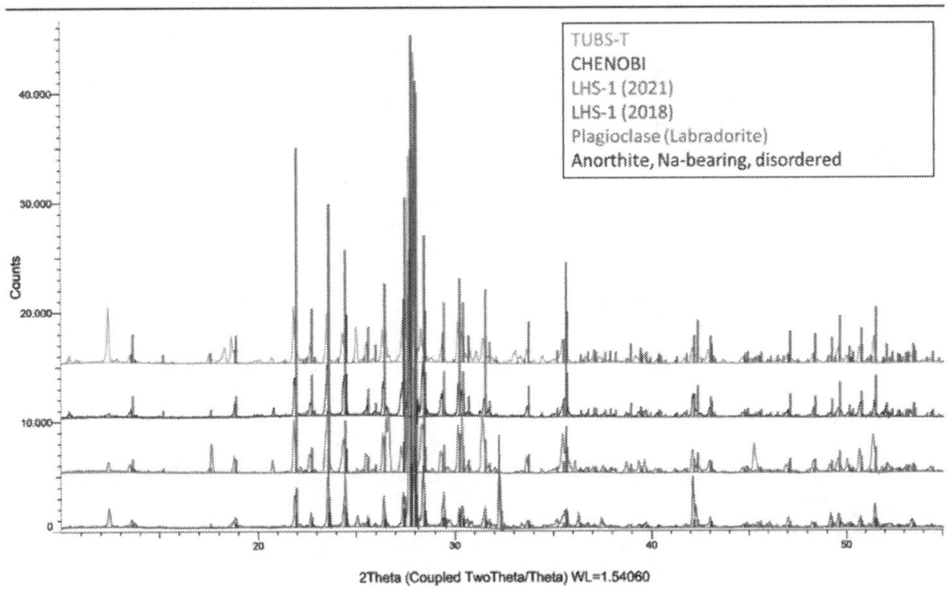

Figure 7. XRD patterns for analysed lunar highland regolith samples; (TUBS-T in orange, Chenobi in brown, LHS-1 2018 sample in pink, LHS-1 2021 sample in red), showing reference peaks for labradorite plagioclase (turquoise) and Na-bearing anorthite (dark blue).

Figure 7 shows the lunar highland simulants analysed, along with the spectra for anorthosite and labradorite. Anorthosite is a critical major component of lunar highland simulants, with other plagioclase minerals also present. Lunar anorthosite is lower in sodic content than that found on earth.

The differences between the 2018 (red) and 2021 (pink) Exolith simulants is important to note. Between these two batches the source of anorthosite changed from the Stillwater mountain to the White mountain in Greenland, and the glass-rich basalt source was also changed.[9] Whilst they are transparent regarding the sources of minerals and have published data referring to the changes made between these batches, the simulants still bear the same name. It is critical that when referring to older literature, this information is borne in mind.

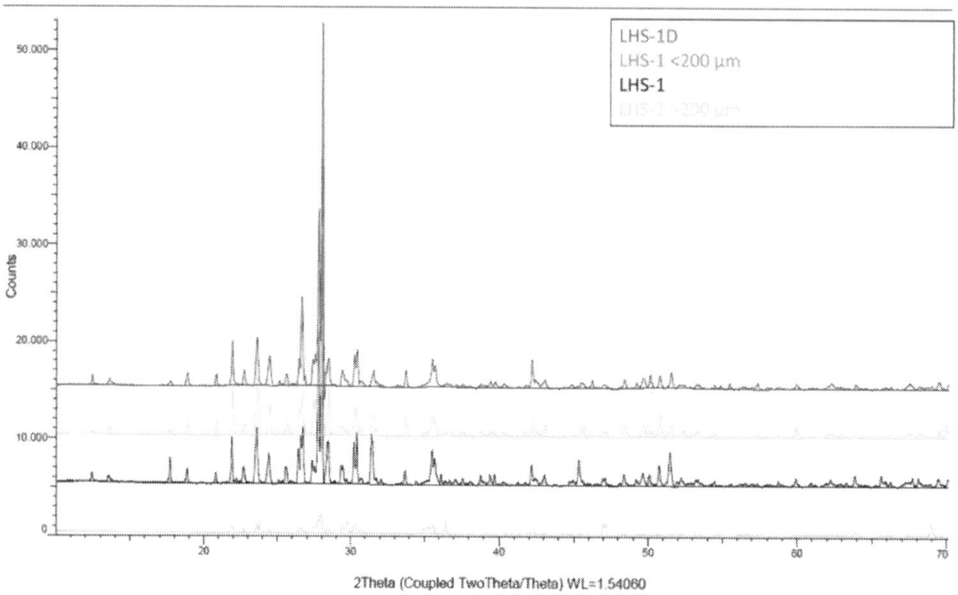

Figure 8. XRD patterns for analysed LHS-1 fractions (>200 µm fraction in light green, as-received in black, <200 µm fraction in aqua) and LHS-1D (dark blue) regolith samples

Figure 8 shows the prepared fractions of LHS-1 (and LHS-1D). It can be seen visually that the separation of the >200 µm fraction of LHS-1 results in the removal of most of the glassy basalt particles. This is not clear at all in these spectra due to the limitations of XRD with regards to amorphous material. Barely any difference is visible despite the huge difference in sample composition. Whilst XRD can in some configurations be used for the semi-quantification of glassy components, this method was set-up to minimise the background scattering signal, thus losing the information from scattering for the glassy component. This set-up of XRD does not allow us to assess the fidelity of glass components to the lunar glass fraction, which is critical for its properties. This limits the usefulness of the technique.

Mare simulants can be expected to be composed of mostly pyroxene, glass-rich basalts, plagioclase, olivine, and ilmenite, depending on their fidelity. The major components of Martian simulants are plagioclase, pyroxene and olivine minerals, as well as a large glassy fraction. The data for JSC-Mars-1A shows a low signal-to-noise ratio, possibly due to the heterogeneity of the grain sizes meaning that large particles disrupted the focal plane of the instrument. Grinding the sample could improve this.

3.4 Interaction with space materials

The coverage of thermal-control surfaces by dust contamination has been reported to cause issues in the past and is expected to be problematic in future. Moreover, dust is also present in cis-lunar space. Some dust adhesion is expected on the orbiting Gateway station due to transfer from lander modules returning with dust contamination that may be dislodged and deposited on other surfaces. With this context in mind thermo-optical testing was carried out to establish a relationship between deposited dust amount and change in absorption and emissivity values. EAC-1A simulant has appropriate thermo-optical properties for its use in this test.

EAC-1A simulant was size-separated to the <100 µm fraction, and deposited on the surface of an OSR with surface coverage varied between samples. Approximate coverages of 10%, 20%, 40%, 55%, 70%, 85% were characterised in the test: with three separate measurable areas per coverage point.

Portable instruments measured infrared emittance and solar absorptance of the bare surface and the dust-contaminated surfaces for each coverage point.

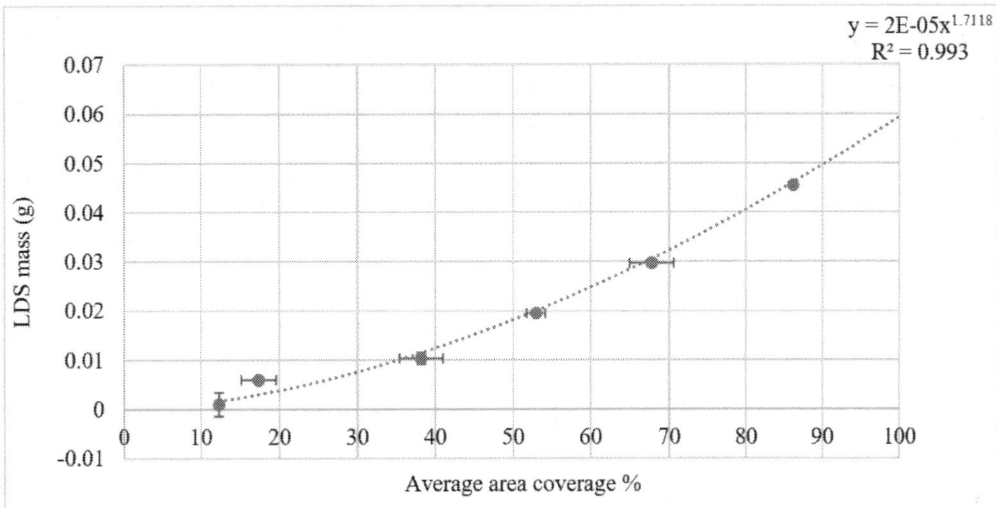

Figure 9. graph showing the mass deposited at each area coverage point.

Figure 9 shows the relationship between deposited mass and average measured area coverage. A polynomial trend is also proposed. As could be expected from the deposition technique, this data points to the stacking of particles and the formation of aggregates in the Z direction. Errors on the graph are from the range in mass measurements made before and after thermo-optical measurements and the range in coverage measured for each area before and after thermo-optical measurements.

Figure 10 shows obtained images of some coverages.

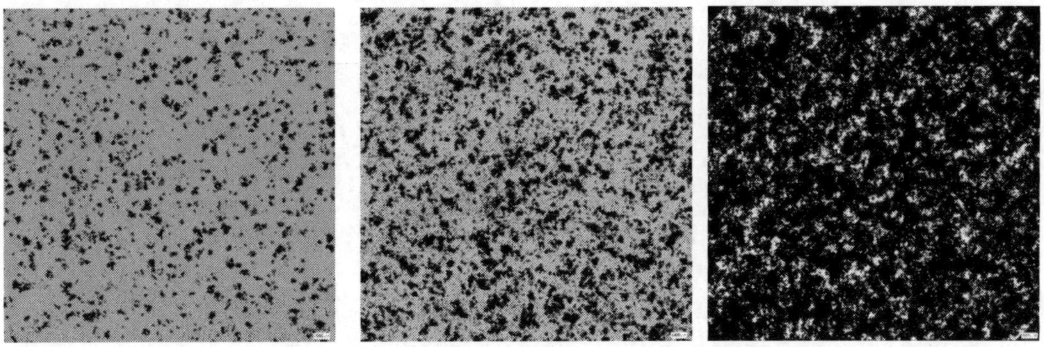

Figure 10. Obtained pre-analysis microscope images at the following coverages a) 10%, b) 40%, and c) 85%. Particles are shown in black; the blue background is not relevant to the analysis.

α_S values obtained on the OSR as a function of the EAC-1A coverage are shown in Figure 11. ε_{IR} values obtained on the OSR as a function of the EAC-1A coverage are shown in Figure 12 and Figure 13. A linear trend is also proposed for each graph. Data for the linear fits are provided in the appendix, including the error of the fit.

The values plotted at 100% coverage (as single points) correspond not to 100% coverage of the OSR but of separately obtained values for the bulk LDS (see Table 2, Table 3), plotted there as a guideline. The error reported for α_S and ε_{IR} is the standard deviation in measurements multiplied by a t value of 2.92 for a 90% confidence interval. Reported coverage value is the average between particle counting measurements carried out before and after thermo-optical measurements. The error reported for the coverage is the difference between those measured coverages, i.e. the range. The absolute instrumental error is not plotted. These values are higher, the measurement uncertainty in the solar absorptance is ± 0.03 (confidence factor = 2) and the estimated uncertainty of the emissivity measurements is ± 0.05.

The ET100 instrument measures reflectance in bands and produces values for emittance at angles of 20° and 60° to the normal. Calculations are subsequently used to produce a value for the hemispherical thermal emittance (HTE). This calculation relies on approximations that differ depending on whether the material is a dielectric or metal surface. Considering the complex nature of the sample, different values could be obtained. Values reported for HTE data were generated using the assumption of a dielectric surface.

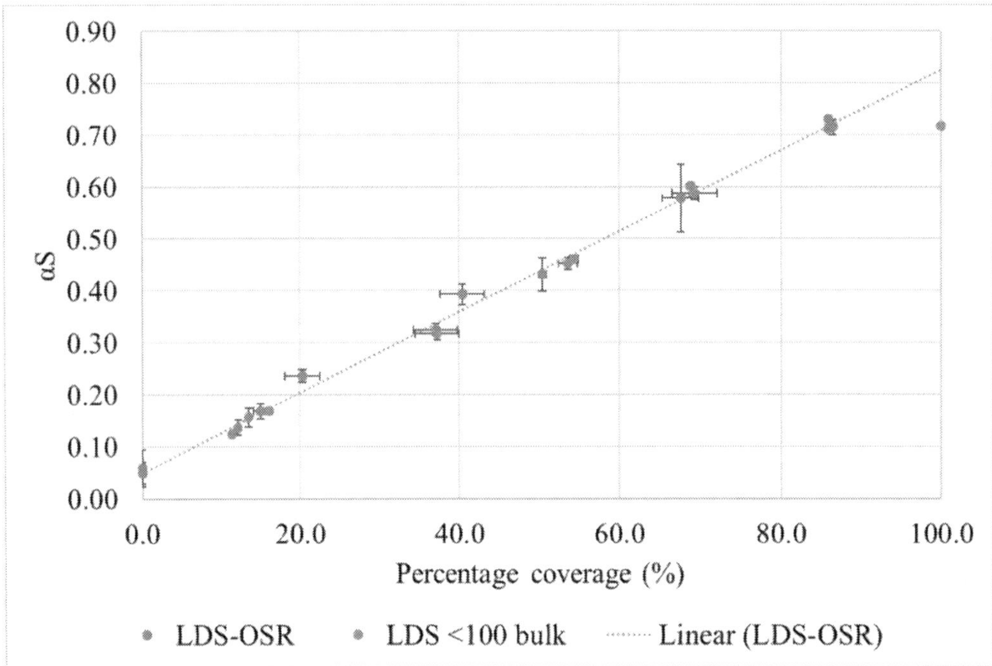

Figure 11. Graph of α_S with surface coverage

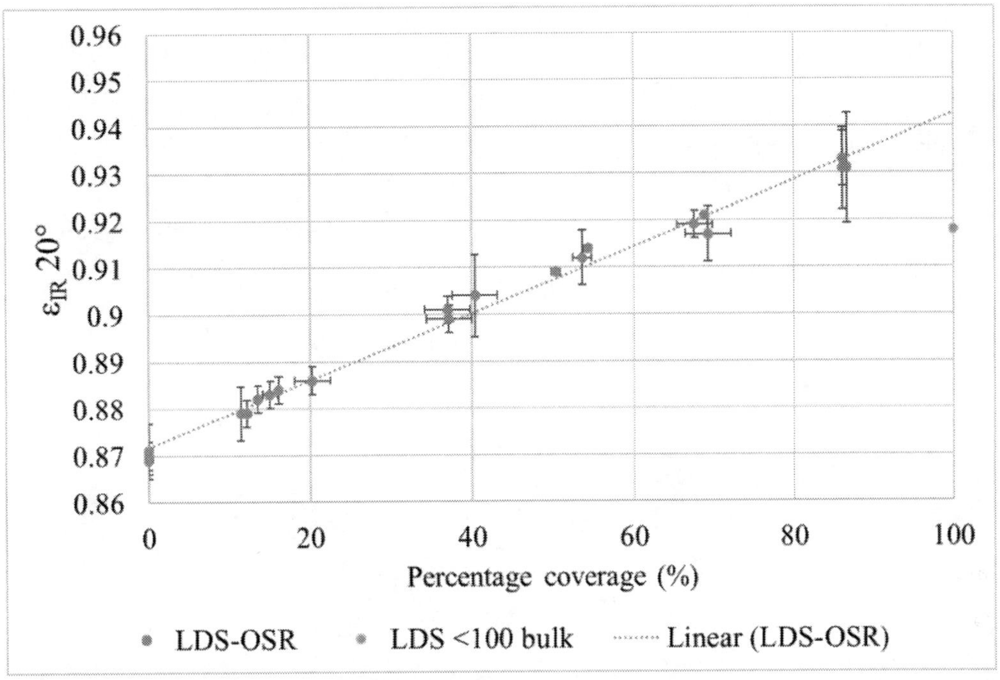

Figure 12. Graph of ε_{IR} at 20° with surface coverage

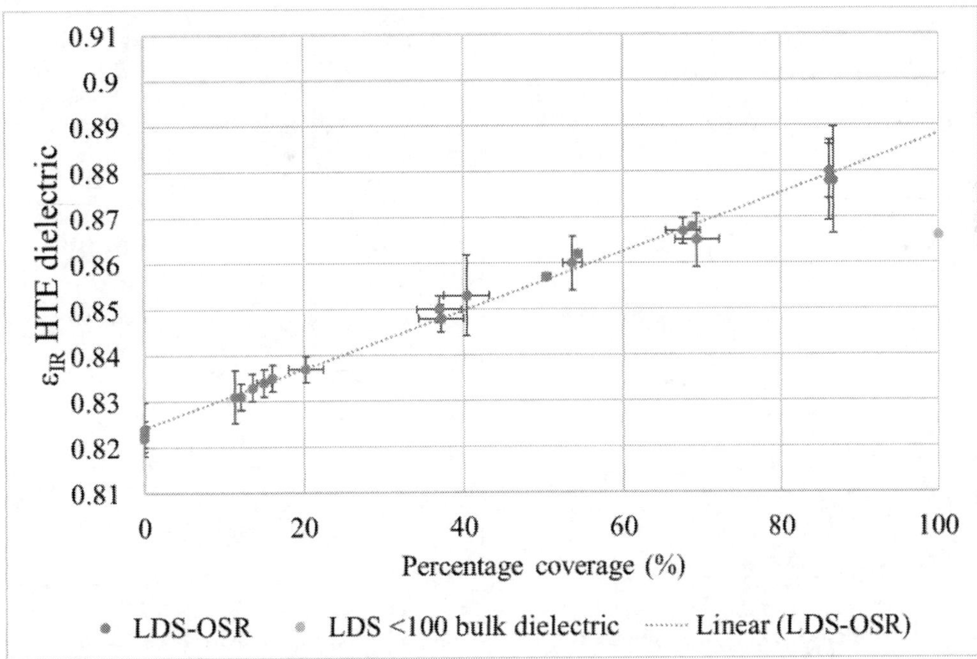

Figure 13. graph of HTE ε_{IR} calculated (assuming a dielectric surface) from values obtained for 20° emissivity.

Values of the ratio of α_S / ε_{IR} can be calculated, as shown in Figure 14.

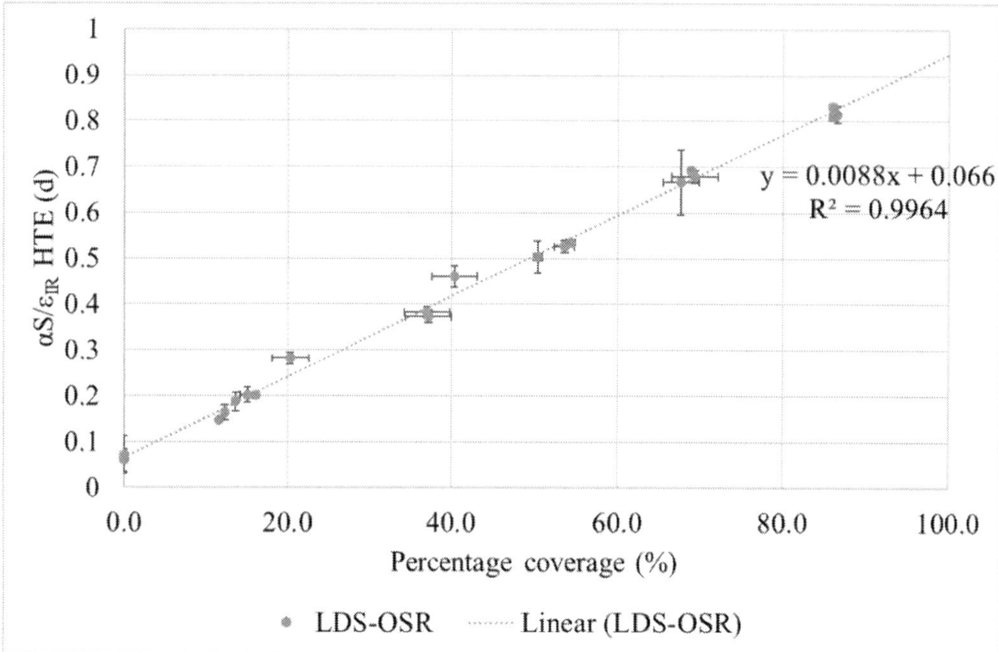

Figure 14. plot of α_S / ε_{IR} HTE ratio with varying percentage coverage

One measure of the distributed dust deposition is mass with coverage. Distributions were generated for each of the coverages:

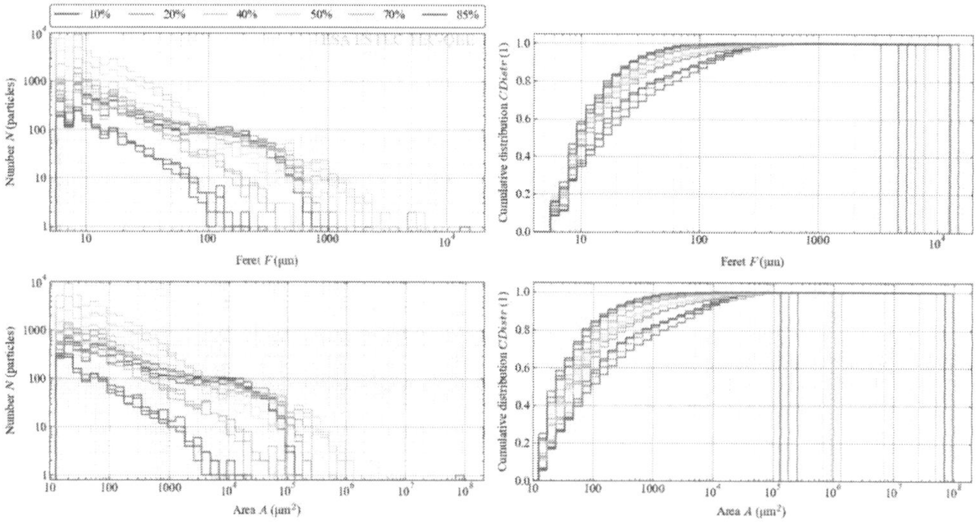

Figure 15. Particle counting distribution characteristics for each level of coverage of each sample

The distribution in values reflects the size of the aggregates formed. Initially, small size particles are more present, then many of these coalesce into larger aggregates following the addition of more dust. As the coverage gets very high, large blocks of space are not separable and are counted by the software as giant particles. For the highest coverage samples, the whole contribution to the area is seen to essentially come from one particle.

The prepared sieved LDS was measured using the same instrumentation as the OSRs, with the difference that 3 mm deep cylinders of LDS were prepared under the same mask placed on top, so as to measure a bulk value. The obtained are presented in Table 2 and Table 3:

Table 2. Values obtained for α_S of the LDS.

Measurement	α_S
Average	0.72
Std dev	0.031
Error	0.06

Table 3. Values of ε_{IR} of the LDS, at 20° from measurements and calculated HTE value.

Measurement	ε_{IR} 20°	ε_{IR} HTE dielectric calculated
Average	0.918	0.866
Std dev	0.002	-
Error	0.003	-

These values obtained using the handheld instrument were lower than expected according to the trend-line for full dust coverage on the OSR. A separate measurement was also carried out as described above in section 3.2 using the quartz window powder cell. The value for α_S was measured to be 0.73, which is a very similar value. Evolution of the thermo-optical properties measured on OSR, from a pristine surface to an EAC1-1A LDS surface coverage up to ~ 85 %, were consistent with a linear increase.

The simple rule of mixing assumes no interaction between the particles on the surface, and models the change in thermo-optical properties as a simple ratio, i.e., a linear combination of both components based on the area of each that is interacting with light. Hence, a linear relationship was expected for this test. However, the value of α_S and ε_{IR} as measured for the bulk would be expected to be higher than the value that was obtained. Using the law of mixtures as a simple approximation, one would expect a value of α_S and ε_{IR} that is equal to that for 100%, which is not shown in our extrapolation of the linear fit. Figure 16 demonstrates the expected theoretical values for α_S (shown as α_S relative), and the deviation from that line.

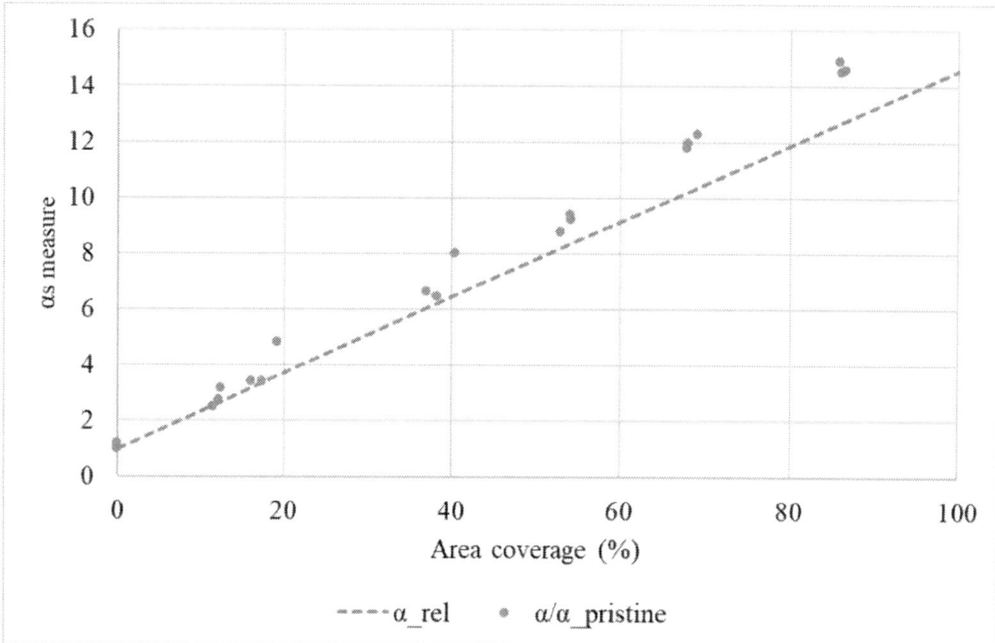

Figure 16. Comparison of theoretical value of α_{rel} (Relative absorptance = $\alpha_{\text{dusted surface}}/\alpha_{\text{pristine surface}}$, blue line), using a value of 0.715 for the α_S of the bulk LDS, obtained using the law of mixing versus obtained experimental values in the form of $\alpha/\alpha_{\text{pristine}}$ (orange points).

Previous literature [28] suggests that glassy or transparent particles would limit the applicability of the law of mixtures, due to the higher absorptivity of bulk LDS versus monolayer coverage. (The exact proportion of transparent particles in this fraction of EAC-1A is unknown.) The opposite phenomenon was measured here, with a lower absorptance of the bulk LDS compared to that expected. Extra testing to obtain a more accurate value for the bulk α_S value did not lead to a revised higher value for α_S in the bulk. The other hypothesis would be that surface coverage was somehow underestimated, considering that the value measured at 85% coverage is much closer to one expected at 100%.

From the above graph, Figure 16, we can see the deviation from theoretical values (line) which becomes more important with higher coverage. The physical parameter with a measured non-linear component is the mass deposition required for coverage, due to the increasing formation of stacks at higher coverage. This was shown in the data for mass deposition with coverage (Figure 9). This is not un-expected due to the method chosen for deposition; as particles fall randomly there is a higher statistical chance with increasing coverage that particles deposited will fall on previously deposited LDS as opposed to on bare OSR surface. We can then derive the hypothesis that the higher value of α_S compared to the law of mixing could result from the non-monolayer nature of the deposition. This is illustrated pictorially in Figure 17.

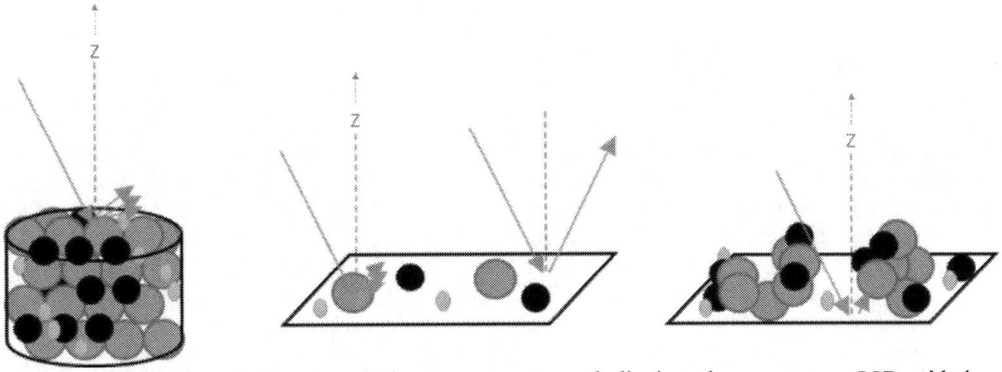

Figure 17. Diagram of thermo-optical measurements on bulk dust, low coverage OSR with low aggregate height, and high coverage OSR with stacking in the Z direction. The spheres represent grains of dust.

As the coverage increases and the height of Z stacks increases also, there is a much higher chance that light (emitted from the instrument at 20°) hitting the OSR will not be able to reflect back into the integrating sphere, due to it being trapped by dust particles which are in the pathway of the light. Thus, the absorption at 85% coverage is already equivalent to that measured for the bulk dust, as the system cannot be treated as a linear combination of surfaces. If one would plot the obtained α_S values with an artificially obtained "experienced coverage", values would be a much closer fit to the aforementioned law of mixing. It can be noted also that the stacking had a greater effect on the value of ε_{IR} than that of α_S (as seen in Figure 11 and Figure 12 where the point for the dust is further below the extrapolated line). That the measured value of ε_{IR} at 85% is even higher than that of the bulk LDS also requires explanation, but could be due to the trapping of diffuse reflections from the surface of the LDS, which in the smooth surface of the bulk measurement would be picked up by the integrating sphere. This question of surface roughness will be further investigated for future deposition tests.

This test measured the impact of the deposition of varying area coverages of LDS EAC-1A sub-100 micron fraction on the thermo-optical properties of OSRs. Solar absorptance and infrared emittance were measured for coverages of 10%, 20%, 40%, 55%, 70% and 85% on 3 areas of an OSR sample. Uniform coverages were obtained with the target size fraction, 90% of the particles used were below 100 μm according to light diffraction analyses. Sample coverage was not monolayer.

A linear increase in solar absorptance was obtained with increasing coverage. The linear fit gave the following equation:

$$\alpha_s = (7.87 \times 10^{-3}) * area\ coverage\ \% + 4.243 \times 10^{-2} \quad (3)$$

with an absolute error of 7×10^{-5} in the slope and 4×10^{-3} in the intercept.

As expected, the increase in ε_{IR} was smaller. A linear relationship was derived also derived from the fit of hemispherical emissivity (dielectric) graph:

$$\varepsilon_{IR} HTE = (6.55 \times 10^{-4}) * area\ coverage\ \% + 8.240 \times 10^{-1} \quad (4)$$

with an absolute error of 1×10^{-5} in the slope and 5×10^{-4} in the intercept.

Some further work is necessary to verify the higher values for α_S and ε_{IR} in the than expected according to the trend measured on OSR substrates, but the current hypothesis is that stacks in the deposition cause the light to interact with a larger volume of LDS than measured using the area coverage. Work on another simulant may also clarify this data.

4. Conclusions

The first step in accurately simulating dusty conditions is to establish the figures of merit for available simulants by measuring them using standardized testing. Reflectance spectra and solar absorptance values were obtained for many as-received simulants using a powder cell holder accessory and performing checks with other diffuse samples. Preliminary thermal conductivity data was obtained for some simulants as a test for future work under more rigorous conditions. Additionally, quality assurance for batch variation should be carried out in order to ensure the reliability of results. To further such an end, XRD data were obtained following a defined process. This data can be reobtained at a further date to check for changes in any simulant obtained for testing.

Testing the impact of regolith simulants on materials was also carried out. EAC-1A simulant was deposited on OSRs and the change in thermo-optical properties with surface area coverage was measured. A linear relationship for the increase in solar absorptance was proposed, giving a value of (7.9×10^{-3}) for the slope with area coverage.

With a more thorough understanding of the simulants available, more advanced tests can be carried out, under representative conditions. Specific facilities have to be dedicated to studies utilising dust, as these tests contaminate the hardware. As there is an increasing need for testing of components, ESA is developing several new facilities for our future research efforts. Firstly, the dust impact (DustI) facility is comprised of various elements which can be combined to investigate the effect of charging and vacuum environment on the dust and exposed materials, including a mobile arm for in-situ manipulations. This facility is under development. A more complex Dust Adhesion Facility is planned for next year in order to address the critical lack of data on dust adhesion. The dust adhesion facility works by using vertical electromagnetic acceleration of a sample plate with controlled pre-deposited dust. The plate hitting a stopper plate transfers the sample momentum to the dust and accelerates it (vertical force). The detachment of dust will happen when the adhesion force is overcome. The aim is to characterise dust adhesion force in vacuum at different temperatures and environments, in order to assess adhesion mechanisms and parameters.

Collaboration on dust contamination is required across multiple disciplines, and we are also developing a European handbook to provide guidelines to ESA's space projects and European industry.

5. Acknowledgements

Support from the following people is acknowledged: Bruno Bras, Szilvia Szmolka, and Ricardo Martins for developing the initial test method for the thermo-optical measurements on OSR surfaces, and Cathal Mooney and Mauricio Portaluppi for their support from the Gateway project. Sarah Rodriguez Castillo for all her help in the thermal lab, particularly the thermal conductivity measurements, and Théo Henry for the linear fits.

6. Appendix

Error analysis:

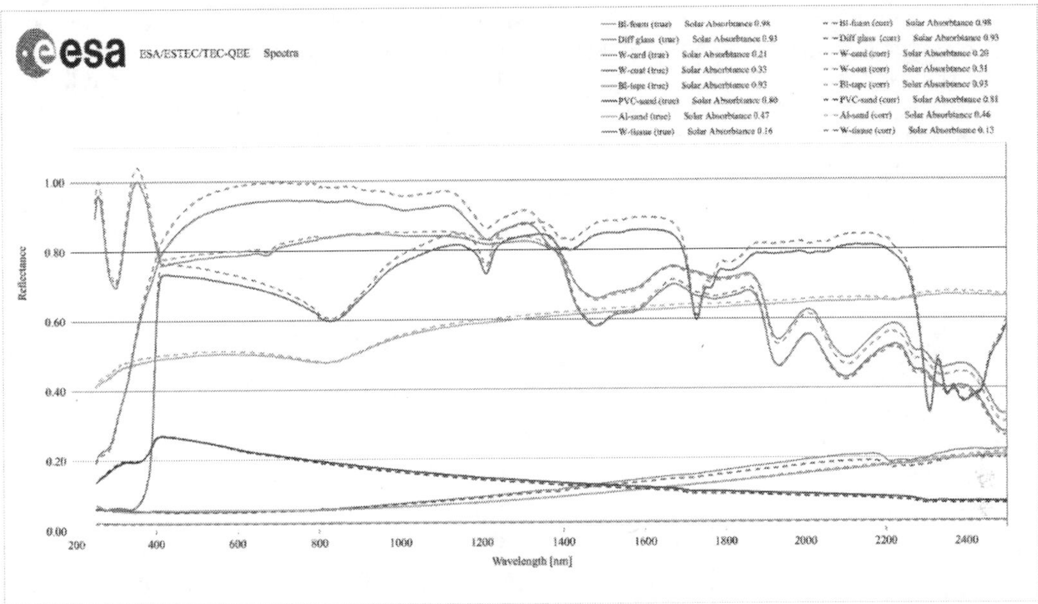

Figure S1. UV-Vis spectra for 8 different solid samples (continuous line) and the corrected spectra for the same sample measured inside the powder holder cell (dashed line). All samples were baseline corrected and a calibration coupon was used to correct instrument response.

Figure S1 shows the UV-Vis spectra collected for 8 different solid samples, as well as the corrected spectra for measurements of those same samples taken through the quartz window of the powder holder cell. Table S1 shows the calculated α_S values for these spectra and the difference between them, as well as the details for each sample.

Table S1. Table showing the values calculated for α_S from measurements of samples without the cell window as well as those values obtained from applying the correction on samples inside the cell and the difference in magnitude between the values.

Sample description	Sample name	αS true	αS corrected	Absolute error in αS
Black foam	Bl-foam	0.98	0.98	0.00
Diffuse reflectance glass with black tape backing	Diff-Glass	0.93	0.93	0.00
White card	W-card	0.21	0.20	0.01
White coating	W-coat	0.33	0.36	0.02
Black tape	Bl-tape	0.93	0.94	0.00
PVC with sandblasted surface	PVC-sand	0.80	0.84	0.01
White tissue (layers)	W-tissue	0.16	0.13	0.03
Aluminium with sandblasted surface	Al-sand	0.47	0.47	0.01

Measurements of reflectance spectra were carried out both including and excluding the collection of the specular reflection, in order to serve as a characterisation of the diffuse component of the sample. This was in order to ascertain whether the diffusivity of the sample had a high impact on the success of the correction, and if so, which values were closest to the powder samples and which error values were therefore more appropriate. This was not shown to impact the error. Reflectance values or α_S values were seen to make more of a difference. (See figure S2 which gives an indication of this effect). It can also be observed visually in figure S1 that the corrected spectra seemed further from the true spectra at higher values of reflectance. Hence the decision was made to carry out separate error approximations for samples based on their corresponding bracket of reflectance or α_S value. Maximum errors were used from each bracket.

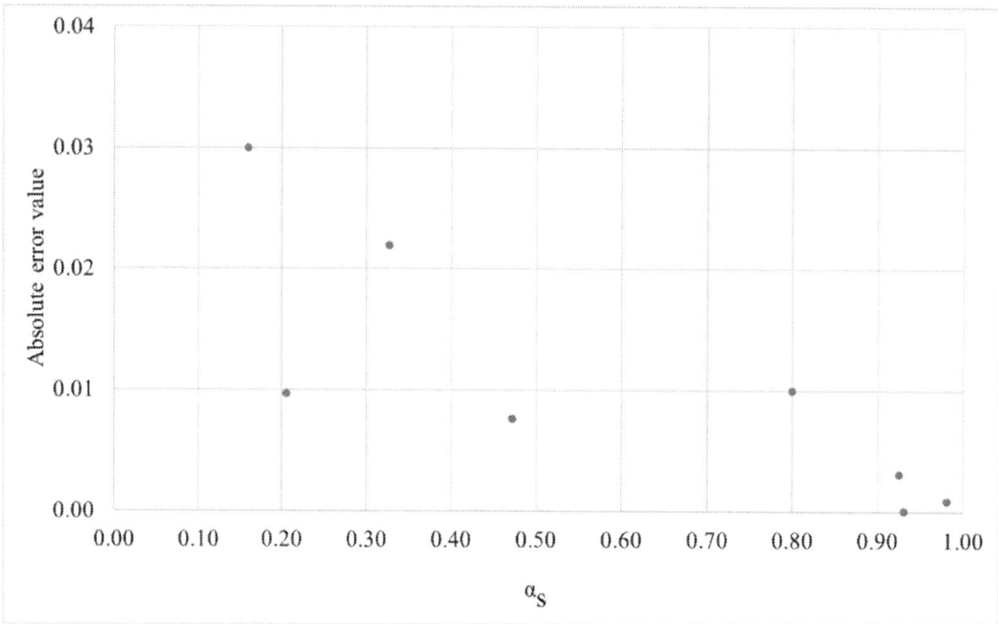

Figure S2. Graph showing the magnitude of the difference between true and corrected α_S value for each sample, plotted against true α_S value

XRD patterns:

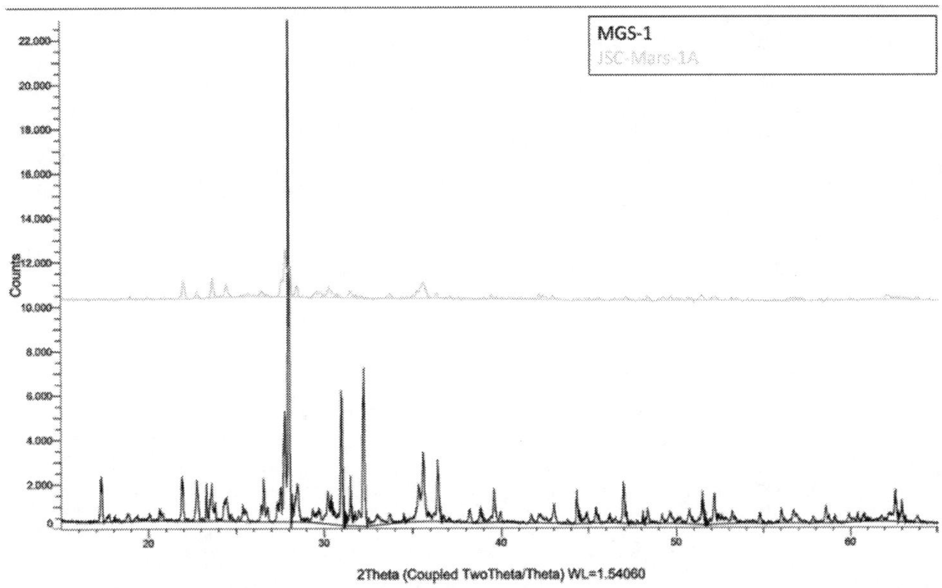

Figure S3. XRD patterns for analysed Martian regolith samples; MGS-1 (black) and JSC-Mars-1A, 90-1000 μm fraction, (green). *JSC-Mars-1A pattern was obtained with 0.5° intervals.*

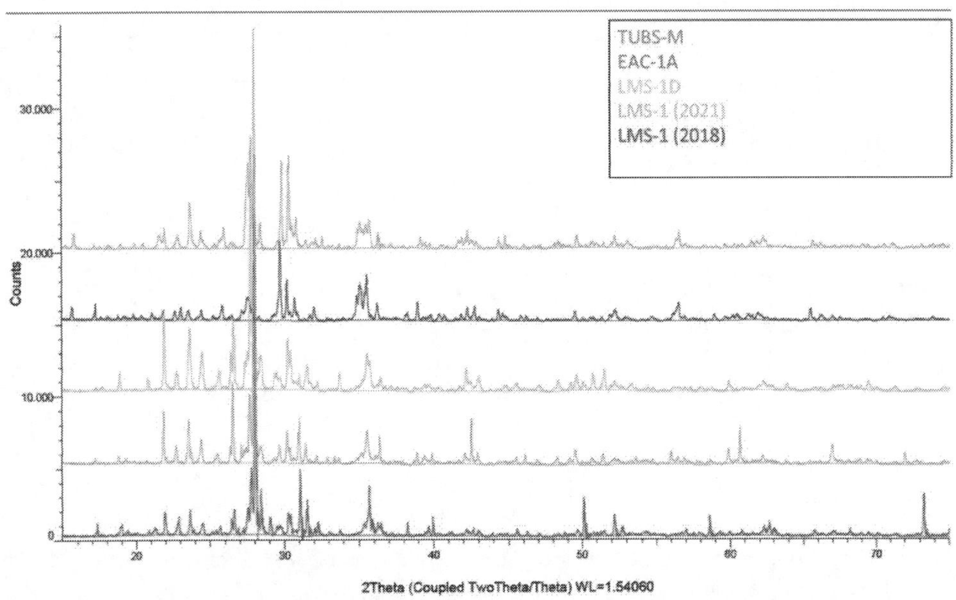

Figure S4. XRD patterns for analysed lunar mare regolith samples; (TUBS-M in light blue, EAC-1A in purple, LMS-1D in pink, LMS-1 2021 in yellow, LMS-1 2018 sample in red)

IOP Conf. Series: Materials Science and Engineering 1287 (2023) 012036 doi:10.1088/1757-899X/1287/1/012036

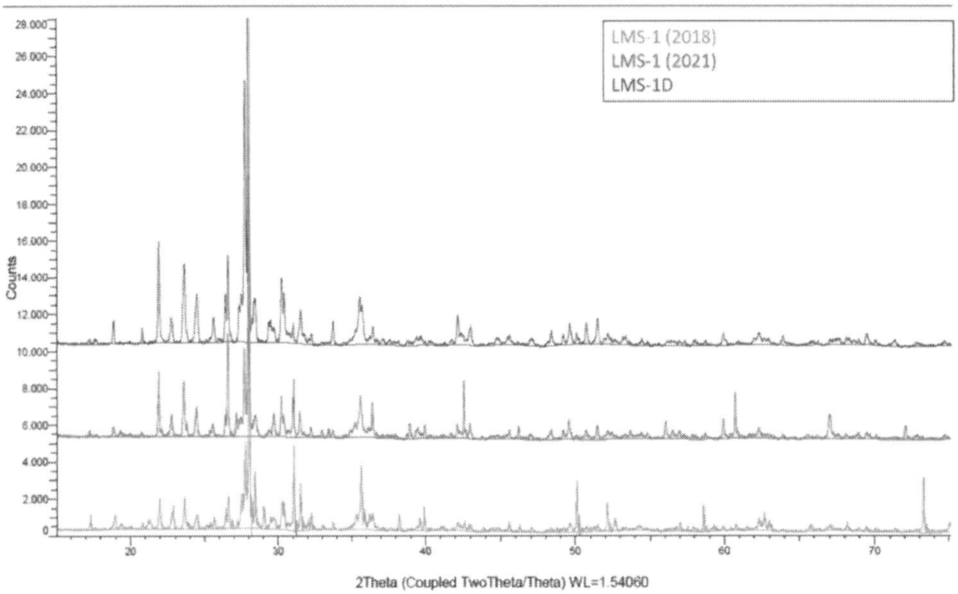

Figure S5. XRD patterns for analysed LMS-1 (2018 sample in orange, 2021 sample in pink) and LMS-1D (brown) regolith samples

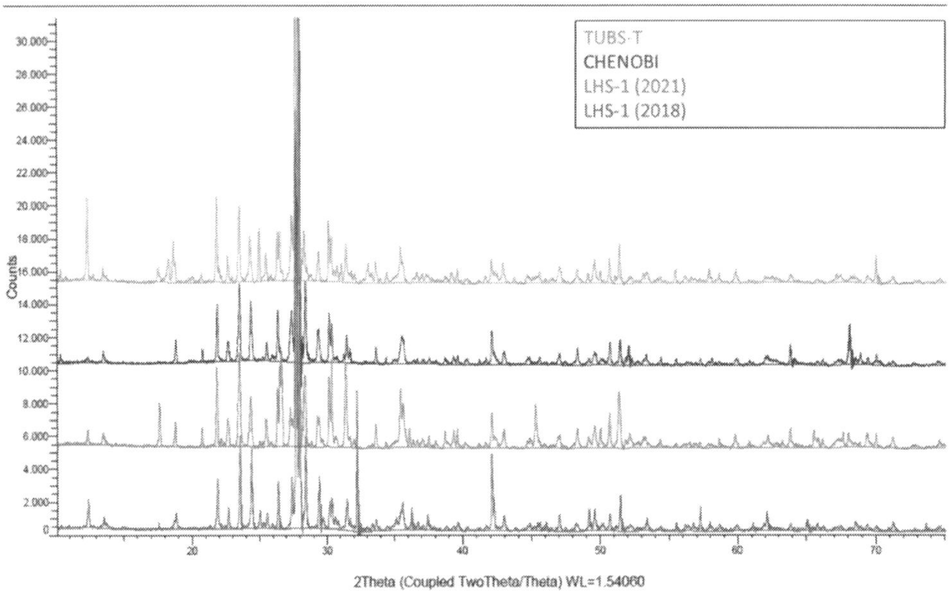

Figure S6. XRD patterns for analysed lunar highland regolith samples; (TUBS-T in orange, Chenobi in brown, LHS-1 2018 sample in pink, LHS-1 2021 sample in red)

Note on HTE calculations: As one mode has to be selected for the measurement, values were obtained using the "metal" setting for the calculation. In order to also provide values for the HTE of the surface when treated as a "dielectric", these values were calculated from the 20° emissivity using theoretical equations from Siegel and Howell, Thermal Radiation Heat Transfer, Taylor and Francis, 2002, chapter 3. [15]

Linear fits: Linear fits were created for each plot using the scaled Levenberg-Marquadt algorithm, taking into account errors in y as calculated from statistics. Where zero standard deviation in values was recorded, an error value of 0.003 was used, corresponding to the minimum value in other points, in order to limit the biasing of the fit towards these points.

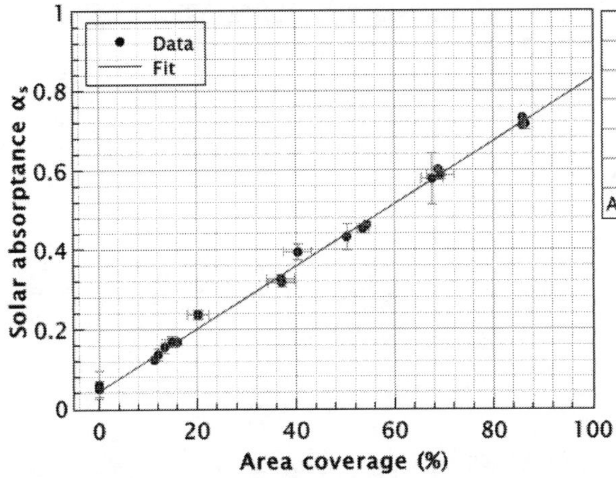

Dataset	Alpha_Alpha
Function	a*x+b
a	7.874e-03 +/- 6.928e-05
b	4.243e-02 +/- 4.010e-03
Chi^2/doF	4.084e+00
R^2	9.985e-01
Adjusted R^2	9.985e-01

Figure S7. Graph of linear fit and associated fit data for solar absorptance measurements with coverage.

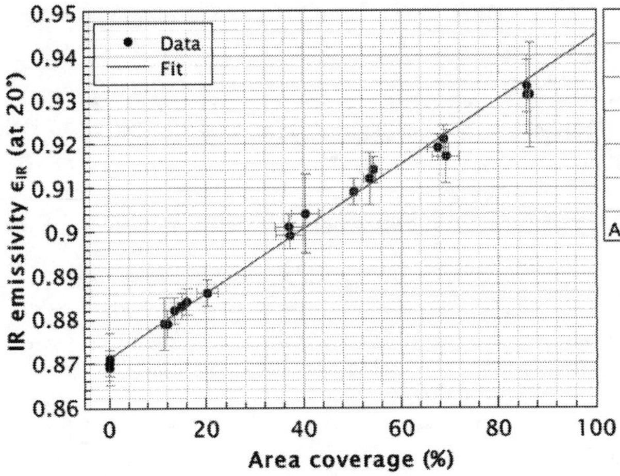

Dataset	e20_E-20
Function	A*x+B
A	7.352e-04 +/- 1.537e-05
B	8.712e-01 +/- 6.254e-04
Chi^2/doF	2.375e-01
R^2	9.918e-01
Adjusted R^2	9.913e-01

Figure S8. Graph of linear fit and associated fit data for emissivity at 20° measurements with coverage.

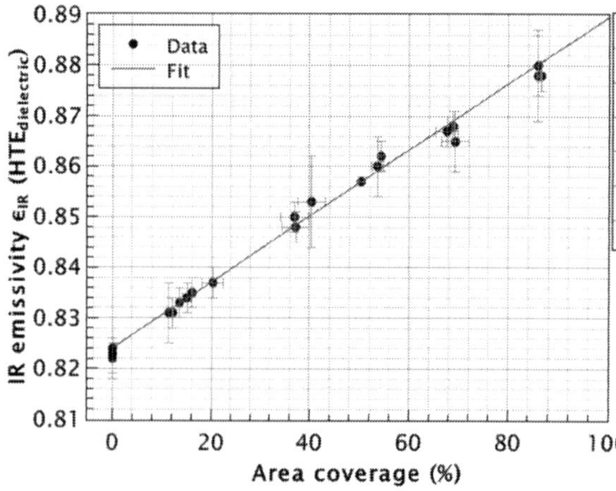

Dataset	eps_HTE_dielec_e-HTE
Function	A*x+B
A	6.547e-04 +/- 1.056e-05
B	8.240e-01 +/- 5.309e-04
Chi^2/doF	1.864e-01
R^2	9.951e-01
Adjusted R^2	9.948e-01

Figure S9. Graph of linear fit and associated fit data for hemispherical emissivity measurements using dielectric assumption with coverage.

7. References

[1] Heiken G H, Vaniman D T and French B M 1991 *Lunar Sourcebook, a user's guide to the Moon*

[2] Gaier J R 2020 The impact of dust on lunar surface equipment during Apollo *The Impact of Lunar Dust on Human Exploration 2141* p 5002

[3] Gaier J R, Siamidis J and Larkin E M 2010 Effect of simulated lunar dust on the properties of thermal control surfaces *Journal of Spacecraft and Rockets* **47**(1) pp 147-152

[4] Walton O R 2007 Adhesion of Lunar Dust NASA/CR-2007-214685

[5] Woods-Robinson R, Siegler M A and Paige D A 2019 A model for the thermophysical properties of lunar regolith at low temperatures *Journal of Geophysical Research: Planets* **124**(7) pp 1989-2011

[6] Wood S E 2020 A mechanistic model for the thermal conductivity of planetary regolith: 1. The effects of particle shape, composition, cohesion, and compression at depth *Icarus* **352** p 113964

[7] Engelschiøn V S, Eriksson S R, Cowley A, Fateri M, Meurisse A, Kueppers U and Sperl M 2020 EAC-1A: A novel large-volume lunar regolith simulant *Scientific reports* **10**(1) pp 1-9

[8] Isachenkov M, Chugunov S, Landsman Z, Akhatov I, Metke A, Tikhonov A and Shishkovsky I 2022 Characterization of novel lunar highland and mare simulants for ISRU research applications *Icarus* **376** p 114873

[9] Stockstill-Cahill K, Blewett D T, Benjamin D, Bussey J, Cahill J T S, Clyde B et al JHU-APL LSII REPORT: 2021 Lunar Simulant Assessment 240-228-0065

[10] Linke S, Windisch L, Kueter N, Wanvik J E, Voss A, Stoll E, Schilde C and Kwade A 2020 TUBS-M and TUBS-T based modular Regolith Simulant System for the support of lunar ISRU activities *Planetary and Space Science* **180** p 104747

[11] Battler M M and Spray J G 2009 The Shawmere anorthosite and OB-1 as lunar highland regolith simulants *Planetary and Space Science* **57**(14-15) pp 2128-2131

[12] Sakatani N, Ogawa K, Arakawa M and Tanaka S 2018 Thermal conductivity of lunar regolith simulant JSC-1A under vacuum *Icarus* **309** pp 13-24

[13] Exolith Lab MGS-1 Mars Global Simulant Fact Sheet 2021 Jan [cited 2023 Jan 27]; Available from: https://exolithsimulants.com/collections/regolith-simulants/products/mgs-1-mars-global-simulant

[14] John K K, Rogers C E 2021 Classifications and requirements for testing systems and hardware to be exposed to dust in planetary environments NASA-STD-1008
[15] Howell J R, Siegel R 2002 *Thermal radiation heat transfer* (4th ed) (Taylor & Francis) p 88
[16] Sakatani N, Ogawa K, Iijima Y I, Arakawa M and Tanaka S 2002 Compressional stress effect on thermal conductivity of powdered materials: Measurements and their implication to lunar regolith *Icarus* **267** pp 1-11
[17] Presley M A and Christensen P R 1997 The effect of bulk density and particle size sorting on the thermal conductivity of particulate materials under Martian atmospheric pressures *Journal of Geophysical Research: Planets* **102**(E4) pp 9221-9229
[18] Hütter E S 2011 Development and testing of thermal sensors for planetary applications PhD thesis University of Graz
[19] Exolith Lab LMS-1 Lunar Mare Simulant Fact Sheet 2021 Jan [cited 2023 Jan 27] Available from: https://exolithsimulants.com/products/lms-1-lunar-mare-simulant?pr_prod_strat=collection_fallback&pr_rec_id=a2f96d085&pr_rec_pid=5644714999966&pr_ref_pid=6687969312946&pr_seq=uniform
[20] Exolith Lab. LMS-1D Dust Simulant Fact Sheet 2021 Jan [cited 2023 Jan 27] Available from: https://exolithsimulants.com/products/lms-1d
[21] Kost P M, Linke S, Gundlach B, Lethuillier A, Baasch J, Stoll E and Blum J 2021 Thermal properties of lunar regolith simulant melting specimen *Acta Astronautica* **187** pp 429-437
[22] Exolith Lab LHS-1 Lunar Highlands Simulant Fact Sheet 2021 [cited 2023 Jan 27] Available from: https://exolithsimulants.com/products/lhs-1-lunar-highlands-simulant
[23] Exolith Lab LHS-1D Dust Simulant Fact Sheet 2021 [cited 2023 Jan 27] Available from: https://exolithsimulants.com/products/lhs-1d-lunar-highland-dust-simulant
[24] Gaier J R, Ellis S and Hanks N 2012 Thermal optical properties of lunar dust simulants *Journal of thermophysics and heat transfer* **26**(4) pp 573-580
[25] Gaier J R 2018 *Optical spectra of lunar dust simulants* (No. NASA/TM-2018-219868)
[26] Goulas A, Binner J G, Harris R A and Friel R J 2017 Assessing extraterrestrial regolith material simulants for in-situ resource utilisation based 3D printing *Applied Materials Today* **6** pp 54-61
[27] Sibille L, Carpenter P, Schlagheck R and French R A 2006 *Lunar regolith simulant materials: recommendations for standardization, production, and usage* (No. NASA/TP-2006-214605)
[28] Gaier J 2010 Effect of Simulant Type on the Absorptance and Emittance of Dusted Thermal Control Surfaces in a Simulated Lunar Environment In *40th International Conference on Environmental Systems* (p 6111)

IOP Conf. Series: Materials Science and Engineering 1287 (2023) 012037 doi:10.1088/1757-899X/1287/1/012037

Monopropellant plume-induced contamination testing for the Europa Lander mission concept

Martin Grabe[1], Leonie J Buntrock[1] and Carlos E Soares[2]

[1] DLR Inst. of Aerodynamics and Flow Technology, Bunsenstr. 10, 37073 Göttingen, Germany
[2] Jet Propulsion Laboratory, California Institute of Technology, Pasadena, CA 91109, USA

E-mail: `martin.grabe@dlr.de`

Abstract. The induced contamination from two hydrazine-propelled thrusters is studied in the high-vacuum plume test facility for chemical thrusters at DLR Göttingen, STG-CT. A number of representative material samples are exposed to the freely expanding pulsed thruster plume for subsequent in-depth analysis at JPL. The plume gas is monitored *in-situ* with a mass spectrometer and quartz crystal microbalances, while non-gaseous ejecta are recorded with a high-speed camera setup. With the detailed chemical analysis ongoing we report on the measurements obtained during the campaign and immediate post-test observations.

1. Introduction

Landings onto 'airless' bodies generate ephemeral 'atmospheres' composed primarily of propulsion byproducts and surface materials above landing sites for at least the duration of powered landing. These spacecraft-induced environments could have contamination impacts to mission science and engineering objectives. The German Aerospace Center (DLR) and NASA Jet Propulsion Laboratory (JPL) have conducted a test program to characterize monopropellant plume induced contamination. These measurements derived from this joint test program are being used for the development of plume-induced contamination models in support of the proposed Europa Lander mission [1, 2]. Two hydrazine-propelled monopropellant thrusters were studied in the high-vacuum plume test facility for chemical thrusters at DLR Göttingen, STG-CT [3], for plume-surface interactions and induced contamination effects. Select representative material samples were exposed to the freely expanding pulsed thruster plumes for subsequent in-depth analytical chemistry analysis at JPL. The plume gas was monitored in-situ with a mass spectrometer and quartz crystal microbalances (QCMs), while non-gaseous ejecta were recorded with high-speed imaging. While the detailed chemical analysis of exposed witness samples is ongoing, we report on the measurements obtained during the test campaign and immediate post-test observations. The JPL/DLR plume induced contamination test program described here is critical to the development of the next generation of plume contamination models in support of technology development maturation for the proposed Europa Lander, and for other future space exploration lander missions.

IOP Conf. Series: Materials Science and Engineering 1287 (2023) 012037 doi:10.1088/1757-899X/1287/1/012037

2. Methods

2.1. Thrusters and Fuel

Two monopropellant hydrazine thrusters from different manufacturers were used in this activity. Though naturally different in design details, both follow a conventional construction, in which the liquid hydrazine (N_2H_4) fuel is injected into a (pre-heated) catalyst bed, where it decomposes into a mixture of mostly hydrogen, nitrogen and ammonia with a net energy release. The mixture of reaction products then expands through a convergent-divergent nozzle, thereby converting the gas enthalpy in the nozzle reservoir into kinetic energy, producing thrust.

Table 1: Theoretical performance (nominal conditions) and history of tested thrusters.

	Thruster A	Thruster B
Nominal thrust	5 N	18 N
Nominal mass flow rate	1.9 g/s	7.8 g/s
χ_{NH_3}	0.184	0.271
χ_{N_2}	0.303	0.288
Cumulative impulse	7 371 Ns	112 931 Ns
Total valve cycles	13 500	99 034

The two thruster specimen, here referred to as Thruster A and Thruster B, differ in nominal mass flow rate and (theoretical) species mole fraction, χ_i, at the nozzle exit, which is relevant for the assessment of their contamination potential, cf. table 1. We do not have information on their detailed operating history, but the cumulative impulse and the total valve cycles are known for each specimen and suggest they have previously been operated primarily in short pulses. While Thruster A is practically new, with 1% and 7% into its specification for cumulative impulse and valve cycles, respectively, Thruster B has served as a qualification unit, experiencing about 80% of the specified valve cycles at 20% specified cumulative impulse.

Table 2: Fuel composition (high-purity grade hydrazine).

	MIL-PRF-26536F	this batch
Hydrazine	> 99 %	99.71 %
Water	< 1 %	0.27 %
Ammonia	< 0.3 %	0.02 %
Particulate	< 1 mg/l	0.4 mg/l
Chloride	< 5 ppm	< 0.1 ppm
Aniline	< 30 ppm	< 0.1 ppm
Iron	< 4 ppm	< 0.1 ppm
Nonvolatile residue	< 10 ppm	4 ppm
Carbon dioxide	< 30 ppm	6 ppm
UDMH/MMH/IPA	< 50 ppm	1 ppm

Reaction control thrusters for space applications typically demand at least monopropellant grade hydrazine according to MIL-PRF-26536, which tolerates up to 0.5% of aniline impurities. The primary impurity is water. The batch from which the fuel tanks for this experimental activity were filled even exceeds the specifications for high-purity grade, as table 2 shows.

Figure 1: DLR high-vacuum plume test facility STG-CT (photo credit: DLR).

2.2. Test facility STG-CT

The high-vacuum plume test facility for chemical thrusters at DLR Göttingen, STG-CT [3], is unique in that it can pump plume gas from reaction control thrusters at the same rate as it is ejected, thus maintaining a background pressure of less than 10^{-5} mbar throughout the thruster operation. The permanently low background pressure permits a free plume expansion, as it would be observed in the vacuum of space, which is a necessary requirement for studying thruster plume induced contamination in a terrestrial setup.

STG-CT's test section is cylindrical, with a diameter of 1.6 m und length 5.05 m. It is shown in figure 1, enclosed by its 3.5 m-diameter stainless-steel vacuum vessel. The test section walls are made entirely of oxygen-free copper and may be cooled to 4.2 K, using liquid helium as a cooling agent. The test section thus acts as a cryo-pump with 30 m^2 surface area, whose ultimate pressure is determined by the vapor pressure of the least condensible plume constituent. In the case of hydrazine thrusters this is hydrogen, whose vapor pressure exceeds 10^{-5} mbar at wall temperatures > 4.7 K. The cryo-pump is capable of handling the power from reaction control thrusters up to about the 20 N class, when they are operated in pulses of a few hundred milliseconds and low duty cycle.

2.3. Instrumentation and setup

The test setup for contamination experiments with both thrusters was almost identical: A thruster test rig contains the required piping, valves, heaters, temperature and pressure sensors to safely operate and monitor the actual thruster unit. It was installed about 1.3 m from the entrance of the test section, facing the other end, with the nozzle axis aligned with the test section axis.

A customized Hiden Analytical HAL PIC 101 quadrupole mass spectrometer with cross-beam electron impact ion source is used to monitor the gas composition on the plume centerline during thruster operation. The sensor head, shrouded and equipped with a skimmer, is placed just over 3 m downstream from the nozzle exit, with the aperture on the plume axis. Sacrificing accuracy for speed, the mass spectrometer is run in RGA mode, resolving only integer values of mass-to-charge ratio in the range $m/z < 45$ amu.

Four QCM Research Mk18 cryo-cooled quartz crystal microbalances (CQCM's) are installed. They passively cool in the cryogenic environment of STG-CT. QCM1 is located about 20 cm below the mass spectrometer aperture, facing the thruster. QCM3 and QCM4 sit next to the

Table 3: Material samples exposed to the thruster plume in high-vacuum.

ID	Name	Specification
A	Aerogel LD	Si, $15\,\mathrm{mg/cm^3}$
B	Aerogel HD	Si, $25\,\mathrm{mg/cm^3}$
C	Ice Simulant	Salt block (NaCl)
D	Kapton®	Black Kapton®
E	Aluminum	6082
F	Copper	CW008A (OFHC)
G	Fused Silica	Corning 7980
H	Gold mirror	protected (SiO_2, TiO_2)
J	Thermal Blanket	Aluminum-coated polyimide
K	StaMet	StaMet (Si/Al-alloy) coated black Kapton®
L	Spectralon®	spectroscopy target, 250 nm to 2000 nm
M	Infragold®	spectroscopy target, $0.7\,\mu$m to $20\,\mu$m

test section wall at 28° off the plume axis and 1.6 m away from the exit plane. QCM2 is mounted onto a remote-controlled movable fixture. During plume exposure, it is located 3.4 m from the nozzle exit plane and 4° off the plume axis. In a second phase of the experiment, in which the QCMs are heated to room temperature for thermogravimetric analysis (TGA), QCM2 moves close to the mass spectrometer (MS) to identify the chemical composition of the desorbing gas.

The nozzle exit plane is monitored with a Phantom v1210 high-speed camera to detect potential non-gaseous ejecta. The camera, equipped with a tele-lens, is mounted on top of the STG-CT vacuum vessel, about 2 m from the nozzle exit, looking down onto the thruster at an angle of about 55°. A high-power LED with a diffusive cover plate, installed right below the nozzle and facing the camera, serves as a background light source for shadowgraph imaging at up to 12 000 frames per second.

2.4. Material samples

In addition to instrumentation for the *in-situ* characterization of the thruster plume effluents, the test setup comprises a number of relevant material samples distributed throughout the test section. An overview of the materials is presented in table 3. With the exception of the two spectroscopy targets (L, M) the samples are affixed to a total of 20 aluminium trays. Each tray comprises a selection of six flat material samples (D ... K) with an exposed surface of 22 mm diameter, and three block materials (A ... C) with an exposed surface of 40 mm × 20 mm and 30 mm depth. The silica aerogel blocks (manufactured by the analytical chemistry group at JPL) are to capture potential fast solid particles in the plume, while the salt block, prepared by the JPL Ocean Worlds Lab, is believed to be a realistic simulant for the Europan ice surface. The sample trays are mounted along the side walls of the test section and to the support of the mass spectrometer, covering an angular range from 3° to 123° off the plume axis.

The spectroscopy targets are grouped in pairs (L and M), located 1.5 m from the nozzle exit at 30° off the plume axis, and 3 m from the nozzle exit at 6° off the axis. They are oriented perpendicular to the thruster streamlines, as are the sample trays.

All material samples were packed and sealed immediately after recovering them from the vacuum chamber after the test. They were then shipped to JPL for detailed inspection of potential plume induced damage and analysis of contamination induced change in surface properties. The analysis is ongoing at the time of writing this paper.

IOP Conf. Series: Materials Science and Engineering 1287 (2023) 012037 doi:10.1088/1757-899X/1287/1/012037

3. Results

For the plume contamination experiments in this activity we operated the thrusters in single pulses with commanded on-times $t_{\mathrm{on}}^{\mathrm{A}} = 1\,\mathrm{s}$ for Thruster A and $t_{\mathrm{on}}^{\mathrm{B}} = 0.4\,\mathrm{s}$ for thruster B, with random duration pauses between 1 min and 25 min. In effect, this implies a variation of catalyst bed temperatures from about 425 K (first pulse in each experiment) to about 650 K. The total number of pulses fired during one experiment was 40 for Thruster A, and 17 for Thruster B. Both thrusters were operated with the nominal feed-line pressure.

3.1. Plume species

While monopropellant hydrazine thrusters appear to be conceptually simple, the detailed chemical processes of catalytic and thermal hydrazine decomposition are actually rather complex [4] and continue to be a subject of research [5]. Engineering textbooks often simplify the decomposition to a two-step scheme of exothermic decomposition of gaseous hydrazine to ammonia and nitrogen, and endothermic decomposition of ammonia to nitrogen and hydrogen:

$$3\,\mathrm{N_2H_4} \rightarrow 4\,\mathrm{NH_3} + \mathrm{N_2} \quad +335\,\mathrm{kJ}, \tag{1}$$

$$4\,\mathrm{NH_3} \rightarrow 2\,\mathrm{N_2} + 6\,\mathrm{H_2} \quad -184\,\mathrm{kJ}, \tag{2}$$

which may be linked through a parameter $x \in [0, 1]$, the degree of ammonia decomposition:

$$3\,\mathrm{N_2H_4} \rightarrow 4(1-x)\,\mathrm{NH_3} + (1+2x)\,\mathrm{N_2} + 6x\,\mathrm{H_2} \quad +(335-184x)\,\mathrm{kJ}. \tag{3}$$

The degree of ammonia decomposition is determined by the thruster design and its operating conditions.

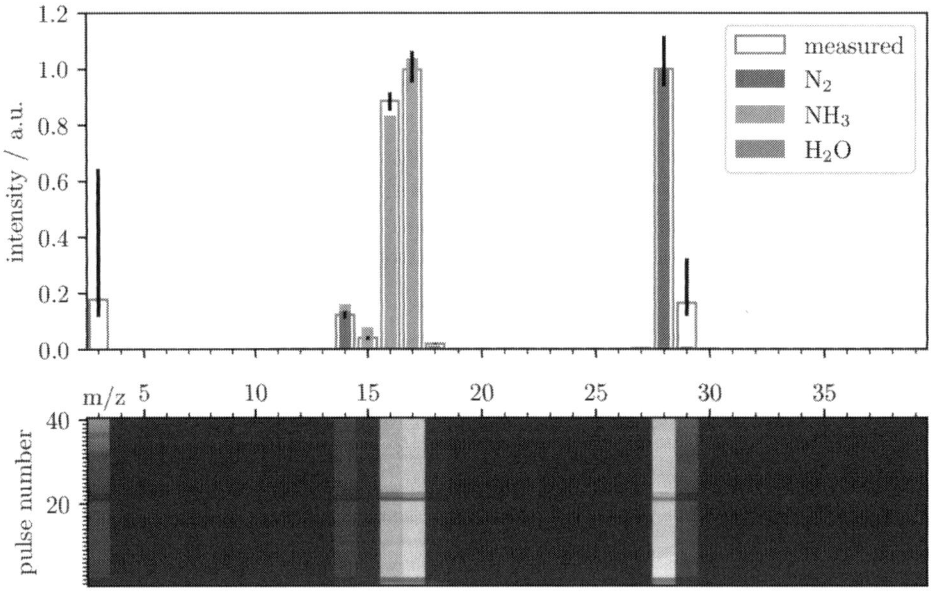

Figure 2: Median mass spectra for Thruster A over all nominal pulses, along with best fits of candidate species [6]. Vertical black lines indicate the variation. Qualitative change of mass spectra with each pulse is shown in the bottom image.

Figure 2 shows the median mass spectrum over all nominal pulses. The top graph shows measured intensities (gray outlines), overlayed in color are scaled reference spectra from the

NIST database [6] for the major species, NH_3 and N_2. The vertical black lines indicate variations from pulse to pulse. Identification of the major species leaves little room for interpretation, even the minute water content (at $m/z = 18$ amu) is picked up. The peak at $m/z = 29$ amu is not part of the reference spectra, and its origin remains subject to speculation. Its intensity appears to gradually increase with the number of pulses fired, as the bottom graph suggests. Observe, that the scanned mass range extends from 3 amu to 39 amu. The lower limit is chosen to avoid overloading the detector with the strong hydrogen signal at 2 amu, which is indirectly picked up by the peak at $m/z = 3$ amu. The bottom graph also shows the adverse impact of a cold thruster structure (pulses 1, 2, 21 and 22 – there was a longer break after pulse 20).

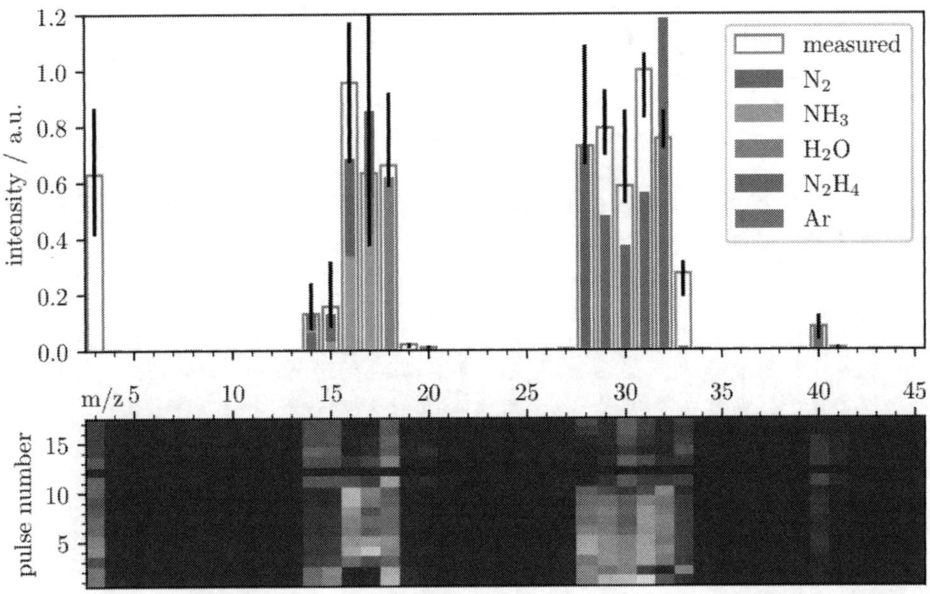

Figure 3: Median mass spectra for Thruster B for pulses 5 to 12, along with best fits of candidate species [6]. Vertical black lines indicate the variation. Qualitative change of mass spectra with each pulse is shown in the bottom image.

The mass spectra measured during operation of Thruster B were expected to be similar to those of Thruster A, perhaps with different proportions of the ammonia and nitrogen peaks. Figure 3 clearly shows a different situation. Pulse to pulse repeatability (bottom) is poor, and thus species identification over a median spectrum, as was done for Thruster A, does not yield meaningful results. The allocation of species from the NIST database is suggestive only. There is, however, a very prominent population of peaks $m/z = 29$ amu ... 32 amu, which in this context must be interpreted as fragments of pure hydrazine. Both, the erratic spectra and the presence of hydrazine in the plume, suggest a problem in the catalyst bed.

The peak identified as argon is real, however. A pneumatically actuated latch valve inside the cryogenic test section exhibited a cold leak during the cool-down phase in preparation for the experiment, and it was decided to keep it open by pressurizing it from an argon gas bottle. Argon would not liquify in the cold environment, is not a native plume constituent and is thus easy to spot in the spectrum.

3.2. *Catalyst bed ejecta*

It is known that monopropellant hydrazine thruster catalyst beds degrade and fragment depending on the operating conditions and employed duty cycles [7]. Voiding and catalyst bed mass loss may occur as a thruster reaches the end of its life [8]. Mass loss can only occur via ejection through the nozzle.

Small particles, compatible with typical mesh sizes of the catalyst bed retainer sieves, were observed when running Thruster A with inert cold gas to check for leaks and proper valve operation during preparation of the test setup. However, no further particulates were observed with the high-speed shadowgraph setup during hot firing.

During operation of Thruster B showers of particulates were repeatedly, but not regularly, observed during hot firing. The vast majority of the particulates appeared during the first 50 ms of a pulse. Limited by the image resolution, the particle diameter could be determined only to be $< 500\,\mu$m. The greatest particle speed observed was 150 m/s, much slower than the speed of the gas at the nozzle exit. All ejected particles were observed within 8° from the nozzle centerline.

Some pulses were monitored with light sources illuminating the vicinity of the nozzle exit from the top, as opposed to using the LED backlight of the shadowgraph setup. From the stray light observed in these images, there appears to be a spray or condensate ejected from thruster B, starting from about 50 ms into the pulse until just the thruster valve closes.

3.3. *Mass deposition on Quartz Crystal Microbalances*

Of the four CQCMs installed in the test section, QCM1 and QCM4 failed during the cooldown of the test section. Among the remaining reliable ones fortunately was QCM2, which is later used for TGA-MS. QCM3, attached to the side of the test section next to QCM4, produced sensible results during the test with Thruster A, but also failed in the Thruster B tests.

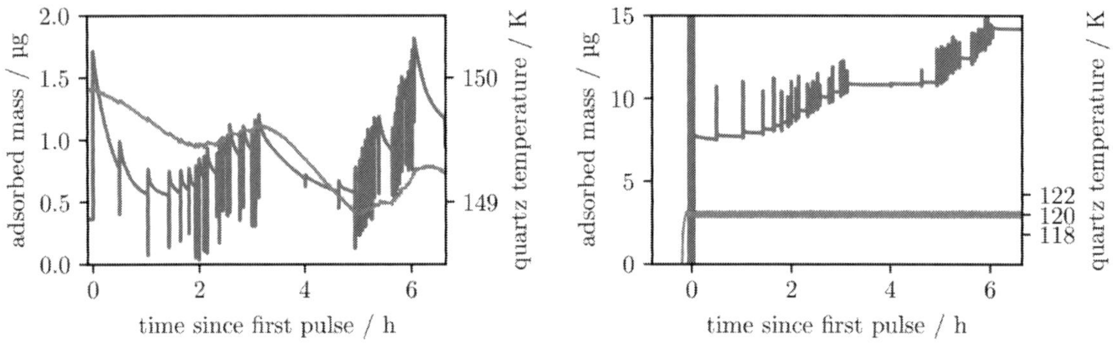

(a) Mass deposition on QCM2.

(b) Mass deposition on QCM3.

Figure 4: Mass deposit recorded by QCM2 and QCM3 during firing of Thruster A.

Figure 4 shows the recorded adsorbed mass (blue curves) for QCM2 and QCM3, along with the temperature of the adsorbing surface (red curve). The temperature for QCM2 is not controlled, but stays between 149 K and 150 K throughout the experiment. One can clearly identify the pulses in the mass trace. and it appears that material is deposited onto the sensor and then evaporates off the surface.

We coarsely aggregate these measurements by dividing the net mass change on the QCMs over the test day by the total thruster on time ($N_{\text{pulse}} \times t_{\text{on}}$). For the tests with Thruster A we thus obtain effective deposition rates of $\phi_{\text{QCM2}} = 1.96 \times 10^{-2}\,\mu\text{g}/(\text{cm}^2\text{s})$ at $\bar{T}_{\text{QCM2}} = 149.6\,\text{K}$, and $\phi_{\text{QCM3}} = 0.254\,\mu\text{g}/(\text{cm}^2\text{s})$ at $\bar{T}_{\text{QCM3}} = 120.0\,\text{K}$.

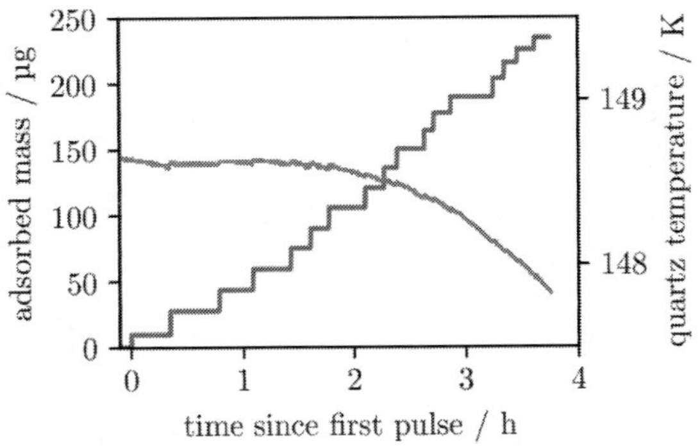

Figure 5: Mass deposit recorded by QCM2 during firing of Thruster B.

The previously noted pathologic performance of Thruster B is manifested also in the measurements of the only remaining operational QCM. As can be seen in figure 5, the recorded deposits onto QCM2 are permanent at that surface temperature, and the adsorbed mass per pulse is substantially greater. The effective deposition rate here is $\phi_{\mathrm{QCM2}} = 33.06\,\mu\mathrm{g}/(\mathrm{cm}^2\mathrm{s})$ at $\bar{T}_{\mathrm{QCM2}} = 148.8\,\mathrm{K}$.

The absolute value of the deposition rates obtained for each thruster depend on the temperature of the adsorbing surface, T_{QCM}, and the chemical composition χ_i of the gas, but also on the location of the sensor with respect to the thruster and the mass flow rate $\dot{m}_{\mathrm{thr}}$ from the thruster. A more meaningful (non-dimensional) value $\hat{\phi}_{\mathrm{QCM}}$ is obtained by scaling the mass flux to the sensor, ϕ_{QCM}, with the mean mass flux at a distance r to the nozzle exit plane:

$$\hat{\phi}_{\mathrm{QCM}}\left(\theta, T_{\mathrm{QCM}}, \chi_i\right) = \frac{4\pi r^2}{\dot{m}_{\mathrm{thr}}} \phi_{\mathrm{QCM}}. \tag{4}$$

In Eq. (4), θ is the angle off the plume axis.

3.4. Thermogravimetric analysis

In a second phase of each test run, after the plume exposure phase, the cryogenic vacuum conditions were maintained and the QCM crystals heated at a rate of no more than 2 K/min. This produced two distinct peaks of evaporation rate for QCM3: 16.06 ng/(cm^{2}s) at 143.63 K and 26.81 ng/(cm^{2}s) at 169.2 K.

QCM2 was moved next to the exit aperture of the mass spectrometer to identify the desorbing species. As the initial temperature of QCM2 was quite high, only a single desorption was observed in each of the test runs. The results of the desorption analysis from QCM2 are summarized in figure 6. The maximum evaporation rate in the Thruster A experiment was found to be 3.13 ng/(cm^{2}s) at 167.5 K. The desorbing species was found to be mainly water, but hydrazine was also detected in an order of magnitude lower concentration. In the Thruster B experiment, the maximum desorption rate was much higher (as expected from the deposition during plume exposure) and occurred at higher temperature (844 ng/(cm^{2}s) at 198.3 K). Here, too, we detect a mixture of hydrazine and water, but hydrazine now is about an order of magnitude more abundant, as anticipated from the analysis of the plume mass spectra, cf. figure 3.

One would expect the traces recorded with the mass spectrum to follow the general trend of the evaporation rate curve. The most plausible explanation for the broad second "peak" in figure 6a or the continued increase of the N_2H_4 concentration in figure 6b is, that the housing of the QCM slowly warmed up, too, as the quartz was heated. Any gas adsorbed to the front face would thus be desorbing and contributing the the mass spectrometer signal, but not be registered by the QCM.

IOP Conf. Series: Materials Science and Engineering 1287 (2023) 012037 doi:10.1088/1757-899X/1287/1/012037

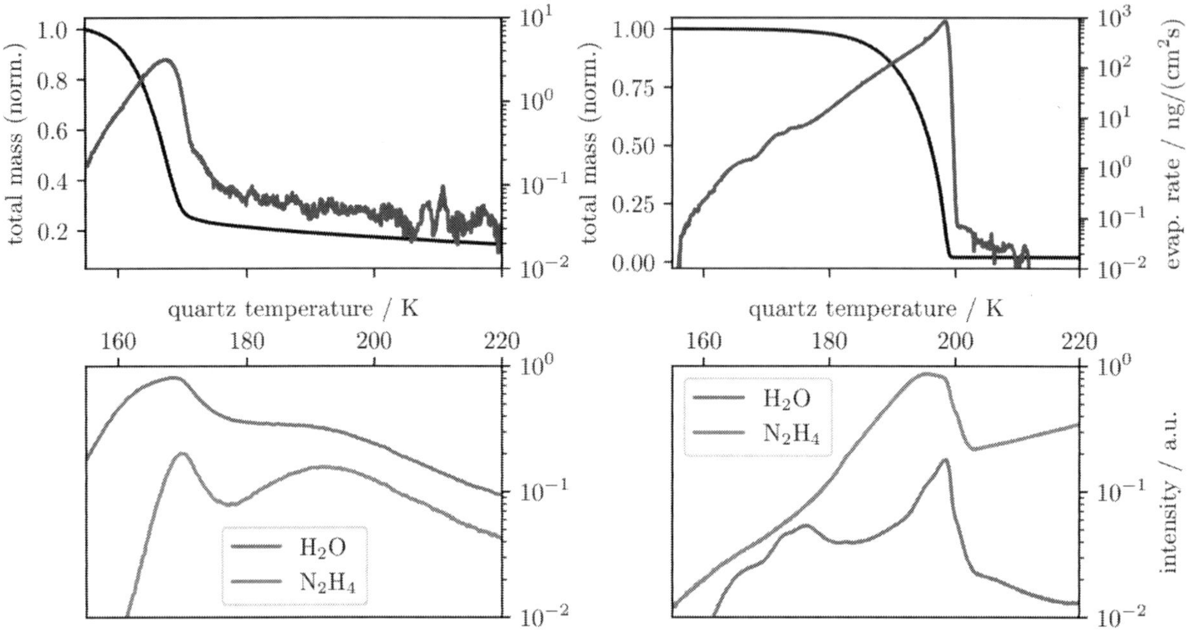

(a) Thruster A; heating rate: 2 K/min. (b) Thruster B; heating rate: 1 K/min.

Figure 6: Temperature dependent desorption from QCM2 (top), and mass spectral analysis of desorbing gas (bottom).

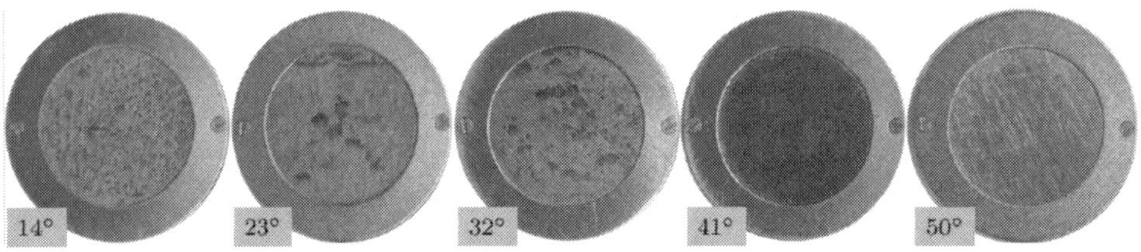

Figure 7: Copper samples after exposure, along with angular position off the plume axis.

3.5. Post-test observations

When recovering the samples after the test with Thruster B, a number of noteworthy observations were possible even without analytical equipment. For one, the copper cryo-pump has attained a greenish tint in a band delimited by angles $\approx 25°$ to $\approx 45°$ off the nozzle centerline. The discoloration is symmetric about the nozzle axis, with no discernible difference in intensity between the top and bottom half of the test section walls.

The copper material samples in that angular range also clearly show features of interaction with what appears to be a liquid substance forming droplets of varying sizes and surface density (figure 7) . The color here is different though, appearing to be more of a dark brown.

4. Discussion

Previous investigation of thruster life (in the spirit of [8]) have focussed primarily on reliability and propulsive performance. As was already pointed out by Williams *et al.* [9], thruster propulsive performance is not necessarily a good indicator of its contamination potential.

We conclude from our study with certainly two extreme cases of new and aged thrusters, that the contamination potential from monopropellant thrusters depends strongly on the the thruster operation history. The deterioration of the catalyst bed is known to be directly dependent on the duty cycle [7], and thus directly (particulates) and indirectly (gas species composition) affects the contamination potential.

For missions equipped with sensitive instruments to detect signs of life, it will be important to know that even for new thrusters residual hydrazine in the plume must be reckoned with. This became apparent only during TGA-MS, as the quadrupole mass spectrometer employed here is not sensitive enough at the required recording speed to detect minute quantities during the thruster pulse. Further data on potential deterioration of surfaces will be available once the chemical analysis is completed.

Future investigations into thruster plume contamination [10] will benefit from experience gained in this activity and improvements derived therefrom. Especially the usefulness of the QCM data can be improved by making better use of the cryo-pump as a heat sink, and thermally strapping the QCMs to the test section walls to capture more plume gas and to yield cleaner TGA-MS results. Designers of spacecraft will benefit from QCM data recorded in the thruster back-flow, i. e. at angles $> 90°$ off the plume centerline.

Acknowledgments

This research described here was carried out at DLR Göttingen under subcontract to the Jet Propulsion Laboratory, California Institute of Technology (1643005).

The authors gratefully acknowledge technical contributions from Günter v. Roden (DLR), Dr. Georg Dettleff (DLR-retired), Dr. Anthony Wong (JPL), Dr. William Hoey (JPL), Dr. John Alred (JPL), as well as contributions from the NASA Goddard Space Flight Center Propulsion Branch (Code 597) and Dr. Alison Rao (Associate Branch Head), which graciously provided thrusters for this test campaign.

References

[1] Soares C E, Hoey W A, Wong A T and Grabe M 2020 Europa Lander Mission Concept Monopropellant Plume-Induced Contamination Testing *Proceedings of the 51st Lunar and Planetary Science Conference* p 2

[2] Hoey W A, Wong A T, Soares C E and Grabe M 2020 Europa Lander mission concept plume-induced contamination: Experimental testing and computational analysis *Systems Contamination: Prediction, Control, and Performance 2020* ed Soares C E, Wooldridge E M and Matheson B A (Online Only, United States: SPIE) p 7 ISBN 978-1-5106-3784-9 978-1-5106-3785-6

[3] DLR Institute for Aerodynamics and Flow Technology 2016 *Journal of Large-Scale Research Facilities* **2** A86

[4] Schmidt E 1984 *Hydrazine and Its Derivatives - Preparation, Properties, Applications* (New York: John Wiley & Sons, Inc.)

[5] Schmidt M W and Gordon M S 2013 *Zeitschrift für Physikalische Chemie* **227** ISSN 0942-9352, 2196-7156

[6] Wallace W E 2022 Mass Spectra *NIST Chemistry WebBook, NIST Standard Reference Database 69* ed Linstrom P J and Mallard W G

[7] Hearn H C 1981 *Journal of Spacecraft and Rockets* **18** 216–221 ISSN 0022-4650, 1533-6794

[8] Wilson M 2005 Demonstration Testing of a Long-Life 5-Lbf (22-N) MR-106L Monopropellant Hydrazine Rocket Engine Assembly *41st AIAA/ASME/SAE/ASEE Joint Propulsion Conference & Exhibit* (Tucson, Arizona: American Institute of Aeronautics and Astronautics) ISBN 978-1-62410-063-5

[9] Williams W D, McCay T D, Powell H M and Lewis J W L 1980 Experimental Study of the Plume Characteristics of a New Monopropellant Hydrazine Thruster Technical Report AEDC-TR-79-54 Arnold Engineering Development Center

[10] Grabe M and Soares C E 2019 Status and Future of Research on Plume Induced Contamination *Proceedings of the 70th International Astronautical Congress* (Washington D.C., USA: IAF) p 15

IOP Conf. Series: Materials Science and Engineering 1287 (2023) 012038 doi:10.1088/1757-899X/1287/1/012038

Modelling of particulate contamination redistribution inside purged cavities

D Faye[1], F Dall'armi[1] and P Bombardier[2]

[1]Centre National d'Etudes Spatiales (CNES), 18 avenue Edouard Belin, 31401 Toulouse cedex 9, France
[2]FAURE QEI, 20 rue Massenet, 38400 Saint-Martin-d'Hères, France

delphine.faye@cnes.fr, p.bombardier@faure-qei.atrix.group

Abstract. Sophisticated instruments on-board most of satellite payloads, especially scientific ones, are sensitive to particulate contamination since their performance can be affected in terms of straylight or surface obscuration. The surface cleanliness levels have to be minimized on ground. However, even if the deposition rate is controlled in cleanroom, the hardware is continuously exposed to particles and fibers, moving in the airflow and settling according to their various sizes and matter compositions. Thus, depending on the initial surface cleanliness level, the redistribution of particulate contamination can be a concern during integration and tests activities until launch, when venting and/or vibrations occur which results in more or less random fallout of particles. Previous studies identified different steps for the transfer mechanism such as detachment, transport and interaction with the surface (with or without bounce). Therefore, it is paramount to understand the physical phenomena of particle resuspension that happens either inside purged cavities of spacecraft or under fairing in order to be able to model them as for molecular contamination. After performing a state of the art of the existing detachment models, a Computational Fluid Dynamics software, scSTREAM, has been chosen among the most relevant ones for a feasibility study to simulate particles redistribution. This tool was particularly interesting due to the possibility to add specific functions for computation. For the purpose, the Rock'n'Roll (R'n'R) model has been implemented to consider particle resuspension. In parallel, some simple test configurations have been defined and a mock-up with several adjacent cavities has been used to compare theory and experiment when a purging under dry nitrogen was applied in the vicinity of a pre-contaminated surface. The relocation of FluoDust particles, as the source of contamination, deposited on a specific substrate has been evaluated thanks to many collector plates disseminated in one or two cavities and the levels quantified. The impact of different test conditions (flow rate, material of the substrate, presence of leakage, initial contamination level…) has been investigated. This paper presents the comparative results that confirmed the redistribution of the particulate contamination with similar trajectories but also some discrepancies due to the limitations of the R'n'R model and depending on experimental parameters such as initial distribution of particles, adherence force to the substrate...

1. Introduction

1.1. Particulate contamination issues for space hardware

By definition according to ECSS-Q-ST-70-01C standard, a contaminant is any unwanted matter on the surface or in the environment of interest that can affect or degrade the relevant performance or life time of a space subsystem (equipment, instrument…).

There are three types of contaminants: chemicals, particles and microorganisms for which requirements have to be specified to maintain the cleanliness of a space hardware at an acceptable level for in-orbit nominal functionality. In this paper, we will focus more particularly on particulate contaminants as undesired foreign solid matter (µm-sized) of different chemical natures with observable length, width and thickness that can be characterized by (1) a size distribution of particles and (2) a number of particles greater than a specific size per unit area.

Two main sources of particulate contamination can be distinguished:

- independent sources:
  - fallout of airborne particles and fibers all along the assembly, integration and storage phases of a satellite,
  - particles generated by human activity such as bacteria, dead cells of the epidermis, hair, fibers, clothing lint. Nevertheless, most of these contaminants can be largely reduced thanks to an appropriate filtration system according to the class of the controlled environment,
  - particles produced during propulsion or attitude control operations,
  - particles resulting from the impact of debris or micrometeorites on materials …
- sources related to the use and handling of materials in subsystems:
  - particles linked to manufacturing processes,
  - flakes of fragile materials such as paints,
  - corrosion products,
  - particles due to atomic oxygen impingement…

Note: This is a non-exhaustive list of contaminants that may be generated on ground or in flight.

By modelling a particle on a surface by a sphere of arbitrary size as shown in Figure 1, two effects can be highlighted: first, degradation by obscuration since the dust will prevent light from illuminating the entire surface; secondly, degradation by scattering causing stray light. Thus, physical properties of contaminated surfaces may be more or less significantly altered. The performance loss will be a function of the amount and size distribution of particles on the surface. Typically, thermal control and optical surfaces, solar arrays, optoelectronic components are the main items impacted by these two effects.

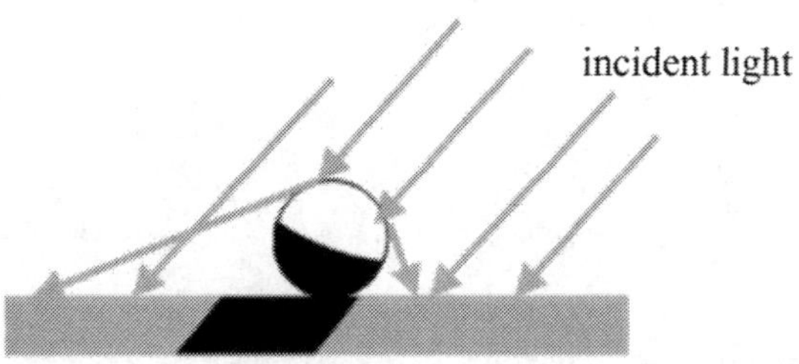

Figure 1. particle effects when illuminating a surface [1]

Moreover, other effects are due to the presence of particles such as failure in mechanisms and microsystems, valve closing anomalies (e.g. in propulsion systems), electrical insulation or short-circuit in electronic devices.

1.2. Purpose of the study

Despite increasingly strict compliance with the cleanliness specifications recommended by space standards and best practices (assembly, integration and test in cleanroom ISO class 8 or 5, preventive actions such as cleaning, use of protective covers, monitoring…), there are still some anomalies resulting in exceeded requirements at different levels of the subsystems assembly until at launch site, due to contamination by undesirable particles, whether environmental or related to the manufacturing steps. Thus, the particulate cleanliness level at beginning of life is often impacted by uncertainties linked to particles deposition and resuspension mechanisms in particular when flushing or venting cavities of space hardware, during pumping down and repressurization of vacuum chambers or during environmental tests (e.g. vibrations or shocks tests) and launch depressurization [2], [3].
Note: The launch budget is currently the biggest contribution to the overall contamination budget.

So this feasibility study was proposed as a first approach using a CNES mock-up to compare simulation and experiments in a simple case of a purged cavity with a pre-contaminated substrate. The aim was to model the redistribution of particles from CFD (Computational Fluid Dynamics) calculations and better understand their transport mechanisms.

2. Redistribution of particles: state of the art

2.1. Particle transport and deposition

Airborne particles are subjected to different physical forces that will favor their deposition on a nearby surface. Each of these forces does not have the same importance depending on the environmental conditions, on the diameter of the particle and its chemical nature (metallic, polymer, mineral) …
As it can be seen in Figure 2, sub-micron particles will rather be subjected to Brownian diffusion which will bring them closer to the surfaces (~3 mm/s for d = 1 nm) with no preferred direction, while the larger particles will be subjected to gravity (~3 mm/s for d = 10 µm), in a proportional manner to the square of their diameter.
Thermophoresis is an additional force to the previous ones, linked to temperature difference between a surface and its environment, which can increase or reduce particle transport velocity.

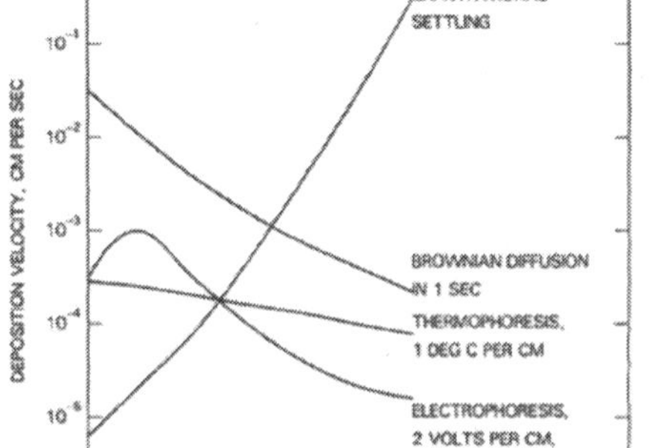

Figure 2. Deposition velocity vs. particle diameter for some mechanisms of particles transport [4]

2.1.1. Particle-Surface Interactions

The particle-surface interaction is described by the sum of adhesion forces existing between two solids (see Figure 3). These forces are generally of electromagnetic origin and then attractive. The main adhesion forces are Van der Waals forces (molecular interactions at a short distance), electrostatic forces (in the presence of surface electric charges) and capillary forces (when a wet film exists between a particle and the surface). The intensity of these forces depends on the characteristics of the particle (material, shape and dimension), of the surface (material, roughness) and of the fluid medium (relative humidity and temperature).

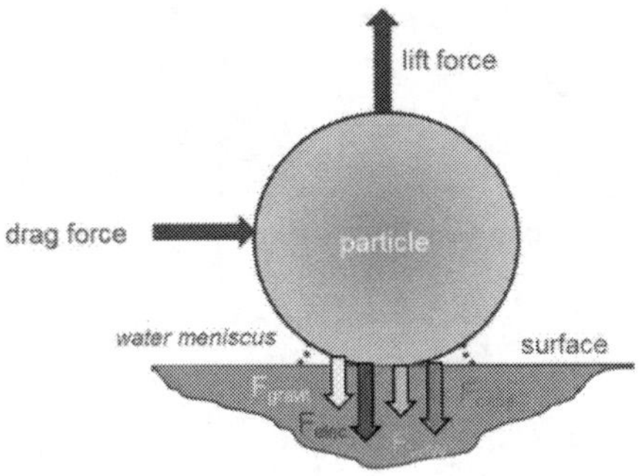

Figure 3. Forces on a particle of diameter < 30 μm in contact with a surface and submitted to airflow

Resuspension (or detachment) forces include aerodynamic forces (Lift and Drag), vibrations, acceleration of the considered environment, electrostatic forces (in case of repulsive charges). Drag forces are related to the friction produced by the movement of air around the particle surface (causing the rolling or sliding of a particle). The lift forces come from the pressure difference created between the side of the particle facing the surface and the opposite side. Thus, the airflow creates an overpressure between the surface and the particle resulting in a force perpendicular to the surface.

When studying the redistribution of particles in a volume, we have to consider that deposition and resuspension take place simultaneously inside this volume, with an intensity that depends on the forces involved at the level of each particle. Moreover, each particle is in a particular configuration since its shape and surface condition are generally sufficient to distinguish it from its neighbors. The chemical nature of the particles is also very important to understand the redistribution because it influences notably the appearance of electrical charges depending on air humidity.

Table 1. Typical adhesion and detachment forces

Adhesion forces ∝ ϕ_p*	**Detachment forces ∝ ϕ_p^2**
Van der Waals (ϕ_p < 50 μm)	Lift forces
Electrostatic	Drag forces
Capillary (Relative Humidity > 60%)	Vibrations

*ϕ_p (diameter of the particle)

Particle resuspension occurs when the balance between adhesion forces (F_A) to surfaces and resuspension forces (F_R) leans in favor of the latter.

Role of surface state:
The surface state (i.e. roughness) defines the effective particle/surface contact surface and also the average distance between the inter-acting molecules (Van der Waals) and electrical contact distance between the two bodies. This surface state also modifies the thickness of the viscous boundary layer at the surface and therefore the air velocities. In addition, particles located on the rough surface and subjected to airflow then find "pivot points" which cause them to rotate rather than slide. This rotation around an asperity on the surface (in the case where the asperity is of significant size compared to the particle diameter) is an effect in favor of resuspension, the adhesion forces then being reduced and the particle being able to acquire a moment (all the more important as the diameter is large).
The Rock'n'Roll (R'n'R) model described in Figure 4 exploits this phenomenon of detachment by rolling to describe the conditions of resuspension.

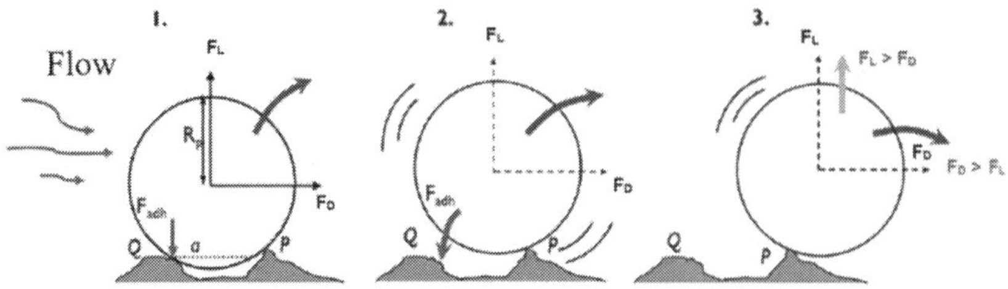

Figure 4. R'n'R model principle [5]

This model uses the moment of forces. The resuspension condition, neglecting the force of gravity, is then written as follows [6]:

$$\frac{F_L}{2} + \frac{R_P}{a} \cdot F_D > F_{adh} \qquad \text{Eq. 1}$$

where:
- F_L is the lift force
- R_p is the particle radius
- F_D is the drag force
- F_{adh} is the adhesion force of the particle at the contact zone Q

The comparison of adhesion ($\propto \phi_p$) and resuspension ($\propto \phi_p^2$) forces shows that the balance between them leans towards resuspension when the particle diameter increases.

It is also necessary to consider the interactions between particles which can modify the shape, size, and density when formation of more or less stable agglomerates.

2.1.2. Particle-Particle Interactions

2.1.2.1 Particle-particle cohesive forces

Particle-particle cohesive forces can bind particles which then form groups of more or less complex geometry and difficult to consider in resuspension calculations. The nature of these forces again depends on the diameter of the particles, their materials and the contact surfaces between particles which will be different according to the shape of each one. The number of "aggregated" particles conditions the complexity of the cohesive forces involved.

2.1.2.2 Particle layer on a surface

When the particles are bound together and deposited on a surface, this forms a particulate layer. The presence of this layer contributes to modifying the adhesion of the external particles. The presence of such a layer is very complex to consider given the difficulty to characterize its properties. It is sometimes linked to a change in surface roughness to take it into account.

2.2. Computational Fluid Dynamics tools

Computational Fluid Dynamics is a branch of fluid mechanics that uses numerical analysis and data structures to analyze and solve problems that involve fluids flow. CFD simulation is a very common tool to model the free-stream flow of a fluid and its interaction with surfaces defined by boundary conditions and many scientific publications deal with the conditions of use to control uncertainties [7], [8].

As an example, thanks to automatic interfacing with 3D Computer-Aided Design (CAD) tool and rapid meshing, CFD is used routinely to predict air movement and distributions of temperature and concentrations in indoor environments but many other fields are also concerned for analysis, design help and prediction, e.g. aerodynamics & aerospace, hypersonics, engine and combustion industrial system design as well as biological engineering, weather simulation. It is particularly helpful where it is difficult to perform experimental measurements.

Specific applications for space industry interest are presented in the two figures below.
Figure 5 shows the airflow velocity vectors in a cleanroom in operation which allows to visualize the air turbulences due to obstacles as furniture or personnel and thus the inherent dead zones. These calculations applied for cleanroom or other controlled environments design are more and more requested to characterize the ventilation efficiency, to study contamination scenarii through aerosol transport and deposition modelling, to predict and assess cleanliness levels from the average air age.

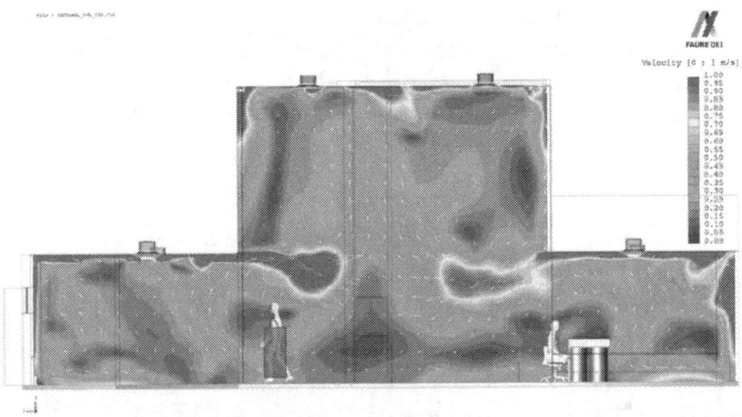

Figure 5. Airflow velocity vectors in INTEGRAL cleanroom at CNES

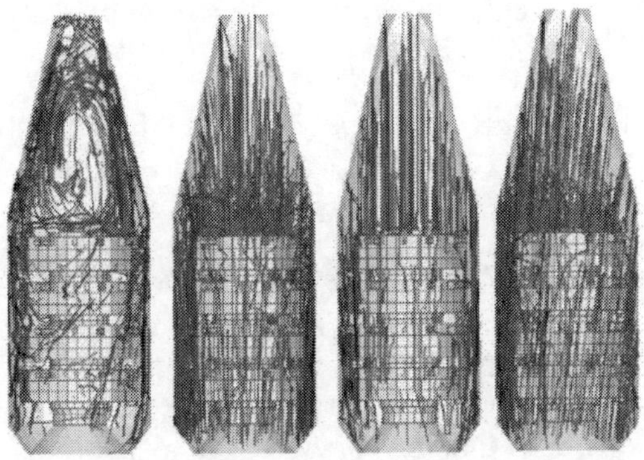

Figure 6. Particle trajectories inside fairing
from left to right: ground, launch, depressurization and engine flameout
colors corresponding to the particle sizes: red: 50 μm, blue: 100 μm, black: 200 μm [9]

Figure 6 gives the results of a NASA (National Aeronautics and Space Administration) study in support of the Magnetospheric MultiScale project regarding the transport of particles inside the fairing of a launcher. The analysis was carried out during the pre-launch, launch and ascent phases in order to obtain quantitative predictions of particulate contamination on the instruments. This showed clearly that depressurization during launch can cause particles to detach not only from fairing surfaces but also from spacecraft surfaces. The redistributed particles may therefore degrade the properties of the new impacted surfaces even very far from the initial source. Then a sensitive surface that met the required cleanliness level at the end of on-ground integration phases may become contaminated after the launch phase. As it is well-known that inert particles of environmental or human origin serve as a support for microorganisms, it could be a real concern for planetary protection annihilating the effects of hardware sterilization and compromising the contamination control. CFD simulations were developed by NASA JPL (Jet Propulsion Laboratory) to take into account these removal and redistribution processes and the results presented at the last COSPAR (Committee on Space Research) symposium [10].

Future works are planned to improve such CFD simulations, showing the importance of studying the particles transport for space applications everywhere resuspension and redistribution can occur: on-ground inside purged or flushed cavities of equipment, at launch and even in flight.

The present study was performed with scSTREAM which is a finite volume general purpose code. scSTREAM uses a structured mesh (hexahedron).
Several RANS (Reynolds Average Navier Stokes equations) turbulence models were tested to assess their capacity to represent high speed airflows colliding with surfaces at different angles. The MP (Modified Production) k-ε model appeared to be the best choice, compared to the RNG (Re-Normalisation Group) k-ε model showing an over-production of turbulence energy.
Different particle models are available and each of them has pros and cons. Figure 7 hereafter shows the main numerical techniques used to mimic particles behavior.

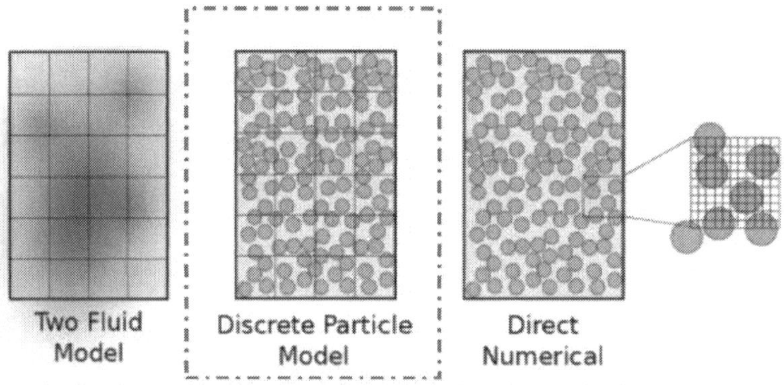

Figure 7. CFD Particle models (Hexagon AB Group, Sweden)

The "Two Fluid Model" doesn't take into account individual particles but replaces them by a gas-like substance with similar behaviour. This so-called "Eulerian" approach is the most common and least computationally demanding. The particles are treated like a second fluid, without being able to distinguish the trajectory of an individual particle. It is therefore a statistical treatment which makes it possible to simulate the concentration of particles by taking into account their characteristics in terms of size and density. In this Eulerian approach, the treatment of resuspension requires implementing a law for the rate of resuspension, then treating dissemination by interaction with air movements. This method has been applied to study the resuspension of particles by a walker [6].
The two other models allow individual particle processing at the expense of computation time. This results in a limitation of the number of particles treated and of their size. Indeed, the calculation of particle interactions such as collisions requires a fine mesh of space and time, all the more so as the particles are small.
The "Discrete Particle" model is generally called DEM (Discrete Element Method). The mesh associated with this method is lightened compared to the DNS (Direct Numerical Simulation) while allowing an individual treatment of the particles. It is however limited to particle sizes greater than 100 μm while the DNS method is too limiting due to the computational cost. Therefore, the Two Fluid Model has been chosen to simulate the transport of particles after resuspension.

3. Redistribution of particles inside a purged cavity: case study

Some simple test configurations have been defined and a mock-up with several adjacent cavities has been used to compare theory and experiment when a purging under dry nitrogen (Alphagaz grade 2 N_2) was applied in the vicinity of a pre-contaminated surface. The relocation of particles deposited on a specific substrate in front of the N_2 jet has been evaluated thanks to many collector plates disseminated in one or two cavities and the levels quantified. The impact of different test conditions (flow rate, material of the substrate, presence of leakage, initial contamination level…) has been investigated [11].

3.1. Ground experiments

3.1.1. Material (see Figure 8)

3.1.1.1 Mock-up

A CNES mock-up, initially designed to study molecular contamination transport, has been modified in order to use it for this study on particles redistribution. This mock-up is composed of 3 separable cavities, formed by assembling polished stainless steel plates, completely hermetic, apart from a circular entrance of 22 mm in diameter in the center of one of the side walls ; a maximum of 2 were used for the experiments.

3.1.1.2 Test particles

Colored calcium carbonate particles of FluoDust (from Fluotechnik) with a mean diameter of 3-4 µm, typically used for filter bag tightness control and contamination simulation, have been chosen as contamination sources, taking into account of their advantage to be fluorescent at 450 nm for the inspection. Due to their hygroscopic properties, the particles were dried at 100°C during 3 hours before dispersion on the test substrate. Anyway, some aggregates still remained.

3.1.1.3 Substrates and collector plates

Several types of substrates have been pre-contaminated: silica optics and polyimide foils from MLI (MultiLayer insulation).
Particle Fallout (PFO) sensors have been used to collect the redistributed particles on the walls of the mock-up. PFO sensors are smooth, non-reflecting glass plates [12].

3.1.2. Variable parameters

Depending on the experiments, several parameters have been modified in order to study the behavior of the particles according to these different cases:

- position of the contaminated substrate,
- nature of the contaminated substrate,
- number of cavities,
- gas flow rate,
- presence or not of a leak,
- presence of obstacles in cavity.

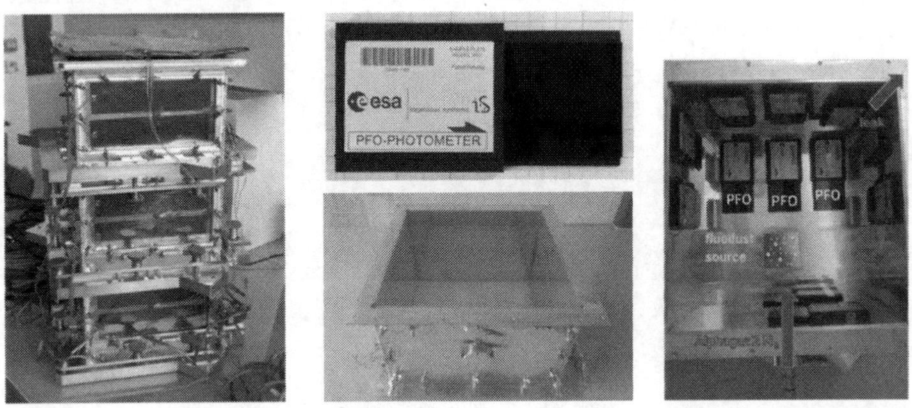

Figure 8. Overview of the CNES mock-up (left), PFO sensor (top center), open bottom cavity with circular entrance on the front plate (bottom center) and the same equipped cavity (right)

3.1.3. Experimental protocol
The figure below describes the main steps of the protocol applied for all experiments, starting by the contamination of the source surface with quantified amounts of FluoDust particles.

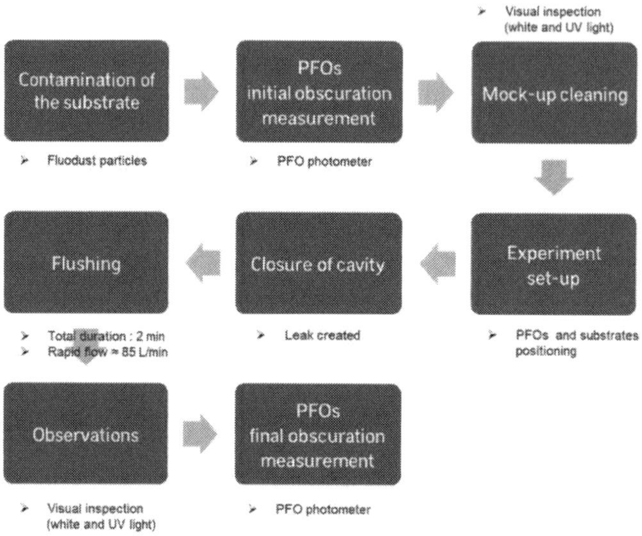

Figure 9. General protocol of redistribution experiments

3.2. CFD simulation

3.2.1. Mock-up geometry and meshing
The mock-up geometry was chosen to be as simple as possible, which simplified the 3D model building and minimized errors. The cavity where dry nitrogen jet is blown has a square ground surface (250 mm side length) and a 158 mm height. This cavity is coupled to a second one just above. A circular opening between them allows to control the experiment chamber volume.
The meshing design is refined where velocity gradients are expected to be high (through volume and surface in front of the N_2 jet). A meshing sensitivity test is undertaken to assess what the optimal mesh size is. The final version is composed of approximately 6.10^6 cells.

3.2.2. Flow simulation
The gas flow corresponds to the mock-up set-up value (85 L/min). The next figures 10 and 11 show the gas velocity distribution and streamlines (Velocity color range is 0 to 3 m/s then 0 to 4 m/s).

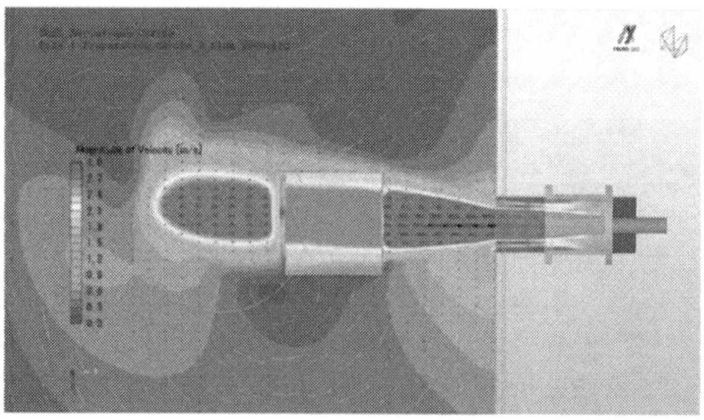

Figure 10. Gas velocity cross section (top view)

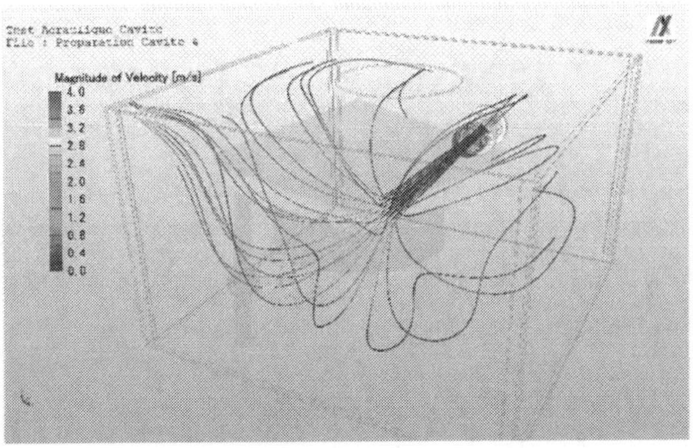

Figure 11. Gas flow streamlines

3.2.3. Particle redistribution simulation

The code integrated into the software is composed of several functions dedicated to the computation of each R'n'R model parameter at every solver iteration and every cell of the mesh, through a multi-dimension table. The final output is the resuspension ratio defining how many particles of each size range present on the surfaces must be resuspended.

The R'n'R model considers the adhesion forces as a distribution. For each particle size, a balance of the forces involved is applied to define their ability to be resuspended. When a particle is considered resuspended, the surface particle concentration is updated and a new particle is released in the volume to follow gas streams until next encounter with a surface (see Figure 12).

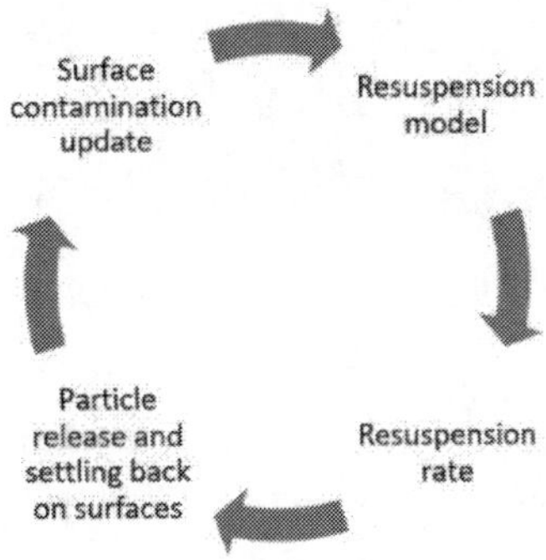

Figure 12. Iterative calculation for resuspension model

The time-step for the transient simulation ranges between 1.10^{-5} et 1.10^{-4} seconds.
User variables allow to define the particle size and range and adhesion forces.

Note: The FluoDust particles were considered as sources of particulate contaminants.
For the simulation, particle sizes of 5, 10, 20 and 30 µm were selected, which corresponds to critical sizes that may damage sensitive surfaces such as optics.

3.2.3.1 Single cavity

To study the particle release and deposition in a single cavity, the mock-up and CFD model are equipped with PFO sensors as shown on Figure 13.

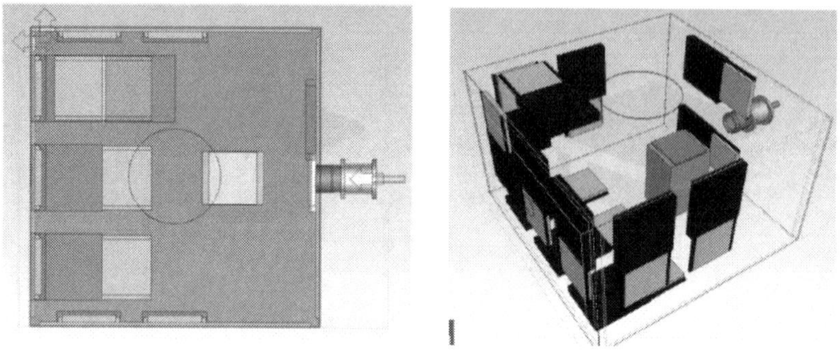

Figure 13. Single cavity equipped with PFO sensors

The grey box in the middle of the cavity is a stand for the contaminated surface (orange) covered with FluoDust particles. Gas output (or leak) is mainly located at a corner of the upper plate (blue arrows, Figure 13 top left).

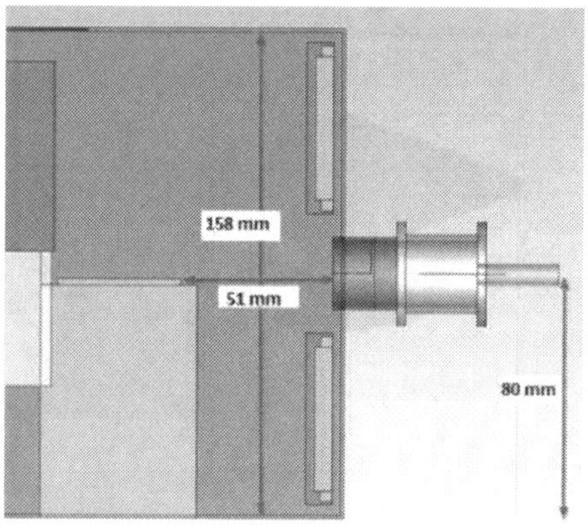

Figure 14. Single cavity side view with input nozzle

3.2.3.2 Double cavity

The double cavity version doubles the volume, the gas output being now on the vertical axis through the centered hole towards the exit located in the upper cavity. The PFO sensors positions are different as it can be seen on Figure 15.

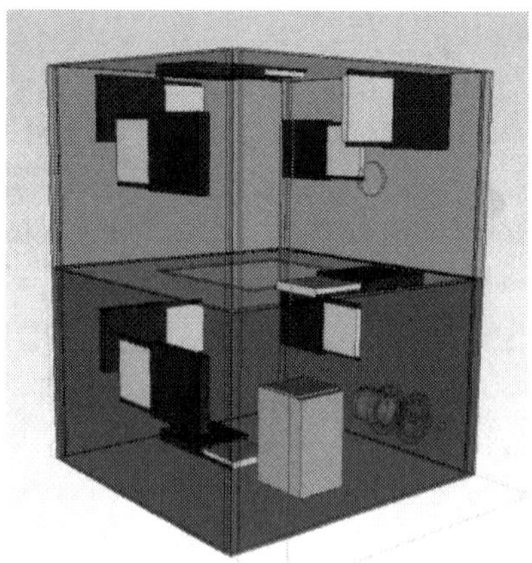

Figure 15. Double cavity model

IOP Conf. Series: Materials Science and Engineering 1287 (2023) 012038 doi:10.1088/1757-899X/1287/1/012038

3.3. Simulation vs. experiment: comparison of relative deposits

Almost twenty different release tests have been performed in the mock-up with various substrates as source surfaces, flow rates, particles concentrations...

Only two compared simulated configurations are presented here. On Figures 16 and 17, the simulation results are normalized with respect to the initial amount of particles on the source surface. In the case of the experiments, the results are displayed as the difference between the obscuration values in ppm on each PFO sensors after and before the test.

These values were measured with a PFO-photometer: the method principle is based on the scattering of light beams caused by the particles on the PFO sensor. The surface used for measuring the contamination is a 15 mm circular area in the middle of the glass plate. The light beams are almost parallel to the surface of the PFO sensor which itself doesn't disperse these beams. The scattered light is then measured and digitally displayed in PFO-units (ppm). The obscuration (i.e. ratio of the area covered by the particles to the total area) is calculated by the following formula:

$$C = n \cdot \frac{d^2}{D^2} \qquad \text{Eq. 2}$$

where:

- C is the obscuration in ppm
- n is the number of particles counted
- d is the particle diameter in µm
- D is the diameter of sensitive zone in mm

For all tests, we observed that the resuspension rate was higher in experimental conditions. This difference can be explained by the difficulty to prepare a contaminant source on the surface without agglomerates and multilayers of particles while the CFD model is monolayer. Aggregates and particles layers showed a higher resuspension rate due to a lower adhesive forces and higher velocity at distance of the surface. Thus, the simulation results hereafter are an extract from the complete simulation performed with the selected particle diameters. Only the particle diameter above 20 µm is considered that corresponds to the lowest limit of the FluoDust particles size distribution (similarly to the experiments since the particles of FluoDust powder were partially agglomerated).

3.3.1. First experiment – Single cavity (source particles on SiO_2 window)

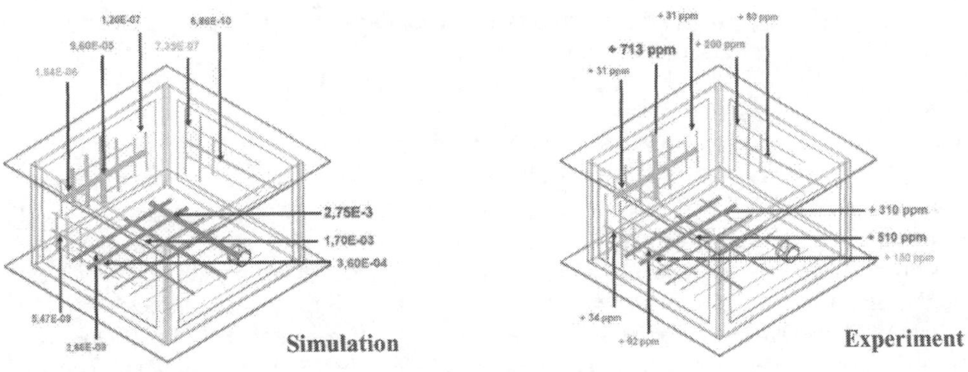

Figure 16. Simulation vs. experiment: comparison of relative deposits in single cavity

It is observed in an identical way that, after a sedimentation time, the particles are deposited mainly on the face which receives the gas flow and on the lower face (no uniformity of the deposit). Side surfaces remain cleaner than front and bottom surfaces. However, there are differences in the particles

IOP Conf. Series: Materials Science and Engineering 1287 (2023) 012038 doi:10.1088/1757-899X/1287/1/012038

distribution. The lower face shows a different distribution with a higher number of particles located at the center in the experiment.

3.3.2. Second experiment – Double cavity (source particles on SiO_2 window)

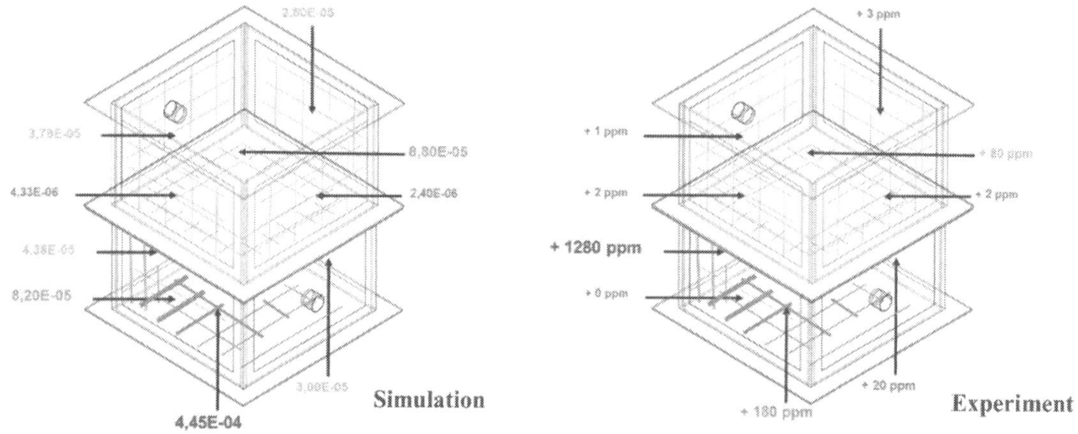

Figure 17. S*imulation vs. experiment: comparison of relative deposits in double* cavity

The redistribution in the first cavity does not change compared to the first experiment. The particle distribution of the second cavity is homogeneous both in simulation and experiment. During the experiment, an important quantity of particles was found on the ceiling of the lowest cavity which is not the case of the simulation. We can distinguish in detailed results that smaller particles are more present in the upper volume than in the lower one (in experimental and simulation results).

Limitations of experimental results:
The main difference is the source of particles. Indeed, the experimental source shows a poor uniformity with aggregates and multilayers of particles that are not considered in the model (complexity of inter-particle forces in cluster).
Moreover, the start of the N_2 jet is simplified with the model since the mechanism for opening the valve is not reproduced. There is a direct transition between "velocity zero" and a given velocity in simulated cases contrary to the short transitional real phenomena.

4. Conclusion and perspectives

Studying how particles are (re)distributed in confined environments is a real topic of interest for several fields and applications: cleanrooms, public health, nuclear energy… Even if it is a complex process, this helps to understand the mechanisms of resuspension of such particles and better predict the surface or airborne contamination levels by modelling.
This feasibility study was dedicated to closed cavities of space instrumentation as an example and allowed the simulation of the resuspension of particles by implementing Rock'n'Roll model in Computational Fluid Dynamics analysis. Experimental tests of resuspension of particles were carried out in parallel. The comparative results confirmed the redistribution of the particulate contamination with similar trajectories. But some discrepancies have to be considered: limitations of the R'n'R model depending on experimental parameters such as initial distribution of the source particles, sizes, shapes, adhesion force to the substrate... So, future works are needed to quantify more accurately the initial deposit of particles and to have a better knowledge of adhesion force in order to improve the simulations. Moreover, the actual code can be adapted for more complex simulations including new inputs such as other particle size ranges, electrostatic forces, acceleration, reduced gravity…

IOP Conf. Series: Materials Science and Engineering 1287 (2023) 012038 doi:10.1088/1757-899X/1287/1/012038

As a perspective, since it does not exist up to now, it should be very useful to develop a modelling tool dedicated to particles transport to be able to quantify contamination levels on space hardware even in orbit, as well as it is performed currently for molecular contamination. In such case, it will be necessary to take into account the pressure change and the density of the residual gas.

References

[1] Tribble A C 2000 *Fundamentals of Contamination Control* SPIE Press p 40-42

[2] Scialdone J et al 1987 Particulate Contaminant Relocation During Shuttle Ascent *Proc. SPIE 0777 Optical System Contamination: Effects, Measurement and Control*

[3] Borson E 2002 Model for Particle Redistribution during Spacecraft Launch *Proc. SPIE 4774 Optical System Contamination: Effects, Measurements and Control*

[4] Cooper D W 1986 Particulate Contamination and Microelectronics Manufacturing: An introduction *Aerosol Science and Technology* **5** p 287-299

[5] Reeks M W and Hall D 2001 Kinetic models for particle resuspension in turbulent flows: theory and measurement *Journal of Aerosol Science*

[6] Mana Z 2015 Study on the resuspension of particles following the walk of an operator *PhD Thesis* Paris Sud University

[7] Nasa Handbook For Models And Simulations: An Implementation Guide For Nasa-Std-7009a 2019

[8] Sørensen D N and Nielsen P V 2003 Quality Control of Computational Fluid Dynamics in Indoor Environments *Indoor Air* **13** p 2-17

[9] Brieda L et al 2010 Analysis of Particulate Contamination During Launch of MMS Mission *Proc. SPIE 7794 Optical System Contamination: Effects, Measurements and Control*

[10] Shallcross G et al 2022 Launch Recontamination: the evaluation of particle adhesion and removal mechanisms in spacecraft payload fairing environments *44th COSPAR Scientific Assembly*

[11] Dall'armi F 2021 CNES internship report

[12] ECSS-Q-ST-70-50, Space product assurance – Particles contamination monitoring for spacecraft systems and cleanrooms 2011

IOP Conf. Series: Materials Science and Engineering 1287 (2023) 012039 doi:10.1088/1757-899X/1287/1/012039

Space polymers erosion in simulated LEO and VLEO environments

Z Iskanderova[1], M Tagawa[2], J Kleiman[3*], S Nishioka[2] and S Horimoto[2]

[1]-AZ Smart Technologies, Toronto, Canada
[2]-Kobe University, Graduate School of Engineering, Kobe, Japan
[3]-Integrity Testing Laboratory Inc., 80 Esna Park Drive, #7-9, Markham, Ontario, Canada, e-mail: jkleiman@itlinc.com (*to whom correspondences should be addressed)

Abstract. The VLEO satellites, that are orbiting at an altitude of 200-300 km and that provide advantages for high-resolution Earth observation and atmospheric science, are becoming important not only for science, but also for agriculture, forestry, and oceans observations, for military, national security and the future of satellite businesses. We evaluated, for the first time, the performance and defined the erosion yields (Ey) for a selected number of main space hydrocarbon, fluorinated, and perfluorinated polymers in the ground-based simulator facility at conditions, imitating the LEO and VLEO environments. We also compared the results with the data for polymers erosion in LEO and with the previously established predictive model, developed for polymers and composites in LEO environment. The obtained data indicated that the predictive model can be extended and used, with some corrections for the predictions of space polymers erosion in the VLEO environment.

1.0 Introduction: Satellites in VLEO - Benefits and Challenges

In general, the Very Low Earth Orbit (VLEO) refers to orbits between 250 and 450 kilometres above the Earth's surface. The possible telecommunication services and applications that could be offered from VLEO, are mostly to be provided by VLEO satellites orbiting at altitudes below 300 km. The VLEO satellites, that are orbiting at an altitude of 200-300 km and that provide advantages for high-resolution Earth observation and atmospheric science, are becoming important not only for science, but also for agriculture, forestry, and oceans observations, for military, national security and the future of satellite businesses [1-3].

A satellite moving through the rarefied gas experiences induced drag, and the atmospheric drag becomes a dominant satellite design consideration. The variations in density and thermospheric winds can create attitude disturbances, and the primary atmospheric component, the highly aggressive atomic oxygen (AO), can erode many of the exposed sensitive materials surfaces. Any mission, operating in VLEO, must address these issues. At VLEO altitudes the residual atmosphere is largely composed of atomic oxygen (AO) that is highly aggressive and can adsorb to and erode exposed surfaces, and of significant amounts of N2.

The atmospheric density and orbital velocity also increase with reducing altitude, resulting in increased flux of AO. The effects of AO adsorption and materials erosion generally increase the momentum accommodation between the spacecraft and the atmospheric flow and, therefore, the drag experienced by the orbiting spacecraft. This, in turn, increases the rate of decay and reduces the useful lifetime, if without propulsive compensation [1-3]. One of the environmental problems on VLEO orbits is an increased, compared to LEO, erosion of materials by dense AO. The N2 that is present in VLEO altitudes in significant amounts also makes its influence, with external satellite materials encountering simultaneous collisions with ~5 eV AO and ~9 eV N2 species. Even while using various protective coatings on polymers, polymer-based compositional materials or composites, and on other sensitive materials, or using some "self-protecting" options, developed originally for LEO applications, the possible surface defects, cracking and spallation may still expose some areas of polymers or polymer-based surfaces to the highly aggressive high AO density VLEO environment.

With a growing interest in the development of VLEO satellites for science, military and business applications, one of the major questions is to obtain reliable data on interaction of sensitive structural materials with VLEO space environment. Another important question is to determine how reliable the materials can be tested in the ground-based simulating facilities, in order to evaluate the materials in question durability and performance expectations in VLEO. It is well-known that the space polymer and polymer-based materials, like some space composites, are extensively used in the satellites, operating on GEO and LEO orbits. And the role of aggressive atomic oxygen interaction with those materials and protection from it in LEO have been studied, and continue to be studied, up to the last decades. The studies for numerous flight-proven and innovative materials have been performed in the ground-based FAO beams simulation facilities. And, starting from MISSE 1 and MISSE 2 in the year 2001, on the numerous MISSE experiments in real LEO space environment, at ISS, that is ~ 450 km over the Earth surface.

Considering the high density of AO in VLEO environment and the presence of other additional environmental components, the goal of the present work was to evaluate for the first time the performance and define the erosion yields (Ey) for a selected number of main space hydrocarbon, fluorinated, and perfluorinated polymers in the ground-based simulator facility at conditions, imitating the LEO and VLEO environments. Another goal was to compare the results with the data for polymers erosion in LEO and with the previously established predictive model, developed for polymers and composites in LEO environment [4,5] and to find out, if the model can be extended and used, with some corrections to be defined and introduced, for the predictions of space polymers erosion in the VLEO environment.

2.0 Experimental - Polymers in Simulated LEO and VLEO Environmental Conditions

2.1 Space Polymer Films Materials Selection

A number of most representative space polymers used as space flight materials that included hydrocarbon, partially fluorinated, and perfluorinated polymer films – Teflon FEP and Teflon PTFE - with erosion rates in LEO covering a range of an order of magnitude have been selected and used for this study. List of these space polymers films (with their Trademarks) is presented below in Table 1.

Table 1 Space polymer films used in this study

Abbreviation	Chemical names	Polymer type	Trademark
PEEK	Poly Ether Ether Ketone	hydrocarbon	Victrex PEEK
PVDF	Poly Vinylidene Fluoride	partially fluorinated	Kynar
FEP	Fluoro Ethylene Propylene	perfluorinated	Teflon FEP
PTFE	Poly Tetra Fluoro-Ethylene	perfluorinated	Teflon PTFE
POM	Poly Oxy-Methylene	hydrocarbon	Acetal
PVF	PolyVinyl Fluoride	partially fluorinated	Tedlar
PET	PolyEthylene Terephthalate	hydrocarbon	Mylar
PI	- Polyimide	hydrocarbon	Kapton 100HN
PI	- Polyimide	hydrocarbon	Kapton 500HN

2.2 Hyperthermal Energy Beam Exposures for LEO and VLEO simulation

A series of exposure experiments were performed at Kobe University using one of the laser-detonation atomic oxygen (AO) beam sources in which hyperthermal AO beams with velocities similar to the orbital velocity of spacecraft are formed. A schematic drawing of the laser-detonation hyperthermal beam system is shown in Figure 1.

The system consists of 3 vacuum chambers; source, drift and TOF. All vacuum chambers are evacuated by turbo molecular pumps and the base pressures are 1E-4, 1E-6 and 1E-6 Pa, respectively. The house-made piezoelectric-driven pulsed supersonic valve (PSV) is used for this experiment. 100% O2 gas was used as a target gas for conventional LEO simulation. In contrast, 50% O2 and 50% Ar premixed gas was used for simulating the VLEO environment that is consisting of AO and N2. The reason for using Ar instead of N2 is to prevent the dissociation reaction of N2 into N in laser plasma which reduces the collision energy of N atoms. A carbon dioxide laser pulse (5-8 J/pulse) is introduced through the ZnSe window and focused by a concave Au mirror. A laser plasma and detonation wave propagating in the nozzle decomposes O2 into AO and accelerates AO into orbital velocity of spacecraft.

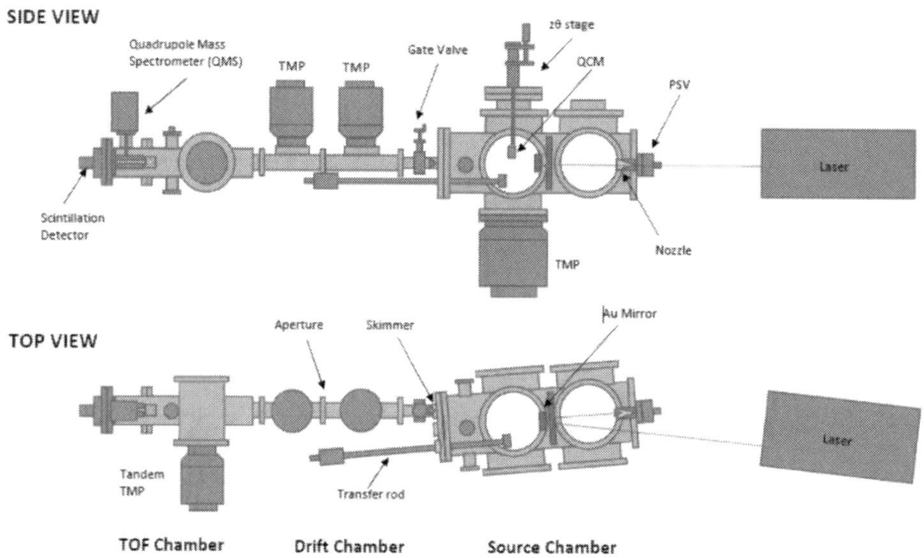

Figure 1. Experimental set-up of the Kobe University's laser-detonation atomic oxygen (AO) beam source (side view and top view)

A centre mirror layout is selected in this experiment in order to minimize the AO fluence variations on the large sample holders. AO flux during the exposure was monitored by polyimide-coated quartz crystal microbalance (QCM). A 3 mm skimmer is inserted between the source and the drift chambers. A quadrupole mass spectrometer and a scintillation detector are installed in the TOF chamber to measure the TOF spectra of the designated component in the beam. The TOF spectra of AO and O2 components in the AO beam for the LEO simulation are shown in Figure 2.

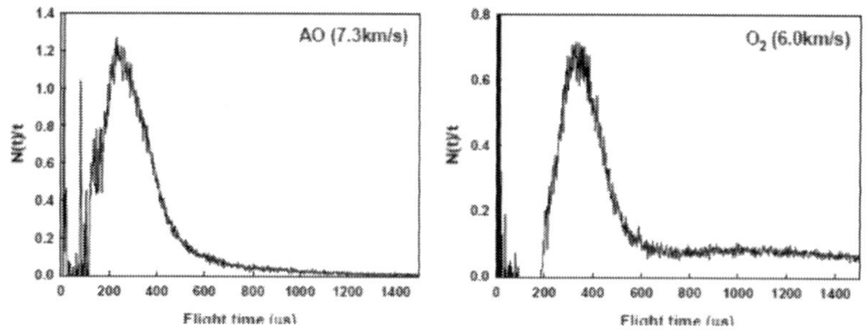

Figure 2. TOF of the AO exposure only. Average velocity of AO and O2 components are 7.3 and 6.0 km/s, respectively.

IOP Conf. Series: Materials Science and Engineering 1287 (2023) 012039 doi:10.1088/1757-899X/1287/1/012039

Average velocity of AO and O_2 components are 7.3 and 6.0 km/s, respectively. Beam composition was analysed from the intensity of the TOF spectra with established relative ionization cross-sections [6]. Fluence of AO on each sample was estimated from the mass loss of Kapton H witness samples located side by side to every sample, for this case - with the established erosion yield of Kapton H in LEO (3.00E-24 cm3/atom) with the equation bellow,

$$F_{AO} = \frac{\Delta m}{A \rho E_y} \quad (1)$$

where F_{AO}: AO fluence (atoms/cm^2), Δm: Mass loss of Kapton (g), A: Exposed area (cm^2), ρ: Kapton' density (g/cm^3) and E_y: Erosion yield (cm^3/atom). VLEO simulation was carried out by the target gas containing 50% O2 and 50% Ar. Figure *3* shows the TOF spectra of AO, O2 and Ar components in the beam used for VLEO simulation.

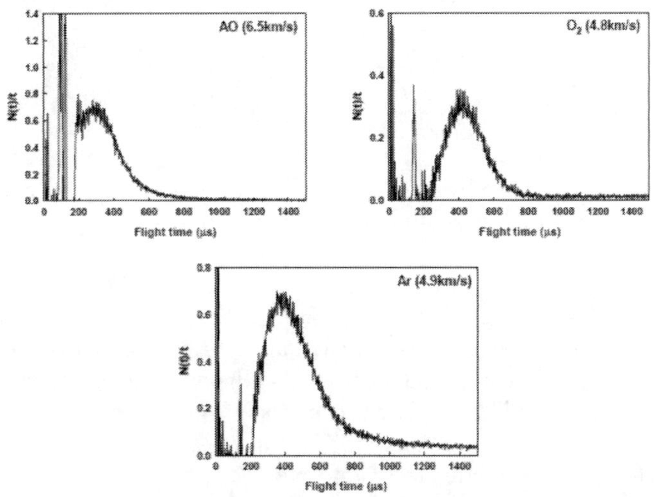

Figure 3. TOF of gas species in AO+Ar exposure. Average velocity of AO, Ar and O_2 components are 6.5 km/s and 4.8 km/s and 4.9 km/s, respectively.

Beam composition is evaluated by the area intensity of each TOF spectra with relative ionization cross-sections [6]. The composition in the beam has been defined to be the following: AO : O2 : Ar = 0.68 : 0.07 : 0.25, that corresponds to approximately 220 km altitude at VLEO (at mean solar activity). Fluence of AO on each sample was also estimated from the mass loss of Kapton witness samples located side by side. However, the established erosion yield of Kapton in LEO (3.00E-24 cm3/atom) should not be used for VLEO simulation, since addition of Ar may enhance the erosion of Kapton H. In the studies of Horimoto et al [7], it was reported, in good agreement from both ground-based simulation and the SLATS VLEO satellite data [8], that addition of Ar enhanced the erosion rate of Kapton proportional to the Ar fraction in the beam. The erosion yield acceleration factor could be expressed by the following equation (1), where the acceleration effect of PI erosion by Ar addition is expressed as:

$$K_e = 1.00 + 0.0267 \cdot P_{Ar} \quad (2)$$

where K_e is the acceleration factor and P_{Ar} is the Ar fraction (%) in the beam. In our experiments, the average Ar fraction was 27%, as derived from the TOF analysis. The erosion yield of Kapton in the AO + Ar exposure is 1.72 times higher than in conventional LEO simulation tests that do not include Ar. Thus, the erosion yield of Kapton polyimide in this VLEO simulation is evaluated to be 5.16×10^{-24} cm^3/atom. By using K_e of Kapton as 5.16×10^{-24}, the equivalent AO fluence at each Kapton sample was calculated, and by using these AO fluences, K_e of other materials have been calculated.

Samples of nine materials listed above in Table 1, prepared as 10 mm discs, were exposed to simulated LEO (AO only) and VLEO (AO + Ar) environments. Highly Oriented Pyrolytic Graphite (HOPG) samples were included as a reference. Erosion rate of the tested samples relative to Kapton H was calculated at both exposures using the mass loss data. For every polymer, 3-5 samples have been exposed on every holder, and the results have been averaged from at least 3-4 weight measurements. In order to minimize the effect of flux variations over sample holders, Kapton witness samples were located close to each sample, with images of both holders with samples shown in Figure 4.

Figure 4 Images of the sample holders for LEO (left) and VLEO (right) exposure tests. Samples marked as 1-1 and 1-2 are HOPG samples exposed for comparison purposes.

It should be emphasized that in our VLEO simulation experiments the effects associated with chemical properties of N_2 during collisions with material surfaces, as in real VLEO flights, are ignored and only the physical effects of the simultaneous N_2 collisions with materials are considered and simulated, when substituting N_2 with Ar. The recent analysis of the world's first real-time material erosion measurement in VLEO flown on SLATS showed good agreement with the ground-based experiments on KaptonH, using this technique [8].

Mass of each sample before and after exposure was measured using a balance with an accuracy of 0.01 mg. The possible influence of moisture absorption was eliminated by measuring the time-dependence of the mass change after venting the sample from vacuum.

3.0 Results and Discussion

The experimental results of the Ey measurements in both simulated LEO and VLEO environments are described below, and the measurements and calculation results are presented in Table 2. The acceleration effect of Ar addition to the AO exposure, illustrated by the ratio of erosion yields obtained in LEO and VLEO experiments, is summarized for all polymers in Table 2 as well.

3.1 LEO environment simulation results

The erosion yield of each sample for LEO simulation was calculated by using the AO fluence and the mass loss of the samples, and the results of Ey (AO) are presented in Table 2. The Ey obtained in the LEO simulation in this study was compared later with the predictive values, based on the model and predictive correlations [4,5] and with the LEO flight results from the (MISSE 2 PEACE) experiment [9]. The erosion yields for the hydrocarbon and fluorinated polymers in this LEO simulation are in reasonable agreement with the predicted values and when compared to MISSE 2 PEACE data, except for Teflon FEP and Teflon PTFE. The erosion yields of both Teflon samples, as obtained in the present work, are ~7.6 and 8.3 times higher than those obtained in MISSE 2 PEACE experiment. I should be noted that such discrepancy for Teflon-based materials between the results of ground-based LEO simulation experiments and actual flight data is quite common [10, 11, 12], with the cause for these differences still not fully clarified.

3.2 VLEO simulation results

The Ey (AO+Ar), obtained in the VLEO simulation with 50% O_2+50% Ar gas as a target gas, are also summarised in Table 2. (Note that Ey of KaptonH is used in this case as 5.16E-24 cm3/atom, as described above). The ratio of the erosion yield in VLEO to that in LEO is 1.72 for Kapton H, and it is in the range from ~1.5 to ~2 for other hydrocarbon and partially fluorinated polymers. For perfluorinated polymers Teflon PTFE and Teflon FEP the erosion yields themselves are significantly higher – see Table 2, and this ratio reached for them ~ 2.4 and ~2.5, values that are slightly higher than for other polymers. (Effects of erosion enhancement for Teflon samples, while considering the influence of additional environmental components in the ground-based space environment simulations of the hyperthermal atomic beams, were proposed to be associated mostly with Collision Enhanced Dissociation (CED) of the polymers, with the formation of volatile degradation products [11, 12].

The results for both Teflon samples indicate that the effect of Ar addition in the beam might be slightly more pronounced for the perfluorinated polymers. Analysing the data, it can be suggested the erosion yields of Teflon FEP and Teflon PTFE can be expected to be higher in VLEO regarding LEO, than for the hydrocarbon and partially fluorinated polymers.

Table 2 Summary of erosion yield values for space polymers in simulated VLEO (AO+Ar) and LEO (AO) and those erosion yield ratios

Sample	Ey by AO+Ar (cm^3/atom)	Ey by AO (cm^3/atom)	Ey (AO+Ar)/ Ey (AO)
PEEK	3.82×10^{-24}	2.06×10^{-24}	1.86
PVDF	2.72×10^{-24}	1.44×10^{-24}	1.89
FEP	3.84×10^{-24}	1.52×10^{-24}	2.52
PTFE	2.85×10^{-24}	1.18×10^{-24}	2.41
Acetal	2.14×10^{-23}	1.02×10^{-23}	2.11
PVF	3.78×10^{-24}	2.43×10^{-24}	1.56
Mylar	5.25×10^{-24}	2.44×10^{-24}	2.15
Kapton 500HN	5.33×10^{-24}	3.00×10^{-24}	1.78
Kapton 100HN	5.33×10^{-24}	3.00×10^{-24}	1.78

It is important to mention that the results of erosion yield determination for all hydrocarbon and partially fluorinated polymer films in the simulated LEO environment, obtained in present work, are in good agreement with the MISSE 2 PEACE flight results in LEO for similar polymer films [9]. The difference of Ey for hydrocarbon polymers is not more than 10-15%, and the values for partially fluorinated polymers are exactly the same as received in [9] for a long-duration LEO flight on MISSE 2, thus validating the methodology and experimental set-up used in the present work. For the perfluorinated polymers (both Teflon PTFE and Teflon FEP) the experimental values for their Ey in ground-based LEO simulation are significantly, up to an order of magnitude, higher then in LEO flights. And, as can be seen from Table 2, similar results have been received in this study for both LEO and VLEO simulation as well. This increase in Ey for Teflon's in the ground-based FAO beam facilities have been known for a number of years, and was explained in [10] for LEO simulation by the presence of higher energy AO species in the FAO beam facilities that can break the higher energy F bonds, and in this way to "assist" in the removal of F atoms and diminishing their "screening" effect on polymer's backbone chains. Other possible reasons have been studied and discussed in [11, 12]. Similar effects were observed also in the present VLEO simulation experiments, as shown in the results presented in Table 2. In regard to these LEO and VLEO simulation studies results, it was interesting to compare the results of this work with the predictive model and correlation, developed earlier in [4, 5] for polymer films in the LEO environment and to find out, if the same trends may be expected to work for VLEO as well, just with experimentally defined correction coefficients.

4.0. Comparison of Data from Present Work with Earlier Predictive Models

Using extensive pool of erosion data from various LEO space flights, a predictive model has been developed at ITL/UTIAS [1995] that allowed calculating and predicting the erosion yield of pure hydrocarbon space polymers in the LEO environment, depending on their chemical composition and structure [4], with high ranking given to this development by professionals in this area [13]. Later, this work was extended to the AO erosion in LEO of partially fluorinated linear chain polymers and the perfluorinated polymers - Teflon PTFE and Teflon FEP [5]. Using this extended predictive model, the fluence-dependent erosion data for the Carbon Fibers Reinforced Plastic (CFRP) composites have been also explained in [5] for both short and long duration exposures in LEO environment flights, especially for the Long-Duration Exposure Facility (LDEF) results [14].

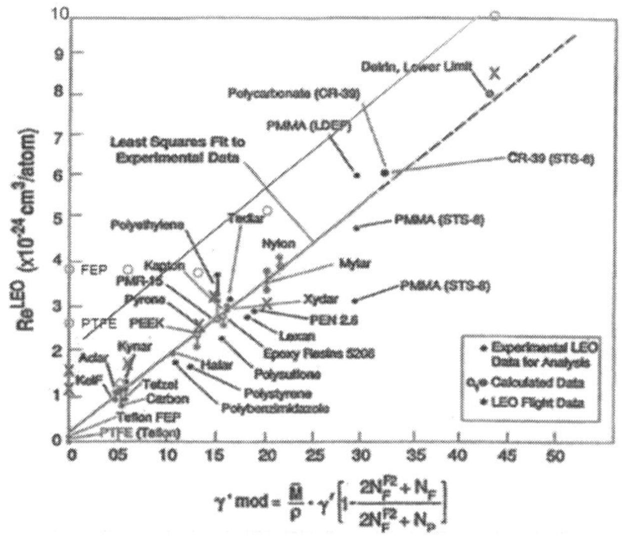

Figure 5 Erosion Yield in LEO for a variety of hydrocarbon and fluorinated polymers versus γ ' mod (for definitions and refs see [5] and refs therein), and the results of this study are added as red crosses (LEO simulation data) and blue open circles (VLEO simulation data).

Figure 5 shows the values of Re (LEO), that is the same as Ey (LEO) in this work, i. e. the LEO erosion yield dependence on the space polymers compositional/structural parameter γ'_{mod} from [5] for a variety of hydrocarbon and fluorinated space polymers (see [4, 5] for details). The erosion yield results from the present VLEO and LEO simulation for the space polymers of this study, tested in the FAO beam facility, are also added to for comparison.In Figure 5 the black solid line represents the extended linear correlation for LEO data, the red crosses are the LEO simulation results of this work (AO only), the blue open circles represent the data of the Ey values of VLEO simulation results of this work (AO + Ar), and the blue line is drawn to indicate the trend (the VLEO simulation results for Teflon FEP and Teflon PTFE are not included). As can be seen from Figure 5, based on the LEO and VLEO simulation results of this study, a similar trend to the one, defined previously for AO erosion in LEO, can be expected for hydrocarbon and fluorinated space polymers at AO erosion in the presence of significant amount of N_2 in the VLEO environment.

The acceleration factor may be expected to be even somewhat higher for perfluorinated space polymers, in particular, for Teflon PTFE and Teflon FEP. However, in both LEO and VLEO simulation in a very reliable FAO beam facility at Kobe University, as well as in other similar ground-based facilities, the erosion yields for Teflons are still consistently significantly higher than in LEO and, as far as we can see, that can be expected for simulated VLEO space environment.

IOP Conf. Series: Materials Science and Engineering 1287 (2023) 012039 doi:10.1088/1757-899X/1287/1/012039

For many years of AO testing experiments of numerous space polymers and composites at ITL, the screening tests by exposure to O-plasma in a O-plasma Asher in pure oxygen plasma demonstrated higher effective erosion yields than our predictions and the ground-based FAO beams facilities testing results or the LEO flights data. For instance, before performing the comparative FAO beam testing of numerous space materials from Russia, Europe, and US in three FAO beam facilities in [15], all of them have been exposed to the O-plasma screening test at ITL in an Asher, using pure oxygen as a working gas. The effective Ey (plasma) erosion yield have been mostly in 1.5 – 2.5 times higher, than in FAO beam facilities. This is in agreement with the results in [9], where a similar comparison was made for numerous MISSE 2 PEACE polymers after testing them in an Asher plasma, using the air as a working gas, with the LEO flight data, and with similar conclusions.

It is interesting to mention that the Ey (plasma) in [9] for the space polymers of this study, excluding the Teflon samples, was higher than the Ey (LEO) at MISSE on average 1.4 - 1.7 times, that is in the range of the ratio Ey(VLEO)/Ey(LEO) obtained in these testing data in the FAO facility (see Table 2) . Slightly higher values of Ey (plasma)/Ey(LEO), around 1.7-1.9, are shown for PMMA, PC, PI CPI, etc. in [9] - but altogether they are still in the range 1.5 - 2, consistent with the results for hydrocarbon and fluorinated space polymers in this study. The reasons and mechanisms for Ey acceleration are different in these two environments - but the final effective results seem to be quite close. It should be noted, however, that for some of the polymers the erosion acceleration in O-plasma may be significantly higher [9]. Based on these results, O-plasma may be considered as a conservative, but more representative testing option of space materials for VLEO, than for LEO. The erosion yield in O-plasma for Teflon PTFA and TeflonFEP is 7-8 times higher then in LEO flights, that is similar to the results in FAO beams facilities. More space polymers in the same conditions and at higher AO+Ar fluences are proposed to test later in ground-based simulation conditions at simulated LEO and VLEO.

5.0 Mechanisms of AO Interaction with Polymers-based Materials and Surface Morphology Changes

Another important question of this study was to compare the surface morphology changes in polymers exposed to the VLEO and LEO simulated environments and what patterns will be formed on the surfaces of polymers exposed to VLEO environment, compared to the well-known, mostly cone-like carpet features formed on surfaces of numerous polymers after LEO flights and FAO beams exposures?

5.1 Analysis of polymer surfaces after AO and AO + Ar exposures

Figure 6 through Figure 11 present the results of the SEM/EDS analysis of a number of materials exposed ether only to AO beam or to a combined AO + Ar beam. Analyzing the data in all figures it can be suggested that in the images of surfaces after AO + Ar exposures the features are finer, i. e. the developed cones are smaller. The Kapton surface developed a very familiar morphology in both LEO and VLEO (Figure *9*). The PTFE sample in both exposures did not develop a cone-like structure (Figure 10). The surfaces looked poked with a large number of small craters (holes). The AO + Ar surface looks smoother with less craters. No obvious features were found in the Teflon FEP samples in either of the two exposures (Figure 11).

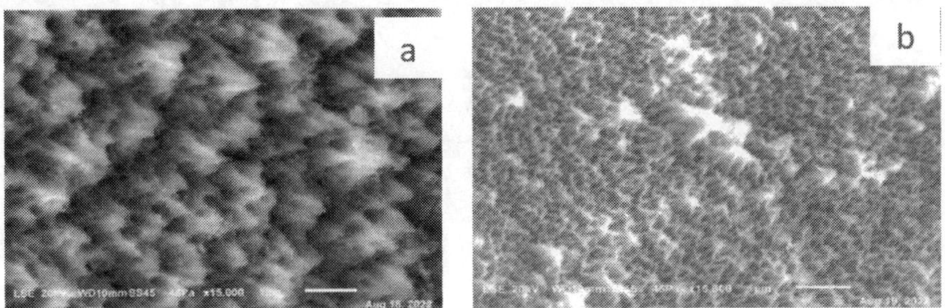

Figure 6. SEM Analysis of Mylar surfaces after AO only (a) and after AO+Ar (b) exposures

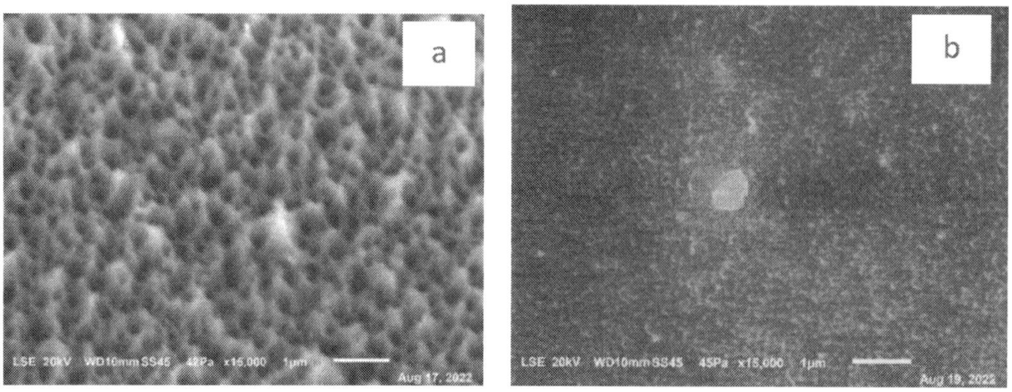

Figure 7. SEM Analysis of PEEK surfaces after AO only (a) and after AO+Ar (b) exposures

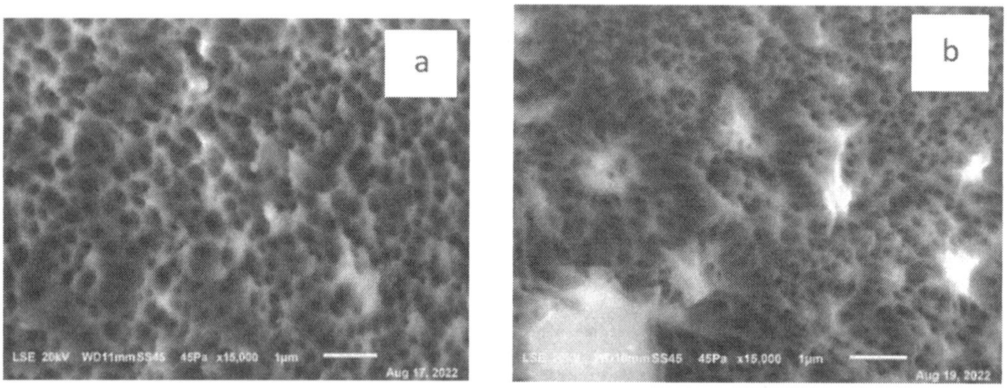

Figure 8. SEM Analysis of PVF surfaces after AO only (a) and after AO+Ar (b) exposures

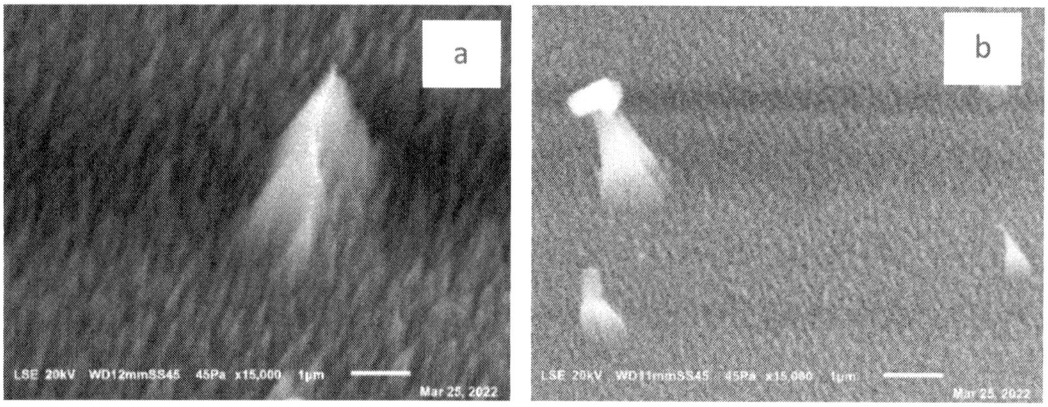

Figure 9. SEM Analysis of Kapton 100HN surfaces after AO only (a) and after AO+Ar (b) exposures

Figure 10. SEM Analysis of PTFE surfaces after AO only (a) and after AO+Ar (b) exposures

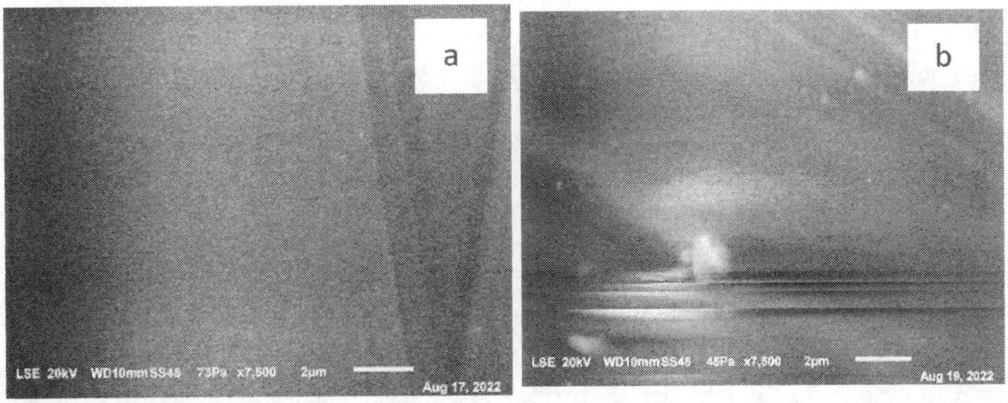

Figure 11. SEM Analysis of Teflon FEP surfaces after AO only (a) and after AO+Ar (b) exposures

5.2 Advanced Protective Coatings and Reducing the Drag in VLEO

Presently, the major solutions to minimise the drag for satellites in VLEO are aligned, mostly, along two directions - by selection and adjustment of the satellite's shapes – see, for example [2], and by using external materials with smooth surfaces that should remain the same during the satellite service life, imposing demanding requirements to the external environmentally sensitive materials that should be well protected and durable in the harsh and aggressive VLEO environment. Different protection schemes should be developed for polymers and composite materials to keep their surfaces smooth and not affected by the higher AO+N_2 erosion rates in VLEO and high AO fluxes.

We are working to explore the use of the innovative surface modification processes developed at ITL for protection of sensitive surfaces in VLEO environment. Superior quality specially doped thin coatings will be deposited on space polymers by Direct Ion Beams Deposition (IBD), for protection and durability enhancement of materials used for VLEO applications. The IBD is a highly versatile technology for deposition of thin superior quality coatings, especially when combined with best selected substrate ion beam pre-treatments and ion beam intermixing and deposition conditions. In a number of high-profile projects, we deposited conformal, highly adhered, high density, pinhole-free, charge dissipative, undoped or specially doped coatings, that met a variety of sophisticated quality and durability requirements in space environments for special applications (see ITL Website www.itlinc.com for examples). Among the flight proven uses, we can mention, for instance, coatings deposited on space polymer films for advanced GEO space antenna applications that were space qualified for GEO environment and that were used on 32 space antennas in 11 space vehicles.

Full protection of the surfaces of thermal control materials in LEO was achieved at a later stage, by applying Si and Si+O specially doped amorphous hydrogenated a:H-DLC super-coatings on Kapton HN, as well as on real thermal control materials such as Kapton HN/Al and TeflonFEP/Al/Inconel [16]. In an 8-hour O-plasma exposure screening test of Kapton HN and other space materials with the successfully doped coating on top, with an effective AO fluence of ~ (1-2) x 10E20 at/cm^2, the coating "converted" to a SiO2 structure, achieving full self-protection and protecting the underlying polymer film. The surfaces of all such coated materials remained smooth and with full restoration of the thermal optical characteristics of pristine materials, and stays unchanged at higher AO fluences.

The obtained results for the innovative TC materials are promising for use in LEO application and provide a hope for future applications in VLEO by reducing the dangers of drag effects that depend on the external satellites materials surface morphology.

6.0 Conclusions

Erosion yield Ey data was obtained for a number of hydrocarbon, fluorinated, and perfluorinated polymers under simulated LEO and VLEO space environments in a fast atomic oxygen beam facility at Kobe University. In tests simulating the VLEO environmental conditions the N_2 gas was substituted for Ar gas to prevent the dissociation reaction of N_2 into N atoms in laser plasma that have lower collision energy.

The results confirmed that the main mechanisms of polymers erosion under FAO in VLEO remain the same as in LEO, just with an acceleration factor for erosion yields Ey between 1.5 - 2 for hydrocarbon and partially fluorinated polymers, and around 2.4 - 2.5 for per-fluorinated polymers Teflon PTFE and Teflon FEP.

The Ey values for both Teflon PTFE and Teflon FEP in VLEO ground-based simulation tests were higher than the data in actual LEO space, similar to the data for these materials from similar ground-based LEO simulation experiments. Despite a number of detailed studies conducted in the last decade, there is no reliable method and facility for ground-based simulation of the Teflon samples erosion in LEO and, as can be expected from these results, in VLEO also, to provide a closer match to data from real LEO and VLEO space environments.

The surface morphology for all polymers, exposed in this study, was carefully analyzed by SEM, and the differences and some similarities have been identified for some of the polymers. It is concluded that for the majority of the polymers, the morphological changes will be developed already at a comparatively low AO fluence at both LEO and VLEO, i. e. after comparatively short duration of the flights in VLEO on unprotected polymers, due to the high AO density.

Innovative specially doped thin super-coatings have been deposited by direct IBD on space polymers at ITL that easily "convert" under AO to a smooth, highly protective surface, with full restoration of original thermal optical properties. The samples are now mounted on MISSE-17 in ram direction, and future exposure is planned for the coated materials at high, VLEO-relevant FAO fluences at Kobe FAO beam facility.

Acknowledgments

Our thanks are going to the ITL team members that contributed to the Project and more specifically to A. Tang for analysing the morphology of the samples and M. Shagalov for his help in organizing the paper materials, and to T. Minton, Colorado, for fruitful discussions and provided important publications.

References

[1] Crisp N 2021 Very Low Earth Orbits — Reducing Orbital Altitude for Lower Cost Earth Observation and Communications Satellites *https://nhcrisp.medium.com/vleo-ea5c5248e857*

[2] Walsh J, Berthoud L and Allen C 2021 Drag Reduction through Shape Optimisation for Satellites in Very Low Earth Orbit *Acta Astronautica* vol 179 pp 105-21 *https://doi.org/10.1016/j.actaastro.2020.09.018*

[3] Nicholas H, Crisp P Roberts P and Livadiotti S 2018 Literature Review of the Benefits and Applications of Very Low Earth Orbits for Earth Observation D5.2 *https://discoverer.space/wp-content/uploads/2020/05/D5_2_Literature_Review.pdf*

[4] Iskanderova Z, Kleiman J, Goudimenko Y and Tennyson R C 1995 Influence of Content and Structure of Hydrocarbon Polymers on Erosion by Atomic Oxygen *JSR* vol 32 #5 pp 878-84

[5] Iskanderova Z, Kleiman J at al 2001 Generalization of the Predictive Model of Erosion of Carbon-Based Materials by Atomic Oxygen *Proc. of ICPMSE-4* Toronto Ed. J. Kleiman and R. Tennyson, (Kluwer Academic Publishers, Dordrecht The Netherlands) vol 4 pp 33 -50

[6] Yokota K, Yasuda S, Mizutani A, Tagawa M 2013 Relative Electron Impact Ionization Probabilities of O, O_2 and Ar Components in the Laser-Detonation Hyperthermal Beams *Japanese Journal of Applied Physics* vol. 52 038002

[7] Horimoto S, Fujita A, Urakawa S, Nishioka S, Ide W, Yokota K and Tagawa M 2021 Ground-based Verification of Promotion Effect by N2 Collision on an Atomic oxygen-induced Material Erosion *Proceedings of the 65th Space Sciences and Technology Conference*, on line, February 26 - March 4, 2021, 4E10, in Japanese.

[8] Yokota K, Tagawa M, Fujimoto Y, Ide W, Kimoto Y, Tsuchiya Y, Goto A, Yukumatsu K, Miyazaki E, Imamura S 2021 Effect of Simultaneous N_2 Collisions on Atomic Oxygen-induced Polyimide Erosion in Sub-low Earth Orbit: Comparison of Laboratory and SLATS Data *CEAS Space Journal* vol 13 No.3 pp 389-97

[9] de Groh K K and Banks B A 2011 Ground-Laboratory to In-Space Atomic Oxygen Correlation for the Polymer Erosion and Contamination Experiment (PEACE) Polymers *NASA/TM—2011-216904*

[10] Tagawa M, Minton T at al 2010 Energy Dependence of Hyperthermal Oxygen Atom Erosion of a Fluorocarbon Polymer: Relevance to Space Environmental Effect *ACS Advanced Materials and Interfaces* vol 2 No.7 pp 1866-71

[11] Zhang J, Lindholm N F, Brunsvold A L, Upadhyaya H P, Minton T K, Tagawa M 2009 Erosion of FEP Teflon and PMMA by VUV *ACS Appl. Mater. Interfaces, Radiation and Hyperthermal O or Ar Atoms, ACS Appl. Mater. Interfaces* 1(3) pp 653-60

[12] Brunsvold A L, Zhang J, Upadhyaya H P and Minton T K2009 Beam-Surface Scattering Studies of the Individual and Combined Effects of VUV Radiation and Hyperthermal O, O2, or Ar on FEP Teflon Surfaces *ACS Appl. Mater. Interfaces* 1 (1) pp 187-96 Publication Date (Web): 24 November 2008

[13] Tribble A C 2003 *The Space Environment: Implications for Spacecraft Design*, (Princeton Univ. Press, Princeton N.J. USA) p 198; LDEF - The Top Ten Results Jan. -Feb. 1995 *Space Flight Environment International Engineering Newsletter (*Silver Spring MD USA; LDEF Corporation) see Ref. 6; Z. Iskanderova, 2003 Personal communication with J. Visentine, Boing at JSC,.

[14] Silverman E M 1995 *Space Environmental Effects on Spacecraft: LEO Materials Selection Guide* vol.1 and vol.2
[15] Iskanderova Z, Kleiman J, Issoupov V, van Eesbeek M, Rohr T, and Minton T 2009 Comparative Testing, Evaluation and Characterization of Selected Advanced Flight-proven Space Materials *Proceedings of 11th Int. Symposium on Materials in a Space Environment, 11th, Int. Symposium of Materials in a Space Environment, France*
[16] Sheldahl Red Book
- https://www.sheldahl.com/sites/default/files/Documents/ShieldingMaterials/RedBook.pdf

IOP Conf. Series: Materials Science and Engineering 1287 (2023) 012040 doi:10.1088/1757-899X/1287/1/012040

Regolith adherence characterization (RAC) experiment on the Moon and it's ground-based simulation: Materials Issues

Jacob Kleiman[1], Zelina Iskanderova[2], Leon Krishtein[3], Richard Ng[1] and Rana Sodhi[4]

[1] Integrity Testing Laboratory, Inc. 80 Esna Park Drive, #7-9, Markham Ontario, L3R 2R7, Canada; e-mail: jkleiman@itlinc.com
[2] AZ SMART Technology, Toronto, Canada
[3] Kvarc Technologies, Mississauga, Ontario, Canada
[4]Ontario Centre for Characterisation of Advanced Materials (OCCAM), Department of Chemical Engineering and Applied Chemistry, University of Toronto, Ontario, Canada

Abstract. The interaction of Lunar dust and other environmental factors, like vacuum, temperatures, solar radiation, ultraviolet irradiation, and electron (e-) and proton (p+) irradiation with structures on the Moon and the outside of the future Gateway Lunar station may lead to permanent change or complete loss of thermal, optical, and other functionalities that could potentially lead to catastrophic failures. Among these factors, Lunar regolith dust is the most aggressive, causing the main problems.

ITL recently initiated a program to evaluate, further develop, and enhance its unique cornerstone Lunar Dust Mitigation Technology (DMT) for sensitive materials on external space structures. In the framework of this program, ITL prepared a set of DMT-treated samples for inclusion in the "Regolith Adherence Characterization (RAC) Payload" funded by NASA and developed and built by Aegis.

Regolith Adherence Characterization (RAC) Payload mission will determine how lunar regolith dust sticks to a range of space materials and coatings (collected from NASA, academia, and industry) exposed to the Moon's environment at different phases of flight caused by 1) landing, and 2) during routine lander operations.

To understand the results of the Lunar exposure experiment, an extensive program was initiated at ITL where a set of experiments on interaction of Lunar dust simulants with samples, similar to the RAC Payload experiments on the Moon, will be conducted in ITL's Vacuum Lunar Dusty Environment Simulator that will be upgraded for this project.

This paper will discuss mainly the material problems and their preparation for experiments. The detailed description of the Vacuum Lunar Dusty Environment Simulator and the results obtained in it will be published elsewhere in upcoming publications.

1. Introduction

Section 1 will briefly review the problems in Lunar exploration and identify the major active and passive mitigation approaches and introduce the reader to the RAC Payload program. In section 2 we shall provide a short review of our past activities in ground-based simulation of interaction of space materials with Lunar dust simulants. Section 3 will describe ITL's effort to prepare the RAC Payload samples and the ground-based simulation of the experiment. Section 4 will describe the rationale behind selecting the materials for the ground-based RAC-imitating experiments. Due to the relevance and the novelty of the discussed in this paper topics, we decided to keep all the discussed subjects in the paper, despite its larger volume.

1.1 The Space and Dusty Planetary Environment on the Moon

The Moon has been a subject of attention and studies, starting, over 50 years ago, with unmanned Lunar missions - the Soviet Union (SU) Lunokhod-1 and Lunokhod-2, and a number of manned Apollo missions by NASA, USA. These have been among the main achievements of humans in space and planetary explorations in the history of mankind. The Lunokhod-1, controlled from Earth, toured the Lunar Mare Imbrium (Sea of Rains) for 11 Lunar days, the equivalent of 322 Earth days, and travelled more than six miles (10 kilometres) across the Lunar surface. The Lunokhod-2 was actively working on the Moon for 4.5 months, until an accident interrupted the communication with it. Both rovers have been able to survive the combined effects of harsh Lunar space environment and planetary dusty environment, getting results of a huge scientific program, with all the information being transferred to Earth [1,2]. Currently a Chinese rover Yutu-2 is operating on the Moon [3].

The Apollo manned missions, Apollo 11-17, and the enormous amount of effort and achievements are well documented, and even today present a unique source of knowledge about the environment and conditions on the Moon. The Apollo program also exposed the unprecedented problems for people to live and work there. Among the Moon's inhospitable environmental conditions one can mention the very thin exosphere that, practically, is a hard vacuum of 10E-14 atm, the space radiation environment that is close to GEO environment, and the temperature extremes that are in the range of -172 °C to + 125 °C, with 13 thermal cycles/year and the Lunar gravity being 6 times less than on the Earth [4,5].

The Lunar surface is covered by a thick layer of sharp, multi-angular, highly erosive, very sticky dark-greyish or blackish (depending on the location on the Moon) regolith dust, that represents the planetary Lunar environment, developed for millions of years by meteors and meteorites striking the basalt Moon surface. Studies of Lunar regolith dust have been performed on dust samples returned from the Moon by USA, SU and recently China. This dust has special chemical composition and particle size distribution, with a large percentage of very small fraction in it [1]. All attempts by astronauts to clean the dusted external surfaces by brushing were not really successful [4]. The dust was disturbed from the Moon surface by landing and lift-off of the Lunar Landers, by movements of humans or rovers on the surface, by drilling for collection of the basalt samples, etc, i. e. by so called anthropogenic factors and activities [4], and was a subject of the astronaut's major concerns. A representative overview of the problems, caused by Lunar regolith dust, especially to external sensitive space materials and structures can be found in [4-8].

It is important to mention that the longest extra-vehicle activities (EVAs) of Apollo astronauts were only about 8 hours long and that all Apollo mission activities have been performed at the Lunar daytime, that is ~ 13 Earth days long. With the NASA's Artemis program plans of returning to the Moon going forward and the extensive plans for future colonization and exploration of the Moon, the dust problems on the Moon became a subject of major concern and numerous extensive studies, with the goal to find and implement the concepts and options of various passive and active Lunar dust mitigation approaches and technologies [7,8].

The present program at ITL is concentrated on studies of dust interaction with and contamination of the external space materials and structures, as a result of human-related activities on the Moon. Unlike the ground-based simulation of LEO and GEO space environment's influence on materials, where often an acceleration factor is introduced, it is not possible to introduce an acceleration factor in studies of interaction of Lunar dusty environments with materials, since the scenarios of human activities on the Moon surface may be drastically different in regard to the total amount of dust disturbed from the surface and the longevity of such disturbances. All these specifics have been addressed in our dusty, vacuum environment simulation studies, by changing the amount of dust simulants reaching the tested materials surfaces, by changing the direction and nature of dust flows, etc. These and other related issues were, very recently, addressed in the first NASA Standard 1080 that was released in August 2021 [9].

1.2 Current concepts and approaches of passive and active Lunar dust mitigation

In Lunar missions of the 1970's, some, either simple or more sophisticated, ways of protecting the hardware and electronics from the Lunar space/planetary environment have been used, with most of them being rather passive. Thus, both Lunokhod's have been fully covered during night-time periods with very tight covers, with the temperature inside kept at 16-18 °C, to avoid malfunctioning problems of the sensitive equipment [1,2]. Significant radiation protection; traveling on the moon at low speed, to minimise dust disturbance from the surface; placing external sensitive materials, structures and some hardware in safe location on the rovers to reduce the dusting; use of special seals, permanent covers and shields; different dust brushing techniques and cleaning by adhesive tapes - all these approaches were tried with various degrees of success and in many cases proved to be not effective enough [4-6].

In light of the intended long-duration manned and unmanned activities on the Moon, significantly stronger efforts are required to develop new and improve old dust mitigation approaches. The main current concepts and approaches in dust mitigation are, generally, divided into passive mitigation

IOP Conf. Series: Materials Science and Engineering 1287 (2023) 012040 doi:10.1088/1757-899X/1287/1/012040

schemes, without permanent energy consumption on the Moon, and active dust mitigation schemes that consume some energy in the process of mitigation.

Among the active Lunar dust mitigation technologies (DMT) are the electrodynamic dust shield, to prevent the accumulation of dust on surfaces and to remove dust already adhered to them; plasma discharging; utilization of an electron beam for surface cleaning; fluidal and mechanical methods; development of special brushes, etc.

Passive DMT's include changing the surface properties of external materials in contact with the dust, using ion implantation, work function matching coatings; deposition of hard, scratch- and wear-resistant, variable transparency thin coatings with charge dissipative properties; laser patterned material surfaces (one type of, so called, multi-hierarchical surface structures); levelling coatings; ion beam textured surfaces (another type of multi-hierarchical surface structures), etc.

In each of the approaches some of the surface properties are improved or changed to provide the dust mitigation effect. However, in order to achieve a full mitigation effect for dust and preserve all other properties of the surfaces like hardness, charge dissipation, thermal optical properties, etc., a new approach that will use a combined, i. e. passive and active mitigation effort is required. We propose to call such an approach Multifunctional Dust Mitigation Technology (MDMT). It is useful to introduce the definition of a MDMT that provides or retains simultaneously all required external functional properties of the surfaces for long-term activities in Lunar space/planetary environment - like mechanical durability, electrical, thermal optical or optical characteristics, long-term space radiation resistance, etc., in addition to the dust mitigation. And a combined use of passive and active dust mitigation technologies looks like the most realistic and promising approach to the complicated problem of Lunar dust mitigation. NASA released a few Requests for Proposals in this regard at 2021 [10]. In a recent overview [11], the various dust mitigation attempts and trials are discussed in detail.

However, the authors overestimate the expected preventive outcome of any passive dust mitigation. From our own and other experimental testing work, in vacuum and in air, on comparative dusting tests of space materials, both for pristine materials surfaces, and with passive dust mitigation properties, some active approaches like shaking or tapping of the samples or providing additional cleaning either by inert gas blows or ultrasonic vibration are still required after dust accumulation stage for comparative removal of accumulated dust.

In order to evaluate the effectiveness of any passive dust mitigation technology, the difference of the dust retention on both pristine and modified surfaces must be evaluated. This can be explained by the existence of not only one adhesion mechanism, that the mitigation method is targeting, but by presence of a few adhesion mechanisms, especially for "rough", multi-angular regolith dust grains on real surfaces, with real surface morphology and roughness, including the possible interlocking effects on existing surface defects and on non-uniform or porous surfaces. Our approach, developed technology and results of passive dust mitigation will be highlighted later in Section 2.

1.3 The RAC Payload NASA mission to the Moon

In preparation for sending humans to Mars for the first time, NASA's Artemis program will send unmanned and manned missions back to the Moon by 2024. Twelve new science and technology payloads have been selected as part of this program that will help studying the Lunar surface. The selected payloads will go to the Moon on future flights through NASA's Commercial Lunar Payload Services (CLPS) project [12] that allows rapid acquisition of Lunar delivery services for payloads that advance capabilities for cutting-edge innovations in science, exploration, and commercial development of the Moon. The first NASA unmanned Regolith Adherence Characterisation (RAC) payload experiment mission is one of the 12 selected projects and it is scheduled to be sent to the Moon in 2023-2024 [13]. While useful data can be generated using ground, high-tech space environment simulator facilities, nothing can be more informative and convincing than the results from real space flight material testing experiments, especially when dealing with planetary environments.

The RAC experiment on the Moon is planned for exposure of a collection of samples of advanced space materials for potential Lunar operation from NASA Centres, industrial companies and academia. A set of samples, directly exposed to the Lunar environment (Figure 1), will be assessed after the Lander landing and another set of similar samples in conditions of routine operations on the Moon, all during

IOP Conf. Series: Materials Science and Engineering 1287 (2023) 012040 doi:10.1088/1757-899X/1287/1/012040

the Moon daytime. ITL, being the only foreign participant on RAC, provided a number of materials for the experiment.

Utilizing Aegis's (Former Alpha Research company to whom the RAC Payload project was awarded by NASA) flight proven MISSE cameras, lights, stepper motors, computers, and software as the basis for a less than 15 kg payload, an experiment was designed to measure the adhesion of regolith to selected space materials using optical methods. Two round-plate optical material disks (OMD) with a variety of materials and coatings will constitute the RAC payload (Figure 2). OMD1 is exposed to the space environment during the flight to the Moon, the decent and landing on the Lunar surface. It is assumed that landing will create a cloud of regolith and some of the regolith will land on OMD1. OMD2, closed during the flight and landing, will open after 24 hours and will be exposed after the cloud from landing settles, to determine the regolith exposure from the Lunar environment during routine lander operations.

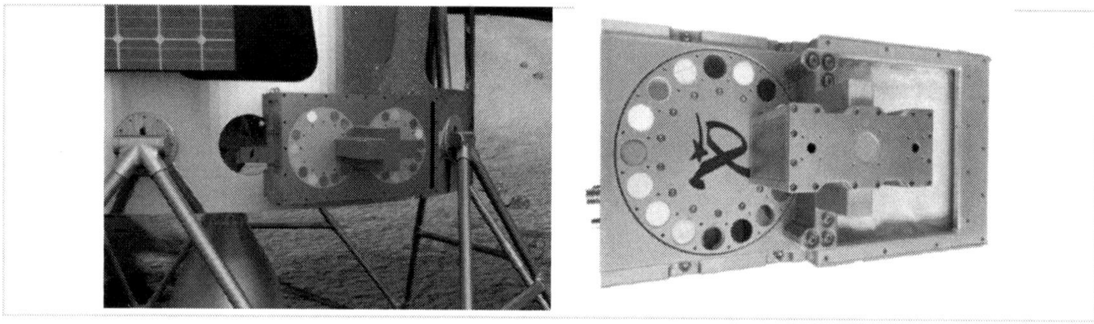

Figure 1 Image of the RAC cassette as integrated on the Intuitive Machines Nova-C Lander (artist view, credit Aegis) after landing (left) and the cassette with the 15 samples that will be exposed to the landing conditions (right). (Credit Aegis).

Each OMD has identical set of materials. Each has a protected light source above and a protected camera system and is mechanized to rotate. In operation, an OMD indexes to each material sample, the light source illuminates the sample, and the light shines and is reflected back in proportion to the accumulated regolith. The OMD then indexes to the next sample, continuing until each sample has been recorded. [12,13].

To understand better, evaluate and explain the results from the RAC Payload experiment on the Moon, ITL initiated a program supported by CSA's LEAP Program [14] to simulate the RAC experiment in a ground-based Dusty Lunar Environment Vacuum Simulator. The experimental set-up, the selection of materials and other aspects of this program will be briefly described in the following sections and in greater details in following publications. The results from this program will be compared with RAC results from the Moon experiment when they will become available.

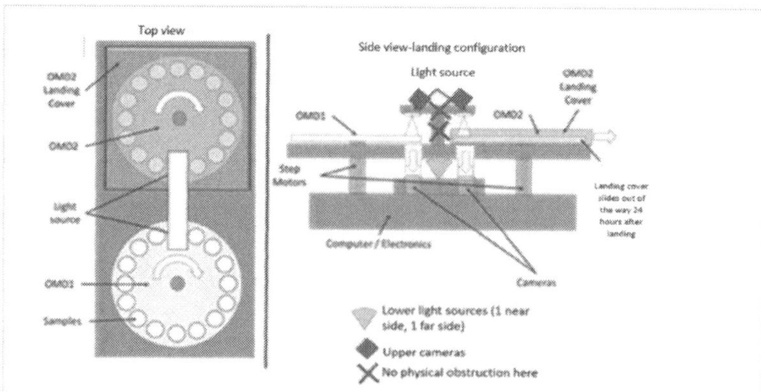

Figure 2 Schematic view of the RAC experiment set-up. (Credit Aegis)

2.0 Ground-based Simulation of Interaction of Space Materials with Lunar Dust Simulants

ITL accumulated extensive experience and knowledge of materials behaviour in LEO, GEO and Lunar environments and the ways of protecting, improving and mitigating the dangers from various factors in these environments. The next few sub-sections will summarize briefly ITL's experience and provide information and rationale on the materials selected for the project.

2.1 Development of Vacuum Lunar Dusty Environment Simulators

ITL Inc., based on previously accumulated experience with design and building of space environment simulators [15], built a series of advanced planetary simulation facilities and associated with them methodologies that addressed the harsh conditions of the Moon in a more efficient way [16-18] (Figure 3). During these projects, a much better understanding of the requirements to dust sources and the interaction of dust with materials was achieved. This experience led to further development of better and more efficient ways of designing the interaction of the dust simulant with samples and more realistic thermal and illumination conditions. A testing protocol was developed at that time that is being used efficiently in all dust exposure experiments [19]. The protocol is being critically assessed presently, in light of the new NASA Standard [9]. An overview of the early activities at ITL in this area can be found in [20].

Based on the feedback from customers developing planetary robotics hardware and other applications, ITL has developed a few basic facility designs to conduct simulated testing in Lunar environment, where dust would be one of the most important environmental factors, to verify the operational capability of different materials, thermal optical surfaces, solar cells and various hardware that may include drive systems, deployment systems, vision systems, mechanical mechanisms, seals, etc.. A dusty Lunar environment vacuum simulator (DLEVAS) at ITL was modified to accommodate the requirements of the RAC Payload experiment. The following modifications were introduced:

a) The improved gravitational and a new horizontal dust source,

b) Special vertical sample holder system to reproduce the conditions of samples in the RAC Payload experiment,

c) The photorecording system, allowing for acquiring and processing the images of the dusty surfaces, similar to the RAC Payload arrangement,

d) The means of inducing and measuring the dust electrical potential and the means of applying the electrical potential to the samples mounted on the sample cassette, if required.

A detailed description of the DLEVAS will be published elsewhere

Figure 3 Images of Dusty Lunar Environment Vacuum Simulators designed and built by ITL.

IOP Conf. Series: Materials Science and Engineering 1287 (2023) 012040 doi:10.1088/1757-899X/1287/1/012040

2.2 Selecting External Sensitive Space Materials for Comparative Testing and Characterization

The first and follow-up studies of interaction of Lunar dust simulants with space materials in vacuum were published for three high-efficiency thermal control space materials, i. e. a white inorganic paint Z93 and two back-metallized Teflon FEP thermal control materials by J. Gaier and his co-workers at NASA GRC [28]. We started our studies of Lunar dust interaction and adhesion by selecting a larger representative set of external sensitive flight-proven space materials, selected from the best manufacturing companies and the world-renown main users worldwide.

From the very beginning it became clear that the black space paints are not a subject of concern, regarding the interaction with dark Lunar dust and dust simulants. Insignificant effects of the dark Lunar dust simulants accumulation have been expected on black paints, that was confirmed experimentally. The same dark regolith dust adhesion and accumulation, however, should and, indeed, significantly influenced the functional performance of white paints and polymer-based TC space materials.

Therefore, for comparative testing, various white space paints originating from a few leading manufacturers in the USA, Europe, and Russia were collected and the accumulation and retention of Lunar dust simulants on them was studied during 2013-2015 period [21,22]. Additionally, a similar study was conducted on pristine and coated space polymer films, space solar cells and solar sell cover glasses [21,22].

Designing the experiments on exposure of samples to dust simulants that are held vertically or horizontally with respect to the dust source in vacuum simulators may present, in many cases, a significant limitation. In real situations, the external sensitive hardware surfaces can be curved, and/or may be positioned at different angles to the incoming dust flux, either by original design or when moving, etc. We designed, therefore, a pyramid-like sample holder, with the samples placed at 45 degrees to the dust flux coming from a gravitational dust source from above (Figure 4). This set-up represents a general case and is accepted in the common materials dust testing standards in air, with the schematic of all forces, applied to the dust grains, presented in [23]. It is emphasized in [23], that both adhesion and dust cohesion forces influence the final results. Two Lunar dust simulants, namely, the JSC-1AF, a bulk Lunar mare regolith simulant developed by NASA, and the OB-1, called Chenobi, with a nano-phase Fe0, developed by CSA, have been used in our testing during the 2013-2015 period (Figure 5).

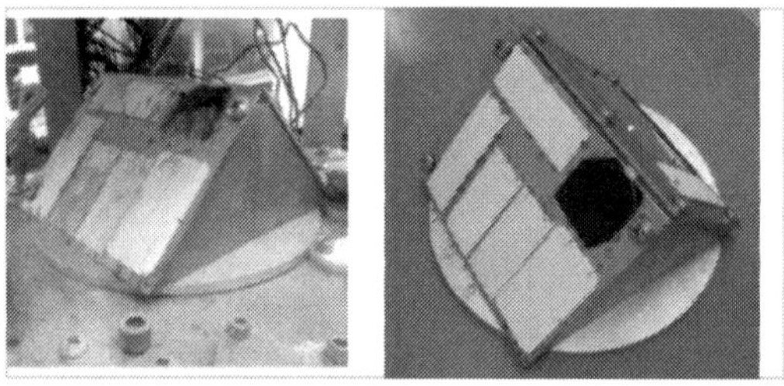

Figure 4 Images of the specially designed pyramid-like sample holder for holding the samples in the DLEVAS system that was used with a vertical gravitational dust source.

Figure 5 Scanning electron microscopy secondary electron images Lunar dust simulants Chenobi and JSC-1F used in 2013-2015 program testing.

We developed a practical test procedure to measure the accumulation and retention of Lunar dust simulants, using the guidelines from the standards of dust adhesion tests on materials in air and considering the experience from the TDS experiment on the Moon in Apollo 14 mission [24].

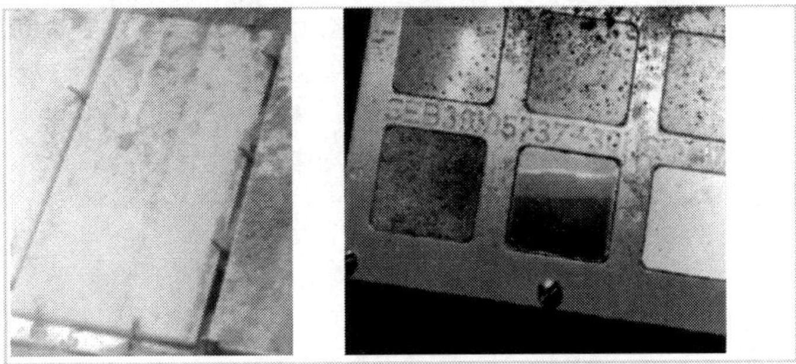

Figure 6 Optical images of ITL's white thermal control painted coupons after an experiment in DLEVAS (left) and, for comparison, images of panels 1 to 6 covered with Lunar dust soil of TDS 1002 flown on Apollo14 Mission [24].

Our procedure consisted of preparation and weighting of the selected samples or structural components and placing then on the pyramid-like sample holder at 45 degrees to the chamber bottom and the dust flux. This configuration allowed us to consider and evaluate both the dust adhesion to sample surfaces and the influence of dust cohesion on the final results of dust adhesion, accumulation and retention. The holder with the samples was placed inside the DLEVAS chamber, the facility was pumped to a vacuum of ~ 10E-6 mbar and samples were left to outgas for a few hours. The selected dust simulant was also loaded into the dust source and baked out. Once the vacuum pressure stabilized after the bake out procedures, the dust source was turned on and the dusting experiment was conducted. After finishing the experiment, the chamber was opened and after a wait of 2 hours, to compensate for sample's mass loss due to outgassing of sample volatiles in vacuum, the samples were removed from the chamber and weighted again (Figure 6), to determine accurately the amount of dust accumulation. Then, inspired by the Apollo14 TDS experiment on the Moon [24], every sample was gently taped (to remove the dust excess), with follow-up repeated weight measurements, to determine the amount of dust retention and retention efficiency on every sample's surface [19].

Exposing the samples to air can, generally, change the dust cohesion [6]. However, for comparative dust adhesion testing of a number of space materials, this phenomenon can be ignored, since the change in dust cohesion, if any, would be the same for all tested materials. The visual observations of samples after taping indicated that only the fine fraction of the dust simulant remained on the surfaces, with the silicate paints being the only exception, where significantly more dust species of various sizes remained on their surfaces. It should be noted that in the described above experiments no special attempts were made either to charge the dust or measure its charge.

A number of external space materials and structures were tested in these studies that included solar arrays elements and thermal control polymer films, advanced space triple-Junction solar cells (EMCORE and AZUR production), space solar cell cover glasses: CMG uncoated (100μ and 200μ) , CMG100/AR, with AR (MgF_2), CMG100/CC/AR, with CC from ITO Conductive (Qioptiq production), advanced white space thermal control paints from USA (mostly from Alion Corp), Russia (Kompozit Corp), and Europe (CNES/MAP) - organic-based, silicone-based, silicate-based: charge dissipative and non-conductive paints, such as white paint samples A276 on aluminium substrates, ECOM-1, ECOM-JC-2, YB-71, Z93C55, DS13N:6N/LO-HP, PSBN, SG121FD; and MAP ATOX K solar reflector (CNES development). Figure 7 shows an example of the sample holder with a number of white paints and a solar cell before and after exposure to a dust simulant in the DLEVAS simulator.

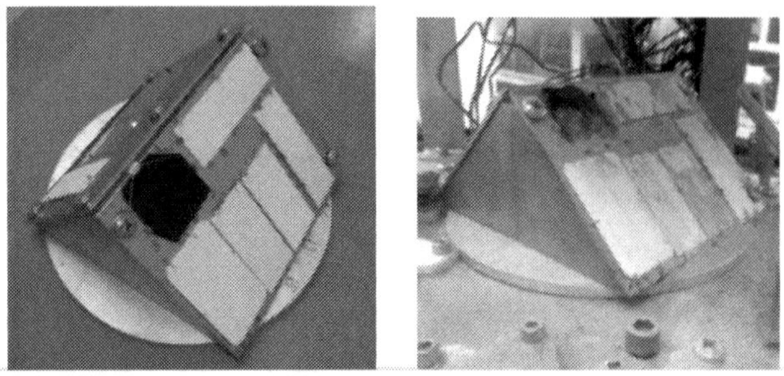

Figure 7 White paints and solar cell samples before (left) and after (right) exposure to Chenobi dust simulant.

A significant number of those paints and polymer films has been previously tested at ITL for durability and performance in LEO and GEO simulated space environments. Protective technologies have been patented for organic paints in LEO applications [25] and polymer films-based space antenna materials in GEO [26,27]. The surface and functional analyses before and after testing were performed that included: Optical Microscopy, Contact Angle Measurements, Scanning Electron Microscopy and Energy Dispersive X-ray microanalysis (SEM/EDS), X-ray Fluorescence Spectroscopy (XRF), and X-ray Photoelectron Spectroscopy (XPS) [14,21,22].

In Lunar applications, in many cases that include but not limited to TC materials, spacesuit surfaces, visors, solar cells and arrays, and space optics, the change of optical and thermal optical properties, due to dust adhesion and accumulation and to mechanical surface damage, such as scratching by sharp angular dust grains, may affect strongly their properties and their functional performance. All such properties should be measured, for comparison, before and after testing.

For comparative testing of materials, the method of calculating the thermal optical (TO) characteristics of dusted samples, based on thermal effects at cooling the dusted samples to cryogenic temperature and heating them up, was used in situ by J. Gaier [28]. However, such an approach would require complicated additional features in the design of the facility.

We used ordinary spectrometers that allowed us to measure the major thermal optical properties, total solar absorptance, α and thermal emittance, ε in accordance with ASTM Standards E408 and E903 [29,30]. To prevent the spectrometers from being contaminated by dust, we did not measure the TO properties right after accumulation, but, instead, conducted the measurements only after taping the samples. Therefore, the conducted measurements of thermal control characteristics had some limitations in this work, indicating only at the changes due to regolith dust retention after taping the dusted samples in air [21, 22]. (Later we developed a special innovative approach to measuring the thermal-optics and/or transmittance on dusted samples both after the dust accumulation and after taping (see Section 3).

We found that, among other factors, the results of dust simulant's adhesion to space materials depend on the surface morphology and roughness that significantly influenced the dust accumulation and retention. These effects, to the best of our knowledge, have not been specifically addressed and/or rarely discussed in previous studies.

In general, we noticed in our testing of different classes of materials in the DLEVAS Simulator that the increase of Lunar dust simulant accumulation and retention (without the dust charge control) was going from less to more as follows: glasses - polymer films – paints; with the highest dust simulant uptake and retention observed on the inorganic silicate paints. The observed inorganic silicate paints behaviour seems to be due to the interlocking effects of dust particles in the porous matrix. This hypothesis was later confirmed in our extensive testing studies of exposure of more silicate paints to high-fidelity Lunar mare simulant LMS-1 that will be described in greater details elsewhere. This observation indicated that despite silicate inorganic paints being radiation and temperature resistant and highly efficient in space applications, and many of them with a significant flight heritage, their use in Lunar applications may be problematic. Experiments with some silicone paints indicated significantly less retention of the dust simulants, compared to silicate paints that looks promising for future work.

However, the dust mitigation for white paints performance on the Moon remains a serious unsolved problem.

2.3 Preliminary Testing of Special Space-qualified Multi-functional Coatings

Early testing and development of special, thin, versatile ion-beam deposited amorphous hydrogenated diamond-like coatings (DLC) like (a:H-DLC) films deposited on space polymers, mostly on KaptonHN, allowed to bring this technology to production and space qualification for space antenna and other space product applications [[27]. The super-coatings demonstrated low temperature dependence of surface resistivity, total RF transmittance (Ka band, 26.5-41 GHz), transparency in the infrared, with the transmittance in the visual depending on the deposition regimes and on the doping elements and conditions, stability in GEO-ground-based simulation tests, a large range of surface resistivity values and an outstanding shelf-life (up to 5-6 years) (Figure 8). Recently, this technology was expanded to some thermal control products on other flexible sensitive substrates, as well as on plastics, quartz, various glass substrates and parts, etc.

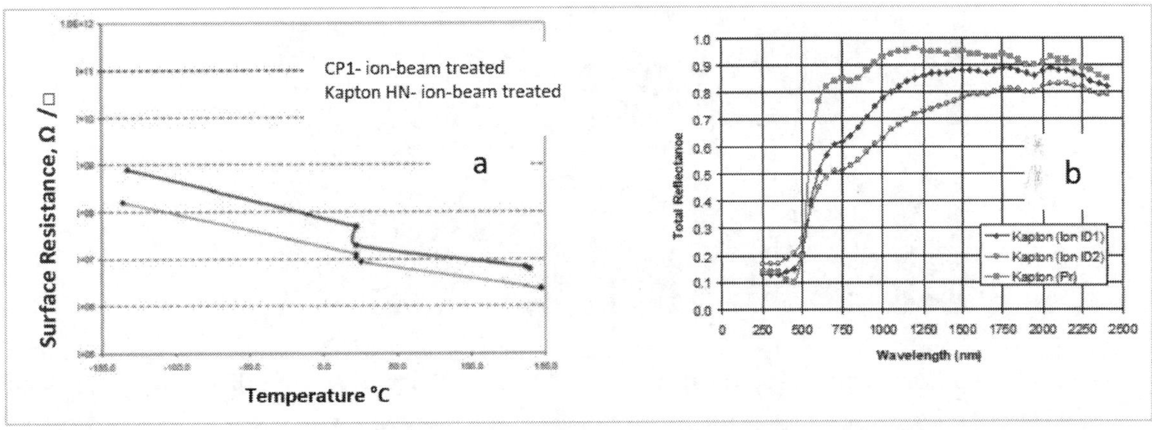

Figure 8 Dependence of surface resistivity SR(T) (a) and Total Reflectance R(W) (b) on the coatings' deposition conditions.

The developed space materials and products with these coating on Kapton HN were extensively characterized and space qualified [26, 27], as well as successfully tested in short- and long-term ground-based GEO environment simulation testing [26,27].

The combination of unique coating properties like the enhanced mechanical protection, adjustable work function and charge dissipation, adjustable in a wide, six orders of magnitude range, looked to us as a good basis for using these coatings as candidates for passive Lunar dust mitigation materials, to prevent the damaging effects of aggressive dust accumulation by reducing the electrostatic adhesion forces. Preliminary testing of the coated polymer films in this study indicated some passive dust mitigation effects, however, it was understood that an extended evaluation of these options requires working with and control of the dust charging and recharging, to verify the degree of influence of electrostatic effects on the dust-surface interaction and adhesion for the coated surfaces.

3.0. Ground-based simulation of the NASA RAC mission to the Moon

3.1 ITL Material's Payload on Lunar RAC Payload Mission

The RAC Payload Program was planned by NASA as a future multi-year program, with more flight opportunities to come up. It was designed with following goals and objectives: to determine how Lunar regolith sticks to a range of materials exposed to the Moon's environment and to determine comparative regolith accumulation rates 1) caused by landing and 2) during routine lander operations - by taking pictures of the samples and sending them back to Earth. The data captured will allow the industry to test, improve, and protect their spacecraft, spacesuits, and habitats from abrasive regolith.

Two types of our innovative materials with expected dust mitigation properties were selected and provided to the RAC Payload experiment. One of the selected samples is a Kapton 500HN space film, treated with our dust mitigation process and technology, providing a thin coating on the surface of the Kapton 500HN that is charge-dissipative, hard and scratch-resistant. The second sample is an innovative static control white space paint with a proprietary fluorinated binder. Figure 9 and Figure 10 show the Kapton 500HN with the coating and the white paint samples, respectively. Figure 11 shows the position of these samples on both RAC Payload cassettes. Figure 12 provides the baseline photos of both sets of ITL samples in the cassettes. The top two photos are of the dust mitigation Kapton-based samples installed in wheel A (A14 location on the wheel) to be exposed the whole mission and wheel B (B14 location on the wheel) to be opened 24 hours after landing. The bottom two photos are of the white paint samples installed in wheel A (A15 location on the wheel) to be exposed the whole mission and wheel B (B15 location on the wheel) to be opened after landing. The images of the LED lights reflected from the samples are seen in the pictures.

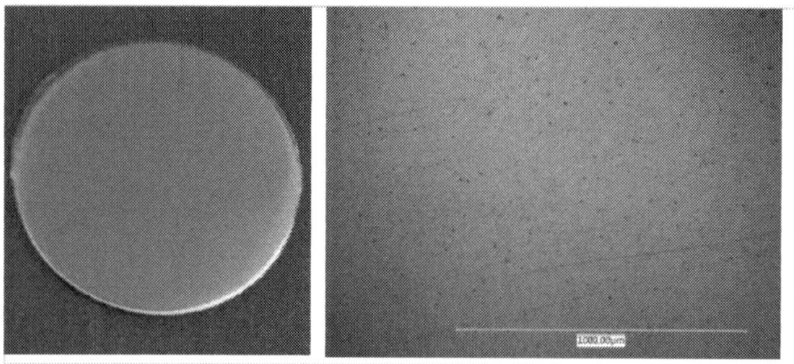

Figure 9 General (left) and surface morphology (right) images of ITL's dust mitigation sample composed of a Kapton 500HN substrate coated with a thin, charge-dissipative, hard, scratch-resistant coating.

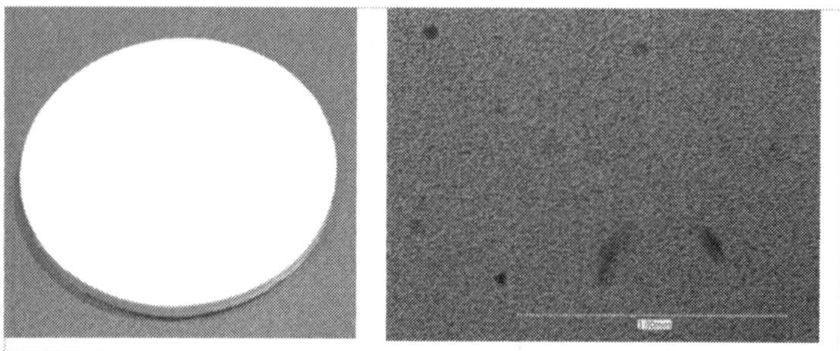

Figure 10 General (left) and surface morphology (right) images of ITL's dust mitigation sample composed innovative static control white space paint with a proprietary fluorinated binder.

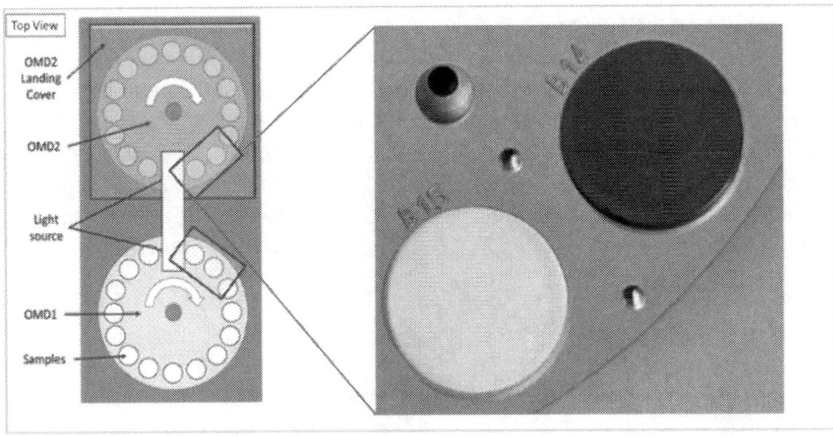

Figure 11 Placement on the RAC cassette of ITL's dust mitigation samples. Right picture shows the expanded view of one of the cassette's locations with two ITL samples installed.

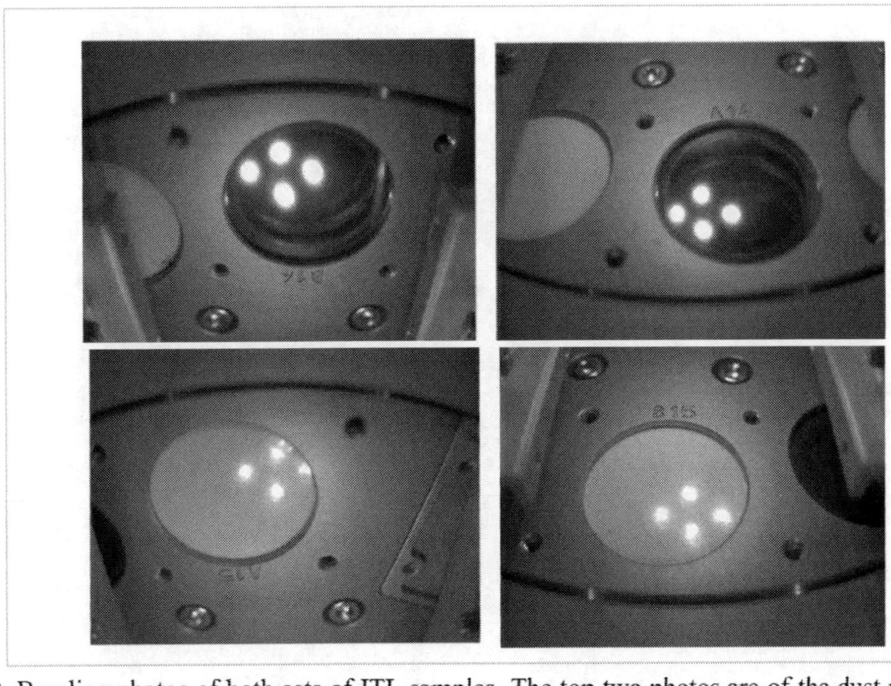

Figure 12. Baseline photos of both sets of ITL samples. The top two photos are of the dust mitigation Kapton-based samples installed in wheel A (A14 location on the wheel) to be exposed the whole mission and wheel B (B14 location on the wheel) to be opened after landing. The bottom two photos are of the white paint samples installed in wheel A (A15 location on the wheel) to be exposed the whole mission and wheel B (B15 location on the wheel) to be opened 24 hours after landing.

3.2 The Ground-based Simulation Experiments

The preliminary studies in the present project of Lunar regolith adhesion and interaction with materials expanded in two directions. In the first one, using the previously developed weight method, studies were conducted in an upgraded vacuum dust simulator facility on further verification of reproducibility of the testing conditions, hardware reliability, variations of dust flux and test duration. The second direction involved the development of prototypes of some real thermal control innovative materials with our super-coatings for space and Lunar exploration, and collection of testing data on various space materials, including the newly developed ones. The high fidelity LMS-1 [33] dust simulant was used in these testing experiments (see section 3.4 for more details on the simulant).

The uptake and retention of the selected regolith simulant on more than 45 main external functional space materials have been preliminary evaluated in the upgraded for the project DLEVAS system. The materials included pristine and coated samples of plastics, pristine and coated samples of quartz, Teflon's with different degrees of surface roughness, to name a few, and Kapton HN with our coatings, undoped and doped by specially selected doping elements. The conducted work and the used protocols for sample preparation and exposure to dust were in agreement with the recent NASA Recommendation of ground-based extensive testing of space materials for Lunar exploration and the Databases of Lunar exploration materials development and testing [34]. Some examples of the materials in this Project are presented in Table 1.

Table 1 Selected list of materials used in the present project

Functionality	Material
Thermal control films	Teflon PTFE - 2mil; Teflon FEP/Ag/Inc- pristine and with our coatings; Teflon FEP/Ag/Inc – textured by specially developed technology; N-DLC/Kapton/Al; Si-DLC/Kapton/Al
Thermal control paints (white)	YB-71; EKOM-1; Alion paints; A-276; Z-93, Z93C55
Charge-dissipating GEO applications	ITO/KaptonHN/Al, (DLC-N)/KaptonHN/Al
Optical	Quartz; Cover glasses for solar cells (CMG100/AR – solar cells cover glass, coated by AR coating of MgF2)
Dust mitigation	Work Function Matching Coating on KaptonHN (Si+O)/ KaptonHN/Al Kapton 100HN; N-DLC/Kapton/Al , Textured surfaces
Fabrics for space suits & inflatables	Ortho fabric (front side) and Ortho fabric (back side)
	Beta cloth; N-DLC/Beta cloth, Ortho fabric, Polyaramid fabric, Gore-Tex,
	Non-woven fabrics
	Kevlar, Nomex
	Aramid fabric
	Gore Tex 604

Additional classes of space materials including advanced space textiles and non-woven fabrics were included as possible candidates for spacesuits, covers and inflatables in these testing experiments. In general, the main trends found earlier, regarding the influence of surface roughness and morphology on the dust retention have been confirmed again. Unexpected results were obtained for Teflon films. While having low surface energy, they accumulated more dust than KaptonHN films, demonstrating also higher retention values. These unexpected results may be caused by the highest tribocharging affinity that Teflon has regarding all other polymer films, with the highest dust adhesion forces, described in [35,36]. It should be noted that the tribocharging of the regolith dust with counterpart surfaces of any moving subjects or objects on the Moon looks as a significant, maybe even, most important, source of the dust charging. These experimental results will be compared to RAC ground-based imitation results that will be obtained in the upgraded for this project simulator that, among other features, will allow to charge the dust and measure it.

A cassette, similar to the RAC cassette shown in Figure 1 and Figure 2, was prepared, with one inch diameter openings to accommodate 15 round samples. We identified a number of important requirements to allow for a successful ground-based simulation of the RAC Payload experiment in the specially upgraded ground-based Vacuum Lunar Dust Simulator. Among them the four most important are as follows:

1. Selection of the material samples for the ground-based experiments, as close as possible, to the actual samples on RAC cassettes and development of appropriate characterization methods.
2. Selection of an appropriate Lunar (mare) dust simulant, with thermal optical characteristics and work function being close to the real Lunar mare regolith dust.
3. To reproduce, as close as possible, the dust flux that will reach the RAC cassette, being disturbed, firstly, by the plume from the rocket engines during the landing, and, secondly, by settlement of dust in the follow up stage of the experiment.
4. To select and use an optical surface imaging method, like in [37], and advanced software for collection and analysis of all the information - pictures, video, etc. - of the pristine and dusted (tested) samples.

The above requirements are further briefly addressed bellow

3.3 Selection and Development of Materials and Characterization Methods for Ground-based RAC Simulation Experiments

3.3.1 Materials. Our efforts to obtain the RAC Payload duplicate samples from RAC Principal Investigators, to include for exposures in our Vacuum Lunar Dusty Simulator, were, unfortunately, unsuccessful. Therefore, using publicly available information, we selected, to the best of our knowledge, from available literature sources, such as [38 – 41] and from corporate and private communications, a number of materials that are representative of the 9 out of the 15 material samples on the RAC payload, with two of them being our own materials.

The selected materials can be divided into two categories, i. e. having surface treatments for potential Lunar dust mitigation, and some already existing space materials with a potential to be used in Moon exploration. Among the latter category is CORIN – an advanced clear polyimide composite material [40], and the Orto fabric - a special woven space fabric from the Company FD Inc (US), used as an outer layer for the NASA spacesuits outside the ISS [41], that is presented on RAC with both sides to be exposed.

Material samples with expected dust mitigation properties that were described in earlier publications [38,39], are represented on the real RAC cassettes by a sample with a Work Function Matching coating (WFMC) [38] and by three space materials, namely, KaptonHN, Teflon FEP, and Ti alloy, from NASA Langley Research Centre that were laser patterned on half of the surfaces to be exposed [39].

Not having the duplicates of these samples, we tried to reproduce the samples of the same three materials with a similar laser patterning. And in the case of WFMC, by depositing one of our super-coatings, with a WF equal to the WF of the LMS-1 simulant, used in these experiments. We, to the best of our knowledge, measured for the first time the WF and the thermal control characteristics of the LMS-1. The WFMC matching samples were prepared by depositing a specially formed Si+O doped DLC-based super-coating on Kapton500HN.

The direct IBD special coatings with various doping elements demonstrated not only an ability to provide a wide range of surface resistivity values, extending over 6 orders of magnitude, but also allowed achieving surfaces with water contact angles ranging from 70 to 92 degrees, i. e. a range of surface energies, and work functions (WF) adjustable in the range of ~3 to ~6.1eV. The measurement of WF was done using the UPS method that is described below. For the ground-based RAC experiments, we selected a sample with WF ~ 5.5 eV, since the WF of the LMS-1 Lunar mare dust simulant has been found to be WF~ 5.52 eV. (Examples of WF measurements for previously used Lunar dust simulants can be found in [42, 43].

In the course of this project, we evaluated a number of potential dust mitigation options and developed a few innovative high efficiency thermal control materials to work on the Moon (see more details in 4.1). We also collected numerous advanced space textile and non-woven fabric samples. Therefore, in the RAC Payload simulation experiments, we used some of these materials to substitute the remaining six samples on the RAC cassette that no information was available to us. The description of these options is presented in Section 4.

3.3.2. Characterization methods used in the RAC simulation experiments: Measurement of Work Function. The work function (WF) was measured using the UPS method. The UPS spectra were obtained using the He lamp attached to a Thermo Fisher Scientific Escalab 250Xi (Thermo Fisher Scientific, East Grinstead UK). The system's calibration was tested by initially taking XPS and UPS spectra of sputter-cleaned Ag. The Ag 3d 5/2 peak binding energy was 368.25 eV with the Fermi Level (FL) at 0.

Work function (WF) for several DLC and doped-DLC coatings were obtained, as well as for the LMS-1 Lunar dust simulant. The latter sample was attached to the sample stub by sprinkling the particles onto C adhesive tape, whereas, for the former, a piece was cut out and attached to the stub. Sample height was determined using the system's video camera focused on the sample surface. For some samples a low count rate was observed which was attributed to charge build up deflecting the photoelectrons. This was alleviated by taping down the edges of the DLC samples with Cu tape.
The spectra were obtained using He (I) (21.22 eV) line at both 0 and -5 eV biases. The latter performed to deconvolute the true sample WF from the internal spectrometer WF and is given by

$$WF = h\upsilon - (E_{cut\text{-}off} - E_f), \quad (2)$$

where hυ is the photon energy (21.22 eV), Ecut-off is the cut-off (corrected for bias) and Ef is the Fermi Level (FL).

3.3.3 Measurement of thermo-optical properties of dusty surfaces. The thermal optical properties of the materials are among the most important characteristics to maintain in the dusty environment of the moon. We adopted the portable spectrophotometers that were developed as part of ITL's project on RAC Characterization (Figure 13) to measure the total solar absorptance, α and the thermal emittance, ε of the samples before and after they were exposed to the dust in the vacuum simulator.

To prevent the dust from contaminating the interiors of the instruments during the measurements, we developed, to the best of our knowledge, a new practical method of solar reflectance and thermal emittance measurement of dust-covered materials and samples. The method employs a specially selected, highly transparent in the wavelength region of interest, adhesive tape with transparent adhesive that is placed over the dusty surface with the adhesive covering the dust particles and securing them in place. The α and the ε measurements then can be performed either by placing the samples on the apertures of the instruments or by positioning the instruments on top of the samples. With some modification to the instruments, this method can be used on the Moon and other dusty planetary environments for direct measurement of dusty surfaces (the Lunar prototypes are being developed now). We compared the measurements for numerous space materials made by the portable spectrophotometers to those made on stationary ones (CARY-5000 system α and Gier-Dunkle system, DB-100 for ε measurements) and confirmed the degree of uncertainty, that has been shown not to exceed ±0.02.

Figure 13 Images of the portable spectrophotometers developed by ITL for measurements of total absorptance, α (the spectrophotometer on the right) and thermal emittance, ε, (the spectrophotometer on the left).

3.4 Selection of Lunar Dust Simulant

Since the landing of the Lunar Lander carrying the RAC payload experiment is planned in the mare region of the Moon, the selection of an appropriate Lunar (mare) dust simulant, with thermal optical characteristics and work function being close to the real Lunar mare regolith dust was very important step. The LMS-1 Lunar mare dust simulant manufactured by Exolith Lab was selected for the ground experiments [33].

Table 2 Basic mineralogy of the LMS-1 simulant [LMS-1].

Component	**Wt.%**
Pyroxene (Bronzite)	32.8
Glass-rich basalt	32.0
Anorthosite	19.8
Olivine	11.1
Ilmenite	4.3

The particle size of the LMS-1 simulant ranges from <0.04 μm to 1000 μm, with the mean particle size reported as 91 μm and the median particle size at 60 μm [33] (Figure 14).

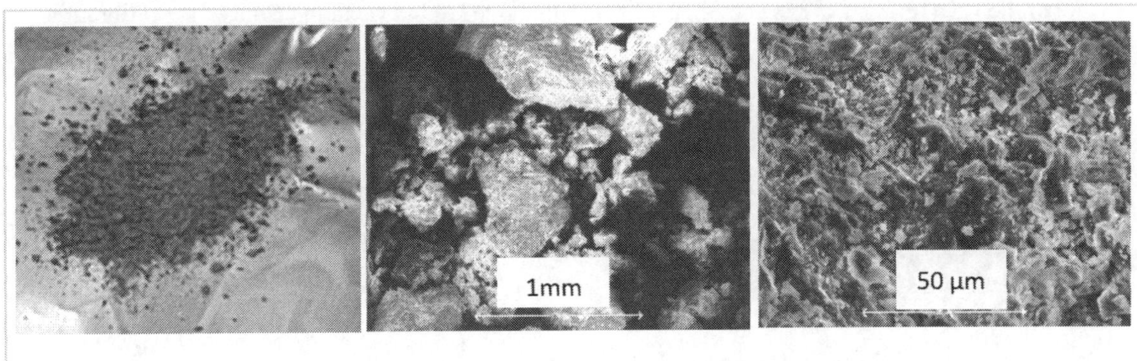

Figure 14 Images of the LMS-1 Lunar dust simulant a) General view; b) and c) scanning electron microscopy (secondary electrons).

To characterize the LMS-1 simulant further, we measured the work function and the thermal optical properties of the simulant (Table 3). The measured value of the WF of LMS-1 simulant is close to the averaged measurements results for WF of other Lunar dust simulants in [42,43]. The thermo-optical characteristics of LMS-1, as measured by our in-house method, are a little bit lower, but close enough to the values of real Lunar mare soil samples, presented in [44]. The WF data and thermal optical characteristics of the LMS-1 confirmed the selection of the LMS-1 as the appropriate Lunar mare dust simulant for our work of simulation on of the RAC Payload experiment on the Moon.

Table 3 Summary of thermal optical and work function values for LMS-1 Lunar dust simulant

Lunar Regolith Simulant	Thermal Optical Properties		WorkFunction,
	Total Asolar absprptance, α	Thermal Emittance, ε	(eV)
LMS-1 [33]	0,86	0,88	5,52

3.5 Selection of Dust Sources

Two different dust sources were designed and will be used in the experiments imitating the RAC Payload experiment on the Moon. The design and performance of these sources will be described in a separate publication that is presently in preparation. Briefly, one dust source (horizontal) will be imitating the arrival of the dust particle at the surfaces of the samples on the RAC Cassette during the landing and the second dust source (vertical) will be used to imitate the settlement of dust in the following days after landing. The main challenge will be in verification of the horizontal dust source and the testing approach to simulate the dust flux reaching the Lunar RAC samples at the Lunar Lander landing. The amount of dust disturbance from Lunar surface by the plume of the Lander's engines and its distribution is a subject of many special studies and modelling and, as written in [9], represent the most difficult case for the simulation of Lunar dust-materials interaction.

3.6 Optical System for Recording the Dust Accumulation on Samples

A special optical camera assembly to measure reflectivity was designed and built to capture the images of the samples positioned vertically in the vacuum Lunar dust simulator. The measurement assembly includes a dust proof housing with a USB b/w camera, white led lighting source, and a laser marker. The brightness of the LEDs and laser are controlled by a micro-controller via a vacuum compatible USB port. To process the acquired images, an image analysis software was developed to calculate the change in the reflectivity of the samples before and after the exposure to the Lunar dust particles. The camera adjusts its signal levels according to the lighting conditions, so, to improve the precision of the measurements, each image to be analysed has 2 areas: a) the area of the laser spot with known intensity, and b) the area of sample's surface with dust particles on it. Change in the reflectivity (i. e. level of dust coverage) is measured in 2 stages. First, the software calculates the ratio between the intensities of the areas (a) and (b) within a single image, and then it compares the calculated ratio among the sequence of images taken as the sample was being exposed to dust. Figure 15 shows the general view of the vacuum-compatible camera that is installed outside the vacuum chamber for calibration.

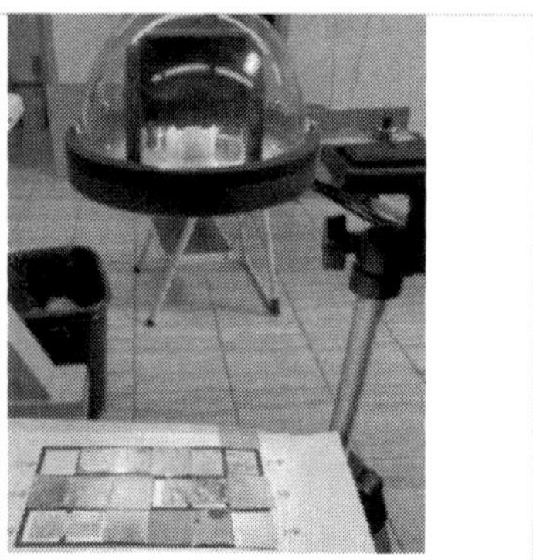

Figure 15 General image of the camera system installed outside the vacuum dust simulator for calibration.

4.0 Selection of Materials for RAC-imitating Ground Testing

4.1 Innovative Thermal Control Materials Development for Space and Lunar Exploration

We explained in the Introduction that for long-term Lunar exploration, the external thermal control space materials should be with static control surface properties and, at the same time, satisfy flexibility requirements and the thermal limits of the Moon environment. The high efficiency materials, like, for example, the ITO/TeflonFEP/Ag/Inconel or ITO/TeflonFEP/Al/Inconel would be highly preferable. However, these products have quite narrow temperature limits, from -60 °C to +70 °C [45] that do not satisfy and are not compatible with the thermal limits of the Lunar environment. Beyond the mentioned above limits, the thermal expansion coefficient's mismatch between the hard and quite brittle ITO and soft TeflonFEP is causing ITO cracking and possible spallation. Even for the widely used ITO/KaptonHN/Al there have been concerns about not enough ITO coating's flexibility when used for curved materials surfaces.

Due to high efficiency and high environmental stability of many current white space inorganic, mostly silicate, paints, using them as thermal reflector materials in Lunar exploration would be the preferred choice. At the same time, we have seen in our numerous space materials testing experiments in the vacuum dusty simulators that the silicate paints, collected from various manufacturers, demonstrated the highest accumulation, retention, and retention efficiency of dust simulants, during and after dusting, due to the porous nature of the silicate binders in the paints. It looks like the dust fills up the pores and stays there, mostly by interlocking, meaning that the use of those white pants on the Moon is at least questionable.

Taking into account the deficiencies discussed above, we extended our work on development of innovative space materials that could replace the ITO/TeflonFEP/Ag/Inconel or ITO/TeflonFEP/Al/Inconel for applications in Lunar environment. Our undoped and specially doped super coatings were deposited first onto front surfaces of back metallized KaptonHN, i. e. KaptonHN/Al (Sheldahl production). Top quality various DLC-like and N-, Si, or Si+O doped DLC thin coatings were deposited at different deposition regimes and substrate temperatures. As a result of this work, uniform, highly flexible, charge-dissipative coatings, with high adhesion, adjustable SR values and with indications of dust mitigation properties have been deposited on KaptonHN/Al samples and then on product prototypes of 0.5m x 1.5m in size that can be used, depending on the type and number of doping elements, for LEO, GEO, or Lunar applications. Some info and data only for N-DLC coating deposition on Kapton/Al in comparison with standard (Sheldahl) ITO/Kapton/Al are presented as an example in Table 4. It can be seen, that the performance efficiency on the innovative thermal control product is even slightly better, than the ITO coated Kapton/Al.

Only a few papers exist in the literature with some examples of DLC successful deposition trials on Teflon PTFE, mostly for medical applications, with no SR values provided there. We also succeeded depositing our coatings on Teflon PTFE films of different thicknesses and with different Teflon PTFE substrate surface finish.

We, however, did not find any data on deposition of DLC coatings on Teflon FEP, that we have been interested in, considering to develop a high efficiency static control thermal control material for potential Moon application. We investigated the possibilities of depositing the DLC coatings on the back-metallized Teflon FEP, which proved to be a real challenge. We developed a technological process of applying a DLC-like coating on Teflon FEP, with SR values in charge dissipative range. The developed technological process required a special innovative surface pre-treatment with follow-up deposition of a thin intermediate layer, on which the specially doped super-coatings were deposited in the required SR range. Data on these materials are also presented in Table 4. The performance efficiency of this product prototype is significantly better, then for the Kapton-based ones.

IOP Conf. Series: Materials Science and Engineering 1287 (2023) 012040 doi:10.1088/1757-899X/1287/1/012040

Table 4 Summary of Thermal Control Materials and Processes for Space Application

Innovative Thermal Control Materials for Space Application	Doping gases	Solar absorptance, (α)	Thermal emittance (ε)	Performance efficiency ξ = (α/ε)	Surface Resistivity (Ohm/sq)
Carbosurf+ technology On Kapton 100HN On Kapton 50E For GEO application and space antennas	N/A	Slightly Changed	Almost unchanged	Slightly Changed	$[3\text{-}4] \times 10^6$ 20×10^6
DLC(N)/KaptonHN/Al	Nitrogen	0.52	0.85	~0.61	10^5 - 10^9 (on request)
KaptonHN/Al, Sheldahl (For comparison)	N/A	0.49	0.81	~0.60	Insulating
ITO/Kapton100/Al; Sheldahl (For comparison)	N/A	0.54	0.81	~0.67	$[2\text{-}10] \times 10^3$
DLC(N)/Teflon FEP/Ag/Inc	Nitrogen	0.23	0.81	~0.28	~ 60×10^6
Textured Teflon FEP/Ag/Inc	N/A	0.12	0.81	0.15	N/A
Photosil –Chemical Process	N/A	Slightly Changed	Slightly Changed	Slightly Changed	N/A

Full surface analysis of all developed innovative materials was conducted, using survey and high-resolution XPS. SR measurements were made right after deposition and then on a monthly basis in lab environment for ~ 8 months (with monitoring still on-going). Abrasion resistance testing and the thermal optical measurements were also performed. Thermal cycling down to Liquid Nitrogen temperatures, with visual appearance and SR measurements have been successfully done on these materials as well. These innovative material samples (Table 4) will be exposed to the LEO space environment in ram and wake directions, depending on the expected environment they will be used, in the upcoming MISSE-17 materials experiment that is scheduled to fly in January 2023.

Another ITL innovative space material product, the Ag back-metallized Teflon FEP, with its front surface textured in the low-energy, high-current ion-beam batch system [46, 47], is already space flight proven, and has a long-time flight and performance history in the LEO environment at the ISS. It was developed to replace the original Ag back-metallized Teflon FEP having mirror-like specular total reflectance properties, with a highly diffuse reflectance surface, while keeping the same total TC characteristics, as the pristine material [46]. Interesting to consider, that in LEO - at ISS, and in GEO - on space antenna sunshields, diffuse reflectance materials are often preferred over materials with specular reflectance properties. It is expected that similar materials will be needed on the Moon as well. In addition, the surface of this textured TeflonFEP is superhydrophobic (water contact angle ~160°), with hierarchical [48] roughness structure (Figure 16). It was suggested in [39] that materials with such surface properties may provide passive dust mitigation properties on the Lunar surface.

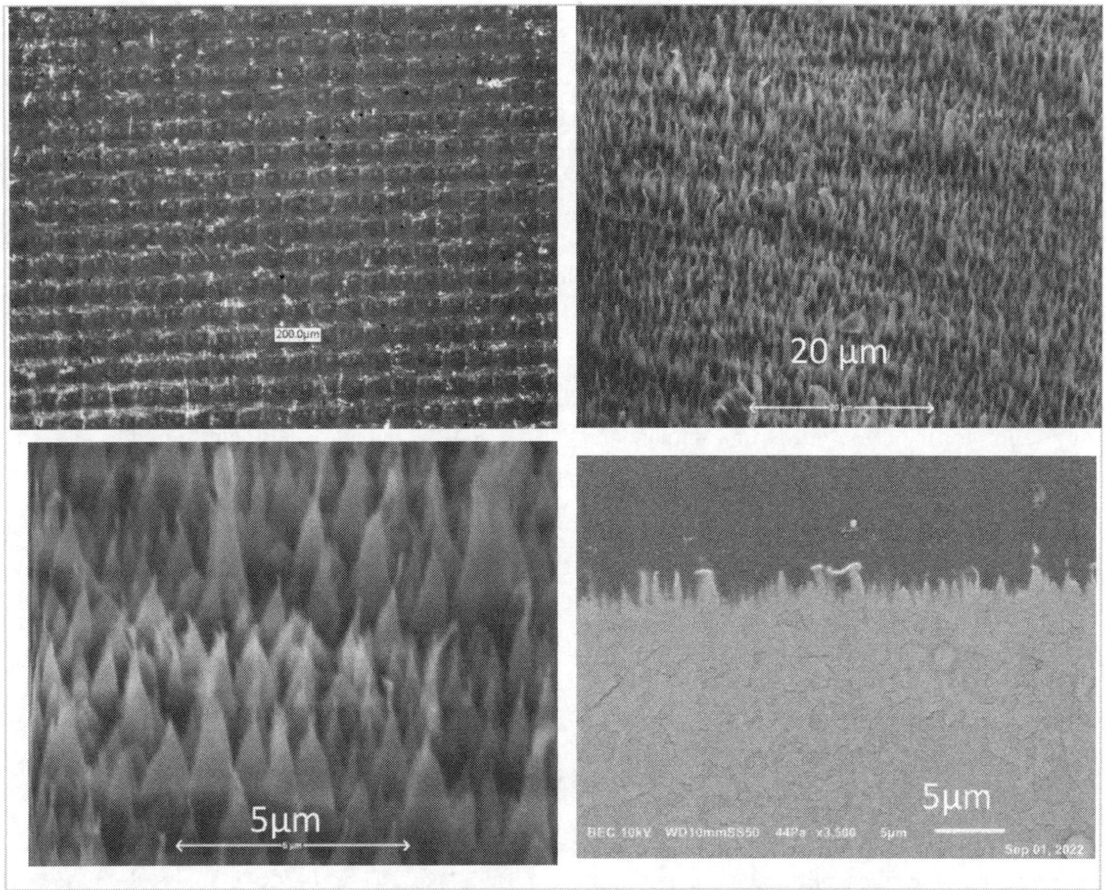

Figure 16 Back metallized Teflon FEP textured in the low-energy high-current ion-beam batch system Superhydrophobic (water contact angle 160°) with hierarchical roughness structure.

For the initial RAC-imitating cassette load, two materials were selected from our innovative materials, as follows:

a. Textured Teflon 500HN/Ag/Inc - for comparison of dust accumulation and adhesion on them;
b. N-DLC/TeflonFEP/Ag/Inc – the innovative high efficiency TC material with our charge dissipative multifunctional super-coating;

A number of other material candidates were selected to fill in the cassette of the RAC-imitating experiment at ITL. Among them we can mention the following:

c. A control Teflon 500HN/Ag/Inc was selected to evaluate the interaction of the LMS-1 simulant with a material surface with a low surface energy but with a very high tribocharging affinity for Teflon, that is known to be the highest among other polymer films.
d. Kevlar space textiles, made of Kevlar 29 and Kevlar 49 - as the main potential components of inflatable materials for space and Lunar exploration,
e. Meta Aramid Nomex textiles, selection from three fabric styles - as potential materials components for dusty environment Lunar spacesuit outer layer;
f. Selection to be made from a few collected non-woven industrial textiles
g. Trials are being planned on transparent levelling special epoxy coatings on Orthofabric (in collaboration with AZ Technology), in order to reduce the Lunar dust accumulation on this space textile. In case of success, those samples may be included in the RAC simulation cassettes.

Lately, it is becoming clear that the passive Lunar dust mitigation options are hard to prove without an active movement of the dusted samples - shaking or tapping them, or trying to blow the dust away with neutral gas pulses, comparing the effects on pristine and surface modified or coated samples. A similar comparison, but using an ultrasound element to remove the dust, was also tried. However, the RAC experiment on the Moon is planned as a passive experiment. Therefore, to extend the experimental program, after making camera pictures of every sample before and after dusting with the Lunar mare dust simulant LMS-1 in the vacuum simulator, we are planning to perform the samples' vibration at various frequency and longevity, making camera pictures afterward, and then trying to blow away the residue dust from the materials' samples. Those actions altogether should help to distinguish the comparative evaluation of adhesion forces and mechanisms for Lunar mare simulant on the imitating RAC cassette samples, the ones reproduced by us, and additional space materials samples, important for Lunar exploration, placed in the RAC imitating cassette.

5 Conclusions

A number of samples with dust mitigation properties were prepared and provided to Aegis Corporation for inclusion in the first unmanned NASA's RAC Payload experiment to be flown to the Moon in 2023-2024. The samples consisted of a Kapton 500HN space film, treated with a potential dust mitigation process and technology, providing a thin hard coating with charge-dissipative, scratch-resistant properties and an innovative static control white space paint with a proprietary fluorinated binder. The coated KaptonHN films are GEO space qualified and space flight proven, and have a flight history in space antennas applications, meaning that they are, practically, space qualified for the whole range of Lunar temperatures and Lunar space environment conditions that are close to those in GEO. The exposure conditions of RAC Payload samples on the Moon will be reproduced, to the best of our abilities in a specially redesigned Lunar Environment Vacuum Simulator. Two different dust sources will be used to reproduce the landing conditions and the stabilized, settled conditions afterwards. A number of important requirements were identified to allow for a successful ground-based simulation of the RAC Payload experiment. Setups were designed to charge the dust and the samples during the experiments and to measure the induced charges.

Since we did not have access to the actual RAC sample duplicates, we selected samples for the ground-based replication of the RAC experiment as close as possible to the RAC materials, based on available in literature information. The selection of other samples was based on our accumulated experience in testing and characterization of passive dust mitigation properties of a large number of pristine and surface-modified materials. The influence of material and surface parameters like charge dissipation, surface work function, surface morphology, multi-hierarchical surface structures, etc. on the adhesion, accumulation and retention of the dust simulants will be studied and compared through optical image matching between the RAC Payload and ground experiments.

The upgraded Lunar Vacuum Dust Simulator will allow evaluation of a proposed concept to combine passive and active Lunar dust mitigation approaches applied to the same samples. To the best of our knowledge, the approach to solve the passive dust mitigation of external materials in a complex way, taking into account all real conditions and requirements affecting the materials on the Moon is offered for the first time.

Acknowledgements

We acknowledge the financial support of Canadian Space Agency (CSA) [20STDPO07]. The acceptance of our samples and help from Aegis Corporation in providing the information on the RAC Payload Lunar mission is acknowledged with thanks. We thank Jean R. Martin from Fabric Development Inc. (USA) for providing a number of modern space textiles samples for testing and evaluation. Our thanks are also going to the ITL staff members that contributed to the previous Lunar dust simulants and space materials testing work and the RAC-related Project, and more specifically, to Dr. S. Horodetsky, Dr. V. Issoupov and Mr. V. Verba that are presently setting up the Lunar Environment Simulator and conducting the RAC-related testing.

References

[1] https://www.nasa.gov/mission_pages/LRO/multimedia/lroimages/lroc-20100318.html,

[2] Gromov V, Kemurdjian A et al 2003 Lunokhod 2 – A Retrospective Glance After 30 years https://www.researchgate.net/publication/234403845_Lunokhod_2_-_A_retrospective_Glance_after_30_Years

[3] https://www.planetary.org/space-missions/change-4 .

[4] Gaier J R 2005 The Effects of Lunar Dust on EVA Systems During the Apollo Missions *Nasa/TM-2005- 213610*,.

[5] Creel R A 2011 Applying Thermal Control Experiences on Apollo Lunar Rover Project to Rovers for Future Space Exploration, Space Systems Engineer Member of the Apollo Lunar Roving Vehicle Team 3rd AIAA Atmospheric Space Environments Conference http://www.hq.nasa.gov/alsj/lrv_thermal_alsj.pdf

[6] Apollo and America's Moon Landing Program - Lunar Dust and Astronaut Spacesuit Contamination, Lessons from the Apollo Moonwalkers 2011(By Progressive Management) https://www.kobo.com/ca/en/ebook/apollo-and-america-s-moon-landing-program-Lunar-dust-and-astronaut-spacesuit-contamination-lessons-from-the-apollo-moonwalkers-evaluation-of-health-risks-to-future-Lunar-explorers

[7] Lunar sourcebook: A user's guide to the moon 1991 (Cambridge University Press, England)

[8] Levine J S, Winterhalter D and Kerschmann R L 2020 The Impact of Lunar Dust on Human Exploration (Cambridge Scholar Publishing, Texas)

[9] NASA-STD-1008, Classifications and Requirements for Testing Systems and Hardware to Be Exposed to Dust in Planetary Environments, https://standards.nasa.gov/sites/default/files/standards/NASA/Baseline/0/2021-08-21_nasa-std-1008-approved.pdf

[10] Active and Passive Dust Mitigation Surfaces (SBIR) https://www.sbir.gov/node/1657747

[11] Budzyń D, Tuohy E at al 2022 Lunar Dust: Its Impact on Hardware and Mitigation Technologies, Proc. of the 46th Aerospace Mechanisms Symposium, Virtual, May 11-13 https://esmats.eu/amspapers/pastpapers/pdfs/2022/budzyn.pdf

[12] Regolith Adherence Characterization (RAC) Payload, AEGIS Publication https://www.nasa.gov/planetarymissions/LSITP.html; Commercial Lunar Payload Services (CLPS)

[13] Goode A Regolith 2021 Adherence Characterization (RAC) and the CLPS payload process https://lsic.jhuapl.edu/uploadedDocs/focus-files/983-DM%20Monthly%20Meeting%20-%202021%2009%20September_Presentation%20-%20Aegis%20Aerospace.pdf.

[14] Kleiman J and Iskanderova Z 2021 Innovative Lunar dust mitigation technologies: performance evaluation on the moon and in ground simulators Canada Reaches for the Moon - 2021: Canadian Lunar Workshop, June 14-16 (CLW-ALC 2021)

[15] Kleiman J, Horodetsky S, Sergeyev V, Issoupov V 2006 Critical Review of The Design of Space Environment Simulators: Lessons Learnt *Proc. of the 10th Int. Symp. on Materials in a Space Environment / 8th Int. Conf. on Protection of Materials and Structure from the LEO Space Environment* (Collioure France) June 19-23

[16] Horodetsky S, Kleiman J, Issoupov V 2010 A Planetary Environmental Simulator/test Facility *Proc. of the 26th Space Simulation Conference (on CD)* (Double Tree Hotel, Annapolis, Maryland), October 18 – 21

[17] Kleiman J, Horodetsky S and Issoupov V 2011 A Planetary Environmental Simulator/test Facility *Proc. of the 10th International Conference on Protection of Materials from Space Environment – ICPMSE-10J*, eds. J. Kleiman, M. Tagawa and Y. Kimoto (Springer Publishers) pp 355-370

[18] Kleiman J, Horodetsky S and Issoupov V 2012 A Planetary Environmental Simulator/test Facility *Proc. of 12th European Conference on Spacecraft Structures, Materials & Environmental Testing,* March 20-23 (Noordwjik The Netherlands)

[19] Materials samples Lunar dust simulants vacuum testing Protocol in Baseline Lunar Dust Simulant Vacuum Testing for Large Surface UUTS 2014 *Integrity Testing Laboratory Report ITL-012914-14027-2M* Submitted under Subcontract #26340 Dust Mitigation Technologies (Moon, Mars) Contract SC-26340/922250)

[20] Kleiman J, Horodetsky S, Issoupov V and Verba V 2014 New Challenges in the Ground-based Technology for Simulated Planetary Environmental Testing Proc. *28th Space Simulation Conference*, November 3-6, 2014 (Westin-BWI in Linthicum, Maryland) pp 213-229,.

[21] Baseline Lunar Dust Simulant Vacuum Testing for Large Surface UUTS 2014 *Integrity Testing Laboratory Report ITL-012914-14027-2M* (Contract SC-26340/922250)

[22] Iskanderova Z, Kleiman J et al 2015 Comparative testing and evaluation of vulnerable external infrastructure materials and components for Lunar applications Presented at ISMSE-13, 22-26 June, (Pau France)

[23] Lee L-H Adhesion and cohesion mechanisms of Lunar dust on the moon's surface 1995 *Journal of Adhesion Science and Technology* 9:8, pp 1103-24

[24] Apollo 14 Preliminary science report 1971 NASA SP 272 TDS Experiment https://www.hq.nasa.gov/alsj/a14/as14psr.pdf pp. 244 – 247

[25] Gudimenko Yu, Kleiman J I, Iskanderova Z A, Tennyson R C and Cool G R 1999 Modification of Subsurface Region of Polymers and Carbon-Based Materials U.S. Patent #5,948,484 issued September 7, 1999

[26] Nömayr Ch, Zimmermann C, Iskanderova Z A and Kleiman J 2015 Method for manufacturing a charge dissipative surface layer European Patent EP 2876649 A1 May 27, 2015

[27] Iskanderova Z, Kleiman J and Bussieres F 2013 Method of making charge dissipative surfaces of polymeric materials with low temperature dependence of surface resistivity and low RF loss US Patent 9174396B2 (2013- 07-03)

[28] Gaier J at al , 2008 The Effect of Simulated Lunar Dust on the Absorptivity, Emissivity, and Operating Temperature on AZ–93 and Ag/FEP Thermal Control Surfaces, NASA/TM—2008-215492 https://ntrs.nasa.gov/api/citations/20090005996/downloads/20090005996.pdf

[29] ASTM Standard E408 – 13 (Reapproved 2019) Standard Test Methods for Total Normal Emittance of Surfaces Using Inspection-Meter Techniques

[30] ASTM Standard E903 – 12, Standard Test Method for Solar Absorptance, Reflectance, and Transmittance of Materials Using Integrating Spheres

[31] Zimon A D 1982 *Adhesion of dust and powders*, Chapter V: "Adhesion of variously shaped particles to rough surfaces", (Consultants Bureau, New York)

[32] Gent A N and Lin C-W 1990 Model Studies of the Effect of Surface Roughness and Mechanical Interlocking on Adhesion, *Journal of Adhesion* 32 pp 113-125

[33] https://exolithsimulants.com/products/lms-1-Lunar-mare-simulant#:~:text=The%20LMS%2D1%20LuLunar Mare (LMS-1) High-Fidelity Moon Dirt Simulantnar%20Mare,mare%20location%20on%20the%20Moon.

[34] Winterhalter D, Levine J S, Kerschmann R L and Brady T K 2020 Lunar Dust and Its Impact on Human Exploration: A NASA Engineering and Safety Center (NESC) Workshop, *NASA/TM−2020-5008219 NESC-RP-19-01469*

[35] Berkebile S P, Gaier G R 2012 Adhesion in a Vacuum Environment and its Implications for Dust Mitigation Techniques on Airless Bodies *NASA/TM—2012-217723*

[36] Lee B W and Orr D E 2023 The Triboelectric Effects Series AlphaLab, Inc. https://www.alphalabinc.com/triboelectric-series/

[37] Benkovski J I, Luedeman W I, Teehan J O, Hörst S M, He Ch and Lorenza R D 2019 Experimental investigation of surface adhesion of Titan analog materials: Mitigation by dust-repellent coatings *Planetary and Space Science*, Vol. 179 December 104721

[38] Gaier J R, Waters D L, Misconin R M, Banks B A and Crowder M 2011 Evaluation of Surface Modification as a Lunar Dust Mitigation Strategy for Thermal Control Surfaces, 41st International

Conference on Environmental Systems
https://scholar.google.ca/scholar?q=Evaluation+of+Surface+Modification+as+a+Lunar+Dust+Mitigation+Strategy+for+Thermal+Control+Surfaces,&hl=en&as_sdt=0&as_vis=1&oi=scholart

[39] Das L, Wohl Ch at al 2022 Evaluation of Materials and Surfaces for Lunar Regolith Adherence Characterization (RAC) Payload Samples, AIAA 2022-1694 Session: Dust Session-Lunar Dust and Mitigation Strategies

[40] Hybrid Plastics CORIN *https://www.knowde.com/stores/hybrid-plastics/products/hybrid-plastics-corin/*

[41] Ortho fabric, Fabric Development Inc, USA, https://www.fabricdevelopment.com/en

[42] Nouzák L, Richterová I, Pavlu J, Nemecek Z and Šafránková J April 2016 Investigations of Photoemission From Lunar Dust Simulant *IEEE Transactions on Plasma Science* Vol 44 No. 4 pp 512-18

[43] Sternovsky Z and Robertson S 2002 Contact charging of Lunar and Martian dust simulants, *J. of Geophysical Research,* Vol. 107, No. E11, 5105 pp 15-1-15-8 *https://www.nasa.gov/centers/johnson/pdf/486012main_Sternovsky.pdf.*

[44] Gaier J at al 2011 Thermal Optical Properties of Lunar Dust Simulants and Their Constituents *NASA/TM—2011-217232*

[45] Sheldahl Red Book - *https://www.sheldahl.com/sites/default/files/Documents/ShieldingMaterials/RedBook.pdf* -

[46] Kleiman J I, Popov O, Tong A and Molenda D 1999 Development of High Diffuse Reflectance Surfaces on Teflon *Proc. Third International Space Conference on Protection of Materials and Structures from the LEO Space Environment* ICPMSE-3, Toronto, 25-26 April, 1996 eds. J.I. Kleiman and R.C. Tennyson (Kluwer Academic Publishers Dordrecht The Netherlands) pp.167 -178

[47] Kleiman J I, Iskanderova Z and Antoniazzi J 2009 Long Term Flight Performance of High Diffuse Reflectance Ag/Teflon Coverings Flown on Canadian Mobile Servicing Station, MSS, Proc.11th ISMSE-11 (Aix-en-Provence, France, CNES-ESA-ONERA,).

[48] https://en.wikipedia.org/wiki/Ultrahydrophobicity#:~:text

[49] Destefanis R, Amerio E, Briccarello M, Belluco M, Faraud M, Tracino M and Lobascio C Space Environment Characterisation of Kevlar®: Good for Bullets, Debris and Radiation Too, *http://esmat.esa.int/materials_news/isme09/pdf/3-ground/s3%20-%20destefanis.pdf* pp 1-9

[50] Destefanis R, et al 2006 Selecting Materials to Protect Inflatable Structures from the Space Environment, *Proc. of the 10th ISMSE & the 8th ICPMSE* (Collioure, France), 19-23 June 2006 (SP-616, September 2006) http://esmat.esa.int/Materials_News/SP-616/session_01/S1_6_Destefanisrev.pdf

IOP Conf. Series: Materials Science and Engineering 1287 (2023) 012041 doi:10.1088/1757-899X/1287/1/012041

Ariane 5 Fairing Preparations for the James Webb Space Telescope (JWST)

E M Stewart[1,4], E M Wooldridge[1], O Schmeitzky[2], J Bonhomme[3], M Madsen[3]

[1] NASA Goddard Space Flight Center, Greenbelt, MD 20771 USA
[2] ESTEC Keplerlaan 1, 2201 AZ Noordwijk, Netherlands
[3] Beyond Gravity Schweiz AG Schaffhauserstrasse 580, 8052 Zürich, Switzerland
[4] Correspondence can be sent to elaine.m.stewart@nasa.gov

Abstract. The James Webb Space Telescope (JWST) is a large, infrared space telescope operating at Lagrange point 2. JWST is a joint effort between NASA, ESA, and CSA and was launched from the Centre Spatial Guyanais (CSG) on an Ariane 5 rocket in December 2021. The Ariane 5 payload faring (PLF), interior and Vehicle Equipment Bay membranes received multiple cleanings, detailed inspections, and verification sampling to achieve necessary cleanliness levels.

Investigation of PLF cleanliness began early in the JWST program to test the Fairing Acoustic Protection (FAP) panels. Initial PLF cleaning was performed at RUAG in Switzerland prior to shipment to the launch site in collaboration with NASA and ESA. The cleaning procedure used at RUAG premises was based on heritage of the PLF cleaning procedure for ESA/Herschel-Planck and NASA/Mars2020 missions on an Atlas V launcher. The Ariane 5 fairing was packaged in a clean manner and inspected and cleaned upon arrival to CSG. Silicon wafers and particle fall out (PFO) plates were used to monitor the internal fairing cleanliness prior to encapsulation. Just after encapsulation, PFO plates and calcium fluoride (CaF2) crystals were placed inside the fairing on a specific tool door for monitoring the cleanliness inside the fairing during the ventilation phase. The fairing itself was specially sealed to protect the inner environment with just a small, doored porthole accessible via diving board for final closeout of the purge interface. Fairing cleanliness was achieved and mitigated contamination distribution upon launch.

IOP Conf. Series: Materials Science and Engineering 1287 (2023) 012041 doi:10.1088/1757-899X/1287/1/012041

1. Introduction

The James Webb Space Telescope (JWST) has four main science objectives to investigate the first light, galaxy change over time, star cycles, and formation of planets as shown in Figure 1.

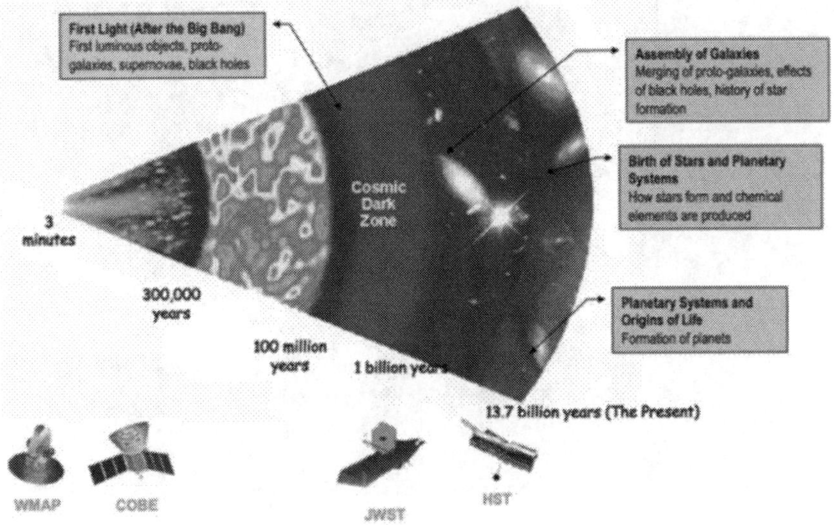

Figure 1. JWST science themes. Credit: NASA

To detect first light, a primary mirror of 6.5 meters observing in the infrared (IR) wavelength range of 0.6 μm to 29 μm was required. To operate at cryogenic temperatures for IR observations, JWST optics needed to be very clean, stringent particulate contamination requirements were set based on optical stray light and throughput analyses. With 30-50 K optics temperatures, there is also sensitivity to molecular and water/ice redistribution during thermal vacuum chamber (TVac) and on-orbit cool-down. To mitigate stray light, the strict particulate requirements were necessary. Because JWST has an open architecture, as shown in Figure 2, the sensitive optical surfaces were exposed for long periods of time throughout assembly, integration, and test (AI&T), during the launch campaign, and finally for launch.

Figure 2. Exposed primary mirror optics during I&T.
Credit: NASA/Chris Gunn

Small particles and Non-Volatile Residue (NVR) are critical to stray light & transmission. In addition, the (NIRSpec) instrument's microshutter array was discovered sensitive to fibers during environmental testing, so it was imperative to implement fiber-free requirements to the extent possible. The other JWST instruments are Mid-Infrared Instrument (MIRI), Near-Infrared Camera (NIRCam), and Near-Infrared Imager and Slitless Spectrograph/Fine Guidance Sensor (NIRISS/FGS).

The contamination requirements for launch campaign through contamination and collision avoidance maneuver (CCAM) were 0.45 Percent Area Coverage (PAC), 4500 ppm and 0.50 mg/.1m^2 NVR (50 angstroms). To meet these requirements, flight hardware was cleaned throughout I&T, the primary mirror was cleaned twice, I&T was conducted in ISO Class 7 or better clean areas with enforced ISO Class 6 cleanroom protocols, extensive in-situ contamination monitoring was performed from mirror segment delivery through fairing encapsulation, and water deposition was controlled through environmental controls monitoring T & RH. Additionally, the Ariane 5 fairing requirements included cleaning with white/ultra violet (UV) light inspections.

2. Pathfinders: Fairing Acoustic Protection (FAP) testing and Herschel-Planck (HP) cleaning and Automatic Transfer Vehicle (ATV)

2.1. ATV-1 Launch Campaign 2008

During the ATV-1 launch campaign, FAP panels met standard visibly clean (VC) at 1 m but not sensitive VC at 30 cm. The requirements were met for ATV-1, but further cleanliness requirements and implementation required for JWST.

IOP Conf. Series: Materials Science and Engineering 1287 (2023) 012041 doi:10.1088/1757-899X/1287/1/012041

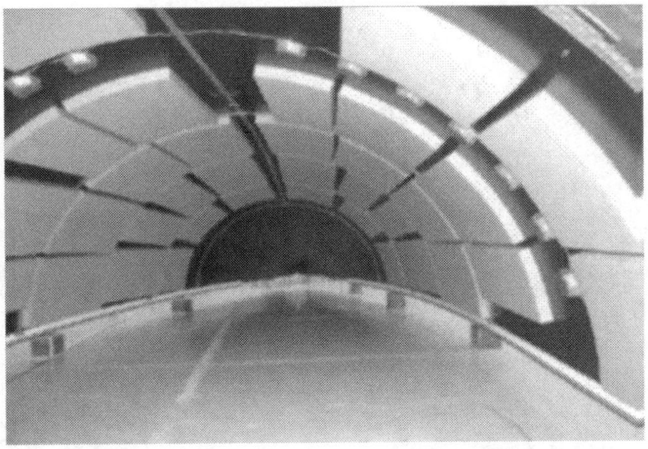

Figure 3. ATV-1 FAP. Credit: ESA

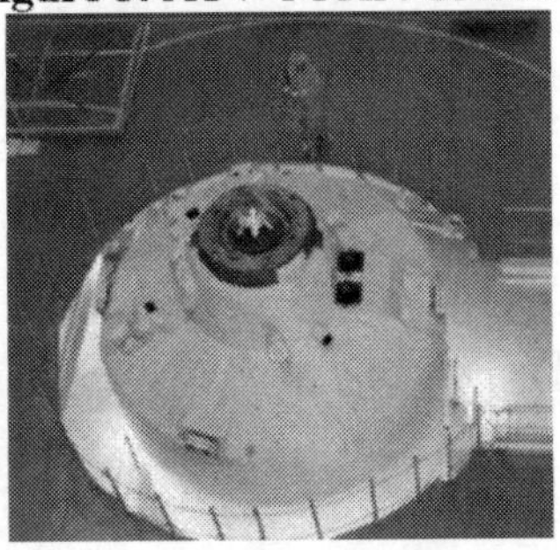

Figure 4. Mating surface for ATV-1 integration on top of launcher. Credit: ESA

Figure 5. ATV-1 launcher accessed from platforms. Credit: ESA

2.2. Heritage from HP telescopes launch 2009 ESTEC FAP Testing

In preparation of the HP launch, A FAP panel was exposed to vibration and depressurization test for assessing of the PAC release. During the test, about 120 ppm of particulate was released by the FAP panel with a view factor 1/1.

Figure 6. Configuration for vibrational testing of FAP panel. Credit: ESA

Figure 7. Acoustic testing of FAP panel. Credit: ESA

Figure 8. Configuration for depressurization testing. Credit: ESA

2.3. Heritage from HP telescopes launch 2009 GSFC FAP Testing

In conjunction with the testing performed at ESTEC, further testing occurred at GSFC. A FAP panel exposed to vibration& depressurization tests, averaged over samples, released:

- 324 ppm (0.0324 PAC) on horizontal samples
- 148 ppm (0.0148 PAC) on vertical samples

The silicon wafer witness samples were configured as shown in Figures 9, 10 and 11 in order to measure particulate fallout during acoustic testing of the panel, and then during rapid depressurization of the same panel. These tests were designed to simulate the launch conditions for HP on the Ariane 5.

Figure 9. Silicon wafers used to monitor FAP panel fallout during acoustic testing. Credit: NASA

IOP Conf. Series: Materials Science and Engineering 1287 (2023) 012041 doi:10.1088/1757-899X/1287/1/012041

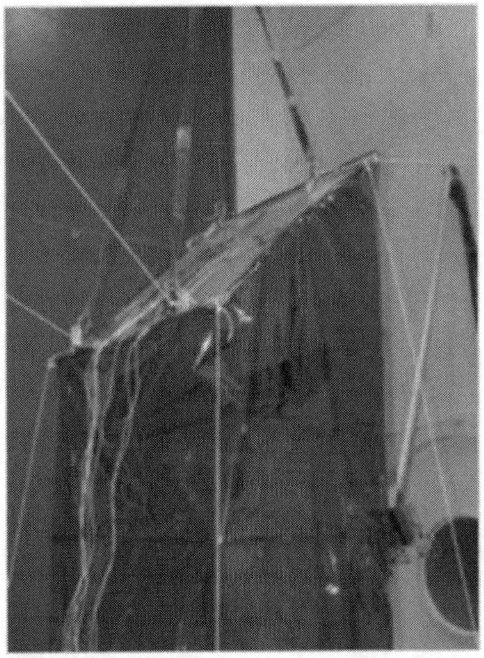

Figure 10. Configuration for acoustic testing of the FAP panel. Credit: NASA

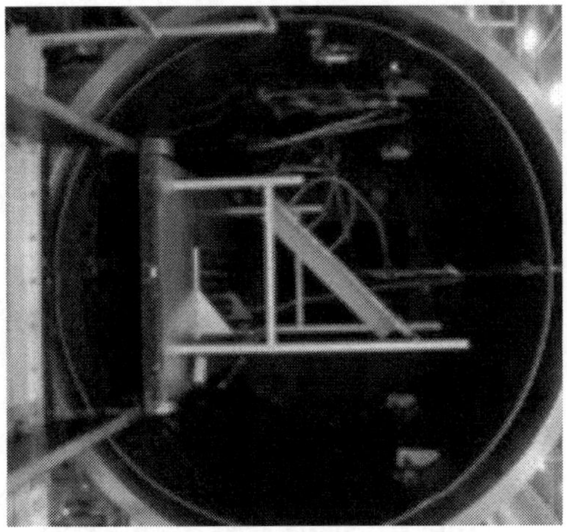

Figure 11. Configuration for rapid depressurization of the FAP panel. Credit: NASA

2.4. Heritage from HP telescope launch in 2009

Before shipment of the fairing to Kourou, a cleanliness inspection with white light for VC highly sensitive criteria was performed. Additional mitigations were performed including: further cleaning of the strongbacks, a dedicated clean blanket (double bagging) and specific HEPA filter implemented on the protective layer.

Figure 12. Bagging over HP fairing for shipment. Credit: ESA-O. Schmeitzky

Figure 13. FAP with bagging integration for protection during shipment. Credit: ESA-O. Schmeitzky

Figure 14. Integration of bagging over HP fairing for shipment. ESA- O. Schmeitzky

HP requirements were met after fine cleaning was performed on the upper part of the fairing with the VC sensitive level but not VCHS+UV required by JWST.

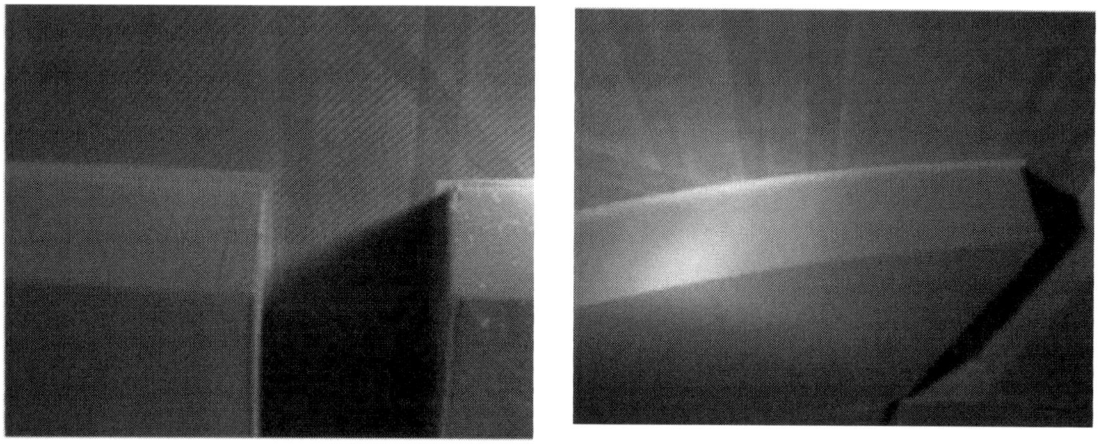

Figure 15. Particulate identified on HP fairing under UV. Credit: ESA

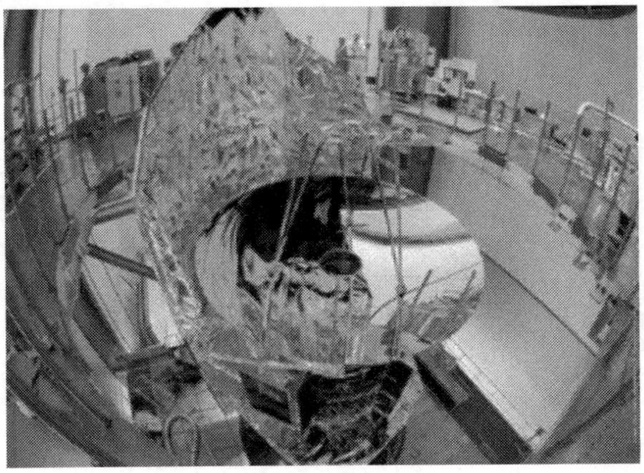

Figure 16. HP integration to Ariane 5 launcher. Credit: ESA

Figure 17. Fairing integration for HP. Credit: ESA

3. Ariane 5 modifications and cleaning at beyond gravity

3.1. Handling point covers

To avoid generating particles and foreign object debris (FOD) during payload fairing (PLF) handling in Kourou, special covers were installed. These covers are installed on the "PLF Handling points" which are used for transportation and encapsulation with different tooling. The covers prevented FOD from entering the PLF when the handling tools are installed.

Figure 18. Filters placed over handling points to block metal particles from coming into the fairing. Credit: Beyondgravity

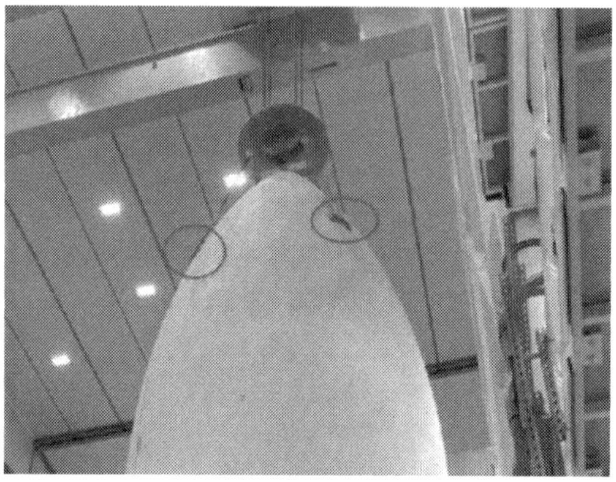

Figure 19. Locations of handling points on fairing for Webb. Credit: Beyondgravity

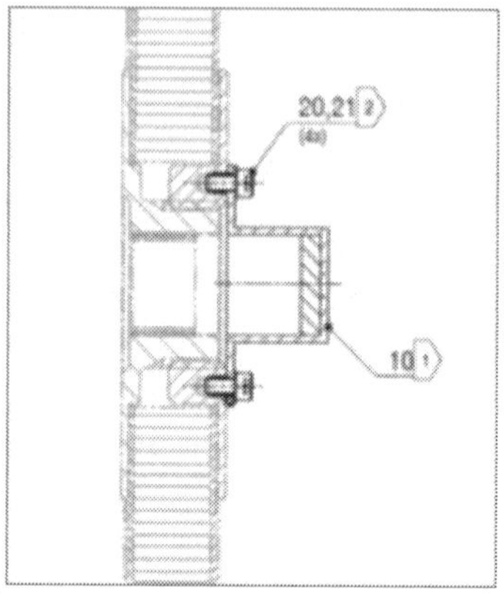

Figure 20. Design of handling points. Credit: Beyondgravity

IOP Conf. Series: Materials Science and Engineering 1287 (2023) 012041 doi:10.1088/1757-899X/1287/1/012041

3.2 Sealing of sandwich panels

The PLF sandwich panels were sealed with adhesive to reduce the residual pressure at PLF separation. Adhesive sealing also helped to maintain the cleanliness of the payload volume by avoiding ingress of air coming from honeycomb venting.

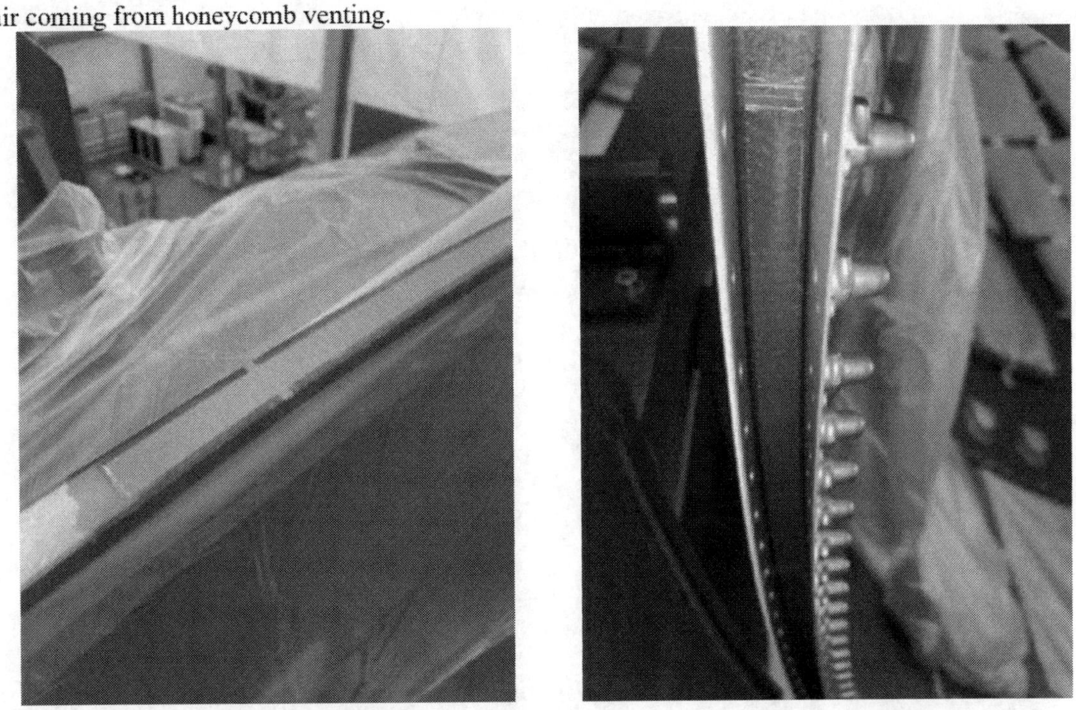

Figure 21. Process of sealing sandwich panels with adhesive. Credit: Beyondgravity

IOP Conf. Series: Materials Science and Engineering 1287 (2023) 012041 doi:10.1088/1757-899X/1287/1/012041

3.3 Improved cleaning of payload fairing (PLF)

Mars 2020 Perseverance mission required cleaning protocols that served as a basis for JWST PLF. Any personnel working on PLF wore cleanroom garments to mitigate contamination. Any facility the PLF was exposed in was cleaned and verified. In addition to the standard white light inspection, a full inspection with UV lights was performed.

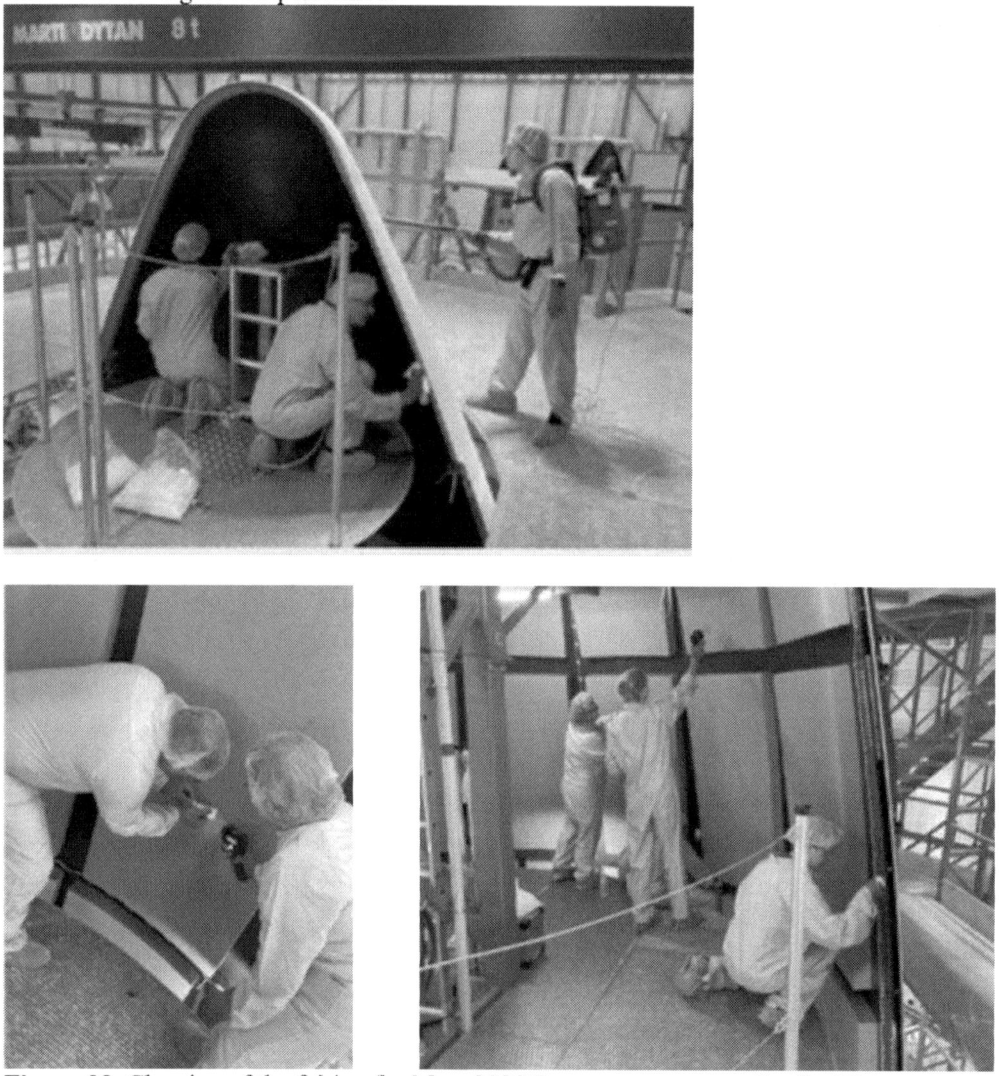

Figure 22. Cleaning of the fairing for Mars 2020 Perseverance rover. Credit: ESA

3.4 Bagging of payload fairing (PLF) for transport

With the heritage bagging designed for ESA Herschel-Planck Mission, each PLF half was bagged with a protective blanketing in its container to reduce any possible risk of contamination during transport. High performance particle air (HEPA) filters were added to ensure the cleanliness and temperature/humidity control of the environment for the transporter container.

Figure 23. Protected fairing with inner bag. Credit: Beyondgravity

Figure 24. HEPA filters to ensure particulate cleanliness during transport. Credit: Beyondgravity

Figure 25. Inner bag protecting fairing for shipping. Credit: Beyondgravity

IOP Conf. Series: Materials Science and Engineering 1287 (2023) 012041 doi:10.1088/1757-899X/1287/1/012041

4. Ariane 5 payload fairing (PLF) cleaning and Centre Spatial Guyanais (CSG)

4.1. Cleanliness preparations in Bâtiment Assemblage Final encapsulation hall (BAF-HE)

The Bâtiment Assemblage Final encapsulation hall (BAF-HE) was vacuumed and wiped using sealed edge wipers, any prohibited surfaces were covered, and white/UV inspections occurred over multiple days prior to the arrival of the A5 fairing. HEPA filters, shown in Figure 26, were turned on after cleaning was completed to ensure the environment JWST would be placed in was to ISO 7 standards. The scaffolding used to access the PLF was cleaned and inspected.

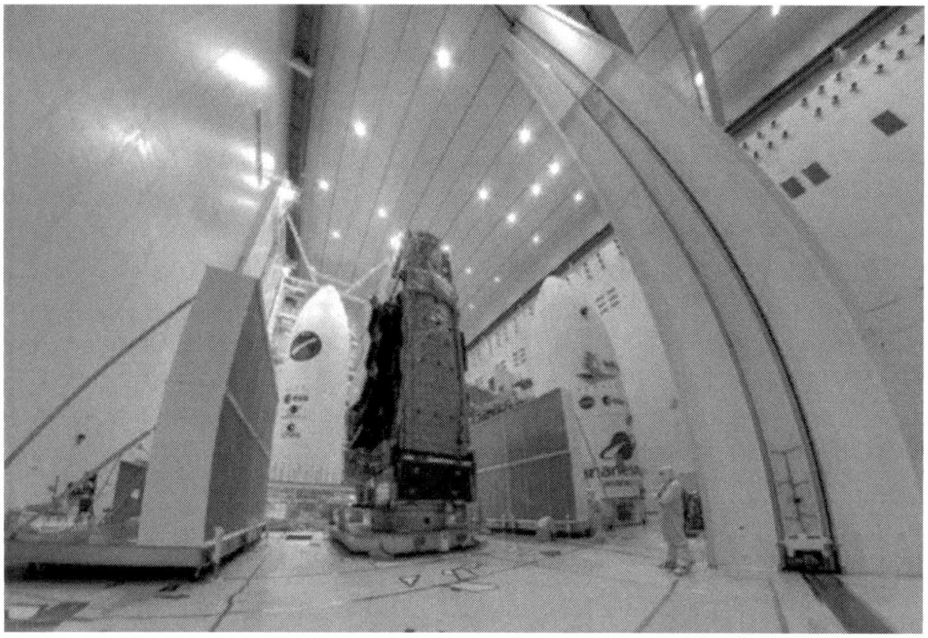

Figure 26. BAF-HE operations with HEPA walls. Credit: ESA-M.Pedoussaut

4.2. Final cleaning at Centre Spatial Guyanais (CSG)

The final cleaning of the PLF during the launch campaign was performed with all stakeholders including beyond gravity, ESA, NASA, and CNES. The PLF was placed over the scaffolding showing in Figure 27 to access the interior. All areas of the PLF able to be accessed were inspected and cleaned. White light and UV, shown in Figure 28, were used during inspections. Metallic surfaces were cleaned using sealed edge wipers with IPA. Particulate identified on the FAP was carefully removed with approved tape. After cleaning, the fairing was sealed out until D-6 lift to launcher. The fallout measured within the fairing environment while sealed was measured as 0.0002 PAC/day (2 ppm).

Figure 27. Ariane 5 fairing for Webb near access scaffolding. Credit: ESA-M.Pedoussaut

Figure 28. Elaine Stewart using UV light to clean the interior FAP of Ariane 5 for Webb. Credit: ESA- O. Schmeitzky

4.3. Particle area coverage (PAC) and molecular organic contamination (MOC) results inside fairing during ventilation phase up to D-2

Figure 29. FAP inside Ariane 5 fairing during lift operations for encapsulation.

Figure 30. Ariane 5 fairing encapsulating Webb observatory.

5. Discussion

On March 11 2022, the Webb team completed the stage of alignment known as "fine phasing." At this key stage, every optical parameter that had been checked and tested was performing at, or above, expectations. The team also found no critical issues and no measurable contamination or blockages to JWST's optical path. The observatory can successfully gather light from distant objects and deliver it to its instruments without issue. The first images were released in July 2022 showing the star and universe cycles with spectrometer measurements of atmospheres. There are frequent releases of new images and data from the Webb space telescope institute (STsci) in Baltimore, MD. Table 1 shows the cleanliness achieved was better than required for End of Life (EOL) for both particulate and molecular.

Table 1. Contamination Levels Achieved

	Particulate Percent Area Coverage (PAC)		Molecular Angstroms (Å)	
	EOL Reqt	Measured	EOL Reqt	Measured
Primary Mirror	1.5	0.754	300	45
Secondary Mirror	0.5	0.149	300	59

IOP Conf. Series: Materials Science and Engineering 1287 (2023) 012041 doi:10.1088/1757-899X/1287/1/012041

References

[1] Stewart M, Wooldridge M and Ward O 2022 *Pred. Contam.* SPIE **122240E**
[2] Abeel C, Wooldridge M, Ward O, Calcabrini and Schmeitzky 2022 *Overview Webb launch campaign* and Ward O 2022 *Pred. Contam.* SPIE **12224–13P**
[3] Wooldridge M and Arenburg 2008 *Contam effects req. dev.* **70690J**
[4] Wooldridge M, Abeel C, Jones B, Goldman M, Lepage D 2022 *Transforming facilities for cleanliness* **1224-12**
[5] Abeel C, Jones B, Wooldridge M, Goldman M, Lepage D 2022 *Evaul. perf. HEPA. Webb* **1224-11P**
[6] Rumler 2008 *Contamination Control Feedback from ATV Launch Campaign*
[7] Rumler and Schmeitzky 2009 *Ariane 5 Cleanliness, Herschel-Planck Launch Campaign Experience*

[8] Van Eesbeck and Schmeitzky 2009 *Herschel-Planck Launch campaign*
[9] Calcabrini and Erin 2021 *JWST-Launch Vehicle Provider-Contamination Control Implementation Plan* **Issue 9-Revision 0**
[10] Lepage D, Abeel C, Stewart M, Rodriguez I, Horton, T and Morales Z 2022 *Transforming any facility for meeting strict cleanliness requirements* **1224-16**
[11] Pinol, Fisher, and Betz 2022 *NASA's Webb Reaches Alignment Milestone, Optics Working Successfully* NASA HQ Press Release

IOP Conf. Series: Materials Science and Engineering 1287 (2023) 012042 doi:10.1088/1757-899X/1287/1/012042

International Space Station Bipropellant Plume Contamination Model update for short thruster pulse widths

Katie Fox[1], **Courtney Steagall**[2], **Taria Usher**[1], **Alexandra Deal**[1], **Erica Worthy**[3]

[1] The Boeing Company, 3700 Bay Area Blvd., Houston, Texas 77058, USA
[2] Jacobs Technology, 2224 Bay Area Blvd., Houston, Texas 77058, USA
[3] NASA Johnson Space Center, 2101 E. NASA Pkwy., Houston, Texas 77058, USA

E-mail: katie.l.fox@boeing.com

Abstract. The International Space Station (ISS) Bipropellant Plume Contamination Model has been a vital tool for characterizing the thruster plume-induced contamination environment at the ISS but was not intended to be used for very short thruster pulse widths. This paper presents an updated model that ensures full start-up and shut-down phases are modeled for each thruster firing, when the majority of liquid phase contaminant mass is released. The updated ISS Bipropellant Plume Contamination Model prevents potential under-prediction of thruster plume-induced contamination due to visiting vehicle proximity operations and provides a way to take advantage of thruster start-up and shut-down performance data gathered during thruster test programs, if available.

1. Introduction

The International Space Station (ISS) Bipropellant Plume Contamination Model developed by ISS Space Environments is a semi-empirical model anchored in flight experiment data and has been a vital tool for characterizing the thruster plume-induced contamination environment at the ISS. [1] The current model utilizes flight experiment data from the Plume Impingement Contamination (PIC) and Shuttle Plume Impingement Flight Experiment (SPIFEX) studies, which include Orbiter 3870 N Primary Reaction Control System (PRCS) and Russian 130 N thrusters, each operating in pulse mode with pulse widths of 80-100 ms. [2]

Contamination due to thruster plumes occurs in the liquid phase (i.e., unburned or partially burned propellant in the plume). Liquid phase releases primarily occur during thruster start-up and shut-down phases, with the steady state phase contributing a small amount relative to the total contaminant mass released. Available PIC and SPIFEX flight experiment data used to develop the current ISS Bipropellant Plume Contamination Model includes a full start-up and shut-down cycle relative to a commanded 80-100 ms pulse width, with the steady state phase contributing a small amount to the total contaminant mass released. However, there is no way to separate how much contaminant mass is released during either the start-up phase (t_{SU}) or steady state phase ($t_{steadystate}$), which are both part of the commanded thruster on-time duration (t), and how much contaminant mass is released during the shut-down phase following the commanded thruster on-time duration (t_{SD}).

The current ISS Bipropellant Plume Contamination Model includes functional dependencies on thruster parameters (e.g., thrust, mass flow rate, specific impulse), distance to receiver surface, and angle with respect to plume centerline. For modeling purposes, contaminant mass released scales linearly with pulse width. This is a conservative approach for pulse widths greater than 80 ms but could under-predict contamination for very short pulse widths (i.e., with little or no steady state phase). As the next generation of crew and cargo visiting vehicles are developed and arrive at ISS, minimum pulse widths of vehicle thrusters used for ISS proximity operations have decreased significantly below 80 ms. An update to the ISS Bipropellant Plume Contamination Model is needed to prevent under-prediction of plume-induced contamination for these very short pulse widths.

2. ISS Bipropellant Plume Contamination Model

The current ISS Bipropellant Plume Contamination Model calculates the contaminant mass flux, $\dot{m}_c$ with units of g/cm^2/s, for a vehicle thruster as a function of thruster parameters (thrust (T), mass flow rate, specific impulse (I_{sp})), distance to receiver surface (r), and angle off plume centerline (θ). These functional dependencies are represented in Equation 1. [1]

$$\dot{m}_c(r,\theta) = f(T) \cdot f(r) \cdot f(\theta) \tag{1}$$

This contaminant mass flux in Equation 1 is then multiplied by the thruster commanded pulse width, or on-time, in seconds to calculate the total contaminant mass released during the thruster firing. Knowing the position of the thruster relative to a receiver surface with a defined surface area, the total mass deposited on the surface can be calculated, or, if desired, the total depth of contamination given a known or assumed contaminant mass density.

The functional dependency on distance to receiver surface is based on mass conservation, where the integrated total mass deposition is constant at any distance from the source. The origin of the plume contamination model is selected to be the nozzle throat where l_{noz} is the nozzle length in cm. The nozzle exit radius, R_{exit} in cm, is selected as the reference distance. The functional dependency on distance r measured from the nozzle exit plane to the receiver surface is then given by Equation 2.

$$f(r) = \left(\frac{r + l_{noz}}{R_{exit}}\right)^{-2} \tag{2}$$

The functional dependency on receiver surface angle off plume centerline θ is based on Trinks functional form for angular distribution of droplets in the plume field, shown in Equation 3. [3] The angular dispersion coefficient θ_0 is defined as 5°. [1]

$$f(\theta) = \left(2e^{-\frac{\theta}{\theta_0}} - e^{-2\frac{\theta}{\theta_0}}\right) \tag{3}$$

The functional dependency on thruster parameters was developed to make the model applicable to all ISS bipropellant thrusters even when experimental data characterizing plume contamination is not available. Engine thrust and the corresponding contaminant mass flux on the plume centerline $\dot{m}_{CL}$ is used to develop this functional model. By focusing on the plume centerline, the angular distribution dependency is removed from the contaminant mass flux equation. Equations 1 and 2 can be combined and solved for $f(T)$, shown in Equation 4.

$$f(T) = \dot{m}_{CL} \cdot \left(\frac{r + l_{noz}}{R_{exit}}\right)^{2} \tag{4}$$

To continue the effort to bridge the gap between all ISS bipropellant thrusters in use and those with available experimental data characterizing the induced contamination environment, the non-dimensional contamination ratio term is introduced. The contamination ratio (CR) term shown in Equation 5 is defined as the contaminant mass flux on the plume centerline $\dot{m}_{CL}$ divided by total propellant mass flow rate $\dot{M}_T$ in g/s and scaled by non-dimensional distance dependency $f(r)$.

$$CR = \frac{\dot{m}_{CL}}{\dot{M}_T / R_{exit}^2} \cdot \left(\frac{r + l_{noz}}{R_{exit}}\right)^2 \quad (5)$$

The contamination ratio value is calculated for available experimental data and plotted with engine steady state thrust T in N. To extend the model to thrusters without experimental data, an exponential curve fit is made separately for both initial contamination deposit results and permanent contamination deposit results, shown in Equation 6, where the contamination ratio curve fit coefficients K and β are provided in Table 1. The CR coefficients were updated in 2008 when the decision was made to apply curve fits to Russian and Orbiter thruster flight test data only, neglecting thruster ground test data included in defining original CR coefficients. [4]

$$CR = KT^{-\beta} \quad (6)$$

Table 1. Contamination Ratio (CR) Curve Fit Coefficients.

	K	β
Initial Deposit	9.84E-03	0.0
Permanent Deposit	0.0022	0.2876

Combining the two forms of the contamination ratio provided in Equations 5 and 6 results in a model for all ISS bipropellant thrusters where all terms will be known except for contaminant mass flux along the thruster centerline, shown in Equation 7.

$$KT^{-\beta} = \frac{\dot{m}_{CL}}{\dot{M}_T / R_{exit}^2} \cdot \left(\frac{r + l_{noz}}{R_{exit}}\right)^2 \quad (7)$$

Solving for the contaminant mass along the thruster centerline makes it possible to ultimately define the functional dependency on thruster parameters, specifically engine size or thrust, shown in Equation 8.

$$f(T) = KT^{-\beta} \cdot \frac{\dot{M}_T}{R_{exit}^2} \quad (8)$$

Finally, the dependencies on thruster parameters (Equation 8), distance to receiver surface (Equation 2), and angle off plume centerline (Equation 3) can be brought together to define the current ISS Bipropellant Plume Contamination Model, given in Equation 9. Depending on the parameters available for a specific thruster, it may be easier to use the ISS Bipropellant Plume Contamination Model with steady state specific impulse, I_{sp}, as an input parameter, in place of total mass flow rate, $\dot{M}_T$; see Equation 10.

$$\dot{m}_c(r, \theta) = KT^{-\beta} \cdot \frac{\dot{M}_T}{R_{exit}^2} \cdot \left(\frac{r + l_{noz}}{R_{exit}}\right)^{-2} \cdot \left(2e^{-\frac{\theta}{\theta_0}} - e^{-2\frac{\theta}{\theta_0}}\right) \quad (9)$$

$$\dot{m}_c(r, \theta) = 101.93 \cdot \frac{KT^{1-\beta}}{I_{sp}} \cdot (r + l_{noz})^{-2} \cdot \left(2e^{-\frac{\theta}{\theta_0}} - e^{-2\frac{\theta}{\theta_0}}\right) \quad (10)$$

The ISS Bipropellant Plume Contamination Model is based on flight experiment data with 80 to 100 ms pulse widths but has been used for thrusters with minimum pulse widths less than 80 ms in the past (e.g., down to 60 ms).

To account for likely higher amounts of contaminant mass leaving the thruster at shorter pulse widths, pulse mode thrust measurements have been used in Equation 6 instead of steady state thrust measurements. The lower thrust value results in a higher contamination ratio term, therefore resulting in higher predicted levels of contamination. The commanded pulse width is unchanged. The plot in Figure 1 shows the contaminant mass released for an example thruster that has a steady state thrust of 350 N and a measured thrust of 225 N at a minimum pulse width of 60 ms, normalized to the intended use of the ISS Bipropellant Plume Contamination Model (e.g., using steady state thrust at an assumed minimum pulse width of 80 ms).

The approach of adjusting the existing ISS Bipropellant Plume Contamination Model using pulse mode rather than steady state thrust measurements has only been used to model thrusters utilizing minimum pulse widths of 60 ms. Implementing this approach is dependent on obtaining reliable pulse mode experimental test data. The adjusted ISS Bipropellant Plume Contamination Model is still not intended to be used at extremely short pulse widths.

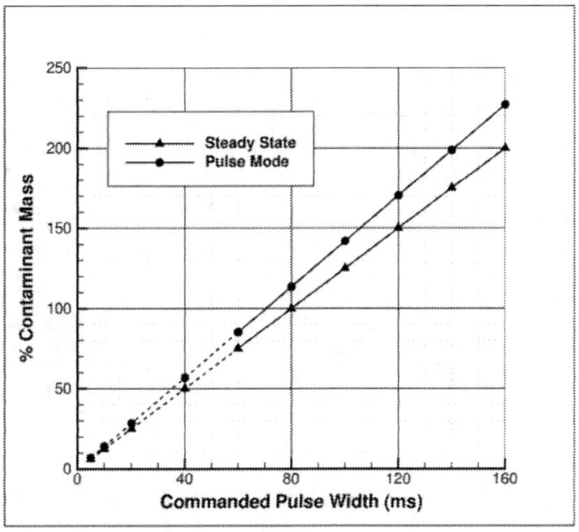

Figure 1. Example thruster plume model using steady state thrust measurement and pulse mode thrust measurement.

3. Updated ISS Bipropellant Plume Contamination Model

ISS Space Environments has added a functional dependency on commanded pulse width to the ISS Bipropellant Plume Contamination Model to address the continued development of visiting vehicle thrusters capable of operating at very short pulse widths, shown in Equation 11. The current ISS Bipropellant Plume Contamination Model is based on empirical data for which the bulk of the contaminant mass is released during 40 ms of combined start-up and shut-down time. This is roughly half of the commanded pulse width used in the experiments. The non-dimensional $f(pw)$ scale factor added to Equation 1 would ensure all contaminant mass released during both start-up and shut-down phases is modeled for each thruster firing, even those with short commanded pulse widths (< 80 ms).

$$\dot{m}_c(r,\theta) = f(T) \cdot f(r) \cdot f(\theta) \cdot f(pw) \tag{11}$$

The scale factor introduced with $f(pw)$ is intended to account for the disproportionate time spent in the start-up and shut-down phases for shorter pulse widths and would be based on both the commanded pulse width and the total start-up plus shut-down time of the thruster. The resulting bipropellant plume model equation would continue to calculate contaminant mass flux, $\dot{m}_c$, using steady state thruster measurements and then must be multiplied by the commanded pulse width or on-time to determine the corresponding contaminant mass deposited on a receiver surface some distance r and angle θ away from the thruster plume centerline.

The functional dependence on pulse width, shown in Equation 12, is a scaled ratio of the thruster start-up time, t_{SU}, thruster shut-down time, t_{SD}, and the commanded thruster pulse width or on-time, t.

$$f(pw) = \begin{cases} 2 \cdot \frac{t_{SU}+t_{SD}}{t_{SU}} & \text{if } t < t_{SU} \\ 2 \cdot \frac{t_{SU}+t_{SD}}{t} & \text{if } t_{SU} \leq t < 2 \cdot (t_{SU} + t_{SD}) \\ 1 & \text{if } t \geq 2 \cdot (t_{SU} + t_{SD}) \end{cases} \tag{12}$$

The constant of 2 is empirically derived from Orbiter PRCS thruster flight experiment data, where the minimum commanded pulse width was $t_{min,PRCS} = 80\,ms$ and the combined start-up and shut-down time was approximately $t_{S,PRCS} = 40\,ms$. This thruster data implies the thruster commanded pulse width must be at least double the total start-up and shut-down time to ensure the contaminant mass resulting from a full startup/shutdown cycle is modeled. [5, 6]

If start-up and shut-down times are not known for a new thruster, the Orbiter PRCS combined start-up and shut-down time of $t_{S,PRCS} = (t_{SU} + t_{SD})_{PRCS} = 40\,ms$ will be used. Steady state thruster parameters (e.g., thrust, mass flow rate, I_{sp}) will continue to be used in this updated ISS Bipropellant Plume Contamination Model. It is not recommended to use pulse mode thruster parameters at minimum pulse widths, in place of steady state thruster parameters as discussed previously, when using the updated ISS Bipropellant Plume Contamination Model with $f(pw)$ term given in Equation 12. The goal of using pulse mode thruster parameters and introducing the functional dependency on pulse width is the same: to ensure the ISS Bipropellant Plume Contamination Model is not underconservative when predicting contaminant mass releases at pulse widths below 80 ms. Using both approaches in the same model may lead to creating an overconservative contamination model for the thruster, and analysis results could introduce unnecessary operational constraints at ISS.

For commanded thruster on-times greater than the required on-time to model a full start-up and shut-down cycle (e.g., $t \geq 2 \cdot (t_{SU} + t_{SD})$), the pulse width functional model will be equal to 1. This is done so contaminant mass levels for longer duration thruster firings are not artificially reduced. Longer pulse widths with more time spent in steady state will produce more contaminant mass than shorter pulse widths and the updated ISS Bipropellant Plume Contamination Model will remain conservative on this point. No test data is available to quantify the ratio of contaminant mass due to start-up and shut-down phases to contaminant mass due to steady state phase for the Orbiter PRCS thrusters. In addition, longer pulse widths are not currently an analysis driver for typical ISS thruster operations.

Situations where the commanded thruster on-time is less than the thruster start-up time ($t < t_{SU}$) are not expected but have been included for completeness. Replacing t with t_{SU} in the denominator assumes the thruster start-up phase can only be as long as the commanded on-time and the shut-down phase will begin once the commanded on-time has elapsed. This also allows for Equation 12 to be used in implementations where the commanded on-time is zero. If the thruster start-up time is not known and Orbiter PRCS thruster total start-up and shut-down time is used as the default, it is recommended to assume a very small t_{SU} value in Equation 12, such as 1 ms or 5 ms. These times are not equal to the Orbiter PRCS thruster start-up time but will help ensure potential contaminant mass levels are not underpredicted when thruster data to support lower contamination levels is not available.

The functional dependence on pulse width, $f(pw)$, can also be interpreted as the ratio of the thruster on-time required to model an entire start-up plus shut-down sequence, t', also described as the reference on-time, and the commanded thruster on-time, t.

$$f(pw) = \frac{t'}{t} \tag{13}$$

where t' is defined as:

$$t' = \begin{cases} 2 \cdot (t_{SU} + t_{SD}) & \text{if } t < 2 \cdot (t_{SU} + t_{SD}) \\ t & \text{if } t \geq 2 \cdot (t_{SU} + t_{SD}) \end{cases} \tag{14}$$

Note the simplified definition of $f(pw)$ is not applicable for commanded thruster on-times less than the thruster start-up time. The full $f(pw)$ expression in Equation 12 should be used for implementation of the updated ISS Bipropellant Plume Contamination Model, particularly to avoid situations where a requested thruster pulse width is equal to zero.

In practice, introducing the functional dependence on pulse width has the effect of forcing thruster on-times used in the ISS Bipropellant Plume Contamination Model to be a constant value once the commanded pulse width is less than the amount of time required to model a full thruster start-up and shut-down sequence. This is demonstrated by the examples in Table 2.

Assume a new bipropellant thruster has a minimum pulse width of 8 ms. In these examples, the contaminant mass release associated with the minimum pulse width is likely occurring over a longer time period than the commanded pulse width (i.e., 8 ms or higher). However, the current ISS Bipropellant Plume Contamination Model would calculate contaminant mass as if the total combined start-up and shut-down time is half the commanded pulse width, or 4 ms. Despite improvements in thruster technology, the actual total start-up and shut-down time is likely much longer and the current ISS Bipropellant Plume Contamination Model would dramatically underpredict contaminant mass associated with a thruster firing.

In the first example shown in Table 2, assume the start-up and shut-down times are unknown for the new thruster. Since the total start-up and shut-down time is unknown, the Orbiter PRCS total start-up and shut-down time of 40 ms will be used. For a commanded pulse width of 8 ms, a scale factor of $f(pw) = 2 \cdot \frac{40\,ms}{8\,ms} = 10$ would ensure a full start-up and shut-down is modeled for the thruster firing. A commanded pulse width of 100 ms is greater than the required thruster on-time to model an entire start-up and shut-down sequence (i.e., $t' = 2 \cdot (t_{SU} + t_{SD}) = 2 \cdot 40\,ms = 80\,ms$) and a scale factor of $f(pw) = 1$ is used.

In the second example shown in Table 2, assume thruster experimental test data is available to demonstrate the total start-up and shut-down time is 20 ms. For this case, applying a scale factor of $f(pw) = 2 \cdot \frac{20\,ms}{8\,ms} = 5$ for a commanded on-time of 8 ms would adjust the contaminant mass flux to account for the disproportionate time spent in the start-up and shut-down phases.

Short-term and long-term options exist for implementing the updated ISS Bipropellant Plume Contamination Model in the existing ISS Space Environments plume contamination analysis process. Visiting vehicle providers typically provide commanded pulse width data during a proximity operations event with ISS in an extensive set of simulated jet firing histories during the ISS integration process. [7] The ISS Space Environments contamination analysis code, NASAN, has historically been modified to accommodate different jet firing history data formats, so the jet firing history data files themselves did not have to be modified. However, updating and verifying NASAN is a time consuming process and with the increase in the number of visiting vehicle providers, each with their own jet firing history file type and data format, it has recently become more efficient to create jet firing history pre-processors in Matlab. Each pre-processor is unique to the visiting vehicle provider and converts the jet firing history files into a generic format readable by NASAN without any code updates. One option for implementing the updated ISS Bipropellant Plume Contamination Model is to modify these jet firing history processors, where the commanded pulse width is replaced with the required or reference pulse width.

Table 2. $f(pw)$ and reference on-time t' for example thrusters with varying assumed total start-up and shut-down times. The thruster in Example 1 assumes a total start-up and shut-down time of 40 ms. The thruster in Example 2 assumes a total start-up and shut-down time of 20 ms. The reference on-time t' is the thruster on-time required in the ISS Bipropellant Plume Contamination Model to model a full start-up and shut-down sequence for the thruster firing.

	Example 1			**Example 2**		
Commanded On-Time t (ms)	Start-up and Shut-down Time $t_{SU} + t_{SD}$ (ms)	$f(pw)$	Reference On-Time t' (ms)	Start-up and Shut-down Time $t_{SU} + t_{SD}$ (ms)	$f(pw)$	Reference On-Time t' (ms)
8	40	10	80	20	5	40
10	40	8	80	20	4	40
15	40	5.33	80	20	2.67	40
20	40	4	80	20	2	40
25	40	3.2	80	20	1.6	40
30	40	2.67	80	20	1.33	40
40	40	2	80	20	1	40
50	40	1.6	80	20	1	50
60	40	1.33	80	20	1	60
70	40	1.14	80	20	1	70
80	40	1	80	20	1	80
90	40	1	90	20	1	90
100	40	1	100	20	1	100
120	40	1	120	20	1	120
140	40	1	140	20	1	140

The long-term implementation option is to update the ISS Bipropellant Plume Contamination Model programmed into NASAN. The user would be required to provide thruster start-up, t_{SU}, and shut-down, t_{SD}, times so the $f(pw)$ term can be calculated. An option should also exist to use Orbiter PRCS thruster parameters as a default. Pulse width data provided in jet firing histories would not have to be altered. The full version of $f(pw)$ in Equation 12 should be used to avoid runtime errors if the commanded on-time is zero.

This ISS Bipropellant Plume Contamination Model update is also applicable when using the plume model for estimating erosion impacts on ISS hardware due to thruster firings at short pulse widths. [4]

An alternative model for thruster plume contamination has been proposed that utilizes ground-based testing for new thrusters. [8] For this model, extensive test data is required to characterize levels of contaminant mass released during the unsteady portions of the pulse (e.g., start-up and shut-down phases), as well as levels of contamination mass released during the steady state portion of the pulse. This test data is typically not available from ISS visiting vehicle providers.

4. Comparing Current and Updated ISS Bipropellant Plume Contamination Models

Developing the plume contamination model for a specific thruster using the ISS Bipropellant Plume Contamination Model must be done in consideration of all available thruster performance data. To demonstrate the advantages and disadvantages of various implementations of both the current and updated ISS Bipropellant Plume Contamination Models, steady state thruster parameters have been manufactured for a small 10 lbf (44.5 N) thruster. Pulse mode thrust and I_{sp} data for a range of pulse widths from 5 ms to 150 ms has also been manufactured for demonstration purposes only.

Five different plume contamination models are considered for each example thruster:

(1) Current ISS Bipropellant Plume Contamination Model utilizing steady state thruster parameters.

(2) Current ISS Bipropellant Plume Contamination Model utilizing minimum pulse width I_{sp} instead of steady state I_{sp}.

(3) Current Bipropellant Plume Contamination Model utilizing I_{sp} versus pulse width data.

(4) Updated ISS Bipropellant Plume Contamination Model with default total start-up and shut-down time of 40 ms.

(5) Updated ISS Bipropellant Plume Contamination Model with an assumed total start-up and shut-down time of 20 ms.

For a comparison of these five potential plume contamination models for the small example thruster, see the contaminant mass versus commanded pulse width data in Figure 2. Since the minimum pulse width is assumed to be much shorter than 80 ms, Model 1 is likely underpredicting contaminant mass released at these short pulse widths. A reflexive correction may be to use Model 2 with thrust and I_{sp} data at the minimum pulse width, which may better predict contaminant mass for very short pulse widths. However, Model 2 is likely severely overpredicting contaminant mass from moderate to long pulse widths that are still used during ISS proximity operations events, which can lead to unrealistic requirement exceedances and unnecessary operational constraints.

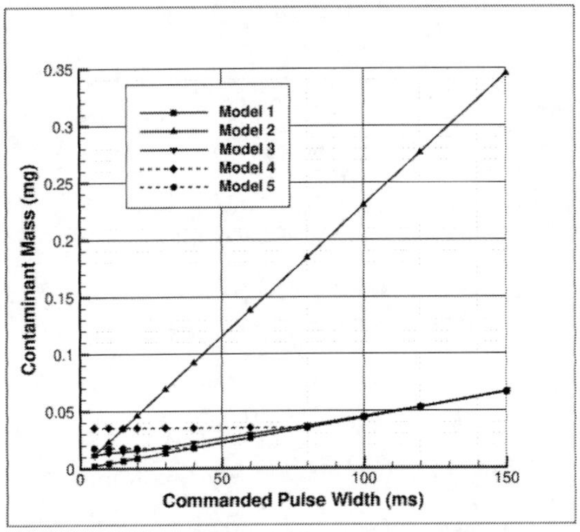

Figure 2. Comparison of five potential plume contamination models for a small (10 lbf) example thruster.

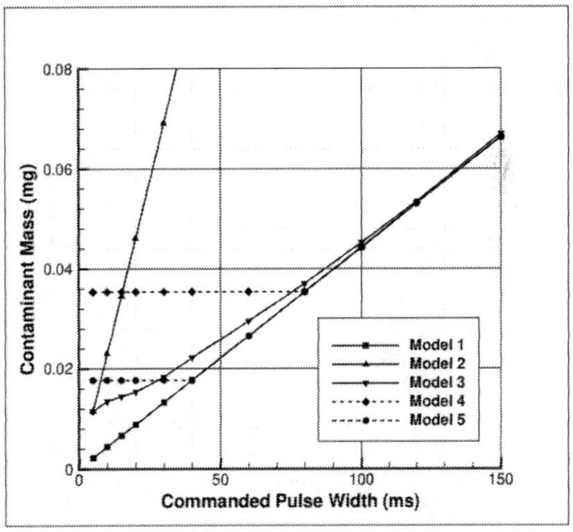

Figure 3. Closer comparison of the potential plume contamination models for a small example thruster.

See Figure 3 for a closer view of Models 3, 4, and 5 for the small thruster. Model 3 demonstrates that when pulse mode I_{sp} data is available for a range of pulse widths, the current ISS Bipropellant Plume Contamination Model can be used with a step function design to likely most accurately predict contaminant mass released from the thruster plume. But again, this data can be difficult to obtain with ground based testing.

The updated ISS Bipropellant Plume Contamination Model with default total start-up and shut-down times used in Model 4 avoids the disadvantages of Model 1 at short pulse widths and of Model 2 at long pulse widths. When test data is available to measure thruster start-up and shut-down time, which is likely consistent across all pulse widths, Model 5 can be used to remove conservatism introduced with Model 4. The results of Model 5 are similar to Model 3, and although Model 3 may be the most accurate, it is likely more difficult to obtain reliable I_{sp} data at a range of pulse widths through a ground test program, versus reliably determining the thruster start-up and shut-down times. This example uses manufactured data to demonstrate the point but these trends are similar to existing thruster data reviewed by ISS Space Environments.

This example shows clearly the plume contamination models to avoid for thrusters operating at short pulse widths (Models 1 and 2) but also shows that the development of a plume contamination model must be made in full consideration of the available thruster data.

5. Conclusion

An update to the Space Environments ISS Bipropellant Plume Contamination Model has been completed to include a functional dependence on thruster pulse width, in addition to functional dependencies on thruster parameters, distance to receiver surface, and angle off plume centerline. This update ensures the portion of the thruster firing spent in start-up and shut-down phases is fully modeled, when the majority of liquid phase contaminant mass is released, even when the commanded pulse widths are very short. This update is necessary for the continued use of the ISS Bipropellant Plume Contamination model for predicting plume-induced contamination and erosion at the International Space Station. The updated model prevents under-prediction of thruster plume-induced contamination due to visiting vehicle proximity operations while providing a way to take advantage of thruster start-up and shut-down data gathered during test programs, if available. Developing the plume model for a thruster using the ISS Bipropellant Plume Contamination Model must be done in consideration of all available performance data.

References

[1] Soares C, Mikatarian R, Barsamian H and Rauer S 2002 International Space Station Bipropellant Plume Contamination Model *Proc. of the 8th AIAA/ASMT Joint Thermophysics and Heat Transfer Conf.* AIAA Paper 2002-3016, St. Louis, MO USA, June 2002

[2] Soares C, Barsamian H and Rauer S 2002 Thruster Plume Induced Contamination Measurements from the PIC and SPIFEX Flight Experiments *Optical System Contamination: Effects, Measurements, and Control VII* **4774** 199-209

[3] Trinks H 1987 Experimental investigation of the exhaust plume flow fields of various small bipropellant and monopropellant thrusters *Proc. of the 22nd AIAA Thermophysics Conf.* AIAA Paper 87-1607, Honolulu, HI USA, June 1987

[4] Soares C, Olsen R, Steagall C, Schmidl W, Myers B, Mikatarian R, Koontz S and Worthy E 2015 Improvements in modeling thruster plume erosion damage to spacecraft surfaces *Proc. of the 13th Int. Symp. on Materials in a Space Environment* Pau, France, June 2015

[5] Kubiak E and Martin M 1983 Minimum impulse limit cycle design to compensate for measurement uncertainties *Journal of Guidance, Control, and Dynamics* **6:6** 432-5

[6] Hattis P 2016 A Short History of the Space Shuttle Orbit Flight Control System Development and Application Evolution *Proc. of the 39th Annual AAS Rocky Mountain Section Guidance and Control Conf.* AAS Paper 16-095 Breckenridge, CO USA, February 2016

[7] Steagall C, Mikatarian R, Huang A, Fox K, Olsen R, Koontz S, Worthy E and Soares C 2018 External contamination integration of visiting vehicles on the International Space Station *Proc. of the 14th Int. Symp. on Materials in a Space Environment* Biarritz, France, October 2018

[8] Larin M, Lumpkin F and Rauer S 2001 Standardization of rocket engine pulse time parameters *Proc. of the 25th AIAA Thermophysics Conf.* AIAA Paper 2001-2817, Anaheim, CA USA, June 2001

AUTHOR INDEX